MECHANICAL AND ELECTRICAL SYSTEMS IN BUILDINGS

MECHANICAL AND ELECTRICAL SYSTEMS IN BUILDINGS

William K. Y. Tao, P.E.

Affiliate Professor
School of Engineering and School of Architecture
Washington University

Richard R. Janis, M. Arch., P.E.

Affiliate Associate Professor
School of Architecture
Washington University

Prentice Hall
Upper Saddle River, New Jersey   Columbus, Ohio

Library of Congress Cataloging-in-Publication Data

Tao, William K. Y.
    Mechanical and electrical systems in buildings / William K. Y.
Tao, Richard R. Janis.
      p. cm.
    Includes index.
    ISBN 0-13-086729-2
    1. Buildings—Mechanical equipment.   2. Buildings—Electric
equipment.   I. Janis, Richard R.   II. Title.
TH6010.T36   1997
696—dc20
                                          95–51717
                                          CIP

Cover photo: Pierre Laclede Building/William Tao & Associates, USA
Editor: Ed Francis
Production Editor: JoEllen Gohr
Test Designer and Production Coordinator: Tally Morgan, WordCrafters Editorial Services, Inc.
Cover/Insert Design Coordinator: Karrie M. Converse
Cover Designer: Russ Maselli
Production Manager: Pamela D. Bennett
Marketing Manager: Danny Hoyt

This book was set in Palatino by Clarinda and was printed and bound by Courier Kendallville, Inc.
The cover was printed by Phoenix Color Corp.

Prentice-Hall International (UK) Limited, *London*
Prentice-Hall of Australia Pty. Limited, *Sidney*
Prentice-Hall Canada Inc., *Toronto*
Prentice-Hall Hispanoamericana, S. A., *Mexico*
Prentice-Hall of India Private Limited, *New Delhi*
Prentice-Hall of Japan, Inc., *Tokyo*
Simon & Schuster Asia Pte. Ltd., *Singapore*
Editora Prentice-Hall do Brasil, Ltda., *Rio de Janeiro*

PREFACE

Throughout their careers, the authors have shared their knowledge by teaching for over 20 years in architectural and engineering schools. In consulting engineering, their expertise covers a spectrum of projects, ranging from single-story to super-high-rise buildings and from simple to high-tech facilities. They have collaborated with hundreds of owners, users, developers, architects, engineers, contractors, and construction managers.

In response to requests by their colleagues, and with feedback from students and professionals in the design and construction fields, the authors have prepared a book that bridges the widening gap between textbooks used in education and problems confronted by professionals in the real world.

There are many well written text and reference books covering building mechanical and electrical (M/E) systems. Some emphasize engineering calculations and construction details; others address simple systems, primarily for residential applications. *Mechanical and Electrical Systems in Buildings* takes another approach, including sufficient engineering calculations and condensed reference data to illustrate M/E principles, but devoting more space to emerging technology and environmental issues. Students and practicing professionals all have ready access to the design and reference data published by leading engineering technical societies, such as ASHRAE and IES. There is no need to duplicate them here.

Through their consulting services, the authors have gathered the most up-to-date information on design direction, on such topics as telecommunications and information technology, alternative M/E equipment and systems, and emerging technology and environmental issues.

Mechanical and Electrical Systems in Buildings, rich with illustrations (many in color), has 17 chapters covering M/E systems and their impact on architectural design, space planning, cost and the environment; HVAC systems from fundamental principles to systems for major structures; plumbing and fire protection; electrical principles, equipment and wiring design; and illumination theory and practice.

A set of questions and exercises is included at the end of each chapter.

These chapters form an introduction to a forthcoming book on applications, including acoustics, vertical transportation, and special building type design considerations.

The authors have debated the merits of using the conventional I–P–S (inch-pound-second) system versus the Système International/SI (meter–gram–second) system for units of quantities. It was decided to use the conventional system as a basis, with SI units as an alternate. Dual units are included for formulas and tables whenever practical.

The authors are indebted to the many reviewers who provided constructive comments and suggestions, and the engineering and technical associates and manufacturers that provided valuable reference data and illustrations.

v

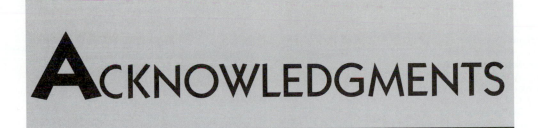

ACKNOWLEDGMENTS

Preparation of *Mechanical and Electrical Systems in Buildings* took several years, and the participation of many. The authors wish to acknowledge those at William Tao & Associates who helped:

Debbie Chollet	Project coordinator
Danniel Ku, P.E.	Chapters on electrical engineering
James Kuba & Steve Andert, P.E.	Chapters on illumination
Tom Uhlmansiek, P.E.	Chapters on HVAC
Karl Normann	Chapters on plumbing
Ralph F. Dames, P.E.	Chapters on fire protection
Lisa J. Sombart, P.E.	Section on automation and controls

In addition we thank Susie Zang, Floy Myers, and Nancy DeNoyer for word processing; Susan Sommers for graphics; Mei Shih and Dennis Harris for CAD work; and Linda Tao for editorial expertise.

Special recognition goes to Rick Oertli and Lynn Gates of Sonacom Company of St. Louis and Peter Tao and Helen Lee of TAO + LEE Associates for the section on architectural design considerations.

Special thanks go to the following organizations for design data:

ASHRAE	American Society for Heating Refrigeration and Air Conditioning Engineers.
IESNA	Illuminating Engineering Society of North America
ASPE	American Society for Plumbing Engineers
NEC	National Electrical Code
NFPA	National Fire Protection Association
TIA/EIA	Telecommunication Industry Association
NSPC	National Standard Plumbing Code/National Association of Plumbing-Heating-Cooling Contractors.

We gratefully acknowledge those product manufacturers that provided valuable photos or illustrations, with special mention of the Trane Company of La Crosse, WI, and General Electric of Nela Park, OH. Others include:

AAF International, Louisville, KY
Advance Transformer Co., Rosemont, IL
Airtherm Mfg. Co., St. Louis, MO
Alfa-Laval Gen. Corp., Richmond, VA
Allen-Bradley Co., Milwaukee, WI
American Insulated Wire Corp., Pawtucket, RI
AMTROL, Inc., West Warwick, RI
Ansul Incorporated, Marinette, WI
ASHRAE, Atlanta, GA
ASPE, Westlake, CA
Atlas/ Soundolier, St. Louis, MO
Aurora Pump, Aurora, IL
Baltimore Aircoil Co., Baltimore, MD
Belden Wire & Cable, Richmond, IN
Ber-Tek, Inc., New Holland, PA
Brasch Mfg. Co., Inc. Maryland Heights, MO
Bryan Steam Corp, Peru, IN
Bussmann, St. Louis, MO
Carnes Company, Inc., Verona, WI
Cerberus Pyrotron, Cedar Knolls, NJ
Challenger Electrical Equipment Corp., Malvern, PA
Chromalox, Pittsburgh, PA
Cleaver Brooks, Milwaukee, WI
Culligan International Co., Northbrook, IL
Cutler-Hammer/Westinghouse, Pittsburgh, PA
Dukane Corp., Saint Charles, IL
Edwards System Technologies, Farmington, CT
Eljer Plumbing, Plano, TX
Elkhart Mfg. Co., Elkhart, IN
Envirovac Inc., Rockford, IL
Fiberstars, Fremont, CA
Figgie Fire Protection System, Charlottesville, VA
Flexonics, Inc., New Braunfels, TX
Fusion Lighting, Inc., Rockville, MD
General Electric, Cleveland, OH
Grinnell Corp., Exeter, NH
Halo/Cooper Ind., Elk Grove Village, IL
IESNA, New York, NY
ITT Bell & Gossett, Morton Grove, IL
Joy Technologies, New Philadelphia, OH
Kidde-Fenwal, Inc., Ashland, MA
Kohler Company, Kohler, WI
Lennox Industries, Richardson, TX
Lightolier, Fall River, MA

MagneTek Drives & Systems, New Berlin, WI
Marley Cooling Towers, Mission, KS
Master Publishers, Inc., Richardson, TX
McQuay International, Minneapolis, MN
Minolta Corp., Ramsey, NJ
Mitel Corporation, Canada K2K 1X3
Mueller Company, Springfield, MO
NFPA, Quincy, MA
NSPC, Falls Church, VA
OSRAM/SYLVANIA, Danvers, MA
Palmer Instruments, Inc., Asheville, NC
Panasonic Co., Secaucus, NJ
PASO Sound Systems, Pelham, NY
Pelli & Associates, New Haven, CT
Peerless Lighting Corp., Berkeley, CA
Philips Lighting Co., Somerset, NJ
Reliable Sprinkler Co., Mt. Vernon, NY
Sams Publishing, Indianapolis, IN
Sloan Valve Co., Franklin Park, IL
SMACNA, Chantilly, VA
Southwire Co., Carrollton, GA
Spirax Sarco, Allenton, PA
Sporlan Valve Co., Washington, MO
Square D Co., Lexington, KY
Star Sprinkler Corp., Milwaukee, WI
State Industries, Ashland City, TN
Sterling Heating Equipment, Westfield, MA
Tate Access Floors, Jessup, MD
Thomas Industries, Tupelo, MS
Titus, Richardson, TX
TIR Systems LTD, Burnaby, BC
Trane Company, La Crosse, WI
United McGill Corp., Groveport, OH
Van-Packer Co., Buda, IL
Victaulic Co. of America, Easton, PA
Viking Corporation, Hastings, MI
Walker Systems, Inc., Parkersburg, WV
Watts Regulator Co., N. Andover, MA
Weil McLain, Michigan City, IN
H.E. Williams, Inc., Carthage, MO
Wiremold Company, West Hartford, CT
York International, York, PA
Zurn Industries, Erie, PA

CONTENTS

CHAPTER 11
ELECTRICAL SYSTEMS AND EQUIPMENT, 319

CHAPTER 12
ELECTRICAL DESIGN AND WIRING CONSIDERATIONS, 353

MECHANICAL AND ELECTRICAL SYSTEMS IN BUILDINGS

1

THE SCOPE AND IMPACT OF MECHANICAL AND ELECTRICAL SYSTEMS

MODERN BUILDINGS ARE NO LONGER JUST SHELters from rain, wind, snow, sun, or other harsh conditions of nature. Rather, they are built to create better, consistent, and productive environments in which to work and to live. Buildings must be designed with features to provide better lighting; comfortable space temperature, humidity, and air quality; convenient power and communication capability; high-quality sanitation; and reliable systems for the protection of life and property. All these desirable features have become a reality with recent advances in building mechanical and electrical (M/E) systems.

These advances have opened the door for a wide range of architectural design innovations in style, form, and scope that are not achievable without the utilization of M/E systems. Block-type buildings without windows, such as department stores, are totally dependent on electrical lighting, ventilation, and space conditioning. High-rise buildings must rely on high-speed vertical transportation and high-pressure water for drinking and cleaning purposes and for protection against fire.

All these benefits, however, are not achieved without penalties. M/E systems demand considerable amounts of floor and ceiling space. Without proper space allocation during the preliminary planning phase of a project, the design process may have to be started over again and often the system performances are compromised. Furthermore, M/E systems add to the cost of construction of a building, in some instances approaching half the total cost. Sophisticated buildings, such as research buildings, hospitals, and computer centers, are just a few examples.

M/E systems require energy to operate them. Energy consumed by occupied buildings, including residential, commercial, institutional, and industrial facilities, account for over 50 percent of all energy usage by an industrialized country. In addition, it accounts for a large portion of the operating costs of such buildings. The high rate and inefficient use of energy by buildings is the major contributing factor to the deterioration of our greater environment.

Properly designed M/E systems utilize space and energy efficiently, thereby reducing building costs and minimizing environmental impacts. This chapter provides an overview of building M/E systems and their impact on space planning, architectural design, construction and operating costs, and the greater environment.

1.1 THE SCOPE OF BUILDING M/E SYSTEMS

The complexity of M/E systems varies with the living standards of the society, climatic conditions of the region, and occupancy and quality of the building. For example, a house located in a mild climate may not require either heating or cooling, regardless of the quality of the house; a warehouse for bulk storage

1

may not require any heating, even in a freezing climate; a modern hospital must have medical gas supply, standby electrical power, and telecommunications systems to meet present health care standards; and a low-cost, small office building may be totally appropriate for window-type air conditioners, whereas an intelligent high-rise office building would most likely be designed with a central HVAC system complete with personal computer (PC) based building automation and management controls.

Building M/E systems may be classified into three major categories as follows:

Mechanical Systems

- *HVAC*   Heating, ventilating, and air conditioning
- *Site utilities*   Water supply, storm water drainage, sanitary disposal, gas supply
- *Plumbing*   Water distribution, water treatment, sanitary facilities, etc.
- *Fire protection*   Water supply, standpipe, fire and smoke detection, annunciation, etc.
- *Special systems*

Electrical Systems

- *Electrical power*   Normal, standby, and emergency power supply and distribution
- *Lighting*   Interior, exterior, and emergency lighting
- *Auxiliary*   Telephone, data, audio/video, sound, security systems, etc.
- *Special systems*

Building Operation Systems

- *Transportation*   Elevators, escalators, moving walkways, etc.
- *Processing*   Products, food service, etc.
- *Automation*   Environmental controls, management, etc.
- *Special systems*

A checklist at the end of the chapter provides a more comprehensive identification of common M/E system features and design criteria that should be considered when initiating a new project. The checklist is an excellent guide for the owner and the architect/engineer to define the scope of services from which to generate a realistic budget, formulate spatial programs, and create architectural concepts.

1.2 THE IMPACT ON SPACE PLANNING

The floor area necessary for M/E systems in a building varies widely, based on the occupancy, climatic

conditions, living standards, and quality and general architectural design of the building. The M/E space affects the gross floor area, footprint (the size and shape of the building's ground floor), floor-to-floor height, geometry, and architectural expression. Reasonable allocations made during the space programming phase allow M/E space to be appropriately sized and strategically located. Space planning for M/E systems is one of the most challenging and least developed procedures in the architectural design process.

Central equipment used for large buildings is usually bulky and tall, requiring floor-to-floor heights of 1½ to 2 times the normal height. Figures 1–1 and 1–2 illustrate a chiller and pump room of a large office-computer building with a floor-to-floor height of 22 ft, about twice the normal height. Figures 1–3 and 1–4 illustrate the complexity of the ductwork, lighting, and wiring of a commercial building, which requires between 2 and 3 ft of ceiling cavity. The ceiling cavity may be reduced in height if the structural beam can be penetrated. Not illustrated, but frequently present in the ceiling cavity, is piping for HVAC and plumbing. These also demand ceiling cavity space and thus affect the floor-to-floor height. Figure 1–5(a) is a photo of the ceiling cavity with the ceiling tiles removed; Figure 1–5(b) shows the finished space; and Figure 1–5(c) illustrates exposed ductwork below the ceiling as a decorative feature of the space. In fact, exposed ductwork, piping, and lighting has been a recent trend in modern architecture.

The purpose of introducing the foregoing complex issues at the beginning of this book is to remind the reader how M/E systems can affect the overall planning of a building. For conceptual planning and budgetary purposes, Table 1–1 lists the range of M/E

TABLE 1–1

Range of M/E floor area required for buildings

Type of Occupancy	Percent of Gross Building Area		
	Low	*Medium*	*High*
Computer centers	10	20	30
Department stores	2	5	7
Hospitals	5	10	15
Hotels	4	7	10
Offices	2	4	6
Research laboratories	5	10	15
Residential, single occupancy	1	2	3
Residential, high rise	1	3	5
Retail, individual stores	1	2	3
Schools, elementary	2	3	4
Schools, secondary	2	4	6
Universities and colleges*	4	6	8

*Buildings other than those used for classrooms follow the space required for specialty-type buildings such as laboratories, computer centers, residences, etc.

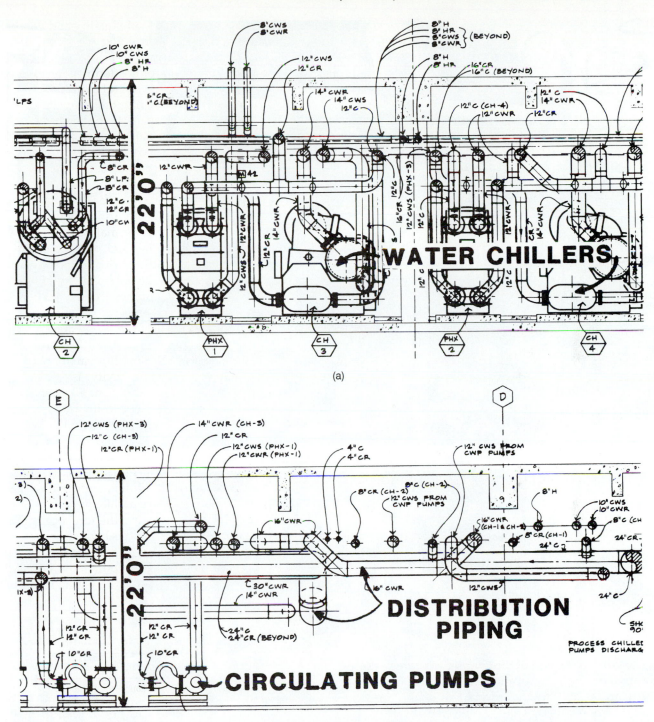

(a)

(b)

■ FIGURE 1–1

(a) Partial layout of mechanical equipment room, showing end elevation of water chillers.
(Reduced scale of drawing: 1/8″ = 1′−0″.) (b) Partial elevation of main distribution
piping and circulation pumps. (Reduced scale of drawing.)

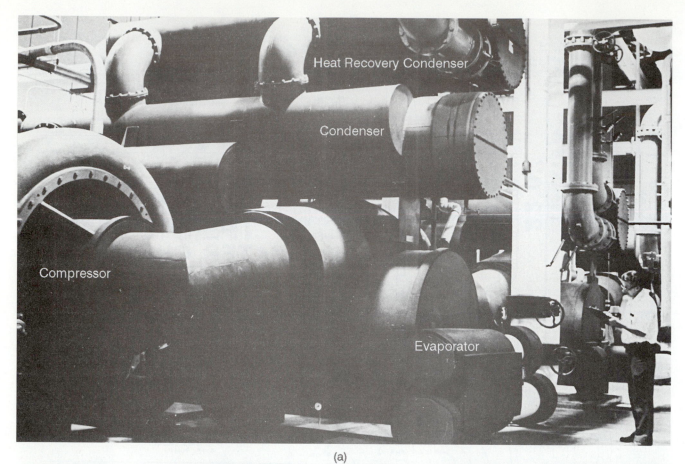

(a)

(b)

■   **FIGURE 1-2**
(a) 1500-ton chillers in a central plant. The maintenance man standing alongside reflects the size of the machine. The ceiling height in this space is 22 feet. (b) Chilled water circulating pumps for the chillers in part (a). Pipe sizes range from 8 to 14 inches in diameter.

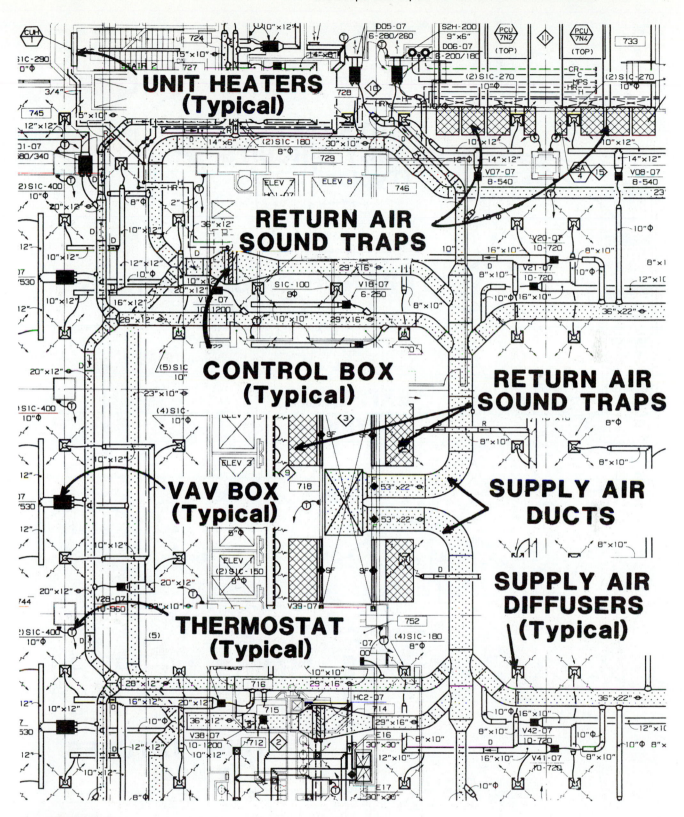

UNIT HEATERS
(Typical)

RETURN AIR SOUND TRAPS

CONTROL BOX (Typical)

RETURN AIR SOUND TRAPS

VAV BOX (Typical)

SUPPLY AIR DUCTS

THERMOSTAT (Typical)

SUPPLY AIR DIFFUSERS (Typical)

■ **FIGURE 1–3**
Partial plan of air distribution ductwork in an office with large computer space, showing the location and size of supply air ducts, diffusers, and return air sound traps. (Reduced scale of drawing.)

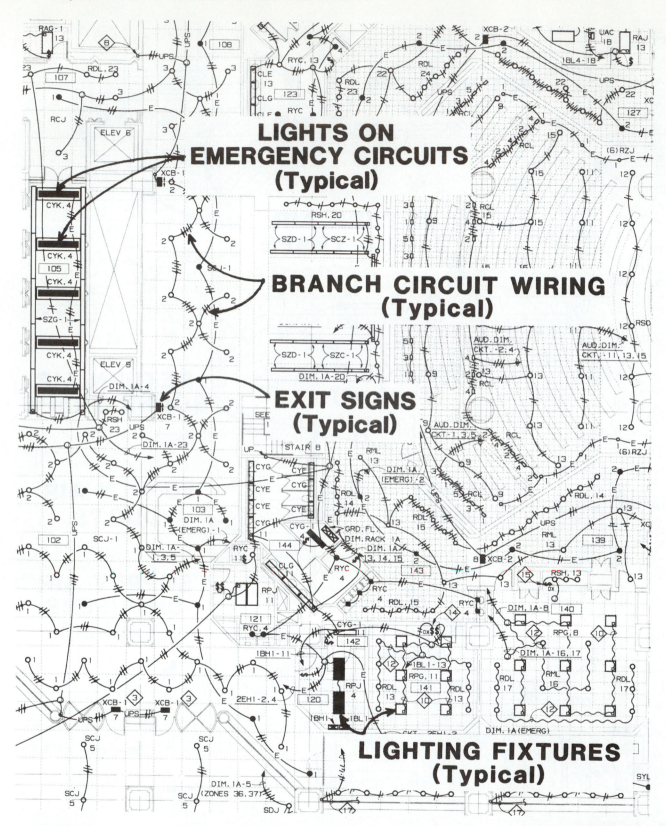

LIGHTS ON EMERGENCY CIRCUITS (Typical)

BRANCH CIRCUIT WIRING (Typical)

EXIT SIGNS (Typical)

LIGHTING FIXTURES (Typical)

■ **FIGURE 1–4**
Partial plan of electrical lighting in an auditorium in a corporate headquarters building, showing lighting fixtures and branch circuit wiring. (Scale of drawing: 1/16" = 1'-0".)

(a)

(b)

(c)

■ **FIGURE 1–5**

(a) Maze of sprinkler piping, air distribution boxes, flexible ducts, supply air diffusers, and fluorescent lights above a typical T-bar acoustic ceiling of a modern office building. (b) Finished space with only diffusers, lights, sprinklers, and smoke detectors shown. (c) Mechanical ducts, lighting fixtures, and illuminated signs are integrated as a dominant architectural design feature in the St. Louis Lambert International Airport concourses.

space requirements, based on building occupancies. Although the ranges are broad, they serve as a basis from which to begin the planning process. The actual space needs are formalized, modified, and refined as the building requirements become better defined.

1.3 THE IMPACT ON ARCHITECTURAL DESIGN

1.3.1 Early Building Forms

Prior to the development of reliable and affordable M/E systems, buildings designed for human occupancy followed a simple rule: every room must have exterior operable windows for the introduction of daylight and for natural ventilation. Accordingly, most buildings are L-, U-, or H-shaped, having either single- or double-loaded corridors. Deep block-type buildings usually have an open interior court for access to daylight and outside air. Figure 1–6 illustrates these basic building forms. It is not difficult to draw the conclusion that buildings of the designs shown in the figure have more exterior wall surface area than the deep block-type design and thus have more heat gain or loss, as well as a higher construction cost.

Great religious buildings, such as cathedrals, are exceptions to the preceding rule. However, they all have extremely high atriums (cathedral domes) to create a good natural draft, heavy exterior walls to retard solar heat gain, and a heavy mass to store thermal energy. Their interiors are usually dimly lit without modern electrical lights. A visit to some famous cathedrals in Europe will easily confirm this fact.

1.3.2 Building Height vs. Space Utilization

It is not difficult to prove that low-rise buildings (six stories or lower) are cheaper to build and more efficient to use than high-rise buildings. However, high-rise buildings (7 to 29 stories), super high-rise buildings (30 to 49 stories), and skyscrapers (50 stories and up) have become increasingly popular in urban settings in recent years, and the trend will continue as long as the availability of urban land is in short supply. No less important is the image associated with high-rise buildings, signifying status, identity, and prestige. These are intangible values to the owners and occupants of the building and are difficult to put into economic terms.

Historically, the height of a building is pretty much limited by a human's ability to walk up and down stairs. Although monuments or bell towers are often built as high as is structurally sound, their stairs are not intended for routine use. The same cannot be said for office-type buildings, however, where stairs are the only means of communication between floors and elevators are a luxury. Modern elevators for

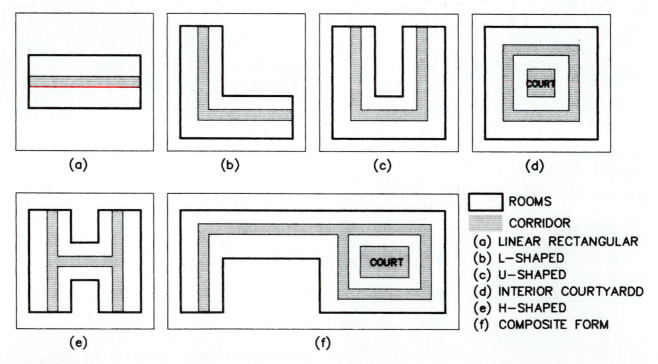

(a)   (b)   (c)   (d)

(e)   (f)

ROOMS
CORRIDOR
(a) LINEAR RECTANGULAR
(b) L—SHAPED
(c) U—SHAPED
(d) INTERIOR COURTYARDD
(e) H—SHAPED
(f) COMPOSITE FORM

■ **FIGURE 1–6**
Common building geometry prior to the development of modern M/E systems.

high-rise buildings are designed to travel between 500 and 2,500 ft (150 and 750 m) per minute. At that speed, it would take less than 30 seconds to reach the top of a 30-story building, excluding the waiting and loading time at each stop.

As the height of the building increases, more floor space is required for stairs, structural elements, elevators, lobbies, M/E system shafts, etc. The result is a reduction in the net usable space on the floor. Measured in terms of the *floor efficiency ratio* (FER), a well-laid floor plan should have an FER between 75 and 85 percent, depending on the selection of M/E systems. As the number of floors is increased, the FER decreases. Figure 1–7 illustrates a typical well-designed floor plan of a 17-story office building, with a compact core plan for elevators, stairs, toilets, and M/E shafts. The FER is in excess of 85 percent.

1.3.3 Building Efficiency Factors

Theoretically, a building is most efficient if 100 percent of the interior space can be utilized for occupancy. This is possible for a small, single-story building with all major M/E equipment located on the roof or at the exteriors. Multistory buildings gradually lose space utilization efficiency due to their need for stairways, elevators, and M/E equipment space. Building efficiency can be expressed by several ratio factors.

Net-to-gross ratio   The net floor area (NFA) of a building is the floor area that can be used by the occupants and is sometimes referred to as "net assignable." NFA is typically defined as the gross floor area (GFA), excluding the area taken by stairs, elevators, lobbies, structural columns, M/E equipment and shafts, etc. The net-to-gross ratio (NGR) is

$$NGR = 100 \times NFA/GFA \qquad (1\text{–}1)$$

where NFA = net floor area, sq ft.
GFA = gross floor area, sq ft.

The available NGR depends on the building's occupancy, M/E systems, and architectural design and usually ranges from 60 to 90 percent. The objective in space planning is to improve the NGR, while maintaining a proper balance between occupant comfort and productivity, M/E system performances, and initial and operating costs.

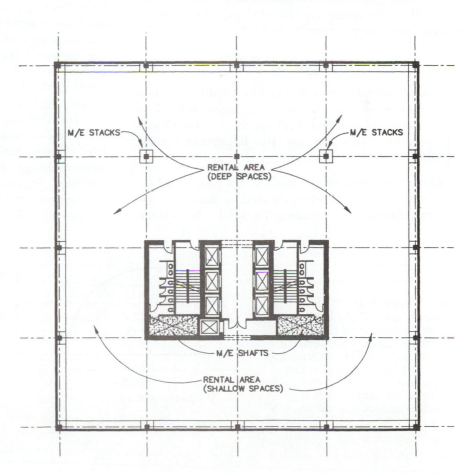

■ **FIGURE 1–7**
A typical rental office floor plan with excellent FER. (Illustrated: Pierre Laclede Building, St. Louis, MO.)

Net rentable area   The net rentable area (NRA) is frequently used by real estate agents when calculating the rental area of a building. The definition of NRA varies widely between regions or countries, depending on local customs and standards. Some calculations of NRA include the proportionate share of public lobbies, part of the exterior walls, and mechanical equipment rooms, while others do not.

Floor efficiency ratio   The FER is frequently used for office buildings to calculate the rentable space on typical rental floors. The FER of a typical floor is calculated as

$$FER = 100 \times (NRA/GFA) \qquad (1\text{--}2)$$

where NRA = net rentable area, sq ft.
GFA = gross floor area, sq ft.

A building with a high FER usually requires more central M/E space and a lower NGR. The trade-off between FER and NGR depends on many factors, such as site constraints, zoning restrictions on the height of a building, space programming, architectural geometry, etc. Normally, the FER of a high-rise office building with centralized M/E systems may be as high as 90 percent. An FER of 85 percent is considered an excellent design.

1.3.4 Geometric Factors

Two additional factors that can be used to evaluate the economics and energy effectiveness related to building geometry and form are the volume-to-surface ratio (VSR) and the area-to-perimeter ratio (APR).

Volume-to-surface ratio   The VSR is a ratio of the total volume of the building divided by the total exterior wall areas of the building. The total volume is the total floor area of the building multiplied by its average floor-to-floor height. Thus, the VSR is actually the floor area per unit area of exterior wall of the building. Naturally, a larger VSR represents a more efficient building geometry.

To be cost effective and energy efficient, the building geometry should minimize exterior surfaces (walls and roof) and maximize interior volume (floor area × height), and thus, the VSR. Mathematically, the optimum VSR of any geometric form is a sphere. However, for occupancy, a hemisphere such as a geodesic dome is the only practical equivalent to a sphere. A dome is ideal for an indoor sports arena, where the height of the dome can be fully utilized. A rectangular building has five exterior surfaces—four walls and one roof. The floor is excluded because it has nearly no energy transfer. Thus, from an energy gain or loss point of view, a rectangular building is a five-sided structure. The optimum rectangular building is a half cube, or semicube, where the height of the building is one-half of its square base. The semispherical and semicubical building configurations are illustrated in Figure 1–8.

It can be shown that any building with height greater than half its base dimension is less energy efficient and more costly to build than one with height less than or equal to half its base dimension. High-rise buildings are in this category, and the taller the building, the less is the energy efficiency. While no building will be built simply to achieve a better VSR, this criterion should not be overlooked when comparing alternative designs of a comparable size building.

The VSR can also be used to evaluate the optimum size of buildings by comparing smaller numbers of large buildings against larger numbers of small buildings. The answer is in favor of the larger buildings. For example, it should be obvious that the cost and energy consumption of (100) 5,000-sq-ft public housing units are considerably higher than those

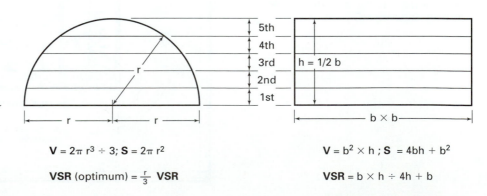

■   **FIGURE 1–8**
The optimum VSR of a building is either a semispherical dome with diminishing upper floor areas or a semicubical building with equal floor areas and height one-half its base dimension.

(*Note:* The optimum VSR is a variable depending on the building size.)

$V = 2\pi\, r^3 \div 3;\; \mathbf{S} = 2\pi\, r^2$

$\mathbf{VSR}\ (\text{optimum}) = \frac{r}{3}\ \mathbf{VSR}$

$V = b^2 \times h;\; \mathbf{S} = 4bh + b^2$

$\mathbf{VSR} = b \times h \div 4h + b$

$\mathbf{VSR}\ (\text{optimum}) = 0.167b$

of (10) 50,000-sq-ft housing units. While other factors, such as security, zoning, community relations, etc., most likely favor the smaller units, an assessment of the VSR is still a valid criterion. (For additional information on the subject, see the reference section at the end of the chapter.)

Area-to-perimeter ratio   The APR is related to the *aspect ratio* (AR) of the building, which is defined as the length (longer dimension) divided by the width of the building, assuming that most upper floors are similar. The APR is the typical floor area divided by the linear perimeter length of the floor, that is, the floor area per linear unit of the perimeter walls. Naturally, the larger the APR, the more is the building energy efficiency. A round or square building should have the optimum APR. On the other hand, a rectangular building with maximum daylight and minimum solar heat gain in the cooling season could override the APR factors. The final design will depend on a computer simulation of the HVAC and lighting load on an hour-by-hour basis. Figure 1–9 illustrates several APR values of a 10,000-sq-ft floor plan.

1.3.5 Impact on Building Geometry

The major influence of M/E systems on modern architecture has been not only in building height, but in architectural style, facade, form, and expression. Architectural and structural system interfacing has long been established in the history of architecture. Typical examples of modern architecture expressed by structural columns and cross-bracing members are the John Hancock Building in Chicago, built during the 1970s, and, more recently, the Bank of China Building in Hong Kong. Architectural and M/E system interfacing became a reality only when the Pompidou Center Art Museum in Paris daringly exposed all M/E system ducts, pipes, and conduits on the exterior of the museum. Since then, the Hong Kong–Shanghai Bank in Hong Kong, built in the 1980s, cre-

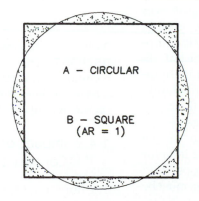

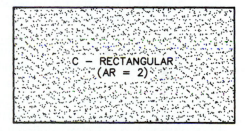

■ FIGURE 1–9

Comparison of the APR values of several geometric configurations.

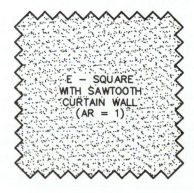

APR OF 10,000 SQ.FT. FLOORS			
DESIGN	AR	DIMENSIONS	APR
A	–	131' DIAM.	28.0
B	1	100' x 100'	25.0
C	2	142' x 71'	23.0
D	8	283' x 35'	15.7
E	1	100' x 100'*	17.7

* The sawtooth fenestration increases the perimeter by 41%

ated another sensation, with exposed structural and M/E elements as the main feature of its architectural design. In fact, most modern buildings are influenced by the presence of M/E systems, as evidenced by the following architectural styles:

1. *Penthouse style.* At the top of the roof is a smaller floor or screened structure to enclose the elevator equipment and elevator overtravel (the extension of the elevator shaft above the last floor to slow down the elevator, should it fail to stop as intended), cooling towers, exhaust fans, and other equipment. This is a fundamental design concept to conceal equipment on the roof and is the minimum building code requirement for many cities and countries.
2. *Flattop style.* The top one or two floors of a high-rise building are usually designed to house central M/E equipment and to conceal upper-level M/E equipment, such as cooling towers, air-handling units, and elevator equipment. The most notable examples of this type of building are the Twin Towers of the World Trade Center in New York.
3. *Intermediate floor bands.* High-rise buildings of over 30 stories are usually designed with one or more intermediate floors to house central M/E equipment. Due to the extra floor-to-floor height required for this equipment, the facade of these intermediate floors is usually treated differently than the adjacent occupied floors. Such floors are easily detected in most high-rise buildings.
4. *Signature buildings.* Postmodern design in the 1990s has deviated from the flattop style in favor of individuality and sculpture-type rooftops. With the sculptured roof design, cooling towers must be concealed in a different manner. They may be located at or near the ground level or may be concealed within the building.

Plates 1–3 (see color insert) show several world famous high-rise buildings and skyscrapers. The Sears Tower in Chicago (Plate 1) has multiple flattop roof lines as the main M/E equipment floors while the M/E floors of Far Eastern Plaza building in Taipei, Taiwan (Plate 2) on the lower, middle and top levels are readily distinguishable from the rest of occupied floors. The M/E floors for the Luijazui financial center in Shanghai, China (Plate 3) are well concealed behind its fine textured façade. Actually, there are principal M/E floors occurring at 15-floor intervals which also accommodate refuge areas during an emergency situation. Due to the unique linear roof line design, the cooling towers can only be located at the ground level.

1.4 THE IMPACT ON CONSTRUCTION COST

1.4.1 Impact of Building Height on Construction Cost

When the building is taller, it will require more time and hoisting equipment to raise the material onto the upper floors. In fact, most construction workers may have to stay on the upper levels during the entire workday, losing productivity. The structural and M/E systems will be more complex. The method of construction will be different. For a building taller than 10 stories, the unit cost per floor area will increase about 10 to 20 percent for the next 5 stories and another 10 to 15 percent for each additional 5 floors. For example, if the unit cost for a 10-story building is $100 per sq ft, then the cost for a 25-story building, using 10 percent as the incremental cost, would likely be calculated as follows:

$$\text{Average Unit Cost} = [(\$100 \times 10) + (\$110 \times 5) \\ + (\$121 \times 5) + (\$133 \times 5)]/25 \\ = \$116 \text{ per sq ft.}$$

Naturally, the incremental values vary with architectural style, construction material, and construction technique. Nevertheless, the general rule can serve as a base for determining the impact of building height on the overall construction cost.

1.4.2 Impact of M/E Systems on Construction Cost

The impact of M/E systems on construction cost varies greatly, depending on the type of building, standard of living of the country, architectural design, and M/E systems selected. The range of M/E systems costs for fully air-conditioned buildings is given in Table 1–2. These values may serve as a general reference from which to modify and to refine the costs throughout the design process.

1.4.3 Impact on Operating Costs

The operating cost of a building includes the cost of routine maintenance, repairs, replacements, and utilities. Most architectural and structural components of a building (except the roof) are normally longlasting, without the need for frequent replacement. This is not the case, however, for most M/E systems, which not only consume energy, but also require ongoing maintenance and repair. Indeed, on a life cycle basis, the cost of owning and operating M/E systems may out-

TABLE 1–2
Range of M/E system cost of buildings

Type of Occupancy	Percent of Total Building Cost		
	Low	Median	High
Computer centers	30	45	60
Department stores	20	25	30
Hospitals	30	40	50
Hotels	20	30	35
Offices	15	30	40
Research laboratories	30	40	50
Residential, single occupancy	10	15	20
Residential, high rise	15	20	25
Retail, individual stores	10	20	25
Schools, elementary	15	20	30
Schools, secondary	15	25	35
Universities and colleges*	20	30	40

*Buildings other than those used for classrooms follow the space required for specialty-type buildings such as laboratories, computer centers, residences, etc.

weigh the initial capital investment of the entire building! Naturally, the importance of efficient M/E systems and management cannot be overemphasized.

1.5 THE IMPACT ON HIGH-RISE BUILDING DESIGN

In high-rise building design, there is no single solution to a problem. Two buildings of similar size and configuration may favor different M/E systems and central plant locations when located on different sites. Climates and economic and cultural backgrounds of the country are other factors that affect the selection of an M/E system.

As indicated in Table 1–1, M/E system equipment takes considerable space. For example, the average M/E floor space in an office building is about 4 percent of the total building gross floor area. In other words, for a 25-story building with a gross floor area of 500,000 sq ft, 20,000 sq ft should be initially allocated as M/E equipment space. This is equivalent to one full floor of the building. Similarly, two floors are needed for M/E equipment in a 50-story building, and four floors in a 100-story building. Needless to say, these are not incidental spaces that can be added at will, except during the initial programming phase of the project. Optimum solutions are the result of close coordination between the architect and the M/E engineers.

Space for M/E equipment may be centralized or decentralized, depending on the system selected. With either plan, there is always the need for on-floor (local) equipment and distribution (shaft) space on every floor. The major difference is that centralized planning concentrates major equipment on one or two floors with smaller on-floor M/E spaces, whereas the decentralized plan is just the opposite.

Without exception, all high-rise buildings have one or more underground levels for utility service, delivery of supplies, fuel storage, etc. Furthermore, underground levels provide a better structural stability for the foundation of the building. More important, underground parking for automobiles is an unavoidable demand by most city codes. Normally, an M/E central plant should be located in an underground level. However, this is not always feasible, since M/E system risers must be close to the core of the building, which frequently hinders access to and from the parking garage. For this and other reasons, the M/E central plant may be located on the rooftop or on other, intermediate levels. Plate 9 shows hoisting the cooling tower to the roof, and the arrangement of M/E equipment in central equipment spaces. Factors affecting the location of a central plant include the following:

- Accessibility for loading/unloading equipment.
- Proximity to the outside air supply and exhaust air discharge.
- Adequacy of floor height.
- Interference with a convenient parking plan.
- Safety. Some equipment, such as boilers, chillers and liquid-filled transformers, contains considerable stored energy or toxic material. This equipment should be confined within fire-proof walls.
- Proximity of system components, such as between the chiller and the condenser with cooling towers.
- Ease of maintenance.
- Vibration and noise from equipment.
- Aesthetics. Last but not least, a central plant may detract from or enhance the architectural concept. A typical example is the Hong Kong–Shanghai Bank, with exposed structural and M/E elements as means of architectural expression.

Table 1–3 lists some of the reasons for locating the central plant on the lower or upper levels.

1.6 ENERGY AND ENERGY CONVERSION

1.6.1 Common Energy Sources

All buildings require electrical power, which is normally supplied by the electrical utility. When utility

TABLE 1–3

Reasons favoring upper or lower central plant locations

In favor of upper levels	In favor of lower levels
▪ When a clean architectural roof line is desired. Elevator overtravel requires that the penthouse protrude over the topmost floor. If the central plant is located at the top, the elevator penthouse can be contained within the same space. ▪ When the cooling tower must be located on the roof. ▪ When a fuel-fired heating plant is used. Vertical shaft space for the flue stacks on all floors is eliminated. ▪ When the outside air quality is a concern. In general, the air quality in urban settings is better at higher elevations than at ground level. ▪ When exhaust/relief air cannot be discharged near ground level, such as exhaust from chemical fume hoods. ▪ When locating the central plant at a lower level will interfere with an underground parking plan. ▪ When the reduction of roof heating/cooling load is significant; the central plant acts as a baffle zone.	▪ When cooling/heating energy is supplied from district sources remote from the building. ▪ When the rooftop is used for other purposes, such as swimming pools, a garden, a restaurant, a club, etc. ▪ When the construction schedule is such that earlier occupancy of the lower floors is desired. A lower floor location for the central plant allows an earlier start and completion of the M/E systems. ▪ When vibration and noise attenuation of the central plant is too costly.

power is not available, or when an on-site electrical power source is required, the supply of fuels such as gas, oil and even coal may need to be considered in the planning process. In addition, HVAC systems also require fuel for heating and cooling. Commonly used energy sources are listed in Table 1–4.

1.6.2 Efficiency of Energy Conversion Process

Energy exists in forms of heat, light, and chemical, sound, mechanical and nuclear energy. According to the law of physics, energy can be neither created nor destroyed, but can be converted from one form to another. Energy conversion is never 100 percent efficient. The loss is usually in the form of low-level (tem-

perature) heat, which is not readily useful. Table 1–4 indicates the normal conversion efficiency of common energy sources through the combustion process. The electric heat pump process utilizes electrical energy to "transport" heat energy from the building exterior to the building interior; it is really not an energy conversion process. The performance (transport efficiency) is called the coefficient of performance (COP), which may vary from 1.5 to 3.0. A COP of 2.0 means that the heat pump process can transport two units of heat energy for each unit of electrical energy consumed.

The efficiency of energy conversion depends on the datum selected for calculations at the building boundary or at the initial energy source, which differ drastically, as demonstrated in Figure 1–10.

TABLE 1–4

Common energy sources and conversion efficiencies

Fuel	Unit of[1] Measure	Nominal Heating Value/Unit, Btu (kJ)	Combustion Efficiency (%)
Natural Gas	Cu ft	1,000 (1,055)	70–85
LP (Propane Gas)	Gallon	93,000 (98,000)	70–85
No. 1 Oil (Diesel)	Gallon	138,000 (146,000)	75–80
No. 5 Oil (Heavy)	Gallon	145,000 (153,000)	72–82
No. 6 Oil (Bunker C)	Gallon	153,000 (161,000)	75–80
Soft Coal (Anthracite)	Pound	13,000 (14,000) 13,700 (14,800)	75–85
Hard Coal (Bituminous)	Pound	12,500 (13,500) 13,200 (14,300)	75–85
Electrical Resistance	kW-hr	3,413 (3,600)	90–95
Electrical Heat Pump	kW-hr	5,100 (10,200) 5,400 (10,800)	1.5–3.0[2]

[1]1 gallon = 3.78 liters; 1 cubic foot = 28.32 liters; 1 pound = 0.454 kilograms
[2]Denotes coefficient of performance (COP) per unit

Plate 1 Sears Tower in Chicago, USA. (Roof: 1450 ft., antenna base: 1515 ft., 110 stories) *Owner:* Sears-Roebuck & Co. *Architect/Engineers:* Skidmore, Owings & Merrill, USA.

Plate 2 Far Eastern Plaza in Taipei, Taiwan. *Owner:* Far Eastern Company, Taiwan. *Architects:* P&T, Hong Kong and C.Y. Lee & Partners, Taiwan. *M/E Engineers:* William Tao & Associates, USA; H.C. Yu & Associates, USA; and Continental Engineering Consultants, Taiwan.

Plate 3 Luijazui Financial Center in Shanghai, China. (1509 ft., 95 stories) *Owner/Developer:* Mori Building, Japan. *Architects:* Kohn Pedersen Fox, USA; Mori Building, Japan; and East China ADRI, China. *M/E Engineers:* Shimizu Company, Japan.

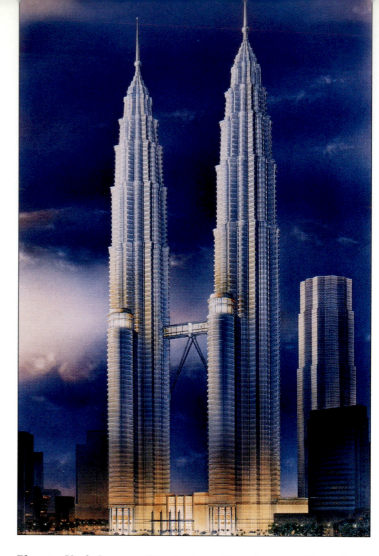

Plate 5 Kuala Lumpur City Center in Kuala Lumpur, Malaysia. (1483 ft., 88 stories) *Architects:* Kuala Lumpur City Center Berhad and Cesar Pelli & Associates, USA. *M/E Engineers:* Flack & Kurtz, USA, and Tenaga Preece, Sdn Bhd, Malaysia.

Plate 4 T. C. Tower in Kao Hsiung, Taiwan. *Owner/Developer:* Tuntex Corporation, Taiwan. *Architects:* Hellmuth, Obata & Kassabaum, USA, and C.Y. Lee & Partners, Taiwan. *M/E Engineers:* William Tao & Associates, USA, and Continental Engineering Consultants, Taiwan.

Plate 6 Pierre Laclede Center in St. Louis, USA. *Owner/Developer:* Connecticut Life Insurance, USA, and Nooney Company, St. Louis, USA. *Architects:* Smith & Entzeroth, USA. *M/E Engineers:* William Tao & Associates, USA. See Plate 9 for M/E equipment photos.

Plate 8 Southwestern Bell Telephone Company in St. Louis, USA. *Architects:* Hellmuth, Obata & Kassabaum, USA. *M/E Engineers:* William Tao & Associates, USA.

Plate 7 New York World Trade Center, USA. (110 stories, 1350 ft.) *Owner:* Port Authority of NY & NJ. *Architects:* Minoru Yamasaki Associates. *Mechanical Engineer:* Jaros Baum Bolles, NY. *Electrical Engineer:* Joseph R. Loring, NY.

Plate 9 M/E equipment and cover photos: Pierre Laclede and Equitable Buildings. *M/E Engineers:* William Tao & Associates, USA.

Plate 10 Gateway Arch and Westward Expansion Museum in St. Louis, USA. *Owner:* National Park Service, USA. *Architect:* Eero Saarinen, USA. *M/E Engineers (museum/arch HVAC upgrade):* William Tao & Associates, USA.

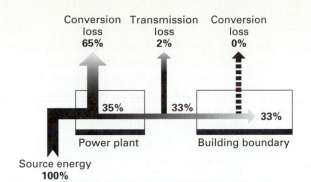

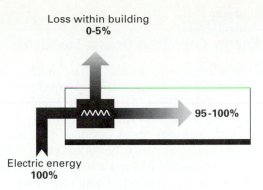

(a) **Energy efficiency of electric heating**
(From energy source)

Energy efficiency of resistance heating
(Within building boundary)

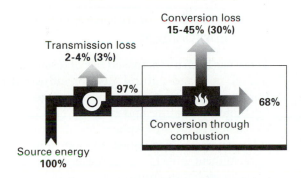

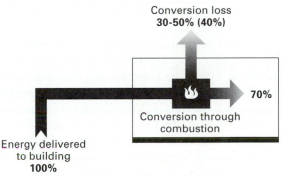

(b) **Energy conversion of combustion heating**
(From energy source)

Energy efficiency of combustion heating
(Within building boundary)

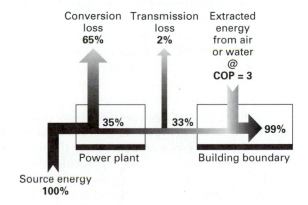

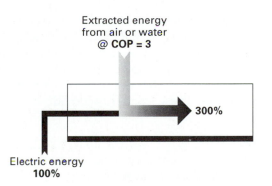

(c) **Energy efficiency of electric heat pump**
(From energy source)

Energy efficiency of heat pump
(Within building boundary)

■ **FIGURE 1–10**

Energy conversion efficiencies of the heating process. The efficiencies differ drastically at the energy source and at the building boundary. (a) *Electric heating:* The conversion is between 95% to 100% efficient since nearly all electrical energy is converted into useful heat within the building. However, for each unit of electrical energy (kilowatt-hour), approximately three units of fuel (coal, gas, or oil) energy is consumed at the power plant. Thus, the overall energy conversion efficiency is only 33%, when using energy source as the datum. (b) *Combustion heating:* To burn fuel in a building heating plant (boiler or heater), the efficiency is only about 70% at the building boundary, but it is nearly the same at the energy source when a small transportation loss is included. Thus, with building boundary as the datum, electrical heating is most efficient, but when using energy source as the datum, combustion heating is more efficient. (c) *Electrical heat pump:* To use electrical energy to pump heat from outside air or underground water to heat the space is called the heat pump process. The COP may be as high as 300% within the building boundary, and nearly 100% at the energy source.

1.6.3 Energy Conscious Design Standards

Prior to the energy crisis of the 1970s, energy consumption in buildings was often more than two times greater than in the 1990s. The annual energy consumption of an office building in the Midwest has typically been reduced from around 150,000 to 200,000 btu per sq ft per year to about 60,000 to 75,000 btu per sq ft per year. With continuously improving building design and technology, energy consumption data will continue to drop. Table 1–5 provides the range of energy consumption data of typical building types (not average figures). These values are primarily based on the author's database and the research findings of the AIA Research Corporation funded by the US Department of Energy. Due to the wide variation of building size, climate, construction methods, architectural styling, and operating schedules, annual energy consumption data spreads widely. Nevertheless, the tabulated data should be useful in identifying relative values. Energy analysis using sophisticated computer simulation programs is necessary to determine the energy consumption for a specific building. These programs are readily available from private and public sources.

The American Society of Heating, Refrigeration and Air-Conditioning Engineers, Inc. (ASHRAE) and the Illuminating Engineering Society of North America (IESNA) have jointly developed energy-efficient design standards known as Standard 90, published in 1975, revised in 1980 and revised in 1996. These standards have been widely adopted as part of the building codes in the United States as well as in a number of other countries. These standards provide three alternative methods of achieving compliance: a prescriptive method, system performance method and building energy cost budget method. The standards are available to design professionals and building owners as a guide to energy-efficient design in new and existing buildings. Building owners and designers should strive to meet or exceed these standards.

1.7 THE IMPACT OF BUILDINGS ON GLOBAL ENVIRONMENT

1.7.1 Impact of Technology on Global Environment

There has been an increased use of fossil fuels, including oil, coal and natural gas, for industrial production, transportation, electrical power generation and for building heating and cooling. This has increased CO_2 levels in the atmosphere and its rate of absorbing solar energy. This phenomenon, known as the "greenhouse effect," is suspected as the primary cause of global warming.

Up to the 1990s, carbon fluorocarbons (CFCs) have been the primary refrigerant for refrigeration and air-conditioning systems. CFCs are also used as cleaning and aerosol agents. Their increased use has caused the destruction of the ozone (O_3) layer in the lower part of the stratosphere, 15–20 miles above the earth. The ozone layer helps shield much of the ultraviolet energy that penetrates down to the earth's surface. With the destruction of the ozone layer, more harmful cancer-causing ultraviolet energy reaches the earth. International agreement has banned the future use of CFCs and promotes the use of alternative chemicals such as hydrochlorofluorocarbons (HCFCs) and hydrofluorocarbons (HFCs) which have lower ozone depletion effect. However, these chemicals are themselves by no means harmless. The most effective way to reduce the environmental damages from man-made pollutants is to minimize the use of energy through more efficient design and controls.

1.7.2 Air Pollutants Due to Building Energy Consumption

According to statistics, the United States consumes more than two trillion kilowatt-hours of electrical energy annually, about one-third by buildings. If the annual energy consumption in buildings is reduced by a mere 20 percent through better design and management, total CO_2 emissions may be reduced by about 150 million tons, the equivalent of taking 30 million

TABLE 1–5

Typical annual energy consumption targets for various building occupancies in the USA

Building Occupancy[a]	Annual Energy Consumption	
	Btu/ft^2-yr	kJ/m^2-yr[b]
Office buildings	60,000	5,300
Multi-family housing	50,000	4,400
Retail stores	60,000	5,300
Shopping centers	70,000	6,200
Hotels/motels	70,000	6,200
Elementary schools	40,000	3,500
Secondary schools	50,000	4,400
Warehouses	35,000	3,100
Assemblies (arenas, stadiums)	60,000	5,300
Health clinics	50,000	4,400
Nursing homes	70,000	6,200
Hospitals	130,000	11,500

[a]Building sites are based on midwestern parts of the USA representing the average condition of the country.
[b]1 Btu/sq ft/yr = 11.36 kJ/m²/yr and rounded.

TABLE 1–6

Air pollutants produced from energy conversion

Energy Converted or Consumed	Air Pollutants Produced, grams (lbs)		
	CO_2	SO_2	NO_x
1 gallon of fuel oil by combustion[a]	10,500 (23.1)	45.0 (0.10)	18.3 (0.04)
1 gallon of gasoline by automobiles[b]	8500 (18.8)	37.0 (0.08)	15.0 (0.03)
1 pound of coal by combustion[c]	1090 (2.4)	9.0 (0.02)	4.4 (0.01)
1 therm of natural gas by combustion[d]	6350 (14.0)	Nil (—)	24.0 (0.05)
1 kW-hr electric energy generated by oil[e]	860 (1.9)	3.7 (0.008)	1.5 (0.003)
1 kW-hr electric energy generated by gas[e]	635 (1.4)	Nil (—)	2.4 (0.005)
1 kW-hr electric energy generated by coal[e]	1090 (2.4)	9.0 (0.02)	4.4 (0.01)

Notes: [a]Calculated by using fuel oil containing 85 percent carbon, and 12 percent hydrogen; and 7.4 lb/gal.
[b]Calculated by using gasoline mixture of C_8H_{18} and (C_nH_{2n+2}) having 84 percent carbon, 15 percent hydrogen and 6.1 lb/gal.
[c]Calculated by using bituminous coal containing 65 percent carbon and 3.8 percent sulfur.
[d]Calculated by using mixture of methane (CH_4) and ethane (C_2H_6) and 100,000 Btu/therm.
[e]Data from Green Light Program, Environmental Protection Agency, USA.

cars off the road. The impact of building energy conservation on the environment could not be more dramatically demonstrated.

Table 1–6 lists air pollutant byproducts from energy conversion processes. For every kilowatt-hour of electrical energy (3413 Btu equivalent) consumed in a building, about 10,000 Btu of fuel is burned at a coal-fired power generation plant, releasing 2.4 pounds (1.09 kg) of CO_2, 0.02 pounds (9 g) of SO_2, and 0.01 pounds (4.4 g) of NO_x.

The amount of air pollutants due to combustion can be calculated by knowing the chemical composition of the fuel and the chemical reactions which occurred during combustion. In the combustion process, the chemical elements of fuel react with oxygen where carbon, having a molecular weight of (12), hydrogen (1), sulfur (32), and nitrogen (14) react with oxygen (16) to form CO_2 (44), SO_2 (64) and NO_x (varies). For every unit weight of carbon combusted, (44/12) or 3.66 unit weights of CO_2 are generated. When the chemical content of a fuel is known, then the amount of air pollutants can be easily calculated. Table 1–6 lists the amount of air pollutants generated due to energy conversion or consumption. The amount is staggering, as shown in the example.

1.7.3 Example:

A large, 600,000-sq-ft office building demands 8 W/sq ft (0.008 kW/sq ft) of electrical power for its heating, air conditioning, lighting, plumbing, fire protection, elevators and other equipment. If the building operates about 4000 hours a year, the annual electrical energy consumed will be:

$$600,000 \text{ sq ft} \times 4,000 \text{ hr/yr} \times 0.008 \text{ kW/sq ft} = 19,200,000 \text{ kW-hr/yr}$$

Assuming that the utility's power-generating plant is oil fired, then from Table 1–3, the amount of air pollutants generated by the power plant that is attributable to electrical usage in the office building is as follows:

- Carbon dioxide (CO_2): 19,200,000 × 1.9 = 36,400,000 lb/yr
- Sulfur dioxide (SO_2): 19,200,000 × 0.02 = 384,000 lb/yr
- Nitrous oxide (NO_x): 19,200,000 × 0.01 = 192,000 lb/yr

Measured by any scale, the amount of pollutants attributable to the operation of this single building is staggering. If a 10 percent energy conservation can be achieved through better design and controls, then the CO_2 emission alone will be reduced by 3,600,000 lb/yr, and the amounts of other pollutants proportionately.

1.8 SYSTEM INTERFACING

The various topics covered in this chapter—from space planning and other architectural concepts to the global environment—are all interrelated. Decision in one area will affect all the others. Figure 1–11 illustrates the need for close coordination and interfacing between the various design disciplines revolving around architectural and engineering decisions.

The design of modern buildings requires a team effort. No single designer or professional knows all of the solutions. The field of building design and construction is analogous to the health care field, where the internist is the prime coordinator of a patient, with the support and consultation of specialists such as a cardiologist, radiologist, ophthalmologist, den-

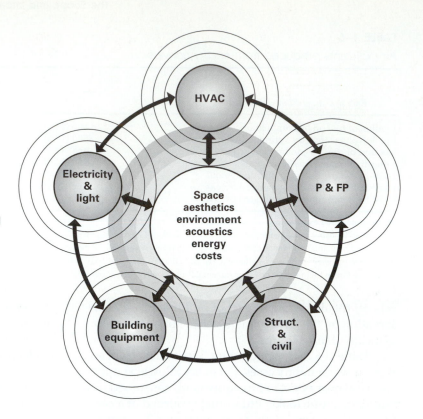

■ FIGURE 1–11
Interfacing is necessary among the design disciplines to achieve a balanced solution as to optimum spatial relations, aesthetics, environmental quality, acoustics, energy efficiency, and cost-effectiveness.

tist, pathologist, etc., as well as the pharmacist, researcher, manufacturer, and hospital administration. In building design and construction, the architect is the prime professional on a project, with the support and consultation of specialists such as civil, structural, mechanical, electrical, lighting, and acoustical engineers as well as contractors and construction managers. Only through a close interface among all professionals can a building achieve its optimum value, balancing functions, cost, and quality.

1.9 ENVIRONMENTALLY RESPONSIVE AND INTEGRATED DESIGNS

With the continued deterioration of our environment and consumption of our limited natural resources, we have pushed the environment to the threshold at which serious design alternatives must be considered and pursued. The nations of the world have gathered together to discuss these alternatives and to develop a global framework for addressing them, with the consensus that it is the duty of everyone to seek the means to conserve our resources by reducing energy consumption.

Historically, for a variety of economic, technological, or logistical reasons, building designs have not addressed environmental issues to their fullest.

However, the general population and various governments are seeking to make a committed change. This change has encouraged design teams, builders and users to be more responsible and flexible when establishing building programs. The result has been the application of "green" issues to buildings, that is, creating environmentally responsive and integrated designs.

1.9.1 The Architectural Façade

A high percentage of the energy consumed in a building is due to cooling loads responding to solar (radiation) or conductive/convective heat gains. These gains are from the occupants and equipment in the building but more so by way of solar gains through the external façade or "skin" of the building.

The conventional wall or "single-skin" façade (Figure 1–12) typically controls the solar gain by controlling the extent of transmittable surfaces, or windows/vision glazing in the building, or by adding reflective coatings and tints to the glass surfaces to reduce solar load. These solutions are effective, but are not optimal since they permit solar radiation to reach and penetrate the glazing. An equal concern from the user's point of view is that, due to the reduced sizes of openings or the quality of coatings and tints, there is a reduction in natural daylighting enter-

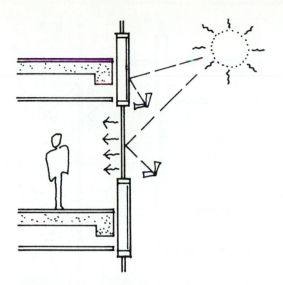

■ **FIGURE 1–12**
Single-skin façade.

ing the building and a reduction in true or clear views to the exterior environment.

The most effective way of controlling solar gain is to prevent the solar radiation from ever touching and heating the glazing surface. If this can be achieved, then less cooling loads (due to reduced heat gains) will be required, and less energy consumption will take place. Typical shading solutions following this approach include the incorporation of external shading devices, with a horizontal orientation for a high-angle sun and a vertical one for a low-angle sun.

To be effective, overhangs should have sufficient depth and cut-off angle. However, excessively deep overhangs or, alternatively, frequently spaced over-

hangs, may be considered visually distracting by causing "tunnel vision" (Figures 1–13 and 1–14).

An additional problem with such external shading solutions is that they can be technically difficult to resolve, especially in buildings such as high rises where wind forces can be extreme, requiring highly engineered and uneconomical viability solutions (Figure 1–15). More recent solutions involve the "cavity" or "double-skin" façade (Figure 1–16). The principle here is that the cavity or zone between the two skins forms a buffer between the conditions of the exterior climate and the interior workplace. This zone can be utilized in a variety of ways to relieve the impact of the outside environment on the internal workplace. The buffering or moderating effect, in turn, leads to a wider range of M/E possibilities for the conditioning of the space—possibilities that cannot be considered in the conventional single-skin façade. Specifications for M/E solutions can be lowered or low tech in nature, leading to a reduction in energy consumption.

The foremost advantage of the double skin is that, regardless of whether a building is a low-rise or high-rise structure, the solar radiation can be effectively prevented from reaching the transmittable surface. The protection of the outer skin permits the use of more conventional shading devices such as blinds. The obvious advantage of the blind is that it is an economical alternative to an external shading device. But equally important is that it offers the user or the tenant choice of lowering or raising the blinds when protection from the sun is not required; by contrast, tinted glasses and external shading devices are permanent.

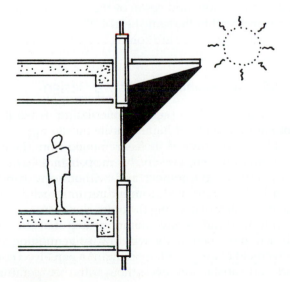

■ **FIGURE 1–13**
External shading.

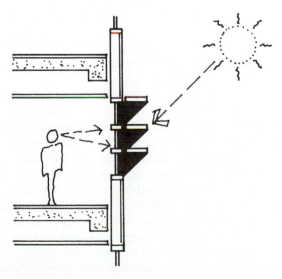

■ **FIGURE 1–14**
External shading.

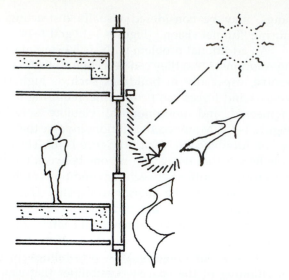

■ FIGURE 1–15
Shading and wind.

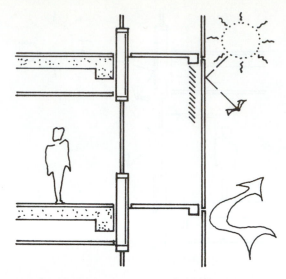

■ FIGURE 1–16
Double-skin façade.

1.9.2 The Technically Integrated Building

In simple terms, a double-skin façade functions as and can be compared to an insulated glass unit (IGU), or double glazing, something that we automatically expect to use in all our designs because of the "thermal buffering" it provides. As with any IGU, heat builds up in the cavity between the skins, so a careful assessment must be made to ensure that there are no excessive temperatures which may lead to failure of the unit or breakage of glass. If one imagines expanding or pulling apart the skins and recognizes this heat buildup and the notion of thermal buffering, then the design opportunities become apparent.

Double-skin designs can take numerous approaches; however, there are two that are typically considered with cavities ranging from 10 to 18 inches (250 to 460 mm) in depths up to 3 feet (915 mm) (Figures 1–17 and 1–18).

First, controlling the temperature of the cavity can reduce the need for mechanical conditioning of the internal space. Blinds prevent the impact of direct solar radiation. Venting the cavity in summer can prevent heat from building up on the inner skin which, through convection, would warm the space and increase cooling loads. Similarly, retaining the heat buildup in the cavity during winter assists in warming the inner skin and often leads to little or no additional heating requirements. This reduces not only operating costs and energy consumption, but also capital costs. Alternatively, it allows the use of unitary electrical heaters in lieu of central heating systems, which require higher capital investment.

The cavity can also be utilized as a "thermal flue," drawing air (naturally or mechanically) from the internal space, which would normally require a ducted or ceiling plenum solution. This use of the cavity can reduce air distribution components (ducts and shafts). However, thermal calculations must be conducted to ensure that the humidity transferred from the internal space will not lead to condensation within the cavity.

Utilizing the flue aspect (this could also be an atrium) as a form of distribution can also lead to lower floor-to-floor heights by reducing distribution zone dimensions in the ceiling; lower heights mean less construction (and costs) or the ability to squeeze more area within the same height. This is particularly relevant where there are zoning height restrictions.

1.9.3 The Socially Interactive Design

In certain locales, there is increasing frustration among occupants of highly mechanically appointed buildings. The occupants feel removed from the natural environment, effectively trapped in a hermetically sealed environment and without any control over it. The cavity wall concept permits such people the possibility of opening the inner skin of the wall, as a window. Studies have demonstrated that the simple action of opening the window for ventilation and movement of air gives the occupant a perceived comfort and satisfaction, even though the temperatures experienced from the cavity may be higher than the conditioned space. By acknowledging this user par-

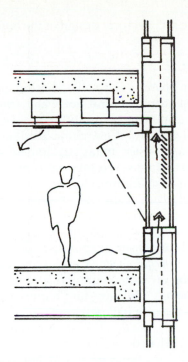

■ **FIGURE 1–17**
Double-skin cavity, 10″–18″.

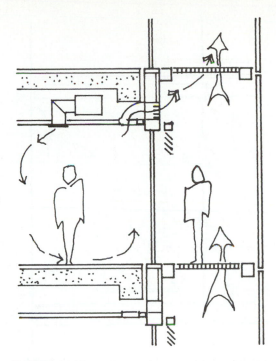

■ **FIGURE 1–18**
Double-skin cavity, (3′-0″).

ticipation factor, it may be possible for designers actually to raise the temperature criterion currently established for internal spaces, again reducing cooling loads and energy consumption. The pressure balance between floors in medium or high-rise buildings must be carefully evaluated to control infiltration, exfiltration, smoke and fire migration (this will be addressed in the HVAC and Fire Protection chapters).

Taking the "user participation factor" further by expanding the cavity into occupiable spaces in the forms of communal atriums or ecocenters (Figure 1–19) provides an alternative means to taking the elevator down 30–40 floors to get a breath of fresh air. These spaces, in turn, can make use of plants to consume CO_2, or can be utilized as chambers or plenums to mix fresh air with exhaust, thereby reducing vertical ductwork distribution and providing better zoning possibilities within the building.

1.9.4 Alternative Technologies

Glass with photovoltaic cells   The glass industry has made great advances in recent years to explore environmentally responsive roles. The integration of photovoltaic cells technology into glazing units is an excellent example of utilizing solar energy by means of the vast surfaces that typically enclose a building structure.

Variable transmittance of glazing material   The development of a glass which is translucent or opaque depending on the intensity of the sun, reducing solar transmittance as well as glare, is in the early research stage.

Radiant ceiling   In certain geographical locations, the use of chilled ceilings may be an alternative solution. This method utilizes the distribution of moderately cool water through pipes or panels much the same way as a heated ceiling. Unlike a fan coil or VAV system which uses low temperature water, this system utilizes water in its natural temperature state without refrigeration.

Wind generation   Wind generators in the form of propeller blades or catenary structures are readily used in open field sites to generate electricity from wind power. However, as technology advances to reduce the size of these devices, it may be feasible to utilize this form of free energy on high-rise buildings where winds are inherently strong at upper elevations.

Thermal storage   In various forms such as chilled water, ice, hot water or liquid refrigerants are increasingly used to reduce the peak building power demand. The stored energy will indirectly reduce the need for the construction of new power plants and

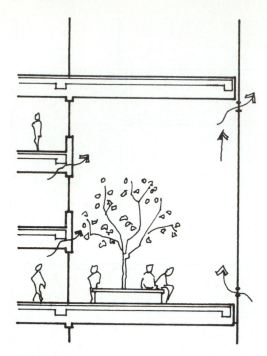

■ **FIGURE 1–19**
Communal atrium concept.

their impact on the greater environment. The methods and techniques of thermal storage are discussed in Chapters 2 to 7.

1.10 CHECKLIST OF BUILDING AND M/E REQUIREMENTS

This section presents a comprehensive checklist that serves to determine the scope of building operational requirements and from which one can determine the scope and criteria of M/E systems. The checklist also will be valuable in formulating the architectural concept, building configuration, space programming, and opportunities of system interfacing. Early identification of these requirements will aid the architect in evaluating construction costs, as well as the allocation of space for M/E equipment, both within and exterior to the building.

1.10.1 Mechanical Systems

Building mechanical systems include heating, ventilating, and air conditioning (HVAC), plumbing and sanitation (P&S), fire protection (FP), and specialty or auxiliary systems. A comprehensive list of services is presented below. Obviously, not all buildings require all services; thus, the list should be tailored to the needs of a specific project. Line items followed with an asterisk (*) are especially important to pin down during the conceptual phase, as they have a major im-

pact on the architectural design of the building's roof lines.

HVAC Systems

- *Energy source* Gas, oil, electrical power, coal, central steam, central hot water, chilled water, etc. (location and capacity)
- *Heating/cooling* Central air handling, direct radiation, in-space unitary equipment, etc.
- *Comfort controls* Number of control zones, humidity, temperature, etc.
- *Central plant* Estimated normal (or base) and standby capacities, etc.
- *Heat rejection (*)* Water cooling tower, air-cooled condenser, evaporative cooler, etc.
- *Location of equipment* Central equipment rooms, on floor, on roof, on ground, etc.
- *Ventilation* Outside air requirements (minimum, high, or 100%)
- *Exhaust (*)* General, food preparation, toxic and special exhaust systems, etc.
- *Automation* Building automation system (BAS), building management system (BMS)

Plumbing and Sanitation Systems

- *Energy source* Electrical power, gas, oil, central steam, hot water, etc.
- *Water supply* Public water, well, etc.; water pressure, capacity available, and location
- *Hot water supply* Hot water heaters or heat exchangers
- *Sewage disposal* Storm, sanitary, or combination sewers, sewage treatment plants, septic tanks, drainage and filtering fields
- *Storm water drainage* Roof, area, and means of discharge; locations
- *Subsoil drainage* Drainpipes, sumps, pumps, and discharge, etc.
- *Building facilities* Plumbing fixtures, water, waste, soil, piping

Fire Protection Systems

- *Energy source* Electrical power, gas, oil
- *Water supply* Flow rate and available pressure at water main, location. Separate service or combination with plumbing water supply
- *Water storage (*)* Lake, pond, storage tanks (locations and capacities)
- *Fire and smoke detection* Thermal and smoke detectors
- *Fire containment* Fire shutters, compartmentalization
- *Smoke containment and evacuation* Smoke exhaust and pressure controls
- *Stairway smoke prevention* Stair pressurization

- *Fire annunciation* Fire alarm, public address, fire department connections
- *Fire extinguishing* Portable extinguishers, automatic sprinklers (water, mist, dry chemical, foam, special gases, etc.)
- *Fire fighting* Fire hose and standpipe systems, Siamese connections
- *Lightning protection* Air terminals, grounding conductors, etc.

1.10.2 Electrical Systems

Building electrical systems include power, lighting, and auxiliary systems. The proliferation of electrical and electronic systems in building applications has greatly expanded the scope of electrical systems and has had a drastic impact on construction costs and the complexity of planning. The general scope of the electrical systems is listed below. The list should be expanded or condensed to fit the needs of a specific project.

Power Systems

- *Normal energy source* Utility power or on-site power (location and capacity); power characteristics (phase and voltage); service entrance (overhead, underground); service requirements (substations, transformer vaults); etc.
- *Emergency power source* Separate utility service or on-site standby generators (location and capacity)
- *Interior power distribution* Primary or secondary voltages, unit substations, distribution panels, etc.
- *On-floor distribution* Floor boxes, under-floor ducts, integrated cellular floors, raised floors, ceiling-cavity conduit network, etc.
- *Emergency power distribution* For critical equipment and emergency lighting loads.
- *Uninterruptible power systems (UPS)* For critical building operations such as computers and communication networks; power storage (battery banks).
- *Power for building systems* HVAC, plumbing, sanitary, fire protection, etc.
- *Power for building operational equipment* Food service, waste disposal, laundry, garage, entertainment equipment, etc.
- *Power for vertical transportation systems* Interface with elevator consultant on power and controls for elevators and escalators

Lighting Systems

- *Basic light source* Incandescent, fluorescent, high intensity discharge (HID), etc.

- *Illumination* Lighting levels, color rendering, controls
- *Lighting fixtures* In offices and other work spaces
- *Architectural lighting* Interface between architect, lighting and/or electrical consultant on public or special spaces
- *Introduction of daylight* Fenestration, skylights, controls, etc.
- *Exit lighting* Exit signs, exitway (evacuation route) lights
- *Exterior lighting* Site, landscape, building façade, aircraft warning lights, etc.

Auxiliary Systems

- *Telephone and telecommunication* Type, number of lines and stations, switchboard (manual, PBX), basic and special features, facsimile, modem, etc.
- *Data distribution systems* Multiple conductor cables, twisted pairs, coaxial cables, Fiber optic cables, wire closets, etc.
- *Public address* Intercom, paging and music systems
- *Audio/video* Radio, TV, and signal distribution systems
- *Satellite dishes (*)* Number, diameter, and orientation
- *Transmission (*)* Transmitter and microwave towers
- *Cable* CCTV distribution systems, locations and interfacing with other auxiliary systems
- *Time and signal* Clock and program systems
- *Fire detection and alarm systems* Interface with fire protection consultants
- *Automatic controls* Interface with HVAC and other building service consultants
- *Security systems* CCTV monitoring, detecting, alarming, controlling, and interface with security consultant
- *Specialty systems* Numerous specialty systems for hospitals, research facilities, computer centers, and industrial, military, or defense facilities, as applicable.

QUESTIONS

1.1 Name some features in modern buildings which are dependent on mechanical and electrical systems.

1.2 Prior to the installation of modern M/E systems, buildings designed for work places, such as offices, are usually limited to a few simple building configurations. Why?

1.3 What is VSR? What is the optimum VSR of a building?

1.4 What are the major categories of building M/E systems?

1.5 What are the major mechanical systems?

1.6 What are the major electrical systems?

1.7 What is the median M/E space to be allowed initially for an office building? Research building? Secondary school?

1.8 Building codes prohibit M/E components, such as ducts, piping, conduits, etc. to penetrate through structural beams. (True) (False)

1.9 What is NRA? How is it defined?

1.10 Central M/E systems tend to improve the FER. (True) (False)

1.11 Why does building design E (a sawtooth curtain wall) in Figure 1–9 have an APR of 17.7, whereas design B (a square) has an APR of 25?

1.12 A rectangular building of low APR is always less energy efficient than a building of higher APR regardless of its orientation and climatalogical conditions (True) (False)

1.13 What is the APR of a 30 ft. $\times$ 30 ft. single-story building?

1.14 There is no strict definition of a high-rise building. The consensus is that buildings seven stories or higher are classified as high-rise buildings. (True) (False)

(Note: Seven stories, or approximately 70 feet to the top floor, were originally considered as reachable by fire truck ladders. With modern fire-fighting equipment, this limitation has long been exceeded.)

1.15 Why is VCR a good measure of construction and energy efficiency for comparing buildings of similar floor areas?

1.16 High-rise buildings have a low VSR and are thus less energy efficient. (True) (False)

1.17 What is the median cost for M/E systems (in percent of total building construction) for an elementary school?

1.18 If the central M/E plant is located on the upper floor of a high-rise building, there will be no need to have any M/E space on the ground or below ground levels. (True) (False)

1.19 Give reasons that favor a central M/E plant on upper levels of a high-rise building.

1.20 What is the heating value of one kW-hr of electrical energy?

1.21 What is the nominal conversion efficiency by burning gas in a boiler?

1.22 An electric heat pump can have an energy efficiency (coefficient of performance, or COP) up to 300%. (True) (False)

1.23 CFC refrigerants are being replaced because they are too expensive. (True) (False)

1.24 How much carbon dioxide will be released into the atmosphere by using 1 gallon of oil in a boiler?

1.25 If a building consumes 40,000 kW-hr of electrical energy per year, how many pounds of CO_2 are released at the coal-fired electrical power generating plant?

1.26 Modern building design process requires close interfacing between the design professionals. Normally, the architect coordinates the team effort. (True) (False)

1.27 Create a comprehensive checklist of M/E scope and criteria for a small office building, say, for 20 occupants. (Note: This is purely a self-test. Do what you think is right at this time, and doublecheck your list when you have completed all the chapters.)

Do the same for a 100-room hotel. (Everyone has stayed in a hotel before. You should be able to generate some basic questions that are relevant to the operation of a hotel). (Again, this is your own self-test.)

REFERENCES

1. Ralph Knowles, *Energy and Form: An Ecological Approach to Urban Growth.* Cambridge, MA: MIT Press, 1974.

2. William Tao and Danniel Ku, *Building Form and Energy.* Paper presented at the National Conference on Energy Performance Standards for Buildings, sponsored by AIA Research Corporation, Department of Energy, and Department of Housing and Urban Development, 1978.

3. Eugene L. Smithart, *CFC's: Today There Are Answers.* La Crosse, Wisconsin: Trane Company, 1992.

4. United States Environmental Protection Agency (EPA), Green Light Program publications. Washington, DC.

5. William Tao, *Energy-Effective Design with the Constraints of Preservation.* Paper presented at the National Conference on Building Redesign and Energy Challenges, sponsored by American Institute of Architects, 1980.

6. William Tao and H. Clay Laird, "Modern Boiler Plant Design." *Heating/Piping/Air Conditioning,* November 1984.

7. William Tao and Richard Janis, "Modern Cooling Plant Design." *Heating/Piping/Air Conditioning,* May 1985.

8. William Tao, "Cooling System Performance." *Heating/Piping/Air Conditioning,* May 1984.

9. William Tao and Danniel Ku, "Design Considerations for Large Electrical Systems," *Specifying Engineer,* June 1986.

10. ASHRAE/IESNA Standard 90.1–1989, Energy Efficient Design of New Buildings except Low-Rise Residential Buildings. Atlanta: 1995.

2

HVAC FUNDAMENTALS

THIS CHAPTER CONTAINS BASIC INFORMATION RE-garding comfort, properties of air, load estimation, and determining the proper flow of heat transfer fluids to satisfy loads. This information is prerequisite for understanding how HVAC systems operate to control the environment.

The next five chapters describe, in turn, the following concepts, subsystems, and equipment used in HVAC systems:

- HVAC delivery
- Cooling production
- Heating production
- Air handling
- Piping systems

Figure 2–1 shows typical equipment used in an HVAC system for a large building. Illustrated are the subsystems described in the five chapters that follow.

2.1 ENVIRONMENTAL COMFORT

2.1.1. Comfort for Occupants

The temperature of a space is not the only factor affecting a person's comfort. Even if the temperature is within an acceptable range, the space may seem warm if the humidity is too high, the airflow is too low, or warmth is being radiated to the occupants.

Conversely, a space may seem cool if the humidity is low, the space is drafty, or warmth is being radiated from the occupants to cold surfaces. Comfort for building occupants is affected by a number of environmental variables, including the following:

- Temperature
- Airflow
- Humidity
- Radiation

Indoor air quality is another aspect of comfort. Good air quality includes the presence of sufficient oxygen and the absence of objectionable impurities such as dust, pollen, odors, and hazardous materials.

Different sets of conditions may be comfortable, depending on the type of activity that goes on in a space. Appropriate conditions for an office would be too warm for a gymnasium and too dry and cool for a natatorium. Expectations must also be considered: Saunas are hot on purpose, and a wide variety of conditions is commonly tolerated in factories. The physical conditions of the occupants, including their age and health, also affect their comfort. Even the seasons affect comfort: Warmer environments are tolerated during summer, cooler in winter, due to clothing and acclimatization.

Economics and concerns about energy conservation are also considered in defining comfort. People will be satisfied with less comfort when faced with a

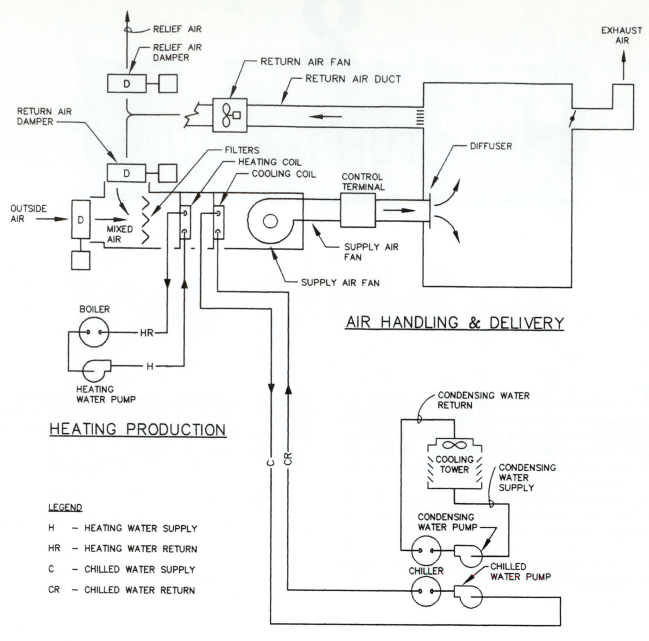

- **FIGURE 2-1**

Components of a large HVAC system. (Based on hot-chilled water system.)

worthy cause or a mandate based on sound business practice.

2.1.2 Temperature and Humidity

Both temperature and humidity affect our sense of comfort. Figure 2–2 shows the acceptable range of each for persons clothed in typical summer and winter clothing at night during sedentary activities. The

lower comfort limit in cold weather is 68°F at about 30 percent relative humidity (RH), and the upper limit in hot weather is 79°F at about 55 percent RH. HVAC systems are generally designed to maintain temperature and RH within a tight range.

An interior design temperature of about 75°F is considered comfortable by most people in general-use spaces as shown in Figure 2–2. During summer, a slightly higher temperature may be appropriate to

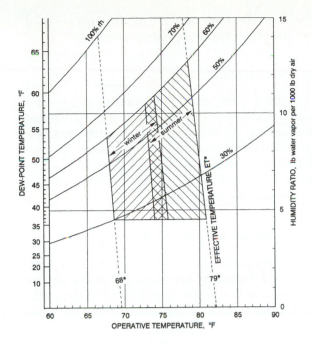

■ **FIGURE 2–2**
Standard effective temperature and comfort zones diagram. (Courtesy: ASHRAE.)

consider in designing air-conditioning systems because of light clothing and acclimatization to warm weather. Conversely, slightly cooler temperatures are acceptable for consideration in the design of heating systems. Most air-conditioning systems are designed to maintain a summer temperature of 72°F–78°F. During winter, heavier clothing and acclimatization to cold weather results in a recommended design temperature of 68°F–72°F for heating systems. These interior design temperatures will be appropriate for the majority of buildings.

Humidity in excess of 60 percent is considered high in general-use spaces. Humidity lower than 25–30 percent can result in uncomfortable drying of breathing passages and problems with static electricity.

2.1.3 Airflow

Systems must be designed for adequate airflow to prevent complaints of "stuffiness" or drafts. The measure of airflow is velocity. Space air velocities less than 10 feet per minute will be stuffy; those more than 50 feet per minute will seem drafty.

2.1.4 Air Quality

Systems must provide sufficient amounts of clean air to keep oxygen levels at an acceptable level and to dilute contaminants generated within occupied spaces. Air should be reasonably free of dust and spaces free of odors or other pollutants that might be hazardous

or objectionable. These conditions are generally achieved through the use of filters and by the introduction of outside air into the system at rates specified in Table 2–8.

2.1.5 Radiant Effects

Even if the temperature, humidity, and airflow in a space are acceptable, the space may be uncomfortable due to radiant effects from cold windows or walls. Systems must therefore compensate for these effects with radiant heat or higher temperatures. Similarly, cooler temperatures or higher air velocities will be needed to offset the effects of warm surfaces. Downdrafts from cold surfaces are also uncomfortable and can be offset by proper placement of heating devices.

2.1.6 Special Considerations

Buildings such as hospitals, computer rooms, and laboratories have special requirements for temperature, humidity, airflow, and air quality. In some instances these requirements are consistent with comfort of the occupants, while in others they are at odds with comfort.

Interior environmental criteria are often set by specifications for equipment used within an occupied space. Computer rooms, for example, are often drafty and cool to suit the environmental requirements of the computing equipment. This will be an uncomfortable environment for operators of the computers, however, and special provisions may be desirable to provide better conditions in certain areas of the room. In a similar manner, materials stored in a warehouse may tolerate cold or hot temperature, but the warehouse employees need a refuge of human comfort.

Economics and expectations of comfort also affect design criteria. Despite the fact that warehouses and factories are occupied by people, it is deemed unnecessary to maintain these buildings at the same interior temperatures as an office building or hospital. The need for energy conservation may also temper expectations of comfort.

2.2 PROPERTIES OF AIR–WATER MIXTURES

The design of environmental control systems relies on an understanding of the properties of air, including temperature and humidity. These properties af-

fect loads on buildings, and HVAC systems are used to alter the properties and produce comfort.

2.2.1 Psychrometry

Psychrometry is the study of properties of air-water mixtures. The psychrometric chart is a convenient source for data on the properties of such mixtures. Figure 2–3 shows how important properties are presented on the psychrometric chart. Figure 2–4 is a complete chart that can be used in analysis of processes associated with HVAC.

2.2.2 Absolute and Relative Humidity

Two basic properties of air-water mixtures are temperature and humidity. The humidity of the air can be expressed in two ways: absolute and relative. Absolute humidity, also known as the humidity ratio

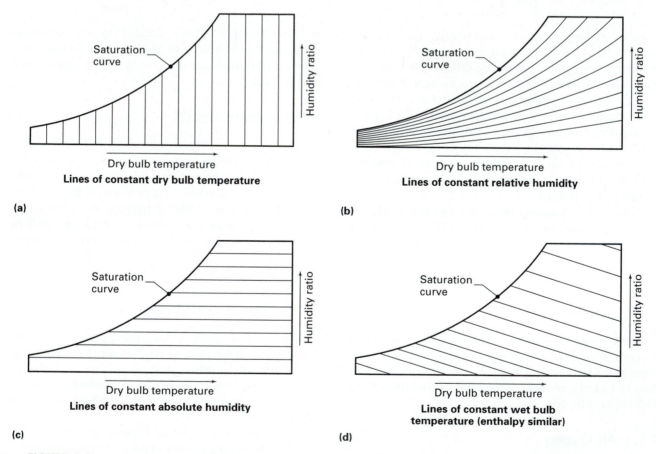

■ **FIGURE 2–3**
Lines representing major properties of air-water mixture on the ASHRA psychrometric chart. (a) Vertical lines: constant dry-bulb (DB) temperature. (b) Curved lines: constant relative humidity (RH). (c) Horizontal lines: constant humidity ratio (W), also commonly referred to as absolute humidity (d) Sloped lines: constant wet-bulb (WB) temperature; lines with same slope: constant enthalpy (h).

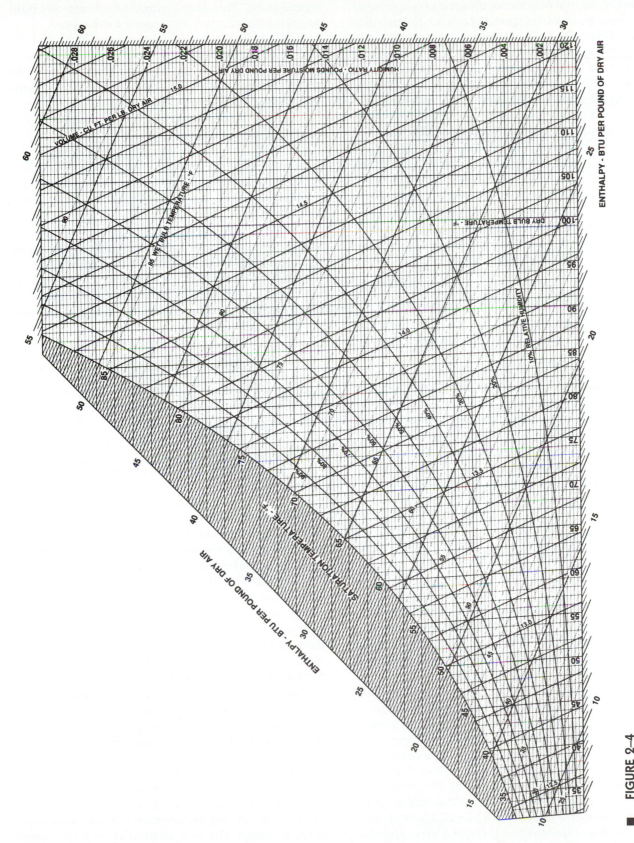

■ **FIGURE 2–4**
The psychrometric chart. (Reprinted with permission of ASHRAE.)

(W), is the amount of water in the air and is measured in grains or pounds per pound of dry air. Relative humidity (RH) is the ratio of the actual water content to the maximum possible moisture content, expressed in percent. If the air is currently holding all the moisture possible, the relative humidity will be 100 percent, and the air is termed *saturated*.

2.2.3 Effect of Temperature on Humidity

The moisture-holding capacity of the air will depend on the air temperature. Warm air will hold more moisture than cold air. For this reason, the same absolute humidity will result in different relative humidities at different temperatures. The psychrometric chart illustrates the relationship among temperature and absolute and relative humidity.

2.2.4 Wet-bulb Temperature

If a wet sock is placed over the bulb of a conventional thermometer, a lower temperature will be recorded due to evaporative cooling. The dryer the air, the more effective will be the evaporative cooling, and the lower will be the temperature measured. If the air is saturated, then there will be no evaporation, and the wet-bulb thermometer will measure the same temperature as a dry-bulb thermometer. By measuring both dry-bulb and wet-bulb temperatures, the temperature and humidity of the air can be determined. A combination of wet- and dry-bulb temperature represents a discrete point on the psychrometric chart.

2.2.5 Sensible, Latent, and Total Heat

Air contains thermal energy in two forms: temperature and water vapor. Water vapor, or humidity, in the air possesses the water's latent heat of vaporization (approximately 1000 Btu/lb of water). Temperature is a measure of sensible heat, while water vapor content is a measure of latent heat. Total heat—the sum of sensible and latent heat—is termed *enthalpy*, symbolized by the Greek letter eta, or H. High temperature or high humidity constitute high energy.

On the psychrometric chart, horizontal movement is associated with sensible heat change (no change in absolute humidity), and vertical movement is associated with latent heat change (no change in temperature). Moving upward or to the right indicates a higher energy level; moving downward or to the left indicates a lower energy level. Lines of constant enthalpy slope upward and to the left at approximately the same slope as lines of constant wet-bulb

temperature. This is no coincidence, for wet-bulb temperature is a good measure of total energy.

Often, changes in air conditions result in changes in both humidity and temperature. The net change in energy level, or enthalpy, can be determined by plotting the initial and final conditions on the psychrometric chart, as shown in Figure 2–5.

2.2.6 Sensible Heating and Cooling

Sensible heating (cooling) occurs when an air-water mixture is raised (lowered) in temperature, but the absolute moisture content remains the same. Each process occurs as air in spaces is warmed or cooled by building loads. Each also occurs in systems as they respond to compensate for loads. For instance, room air might be cooled at an outside wall during cold winter weather. To compensate, a heater at the base of the wall might warm the air. Sensible heating or cooling is represented by a horizontal movement along the psychrometric chart.

2.2.7 Processes Involving Latent Heat

Heating and cooling represent a transfer of sensible heat; humidification and dehumidification represent a transfer of latent heat. The amount of moisture liberated or absorbed by air is measured by its initial and final absolute humidities.

Humidification can be accomplished either by adding dry steam to or by evaporating moisture into the air. If dry steam is added, the air will have a higher energy level, taking on the latent heat of the steam. (There will also be a slight increase in temperature due to the sensible heat of the steam, but the effect is small and generally ignored in practice.) On the psychrometric chart, this process is represented by a vertical movement.

If water is evaporated into the air, the final energy level will not change. The air will cool to provide the heat required for vaporization of the water. The sensible heat loss will equal the latent heat gain, resulting in constant enthalpy. This process is called *adiabatic* saturation. (No energy is added or removed.) Evaporative humidification is accompanied by evaporative cooling and is represented on the psychrometric chart by an upward movement along a line of constant enthalpy, coincidentally a line of constant wet-bulb temperature.

Cooling is a method for dehumidifying air. If moist air is cooled to the saturation curve, further cooling will not only reduce temperature, but also remove moisture. The temperature at which moisture will begin to condense is termed the *dew point*. Liquid

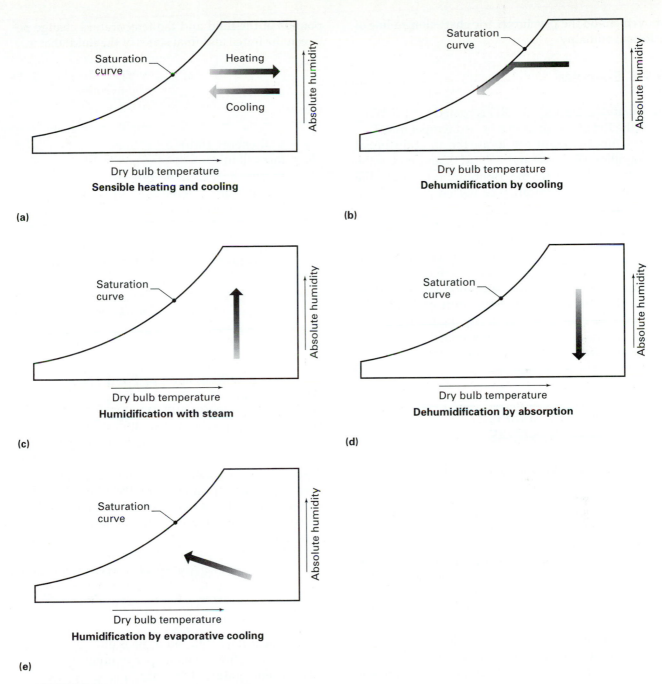

Saturation
curve

■   **FIGURE 2–5**
Basic psychrometric process. (a) Sensible heating and cooling—horizontally.
(b) Dehumidification by cooling—follows saturation curve. (c) Humidification with
steam—up vertically. (d) Dehumidification by absorption—down vertically.
(e) Humidification by evaporative cooling—follows the wet-bulb line.

moisture removed from the air by this process is termed *condensate.* The air that results from the process will be both cooler and less humid than it was initially.

Dehumidification can also be accomplished by absorption. Some substances are *hygroscopic,* meaning that they absorb moisture. Hygroscopic substances, or desiccants, such as silica gel and lithium bromide, are used in certain applications to absorb moisture from the air. As moisture is condensed in the desiccant, its latent heat is liberated, resulting in heating of the air. Absorption is represented by a downward movement on the psychrometric chart along a line of constant enthalpy.

2.2.8 Examples

1. Air at 70°F DB and 75% RH is heated to 84°F. What is the RH of the air at this heated temperature?

    Ans. In Figure 2–4, locate the air at the initial condition (70°F DB and 75% RH), and follow the horizontal line to the right until it meets the 84°F DB line (vertical). The heated air is now at 47% RH.

2. Air at 90°F DB and 70% RH is being cooled down to 75°F. What is the relative humidity?

    Ans. In Figure 2–4, from the intersecting point of 90°F DB and 70% RH, draw a line to the left. This line will meet the saturation curve at 79°F, which is the dew point temperature of the air. The air is then cooled further, following the saturation curve until it stops at 75°F. Between 79°F and 75°F, the air is saturated, and moisture condenses out of it. The air now has 100% RH.

2.3 ENERGY TRANSPORT IN HVAC SYSTEMS

2.3.1 Heat Transport by Fluid Flow

Figure 2–1 shows that HVAC systems use fluids to transport heat and cold to satisfy loads and maintain comfort. Such fluids include air, water, steam, and refrigerant. Equations are developed in this section that can be used to determine heat transport based on the flow rate and the initial and final conditions of the fluid. These equations can also be used to determine flow rates or conditions, based on requirements for heat transport.

Heat is measured in British thermal units, or Btus. A Btu is the amount of heat required to raise 1 lb of water 1°F. The rate of heat flow is measured in Btus per hour, or Btuh. Fluids are used to transport heat in HVAC systems.

2.3.2 Heat Transport by Sensible Heating and Cooling

The natural property of a fluid that affects heat transfer is called *specific heat*, which is the amount of energy required to raise 1 lb of a substance 1°F. The specific heat of water is 1.0; that of air is 0.24. The heat liberated from a quantity of fluid is equal to the specific heat of the fluid multiplied by the number of pounds of the fluid and the temperature change between the initial and final states of the fluid; that is,

$$q = m \times C \times TD \tag{2–1}$$

where

q = heat energy (Btu)
m = mass (lbm)
C = specific heat (Btu/lb·°F))
TD = temperature difference, °F (final minus initial temperature)

HVAC equipment loads, equipment capacity, and output are expressed as quantities per unit time, or rates:

$$Q = M \times C \times TD \tag{2–2}$$

where

Q = heat flow (Btu/hour, or Btuh)
M = mass flow (lbm/hour)
C = specific heat (Btu/(lb·°F))
TD = temperature difference, °F

Heat transfer in water Mass flow is quantified in units of gallons per minute, or GPM. Knowing that 1 gallon of water has a mass of 8.35 lbs and that there are 60 minutes in 1 hour, the following equation can be derived:

$$Q = 500 \times GPM \times TD \tag{2–3}$$

where

Q = heat flow (Btu/hour, or Btuh)
GPM = water flow (gallons per minute)
TD = temperature difference, °F

Heat transfer in air Mass flow is quantified in units of cubic feet per minute, or CFM. An equation for heat transfer in air can be derived given that the density of air at standard pressure is .076 lb per cubic foot and that the specific heat of air is 0.24 Btu/lb·°F:

$$Q = 1.1 \times CFM \times TD \tag{2–4}$$

where

CFM = airflow (cubic feet per minute)

2.3.3 Heat Transport by Fluid Phase Change

Heat transfer in steam   In steam, heat is liberated by a change of phase from vapor to liquid. One lb of steam liberates approximately 1000 Btu as it condenses. For steam, heat flow is approximately by the equation

$$Q = 1000 \times SFR \qquad (2\text{–}5)$$

where

SFR = steam flow rate (lb per hour)

Heat transfer in refrigerants   Refrigerants absorb heat by changing phase from liquid to gas. The heat absorbed will be equal to the latent heat of vaporization, measured in Btu per lb, times the refrigerant flow rate, measured in lb per hour.

2.3.4 Selecting Fluid Flow Rates for HVAC Systems

HVAC systems and subsystems are designed to satisfy heat loads by using heat transport fluids. Fans, pumps, boilers, and distribution elements are sized according to flow requirements, which must be determined by the HVAC designer. The first step is to estimate building heat loads. Methods for estimating loads are presented later in the chapter. Once they are estimated, the HVAC designer must decide the proper combination of flow and conditions for fluids used to transfer heat and compensate for loads. Initial and final conditions are generally selected on the basis of accepted general practice found to achieve acceptable results. The equations developed above can be used to calculate flow, given the heat transfer requirement and the initial and final conditions of the fluid.

Water flow   For devices using hot water for heating, a supply temperature of 160°F might be chosen and the load equipment selected to allow a 20° drop in water temperature, resulting in a 140° return temperature. Once this decision is made, the required water flow rate can be calculated as

$$GPM = Q/(500 \times TD) \qquad (2\text{–}6)$$

Similarly, chilled water can be used for cooling. Chilled water supply temperatures between 40° and 50°F are common for building HVAC applications, and systems are designed for water temperature rises ranging from 10° to 15°F. The same equation applies for determining the required chilled water flow rate.

Airflow   For systems using warm air for heating, supply temperatures between 105° and 140°F will be appropriate to maintain a space at, say, 75°F. Given a space temperature and a selected supply temperature, the required airflow rate can be calculated as

$$CFM = Q/(1.1 \times TD) \qquad (2\text{–}7)$$

Systems using chilled air for cooling generally employ supply air temperatures between 50° and 60°F. The preceding equation can be used to determine the airflow rate required to satisfy the sensible portion of cooling loads. Once the airflow rate is determined, the humidity can be checked by plotting on a psychrometric chart.

Steam flow   The rate of steam flow (lb per hour) required to satisfy a given heating load is determined by the equation

$$Steam\ flow = Q/1000 \qquad (2\text{–}8)$$

Refrigerant flow   The rate of refrigerant flow for cooling is determined by dividing the cooling load by the latent heat of vaporization.

The foregoing concepts and equations are used to estimate theoretical fluid flow rates required to meet a given load. The resulting estimates are the basis for sizing the piping and duct systems, along with the pumps and fans required to transport heat and cold.

2.4 HVAC LOAD ESTIMATION

2.4.1 Nature of HVAC Loads

The design of HVAC systems starts with an estimation of the loads the system must satisfy. Heating loads determine how much heat is lost and must be made up by the system. Cooling loads determine how much heat is gained and must be removed. Humidity must also be considered. Internal and external sources of moisture gains and losses need to be compensated in maintaining proper humidity levels. Estimates of these quantities involve heat transfer and conversion into and within the building.

2.4.2 Methods for Estimating Loads

Many methods are available for calculating heating and cooling loads for buildings. All, however, should be considered estimates, the precision of which depends on how the method accommodates the non-homogeneous qualities of building assemblies and contents and the non-steady-state nature of building loads. Heat transfer in building systems is a dynamic process, with ever-changing loads from outside and within the building.

2.4.3 Accuracy and Precision

Components of the building envelope are not the ideal elements described in elementary heat transfer texts. Building elements and contents are complex assemblies with both series and parallel paths for heat flow. They absorb and release heat at various rates in response to loads that change over time, and there is no method that will model the process with 100-percent certainty.

Appreciating the relative accuracy of calculations is especially important, since loads must be estimated to select HVAC system components before the building itself is fully designed, much less constructed. To accommodate the variations that can be expected during building design and construction, sizing equipment and systems must be somewhat conservative. It would be impractical to design and construct the building and then measure load, and size the HVAC system. This is not to say that precision is unwarranted, and some calculational methods are more precise than others. Generally, however, the more precise the method, the more laborious it is.

A few buildings are designed to meet an outside temperature criterion corresponding to a median of extremes. For heating systems, this is the mean of the coldest recorded temperatures. The median value has as many annual extremes above as below it. This is a stringent criterion that may be appropriate when system performance is critical, during the coldest temperatures. For instance, performance is important in a hospital due to the condition of the occupants. Also, the occupants will be present during the early morning hours, when the lowest outside temperatures generally occur.

For most buildings, criteria need not be so stringent. If the inside temperature of an office building falls a few degrees lower than the intent of the design, no great harm results. In addition, the schedule of office occupancy is such that the lowest outside temperature is not coincident with many people being in the building. For most buildings, outside design temperatures are selected on the basis of a percent concept.

For heating load calculations, the percent concept assumes a heating season of December through February, which equals 2160 hours per year. Design conditions are tabulated according to percentages which imply that, statistically, the weather will be colder than the listed condition only for the specified percentage of 2160 hours per year. For example, for a 97.5 percent value listed as 6°F, statistically, the temperature can be expected to be below 6°F for 2.5 percent of 2160 hours, or approximately 50 hours per year. Data are listed for criteria of 99 percent and 97.5 percent. The former criterion is more stringent than the latter.

Cooling load criteria are similar, but use different percentage values, including 1 percent, 2.5 percent, and 5 percent. If a 5 percent design criterion is used, we can expect that, on average, 5 percent of the summer will be warmer than anticipated by the load calculation, and analogously for 2.5 percent and 1 percent. Summer is defined as June through September, a total of 2928 hours per year.

Extremes of hot weather are expressed as the dry-bulb temperature and the mean coincident wet-bulb temperature. The wet-bulb temperature is also listed in the table according to percent of occurrence. This information is provided for specifying equipment for which performance is sensitive to the wet-bulb temperature. Such equipment includes cooling towers, evaporative condensers, and evaporative coolers.

Accepted methods for calculating heating and cooling loads are documented in the ASHRAE *Handbook of Fundamentals,* which is revised periodically. ASHRAE methods have become increasingly precise and are practical only with the use of computer programs. In what follows, we describe the basic principles of calculating heating and cooling loads using a simple method that can be performed with a pencil, calculator, and tables. In most cases, the results will be conservative. Greater precision can be achieved by computer analysis using more complex methods. However, precision should not be confused with ultimate accuracy. Safety factors and allowances for unknown developments need to be applied in sizing HVAC systems and components.

2.4.4 Critical Conditions for Design

The designer must select an appropriate set of conditions for the load calculation. Relevant conditions include the outside weather, solar effects, the inside temperature and humidity, the status of building operations, and many other factors.

For heating, the critical design condition will occur during cold weather, at a time when there is little or no assistance from radiant solar energy or internal heat gains from lights, appliances, or people. The selection of an appropriately cold outside air temperature for design is an important decision for the designer.

For cooling load calculations, the critical design condition will be the peak coincident occurrence of heat, humidity, solar effects, and internal heat gains from equipment, lights, and people. The position of the sun varies by season and through the day, as does the weather. Building operations also vary. Sometimes, several estimates must be performed for different times to determine the highest combination of individual loads.

2.4.5 Temperature Criteria

The inside temperature and humidity are set by criteria based on expectations of comfort, as discussed earlier.

Outside weather conditions affect heating and air-conditioning loads from infiltration of air and conduction of heat through the building envelope. Anticipated extremes of temperature and humidity will be the basis for design load calculations here. Statistical data compiled for locations throughout the world are used by HVAC designers.

Table 2–1 shows design temperatures for selected locations throughout the United States and abroad. The table includes data not only on weather, but also on latitude, longitude, and relative wind severity, as well as other useful information. Latitude is important in estimating solar loads. Wind conditions are important for judging the severity of infiltration through a building envelope.

The outside temperature criteria used to calculate the required capacity of heating and cooling systems depend on the nature of the building. If it is essential that the system be capable of always meeting demand, the designer might assume the coldest re-

corded temperature. Buildings are seldom designed according to that criterion, however.

2.5 CALCULATING HEATING LOADS

The first step in the design of HVAC systems is to calculate loads. Heat losses include conducted loads through building envelope elements (walls, windows, roof, etc.) and outside air loads from leakage (infiltration and exfiltration) and ventilation air. (See Figure 2–6.) The amount of ventilation air required depends on the type of occupancy and the number of occupants contemplated. See Table 2–8 for ASHRAE-recommended occupant density and ventilation air quantity.

2.5.1 Manual vs. Computer Calculations

Building load calculations are done almost exclusively by computer programs. Many good programs are available. Most use data, algorithms, and methods developed by ASHRAE. Manual load calculations should be considered only for preliminary design or simple buildings. For that purpose, the simplified methods presented here will tend to be appropriately conservative.

2.5.2 Conduction

Heat transfer by conduction is proportional to the temperature difference between the warm and cold sides of the building envelope element. Inside temperature will depend on design criteria for either personal comfort or manufacturing process, whichever is

TABLE 2–1

Climatic conditions in the United States of America
(reprinted with permission from ASHRAE)

Col. 1	Col. 2		Winter, °F Col. 3		Main	Summer, °F Col. 4 Design Dry Bulb and Coincident	Wet Bulb		Design Wet Bulb Col. 5			Temp, °F Col. 6 Median of Annual Extremes	
State and Station	Lat.		Design Dry Bulb										
	°	′	99%	97.5%	1%	2.5%	5%	1%	2.5%	5%	Max.	Min.
ALABAMA												
Birmingham AP	33	34	17	21	96/74	94/75	92/74	78	77	76	98.5	12.9
CALIFORNIA												
Los Angeles CO (S)	34	3	37	40	93/70	89/70	86/69	72	71	70	98.1	35.9
Oakland AP	37	49	34	36	85/64	80/63	75/62	66	64	63	93.0	31.8
MISSOURI												
St. Louis AP	38	45	2	6	97/75	94/75	91/74	78	77	76		

(For convenience, only a portion of Table 2–1 is shown here. The full table is located at the end of the chapter.)

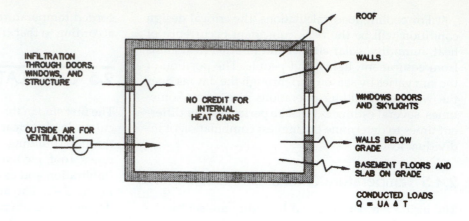

■ **FIGURE 2–6**
Components of building heating load.

applicable, and weather conditions—such as outside air temperature, wind velocity and humidity—selected for design.

Not only is conduction proportional to the difference between the outside and inside temperatures, but it is also proportional to the area through which the heat is transferred; that is, twice the wall begets twice the heat transfer. Conduction also depends on the insulating quality of the wall, which is measured by resistance to heat transfer, or the R-value. Envelope elements are made up of several layers, and heat must flow through each layer in sequence. The insulating value of a total assembly is the sum of the R-values of each component. The higher the resistance, the lower is the heat transfer. Heat transfer, temperature difference, area, and resistance are related by the equation

$$Q = A \times TD/R \qquad (2\text{–}9)$$

where

Q = heat transfer (Btuh)
A = area of assembly (sq ft)
R = resistance (h·ft^2·°F/Btu)

The susceptibility of an assembly to heat flow is called the U-factor, or U, which is mathematically the reciprocal of R. Thus, heat transfer is inversely proportional to R and directly proportional to U. Using U, the preceding equation becomes

$$Q = U \times A \times TD \qquad (2\text{–}10)$$

where

U = U-factor (Btu/h·ft^2·°F)

This equation is used in heating load calculations to estimate conduction heat loss through a wall, roof, or window. For air-conditioning load calculations, different equations must be used that consider the ef-

fects of the heat of the sun, as well as the outside air temperature.

2.5.3 Estimating U-factors for Building Assemblies

Resistance to heat flow and the U-factor for a wall or roof can be calculated using thermal properties for the elements that comprise the assembly. The sum of the resistances of individual layers of the assembly will be the total resistance for the assembly.

Thermal properties of commonly used building materials are listed in Table 2–2. For some materials, resistance is tabulated by the inch. For example, concrete has a resistance of 0.08 per inch. Thus, 8 inches of concrete has a thermal resistance of 8 × 0.08, or 0.64.

For certain commonly used modules, resistances are tabulated for the module. An 8-in concrete masonry unit has a resistance of 1.11. Some insulating materials are specified according to their R-values, such as an R-13 fiberglass batt.

Air spaces contained in building assemblies offer significant resistance to heat flow. Table 2–3 can be used to estimate this resistance. Note that the resistance of an air space depends not only on the thickness of the space and orientation of the heat flow, but also on the emissivity of surfaces facing the air space. Emissivity is a measure of a surface's ability to reflect and absorb radiant heat. A film of air clings to the surface and has a resistance to heat flow that depends on the thickness of the film. In still air, the film will be thick. If wind is present, its thickness will be less. The direction of heat flow (up, down, or horizontal) will also affect the resistance of the air film. Resistances of air films are listed in Table 2–4.

Some assemblies have a nonuniform construction, such as stud walls with insulation between framing. The U-factor of the assembly will be the average based on areas of the different constructions.

TABLE 2–2

Thermal properties of typical building and insulating materials
(reprinted with permission from ASHRAE)

Description	Density lb/ft^3	Conductivity λ $Btu \cdot in/h \cdot ft^2 \cdot F$	Conductance (C) $Btu/h \cdot ft^2 \cdot F$	Resistance (R) Per inch thickness $(1/\lambda)$ $h \cdot ft^2 \cdot F/Btu$	Resistance (R) For thickness listed $(1/C)$ $h \cdot ft^2 \cdot F/Btu$
BUILDING BOARD					
Boards, Panels, Subflooring, Sheathing					
Woodboard Panel Products					
Asbestos-cement board	120	4.0	—	0.25	—
Asbestos-cement board .... 0.125 in	120	—	33.00	—	0.03
Asbestos-cement board .... 0.25 in	120	—	16.50	—	0.06
Gypsum or plaster board .... 0.375 in	50	—	3.10	—	0.32
Gypsum or plaster board .... 0.5 in	50	—	2.22	—	0.45
Gypsum or plaster board .... 0.625 in	50	—	1.78	—	0.56
Plywood (Douglas fir)...	34	0.80	—	1.25	—
Plywood (Douglas fir).. 0.25 in	34	—	3.20	—	0.31
Plywood (Douglas fir)....................................... 0.375 in	34	—	2.13	—	0.47
Plywood (Douglas fir).. 0.5 in	34	—	1.60	—	0.62
Plywood (Douglas fir)....................................... 0.625 in	34	—	1.29	—	0.77

(For convenience, only a portion of Table 2–2 is shown here. The full table is located at the end of the chapter.)

TABLE 2–3

Thermal resistances of air spaces (h-sq ft-F/Btu)
(reprinted with permission from ASHRAE)

Position of Air Space	Direction of Heat Flow	Thickness of Air Space (inches) 0.5	0.75	1.5	3.5
Horizontal	Up	0.91	0.93	0.97	1.03
45° Slope	Up	1.02	1.00	1.04	1.06
Vertical	Horizontal	1.13	1.18	1.12	1.14
45° Slope	Down	1.15	1.26	1.27	1.27
Horizontal	Down	1.15	1.30	1.49	1.62

Values are listed for 0° mean temperature, 20° temperature difference, and effective space emittance of 0.82.

Figures 2–7 through 2–10 illustrate the procedure for calculating the total assembly resistance and *U*-factor.

2.5.4 Infiltration

The HVAC system must have sufficient capacity to heat or cool air infiltration at windows and entries and through loose construction. Infiltration loads will depend on the amount of outside air leakage and the difference in conditions between the outside and inside air. For heating load calculations, the following equation, derived earlier in the chapter, is used:

$$Q = 1.1 \times CFM \times TD \qquad (2\text{–}11)$$

If the space must be humidified, the humidification load can be determined by comparing the absolute

TABLE 2–4

Surface conductances (Btu/h-sq ft-F) and resistances (h-sq ft-F/Btu) for air
(reprinted with permission from ASHRAE)

Position of Surface	Direction of Heat Flow	Surface Emittance Nonreflective h	Surface Emittance Nonreflective R
STILL AIR			
Horizontal	Upward	1.63	0.61
Sloping—45°	Upward	1.60	0.62
Vertical..............................	Horizontal	1.46	0.68
Sloping—45°	Downward	1.32	0.76
Horizontal	Downward	1.08	0.92
MOVING AIR		h	R
(Any position)			
15-mph wind (for winter)	Any	6.00	0.17
7.5-mph wind (for summer)	Any	4.00	0.25

humidities of the outside air and the space. Absolute humidity (*W*) is expressed in terms of grains of water per pound of dry air. Knowing that there are 7000 grains per pound and that the density of dry air is approximately 0.076 pound per cubic foot, we can derive the following equation for latent heat:

$$Q = 0.68 \times CFM \times (W_{final} - W_{initial}) \qquad (2\text{–}12)$$

where

W = absolute humidity (grains per lb dry air)

These coefficients are expressed in Btu per (hour) (square foot) (degree Fahrenheit difference in temperature between the air on the two sides), and are based on an outside wind velocity of 15 mph

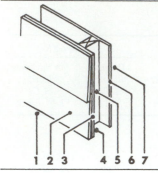

	Replace Air Space with 3.5-in. R-11 Blanket Insulation (New Item 4)				
		1		**2**	
		Resistance (R)			
Construction		Between Framing	At Framing	Between Framing	At Framing
1. Outside surface (15 mph wind)		0.17	0.17	0.17	0.17
2. Siding, wood, 0.5 in.× 8 in. lapped (average)		0.81	0.81	0.81	0.81
3. Sheathing, 0.5-in. vegetable fiber board		1.32	1.32	1.32	1.32
4. Nonreflective air space, 3.5 in. (50 F mean; 10 deg F temperature difference)		1.01	—	11.00	—
5. Nominal 2-in. × 4-in. wood stud		—	4.35	—	4.35
6. Gypsum wallboard, 0.5 in.		0.45	0.45	0.45	0.45
7. Inside surface (still air)		0.68	0.68	0.68	0.68
Total Thermal Resistance (R)		R_i=4.44	R_s=7.78	R_i=14.43	R_s=7.78

Construction No. 1: $U_i = 1/4.44 = 0.225$; $U_s = 1/7.81 = 0.128$. With 15% framing (typical of 2-in. × 4-in. studs @ 16-in. o.c.), $U_{av} = 0.8(0.225) + 0.15(0.128) = 0.199$ (See Eq 9)
Construction No. 2: $U_i = 1/14.43 = 0.069$; $U_s = 0.128$. With framing unchanged, $U_{av} = 0.8(0.069) + 0.2(0.128) = 0.081$

■ **FIGURE 2–7**
Sample calculation of U-factor for frame walls.

Coefficients are expressed in Btu per (hour) (square foot) (degree Fahrenheit difference in temperature between the air on the two sides), and are based on an outside wind velocity of 15 mph

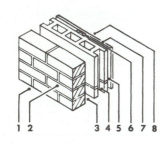

	Replace Furring Strips and Gypsum Wallboard with 0.625-in. Plaster (Sand Aggregate) Applied Directly to Concrete Block-Fill 2.5-in. Air Space with Vermiculite Insulation, 7-8.2 lb/ft³ (New Items 3 and 7)			
		1		**2**
		Resistance (R)		
Construction		Between Furring	At Furring	
1. Outside surface (15 mph wind)		0.17	0.17	0.17
2. Common brick, 4 in.		0.80	0.80	0.80
3. Nonreflective air space, 2.5 in. (30 F mean; 10 deg F temperature difference)		1.10*	1.10*	5.32**
4. Concrete block, three-oval core, stone and gravel aggregate, 4 in.		0.71	0.71	0.71
5. Nonreflective air space 0.75 in. (50 F mean; 10 deg F temperature difference)		1.01	—	—
6. Nominal 1-in. × 3-in. vertical furring		—	0.94	—
7. Gypsum wallboard, 0.5 in.		0.45	0.45	0.11
8. Inside surface (still air)		0.68	0.68	0.68
Total Thermal Resistance (R)		$R_i = 4.92$	$R_s = 4.85$	$R_i = R_s = 7.79$

Construction No. 1: $U_i = 1/4.92 = 0.203$; $U_s = 1/4.85 = 0.206$. With 20% framing (typical of 1-in. × 3-in. vertical furring on masonry @16-in. (o.c.), $U_{av} = 0.8(0.203) + 0.2(0.206) = 0.204$
Construction No. 2: $U_i = U_s = U_{av} = 1.79 = 0.128$

■ **FIGURE 2–8**
Sample calculation of U-factor for masonry cavity walls.

Tables 2–5 and 2–6 offer guidelines for estimating the amount of air leakage through construction, in units of cubic feet per hour per square foot of wall surface and linear feet of crack for fenestrations. Values are listed for various constructions and pressure differences between the inside and outside. A detailed estimate can be prepared assuming the type of construction areas and length of crack for sources of building leakage. This procedure is known as the *crack method* for estimating infiltration.

Most designers prefer to use a simpler procedure called the *air change method*. Airflow into a space can be measured in air changes per hour. An air change is equal to the volume of the space. For estimating infil-

tration, air change rates are assumed on the basis of one's judgment regarding the tightness of the space. Air change rates used for design will vary from 1/2 air change per hour for perimeter rooms of a tight building to 4 air changes per hour for a loosely constructed building. Exceptional buildings might warrant higher or lower estimates. Guidelines for estimating air change rates in residential construction are contained in Table 2–7.

2.5.5 Ventilation

Outside air is introduced by the HVAC system to dilute building air contaminants and to make up for ex-

Coefficients are expressed in Btu per (hour) (square foot) (degree Fahrenheit difference between the air on the two sides), and are based on still air (no wind) on both sides

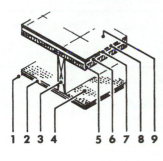

	Assume Unheated Attic Space above Heated Room with Heat Flow Up—Remove Tile, Felt, Plywood, Subfloor and Air Space—Replace with R-19 Blanket Insulation (New Item 4)			
Heated Room Below Unheated Space	**1**		**2**	
	Resistance (R)			
Construction (Heat Flow Up)	Between Floor Joists	At Floor Joists	Between Floor Joists	At Floor Joists
1. Bottom surface (still air)	0.61	0.61	0.61	0.61
2. Metal lath and lightweight aggregate, plaster, 0.75 in.	0.47	0.47	0.47	0.47
3. Nominal 2-in. × 8-in. floor joist	—	9.06	—	9.06
4. Nonreflective airspace, 7.25-in. (50 F mean; 10 deg F temperature difference)	0.93*	—	19.00	—
5. Wood subfloor, 0.75 in.	0.94	0.94	—	—
6. Plywood, 0.625 in.	0.77	0.77	—	—
7. Felt building membrane	0.06	0.06	—	—
8. Tile	0.05	0.05	—	—
9. Top surface (still air)	0.61	0.61	0.61	0.61
Total Thermal Resistance (R)	R_i= 4.44	R_s= 12.57	R_i= 20.69	R_s=10.75

Construction No. 1: U_i= 1/4.45= 0.225; U_s= 1/12.58= 0.079. With 10% framing (typical of 2-in. joists @ 16-in. o.c.), U_{av} = 0.9 (0.225) + 0.1 (0.079)= 0.210

Construction No. 2: U_i = 1/20.69 = 0.048; U_s = 1/10.75 = 0.093. With framing unchanged, U_{av} = 0.9 (0.048) + 0.1 (0.093) = 0.053

■ **FIGURE 2–9**

Sample calculation of U-factor for frame construction ceiling and floor.

These Coefficients are expressed in Btu per (hour) (square foot) (degree Fahrenheit difference in temperature between the air on the two sides), and are based upon an outside wind velocity of 15 mph

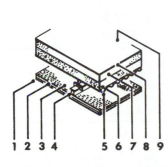

	Add Rigid Roof Deck Insulation, $C = 0.24$ ($R = 1/C = 4.17$) (New Item 7)	
Construction (Heat Flow Up)	1	2
1. Inside surface (still air)	0.61	0.61
1. Metal lath and lightweight aggregate plaster, 0.75 in.	0.47	0.47
3. Nonreflective air space, greater than 3.5 in. (50 F mean; 10 deg F temperature difference)	0.93*	0.93*
4. Metal ceiling suspension system with metal hanger rods	0**	0**
5. Corrugated metal deck	0	0
6. Concrete slab, lightweight aggregate, 2 in. (30 lb/ft³)	2.22	2.22
7. Rigid roof deck insulation (none)	—	4.17
8. Built-up roofing, 0.375 in.	0.33	0.33
9. Outside surface (15 mph wind)	0.17	0.17
Total Thermal Resistance (R)......................	4.73	8.90

Construction No. 1: U_{av} = 1/4.73 = 0.211
Construction No. 2: U_{av} = 1/8.90= 0.112

■ **FIGURE 2–10**

Sample calculation of U-factor for flat masonry roofs with built-up roofing, with and without suspended ceiling (winter conditions, upward flow).

haust. During cold weather, the air must be heated to the temperature of the space. The ventilation load is calculated by the same equations used for infiltration.

Determining the proper amount of outside air requires analyses of the building's exhaust systems and fresh air requirements for occupancy and a consideration of excess air to pressurize the building slightly and prevent undue infiltration.

Minimum values for outside air are specified by code to maintain acceptable indoor air quality. Model codes generally incorporate recommendations of

ASHRAE Standard 62. Code requirements for outside air have varied significantly over the years due to concerns about energy conservation and recent attention to "sick building syndrome." Table 2–8 shows the consensus standard for outside air rates as of 1992. Rates are specified according to usage of space.

2.5.6 Miscellaneous Loads

In addition to conduction, infiltration, and ventilation, heating loads should take into account miscella-

TABLE 2–5

Air leakage through walls (cfh per sq ft)
(reprinted with permission from ASHRAE)

Type of Wall	Pressure Difference, in. Water			
	0.05	0.10	0.20	0.30
Brick Wall: 8.5 in Plain	5	9	16	24
Plastered Two coats on brick	0.05	0.08	0.14	0.2
Brick Wall: 13 in Plain	5	8	14	20
Plastered Two coats on brick	0.01	0.04	0.05	0.09
Plastered, Furring, Lath Two coats gypsum Plaster	0.03	0.24	0.46	0.66
Frame Wall:				
Bevel siding painted or cedar singles, sheathing, building paper, wood lath, and three coats gypsum plaster	0.09	0.15	0.22	0.29

TABLE 2–6

Infiltration through double-hung wood windows
(cfh per foot of crack)
(reprinted with permission from ASHRAE)

Type of Window	Pressure Difference, in. Water		
	0.10	0.20	0.30
A. Wood Double-Hung Window (Locked)			
1. Nonweatherstripped, loose fit or weatherstripped, loose fit	77	122	150
2. Nonweatherstripped, average fit	27	43	57
3. Weatherstripped, average fit	14	23	30
B. Frame-Wall Leakage (Leakage is that passing between the frame of a wood double-hung window and the wall)			
1. Around frame in masonry wall, not caulked	17	26	34
2. Around frame in masonry wall, caulked	3	5	6
3. Around frame in wood frame wall	13	21	29

TABLE 2–7

Infiltration air changes per hour occurring under average conditions in residences
(reprinted with permission from ASHRAE)

Kind of room	Single Glass, No Weatherstrip	Storm Sash or Weatherstripped
No windows or exterior doors	0.5	0.3
Windows or exterior doors on one side	1	0.7
Windows or exterior doors on two sides	1.5	1
Windows or exterior doors on three sides	2	1.3
Entrance halls	2	1.3

TABLE 2–8

Outdoor air requirements for ventilation
(reprinted with permission from ASHRAE)

Application	Estimated Maximum Occupants per 1000 sq ft	Outdoor Air Requirements		
		CFM/ person	CFM/ sq ft	CFM/ room
Dry Cleaners, Laundries (a)				
Commercial laundry	10	25		
Commercial dry cleaner	30	30		
Storage, pick up	30	35		
Coin-operated laundries	20	15		
Coin-operated dry cleaner	20	15		

(For convenience, only a portion of Table 2–8 is shown here. The full table is located at the end of the chapter.)

TABLE 2–9

Below-grade heat losses for basement walls and floors
(reprinted with permission from ASHRAE)

Ground Water Temperature	Basement Floor Loss, (a) Btu/sq ft	Below-Grade Wall Loss, (b) Btu/sq ft
40	3.0	6.0
50	2.0	4.0
60	1.0	2.0

(a) Based on basement temperature of 70°F and U of 0.10.
(b) Assumed twice basement floor loss.

TABLE 2–10

Heat loss of concrete floors at or near grade (Btuh/ft slab edge)
(reprinted with permission from ASHRAE)

Outdoor Design Temperature, °F	Unheated (Heated) Slab; Resistance of Insulation		
	5.0	3.3	2.5
5 to −5	23 (31)	35 (47)	46 (62)
−5 to −15	26 (36)	39 (54)	53 (72)
−15 to −25	29 (40)	44 (60)	59 (80)

neous factors such as losses through walls below grade and slabs on grade. Guidelines are shown in Tables 2–9 and 2–10.

2.5.7 Heating Load Problem 2–1

This sample problem demonstrates methods for calculating heating and humidification loads for a small office building defined in Figure 2–11. The criteria and physical properties of the building are as follows:

- Design conditions—indoor (72°F, 20% RH), outdoor (8°F, near 0% RH)
- Infiltration—2 air changes per hour
- Ventilation—500 CFM

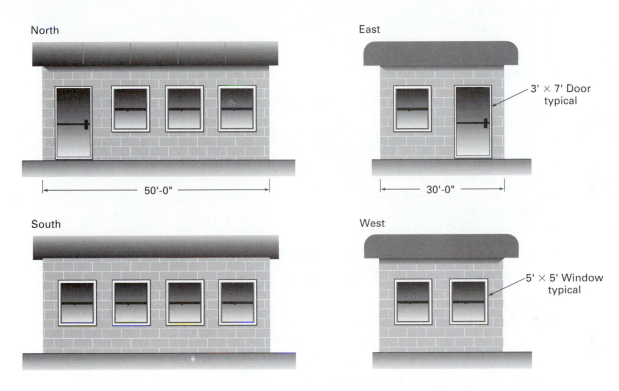

Elevations

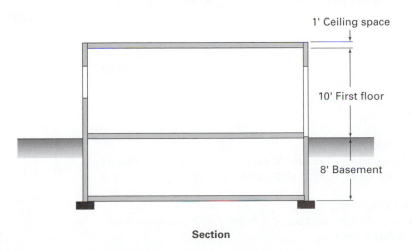

Section

■ **FIGURE 2–11**
Elevations and section of the building, problem 2–1. U-factors are as follows: walls, 0.062; roof, 0.16; windows, 0.58; and doors, 0.64.

Calculations for Problem 2–1

LOAD COMPONENTS			LOAD (Btuh)

Roof

$Q = U \times A \times TD$

$Q = 0.16 \times (30 \times 50) \times (72 - 8)$ — 15,360

Walls

$Q = U \times A \times TD$

North	$Q = 0.062 \times (11 \times 50) \times (72 - 8)$	2182
South	$Q = 0.062 \times (11 \times 50) \times (72 - 8)$	2182
East	$Q = 0.062 \times (11 \times 30) \times (72 - 8)$	1309
West	$Q = 0.062 \times (11 \times 30) \times (72 - 8)$	1309

Doors

$Q = U \times A \times TD$

North	$Q = 0.64 \times (3 \times 7) \times (72 - 8)$	860
East	$Q = 0.64 \times (3 \times 7) \times (72 - 8)$	860

Windows

$Q = U \times A \times TD$

North	$Q = .58 \times (3 \times 5 \times 5) \times (72 - 8)$	2784
South	$Q = .58 \times (4 \times 5 \times 5) \times (72 - 8)$	3712
East	$Q = .58 \times (1 \times 5 \times 5) \times (72 - 8)$	928
West	$Q = .58 \times (2 \times 5 \times 5) \times (72 - 8)$	1856

Basement floor

$Q = Btuh / ft^2 \times area$

$Q = 3 \times (30 \times 50)$ — 4500

Basement walls

$Q = Btuh / ft^2 \times area$

$Q = 1.5 \times (8 \times (30 + 50 + 30 + 50))$ — 1920

Infiltration, sensible only

$Q = 1.1 \times CFM \times TD$

CFM = (Air exchanges per hour $\times$ volume) / 60 minutes per hour

$Q = 1.1 \times ((2 \times (30' \times 50' \times 10'))/60) \times (72 - 8)$ — 35,200

Ventilation, sensible only

$Q = 1.1 \times CFM \times TD$

CFM = (Air exchanges per hour $\times$ volume) / 60 minutes per hour

$Q = 1.1 \times 500 \times (72 - 8)$ — 35,200

Total Heat Loss = 110,162 Btuh

HUMIDIFICATION (Optional)

Infiltration air

$Q = .68 \times CFM \times (W_{in} - W_{out})$

CFM = (Air exchanges per hour $\times$ volume) / 60 minutes per hour

$Q = 0.68 \times ((2 \times (30' \times 50' \times 10'))/60) \times (36 - 0)$ — 12,240

Ventilation air

$Q = .68 \times CFM \times (W_{in} - W_{out})$

CFM = (Air exchanges per hour $\times$ volume) / 60 minutes per hour

$Q = 0.68 \times 500 \times (36 - 0)$ — 12,240

Total Humidification of Outside Air = 24,480 Btuh

2.6 CALCULATING COOLING LOADS

The first step in sizing air conditioning equipment is to calculate loads. Heat gains include conduction, solar effects, outside air loads, and internal heat loads, as illustrated in Figure 2–12.

2.6.1 Conduction through Walls and Roofs

In estimating cooling loads, a simple temperature difference between inside and outside air will not account for solar heat. The outside surface of a wall or roof may be much warmer than the surrounding air, due to solar effects. Accordingly, conduction through walls and roofs is estimated by equations using a quantity that considers such effects. One quantity that is used is the total equivalent temperature difference, or TETD. The value for TETD will vary with the orientation, time of day, absorption property of the surface, and thermal mass of the building assembly, which affects the timing of heat entry to the interior. Values for TETD are shown in Table 2–11 for roofs and Table 2–12 for walls. Wall construction is defined in Table 2–13. (*Note:* These tables are given in their entirety at the end of the chapter.)

For walls and roofs, the cooling load is calculated by using the equation

$$Q = U \times A \times \text{TETD} \qquad (2\text{--}13)$$

where

TETD = total equivalent temperature difference (°F)

2.6.2 Conducted and Solar Heat through Glazing

Solar effects must also be considered in estimating heat gains through windows and skylights. These heat loads are considered in two parts—simple conduction and solar transmission—as shown in Figure 2–13. Equation 2–14 applies:

$$Q = U \times A \times \text{TD} + \text{SC} \times A \times \text{SHGF} \qquad (2\text{--}14)$$

where

SHGF= solar heat gain factor (Btuh/sq ft)
  SC = shading coefficient (dimensionless)

The solar heat gain factor (SHGF) represents the amount of solar heat that will enter a single-pane, clear window at a given time of year and time of day, facing the specified orientation. Values of SHGF are shown in Table 2–14 for 40° N latitude. Tables for other latitudes are included in the ASHRAE *Handbook of Fundamentals*. The tables also include values for the azimuth and altitude of the sun, which are important in accounting for the geometry of shading from fins, overhangs, and adjacent structures.

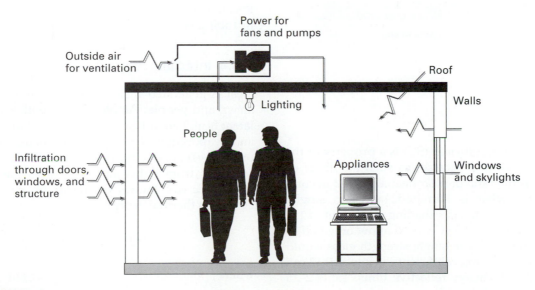

■ **FIGURE 2–12**
Components of building cooling loads.

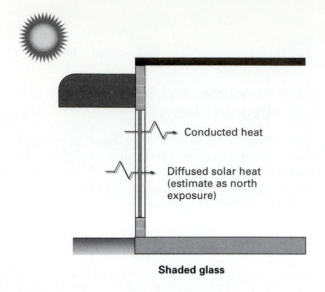

Conducted heat

Diffused solar heat
(estimate as north
exposure)

Shaded glass

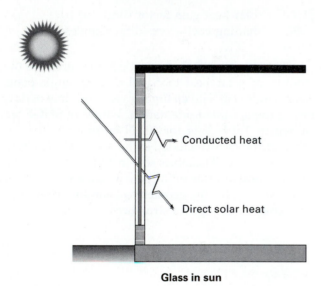

Conducted heat

Direct solar heat

Glass in sun

■ **FIGURE 2–13**
Heat gain through glazing.

The shading coefficient (SC) is a property of the glazing material and accessories such as blinds or drapes. SC is the ratio of solar heat admitted in comparison with what is admitted by clear, single-strength glass, which has a shading coefficient of 1.0. Thus, a glazing with 0.5 SC would allow only half as much solar heat as clear, single-strength glass would into the space. Shading coefficients as low as 0.12 can be achieved with heavy reflective films. Table 2–15 gives shading coefficients for commonly used glazing materials.

2.6.3 Infiltration and Ventilation

Generally, the amount of air infiltration during hot weather will be much lower than during cold weather. This is because milder winds and lower temperature differentials cause less of a chimney effect. Accordingly, for summer air change rates should be estimated lower than for winter.

Infiltration loads have two components: sensible and latent. The equations governing such loads are identical to those cited earlier for heating load calculations:

$$Q_{sensible} = 1.1 \times CFM \times TD \qquad (2\text{--}15)$$

$$Q_{latent} = 0.68 \times CFM \times (W_{final} - W_{initial}) \qquad (2\text{--}16)$$

Total heat, termed enthalpy (H), is the sum of sensible and latent heat. The total heat of air at various conditions of temperature and humidity can be taken from a psychrometric chart or tables, and the following equation can be used to determine energy flow:

$$Q = 4.5 \times CFM \times \Delta H \qquad (2\text{--}17)$$

where

ΔH = change in enthalpy (Btu/lb)

Outside air introduced for ventilation by the air-conditioning equipment will result in sensible and latent loads calculated according to the same equations. Recommended minimum ventilation rates are shown in Table 2–8.

2.6.4 Internal Heat Gains

Heat is generated inside buildings by lights, appliances, and people. People liberate both sensible and latent heat. Latent heat results from exhaled moisture and evaporation of perspiration. Loads will depend on the level of activity, as shown in Table 2–16.

Heat from lights and appliances can be calculated once one knows the factor for conversion of electric to thermal energy:

$$Q = 3.41 \times P \qquad (2\text{--}18)$$

where

P = power input to light fixture or appliance (watts)

TABLE 2–14

Solar position and intensity and solar heat gain factors for 40° north latitude
(reprinted with permission from ASHRAE)

Date	Solar Time, A.M.	Solar Position Altitude	Solar Position Azimuth	Direct Normal Irradiation, Btuh/sq ft	Solar Heat Gain Factors, Btuh/sq ft N	NE	E	SE	S	SW	W	NW	Horizontal	Solar Time, P.M.
Jan 21	8	8.1	55.3	141	5	17	111	133	75	5	5	5	13	4
	9	16.8	44.0	238	11	12	154	224	160	13	11	11	54	3
	10	23.8	30.9	274	16	16	123	241	213	51	16	16	96	2
	11	28.4	16.0	289	18	18	61	222	244	118	18	18	123	1
	12	30.0	0.0	293	19	19	20	179	254	179	20	19	133	12
			Half-day Totals		59	68	449	903	815	271	59	59	353	
Feb 21	7	4.3	72.1	55	1	22	50	47	13	1	1	1	3	5
	8	14.8	61.6	219	10	50	183	199	94	10	10	10	43	4
	9	24.3	49.7	271	16	22	186	245	157	17	16	16	98	3
	10	32.1	35.4	293	20	21	142	247	203	38	20	20	143	2
			Half-day Totals		81	144	634	1035	813	250	81	81	546	
Mar 21	7	11.4	80.2	171	8	93	163	135	21	8	8	8	26	5
	8	22.5	69.6	250	15	91	218	211	73	15	15	15	85	4
	9	32.8	57.3	281	21	46	203	236	128	21	21	21	143	3
	10	41.6	41.9	297	25	26	153	229	171	28	25	25	186	2
	11	47.7	22.6	304	28	28	78	198	197	77	28	28	213	1
	12	50.0	0.0	306	28	28	30	145	206	145	30	28	223	12
			Half-day Totals		112	310	849	1100	692	218	112	112	764	
Apr 21	6	7.4	98.9	89	11	72	88	52	5	4	4	4	11	6
	7	18.9	89.5	207	16	141	201	143	16	14	14	14	61	5
	8	30.3	79.3	253	22	128	225	189	41	21	21	21	124	4
	9	41.3	67.2	275	26	80	203	204	83	26	26	26	177	3
	10	51.2	51.4	286	30	37	153	194	121	32	30	30	218	2
	11	58.7	29.2	292	33	34	81	161	146	52	33	33	244	1
	12	61.6	0.0	294	33	33	36	108	155	108	36	33	253	12
			Half-day Totals		153	509	969	1003	489	196	146	145	962	

(For convenience, only a portion of Table 2–14 is shown here. The full table is located at the end of the chapter.)

TABLE 2–15

Shading coefficient for single glass and insulating glass
(reprinted with permission from ASHRAE)

Type of Glass	Normal Thickness	Shading Coefficient (a)
Single Glass		
Regular Sheet	1/8	1.00
Regular Plate/Float	1/4	0.95
	3/8	0.91
	1/2	0.88
Grey Sheet	1/8	0.78
	1/4	0.86
Heat-Absorbing Plate/Float	3/16	0.72
	1/4	0.70
Insulating Glass		
Regular Sheet Out, Regular Sheet In	1/8	0.90
Regular Plate/Float Out, Regular Plate/Float In	1/4	0.83
Heat-Absorbing Plate/Float Out, Regular Plate/Float In	1/4	0.06

(a) Wind velocity 7.5 mph.

Often, a precise figure for the heat from building lighting and appliances is unavailable at the time the air-conditioning system is being designed. In that case, an allowance is assumed in watts per sq ft of building or room area. Guidelines for estimating electrical loads for common types of buildings are given in Chapter 12.

2.6.5 Loads in Return Air Plenums

If lighting is recessed in a ceiling that is used as a return air plenum, heat from the back side of the fixtures will not enter the occupied space. Heat to the plenum will still need to be removed by the air-conditioning system, but will not affect the amount of supply air delivered to cool the occupied space. Similar adjustments should be made for roof loads that occur above return air plenum ceilings. The actual load to the space should be calculated by considering conduction from the plenum through the ceiling.

TABLE 2–16

Rates of heat gain from occupants of conditioned space[a]
(reprinted with permission from ASHRAE)

Degree of Activity	Typical Application	Total Heat Adults, Male, Btu/hr	Total Heat Adjusted,[b] Btu/hr	Sensible Heat, Btu/hr	Latent Heat, Btu/hr
Seated at rest	Theater—matinee	390	330	225	105
	Theater—evening	390	350	245	105
Seated, very light work	Offices, hotels, apartments	450	400	245	155
Moderately active office work	Offices, hotels, apartments	475	450	250	200
Standing, light work; or walking slowly	Department store, retail store, dime store	550	450	250	200
Walking; seated Standing; walking slowly	Drugstore, bank	550	500	250	250
Sedentary work	Restaurant[c]	490	550	275	275
Light bench work	Factory	800	750	275	475
Moderate dancing	Dance hall	900	850	305	545
Walking 3 mph; moderately heavy work	Factory	1000	1000	375	625
Bowling[d] Heavy work	Bowling alley Factory	1500	1450	580	870

[a]*Note:* Tabulated values are based on 75°F room dry-bulb temperature. For 80°F room dry bulb, the total heat remains the same, but the sensible heat values should be decreased by approximately 20 percent and the latent heat values increased accordingly.

[b]*Adjusted total heat gain* is based on normal percentage of men, women, and children for the application listed, with the postulate that the gain from an adult female is 85 percent of that for an adult male and that the gain from a child is 75 percent of that for an adult male.

[c]Adjusted total heat value for *sedentary work, restaurant,* includes 60 Btu per hour for food per individual (30 Btu sensible and 30 Btu latent).

[d]For *bowling,* figure one person per alley actually bowling and all others sitting (400 Btu per hour) or standing (550 Btu per hour).

Space loads and plenum loads can be distinguished by comparing the two cases shown in Figure 2–14.

2.6.6 Sample Problems 2–2 and 2–3

The following problems illustrate the calculation of cooling loads for a small office building, shown in Figure 2–15. Criteria and physical properties of the building are as follows:

- Design conditions—inside (78°F, 50% RH); outside (95°F DB and 78°F WB).
- Construction—wall, type F; *U*-factor, 0.103
  roof—2″ insulation over metal deck; *U*-factor, 0.125
  windows—*U*-factor, 0.58; shading coefficient, 0.65
  doors—*U*-factor, 0.64
  ceiling (Problem 2–2)—*U*-factor, 0.30
- Outside air—infiltration, 1/2 air exchange per hour; ventilation, 500 CFM.

- Lighting—30 fluorescent fixtures with (4) 40-watt lamps each; ballast factor, 1.2: all heat to occupied space in Problem 2–1 and 50% in Problem 2–2.
- Appliances—allowance of 12.5 watts per square foot.
- Occupants—(10) adults, general office work.

Loads are calculated for two cases: In Problem 2.2, the occupied space extends to the underside of the roof, and lighting is suspended in the space. Problem 2.3 includes a dropped ceiling used as the underside of a return air plenum. Lighting fixtures are recessed into the plenum.

Completing the problems shows that the overall air-conditioning load will be about the same, regardless of whether or not there is a return air plenum. However, the load in the occupied space will be considerably lower if the design includes a return air plenum. This will have an important effect on the amount of supply air required for cooling.

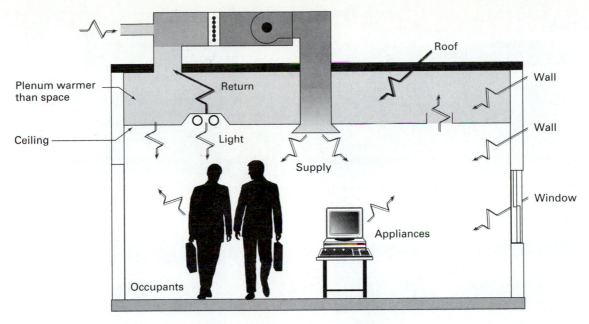

(a)   **Room with return air plenum**

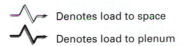

⟿ Denotes load to space

⟿ Denotes load to plenum

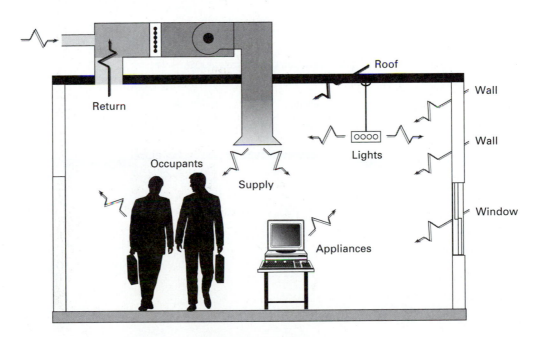

(b)   **Room with direct return**

■ **FIGURE 2–14**

Effect of return air plenum on load to conditioned space. (a) Room with return air plenum, (b) Room with direct return.

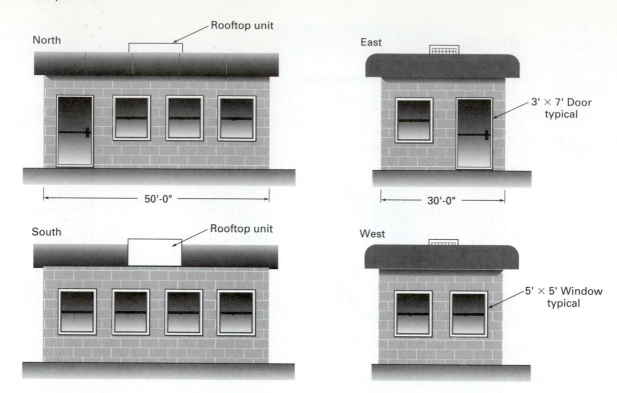

North

Rooftop unit

East

3' × 7' Door typical

50'-0"

30'-0"

South

Rooftop unit

West

5' × 5' Window typical

Elevations

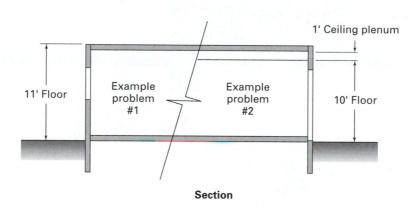

1' Ceiling plenum

11' Floor

Example problem #1

Example problem #2

10' Floor

Section

■ **FIGURE 2–15**
A small office building.

Answer to Problem 2–2.
Cooling Load for a Simple Office Building

		Heat gains	
		Sensible	**Latent**

HEAT GAINS TO SPACE

Roof $Q = U \times A \times$ **TETD**
$Q = 0.16 \times (30 \times 50) \times 81$ — 19,440 — —

Walls in space (below plenum)
$Q = U \times A \times$ **TETD**
$Q = 0.103 \times (11 \times 50) \times 15$ — 850 — —
$Q = 0.103 \times (11 \times 50) \times 26$ — 1473 — —
$Q = 0.103 \times (11 \times 30) \times 37$ — 1258 — —
$Q = 0.103 \times (11 \times 30) \times 19$ — 646 — —

Doors $Q = U \times A \times$ **TETD**
North $Q = 0.64 \times (3 \times 7) \times 21$ — 282 — —
East $Q = 0.64 \times (3 \times 7) \times 28$ — 376 — —

Windows $Q = (U \times A \times$ **TD**$) + (A \times$ **SC** $\times$ **SHGF**$)$
North $Q = (.56 \times (3 \times 5 \times 5) \times (95 - 78)) + ((3 \times 5 \times 5) \times 0.65 \times 28)$ — 2079 — —
South $Q = (.56 \times (4 \times 5 \times 5) \times (95 - 78)) + ((4 \times 5 \times 5) \times 0.65 \times 29)$ — 2837 — —
East $Q = (.56 \times (1 \times 5 \times 5) \times (95 - 78)) + ((1 \times 5 \times 5) \times 0.65 \times 26)$ — 661 — —
West $Q = (.56 \times (2 \times 5 \times 5) \times (95 - 78)) + ((2 \times 5 \times 5) \times 0.65 \times 216)$ — 7496 — —

Lighting $Q =$ **Wattage** $\times$ **3.41** $\times$ **portion to space**
Wattage = Fixtures $\times$ lamps/fix. $\times$ watts/lamp $\times$ ballast factor
$Q = 30 \times 4 \times 40 \times 1.2 \times 3.41 \times 1.0$ — 19,642 — —

Appliances $Q =$ **Wattage** $\times$ **3.41**
Wattage = Watts/SF $\times$ area
$Q = 1.5 \times (30 \times 50) \times 3.41$ — 7673 — —

Occupants, sensible load $Q =$ **Btuh, sensible/occupant** $\times$ **no. occupants**
$Q = 250 \times 10$ — 2500 — —

Occupants, latent load $Q =$ **Btuh, latent/occupant** $\times$ **no. occupants**
$Q = 250 \times 10$ — — 2500

Infiltration, sensible load $Q =$ **1.1** $\times$ **CFM** $\times$ **TD**
CFM = (Air exchanges per hour $\times$ volume) / 60 minutes per hour
$Q = 1.1 \times ((0.5 \times (30' \times 50' \times 11'))/60) \times (95 - 78)$ — 2338 — —

Infiltration, latent load $Q =$ **.68** $\times$ **CFM** $\times (W_{out} - W_{in})$
CFM = (Air exchanges per hour $\times$ volume) / 60 minutes per hour
$Q = 0.68 \times ((0.5 \times (30' \times 50' \times 11'))/60) \times (120 - 72)$ — — 4080

Subtotals, space heat gains* = **69,551** **6,580**

			Heat gains	
			Sensible	Latent

HEAT GAINS FROM OUTSIDE AIR FOR VENTILATION

Sensible load

$Q = 1.1 \times CFM \times TD$

$Q = 1.1 \times 500 \times (95 - 78)$ — 9350 —

Latent load

$Q = 0.68 \times CFM \times (W_{out} - W_{in})$

$Q = 0.68 \times 500 \times (120 - 72)$ — — 16,320

FAN HEAT**

$Q =$ **(Horsepower × 2545 Btuh/brake horsepower) / motor efficiency**

$Q = (2 \times 2545)/.95$ — 5358 —

LOAD SUMMARY

	Sensible	Latent
Space	69,551	6,580
Outside air for ventilation	9,350	16,320
Fan heat	5,358	—
Total air-cooling load (Btuh)	**84,259**	**22,900**
	(9 tons)[†]	

*At this point, the space sensible heat gains would be used to calculate the required airflow rate for cooling. The equation is

CFM supply = Space sensible load/(1.1 × (Tspace − Tsupply)) = 69,551/(1.1 × (78 − 55)) = 2750

**The fan horsepower would be obtained from catalogued data after a preliminary selection of the air-conditioning unit.

[†]Refrigeration system and equipment are frequently expressed in the term of *tons* which originated from the ice making industry. One ton of refrigeration is equal to the capacity to make one ton (2000 lbs.) of ice in 24 hours. The total heat to be removed from 2000 pounds of water at 32°F is 288,000 Btu in 24 hours, or 12,000 Btu/hr.

Answer to Problem 2–3.
Cooling Load for a Simple Office Building with Return Air Plenum

		Heat gains	
		Sensible	**Latent**

HEAT GAINS TO SPACE

Walls in space (below plenum)

		Sensible	Latent
	$Q = U \times A \times \text{TETD}$		
North	$Q = 0.103 \times (10 \times 50) \times 15$	773	
South	$Q = 0.103 \times (10 \times 50) \times 26$	1339	
East	$Q = 0.103 \times (10 \times 30) \times 37$	1143	
West	$Q = 0.103 \times (10 \times 30) \times 19$	587	

Doors

		Sensible	Latent
	$Q = U \times A \times \text{TETD}$		
North	$Q = 0.64 \times (3 \times 7) \times 21$	282	—
East	$Q = 0.64 \times (3 \times 7) \times 28$	376	—

Windows

		Sensible	Latent
	$Q = (U \times A \times \text{TD}) + (A \times SC \times SHGF)$		
North	$Q = (.56 \times (3 \times 5 \times 5) \times (95 - 78)) + ((3 \times 5 \times 5) \times 0.65 \times 28)$	2079	—
South	$Q = (.56 \times (4 \times 5 \times 5) \times (95 - 78)) + ((4 \times 5 \times 5) \times 0.65 \times 29)$	2837	—
East	$Q = (.56 \times (1 \times 5 \times 5) \times (95 - 78)) + ((1 \times 5 \times 5) \times 0.65 \times 26)$	661	—
West	$Q = (.56 \times (2 \times 5 \times 5) \times (95 - 78)) + ((2 \times 5 \times 5) \times 0.65 \times 216)$	7496	—

Lighting to space

$Q = \text{Wattage} \times 3.41 \times \text{portion to space}$
Wattage = fixtures × no. lamps/fixture × watts/lamp × ballast factor
$Q \times (30 \times 4 \times 40 \times 1.2) \times 3.41 \times 0.5$ — Sensible: 9821 — Latent: —

Appliances

$Q = \text{Wattage} \times 3.41$
Wattage = Watts/SF × area
$Q = (1.5 \times (30 \times 50)) \times 3.41$ — Sensible: 7673 — Latent: —

Occupants, sensible load

$Q = \text{Btuh, sensible/occupant} \times \text{no. occupants}$
$Q = 250 \times 10$ — Sensible: 2500 — Latent: —

Occupants, latent load

$Q = \text{Btuh, latent/occupant} \times \text{no. occupants}$
$Q = 250 \times 10$ — Sensible: — — Latent: 2500

Infiltration, sensible load

$Q = 1.1 \times \text{CFM} \times \text{TD}$
CFM = (Air exchanges per hour × volume) / 60 minutes per hour
$Q = 1.1 \times ((0.5 \times (30' \times 50' \times 10'))/60) \times (95 - 78)$ — Sensible: 2338 — Latent: —

Infiltration, latent load

$Q = .68 \times \text{CFM} \times (W_{out} - W_{in})$
CFM = (Air exchanges per hour × volume) / 60 minutes per hour
$Q = 0.68 \times ((0.5 \times (30' \times 50' \times 10'))/60 \times (120 - 72)$ — Sensible: — — Latent: 4080

Space gain from plenum (assure plenum temperature 10° higher than space)

$Q = U \times A \times \text{TD}$
$Q = 0.30 \times ((30 \times 50) - (30 \times 2 \times 4)) \times (10)$ — Sensible: 3780 — Latent: —

Subtotals, space heat gains = **43,684*** **6580**

HEAT GAINS TO RETURN AIR PLENUM**

		Sensible	Latent
		Heat gains	

Roof
$Q == U \times A \times \text{TETD}$
$Q = 0.16 \times (30 \times 50) \times 71$ — 17,040 —

Walls (above ceiling) $Q = U \times A \times \text{TETD}$

		Sensible	Latent
North	$Q = 0.103 \times (1 \times 50) \times 15$	77	—
South	$Q = 0.103 \times (1 \times 50) \times 26$	134	—
East	$Q = 0.103 \times (1 \times 30) \times 37$	114	—
West	$Q = 0.103 \times (1 \times 30) \times 19$	59	—

Lighting to plenum $Q = \text{Wattage} \times 3.41 \times \text{portion to plenum}$
Wattage = Fixtures × no. lamps/fixture × watts/lamp × ballast factor
$Q = (30 \times 4 \times 40 \times 1.2) \times 3.41 \times 0.5$ 9821 —

Subtotal, plenum sensible heat gain **27,245** —

FAN HEAT***

$Q = (\text{Horsepower} \times 2545 \text{ Btuh/brake horsepower})/\text{motor efficiency}$
$Q = (1.5 \times 2545)/.95$ 4018 —

HEAT GAIN FROM OUTSIDE AIR FOR VENTILATION

Sensible
$Q = 1.1 \times \text{CFM} \times \text{TD}$
$Q = 1.1 \times 500 \times (95 - 78)$ 9350 —

Latent
$Q = 0.68 \times s \text{ CFM} \times (W_{\text{out}} - W_{\text{in}})$
$Q = 0.68 \times 500 \times (120 - 72)$ — 16,320

LOAD SUMMARY

	Sensible	Latent
Space	43,684	6580
Plenum	27,245	—
Fan Heat	4018	—
Outside Air Heat Load	9350	16,320
Total air-cooling load (Btuh) =	**84,297**	**22,900**

*At this point, the space sensible heat gains would be used to calculate the required airflow rate for cooling. The equation is

$$\text{CFM supply} = \text{space sensible load} / (1.1 \times (T_{\text{space}} - T_{\text{supply}})) = 43,684/(1.1 \times (78 - 55)) = 1727$$

**The roof TETD is adjusted to account for a higher plenum temperature.
***The fan horsepower would be obtained from catalogued data after a preliminary selection of the air-conditioning unit.

2.7 REFERENCE TABLES AND FIGURES

This section contains tables and figures that are too lengthy to be included in the body of the text. The text contains a sample of them sufficient for the discussion at hand. Unless otherwise indicated, all of the tables and figures are reproduced with permission from publications of the American Society of Heating, Ventilating and Air Conditioning Engineers, Inc. (ASHRAE).

Additional data on weather, ventilating air, thermal and solar properties can be found from ASHRAE handbooks, Atlanta, GA.

TABLE 2–1A

Climatic Conditions for the United States
(condensed from ASHRAE *Handbook of Fundamentals*)

					Winter,[b] °F			Summer,[c] °F						
Col. 1	Col. 2		Col. 3		Col. 4	Col. 5		Col. 6			Col. 7	Col. 8		
State and Station[a]	Lat.		Long.		Elev.	Design Dry Bulb		Design Dry Bulb and			Mean Daily Range	Design Wet Bulb		
								Mean Coincident		Wet Bulb				
	°	′	°	′	Feet	99%	97.5%	1%	2.5%	5%		1%	2.5%	5%
ALABAMA														
Birmingham AP	33	34	86	45	620	17	21	96/74	94/75	92/74	21	78	77	76
Mobile Co	30	40	88	15	211	25	29	95/77	93/77	91/76	16	80	79	78
ALASKA														
Anchorage AP	61	10	150	01	114	−23	−18	71/59	68/58	66/56	15	60	59	57
Fairbanks AP	64	49	147	52	436	−51	−47	82/62	78/60	75/59	24	64	62	60
ARIZONA														
Phoenix AP	33	26	112	01	1112	31	34	109/71	107/71	105/71	27	76	75	75
Tucson AP	32	07	110	56	2558	28	32	104/66	102/66	100/66	26	72	71	71
ARKANSAS														
Little Rock AP	34	44	92	14	257	15	20	99/76	96/77	94/77	22	80	79	78
Pine Bluff AP	34	18	92	05	241	16	22	100/78	97/77	95/78	22	81	80	80
CALIFORNIA														
Los Angeles AP	33	56	118	24	97	41	43	83/68	80/68	77/67	15	70	69	68
San Francisco AP	37	37	122	23	8	35	38	82/64	77/63	73/62	20	65	64	62
COLORADO														
Denver AP	39	45	104	52	5283	−5	1	93/59	91/59	89/59	28	64	63	62
Pueblo AP	38	18	104	29	4641	−7	0	97/61	95/61	92/61	31	67	66	65
CONNECTICUT														
Bridgeport AP	41	11	73	11	25	6	9	86/73	84/71	81/70	18	75	74	73
Norwalk	41	07	73	25	37	6	9	86/73	84/71	81/70	19	75	74	73
DELAWARE														
Dover AFB	39	08	75	28	28	11	15	92/75	90/75	87/74	18	79	77	76
Wilmington AP	39	40	75	36	74	10	14	92/74	89/74	87/73	20	77	76	75
FLORIDA														
Miami AP	25	48	80	16	7	44	47	91/77	90/77	89/77	15	79	79	78
Orlando AP	28	33	81	23	100	35	38	94/76	93/76	91/76	17	79	78	78
GEORGIA														
Atlanta AP	33	39	84	26	1010	17	22	94/74	92/74	90/73	19	77	76	75
Gainesville	34	11	83	41	50	24	27	96/77	93/77	91/77	20	80	79	78
HAWAII														
Hilo AP	19	43	155	05	36	61	62	84/73	83/72	82/72	15	75	74	74
Honolulu AP	21	20	157	55	13	62	63	87/73	86/73	85/72	12	76	75	74
IDAHO														
Boise AP	43	34	116	13	2838	3	10	96/65	94/64	91/64	31	68	66	65
Burley	42	32	113	46	4156	−3	2	99/62	95/61	92/66	35	64	63	61
ILLINOIS														
Champaign/Urbana	40	02	88	17	777	−3	2	95/75	92/74	90/73	21	78	77	75
Chicago, Midway AP	41	47	87	45	607	−5	0	94/74	91/73	88/72	20	77	75	74

[a] AP, AFB, following the station name designates airport or military airbase temperature observations. Co designates office locations within an urban area that are affected by the surrounding area. Undesignated stations are semirural and may be compared to airport data.

[b] Winter design data are based on the 3-month period, December through February.

[c] Summer design data are based on the 4-month period, June through September.

[d] Mean wind speeds occurring coincidentally with the 99.5% dry-bulb winter design temperature.

TABLE 2–1A   *(continued)*

Col. 1	Col. 2		Col. 3		Col. 4	Winter,[b] °F Col. 5		Summer,[c] °F Col. 6			Col. 7	Col. 8		
State and Station[a]	Lat.		Long.		Elev.	Design Dry Bulb		Mean	Design Dry Bulb and Coincident Wet Bulb		Mean Daily Range	Design Wet Bulb		
	°	′	°	′	Feet	99%	97.5%	1%	2.5%	5%		1%	2.5%	5%
INDIANA														
Huntington	40	53	85	30	802	−4	1	92/73	89/72	87/72	23	77	75	74
Indianapolis AP	39	44	86	17	792	−2	2	92/74	90/74	87/73	22	78	76	75
IOWA														
Iowa City	41	38	91	33	661	−11	−6	92/76	89/76	87/74	22	80	78	76
Sioux City AP	42	24	96	23	1095	−11	−7	95/74	92/74	89/73	24	78	77	75
KANSAS														
Dodge City AP	37	46	99	58	2582	0	5	100/69	97/69	95/69	25	74	73	71
El Dorado	37	49	96	50	1282	3	7	101/72	98/73	96/73	24	77	76	75
KENTUCKY														
Lexington AP	38	02	84	36	966	3	8	93/73	91/73	88/72	22	77	76	75
Louisville AP	38	11	85	44	477	5	10	95/74	93/74	90/74	23	79	77	76
LOUISIANA														
Alexandria AP	31	24	92	18	92	23	27	95/77	94/77	92/77	20	80	79	78
Baton Rouge AP	30	32	91	09	64	25	29	95/77	93/77	92/77	19	80	80	79
MAINE														
Augusta AP	44	19	69	48	353	−7	−3	88/73	85/70	82/68	22	74	72	70
Bangor, Dow AFB	44	48	68	50	192	−11	−6	86/70	83/68	80/67	22	73	71	69
MARYLAND														
Baltimore AP	39	11	76	40	148	10	13	94/75	91/75	89/74	21	78	77	76
Salisbury	38	20	75	30	59	12	16	93/75	91/75	88/74	18	79	77	76
MASSACHUSETTS														
Boston AP	42	22	71	02	15	6	9	91/73	88/71	85/70	16	75	74	72
Clinton	42	24	71	41	398	−2	2	90/72	87/71	84/69	17	75	73	72
MICHIGAN														
Battle Creek AP	42	19	85	15	941	1	5	92/74	88/72	85/70	23	76	74	73
Detroit	42	25	83	01	619	3	6	91/73	88/72	86/71	20	76	74	73
MINNESOTA														
Minneapolis/St. Paul AP	44	53	93	13	834	−16	−12	92/75	89/73	86/71	22	77	75	73
Rochester AP	43	55	92	30	1297	−17	−12	90/74	87/72	84/71	24	77	75	73
MISSISSIPPI														
Biloxi, Keesler AFB	30	25	88	55	26	28	31	94/79	92/79	90/78	16	82	81	80
Jackson AP	32	19	90	05	310	21	25	97/76	95/76	93/76	21	79	78	78
MISSOURI														
Kansas City AP	39	07	94	35	791	2	6	99/75	96/74	93/74	20	78	77	76
St. Louis AP	38	45	90	23	535	2	6	97/75	94/75	91/74	21	78	77	76
MONTANA														
Glasgow AP	48	25	106	32	2760	−22	−18	92/64	89/63	85/62	29	68	66	64
Great Falls AP	47	29	111	22	3662	−21	−15	91/60	88/60	85/59	28	64	62	60
NEBRASKA														
Lincoln Co	40	51	96	45	1180	−5	−2	99/75	95/74	92/74	24	78	77	76
Omaha AP	41	18	95	54	977	−8	−3	94/76	91/75	88/74	22	78	77	75
NEVADA														
Las Vegas AP	36	05	115	10	2178	25	28	108/66	106/65	104/65	30	71	70	69
Reno AP	39	30	119	47	4404	5	10	95/61	92/60	90/59	45	64	62	61
NEW HAMPSHIRE														
Concord AP	43	12	71	30	342	−8	−3	90/72	87/70	84/69	26	74	73	71
Manchester, Grenier AFB	42	56	71	26	233	−8	−3	91/72	88/71	85/70	24	75	74	72
NEW JERSEY														
Newark AP	40	42	74	10	7	10	14	94/74	91/73	88/72	20	77	76	75
New Brunswick	40	29	74	26	125	6	10	92/74	89/73	86/72	19	77	76	75
NEW MEXICO														
Albuquerque AP	35	03	106	37	5311	12	16	96/61	94/61	92/61	27	66	65	64
Farmington AP	36	44	108	14	5503	1	6	95/63	93/62	91/61	30	67	65	64
NEW YORK														
NYC-Kennedy AP	40	39	3	47	13	12	15	90/73	84/72	84/71	16	76	75	74
Rochester AP	43	07	77	40	547	1	5	91/73	88/71	85/70	22	75	73	72
NORTH CAROLINA														
Asheville AP	35	26	82	32	2140	10	14	89/73	87/72	85/71	21	75	74	72
Charlotte AP	35	13	80	56	736	18	22	95/74	93/74	91/74	20	77	76	76

					Winter,[b] °F				Summer,[c] °F						
Col. 1	Col. 2		Col. 3		Col. 4	Col. 5		Col. 6				Col. 7	Col. 8		
State and Station[a]	Lat.		Long.		Elev.	Design Dry Bulb		Mean	Design Dry Bulb and Coincident		Wet Bulb	Mean Daily Range	Design Wet Bulb		
	°	′	°	′	Feet	99%	97.5%	1%	2.5%		5%		1%	2.5%	5%
NORTH DAKOTA															
Bismarck AP	46	46	100	45	1647	−23	−19	95/68	91/68		88/67	27	73	71	70
Fargo AP	46	54	96	48	896	−22	−18	92/73	89/71		85/69	25	76	74	72
OHIO															
Cleveland AP	41	24	81	51	777	1	5	91/73	88/72		86/71	22	76	74	73
Columbus AP	40	00	82	53	812	0	5	92/73	90/73		87/72	24	77	75	74
OKLAHOMA															
Norman	35	15	97	29	1181	9	13	99/74	96/74		94/74	24	77	76	75
Oklahoma City AP	35	24	97	36	1285	9	13	100/74	97/74		95/73	23	78	77	76
OREGON															
Eugene AP	44	07	123	13	359	17	22	92/67	89/66		86/65	31	69	67	66
Portland AP	45	36	122	36	21	17	23	89/68	85/67		81/65	23	69	67	66
PENNSYLVANIA															
Philadelphia AP	39	53	75	15	5	10	14	93/75	90/74		87/72	21	77	76	75
Pittsburgh AP	40	30	80	13	1137	1	5	89/72	86/71		84/70	22	74	73	72
RHODE ISLAND															
Newport	41	30	71	20	10	5	9	88/73	85/72		82/70	16	76	75	73
Providence AP	41	44	71	26	51	5	9	89/73	86/72		83/70	19	75	74	73
SOUTH CAROLINA															
Greenville AP	34	54	82	13	957	18	22	93/74	91/74		89/74	21	77	76	75
Greenwood	34	10	82	07	620	18	22	95/75	93/74		91/74	21	78	77	76
SOUTH DAKOTA															
Aberdeen AP	45	27	98	26	1296	−19	−15	94/73	91/72		88/70	27	77	75	73
Rapid City AP	44	03	103	04	3162	−11	−7	95/66	92/65		89/65	28	71	69	67
TENNESSEE															
Greeneville	36	04	82	50	1319	11	16	92/73	90/72		88/72	22	76	75	74
Jackson AP	35	36	88	55	423	11	16	98/76	95/75		92/75	21	79	78	77
TEXAS															
Dallas Ap	32	51	96	51	481	18	22	102/75	100/75		97/75	20	78	78	77
Houston AP	29	58	95	21	96	27	32	96/77	94/77		92/77	18	80	79	79
UTAH															
Salt Lake City AP	40	46	111	58	4220	3	8	97/62	95/62		92/61	32	66	65	64
Vernal AP	40	27	109	31	5280	−5	0	91/61	89/60		86/59	32	64	63	62
VERMONT															
Barre	44	12	72	31	600	−16	−11	84/71	81/69		78/68	23	73	71	70
Burlington AP	44	28	73	09	332	−12	−7	88/72	85/70		82/69	23	74	72	71
VIRGINIA															
Richmond AP	37	30	77	20	164	14	17	95/76	92/76		90/75	21	79	78	77
Roanoke AP	37	19	79	58	1193	12	16	93/72	91/72		88/71	23	75	74	73
WASHINGTON															
Port Angeles	48	07	123	26	99	24	27	72/62	69/61		67/60	18	64	62	61
Seattle-Boeing Field	47	32	122	18	23	21	26	84/68	81/66		77/65	24	69	67	65
WEST VIRGINIA															
Charleston AP	38	22	81	36	939	7	11	92/74	90/73		87/72	20	76	75	74
Wheeling	40	07	80	42	665	1	5	89/72	86/71		84/70	21	74	73	72
WISCONSIN															
La Crosse AP	43	52	91	15	651	−13	−9	91/75	88/73		85/72	22	77	75	74
Milwaukee AP	42	57	87	54	672	−8	−4	90/74	87/73		84/71	21	76	74	73
WYOMING															
Casper AP	42	55	106	28	5338	−11	−5	92/58	90/57		87/57	31	63	61	60
Cheyenne	41	09	104	49	6126	−9	−1	89/58	86/58		84/57	30	63	62	60

TABLE 2–1B

Climatic conditions of Canada
(condensed from ASHRAE *Handbook of Fundamentals*)

							Winter,[b] °F					Summer,[c] °F			
Col. 1	Col. 2		Col. 3		Col. 4	Col. 5		Col. 6				Col. 7	Col. 8		
						Design Dry Bulb		Design Dry Bulb and					Design Wet Bulb		
State and Station[a]	Lat.		Long.		Elev.			Mean	Coincident		Wet Bulb	Mean Daily Range			
	°	′	°	′	Feet	99%	97.5%	1%	2.5%		5%		1%	2.5%	5%
Calgary AP	51	06	114	01	3540	−27	−23	84/63	81/61		79/60	25	65	63	62
Prince George AP	53	53	122	41	2218	−33	−28	84/64	80/62		77/61	26	66	64	62
Winnipeg AP	49	54	97	14	786	−30	−27	89/73	86/71		84/70	22	75	73	71
Edmundston Co	47	22	68	20	500	−21	−16	87/70	83/68		80/67	21	73	71	69
St John's AP	47	37	52	45	463	3	7	77/66	75/65		73/64	18	69	67	66
Fort Smith AP	60	01	111	58	665	−49	−45	85/66	81/64		78/63	24	68	66	65
Halifax AP	44	39	63	34	83	1	5	79/66	76/65		74/64	16	69	67	66
Ottawa AP	45	19	75	40	413	−17	−13	90/72	87/71		84/70	21	75	73	72
Charlottetown AP	46	17	63	08	186	−7	−4	80/69	78/68		76/67	16	71	70	68
Montreal AP	45	28	73	45	98	−16	−10	88/73	85/72		83/71	17	75	74	72
Prince Albert AP	53	13	105	41	1414	−42	−35	87/67	84/66		81/65	25	70	68	67

TABLE 2–1C

Climatic conditions of the world (other than USA and Canada)
(condensed from ASHRAE *Handbook of Fundamentals*)

							Winter,[b] °F			Summer,[c] °F					
Col. 1	Col. 2				Col. 3		Col. 4		Col. 5			Col. 6	Col. 7		
									Design Dry-Bulb				Design Wet-Bulb		
Country and Station[a]	Lat.		Long.		Elevation, ft	Mean of Annual Extremes						Mean Daily Range			
	°	′	°	′			99%	97.5%	1%	2.5%	5%		1%	2.5%	5%
ARGENTINA															
Buenos Aires	34	35S	58	29W	89	27	32	34	91	89	86	22	77	76	75
AUSTRALIA															
Melbourne	37	49S	144	58E	114	31	35	38	95	91	86	21	71	69	68
Sydney	33	52S	151	12E	138	38	40	42	89	84	80	13	74	73	72
AUSTRIA															
Vienna	48	15N	16	22E	644	−2	6	11	88	86	83	16	71	69	67
BELGIUM															
Brussels	50	48N	4	21E	328	13	15	19	83	79	77	19	70	68	67
BRAZIL															
Rio de Janeiro	22	55S	43	12W	201	56	58	60	94	92	90	11	80	79	78
BULGARIA															
Sofia	42	42N	23	20E	1805	−2	3	8	89	86	84	26	71	70	69
CHILE															
Santiago	33	24S	70	47W	1555	27	30	32	90	88	86	37	68	67	66
CHINA															
Chungking	29	33N	106	33E	755	34	37	39	99	97	95	18	81	80	79
Shanghai	31	12N	121	26E	23	16	23	26	94	92	90	16	81	81	80
COMMONWEALTH OF INDEPENDENT STATES (formerly SOVIET UNION)															
Moscow	55	46N	37	40E	505	−19	−11	−6	84	81	78	21	69	67	65
St. Petersburg (Leningrad)	59	56N	30	16E	16	−14	−9	−5	78	75	72	15	65	64	63
CUBA															
Havana	23	08N	82	21W	80	54	59	62	92	91	89	14	81	81	80
CZECHOSLOVAKIA															
Prague	50	05N	14	25E	662	3	4	9	88	85	83	16	66	65	64
DENMARK															
Copenhagen	55	41N	12	33E	43	11	16	19	79	76	74	17	68	66	64

Col. 1	Lat.		Long.		Col. 3 Eleva-tion, ft	Mean of Annual Extremes	Winter,[b] °F Col. 4		Summer,[c] °F Design Dry-Bulb Col. 5			Col. 6 Mean Daily Range	Design Wet-Bulb Col. 7		
Country and Station[a]	°	′	°	′			99%	97.5%	1%	2.5%	5%		1%	2.5%	5%
EGYPT															
Cairo	29	52N	31	20E	381	39	45	46	102	100	98	26	76	75	74
FINLAND															
Helsinki	60	10N	24	57E	30	−11	−7	−1	77	74	72	14	66	65	63
FRANCE															
Paris	48	49N	2	29E	164	16	22	25	89	86	83	21	70	68	67
GERMANY															
Berlin	52	27N	13	18E	187	6	7	12	84	81	78	19	68	67	66
HONG KONG															
Hong Kong	22	18N	114	10E	109	43	48	50	92	91	90	10	81	80	80
INDIA															
New Delhi	28	35N	77	12E	703	35	39	41	110	107	105	26	83	82	82
INDONESIA															
Djakarta	6	11S	106	50E	26	69	71	72	90	89	88	14	80	79	78
ISRAEL															
Jerusalem	31	47N	35	13E	2485	31	36	38	95	94	92	24	70	69	69
ITALY															
Milan	45	27N	9	17E	341	12	18	22	89	87	84	20	76	75	74
JAPAN															
Tokyo	35	41N	139	46E	19	21	26	28	91	89	87	14	81	80	79
KOREA															
Pyongyang	39	02N	125	41E	186	−10	−2	3	89	87	85	21	77	76	76
Seoul	37	34N	126	58E	285	−1	7	9	91	89	87	16	81	79	78
MALAYSIA															
Kuala Lumpur	3	07N	101	42E	127	67	70	71	94	93	92	20	82	82	81
MEXICO															
Mexico City	19	24N	99	12W	7575	33	37	39	83	81	79	25	61	60	59
Monterrey	25	40N	100	18W	1732	31	38	41	98	95	93	20	79	78	77
NETHERLANDS															
Amsterdam	52	23N	4	55E	5	17	20	23	79	76	73	10	65	64	63
NEW ZEALAND															
Auckland	36	51S	174	46E	140	37	40	42	78	77	76	14	67	66	65
NORWAY															
Bergen	60	24N	5	19E	141	14	17	20	75	74	73	21	67	66	65
Oslo	59	56N	10	44E	308	−2	0	4	79	77	74	17	67	66	64
PHILIPPINES															
Manila	14	35N	120	59E	47	69	73	74	94	92	91	20	82	81	81
POLAND															
Krakow	50	04N	19	57E	723	−2	2	6	84	81	78	19	68	67	66
Warsaw	52	13N	21	02E	394	−3	3	8	84	81	78	19	71	70	68
SAUDI ARABIA															
Dhahran	26	17N	50	09E	80	39	45	48	111	110	108	32	86	85	84
SINGAPORE															
Singapore	1	18N	103	50E	33	69	71	72	92	91	90	14	82	81	80
SOUTH AFRICA															
Cape Town	33	56S	18	29E	55	36	40	42	93	90	86	20	72	71	70
SPAIN															
Barcelona	41	24N	2	09E	312	31	33	36	88	86	84	13	75	74	73
SWEDEN															
Stockholm	59	21N	18	04E	146	3	5	8	78	74	72	15	64	62	60
SWITZERLAND															
Zurich	47	23N	8	33E	1617	4	9	14	84	81	78	21	68	67	66
TAIWAN															
Tainan	22	57N	120	12E	70	40	46	49	92	91	90	14	84	83	82
Taipei	25	02N	121	31E	30	41	44	47	94	92	90	16	83	82	81
THAILAND															
Bangkok	13	44N	100	30E	39	57	61	63	97	95	93	18	82	82	81
TURKEY															
Istanbul	40	58N	28	50E	59	23	28	30	91	88	86	16	75	74	73
UNITED KINGDOM															
London	51	29N	0	00	149	20	24	26	82	79	76	16	68	66	65

TABLE 2-2

Thermal properties of typical building and insulating materials

Description	Density lb/ft³	Conductivity λ Btu·in./h·ft²·F	Conductance (C) Btu/h·ft²·F	Resistance (R) Per inch thickness (1/λ) h·ft²·F/Btu	For thickness listed (1/C) h·ft²·F/Btu
BUILDING BOARD					
Boards, Panels, Subflooring, Sheathing					
Woodboard Panel Products					
Asbestos-cement board	120	4.0	—	0.25	—
Asbestos-cement board ...0.25 in.	120	—	16.50	—	0.06
Gypsum or plaster board ...0.5 in.	50	—	2.22	—	0.45
Plywood (Douglas fir)	34	0.80	—	1.25	—
Plywood (Douglas fir) ...0.25 in.	34	—	3.20	—	0.31
Plywood (Douglas fir) ...0.5 in.	34	—	1.60	—	0.62
Vegetable fiberboard					
Sheathing, regular density ...0.5 in.	18	—	0.76	—	1.32
Nail-base sheathing ...0.5 in.	25	—	0.88	—	1.14
Shingle backer ...0.375 in.	18	—	1.06	—	0.94
Sound-deadening board ...0.5 in.	15	—	0.74	—	1.35
Tile and lay-in panels, plain or acoustic	18	0.40	—	2.50	—
Laminated paperboard	30	0.50	—	2.00	—
Hardboard					
Medium density	50	0.73	—	1.37	—
High density, std. tempered	63	1.00	—	1.00	—
Particle board					
Low density	37	0.54	—	1.85	—
Medium density	50	0.94	—	1.06	—
Wood subfloor ...0.75 in.		—	1.06	—	0.94
BUILDING MEMBRANE					
Vapor—permeable felt	—	—	16.70	—	0.06
Vapor—seal, plastic film	—	—	—	—	Negl.
FINISH FLOORING MATERIALS					
Carpet and fibrous pad	—	—	0.48	—	2.08
Carpet and rubber pad	—	—	0.81	—	1.23
Terrazzo ...1 in.	—	—	12.50	—	0.08
Tile—asphalt, linoleum, vinyl, rubber	—	—	20.00	—	0.05
Wood, hardwood finish ...0.75 in.	—	—	1.47	—	0.68
INSULATING MATERIALS					
Blanket and Batt					
Mineral fiber, fibrous form processed from rock, slag, or glass approx. 3.5 in.	0.3–2.0	—	0.077	—	13
Board and Slabs					
Cellular glass	8.5	0.35	—	2.86	—
Glass fiber, organic bonded	4–9	0.25	—	4.00	—
Expanded rubber (rigid)	4.5	0.22	—	4.55	—
Expanded polystyrene, extruded					
Cut cell surface	1.8	0.25	—	4.00	—
Cellular polyurethane (R-11 exp.)(unfaced)	1.5	0.16	—	6.25	—
Cellular polyisocyanurate (R-11 exp.) (foil faced, glass fiber-reinforced core)	2.0	0.14	—	7.20	—
Nominal 1.0 in.		—	0.139	—	7.2
Mineral fiber with resin binder	15.0	0.29	—	3.45	—
Mineral fiberboard, wet felted					
Core or roof insulation	16–17	0.34	—	2.94	—
Acoustical tile	18.0	0.35	—	2.86	—
Mineral fiberboard, wet molded					
Acoustical tile	23.0	0.42	—	2.38	—
Cement fiber slabs (shredded wood with Portland cement binder)	25–27.0	0.50–0.53	—	2.0–1.89	—

Description	Density lb/ft^3	Conductivity λ Btu·in./h·ft^2·F	Conductance (C) Btu/h·ft^2·F	Resistance (R) Per inch thickness (1/λ) h·ft^2·F/Btu	Resistance (R) For thickness listed (1/C) h·ft^2·F/Btu
LOOSE FILL					
Wood fiber, softwoods	2.0–3.5	0.30	—	3.33	—
Perlite, expanded	2.0–4.1	0.27–0.31	—	3.7–3.3	—
	4.1–7.4	0.31–0.36	—	3.3–2.8	—
	7.4–11.0	0.36–0.42	—	2.8–2.4	—
Mineral fiber (rock, slag, or glass)					
approx. 3.75–5 in.	0.6–2.0	—	—		11.0
approx. 6.5–8.75 in.	0.6–2.0	—	—		19.0
Mineral fiber (rock, slag, or glass)					
approx. 3.5 in. (closed sidewall application)	2.0–3.5	—	—	—	12.0–14.0
Vermiculite, exfoliated	7.0–8.2	0.47	—	2.13	—
FIELD APPLIED					
Polyurethane foam	1.5–2.5	0.16–0.18	—	6.25–5.26	—
Spray cellulosic fiber base	2.0–6.0	0.24–0.30	—	3.33–4.17	—
PLASTERING MATERIALS					
Cement plaster, sand aggregate	116	5.0	—	0.20	—
Sand aggregate ... 0.375 in.	—	—	13.3	—	0.08
Gypsum plaster:					
Lightweight aggregate ... 0.5 in.	45	—	3.12	—	0.32
Lightweight aggregate on metal lath ... 0.75 in.	—	—	2.13	—	0.47
PLASTERING MATERIALS					
Sand aggregate	105	5.6	—	0.18	—
Sand aggregate ... 0.5 in.	105	—	11.10	—	0.09
MASONRY MATERIALS					
Concretes					
Cement mortar	116	5.0	—	0.20	—
Lightweight aggregates including	120	5.2	—	0.19	—
slags; cinders; pumice; vermiculite.	80	2.5	—	0.40	—
Perlite, expanded	40	0.93	—	1.08	—
Sand and gravel or stone aggregate					
(oven dried)	140	9.0	—	0.11	—
Sand and gravel or stone aggregate					
(not dried)	140	12.0	—	0.08	—
Stucco	116	5.0	—	0.20	—
MASONRY UNITS					
Brick, common	120	5.0	—	0.20	—
Brick, face	130	9.0	—	0.11	—
Clay tile, hollow:					
1 cell deep ... 3 in.	—	—	1.25	—	0.80
2 cells deep ... 6 in.	—	—	0.66	—	1.52
3 cells deep ... 12 in.	—	—	0.40	—	2.50
Concrete blocks, three oval core:					
Sand and gravel aggregate ... 4 in.	—	—	1.40	—	0.71
... 8 in.	—	—	0.90	—	1.11
Cinder aggregate ... 4 in.	—	—	0.90	—	1.11
... 8 in.	—	—	0.58	—	1.72
Lightweight aggregate					
(expanded shale, clay, slate ... 4 in.	—	—	0.67	—	1.50
or slag; pumice): ... 8 in.	—	—	0.50	—	2.00
Stone, lime, or sand	—	12.50	—	0.08	—
ROOFING					
Asbestos-cement shingles	120	—	4.76	—	0.21
Asphalt roll roofing	70	—	6.50	—	0.15
Asphalt shingles	70	—	2.27	—	0.44
Built-up roofing ... 0.375 in.	70	—	3.00	—	0.33
Slate ... 0.5 in.	—	—	20.00	—	0.05

TABLE 2–2 *(continued)*

Description	Density lb/ft³	Conductivity λ Btu·in./h·ft²·F	Conductance (C) Btu/h·ft²·F	Resistance (R)	
				Per inch thickness (1/λ) h·ft²·F/Btu	For thickness listed (1/C) h·ft²·F/Btu
SIDING MATERIALS (on flat surface)					
Shingles					
Asbestos-cement	120	—	4.75	—	0.21
Wood, 16 in., 7.5-in. exposure	—	—	1.15	—	0.87
Wood, plus insul. backer board, 0.3125 in.	—	—	0.71	—	1.40
Siding					
Asbestos-cement, 0.25 in., lapped	—	—	4.76	—	0.21
Asphalt roll siding	—	—	6.50	—	0.15
Asphalt insulating siding (0.5 in. bed.)	—	—	0.69	—	1.46
Wood, bevel, 0.5 · 8 in. lapped	—	—	1.23	—	0.81
Aluminum or steel, over sheathing					
Hollow backed	—	—	1.61	—	0.61
Insulating-board backed nominal 0.375 in.	—	—	0.55	—	1.82
Architectural glass	—	—	10.00	—	0.10
WOODS (12% moisture content)					
Hardwoods					
Oak	41.2–46.8	1.12–1.25	—	0.89–0.80	—
Maple	39.8–44.0	1.09–1.19	—	0.94–0.88	—
Softwoods					
Southern pine	35.6–41.2	1.00–1.12	—	1.00–0.89	—
Douglas fir-larch	33.5–36.3	0.95–1.01	—	1.06–0.99	—
California redwood	24.5–28.0	0.74–0.82	—	1.35–1.22	—

TABLE 2–8

Outdoor air requirements for ventilation
(reprinted with permission from ASHRAE)

Application	Estimated Maximum Occupants per 1000 sq ft	Outdoor Air Requirements		
		cfm/ person	cfm/ sq ft	cfm/ room
Dry Cleaners, Laundries (a)				
Commercial laundry	10	25		
Commercial dry cleaner	30	30		
Food and Beverage Service				
Dining rooms	70	20		
Cafeteria, fast food	100	20		
Hotels, Motels, Resorts, Dormitories				
Bedrooms				30
Living rooms				30
Baths (c)				35
Conference rooms	50	20		
Assembly rooms	120	15		
Offices				
Office space (d)	7	20		
Reception areas	60	15		
Conference rooms (b)	50	20		
Public Spaces				
Corridors and utilities			0.05	
Public restrooms, cfm/wc or cfm/urinal (i)		50		
Locker and dressing rooms			0.5	

Application	Estimated Maximum Occupants per 1000 sq ft	Outdoor Air Requirements		
		cfm/ person	cfm/ sq ft	cfm/ room
Retail Stores, Sales Floors, and Showroom Floors				
Basement and street	30		0.30	
Upper floors	20		0.20	
Malls and arcades	20		0.20	
Warehouses	5		0.05	
Specialty Shops				
Beauty	25	25		
Clothiers, furniture			0.30	
Supermarkets	8	15		
Pet shops			1.00	
Sports and Amusement				
Spectator areas	150	15		
Ice arenas (playing areas)			0.50	
Swimming pools (pool and deck area) (f)			0.50	
Playing floors (gymnasium)	30	20		
Theaters (g)				
Lobbies	150	20		
Auditoriums	150	15		
Stages, studios	70	15		
Transportation (h)				
Waiting rooms	100	15		
Vehicles	150	15		
Workrooms	10	15		
Education				
Classroom	50	15		
Laboratories (i)	30	20		
Libraries	20	15		
Locker rooms			0.50	
Auditoriums	150	15		
Hospitals, Nursing and Convalescent Homes				
Patient rooms (j)	10	25		
Operating rooms	20	30		
Autopsy rooms (k)			0.50	
Physical therapy	20	15		

a. Dry-cleaning processes may require more air.

b. Supplementary smoke-removal equipment may be required.

c. Installed capacity for intermittent use.

d. Some office equipment may require local exhaust.

e. Normally supplied by transfer air. Local mechanical exhaust with no recirculation recommended.

f. Higher values may be required for humidity control.

g. Special ventilation will be needed to eliminate special stage effects (e.g., dry ice vapors, mists, etc.)

h. Ventilation within vehicles may require special considerations.

i. Special contaminant control systems may be required for processes or functions, including laboratory animal occupancy.

j. Special requirements or codes and pressure relationships may determine minimum ventilation rates and filter efficiency. Procedures generating contaminants may require higher rates.

k. Air shall not be recirculated into other spaces.

TABLE 2-11

Total equivalent temperature differentials for calculating heat gain through flat roofs

Description of Roof Construction	Wt, lb per sq ft	U value Btu/(hr) (ft²)(F°)	A.M. 8 D	8 L	10 D	10 L	12 D	12 L	P.M. 2 D	2 L	4 D	4 L	6 D	6 L	8 D	8 L	10 D	10 L	12 D	12 L
Light Construction Roofs—Exposed to Sun																			
1" insulation + steel siding	7.4	0.213	28	11	65	31	90	48	95	53	78	45	43	27	8	6	1	1	−3	−3
2" insulation + steel siding	7.8	0.125	24	8	61	29	88	46	96	53	81	46	48	30	10	8	2	2	−3	−3
1" insulation + 1" wood	8.4	0.206	12	2	47	21	77	39	92	50	86	48	61	36	25	16	7	5	0	−1
2" insulation + 1" wood	8.5	0.122	8	0	41	18	72	36	90	48	88	40	65	38	30	19	9	7	1	0
1" insulation + 2.5" wood	12.7	0.193	2	−2	23	8	48	23	70	36	79	42	71	40	50	29	29	17	15	9
2" insulation + 2.5" wood	13.1	0.117	1	−2	19	6	43	20	65	33	76	41	72	40	53	31	33	20	18	11
Medium Construction Roofs—Exposed to Sun																			
1" insulation + 4" wood	17.3	0.183	5	0	14	5	31	14	49	24	62	32	65	35	56	31	41	24	29	17
2" insulation + 4" wood	17.8	0.113	6	1	13	4	28	12	45	22	58	30	63	34	56	31	43	25	32	18
1" insulation + 2" h.w. concrete	28.3	0.206	4	−1	27	11	54	26	74	39	81	44	70	40	45	27	24	15	12	7
2" insulation + 2" h.w. concrete	28.8	0.122	2	−2	23	9	49	23	70	36	79	43	71	40	49	29	28	17	15	9
4" l.w. concrete	17.8	0.213	1	−3	28	11	59	28	82	43	88	48	74	42	44	27	19	12	6	4
6" l.w. concrete	24.5	0.157	−2	−4	9	2	31	13	55	27	72	38	76	41	64	36	42	25	25	15
8" l.w. concrete	31.2	0.125	6	2	6	1	16	6	32	14	49	24	61	32	63	34	55	31	41	24
Heavy Construction Roofs—Exposed to Sun																			
1" insulation + 4" h.w. concrete	51.6	0.199	7	1	17	6	33	15	50	25	61	32	63	34	53	30	40	23	28	16
2" insulation + 4" h.w. concrete	52.1	0.120	7	2	15	6	30	13	46	23	58	30	61	33	54	30	41	23	31	17
1" insulation + 6" h.w. concrete	75.0	0.193	13	6	17	7	26	12	38	18	48	25	53	28	51	27	43	24	35	19
2" insulation + 6" h.w. concrete	75.4	0.117	15	7	17	7	25	11	36	17	46	23	51	27	50	27	43	24	36	20

1. *Application.* These values may be used for all normal air conditioning estimates, usually without correction in latitude 0° to 50° north or south when the load is calculated for the hottest weather.

2. *Corrections.* The values in the table were calculated for an inside temperature of 75°F and an outdoor maximum temperature of 95°F with an outdoor daily range of 21°F. The table remains approximately correct for other outdoor maximums (93–102°F) and other outdoor daily ranges (16–34°F), provided that the outdoor daily average temperature remains approximately 85°F.

3. *Attics or other spaces between the roof and ceiling.* If the ceiling is insulated and a fan is used for positive ventilation in the space between the ceiling and roof, the total temperature differential for calculating the room load may be decreased by 25 percent.

   If the attic space contains a return duct or other air plenum, care should be taken in determining the portion of the heat gain that reaches the ceiling.

4. *Light Colors.* Credit should not be taken for light-colored roofs, except where the permanence of light color is established by experience, as in rural areas or where there is little smoke.

Note: h.w. = heavy weight
   l.w. = light weight

TABLE 2–12

Total equivalent temperature differentials for calculating heat gain through sunlit walls

Sun Time — *A.M.* (8, 10, 12) / *P.M.* (2, 4, 6, 8, 10, 12) — Exterior color of wall: D = dark, L = light

Groups A, B, C

North Latitude Wall Facing	8 D	8 L	10 D	10 L	12 D	12 L	2 D	2 L	4 D	4 L	6 D	6 L	8 D	8 L	10 D	10 L	12 D	12 L
Group A																		
NE	27	16	31	18	26	17	24	17	24	18	23	17	20	15	17	13	15	11
E	32	18	41	24	37	22	29	30	28	20	26	19	23	16	20	14	18	13
SE	25	15	36	21	38	23	33	21	28	20	26	18	22	16	19	14	18	12
S	14	9	20	13	28	18	33	22	31	21	25	18	20	15	17	13	15	11
SW	17	11	20	13	24	16	34	22	42	27	41	26	28	20	20	15	17	13
W	17	11	20	13	24	16	30	20	42	27	48	30	33	22	22	16	19	14
NW	14	9	17	11	21	14	23	17	31	21	38	25	26	18	18	14	16	12
N	14	9	15	10	17	12	20	15	21	16	21	16	20	15	17	13	15	11
Group B																		
NE	12	7	27	14	31	17	30	19	31	21	30	22	27	20	21	17	16	13
E	14	8	34	18	45	24	43	25	39	25	35	24	30	22	23	18	17	14
SE	9	5	25	13	39	21	44	26	41	26	37	25	31	23	24	18	17	14
S	4	3	7	4	18	11	32	19	41	26	39	27	33	24	25	17	18	15
SW	5	3	7	4	11	7	23	15	26	18	34	22	51	33	38	25	26	19
W	6	4	7	4	11	7	18	12	23	16	34	23	59	37	43	28	30	20
NW	5	3	6	3	11	7	17	12	18	13	27	19	47	31	36	24	25	18
N	6	4	9	5	12	8	18	12	17	13	20	16	27	21	22	17	16	14
Group C																		
NE	9	6	19	10	26	15	28	17	29	18	29	20	28	20	24	19	20	16
E	10	7	22	12	36	19	40	23	39	23	36	24	33	23	28	20	22	17
SE	8	6	16	9	29	16	38	21	39	24	37	24	34	23	28	21	23	17
S	7	5	7	4	12	7	22	14	32	20	36	24	34	24	29	21	23	17
SW	9	6	8	5	16	6	28	10	28	18	42	26	48	30	42	28	33	22
W	10	7	9	5	14	6	24	9	24	16	40	25	52	32	47	30	37	24
NW	8	6	9	5	13	6	19	9	19	14	30	20	40	27	38	26	30	21
N	7	5	8	5	14	7	18	9	13	13	22	16	25	19	23	18	19	16

Groups G, H, I

North Latitude Wall Facing	8 D	8 L	10 D	10 L	12 D	12 L	2 D	2 L	4 D	4 L	6 D	6 L	8 D	8 L	10 D	10 L	12 D	12 L
Group G																		
NE	11	9	10	15	20	12	24	14	25	16	26	17	27	18	26	18	23	17
E	13	9	17	11	26	15	32	18	34	20	34	21	33	22	31	21	27	19
SE	13	9	14	9	21	12	28	16	33	19	34	21	33	22	31	21	27	19
S	12	9	10	7	11	8	16	10	23	15	29	18	30	20	29	20	26	19
SW	16	11	13	9	13	8	14	9	20	13	29	19	37	24	39	25	35	23
W	18	12	15	10	14	9	14	9	18	12	27	18	38	24	42	26	38	25
NW	14	10	12	8	12	8	13	9	16	11	21	15	29	20	33	22	31	21
N	10	8	10	7	10	7	12	8	15	10	18	13	20	15	21	16	16	14
Group H																		
NE	15	11	16	11	18	12	20	13	22	14	24	15	25	16	25	17	24	17
E	18	13	18	12	22	14	26	16	29	17	30	19	31	20	30	20	29	19
SE	18	14	17	12	19	12	23	14	27	16	29	18	30	20	30	20	28	19
S	16	12	14	10	14	10	15	10	19	12	23	15	25	18	26	18	26	18
SW	22	14	19	12	17	11	16	11	18	12	23	15	29	18	32	21	32	21
W	23	15	20	13	18	12	17	11	18	12	22	15	29	18	33	21	34	22
NW	19	13	17	11	15	10	15	10	16	11	18	12	23	15	26	18	27	19
N	13	10	12	9	11	9	12	9	13	10	15	11	17	13	18	14	19	14
Group I																		
NE	16	11	18	12	20	13	22	14	23	15	24	16	24	16	23	16	22	16
E	19	13	21	14	25	16	29	17	30	18	30	19	29	19	28	18	26	18
SE	19	13	19	13	22	14	26	16	28	18	29	18	29	19	28	18	26	18
S	16	12	15	11	16	11	18	12	21	14	24	16	25	17	25	17	23	16
SW	20	14	19	13	18	12	19	13	22	14	27	17	31	20	32	21	30	20
W	22	14	20	13	19	13	20	14	22	14	26	17	31	20	33	21	32	21
NW	18	12	16	11	16	11	17	12	18	12	21	14	25	17	27	18	26	18
N	13	10	12	9	13	9	13	10	15	11	16	12	18	13	18	13	18	14

TABLE 2–12 (*continued*)

Sun Time — Groups D, E, F

Exterior color of wall—D = dark, L = light

North Latitude Wall Facing	A.M. 8 D	L	A.M. 10 D	L	A.M. 12 D	L	P.M. 2 D	L	P.M. 4 D	L	P.M. 6 D	L	P.M. 8 D	L	P.M. 10 D	L	P.M. 12 D	L
Group D																		
NE	8	5	19	10	28	15	30	17	30	19	21	21	24	19	23	19	19	16
E	9	6	23	12	38	20	37	24	37	24	24	24	27	23	29	20	21	17
SE	7	5	16	9	30	16	40	22	41	25	34	25	28	24	28	22	22	17
S	5	4	6	4	12	7	23	14	34	21	35	25	29	29	23	23	23	17
SW	8	5	7	4	9	6	16	10	30	19	51	28	43	28	33	33	33	22
W	8	6	7	5	9	6	14	9	25	16	55	27	49	31	37	37	37	25
NW	7	5	7	4	9	6	13	9	20	14	42	21	40	27	31	31	31	21
N	6	4	8	5	10	6	10	10	23	14	25	17	24	19	19	19	19	16
Group E																		
NE	10	6	23	12	30	16	30	18	30	20	21	21	23	23	18	19	14	14
E	11	6	28	15	42	22	39	24	39	24	24	24	25	25	19	20	15	15
SE	8	5	20	11	35	19	41	24	41	25	25	25	26	26	20	20	16	16
S	4	3	6	4	15	9	38	17	38	24	26	26	27	27	20	20	16	16
SW	6	4	7	4	10	6	35	12	35	22	49	28	52	30	27	34	21	21
W	7	5	7	4	10	6	30	11	30	20	48	31	57	34	30	37	23	23
NW	6	4	6	4	10	6	23	10	23	16	36	24	45	28	26	31	20	20
N	6	4	8	5	11	7	21	11	21	15	24	18	26	20	18	20	16	16
Group F																		
NE	9	7	14	9	21	12	27	15	27	17	29	17	20	20	19	22	17	17
E	10	8	17	10	28	15	37	19	37	22	37	22	23	23	22	26	19	19
SE	10	7	13	8	22	13	36	17	37	21	37	21	23	23	22	27	19	19
S	9	7	7	5	10	6	26	10	32	16	32	16	22	31	22	27	19	19
SW	12	9	10	6	9	6	22	8	33	14	42	21	52	30	27	37	25	25
W	14	9	11	7	10	6	19	8	31	12	43	20	57	34	29	41	27	27
NW	12	8	9	6	9	6	16	8	24	11	33	16	45	28	24	33	23	23
N	8	7	7	6	11	6	15	8	19	11	22	14	26	20	18	23	17	17

Sun Time — Groups J, K, L

Exterior color of wall—D = dark, L = light

North Latitude Wall Facing	A.M. 8 D	L	A.M. 10 D	L	A.M. 12 D	L	P.M. 2 D	L	P.M. 4 D	L	P.M. 6 D	L	P.M. 8 D	L	P.M. 10 D	L	P.M. 12 D	L
Group J																		
NE	18	13	17	12	18	12	19	13	21	13	22	14	23	15	23	16	23	16
E	22	15	20	14	21	14	24	16	26	16	28	17	29	18	29	19	20	19
SE	21	15	20	14	20	13	21	14	24	15	25	16	28	17	28	17	28	18
S	19	14	17	12	16	11	17	11	17	12	20	13	22	15	24	16	24	16
SW	24	16	22	15	20	13	19	13	19	13	21	14	24	16	28	18	30	19
W	26	17	24	16	22	14	20	13	20	13	21	14	24	16	28	18	31	20
NW	21	15	19	13	18	12	17	11	17	11	17	12	19	13	22	15	25	17
N	15	11	14	11	13	10	13	9	13	10	14	10	15	11	17	12	17	13
Group K																		
NE	19	14	19	13	19	13	20	13	20	14	21	14	22	15	22	15	22	15
E	23	16	22	15	23	15	26	16	26	16	27	17	27	17	28	18	27	18
SE	23	15	22	15	22	14	24	15	24	15	26	16	26	17	27	17	27	17
S	20	14	19	13	18	12	18	12	18	13	21	14	21	14	22	15	23	15
SW	25	16	23	15	22	15	21	14	21	14	22	15	24	16	26	17	27	18
W	26	17	24	16	23	15	22	15	22	14	23	15	24	16	27	17	28	18
NW	21	15	20	13	19	13	18	13	18	12	19	13	20	14	22	15	23	16
N	15	11	14	11	14	10	14	10	14	10	14	11	15	11	16	12	16	12
Group L																		
NE	18	13	18	13	19	13	20	13	21	14	22	15	23	15	23	16	22	15
E	22	15	22	14	23	15	25	16	27	17	28	18	28	18	28	18	27	18
SE	21	14	21	13	22	14	23	15	25	16	27	17	27	17	27	17	26	18
S	19	13	17	12	17	12	18	12	19	13	21	14	23	15	23	16	23	16
SW	23	15	22	14	21	14	20	13	20	14	21	15	26	17	28	18	28	18
W	25	16	23	15	21	14	21	14	21	14	20	15	26	17	29	19	30	19
NW	20	14	19	13	18	12	18	12	18	12	16	13	21	15	23	16	24	16
N	14	11	14	10	13	10	13	10	14	10	14	11	16	12	17	13	17	13

1. *Application.* These values may be used for all normal air conditioning estimates, usually without correction when the load is calculated for the hottest weather.

2. *Corrections.* The values in the table were calculated for an inside temperature of 75°F and an outdoor maximum temperature of 95°F with an outdoor daily range of 21°F. The table remains approximately correct for other outdoor maximums (93–102°F) and other outdoor daily ranges (16–34°F), provided that the outdoor daily average temperature remains approximately 85°F.

3. *Color of exterior surface of wall.* Use temperature differentials for light walls only when the permanence of the light wall is established by experience. For cream colors, use the values for light walls. For medium colors, interpolate halfway between the dark and light values. Medium colors are medium blue, medium green, bright red, light brown, unpainted wood, natural color concrete, etc. Dark blue, red, brown, green, etc., are considered dark colors.

TABLE 2–13

Description of wall construction

Group	Components	Wt. lb per sq ft	U Value
A	1″ stucco + 4″ l.w. concrete block + air space	28.6	0.267
	1″ stucco + air space + 2″ insulation	16.3	0.106
B	1″ stucco + 4″ common brick	55.9	0.393
	1″ stucco + 4″ h.w. concrete	62.5	0.481
C	4″ face brick + 4″ l.w. concrete block + 1″ insulation	62.5	0.158
	1″ stucco + 4″ h.w. concrete + 2″ insulation	62.9	0.114
D	1″ stucco + 8″ l.w. concrete block + 1″ insulation	41.4	0.141
	1″ stucco + 2″ insulation + 4″ h.w. concrete block	36.6	0.111
E	4″ face brick + 4″ l.w. concrete block	62.2	0.333
	1″ stucco + 8″ h.w. concrete block	56.6	0.349
F	4″ face brick + 4″ common brick	80.5	0.360
	4″ face brick + 2″ insulation + 4″ l.w. concrete block	62.5	0.103
G	1″ stucco + 8″ clay tile + 1″ insulation	62.8	0.141
	1″ stucco + 2″ insulation + 4″ comon brick	56.2	0.108
H	4″ face brick + 8″ clay tile + 1″ insulation	96.4	0.137
	4″ face brick + 8″ common brick	129.6	0.280
	1″ stucco + 12″ h.w. concrete	155.9	0.365
	4″ face brick + 2″ insulation + 4″ common brick	89.8	0.106
	4″ face brick + 2″ insulation + 4″ h.w. concrete	96.5	0.111
	4″ face brick + 2″ insulation + 8″ h.w. concrete block	90.6	0.102
I	1″ stucco + 8″ clay tile + air space	62.6	0.209
	4″ face brick + air space + 4″ h.w. concrete block	69.9	0.282
J	face brick + 8″ common brick + 1″ insulation	129.8	0.145
	4″ face brick + 2″ insulation + 8″ clay tile	96.5	0.094
	1″ stucco + 2″ insulation + 8″ common brick	96.3	0.100
K	4″ face brick + air space + 8″ clay tile	96.2	0.200
	4″ face brick + 2″ insulation + 8″ common brick	129.9	0.098
	4″ face brick + 2″ insulation + 8″ h.w. concrete	143.3	0.107
L	4″ face brick + 8″ clay tile + air space	96.2	0.200
	4″ face brick + air space + 4″ common brick	89.5	0.265
	4″face brick + air space + 4″ h.w. concrete	96.2	0.301
	4″ face brick + air space + 8″ h.w. concrete block	90.2	0.246
	1″ stucco + 2″ insulation + 12″ h.w. concrete	156.3	0.106

TABLE 2–14

Solar position and intensity; solar heat gain factors for 40° north latitude

Date	Solar Time, A.M.	Altitude	Azimuth	Direct Normal Irradiation, Btuh/sq ft	N	NE	E	SE	S	SW	W	NW	Horizontal	Solar Time, P.M.
Jan 21	8	8.1	55.3	141	5	17	111	133	75	5	5	5	13	4
	9	16.8	44.0	238	11	12	154	224	160	13	11	11	54	3
	10	23.8	30.9	274	16	16	123	241	213	51	16	16	96	2
	11	28.4	16.0	289	18	18	61	222	244	118	18	18	123	1
	12	30.0	0.0	293	19	19	20	179	254	179	20	19	133	12
	Half-day Totals				59	68	449	903	815	271	59	59	353	
Feb 21	7	4.3	72.1	55	1	22	50	47	13	1	1	1	3	5
	8	14.8	61.6	219	10	50	183	199	94	10	10	10	43	4
	9	24.3	49.7	271	16	22	186	245	157	17	16	16	98	3
	10	32.1	35.4	293	20	21	142	247	203	38	20	20	143	2
	11	37.3	18.6	303	23	23	71	219	231	103	23	23	171	1
	12	39.2	0.0	306	24	24	25	170	241	170	25	24	180	12
	Half-day Totals				81	144	634	1035	813	250	81	81	546	

TABLE 2–14   (continued)

Date	Solar Time, A.M.	Solar Position Altitude	Solar Position Azimuth	Direct Normal Irradiation, Btuh/sq ft	N	NE	E	SE	S	SW	W	NW	Horizontal	Solar Time, P.M.
Mar 21	7	11.4	80.2	171	8	93	163	135	21	8	8	8	26	5
	8	22.5	69.6	250	15	91	218	211	73	15	15	15	85	4
	9	32.8	57.3	281	21	46	203	236	128	21	21	21	143	3
	10	41.6	41.9	297	25	26	153	229	171	28	25	25	186	2
	11	47.7	22.6	304	28	28	78	198	197	77	28	28	213	1
	12	50.0	0.0	306	28	28	30	145	206	145	30	28	223	12
		Half-day Totals			112	310	849	1100	692	218	112	112	764	
Apr 21	6	7.4	98.9	89	11	72	88	52	5	4	4	4	11	6
	7	18.9	89.5	207	16	141	201	143	16	14	14	14	61	5
	8	30.3	79.3	253	22	128	225	189	41	21	21	21	124	4
	9	41.3	67.2	275	26	80	203	204	83	26	26	26	177	3
	10	51.2	51.4	286	30	37	153	194	121	32	30	30	218	2
	11	58.7	29.2	292	33	34	81	161	146	52	33	33	244	1
	12	61.6	0.0	294	33	33	36	108	155	108	36	33	253	12
		Half-day Totals			153	509	969	1003	489	196	146	145	962	
May 21	5	1.9	114.7	1	0	0	0	0	0	0	0	0	0	7
	6	12.7	105.6	143	35	128	141	71	10	10	10	10	30	6
	7	24.0	96.6	216	28	165	209	131	20	18	18	18	87	5
	8	35.4	87.2	249	27	149	220	164	29	25	25	25	146	4
	9	46.8	76.0	267	31	105	197	175	53	30	30	30	196	3
	10	57.5	60.9	277	34	54	148	163	83	35	34	34	234	2
	11	66.2	37.1	282	36	38	81	130	105	42	36	36	258	1
	12	70.0	0.0	284	37	37	40	82	112	82	40	37	265	12
		Half-day Totals			203	643	1002	874	356	194	171	170	1083	
June 21	5	4.2	117.3	21	10	21	20	6	1	1	1	1	2	7
	6	14.8	108.4	154	47	142	151	70	12	12	12	12	39	6
	7	26.0	99.7	215	37	172	207	122	21	20	20	20	97	5
	8	37.4	90.7	246	29	156	215	152	29	26	26	26	153	4
	9	48.8	80.2	262	33	113	192	161	45	31	31	31	201	3
	10	59.8	65.8	272	35	62	145	148	69	36	35	35	237	2
	11	69.2	41.9	276	37	40	80	116	88	41	37	37	260	1
	12	73.5	0.0	278	38	38	41	71	95	71	41	38	267	12
		Half-day Totals			242	714	1019	810	311	197	181	180	1121	
July 21	5	2.3	115.2	2	0	2	1	0	0	0	0	0	0	7
	6	13.1	106.1	137	37	125	137	68	10	10	10	10	31	6
	7	24.3	97.2	208	30	163	204	127	20	19	19	19	88	5
	8	35.8	87.8	241	28	148	216	160	29	26	26	26	145	4
	9	47.2	76.7	259	32	106	194	170	52	31	31	31	194	3
	10	57.9	61.7	269	35	56	146	159	80	36	35	35	231	2
	11	66.7	37.9	274	37	39	81	127	102	42	37	37	255	1
	12	70.6	0.0	276	38	38	41	80	109	80	41	38	282	12
		Half-day Totals			116	300	803	1045	672	221	117	116	738	
Aug 21	6	7.9	99.5	80	12	67	82	48	5	5	5	5	11	6
	7	19.3	90.0	191	17	135	191	135	17	15	15	15	62	5
	8	30.7	79.9	236	23	126	216	180	40	22	22	22	122	4
	9	41.8	67.9	259	28	82	197	196	79	28	28	28	174	3
	10	51.7	52.1	271	32	40	149	187	116	34	32	32	213	2
	11	50.3	29.7	277	34	35	81	156	140	52	34	34	238	1
	12	62.3	0.0	279	35	35	38	105	149	105	38	35	247	12
		Half-day Totals			161	503	936	961	471	202	154	153	945	
Sep 21	7	11.4	80.2	149	8	84	146	121	21	8	8	8	25	5
	8	22.5	69.6	230	16	87	205	199	71	16	16	16	82	4
	9	32.8	57.3	263	22	47	195	226	124	23	22	22	138	3
	10	41.6	41.9	279	26	28	148	221	165	30	26	26	180	2
	11	47.7	22.6	287	29	29	77	192	191	77	29	29	206	1
	12	50.0	0.0	290	30	30	32	141	200	141	32	30	215	12

Date	Solar Time, A.M.	Solar Position		Direct Normal Irradiation, Btuh/sq ft	Solar Heat Gain Factors, Btuh/sq ft									Solar Time, P.M.
		Altitude	Azimuth		N	NE	E	SE	S	SW	W	NW	Horizontal	
	Half-day Totals				116	300	803	1045	672	221	117	116	738	
Oct 21	7	4.5	72.3	48	1	20	45	41	12	1	1	1	3	5
	8	15.0	61.9	203	10	49	173	187	88	10	10	10	43	4
	9	24.5	49.8	257	17	23	180	235	151	18	17	17	96	3
	10	32.4	35.6	280	21	22	139	238	196	38	21	21	140	2
	11	37.6	18.7	290	23	23	70	212	224	100	23	23	167	1
	12	39.5	0.0	293	24	24	26	165	234	165	26	24	177	12
	Half-day Totals				83	143	610	989	783	245	84	83	535	
Nov 21	8	8.2	55.4	136	5	17	107	128	72	5	5	5	14	4
	9	17.0	44.1	232	12	13	151	219	156	13	12	12	54	3
	10	24.0	31.0	267	16	16	122	237	209	50	16	16	96	2
	11	28.6	16.1	283	19	19	61	218	240	116	19	19	123	1
	12	30.2	0.0	287	19	19	21	176	250	176	21	19	132	12
	Half-day Totals				61	71	442	884	798	267	62	61	353	
Dec 21	8	5,5	53.0	88	2	7	67	83	49	3	2	2	6	4
	9	14.0	41.9	217	9	10	135	205	151	12	9	9	39	3
	10	20.7	29.4	261	14	14	113	232	210	55	14	14	77	2
	11	25.0	15.2	279	16	16	56	217	242	120	16	16	103	1
	12	26.6	0.0	284	17	17	18	177	253	177	18	17	113	12
	Half-day Totals				49	54	380	831	781	273	50	49	282	↑
					N	NW	W	SW	S	SE	E	NE	HOR.	←P.M.

Data shown in table is for 40 degree north latitude. Solar intensity is given in the Direct Normal column. Solar heat gain factor is the peak heat gain of the hour calculated with clearness factor = 1.0 and ground reflectance = 0.20. Half day total is the average heat gain based on Simpson's rule on 10-minute interval. See ASHRAE Handbook of Fundamentals for data of other latitudes (16, 24, 32, 48, 56, and 64).

QUESTIONS

2.1 State the factors that affect environmental comfort and their general ranges of values, where applicable.

2.2 Briefly describe the difference between sensible and latent heat.

2.3 For air at 75°F, 50% relative humidity, what will happen to relative humidity if the air is cooled to 65°F? To 50°F?

2.4 How much water is present in 1 lb of air at 75°F, 50% relative humidity? How much water is present in 1 lb of air at 95°F DB (dry bulb)? 78°F WB (wet bulb)?

2.5 What is the rate of removal of moisture required to reduce a 100-cfm airstream from the higher to the lower condition in problem 2.4?

2.6 If 20 lb/hr of water is allowed to evaporate into an airstream, what will be the effect on the dry-bulb temperature? What will be the effect on the wet-bulb temperature?

2.7 How much heat is required to warm 100 gallons of water from 60°F to 130°F?

2.8 How much heat is liberated when 500 gallons of water cools from 160°F to 140°F?

2.9 What is the heat liberation rate for a 300-gpm water flow cooling from 140°F to 110°F?

2.10 A heating system load is 200,000 Btuh. How much heating water flow is required to satisfy the load if the system is designed for a 20°F temperature drop? How much flow is required for a 30°F drop?

2.11 How much steam flow would be required for a 200,000 Btuh heating load?

2.12 A cooling system load is 36,000 Btuh sensible. How much chilled air is required to satisfy the load if the system is designed for a 20°F temperature rise? How much flow is required for a 15°F rise?

2.13 Your client desires an interior temperature of 72°F for the design of his HVAC system in St. Louis (close to the airport weather station). Assuming that a 97.5% design will be satisfactory, what is the design temperature difference?

2.14 Your client desires an interior temperature of 78°F for the design of her HVAC system in St. Louis (close to the airport weather station). Assuming that a 5% design will be satisfactory, what is the design temperature difference?

2.15 What will be the *U*-factor for a wall constructed as follows?
- 6"-thick concrete wall.
- The wall is finished on the interior with 1/2" drywall, adhesively applied to the surface.
- The drywall is applied over 1.5" furring.
- The furring space is filled with insulation at R-4 per inch.

2.16 What will be the heating load for a wall 10' high by 50' long that is constructed as per the last description in Question 15 under the conditions specified in Question 13?

2.17 What will be the cooling load if the wall in Question 16 faces south? North? West?

2.18 What will be the heating load for a 5'-by-5' window constructed of single-pane clear glass according to the criteria in Question 13? Assume U = 1.05. Double-pane clear glass? Assume U = 0.55.

2.19 What will be the July cooling load for a 5' × 5' west-facing window constructed of single-pane clear glass according to the criteria in Question 13? Double-pane clear glass? Double-pane tinted glass with a shading coefficient of 0.75?

2.20 What internal heat gain will result from each of the following in an office?
- Five people, moderate activity
- Two copy machines, 500 watts each
- Six suspended light fixtures, two 60-watt lamps each

2.21 What minimum outside air quantity would be required for the room in Question 20?

2.22 What heating load would result from the minimum outside air in Question 21 under the criteria of Question 13? What cooling load?

3

HVAC DELIVERY SYSTEMS

COMPLETE HVAC SYSTEMS ARE COMPRISED OF SUB-systems that produce heating and cooling, move heat transfer fluids, and control delivery to a space to maintain stable conditions. The production of heating and cooling is performed by head-end equipment, including refrigeration devices, furnaces, and boilers. The movement of heat transfer fluids is performed by air-handling equipment, ductwork, grilles, and diffusers for air and by pumps and piping systems for water. Delivery is performed through concepts and equipment described in this chapter.

Delivery methods vary greatly in their ability to maintain space conditions, in the complexity of their operation and maintenance, and in their energy consumption. Selecting appropriate delivery systems is critical to the successful performance of systems.

3.1 CONTROL OF HEATING AND COOLING

Heating and cooling loads vary with time; therefore, the amount of heating or cooling supplied to a space must vary to keep the temperature constant within certain limits. Heating and cooling must be controlled to match the load and vary the flow of energy. Load conditions are measured at a control device, generally a space thermostat.

Spaces with similar load characteristics will require either heating or cooling, but not both at the

same time. Thus, HVAC services can be delivered by single packages or systems capable of nonsimultaneous heating and cooling. A single system will suffice, as in the case of a residential combination furnace and air conditioner, or a rooftop heating and cooling unit for a small office building.

Large buildings are comprised of interior and perimeter spaces. Interior spaces experience heat gains year round due to lights, appliances, and people and have no exterior exposure from which to lose heat during winter. Accordingly, interior spaces need air conditioning in both summer and winter. By contrast, perimeter spaces have walls and windows exposed to the outside. Hence, they will need heating during cold weather if the interior gains are not sufficient to compensate for heat losses to the outside.

Large buildings can be served by multiple small systems capable of providing either heating or cooling to individual spaces. They can also be served by larger central systems capable of simultaneously delivering heating and cooling services as required by individual spaces.

Central systems most often use air-handling units with coils to raise or lower the temperature of air supplied to spaces. The air is delivered by fans, either directly to the space or through a distribution system with ducts. When the air-handling unit serves multiple spaces with different load conditions, there must be means for providing various degrees of heating and/or cooling to individual areas.

Air systems offer several methods of control to vary the amount of heating or cooling supply:

- varying the temperature of the air supplied, while holding the flow constant
- varying the flow of warm or cold air supplied, while holding the temperature constant
- varying both the temperature and the flow of air supplied

These functions may be accommodated within the air-handling unit or by control devices located downstream of the unit.

Air systems are sometimes combined with convection and radiation devices for heating. These are generally controlled by turning them on and off, as in the case of electric heat, or by modulating the temperature or flow of the heat source, as in the case of hot water or steam heat.

3.2 ZONING

In HVAC terminology, a *zone* is an area or temperature that is controlled by a single thermostat. For instance, a house with one furnace controlled by a single thermostat would be termed a single-zone system. If a larger house were to have two furnaces, each controlled by a separate thermostat, the system would have two zones. Zones are not to be confused with rooms. Several rooms, or even an entire building, might be controlled as a single zone. Complex buildings require many zones of control to accommodate load variations among their spaces.

Many systems use air terminal devices to serve individual spaces or groups of spaces. In general, each terminal will correspond to a zone of control. Terminals can be physically located as shown in Figure 3–1.

3.3 CONTROLS AND AUTOMATION

3.3.1 Definition

Controls and automation provide the intelligence of mechanical and electrical systems. Automation is the function of having the equipment react automatically, without any operator intervention, to satisfy preset conditions. Control occurs when a signal to the equipment causes the movement or adjustment of a component to produce the desired result. Equipment is selected predominantly to provide adequate capacity for the needs of the building under the design conditions, whereas controls and automation make the equipment operate under all anticipated conditions. Once the equipment is selected, the design of the controls and automation system begins in earnest.

3.3.2 Basic Control Systems and Devices

All HVAC systems require some form of controls, either manual or automatic. Automatic controls enable the equipment or the entire system to operate more

■ **FIGURE 3–1**
Typical placement of air terminals. (a) Sill line—applicable to fan-coil units, unit ventilators, induction units, water source heat pumps, PTACs, etc. (b) Ceiling suspended—applicable to all systems. (c) Concealed in soffit—applicable to fan-coil units and water source heat pumps. (d) Concealed in ceiling—applicable to all systems.

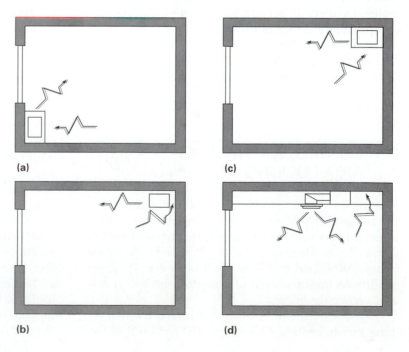

(a)

(c)

(b)

(d)

precisely and reliably to provide comfort, safety, and energy efficiency. They accomplish their task by controlling one or more of the following properties of the transporting medium, such as air or water, and the related equipment:

- Temperature—with sensors set for an operating temperature, a differential, or temperature limits
- Pressure—with sensors set for an operating pressure, a differential, or pressure limits
- Flow rate—with sensors set for an operating rate, a differential, or flow rate limits
- Humidity—with sensors set for an operating level, a differential, or humidity limits
- Speed—with sensors to control the equipment so that it is either on or off or has variable or multiple speeds
- Time—with a clock or a program to control the duration of operation of the equipment

Basic control systems   Control systems may be electric, electronic, pneumatic, direct digital control, or a combination of these.

Electric controls use line voltage (120 volts) or low voltage (12 to 24 volts) to perform the basic control functions. A low-voltage control system is more sensitive and is therefore preferred.

Pneumatic controls use 5- to 30-psi compressed air and receiver controllers with force-balance mechanisms. The receiver controllers constantly adjust output air pressure to actuators in response to input pressure from sensors to produce the desired result. Electronic control uses a similar form of controller, except the signals are electronic rather than pneumatic. These basic systems require manual calibration and adjustment.

More sophisticated than electric or pneumatic control systems, and increasingly more common, are direct digital control systems, or DDC systems. As with pneumatic or electric controls, an input signal to a controller results in an output to the appropriate device. The major difference is in the controller: Instead of physically adjusting the controller components to result in the same reaction again and again, DDC system controllers contain microprocessors that are programmed to interpret the input signal, process the data in resident programs, and intelligently decide the appropriate response. These systems have all of the features of pneumatic and electronic controls and can, additionally, anticipate needs based on recorded trends of previous operation of the equipment.

Many buildings have hybrid systems of automation and controls that combine DDC controllers with pneumatic or electronic sensors and actuators.

These systems are typically less expensive than full DDC, are generally more familiar to most facility operating staff, and have been proven reliable through the past experience of many actual installations. Figure 3–2 shows the typical hierarchy of a distributed control system.

Basic control devices

Control devices include sensors, controllers, and actuators. Sensors measure the monitored or controlled variable. The signal from the sensors is input to a controller for processing and decision making. The controller determines if a signal should be sent to a monitoring station or to an actuator. The controller also determines if the output signal is two-position or proportional, or direct- or reverse-acting. Actuators manipulate the equipment to meet the desired set point of the controlled variable.

Two-position signals are used to indicate the operating status of the equipment, such as on or off, normal or alarm, open or closed, or uses output signals to start, stop, open, or close the controlled equipment. Proportional signals are used to monitor and control variables that change continuously, such as temperature, pressure, or flow. Proportional signals can provide multiple levels of control and alarm through the use of a single sensor or actuator.

Thermostats are devices that sense and respond to temperature combining the functions of the sensor and controller. Room (space) thermostats may be designed for heating or cooling alone or for heating and cooling in combination. The simplest combination thermostat normally contains a mode selection switch. If this switch is set on the heating mode, the thermostat allows more or less heating energy to be added to the space until a certain level is reached. If the space temperature is already higher than the thermostat setting, due to an uncontrollable heat source, such as solar heat gain through the windows, then lowering the thermostat set point will not bring down the space temperature. The reverse will be true when the mode switch is set in the cooling position. These simple facts are often misunderstood by most users or occupants of buildings.

More sophisticated, programmable electronic thermostats can be seasonally switched if they have 365-day programs. This, of course, requires the programmer to anticipate the beginning of the heating and cooling seasons in order to identify when the switch from heating to cooling should occur. A warm winter day or a cold spring or fall day can contribute to occupants' discomfort with such a system. DDC thermostats are capable of being tied into distributed communication networks and can share information,

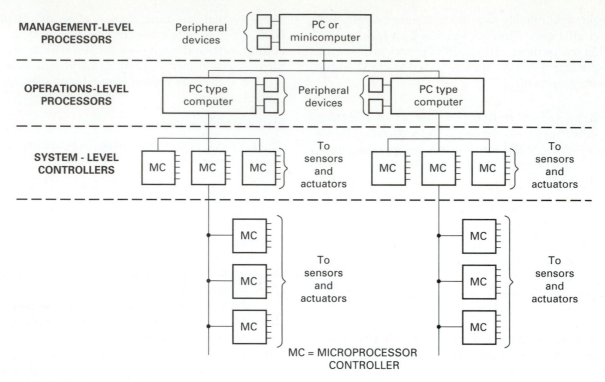

MC = MICROPROCESSOR CONTROLLER

■ **FIGURE 3–2**
Hierarchical control system configuration.

such as what the outdoor temperature is, to "decide" whether the HVAC system should be in heating or cooling mode, without any intervention by an operator.

Some thermostats can automatically change over from heating to cooling or vice versa. However, such changeovers usually require a more sophisticated HVAC system. A simple heating system using a thermostat is shown in Figure 3–3 (a). Adding other control components to a simple heating system can enhance the system control. An example of an expanded heating control arrangement is shown in Figure 3–3 (b).

Humidistats are devices that sense and respond to humidity—either relative or absolute. A space humidity sensor would send the appropriate signal to the controller to determine whether humidity should be added to or removed from the supply air. In order to control the space humidity, the airflow can have humidity added by a humidifier or removed by overcooling the air to remove excess humidity and then reheating the air to meet the requirements of the thermostat. All of these functions are performed through controls, without any action by the occupants other than adjusting the set point of the humidistat to the appropriate level for space functions.

Other control sensors include pressure switches and transmitters, which respond to pressures;

flow switches and transmitters, which respond to rates of flow; speed switches, which respond to flow, pressure, or a program in order to control the speed of the equipment; and position switches, which respond to signals to open, close, or modulate dampers, valves, etc.

Controllers provide the decision-making function of the control system. Controllers are available for all types of control systems: electric, pneumatic, electronic and DDC. Output signals from these devices are typically two-position, proportional, direct acting or reverse acting. DDC controllers are available as preprogrammed or fully programmable. Various levels of control that are identified in Figure 3–2 are provided at the controller. Zone level controllers are used to operate terminal units or small unitary equipment. Many control system manufacturers provide dedicated controllers for specific types of zone equipment. Preprogrammed controllers for unitary equipment such as heat pumps or VAV terminal units can be installed very inexpensively. System-level controllers are typically provided without any programming because of the uniqueness of each installation. These controllers are provided with additional input and output capabilities over the zone controllers. Also, the signals are usually "universal," which results in a completely flexible system for adding or modifying the inputs and outputs.

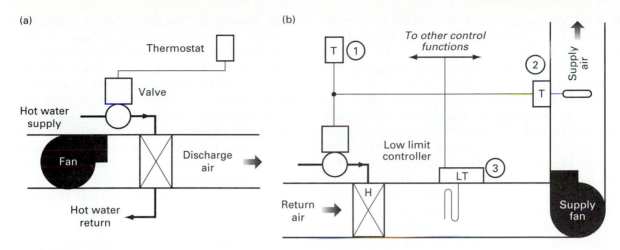

(a)   (b)

Thermostat

Valve

Hot water supply

Fan

Discharge air

Hot water return

To other control functions

T ①

② Supply air

T

Low limit controller

LT ③

H

Return air

Supply fan

■ **FIGURE 3–3**

(a) Shows an air handling system with a hot water coil. The thermostat modulates a 2-way valve to control the flow of hot water through the coil for heating the discharge air. (Courtesy: Honeywell, Minneapolis, MN.) (b) Schematic of an expanded heating control system where space thermostat (1) modulates a 2-way coil valve to maintain space temperature. A low limit discharge temperature controller (2) overrides the thermostat to prevent the discharge air temperature from falling below a preset minimum temperature. A separate controller (3) may be installed to initiate a sequence of operations, such as to close a damper, to turn off the fan, to open the valves, etc. (Courtesy: Honeywell, Minneapolis, MN.)

Actuators provide the physical control of the equipment, typically dampers and valves. Movement of the actuator results in a proportional response of the controlled equipment. Actuators are provided as normally open or normally closed. Normally open actuators return to the open position if the control signal is off; normally closed actuators return to the closed position.

Control valves are used to control the flow of fluids. The valve can be either direct-acting or reverse-acting. A direct-acting valve allows flow with the stem up, while a reverse-acting valve shuts off flow with the stem up. The combination of valve body and actuator (called the valve assembly) determines the valve stem position. Figure 3–4 (left) shows a direct-acting two-way valve; Figure 3–4 (right) shows a reverse-acting two-way valve. A two-way valve is a valve with one inlet and one outlet port; a

three-way valve may have either one inlet and two outlets or two inlets and one outlet port.

3.3.3 Energy Management

Energy management by automation and control systems became prominent in the 1970s. Automation and controls continue to play a major role in managing a building's energy costs. As a rule, the more segregated the equipment (numerous zones, multiple thermostats, etc.), the less control technology is required to make the systems perform at their most efficient energy level. In order to effectively manage energy usage, the controllers of the energy management system must be able to "speak" to one another. Addition of a local area network for dedicated communication between all controllers is the most common way to have all controllers communicate. Communi-

■ **FIGURE 3–4**

Two-way valves: the selection of valve body whether direct-acting or reverse-acting depends on the combination of valve body and actuator.

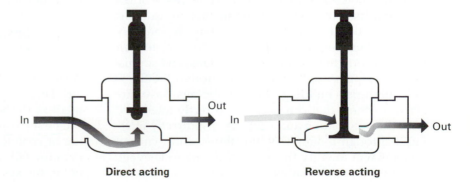

In   Out

In   Out

Direct acting   **Reverse acting**

cation by telephone lines is a method more commonly used to communicate between buildings or sites. A common headend computer is the interface for an operator to monitor, control, program, and generate reports.

The most basic energy management option is to shut down unnecessary equipment. Scheduling equipment on and off is more feasible with multiple HVAC systems because it can be done without disturbing areas served by other systems. On the other hand, large central HVAC systems can provide the needed service to all areas covered by multiple systems. With a more sophisticated automation and control system, zoning can be achieved through the automation and control equipment and, with proper programming, can result in similar energy savings.

Operating equipment at optimum conditions is another way to save energy with controls. Evaluation of loads and conditions leads to a decision about what to operate at that time. An excellent example is operating multiples of a single equipment type. The energy management system can determine if it is more economical to operate one at high loading, two at low loading, or some other combination that results in minimum energy consumption.

Energy can also be saved through monitoring demand and shutting down equipment to prevent consumption from exceeding a preset limit. This requires an interface with the main electrical service and individual panels or circuits to achieve the desired results.

Another method of providing energy management is to record equipment trends. When a mechanical or electrical system begins operating outside its normal range, a typical indicator is increased energy usage. By tracking the energy usage of the equipment, it is possible to identify when operations are outside the normal range and perform the required maintenance.

3.3.4 Life Safety

Although life safety is not a required part of automation and controls systems, one must be aware of the potential uses of such systems in that regard. The most significant example is providing an interface with the building fire alarm system to activate stair pressurization fans, smoke exhaust fans, and sophisticated air pressure "sandwiches" around fire floors in a high-rise building. Most buildings have fire alarm systems that can be either directly interfaced with the automation and controls or indirectly interfaced through a simple relay cabinet. Most current systems will identify the exact location of the event. Based on that information, the automation and con-

trols system can "decide" whether it is necessary to activate any pre-defined sequences for smoke removal and containment and, if so, which areas will be affected. "Sandwiching" of fire floors requires information on the exact location of the event and floor-to-floor control of the HVAC systems through equipment zoning or zone segregation by dampers. This arrangement creates a negative pressure on the fire floor and a positive pressure on the floors immediately above and below the fire floor. The intent is to allow occupants to evacuate the fire floor and to minimize the opportunity for smoke to migrate to surrounding floors. Other typical interfaces with the fire alarm system include the operation of a smoke exhaust fan and a stair pressurization fan. Also, a shutdown of equipment due to smoke or fire can be monitored by both the fire alarm and automation and controls systems. Such monitoring allows HVAC operators to know whether the abnormal off condition is due merely to a mechanical failure or to a potentially life-threatening situation.

3.3.5 Equipment Safety

Equipment safety is typically provided by monitoring specific components for their hours of operation, temperature, and other parameters and then providing the appropriate maintenance response. Additional safety features of the control system include devices in air systems that prevent their operation if damage could occur. Typical examples of these devices are "freezestats" to protect water coils from low temperatures, and smoke detectors or "firestats," to protect equipment from operating if a fire occurs in the motor.

3.4 COMMONLY USED SYSTEMS FOR ZONE CONTROL

A wide variety of systems is available for building applications. They differ with respect to comfort, cost, energy efficiency, maintainability, and flexibility in terms of altering them.

3.4.1 Constant Temperature, Variable Volume (On-Off)

A simple residential system will provide hot or cold air. The supply temperature in either the cooling mode or the heating mode is fairly constant. Controls can be set to allow the fan to operate only when heating or cooling is being furnished. As the space temperature falls below or rises above the thermostat set point, the system is activated, and hot or cold air is supplied to the space at a constant rate until the space warms or cools and trips the thermostat. The volume of air supplied to the space will depend on how long the fan is activated. When the proper amount of warm or cold air has been supplied, the system deactivates.

3.4.2 Single-Zone Constant Air Volume

In commercial buildings, fans are generally required to run continuously to provide ventilation while the building is occupied. This mode of operation is called

constant volume. Single-zone constant volume is the simplest of systems. Air is supplied at a constant rate to the space. The air-handling unit includes a cooling coil and/or a heating coil that varies the temperature of the air in response to a space thermostat. Heating and cooling may be on-off or proportional. (See Figure 3–5).

If the space is composed of rooms or areas with different load characteristics, single-zone constant volume will not be a good choice. Good control can be achieved at the location of the thermostat, but the temperature will not be controlled elsewhere. In gen-

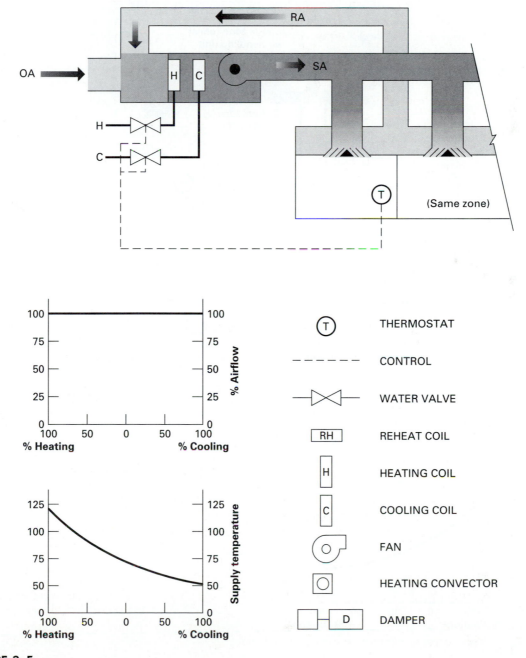

■ FIGURE 3–5
Single-zone constant air volume system.

eral, single-zone constant volume is satisfactory only for small simple buildings comprised of spaces with similar load characteristics or for large open spaces. However, single zone constant volume systems used in multiples can provide satisfactory control for large or complex buildings.

3.4.3 Single-Zone Reheat

Single-zone reheat is a constant-volume system used for air conditioning when humidity control is espe-

cially important. Similar to the single-zone constant volume system, the single-zone reheat system is satisfactory only for a single spaces with similar load characteristics or for large open spaces. The single-zone reheat system uses a cooling coil to cool and dehumidify air, along with a reheat coil for temperature control. Humidifiers may also be present. (See Figure 3–6.)

A space humidistat may be used to adjust the air temperature to condense enough moisture so that the supply air is sufficiently dry for maintaining proper humidity levels in the space. Since the cooling coil is

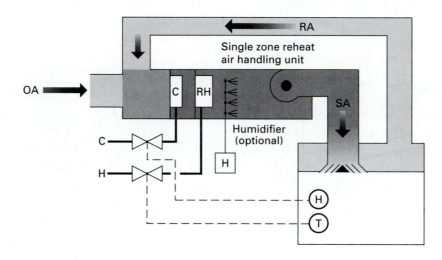

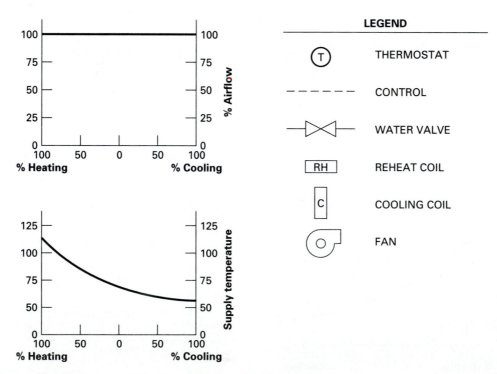

■ **FIGURE 3–6**
Single-zone constant-volume reheat system.

controlled in order to maintain the humidity level, the air temperature would generally be too cold to maintain a proper space temperature. Therefore, a heating coil is placed downstream of the cooling coil to reheat the air. The heating coil is controlled by the space thermostat.

Single-zone reheat systems are characterized by high energy usage. Air may be overcooled to maintain humidity and reheated to compensate for the overcooling. These systems are used only for special applications, such as hospital operating rooms and computer rooms.

3.4.4 Constant-Volume Terminal Reheat, Multiple Zones

Constant-volume terminal reheat systems (Figure 3–7) are similar in principle to single-zone reheat systems, except that multiple zones of temperature control can be achieved. The air-handling unit contains a cooling coil that chills all the air supplied to the various zones. One or more main trunk ducts are used to distribute air throughout the area served, and a terminal box containing a reheat coil is installed for each zone. The cooling coil chills air to the same temperature for all zones. The reheat coils respond to their respective thermostats to maintain temperature control in each zone.

Terminal reheat systems are capable of serving multiple zones from a single air-handling unit. Reheat terminals, often called boxes, are modular devices that are available in sizes from approximately 50 cfm to 3000 cfm. Other types of terminal are available in similar sizes. These sizes are adequate to satisfy spaces ranging from individual small offices to large open areas up to approximately 5000 sq ft. A typical reheat box with hot water and electric coils is shown in Figure 3–8. Reheat coils are normally installed above the ceiling. Hot water coils require piping and accessories, which may be prone to leakage.

Constant-volume terminal reheat systems are flexible enough that one may add or rearrange reheat boxes on their main trunk as space is reconfigured. Excellent humidity control is another of their characteristics. These advantages led terminal reheat systems to be used extensively in commercial and institutional buildings designed from the 1950s through the early 1970s. During the mid-1970s, reheat systems lost favor due to a concern for energy conservation and operating costs.

3.4.5 Constant-Volume Dual Duct

Constant-volume dual duct systems (Figure 3–9) send warm and chilled air through a pair of main trunk ducts to the areas being served. At the air-handling unit, a single fan blows air through a heating coil and a cooling coil into a warm-air trunk duct and a chilled-air trunk duct. Constant-volume dual duct systems are capable of controlling multiple zones, with each zone served by an individual mixing box. (See Figure 3–10.)

Mixing boxes are controlled with dampers to vary the quantity of warm and chilled air in response to the zone thermostat. A typical box is shown in Figure 3–7. If a zone needs more cooling, the dampers will be positioned to deliver more chilled air and less warm air. Conversely, the dampers will be positioned to deliver more warm air and less chilled air if a space needs heating or less cooling. The total airflow, consisting of warm and chilled air, stays fairly constant, since each damper closes as the other opens.

During warm weather, the cooling coil operates to provide air conditioning, but the heating coil can be inactive, as there will be no spaces requiring heating. The return air will be sufficiently warm to mix with chilled air for space temperature control. During cold weather, the heating coil must be activated to provide warmer air for heating. At the same time, interior spaces will require cooling. When cooling and heating are operated simultaneously, the mixing of warm and cold air wastes energy, a problem similar to that caused by reheating.

The constant-volume dual-duct system has the flexibility to add or rearrange mixing boxes on the main trunks as space is reconfigured. It also has the advantage of being an all-air system. Unlike the constant-volume terminal reheat system, there are no coils installed over occupied spaces, thus reducing maintenance and avoiding the possibility of water leaks.

3.4.6 Multizone

Multizone systems are very similar in operating concept to dual-duct systems. Warm and cold air are mixed to produce the right air temperature for conditioning according to signals from the zone thermostats. Unlike dual-duct systems, however, in multizone systems the mixing is done by dampers located at the air-handling unit rather than at terminals located near the spaces served. Figures 3–11 and 3–12 show the operating concept and configuration of a multizone air-handling unit. An individual duct is run from the air-handling unit to each zone of control. A single multizone unit can serve up to approximately 12 zones.

Similar to dual-duct systems, multizone systems waste energy due to mixing warm and cold air. In addition, multizone systems have the disadvantage

(Text continues, page 81.)

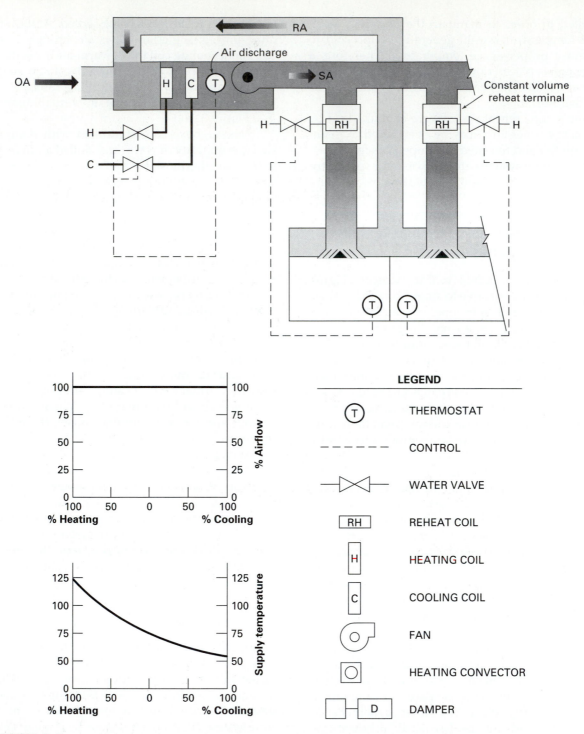

LEGEND

(T)	THERMOSTAT
– – – – –	CONTROL
⋈	WATER VALVE
RH	REHEAT COIL
H	HEATING COIL
C	COOLING COIL
	FAN
◯	HEATING CONVECTOR
D	DAMPER

■ **FIGURE 3–7**
Constant-volume terminal reheat system.

■ FIGURE 3–8
A typical single-duct terminal with hot water reheat coil. The coil may also be electric. (Courtesy: Carnes Company, Inc., Verona, WI.)

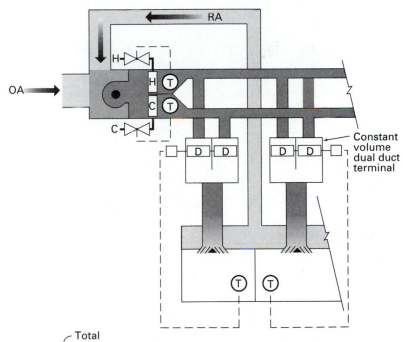

■ FIGURE 3–9
Constant-volume dual-duct system.

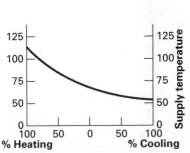

LEGEND

Ⓣ	THERMOSTAT
- - - - -	CONTROL
⋈	WATER VALVE
RH	REHEAT COIL
H	HEATING COIL
C	COOLING COIL
⊙	FAN
◎	HEATING CONVECTOR
□—D	DAMPER

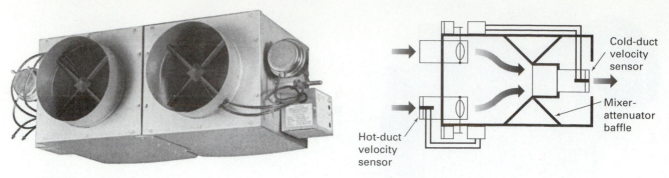

■ **FIGURE 3–10**
Typical dual-duct terminal and its internal section. (Courtesy: Titus, Richardson, TX.)

Cold-duct velocity sensor

Mixer-attenuator baffle

Hot-duct velocity sensor

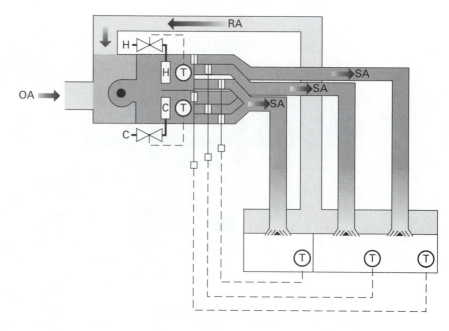

■ **FIGURE 3–11**
Constant-volume multizone system.
(Sketch shown is of a triple-zone, double-deck multizone unit.)

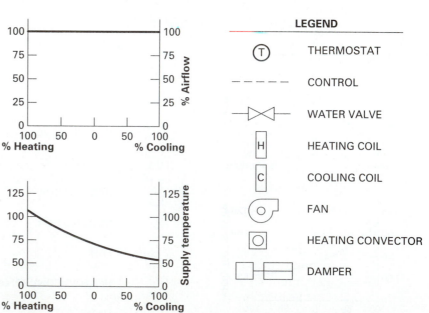

LEGEND	
Ⓣ	THERMOSTAT
– – – – –	CONTROL
⋈	WATER VALVE
H	HEATING COIL
C	COOLING COIL
Ⓞ	FAN
⊙	HEATING CONVECTOR
▭	DAMPER

(a)

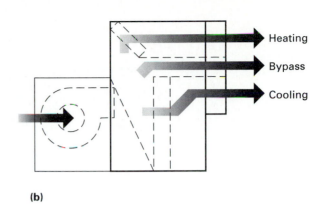

Heating

Bypass

Cooling

(b)

■ **FIGURE 3–12**

(a) Photo of a triple-deck multizone unit. (b) Schematic of the triple-deck unit. The three decks are the standard hot and cold decks plus a bypass deck. This configuration offers significant energy conservation opportunities by allowing return or outside air to bypass both coils, and the thermal inefficiency of mixing heated and cooled air is eliminated. (Courtesy: McQuay International, Minneapolis, MN.)

of being inflexible with regard to change. Rearranging or adding zones will involve considerable ductwork and modification of the unit. Multizone does, however, have the advantage of centralizing controls within the equipment room for easier maintenance and less intrusion into occupied spaces. This is important where security is an issue or where ceiling terminals might be difficult to access.

3.4.7 Single-Zone Variable Air Volume

Single-zone variable air volume (VAV) systems gained popularity during the mid-1970s as an energy-efficient alternative to constant-volume terminal reheat and constant-volume dual-duct systems. VAV systems used for cooling vary the quantity of air supplied to the space in response to the space thermostat. The air-handling unit supplies chilled air. If the cooling load is high, the chilled airflow will be high. As the load diminishes, the airflow is reduced accordingly. No energy is wasted due to reheating or mixing of warm and chilled air. In addition, varying the air quantities offers energy savings in terms of operation of the fan in comparison with constant-volume systems.

Single-zone VAV systems use a cooling coil to produce chilled air at a constant temperature. Airflow is increased or decreased in response to the space thermostat by adjusting dampers or the speed of the fan. Such a unit might also be equipped with an additional coil for heating. Because the unit can serve just one zone, it is suitable only for small, simple buildings or, in multiples, for large buildings. (See Figure 3–13.)

3.4.8 Multiple-Zone Variable Air Volume

Multiple-zone VAV systems use a VAV air-handling unit to supply chilled air into a main trunk duct-feeding multiple VAV terminals serving individual zones of control. VAV terminals, or boxes, contain dampers that vary the airflow to individual zones of control in response to their thermostats. A typical box is shown in Figure 3–14. This form of VAV system can serve large, complex buildings, offering the flexibility of being able to add or rearrange zones by adding or rearranging terminals on the main trunk ducts.

VAV terminals are inexpensive and energy efficient. Despite complaints that they produce stuffiness at low airflow, they are used extensively in commercial and institutional buildings. VAV terminals are designed to provide only cooling. For buildings with heating loads, they must be used in combination with other devices to provide heat. Figure 3–15 shows the operating concept of a typical multizone VAV terminal.

Interior spaces of large buildings need cooling year round, and VAV systems often meet this need. Perimeter spaces need heating during cold weather and can be served by several means, including convectors, radiant panels, or air terminals with heating capability as well as cooling.

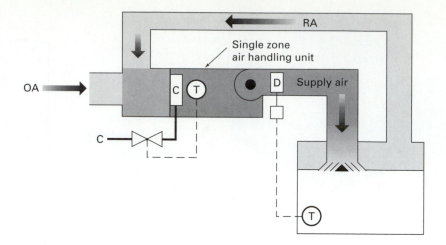

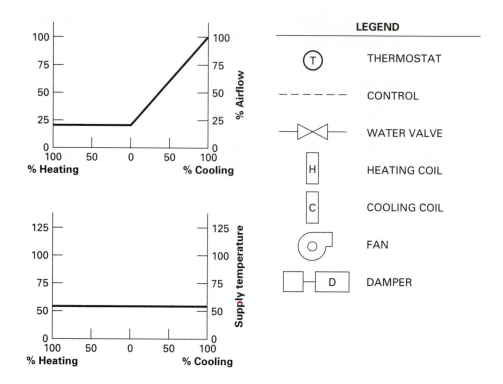

■ **FIGURE 3–13**
Single-zone variable air volume system.

LEGEND

(T)	THERMOSTAT
- - - - -	CONTROL
▷◁	WATER VALVE
H	HEATING COIL
C	COOLING COIL
(O)	FAN
□—D	DAMPER

Perimeter convectors are often used as a source of heat in conjunction with VAV terminals for cooling. Perimeter convectors use electric resistance elements or hot water piping equipped with fins to enhance heat transfer. Often called *finned tube,* this equipment is placed in a low, linear enclosure called a baseboard or in a taller configuration called a sill line. Baseboard or sill line convectors are excellent at preventing downdrafts under windows. They have the disadvantage, however, of taking up several inches of perimeter floor space and sometimes impairing the flexibility of furniture arrangement. (See Figures 3–16 and 3–17.) A combination system, including perimeter convectors with VAV, is shown in Figure 3–18.

Radiant panels are generally located at the ceiling and use electric resistance elements or hot water piping to warm their surface and radiate infrared heat downward. Infrared heat offers little protection against downdrafts and is directional. Areas under desks and tables may not be adequately exposed to the heating effect.

The top story of a building poses another perimeter problem: heat lost through the roof. Convectors or unit heaters can be installed to prevent cold air

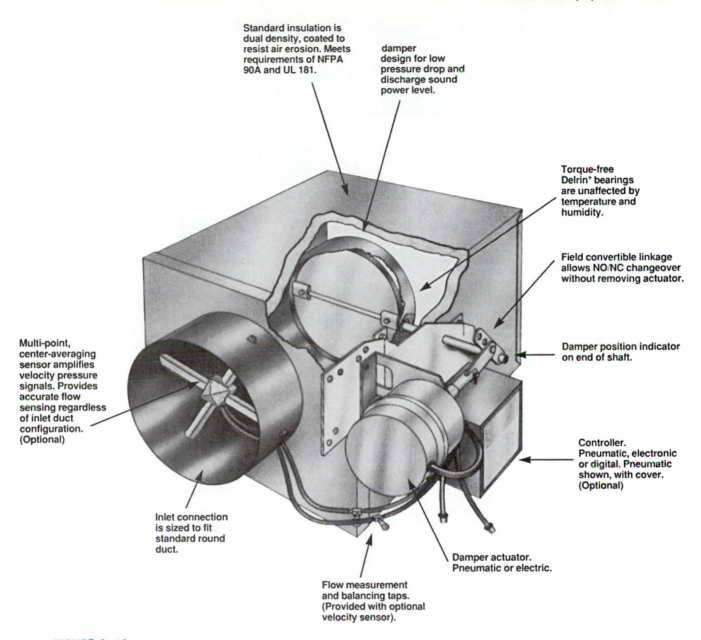

Standard insulation is dual density, coated to resist air erosion. Meets requirements of NFPA 90A and UL 181.

damper design for low pressure drop and discharge sound power level.

Torque-free Delrin® bearings are unaffected by temperature and humidity.

Field convertible linkage allows NO/NC changeover without removing actuator.

Multi-point, center-averaging sensor amplifies velocity pressure signals. Provides accurate flow sensing regardless of inlet duct configuration. (Optional)

Damper position indicator on end of shaft.

Controller. Pneumatic, electronic or digital. Pneumatic shown, with cover. (Optional)

Inlet connection is sized to fit standard round duct.

Damper actuator. Pneumatic or electric.

Flow measurement and balancing taps. (Provided with optional velocity sensor).

■ **FIGURE 3–14**
Variable air volume terminal or box. Air terminals come in a number of designs, including constant air volume (CAV) or variable air volume (VAV) and single duct or dual duct, and the airflow rate may be dependent on or independent of the system pressure. Control power may be electric, pneumatic, electronic, or direct digital. The flow of air to a room or a zone is in response to a room thermostat or other signaling devices. The illustration shown is a pneumatic-controlled single-duct terminal. (Courtesy: Titus, Richardson, TX.)

from entering occupied spaces served by cooling-only systems.

3.4.9 Variable Air Volume Reheat Terminals

VAV reheat terminals are equipped with electric or hot water heating coils (see Figure 3–19) and are similar to constant-volume reheat terminals except for control of the airflow. While the space requires cooling, the VAV reheat terminal supplies chilled air. As the cooling load diminishes, the amount of chilled air is reduced accordingly, to a preset minimum quantity. When heating is required, the heating coil warms the minimum quantity of chilled

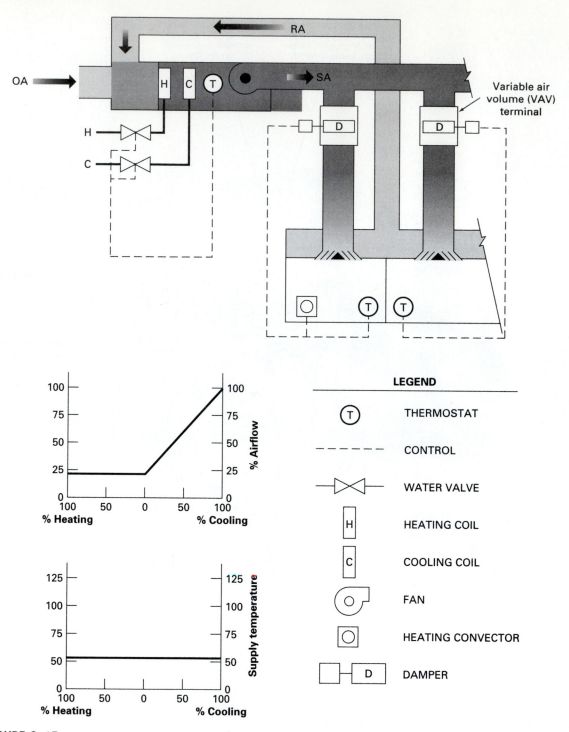

LEGEND

(T)	THERMOSTAT	
– – – –	CONTROL	
⋈	WATER VALVE	
H	HEATING COIL	
C	COOLING COIL	
(○)	FAN	
▢	HEATING CONVECTOR	
▢—D	DAMPER	

■ **FIGURE 3–15**
Multiple-zone VAV system.

air to a temperature that is suitable for heating the space. The energy waste due to reheat is much less than in constant-volume reheating. Figure 3–20 illustrates the operation of a VAV reheat terminal.

VAV reheat terminals are equipped with electric or hot water heating coils, and VAV reheat terminals are often used for perimeter heating in combination with simple VAV terminals for the interior (Figure

■ FIGURE 3–16
A typical convector installed at the base of a perimeter wall to overcome the downdraft during cold weather.

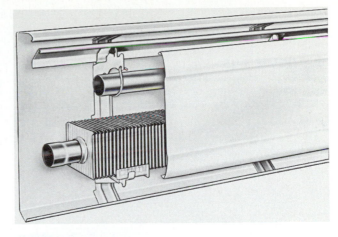

■ FIGURE 3–17
Cutaway view of the hot water convector showing the finned tube and supply pipe. Convectors may use steam or electric power and come in various designs and heights. (Courtesy: Sterling Co., Westfield, MA.)

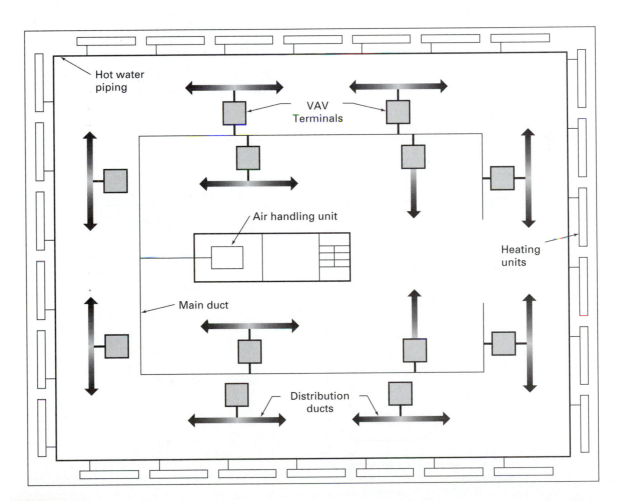

■ FIGURE 3–18
A multiple VAV interior and hot water convector exterior zone system.

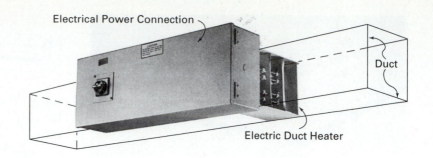

■ **FIGURE 3–19**
Construction of an electrical reheat coil in the ductwork. The coil may be a hot water or an electric coil. (Courtesy: Titus, Richardson, TX.)

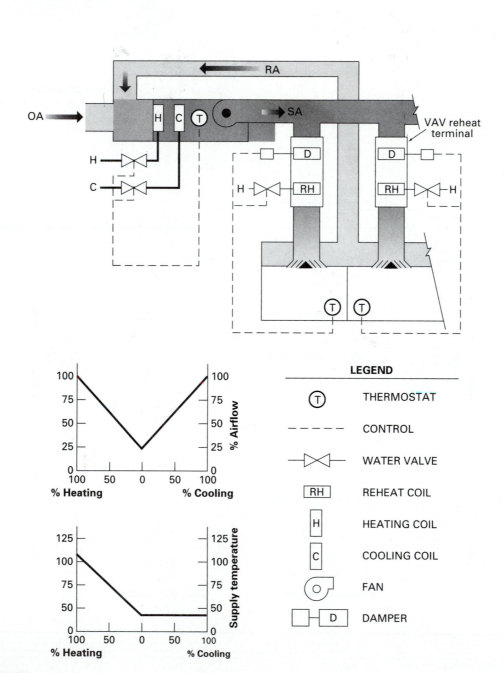

LEGEND

(T)	THERMOSTAT
– – – – –	CONTROL
▷◁	WATER VALVE
RH	REHEAT COIL
H	HEATING COIL
C	COOLING COIL
⊙	FAN
□–D	DAMPER

■ **FIGURE 3–20**
Variable air volume reheat system.

3–17). VAV reheating uses more energy than VAV with perimeter convectors and offers little resistance to downdrafts. On the other hand, VAV reheating is generally less expensive and avoids locating equipment on the floor, improving the flexibility of space usage.

3.4.10 Variable Air Volume Dual-Duct Terminals

VAV dual-duct terminals are arranged for operation at a variable rather than a constant air volume in order to conserve energy. They are served from a warm air duct and a chilled air duct. While the space requires cooling, the VAV dual-duct terminal supplies chilled air through a damper from the chilled air duct. As the cooling load diminishes, the amount of chilled air is reduced accordingly, while the damper from the warm air duct remains closed. When heating is required, the chilled air damper will be at or near the closed position, and the warm air damper will start to open and increase the flow of warm air as the heating load increases. By this mode of operation, there is minimal overuse of energy due to mixing of warm and cold air at the terminal. Figure 3–21 illustrates the operation of a VAV dual-duct terminal.

VAV dual-duct terminals are another common choice for perimeter heating with an interior system using cooling-only VAV terminals. The central air-handling unit is arranged to provide both warm and chilled air. VAV terminals serving interior spaces are connected to the chilled air supply. Dual-duct terminals are connected to both the warm and the chilled air supply.

VAV dual-duct terminal systems are relatively energy efficient and have several advantages over VAV with perimeter convectors and VAV reheat. VAV dual duct does not require any electric or hot water devices to be maintained in occupied spaces at the perimeter; all heating equipment is confined to the mechanical room.

3.4.11 Variable Air Volume Multizone

VAV multizone is similar in operating concept to VAV dual duct. Separate dampers are used for the warm and cold portions of the mixing section. (See Figure 3–22.) The cold air damper is almost closed before the warm air is allowed to open. This prevents energy waste due to mixing.

3.4.12 Fan Terminal Units

Fan terminal units are designed to overcome several of the complaints about the performance of VAV terminals. The nature of VAV is to decrease the flow of cooling air under light load conditions. This can create stuffiness in occupied spaces. The fan terminal unit contains a VAV damper that responds to the space thermostat. Chilled air through the VAV damper is mixed with return air from above the ceiling and delivered to the space via a small constant-volume fan within the unit. Since the fans within the fan terminal units have only a short length for distribution to the space, their pressure requirements are low, and energy consumption is accordingly modest. The central air-handling unit takes full advantage of reduced chilled air requirements at partial load, and overall energy savings are achieved at the central fan. Fan terminal units can also be equipped with heating coils and are sometimes used in combination with VAV systems. Figure 3–23 shows a typical fan terminal unit; its operation is diagrammed in Figure 3–24.

Figures 3–18 and 3–25 illustrate two of many possible choices of heating and cooling systems for a typical building. The former system, a single-duct VAV interior system with a hot-water convector exterior system, is sometimes described as a "split" system which incorporates both air and water (or steam) as the means of energy distribution. The latter system, a single-duct VAV interior system with dual-duct VAV for the exterior zones, is sometimes described as an "all air" system which incorporates air only as the means of energy distribution. The application of each system depends on a number of factors, such as the availability of hot water or steam, relative cost of energy, building construction, floor-to-floor height, depth of ceiling cavity, depth of structural members, aesthetics, and the number of exterior zones desired. Although no generalized conclusion can be made, the all-air system is usually lower in cost, but requires more space, both for the high-velocity and the low-velocity ducts. In specific building applications, the increased duct sizes and number of terminal units may preclude the use of an all-air system regardless of the economics.

3.4.13 Fan Coils

Fan coils are another system that can be used to provide multiple zones of control. A fan coil consists of a filter, a cooling and/or heating coil, a fan, and controls. Units can be located above the ceiling, in wall cabinets, or in soffits. Typical fan coil units are shown in Figure 3–26. Fan coils can also be mounted above ceilings or in soffits. Each fan coil is capable of control by an independent thermostat.

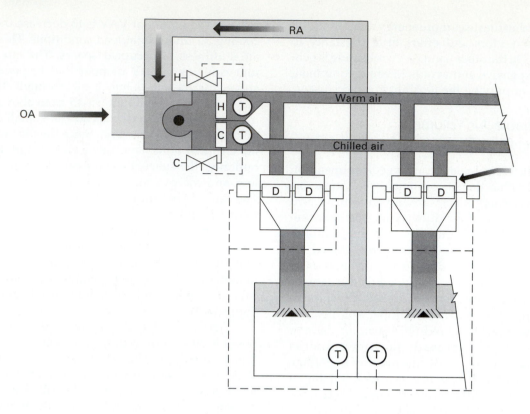

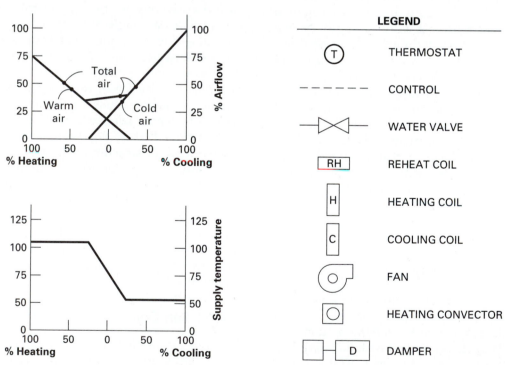

■ **FIGURE 3–21**
Variable air volume dual-duct system.

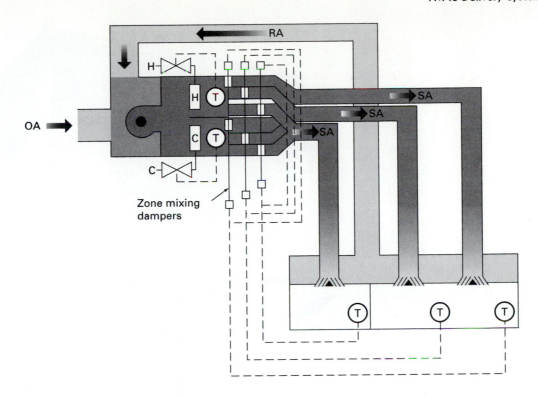

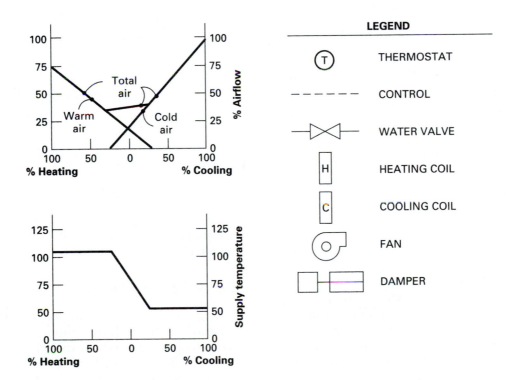

■ **FIGURE 3–22**
Variable air volume multizone system.

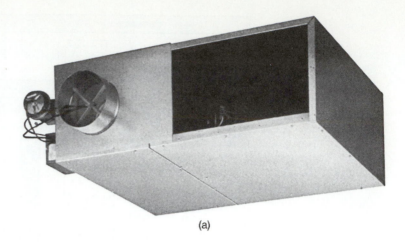

(a)

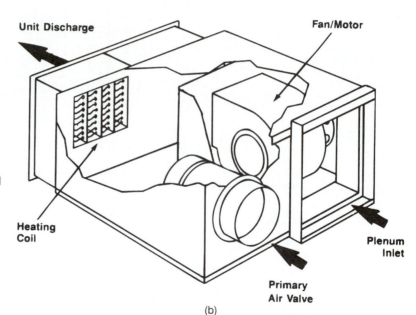

(b)

■ FIGURE 3–23
(a) Typical fan terminal, which contains a fan drawing air from the space or the return air plenum, a primary air duct connection (behind photo, not visible), and an air outlet. (b) Side and top view of a fan terminal box with arrows indicating the mixing of primary air and induced air. (Courtesy: Titus, Richardson, TX.)

Fan coils can be two pipe or four pipe in design, as shown in Figure 3–27. Two-pipe fan coils are served by a set of supply and return pipes that can carry either hot or chilled water; thus, a particular unit is capable of heating or cooling, depending on which water service is available. The piping may be zoned by its exposure to the sun, with some portions of the building capable of cooling while the rest is heating. Obviously, some compromises in comfort can be expected with a two-pipe system.

A four-pipe fan coil system has two sets of piping—one set of supply and return for chilled water and another set for hot water. The four-pipe system is more flexible than the two-pipe system, but also more expensive to install.

Fan coil units can be used at building perimeters in combination with VAV cooling-only systems serving interior spaces.

3.4.14 Package Terminal Air Conditioners

Package terminal air conditioners (PTACs) are window or through wall units containing their own compressors and air-cooled condensers. (See Figure 3–28.) Small package rooftop units with limited ductwork can also be classified as PTACs. PTACs can be installed with or without electric heat and may be designed as heat pumps for more economical operation. PTACs are limited to rooms with outside exposure or spaces with suitable adjacent places to reject heat from refrigeration. PTACs tend to be noisy

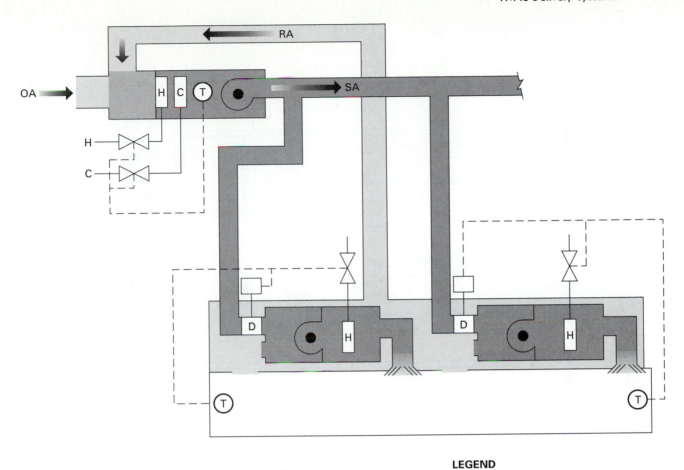

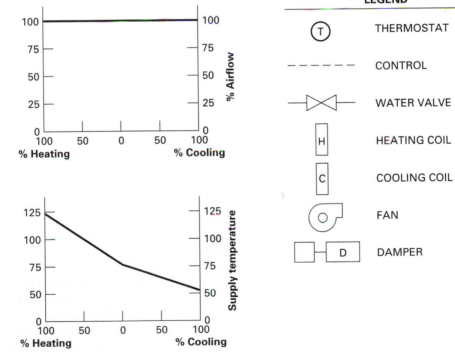

LEGEND

(T)	THERMOSTAT
— — — —	CONTROL
⋈	WATER VALVE
H	HEATING COIL
C	COOLING COIL
⊙	FAN
D	DAMPER

■ **FIGURE 3–24**
Fan terminal units.

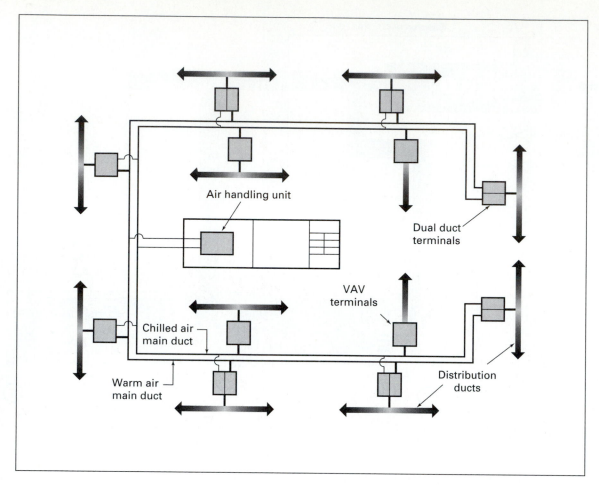

■ FIGURE 3–25

A typical layout of an air distribution system with single-duct VAV for the interior zones and dual-duct constant or variable air volume for the exterior zones.

and require high maintenance as their compressors age.

Operating as a heat pump, the unit can reverse the refrigeration process to move heat from outdoors to indoors, thus acting as a heater instead of an air conditioner.

3.4.15 Water Source Heat Pumps

Water source heat pumps are similar to PTAC heat pumps, except that their condensers are water cooled rather than air cooled. Water source heat pumps do not need access to outside air for condensing, so they can be located at the interior of the building. Small units are generally located above ceilings; larger units can be located in equipment rooms.

Condensing water for water source heat pumps is produced by a central evaporative fluid cooler and circulated through the heat pump condensers to absorb heat generated by the air-conditioning process. A boiler is also included to provide heat for units operating in the heating mode by reversal of the refrigeration process. During winter, units serving interior spaces function in the cooling mode, and units serving perimeter spaces operate in the heating mode. Energy rejected from the cooling units will be used to supplement heat from the boiler.

Water-cooled condensers are much more effective at heat transfer than are air-cooled condensers. In addition, with proper control of the boiler and evaporative cooler refrigerant, condensing temperatures will be more moderate than with outside air. The result is that the compressors need not work as hard, and water source heat pumps are more energy efficient and have a longer life than air-cooled PTACs.

(a)

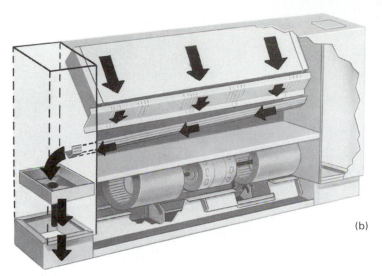

(b)

■ **FIGURE 3–26**

(a) Fan coil. A close-up view of the valve assembly showing control and shutoff valves. The small pan under the valves is the auxiliary drain pan, which receives collected condensate from the drain pan under the coil and condensate occurring on the valve assembly. Drain pans require periodic cleaning to avoid clogging. (b) A cutaway view of floor-mounted fan coil units with enclosed coil. The coil is totally enclosed for effective air distribution. (Courtesy: Airtherm Manufacturing Co., St. Louis, MO.) (c) Photos showing ceiling surface and ceiling recess mounted fan coil units. (Courtesy: Airtherm Manufacturing Co., St. Louis, Mo.)

(c)

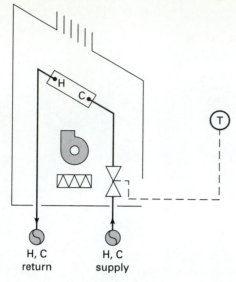

Two-pipe system

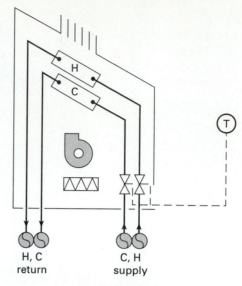

Four-pipe system

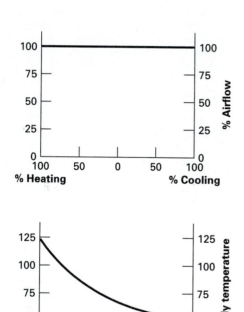

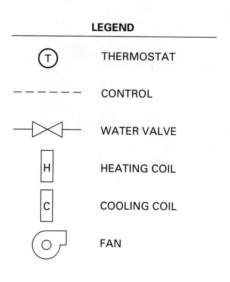

LEGEND

(T)	THERMOSTAT
–––––––	CONTROL
⋈	WATER VALVE
H	HEATING COIL
C	COOLING COIL
(O)	FAN

■ **FIGURE 3–27**
Fan coil systems.

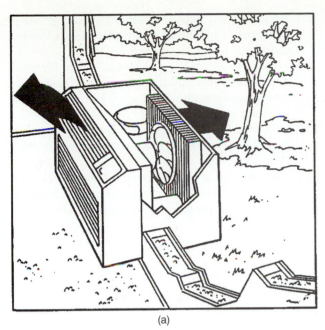

(a)

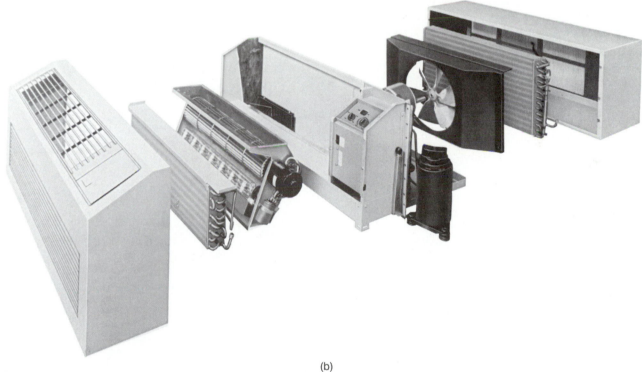

(b)

■ **FIGURE 3–28**

(a) A cutaway view of a package terminal air conditioner (PTAC) shows two sections: the indoor section and the outdoor section. The two sections should be completely insulated and separated from each other to increase efficiency and decrease noise. The indoor section contains the evaporator coil, fan, and controls. The outdoor section contains the compressor, condenser coil, and fan. A PTAC unit may also be designed to operate on the reverse refrigeration cycle as an air-to-air heat pump. (b) An exploded view of the PTAC showing all components. From left to right: indoor cabinet, evaporator coil, electric heating coil (optional), separating partition, refrigerant compressor, condenser fan, condenser coil, and outdoor cabinet. (Courtesy: Trane Company, La Crosse, WI.)

QUESTIONS

3.1 Describe the basic difference between interior and perimeter space with respect to HVAC loads.

3.2 What principles are used to vary the amount of HVAC service to a space with varying loads?

3.3 What is an HVAC zone, and how might a zone differ from a room?

3.4 Which HVAC delivery systems are best for preventing high humidity in spaces? Why?

3.5 Which HVAC delivery systems are generally avoided due to their high energy usage?

3.6 Name the HVAC delivery systems that are most flexible for adding or rearranging zones. Why are they effective in this respect?

3.7 Which HVAC delivery systems are most energy efficient? Why?

3.8 Which HVAC delivery systems would be most appropriate in a building in which risk of water damage is a serious concern?

3.9 Which HVAC delivery systems would be most appropriate in a building in which intrusion for maintenance within occupied spaces must be kept at a minimum?

3.10 Of all the HVAC delivery systems discussed in this chapter, which is the most energy efficient? Why?

3.11 Describe the major difference between electric or pneumatic control and direct digital control.

3.12 When is it appropriate to use two-position control?

3.13 When is it appropriate to use proportional control?

4

Cooling Production Equipment and Systems

Cooling systems may be designed using any of several methods for producing and distributing cooling, including vapor compression and absorption refrigeration cycles, and direct expansion and chilled water distribution systems. Large buildings with central chilled water plants require considerable attention to the placement and design of the major equipment room. Guidelines are given along with examples at the end of the chapter.

4.1 REFRIGERATION CYCLES

Refrigeration is used to produce cooling. In doing so, it moves heat in a manner unlike any natural mode of heat transfer. Conduction, convection, and radiation move heat from high-temperature sources to low-temperature destinations—like a ball rolling downhill. Refrigeration does the opposite, moving heat from low-temperature sources to high-temperature destinations—like rolling a ball uphill. The ball rolls downhill naturally, but requires pushing to go uphill. Similarly, refrigeration requires energy to counter the natural flow of heat.

4.1.1 Vapor Compression Cycle

The refrigeration process most often used in HVAC systems is called the *vapor compression cycle*. The process involves boiling and condensing fluids called refrigerants at temperatures that will produce a cooling effect. We are familiar with the process in other contexts. Consider a pot of water boiling on a stove. As the hot gas of the flame passes by the pot, heat is absorbed by the pot and used to boil the water. The water absorbs heat and boils, similar to a refrigerant; the gas, losing its heat, is cooled, similar to air across a cooling coil (see Figure 4–1).

Refrigeration involves different substances at different temperatures. Warm air passes across a cooling coil filled with a refrigerant. The heat of the warm air is absorbed to boil the refrigerant. The air is cooled and the refrigerant is boiled, or evaporated. The basic difference between boiling water on a stove and boiling a refrigerant for air conditioning is the temperatures at which they boil. Boiling water at 212°F can absorb heat and cool a gas stream starting at 1000°F. Boiling a refrigerant at 45°F can absorb heat and cool an airstream starting at 80°F. The heat absorbed from the air boils the refrigerant, making it a gas.

The vapor compression cycle returns the refrigerant gas to its liquid state to be boiled again in the coil. Returning gas to a liquid phase is termed *condensing*, which is the opposite of boiling. At atmospheric pressure, steam absorbs heat and boils at 212°F and can be condensed back to water by removing heat at 212°F. The boiling or condensing temperature of a fluid depends on pressure: The higher the pressure, the higher will be the boiling or condensing temperature, and vice versa. Refrigerants that boil at

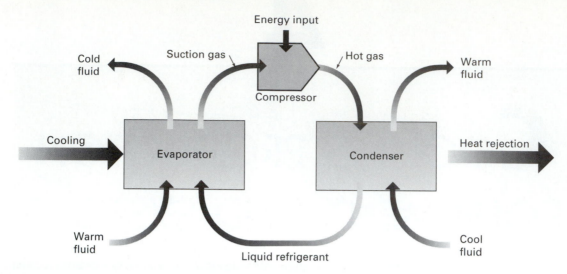

■ FIGURE 4–1
Vapor compression cycle.

low temperature and pressure to produce cooling can be made to condense at high temperature if the pressure is raised.

In the vapor compression refrigeration cycle, the compressor is used to reduce the pressure of the refrigerant in the cooling device. Refrigerant gas drawn by the suction of the compressor has its pressure raised at the discharge of the compressor. The refrigerant has a higher boiling point at higher pressure and can be condensed, releasing heat to another fluid at a temperature higher than that of the fluid which was cooled.

Liquid refrigerant from the condenser is still at high pressure. The pressure is reduced at the inlet of the cooler by an expansion valve or other throttling device, similar to a faucet, which allows a metered amount of fluid to pass from the high-pressure portion of the system to the low-pressure portion.

Temperatures for the process vary with specific applications. The refrigerant on the suction side of the compressor might be 40°F, which is cold enough to reduce 80°F air to 55°F. The 55°F air is suitable for cooling the space. At the discharge of the compressor, the boiling point of the refrigerant might be 130°F due to higher pressure, which is hot enough to release heat to 100°F outside air. Releasing heat from the compressed gas allows it to condense. Once condensed, the liquid refrigerant can be cycled back to the evaporator and used again for cooling.

4.1.2 Absorption Refrigeration Cycle

The absorption refrigeration cycle uses water as a refrigerant. As with all refrigerants, the temperature of

boiling and condensation will depend on pressure. At atmospheric pressure, water boils and condenses at 212°F. If the pressure is reduced to a low vacuum (measured in inches of mercury), water will boil and condense at temperatures low enough for producing cooling. At 5–7 mm Hg vacuum, water boils at 40°F and can be used as a refrigerant to produce chilled water or chilled air within the temperature ranges common to air-conditioning systems.

As in the vapor compression cycle, the refrigerant is boiled in an evaporator that absorbs heat from the fluid being cooled. The water vapor produced in the evaporator is conducted to a chamber that contains a strong solution of a hygroscopic chemical such as lithium bromide. The solution absorbs the water vapor, which then becomes liquid. This portion of the process is similar to condensation in the vapor compression cycle.

As the solution becomes more dilute, it must be regenerated by removing water. This is done in a concentrator chamber, which uses heat to boil off excess water and strengthen the solution. The source of the heat can be piped steam, hot water, or direct firing with gas or oil. The concentrated solution must then be cooled for subsequent return to the absorber chamber. The excess water that is driven off must be condensed for return to the evaporator. These cooling processes are generally accomplished using water from a cooling tower (see Figure 4–2).

4.1.3 Coefficient of Performance

The energy efficiency of refrigeration processes is measured as the coefficient of performance or COP,

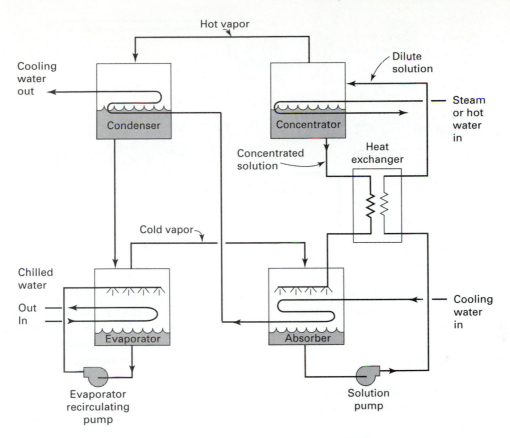

■ **FIGURE 4–2**

Operation of an absorption cycle for generation of chilled water.

which is the ratio of the cooling effect divided by the power used to accomplish the process. The cooling effect is measured in Btuh. Several forms of power might be used to accomplish cooling. Their units must be converted to Btuh for calculating the COP (see Figure 4–3).

Absorption refrigeration is less efficient than the vapor compression cycle. The COP of the absorption cycle ranges from 0.5 to 1.0, in comparison to values of 2.5 to almost 7 for vapor compression. Most of the energy used in the absorption process is heat, with a small amount of electricity used for pumps and accessories. Most vapor compression cycle machines rely solely on electricity as a source of energy for refrigeration. In general, electricity is a more expensive form of energy than heat produced by burning fuels. However, its COP is higher, and either energy form may be an appropriate selection, depending on energy rates and forecasts.

Under some circumstances, absorption refrigeration can be economical despite its poor COP. Absorption is also the logical choice of cooling for sys-

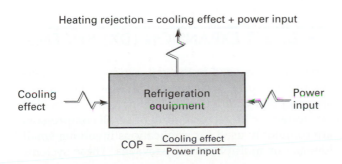

■ **FIGURE 4–3**

Definition of coefficient of performance: COP = cooling effect ÷ power input in constant units. E.g., cooling effect = 10,000 Btuh; power input = 1 kW = 3414 Btuh; COP = 10,000/3414 = 2.6.

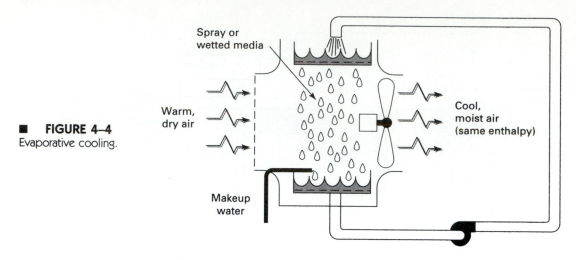

■ FIGURE 4–4
Evaporative cooling.

Spray or
wetted media

Warm,
dry air

Cool,
moist air
(same enthalpy)

Makeup
water

tems that rely on waste heat from industrial processes or heat that is a by-product of turbines or engines used to generate electricity.

4.1.4 Evaporative Cooling

Under favorable outdoor conditions, evaporative cooling is an economical alternative or supplement to vapor compression or absorption refrigeration. Evaporative air coolers, common in arid regions, pass outside air through a water spray or wetted medium. The air is cooled and humidified for direct use in interior spaces or indirect use to cool another airstream via an air-to-air heat exchanger. The indirect method is necessary if the addition of humid outside air to the space is a problem. Figure 4–4 shows how evaporative cooling can be used directly and indirectly for air conditioning. Evaporative cooling is also a part of many systems that use vapor compression or absorption refrigeration. A typical large system will include a cooling tower to produce condensing water during warm weather (see Figure 4–4).

Evaporative cooling can also be used for making chilled water during cool weather. Many buildings and processes need air conditioning year round, regardless of weather. During cool weather, the tower will have sufficient capacity to generate water temperatures that are cold enough for air conditioning. The water can be filtered and used directly in the chilled water system, or it can be used with a heat exchanger to transfer cold to the chilled water system. The tower water system contains water that is exposed to oxygen and dirt, which can result in corrosion and fouling if not properly tended. The use of a heat exchanger to keep this water out of the entire building system is more expensive, but reduces the concern about damage. Cooling can also be generated

by the direct migration of the refrigerant from the condenser coil to the evaporator coil when the temperature in the condenser is colder than the evaporator during cold weather. Figure 4–5 shows the various methods for producing chilled water by the evaporative cooling process.

4.2 COOLING PRODUCTION EQUIPMENT

The selection of a cooling production and distribution system will depend on the size of the project, the budget, life cycle cost concerns, and operating cost concerns, including utilities and maintenance. There are two basic options: direct expansion (DX) and chilled water.

Cooling produced by refrigeration must be transferred to a space for air conditioning. The transfer is accomplished at a cooling coil, which is constructed of tubes surrounded by fins. A chilled fluid passes through the tubes, and cooling is conveyed to the airstream, which flows over the fins. The air-conditioning coil may use refrigerant directly or use chilled water. The latter is produced by refrigerant in a water chiller and distributed to the coil or coils of the system.

4.3 DIRECT EXPANSION (DX) SYSTEMS

If refrigerant is used in the coil, the system is generally termed *direct expansion*, or DX. Most DX systems are equipped with air-cooled condensers. Air-cooled DX systems with reciprocating or scroll compressors are commonly used for applications involving small tonnage or multiple small packages. These systems

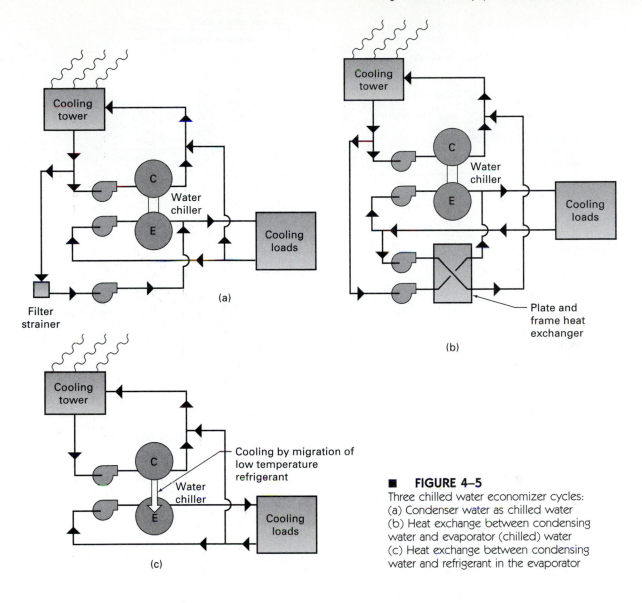

■ **FIGURE 4–5**

Three chilled water economizer cycles:
(a) Condenser water as chilled water
(b) Heat exchange between condensing
water and evaporator (chilled) water
(c) Heat exchange between condensing
water and refrigerant in the evaporator

can be very economical in first cost; however, their power demand is high due to high condensing temperatures when reciprocating compressors are used. Scroll compressors are a more energy-efficient option for small systems. Larger DX systems utilizing helical or centrifugal compressors with water-cooled condensing are even more energy efficient.

Basic DX system components are shown in Figure 4–6. DX refrigerant coils are used in most small air-conditioning equipment and can be used in large equipment. If the compressor is included with the equipment, the resulting device is termed *unitary*. Unitary devices include small-package terminal units such as window air conditioners and larger units such as rooftop air conditioners.

If the compressor is remote, and refrigerant is piped to the unit, the resulting system is termed a *split system*. Most residences and many light commercial facilities use split systems. The cooling coil or evaporator coil is contained in an air-handling unit or attached to a furnace. The compressor is placed in a package outside at grade or on the roof. The condenser is generally in the same package, and the entire assembly is called a *condensing unit*.

Some systems include the compressor with the air-handling equipment and use a remote condenser. This relationship between the coil and the compressor would be termed unitary, rather than split, despite the fact that the components are separated. The terminology is illustrated in Figure 4–7.

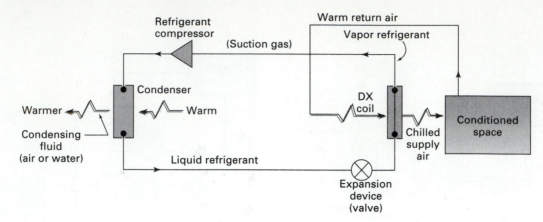

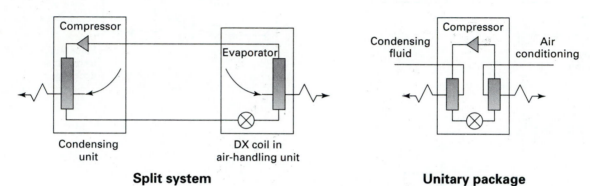

■ **FIGURE 4–6**
Basic DX system components.

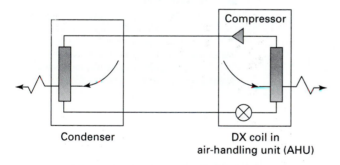

Split system

Unitary package

Unitary with remote condenser

■ **FIGURE 4–7**
Alternative DX system configurations (fans not shown).

DX systems generally use compressor equipment at each air-conditioning system. If the associated condensers are air cooled, the systems must be in close proximity to the outdoors. Otherwise, long runs of refrigerant piping will create pressure drops, large volumes of expensive refrigerant, and the potential for costly leaks. For this reason, DX systems with air-cooled condensers are generally used only for small air-conditioning applications, such as in residences. Larger DX equipment is practical only if it can be conveniently located outside, such as rooftop units for low-rise commercial buildings.

DX systems are frequently utilized for individual tenants in large building complexes, such as shopping malls, apartments, and low-rise buildings. The DX system is used to reduce the building owner's responsibility for maintenance. DX systems (window air conditioners, heat pumps, or through-the-wall units) require openings in the windows or walls, which is not aesthetically desirable.

For massive buildings or high-rise projects, it may be impractical to access the outdoors for air-cooled condensers. Most commonly, the solution for massive buildings and high-rise projects is to distribute cooling by using chilled water. The chilled water is produced centrally and distributed to various air-conditioning units dispersed throughout the building.

There are other alternatives. For example, unitary DX equipment can utilize water-cooled condensers. Cooling water is furnished by a central cooling tower. For this type of system, the condenser is a refrigerant-to-water heat exchanger, and condensing water is distributed from a central source. This concept avoids the necessity for long runs of refrigerant when DX systems have poor access to the outdoors.

4.4 CHILLED-WATER SYSTEMS

Water is a low-cost thermal transfer medium that can be pumped from a central cooling plant to dispersed air-conditioning equipment in a large facility. Water chillers may utilize vapor compression or absorption cycle refrigeration, and condensing may be accomplished by air cooling or water cooling. Basic components of chilled-water systems are illustrated in Figure 4–8.

4.4.1 Chilled Water vs. DX

Based on total refrigeration capacity, chilled-water systems are generally more expensive than DX, but have advantages that may be important for individual projects. Unlike most DX systems, water chillers can serve any number of individual air-conditioning systems. Since not all systems will experience peak load at the same time, the central chilled-water system can be sized for the net demand load rather than for the accumulated peak design loads. Typically, diversity factors for commercial and institutional buildings vary between 70 and 90 percent of combined peak loads. This can result in substantial equipment savings through the use of chilled-water systems. For industrial applications, the diversity factor may be as low as 30 to 50 percent.

Chilled-water systems can be quieter and less visible than DX systems. Water chillers and heat

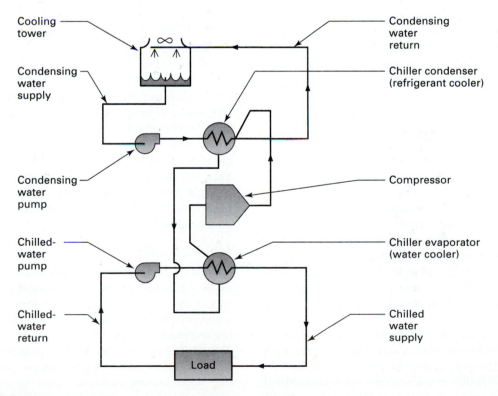

■ **FIGURE 4–8**
Basic chilled water system.

rejection equipment are generally located in a mechanical area remote from the air-handling equipment. This confines the noisiest and most visually objectionable components of the system to a space that can be isolated from occupied areas of the building.

Controlling leaks of refrigerant is an important environmental and safety issue. Chilled-water systems confine the refrigerant to fewer pieces of equipment, so it is less subject to leaks. DX systems, which generally have extensive external refrigerant piping and valves between the compressor, condenser, and evaporator, are more susceptible to leaks. This enhances the reliability of chilled-water refrigeration, as well as reducing the likelihood of costly recharging of the refrigerant.

Chilled-water systems are more flexible than DX systems. Chilled-water distribution can be considered an in-building utility, which can be easily tapped for the addition or rearrangement of air-conditioning systems. This is extremely important for facilities that undergo frequent remodeling due to a changing mission. High-technology buildings in health care, research, and manufacturing will greatly benefit from the flexibility that is inherent in chilled-water systems.

Centralized chilled-water plants are easier to maintain than decentralized or distributed plants, due to their good accessibility and proximity to maintenance personnel. This is especially important if security is an issue. With a centralized plant, maintenance personnel and outside contractors can be kept away from spaces that have limited access because of safety and security concerns.

4.4.2 Vapor Compression Chillers

Chillers may operate on either the vapor compression or absorption cycle. Characteristics of water chillers are summarized in Table 4–1.

When a liquid refrigerant is evaporated or vaporized, it absorbs heat from the surrounding medium. Conversely, when a vapor refrigerant is condensed, it releases heat. Through a closed-circuit mechanical system, the heat in the building (cooling load) is absorbed by the refrigerant in the evaporator, transported by the compressor, and rejected by the condenser.

The evaporator and condenser are usually of the shell-and-tube design, with water circulating through the tubes. In the evaporator, heat to vaporize the liquid refrigerant is absorbed from the circulated chilled water. In the condenser, heat to condense the gaseous refrigerant is absorbed by water from a cooling tower

or other source. Air-cooled condensers are sometimes used if water is a scarce commodity or if energy efficiency is not a high priority, or if the ambient air temperature is low.

Chillers operating on the vapor compression cycle are classified according to the type of compressor they use, including reciprocating, rotary (scroll or helical), and centrifugal compressors. Chillers of these basic types are illustrated in Figures 4–9 through 4–11.

Due to concern about global warming and ozone depletion, certain refrigerants are banned. Table 4–1 has been updated to include some newer refrigerants.

Reciprocating chillers

Reciprocating chillers use positive displacement compressors, which may contain two or more cylinders. The compressors operate at pressures above atmospheric pressure. Due to inherently high losses associated with reciprocating motion, this type of compressor is generally noisier and uses more power than rotary-type compressors.

Capacity control is achieved by cycling compressors or deactivating cylinders in steps. Cycling and step control will result in varying temperatures of the supply water, and reciprocating chillers are not appropriate if tight temperature control is required.

Rotary helical chillers

Rotary helical chillers use screw-type compressors, which consist of two intermeshing helical grooved rotors. The operation of this type of compressor is relatively vibration free. The operating pressure is usually above atmospheric pressure. Capacity control is achieved by altering the length of the rotors available for service, using a slide valve. Unlike reciprocating chillers, rotary helical chillers offer modulation rather than stepped control. Their refrigerants can operate at high condensing temperatures (up to 150°F). Accordingly, helical chillers are well suited for heat recovery applications.

Centrifugal chillers

Centrifugal chillers use turbocompressors with impellers that operate on a centrifugal principle. Volumetric capacities are higher than those of equivalent-sized reciprocating compressors because the flow of the refrigerant is continuous rather than intermittent. There is very little vibration and only minimum wear, due to the nonpulsating gas flow and minimum number of contacting surfaces. Capacity control is most often achieved by modulation of a refrigerant flow device at the inlet of the compressor.

TABLE 4-1

Characteristics of water chillers, typical refrigerants, operating pressures, and COP

Type	Speed, rpm	Capacity range, tons	Typical Type	Evaporator pressure, psia	Condensing pressure, psia	Condensing media	Theoretical performance (COP)	Applications
Reciprocating								
Semihermetic	1750	20-200	HCFC-22	43	173	Water	4.7	Commercial/industrial Cooling systems up to 200 tons with water or air cooled condensers
Open	1750	20-150	HCFC-22	43	173	Water	4.7	
Semihermetic	1750	20-200	HCFC-22	43	173	Air	4.7	
Open	1750	20-100	HFC-134a	24	112	Air	4.4	
Rotary								
Hermetic	3500	50-460	HCFC-22	43	173	Water	4.7	Commercial cooling systems
Open	3500	120-750	HCFC-22	43	173	Water	4.7	Industrial cooling systems
Centrifugal								
Direct driven hermetic	3500	100-2000	HCFC 123	-2.3	16	Water	4.8	Commercial/industrial cooling systems over 100 tons
Open gear drive	5000 to 7500	150-8000	HCFC 123 / HFC-134a	-2.3 / 24	16 / 112	Water	4.8 / 4.4	Commercial/industrial cooling systems over 150 tons
Hermetic gear drive	6000 to 20,000	100-2000	HCFC 123 / HFC 134a	-2.3 / 24	16 / 112	Water	4.8 / 4.4	Commercial/industrial cooling systems over 100 tons
Absorption								
Direct fired	—	100-1500	Water	6 mm Hg absolute	50 mm Hg absolute	Water	1.0	Commercial/industrial cooling; some units can be equipped with heat exchangers for generating or heating hot water
Indirect fired (Steam, hot water)	—	75-2000	Water	7 mm Hg absolute	7.5 mm Hg. absolute	Water	0.8	Commercial/industrial cooling; where waste heat or steam is available

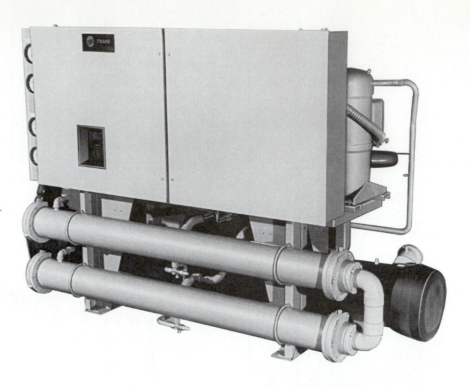

■ **FIGURE 4–9**
A packaged chiller with rotary scroll compressor, evaporator, and condenser assembly. Capacity range is 20 to 60 tons. (Courtesy: The Trane Company, LaCrosse, WI.)

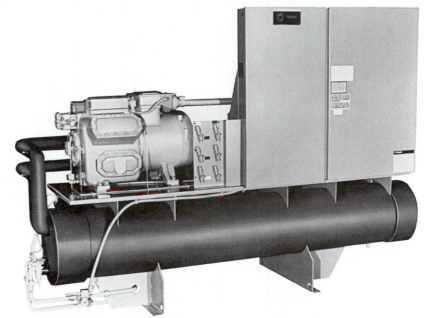

■ **FIGURE 4–10**
A packaged chiller assembly with reciprocating compressor. Capacity range is 70 to 120 tons. (Courtesy: The Trane Company, LaCrosse, WI.)

4.4.3 Design Variations

Chillers have several design variations:

- Open vs. hermetic
- Direct vs. gear drive
- Refrigerant operating pressure

Open chillers differ from hermetic chillers in the placement of the compressor motor. The motor of an open chiller is externally coupled to the compressor, while the motor of a hermetic unit is housed with the compressor in a common enclosure and cooled by the surrounding refrigerant.

Direct-drive machines are usually two- or three-stage machines operating at 3600 rpm. These units have fewer moving parts than gear-driven units, which operate through gear trains at speeds ranging from 6000 to 20,000 rpm. High-speed gear-driven units have been improved over the years, and many

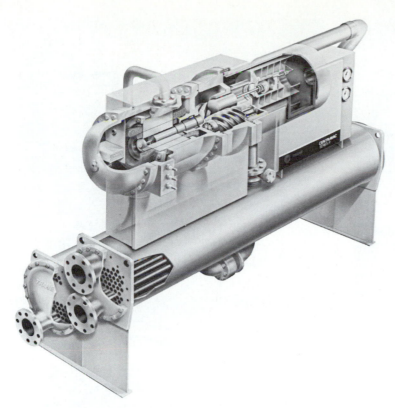

■ FIGURE 4–11
A packaged chiller with cutaway view of the helical-rotary compressor. Capacity range is from 100 to 450 tons. Large chillers up to 5000 tons are made with multistage centrifugal compressors. (All chiller illustrations courtesy of The Trane Company, LaCrosse, WI.)

of the early objections to them, such as their reliability, shaft seal leaks, etc., have been overcome. Gear-driven machines offer the economic advantage of providing a series of available capacities with fewer basic compressor sizes.

Centrifugal chillers may use refrigerants that operate at pressures above or below atmospheric pressure. Chillers that use refrigerants operating below atmospheric pressure are susceptible to air leaks into the refrigerant circuit. Air in the system can cause serious surging of the compressor and loss of capacity. For this reason, there must be a provision for air purging. A small amount of refrigerant is lost along with the air. Due to concerns about CFCs in the environment, purge units on chillers must be designed for very low refrigerant losses.

With refrigerants that operate above atmospheric pressure, leaks can occur. Even so, there is no opportunity for air to leak into a unit. Therefore, an air purge system is not required. It is necessary, however, to have a built-in cycle arranged to store the refrigerant in separate pressurized drums whenever the chiller must be evacuated for maintenance.

4.4.4 Absorption Chillers

Chillers based on absorption refrigeration utilize water as the refrigerant. In lieu of using a compressor to facilitate alternate evaporating and condensing, as in the vapor compression cycle, the absorption cycle depends on operating the unit at near perfect vacuum conditions and on the affinity lithium bromide has for water. Absorption equipment is factory evacuated and sealed to preserve the required vacuum for proper operation.

Depending on the source of applied heat to regenerate the lithium bromide solution, absorption chillers may be classified as indirect or direct fired. Figure 4–12 shows typical absorption chillers. Figure 4–13 shows the operating features of a single-stage absorption chiller.

Indirect-fired chillers are available up to 2000 tons for use with 5- to 150-psi steam or with 150° to 400°F hot water as the energy source. Direct-fired chillers, available up to 1500 tons, use natural gas combustion or hot-process waste gas as the energy source.

The cooling tower capacity for absorption refrigeration must be adequate to reject the heat absorbed as a result of the refrigeration effect, plus the heat added to the system for the purpose of regeneration. The rejected heat is approximately double that of a vapor compression cycle.

An absorption chiller operates in an extremely high vacuum. Proper operation is dependent on a very tight system to prevent the infiltration of air into

(a)

(b)

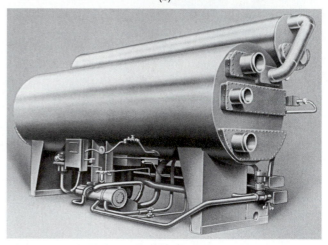

(c)

■ **FIGURE 4–12**
(a) Direct-fired absorption chiller. (b) Single-stage indirect-fired absorption chiller.
(c) Two-stage indirect absorption chiller. Two-stage absorption chillers have a higher
COP than does the single-stage design. (Courtesy of The Trane Company, LaCrosse, WI.)

the unit. Special care and procedures are implemented during the manufacture and operation of such a chiller.

Lithium bromide is a salt and will crystallize when in concentrated form and when subcooled. Crystallization will occur whenever the source of heat is interrupted during a normal operating cycle or at very low load conditions. Safety controls are normally incorporated to prevent crystallization. The plant operator should be aware of the abnormal conditions that cause crystallization. Once a unit has crystallized, elaborate procedures are required for decrystallization, and downtime may be as much as two or three days.

Lithium bromide is highly corrosive to some materials, so precaution must be exercised in the design and operation of absorption units. Some manufacturers depend entirely on corrosion-resistant materials. Others depend on both corrosion-resistant materials and an inhibitor. When an inhibited solution is required, it is important that the concentration of the inhibitor be periodically checked and maintained.

The COP of an absorption chiller is considerably lower than that of a vapor compression type of chiller. The COP varies from 0.5 for small, single-stage, indirect-fired absorption chillers to 0.8 for large, two-stage, indirect-fired absorption chillers. A two-stage, direct-fired unit has a COP of approxi-

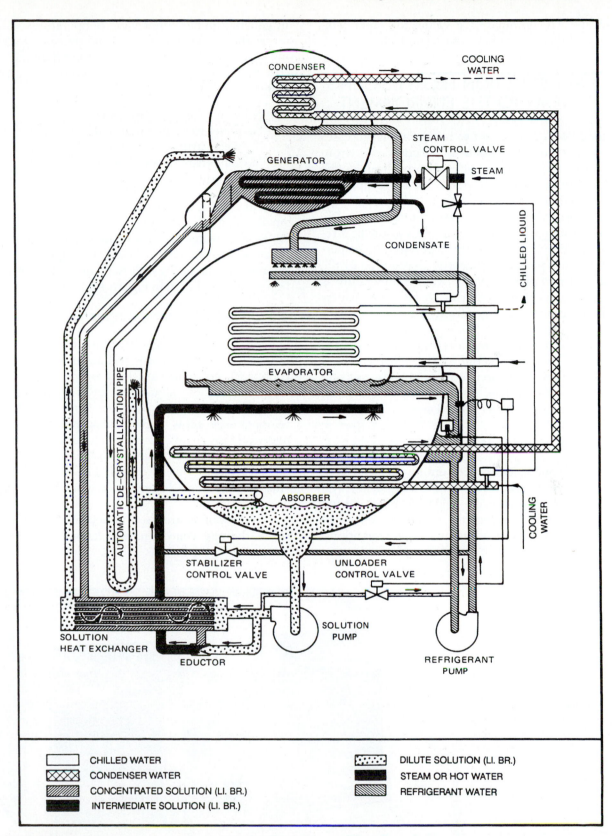

CONDENSER

COOLING WATER

GENERATOR

STEAM CONTROL VALVE

STEAM

CONDENSATE

CHILLED LIQUID

EVAPORATOR

AUTOMATIC DE–CRYSTALLIZATION PIPE

ABSORBER

COOLING WATER

STABILIZER CONTROL VALVE

UNLOADER CONTROL VALVE

SOLUTION PUMP

REFRIGERANT PUMP

SOLUTION HEAT EXCHANGER

EDUCTOR

	CHILLED WATER		DILUTE SOLUTION (LI. BR.)
	CONDENSER WATER		STEAM OR HOT WATER
	CONCENTRATED SOLUTION (LI. BR.)		REFRIGERANT WATER
	INTERMEDIATE SOLUTION (LI. BR.)		

■ **FIGURE 4–13**
Operation of a single-stage indirect-absorption chiller. (Courtesy of York, York, PA.)

mately 1.0. The nominal COP values for vapor compression chillers range between 2.5 and 7.0.

4.5 HEAT REJECTION FROM COOLING SYSTEMS TO THE ENVIRONMENT

Air-conditioning systems use refrigeration processes to move heat from the indoor to the outdoor environment. The refrigeration cycle absorbs heat by the evaporation of liquid refrigerant in the evaporator (indoor coil), and rejects heat by the condensation of vapor refrigerant in the condenser (outdoor coil). The condenser may use either water or air as the heat rejection medium. The lower the heat rejection temperature, the less power (and energy) is required by the refrigerant compressor. Since water temperature is normally cooler than air during the summer, water-cooled condensers are generally more energy efficient.

The heat rejected from a refrigeration system is the sum of the actual cooling load and the energy added by the refrigeration equipment. Heat rejection systems may be grouped into two major classes: air-cooled systems and water-cooled systems.

4.5.1 Air-Cooled Systems

Air-cooled condensers utilize air to cool and condense refrigerant gas to the liquid state. These systems normally require 600 to 900 CFM of air per ton of refrigeration. Condensing temperatures range from 115° to 140°F, depending on climatic conditions.

In general, higher condensing temperatures will increase the power requirement and decrease the cooling system performance. Despite their higher energy cost, air-cooled systems offer several advantages over water-cooled systems:

- No water-freezing problem.
- No water treatment required.
- No water drift concern.
- No condensing water pump required.
- Lower initial cost.

Air cooling of condensers may be appropriate for small systems that do not warrant the complexity and expense of water systems or in areas where energy is more available than water. A typical rooftop-mounted air handling unit with air-cooled refrigeration system is shown in Figure 4–14.

4.5.2 Water-Cooled Systems

Sources of condensing water include spray ponds, domestic supply systems, wells, surface water, and cooling towers, which are the most common solution. Water from these sources is lower in temperature than air is during hot weather. Lower condensing temperatures allow higher refrigeration efficiency, and water is generally preferred to air on this basis.

Using domestic water in condensers is wasteful and expensive. This method is undertaken only for small systems as a convenience when other options are not practical.

Spray ponds are used in lieu of cooling towers when the site conditions are favorable. However,

■ **FIGURE 4–14**
A typical rooftop model of an air-cooled cooling and heating unit consisting of all indoor and outdoor components. (Courtesy of The Trane Company, LaCrosse, WI.)

spray ponds are more costly than cooling towers and require large surface areas (approximately 20 to 30 sq ft per ton). Ponds can be aesthetically pleasing when properly designed, but maintenance is high.

Wells and river water are viable sources of cooling water when they are readily available. Shipboard installations offer a good example of this application. Obviously, the cost of cooling towers and similar devices is eliminated. It is necessary, however, to provide more elaborate water treatment and filtration facilities than with conventional heat rejection methods.

4.5.3 Cooling Towers

Used in most applications, cooling towers produce water at an appropriate temperature for condensing by evaporation. A portion of the circulated water is evaporated in the cooling tower to cool the remainder, thus minimizing the use of water and any associated expense for processing. Approximately 10 percent of the circulation rate is evaporated and lost to bleed-off in the tower. New water is introduced to make up for that which is lost. Evaporation is achieved by flowing the water over a fill ma-

terial in the tower, which is designed to effect good contact between water and air.

The typical cooling tower is capable of producing water within 5°F of the ambient wet-bulb temperature without becoming prohibitively expensive.

Selecting a tower

Selection of the right cooling tower depends on many factors, including the available space, access for installation and possible replacement of the tower, the initial budget, and life cycle cost expectations, particularly energy, maintenance, and replacement costs.

Performance characteristics and applications of mechanical heat rejection equipment are given in Table 4–2. The indicated data are useful for preliminary evaluation, but should not be used for sizing equipment, due to substantial variations among manufacturers.

Airflow

There are three basic operating concepts for moving air through a cooling tower: gravity draft, induced draft, and forced draft. Some larger towers used primarily in utility power plants employ a chimney effect to make the air flow past the water. These towers,

TABLE 4–2

Characteristics of heat rejection equipment

Type	Capacity range, tons	Airflow CFM per ton	KW input per MMBtuh	Applications
Air cooled				
Air-cooled condensers (propeller fans)	3–150 (1)	500–700	9.0–12.0	Commercial cooling systems where towers are not practical and for year-round systems where freezing of cooling towers is difficult to control, as in extreme climates.
Dry coolers (propeller fans)	3–65 (2)	1000–1400	10.0–13.0	Can be used to cool condenser water or to directly cool chilled water in northern climates: not economical in southern climates.
Water cooled				
Cooling towers				
Packaged induced draft (propeller fan)	5–1000 (3)	200–250	2.0–3.0	Ideal for small and medium-size cooling plants. Somewhat less expensive than built-up towers, but usually has a shorter life span.
Packaged forced draft (centrifugal fan)	10–400 (3)	200–250	4.0–6.0	Applicable to commercial and industrial cooling systems; especially adapted to indoor and restricted outdoor installations; very compact units, relatively quiet.
Field-erected induced draft (propeller fans)	200–1500 (3)	200–250	1.5–2.0	For use with large water-cooled systems. Towers can be built up to 20,000 tons per cell; however, the most commonly used preengineered sizes range from 200–1500 tons.
Water spray type of fluid coolers (centrifugal fans)	5–150 (3)	500–700	14–18	Minimizes water treatment requirements. Eliminates condenser freeze protection when used with water glycol solution. Can be used for direct chilled-water cooling during cold weather.

Notes:
(1) Larger units available when factory coupled to air-cooled chillers. Data based on 95°F ambient, 115°F condensing temperature.
(2) Data based on cooling condenser containing water glycol solution from 100° to 110°F at 95°F ambient.
(3) Based on one cell; multiple cells available. Data based on cooling from 95° to 85°F at 78°F WB.

generally hyperbolic in configuration, are bulky and tall, and thus are not used in air-conditioning applications.

Towers used in building applications generally have fans. The fans can be located to blow air through the fill, which is termed *forced draft;* or they can be located to draw air, termed *induced draft.* Air can enter on one side of the tower, termed *single inlet,* or two sides, termed *double inlet.* Air discharge can be vertical or horizontal.

Airflow may be horizontal or vertical through the fill. Horizontal airflow results in a cross flow of air

and water; vertical airflow results in a counterflow. (See Figure 4–15.)

Placement of cooling tower

Free circulation of air is needed for efficient operation of the tower. Discharge from the tower should be unimpeded, and any obstructions at the inlet should be held at a distance. Inlets are also important. A general rule for induced-draft towers is to allow a clear air passage equal to the tower intake height and for forced-draft towers to allow air passage space equal to 1.5 times the tower width. Violations of these guide-

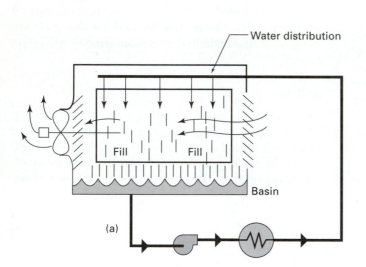

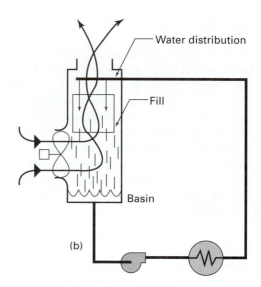

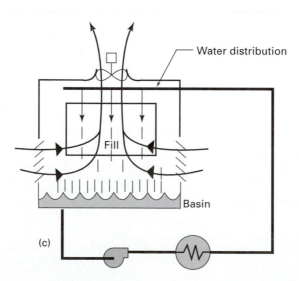

■ **FIGURE 4–15**

Typical cooling towers for building applications. (a) Induced-draft cross flow (draw-through horizontal discharge design). (b) Forced-draft counterflow (single inlet and vertical discharge design). (c) Induced-draft counterflow (double inlet and vertical discharge design).

lines can cause recirculation of humid discharge into the inlet or impede the airflow. Both effects will degrade the tower's performance, resulting in the need for a larger tower or more powerful fans. Airflow at the inlet must be considered in the placement of towers and design of visual screening elements.

Discharge air from towers is humid and can contain droplets of water with high concentrations of minerals and water treatment chemicals. Surfaces exposed to tower water carry-over may become corroded, soiled, or discolored. Wind can blow humid air and water droplets horizontally from the tower discharge or through the tower body. This phenomenon is most likely to happen during conditions of light load when the tower fans are off or running slowly during cool weather. Under these conditions, humid air from the tower may also condense into fog or cause ice on paved adjacent surfaces. All of these possibilities must be considered in locating a cooling tower. (See Figure 4–16.)

Configurations and materials

Towers can be either packaged, or erected in the field. Package tower cells are delivered preassembled to the job site. Package towers are used for small applications, and cells are available up to approximately 900 tons. A large tower might be assembled from a number of single-cell packages or it might be erected in the field if the packaged cells are too large for easy transport to the job site.

Several options are available for the tower's frame and fill material. The framing material of package towers can be fiberglass, metal, wood, or concrete. Field-erected towers are generally wood framed, built by specialty carpenters. Fill material can be corrugated plastic, wood lath, or ceramic tile. Basins can be plastic, metal, or wood.

Most small cooling towers are steel-framed package units with plastic fill. Despite coatings to protect against corrosion, metal framing will ultimately corrode in the humid environment of a cooling tower. Fiberglass, wood, or concrete framing is a higher cost option with a longer life. Plastic fill is most economical, but it is fragile and ultimately needs replacement. Wood or ceramic tile fill has a longer life at a higher initial cost.

Figure 4–17 shows a variety of cooling towers used with HVAC systems.

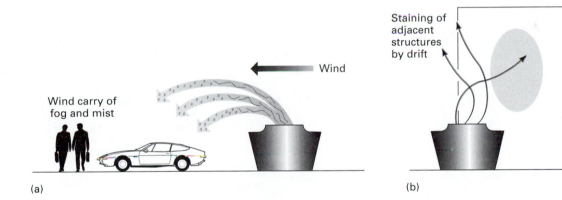

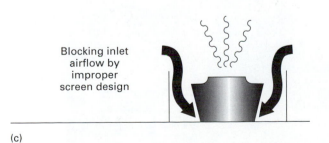

■   FIGURE 4–16
Factors to consider in locating cooling towers. (a) Check for wind directions and drift.
(b) Avoid a blocked air intake. (c) Be concerned about staining adjacent structures.

(a)

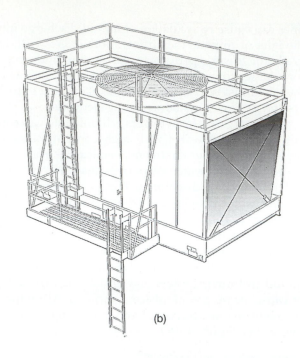

(b)

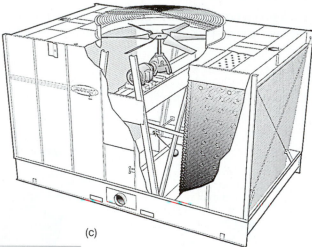

(c)

■ **FIGURE 4–17**

(a) A typical draw-through, horizontal flow tower with fiberglass construction. The flow rate is from a few GPM to several thousand GPM. (b) Large towers require a service platform and safety guard rails. (c) A typical double-flow vertical discharge tower with sizes from 500 to 30,000 GPM. (d) Large custom-designed masonry towers may be constructed of brick, stone, metal, or other materials to match the building material and can also be constructed as a part of the building. (All illustrations courtesy of Marley Company, Mission, KS.)

(d)

4.6 CHILLED-WATER PLANT DESIGN

4.6.1 Basic Configurations

Most large cooling systems are served by chilled water. The following chilled-water plant configuration concepts are illustrated in Figure 4–18:

- Central plant
- Distributed water chillers
- Single- and double-pipe chilled water loop
- District plant

Each design offers solutions to specific problems.

Central plant

Compared with multiple plants, a central plant offers the advantage of lower space requirements and fewer auxiliary systems. The greatest advantage is in the size of equipment that can take advantage of the diversified load, which may be as low as 50 percent of the loads of individual zones.

Central plants also offer operating flexibility, inexpensive redundancy, and centralized maintenance. In addition, the expense of special design provisions is confined to a single location. Among such provisions are sound treatments, ventilation, access, and a consideration of fogging or drift from cooling towers. The central plant is generally the least expensive and best overall solution for a single building or closely coupled building complex.

Distributed water chillers

When growth of the cooling load is phased in or the load is physically separated, distributed plants are appropriate to serve separate areas. Plants can be added as needed, with little commitment of initial investment in space or other provisions.

Distributed chillers will result in smaller piping than central plants will, since less capacity needs to be delivered from a given location. However, the sum of multiple plant capacities is usually substantially greater than the capacity of a central plant, as a diversity of loads can be applied to the individual plants.

Another application of distributed chiller plants is in high-rise buildings over 50 stories in height when two or more vertical zones are necessary to limit static pressure in chilled-water piping systems.

Chilled-water loop

The loop consists of common distribution piping that connects a number of plants. The scheme is applicable to single buildings or a group of buildings within reasonable distance of one another. The loop will al-

low equipment to operate as if it were all installed in a single plant.

The loop shares some of the advantages of the central plant. Reliability is enhanced, since a chiller in one location can act as a standby for chillers in other locations. Economy of operation can result from the ability to shift the load from one plant to another to take advantage of the most efficient equipment. The life of the equipment can be extended by concentrating the load in the fewest number of machines possible and minimizing the operation of the remaining equipment. Because maintenance is often scheduled on the basis of elapsed run time, further economies can result as well.

A loop can be used to supply chilled water to buildings that lack physical space for chilled water plants. This makes the concept ideal for upgrading older campus buildings.

Loops can be constructed with single- or double-pipe distribution. Single-pipe loops have a distribution pipe, generally underground, that passes by each building served by the system. Some buildings will have water chillers, and some will not. The water chillers serve to reduce the temperature of the water in the loop. Chilled water used by the buildings is returned to the loop at a higher temperature. Ideally, the chillers and loads are distributed fairly evenly so that adequately chilled water is available at all locations. Figure 4–18 shows typical piping arrangements for buildings on a single-pipe loop.

With a single-pipe loop, simply having enough capacity is not sufficient to serve the buildings; the capacity needs to be located properly also. Double-pipe chilled-water loops use separate supply and return piping so that the same water temperature is available to all buildings. Because of the extra piping, a double-pipe loop will generally be more expensive to install, but its performance is superior to that of the single-pipe loop.

District plants

A large central chilled-water plant may serve a campus, district, or community. Such plants rely on economies of scale to provide low cost per unit of capacity. These savings are used to offset the expense of distribution piping. The district plant will generally occupy a structure and site built expressly for this purpose; therefore, the design can be optimized with respect to spacing of columns, room height, and location.

The district plant concept centralizes maintenance, keeps noise and cooling tower moisture away from other buildings, and can accommodate improvements without intruding on functions of other buildings. If chilled-water generation is combined

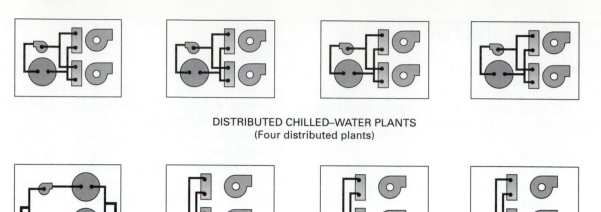

DISTRIBUTED CHILLED–WATER PLANTS
(Four distributed plants)

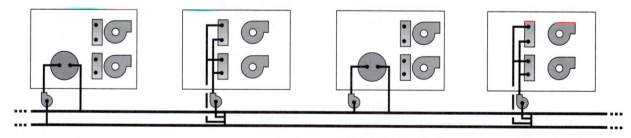

CENTRAL CHILLED–WATER PLANT
(DISTRICT PLANTS SIMILAR)
(One central plant, three distributed)

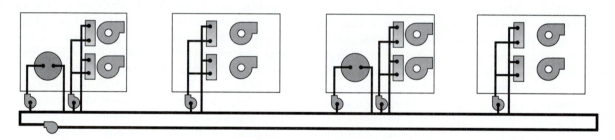

SINGLE-PIPE CHILLED–WATER LOOP
(Two central plants, four distributed)

DOUBLE-PIPE CHILLED–WATER LOOP
(Two central plants, four distributed)

■ **FIGURE 4–18**
Chilled-water plant options.

with power or heat generation, a district plant is ideal for incorporating total energy and reclaiming heat.

4.6.2 Sizing Equipment

Several principles must be considered when sizing equipment. Among these are growth considerations, redundancy, partial redundancy, and capacity modulation.

Cooling requirements grow as buildings are more fully utilized and expanded. Allowance for future growth in the size of equipment and the provision of space for future units should be considered.

Redundancy of cooling service is important for critical applications such as computer space, research laboratories, and portions of health care facilities. A single unit is obviously less costly than multiple units. However, if a single unit is installed, there will be a total loss of service if the unit fails. If continuity of service is critical, some degree of redundancy must be considered. To ensure full load service, a standby chiller must be provided with a capacity equal to the capacity of the larger chiller in the bank of chillers.

The cooling environment in most commercial or office spaces can be compromised for short durations. Partial redundancy can be achieved economically by spreading the design load among two or more units. For example, using two chillers, each sized at half the total load, will ensure 50-percent capacity in the event of one unit's failure. Installing three and four units for the total load would ensure 66- and 75-percent capacity, respectively, if one unit fails. Each unit may be slightly oversized at only nominal increase in cost. This will result in added spare capacity for the future, as well as additional redundancy in the event of equipment failure.

Turndown capability to meet the varying demand loads is another sizing consideration. Modern large chillers can generally operate down to approximately 15 percent of the load without cycling off. If there are periods when the plant must operate at light load, then the chillers must be selected accordingly. If the machines are too large, they will cycle on and off at light load, and the temperature of the chilled water will fluctuate. Also, the machines' lives may be shortened by frequent starting.

Chiller units need not be identical in size. A smaller chiller in the group is often useful for light load conditions during colder weather, on weekends, and at night.

4.6.3 Chiller Circuiting Arrangements

Multiple water chillers can be circuited in series, in parallel, or in a combined series/parallel arrangement, as shown in Figure 4–19.

Series arrangement

A series arrangement reduces water temperature successively through two chillers. The discharge chiller will produce low-temperature leaving water, and the inlet chiller will produce water at a temperature between the return and supply conditions. The advantage of this arrangement is that only the second chiller suffers the performance penalty of low-temperature discharge. On the other hand, all the chilled water is pumped through both chillers. This results in

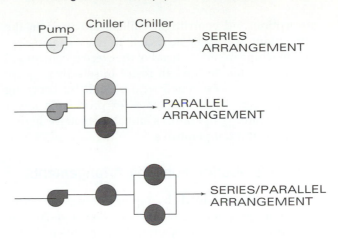

■ **FIGURE 4–19**
Chiller piping arrangements (series, parallel, and series/parallel).

higher pumping energy than for parallel arrangements. In addition, water flow will be maintained through both chillers regardless of their operating status. This can accelerate the erosion of tubes in the evaporator, however.

The series arrangement can be very economical in its initial cost, since piping is simple and only one pump is required for each set of chillers in series. However, a standby pump and bypass piping are recommended to ensure continuity of operation if one machine requires service. Once these items are considered, the parallel arrangement can be just as economical to install.

Parallel arrangement

Chillers in parallel should each have an associated chilled water pump. The use of a single pump for multiple chillers in parallel will allow mixing of return water with supply water, unless all of the chillers are operating.

Any number of chillers can be operated in parallel. Unlike the series arrangement, each will be required to produce the same leaving water temperature. The compressor power will be slightly higher; however, the pumping power will be lower.

The flexibility and reliability of the parallel arrangement are excellent. When the load dictates, chillers and their respective pumps can be shut down. In addition, some machines can be withdrawn from service for maintenance without affecting the operation of others.

Combination series/parallel arrangement

A combination series/parallel arrangement may be appropriate for certain applications. For instance, if absorption and centrifugal chillers are used in the same plant, it may be advisable to place the absorption machine in series ahead of the electric centrifugal machines. This design can result in selecting equipment that is more economical overall by choosing electrical centrifugal machines for producing lower leaving water temperatures and absorbers for higher leaving water temperatures.

4.6.4 Distribution Pumping Arrangements

Chillers can serve loads by piping arranged in three basic configurations, as shown in Figure 4–20. The simplest configuration is a unit loop in which chilled water is pumped through the chiller directly to the load.

Unit loop

To avoid freezing of tubes, the water flow through the chiller cannot be reduced significantly. Therefore, the unit loop arrangement is generally designed for a flow at constant volume. This will require three-way valves at loads or two-way valves at loads and a single valve to bypass unused water. If multiple chillers are arranged in a unit loop configuration, the flow from the chiller plant will vary in steps, depending on the number of chillers required to satisfy the load.

The unit loop is lowest in initial cost and can be economical to operate for applications with minimal distribution piping and low pumping power requirements. If distribution is extensive, pumping costs will be high due to the requirement of a constant flow.

Primary loop

The addition of a primary loop can save pumping energy. The unit loop acts to maintain a constant flow through individual chillers, and the primary pumps draw and return water from the unit loop as the load dictates. For energy savings to occur, the primary loop must vary the flow according to the load. Most buildings exhibit considerable load variations due to occupancy and weather. Therefore, a primary loop distribution has a large potential for saving energy and is commonly used, especially in large applications.

If the distribution is extensive, there may be considerable differences in the available head between the near and far ends of the system. Selecting control valves may then prove difficult. In addition, the head requirement for the primary pumps may be excessive. When this is the case, secondary loops may be added to the system.

Secondary loops

A secondary loop pump withdraws water from a bridge circuit between primary supply and return piping. Secondary loops are generally controlled for variable flow to save pumping energy. Secondary pumping systems can serve entire buildings from a campus primary distribution system. Space and maintenance requirements for secondary pump stations within the building are minimal.

4.6.5 Energy Features

Heat recovery

Chillers move heat from one fluid stream to another. If the heat can be used instead of rejected, energy will be saved. This is the principle of heat recovery. There are many potential applications of heat recovery in process and HVAC chilled-water systems.

Most large buildings require cooling year round due to heat gains from lights, people, and appliances in the interior of the building. During cold weather, air conditioning is provided either by cooling with outside air or by refrigeration. If refrigeration is used, the warm condensing water can serve to heat the building, to heat hot water, or to preheat outside air.

Virtually any building with internal heat gains and simultaneous heating requirements will be a candidate for reclaiming heat. Data-processing equipment is a strong heat source due to its high power usage and continuous operation. Other process cooling requirements should also be considered.

A heat-reclaiming chiller generally has one condenser that uses cooling tower water and an additional closed-circuit condenser for the reclaimed heating water, as shown in Figure 4–21. The available hot water temperature is a limiting factor in reclaiming heat.

Centrifugal equipment designed for HVAC duty is generally selected to be capable of generating hot water between 105° and 110°F. Reciprocating and rotary helical machines are capable of temperatures up to 140°F, but are limited in size.

Higher temperatures may be attained by cascading chillers. The heating-water temperatures of the second chiller can be much higher than for a single chiller. The equipment is more costly, and electric power requirements can be excessive. The best economy is generally achieved by designing heating equipment to utilize lower temperature water.

Process cooling often does not require the low evaporator temperatures normally encountered in air-conditioning applications. At higher evaporator

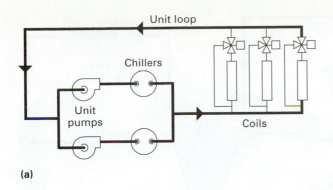

(a)

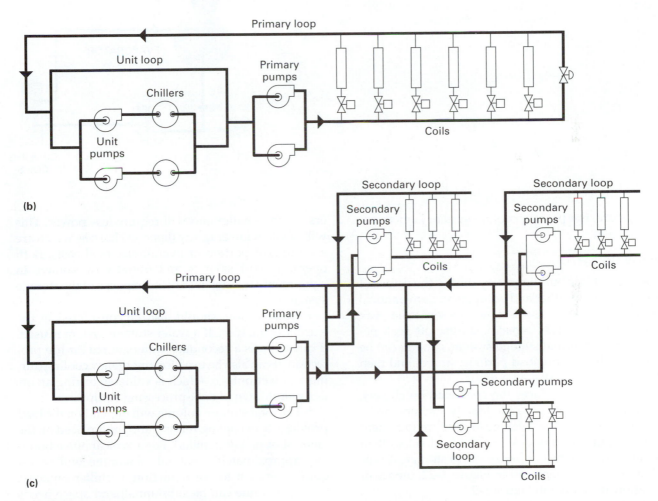

(b)

(c)

■ **FIGURE 4–20**

Alternative configurations of chilled-water distribution systems. (a) Unit loop distribution for small systems. (b) Unit loop with primary distribution for large single buildings. (c) Unit loop with primary and secondary distribution, most often used for groups of buildings such as college and corporate campuses, large retail malls, and industrial complexes.

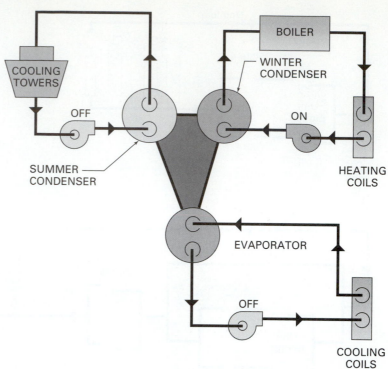

■ **FIGURE 4–21**
Operation of a heat-reclaiming chiller with winter condenser as preheat to the hot water boiler (heater). The winter condenser may be a separate shell-and-tube heat exchanger or a second tube bundle within the summer condenser shell.

temperature, the process cooling system is more energy efficient.

Thermal storage

Cooling loads for buildings vary during the course of the day due to weather and occupancy. Conventional chilled-water plants are designed to serve the peak load and have excess capacity during off-peak periods. With a thermal storage system, cooling can be generated during low load periods, stored, and then retrieved during peak load periods.

Since storage reduces the peak load on chillers, their size can also be reduced. Ideally, chillers can be sized to produce the average output requirement over the course of a day. When the load is less than average, the chillers will contribute to storage. A typical application is analyzed in Figure 4–22; modes of operation are shown in Figure 4–23.

Water can be stored by chilling it or freezing it so that it becomes ice. Water systems have the advantage of using conventional refrigeration equipment, which will chill water to approximately 42°F. The water temperature will generally rise in the system by 15° to 18°F. The required water volume will range between 70 and 100 gal per ton-hr of stored cooling, depending on the rise in temperature and the efficiency of heat retrieval.

If a storage system is installed, chillers, unit pumps, condensing-water pumps, and cooling tow-

ers will be smaller and will require less power. This will result in lower utility demand charges, which are a significant portion of overall electrical costs. Both operating cost savings and initial cost savings in equipment can often justify the expense of the storage system.

Other factors should also be considered in the feasibility analysis. If a water storage system is used, it can serve as a secondary water source for fire protection. It can also be used for emergency cooling during a power outage, which is valuable during an orderly shutdown of data-processing equipment.

The total storage volume will depend on the load profile, rate of cooling release, and limit placed on the amount of power demand. For a typical office building, approximately 1000 gal of storage will be required for each ton of reduction in chiller capacity. Storage volume can be substantial, and space needs should be considered early in the design process. Water can be stored in closed tanks or in open vats. Vats are generally less expensive for large systems and can often be integrated with construction of the foundation to reduce costs further.

Ice storage

Storing ice is an alternative to storing water. Ice is generated by low-temperature brine or direct refrigerant heat exchange and stored in either open vats or enclosed housings. Chilled water is circulated

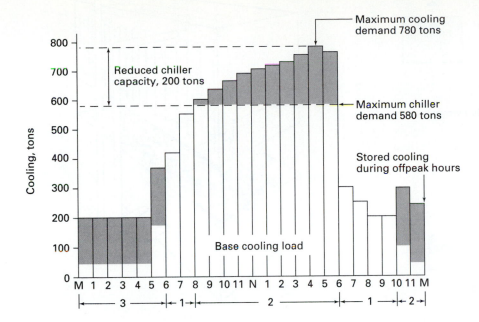

■ **FIGURE 4–22**
Load profile of a 780-ton peak-load cooling system. With thermal storage, it is possible to reduce the capacity of the chiller to 580 tons. During peak hours, the chiller will be supplemented by the stored chilled water that was generated by using the chiller during off-peak and evening hours. From the load profile between 8 A.M. and 6 P.M., it is easy to determine that the chilled-water storage capacity should be a minimum of 1,500 ton-hours. (Reprinted with permission from *HPAC Magazine,* May 1985.)

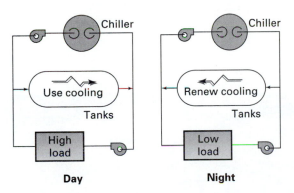

■ **FIGURE 4–23**
Chilled-water storage operations, day (heavy load) cycle and night (light load) cycle.

through the ice or closed storage modules, causing the ice to melt and release its latent cooling effect. The resulting chilled water will be colder than is normally used in conventional systems. It can either be blended into a higher temperature circulating system or used to produce colder-than-normal supply air. The advantage of colder air is reduced flow requirements for delivering a given amount of cooling. This feature offers cost savings from smaller fans and ductwork. Energy is also saved by moving less air.

Ice systems can be used for applications requiring low humidity. Colder water temperatures associated with ice systems allow coil leaving-air dew points as low as 40°F, which corresponds to approximately 30 percent relative humidity at normal room temperatures. Low humidity is necessary in some applications to preserve artifacts and control the growth of mold.

Ice storage systems are more compact than water storage systems. Theoretically, ice requires about one-tenth the volume of water to be stored (latent heat of ice = 144 Btu per lb). On the other hand, refrigeration systems must operate at much lower temperatures than for conventional building applications. This increases power requirements by 30 to 80 percent. Nonetheless, ice storage systems have proved feasible in certain applications and are available as packaged systems from a number of manufacturers. Two types of ice storage systems are shown in Figures 4–24 and 4–25.

Cooling without compressors

Outside air economizers are most commonly used to produce cooling for air conditioning needs during mild and cold weather. However, air economizers may not be desirable where the air quality is poor, where loss of humidity is a concern, or where intake of snow is a problem. There are various means of attaining chilled water without the use of a refrigerant compressor. These systems find applications at outdoor temperatures below 40°F. They can be designed using methods involving evaporative cooling. (See Figure 4–5.) Among such systems are the following:

- Cooling towers can be controlled to produce water as low as 45°F. With appropriate filtration, the water can be introduced directly into the chilled-water system.
- In lieu of the direct use of tower water, plate heat exchangers can be used to transfer cold into the chilled-water system.

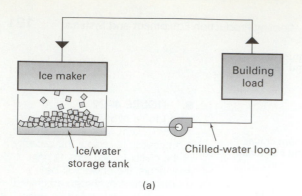

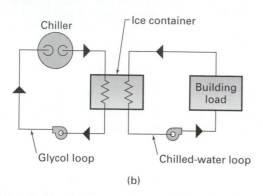

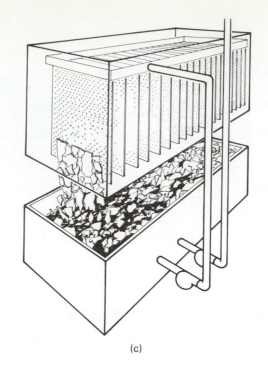

(a)

Chiller

Ice container

Building load

Glycol loop

Chilled-water loop

(b)

(c)

■ **FIGURE 4–24**

(a) Schematic of an open ice storage system. (b) Schematic of a closed ice storage system. (c) An actual product design showing ice maker on top and ice/water tank below. (Courtesy: Mueller Co., Springfield, MO.)

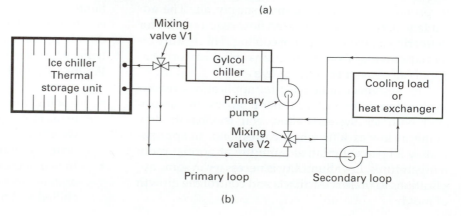

(a)

■ **FIGURE 4–25**

(a) A multiple tank closed ice system. (Courtesy: Trane Company, LaCrosse, WI.) (b) A primary-secondary ice storage system. (Courtesy: Baltimore Air Coil Company, Baltimore, MD.)

(b)

- Fluid coolers operate on the principle of spraying evaporatively cooled water over heat exchange surfaces through which chilled water (a glycol solution) is circulated through the system.
- A chiller-free cooling feature uses the natural pressure differential of the refrigerant within the chiller that is attainable by the use of low-temperature condenser water to make chilled water. In this mode of operating, chilled water can be produced without operating a compressor.

4.6.6 Plant Layout

The layout of equipment in a cooling plant shares equal importance with the selection of a system and the specification of equipment. Following are the salient factors, with several examples of good design.

Arrangement of equipment   The orientation and relative position of equipment are important to achieve a workable and efficient plant. The major criteria are as follows:

- Adequate aisle and circulation space for service and maintenance of equipment must be provided.
- Code clearance from electrical gear must be maintained, i.e., 3 ft minimum in front of wall-mounted panels and 3 to 5 ft at the front and rear of freestanding switchgear. The minimum clearance also depends on the voltage class.
- Means of removing and replacing equipment must be provided.
- Vertical clearance must be maintained. Some equipment may be 10 to 12 ft in height; associated pipe and valving may be at higher elevations. Access platforms, ladders, etc., should be provided.
- Equipment should be arranged in an orderly fashion to allow efficient piping, which is essential to low cost and good maintenance.
- Efficient pipe routing is largely dependent on the arrangement of equipment. Associated pieces of equipment should be grouped so as to allow short, direct runs of piping. Also, piping should be grouped as much as possible to result in runs of parallel piping at one or more elevations. This will allow the use of trapeze-type hangers and will result in a neat, orderly arrangement of piping. Vertical piping should be arranged to clear tube pulling and maintenance spaces.
- Heat rejection equipment, including cooling towers and air-cooled condensers, requires ready access to a source of air and is therefore usually located on the roof or on a site adjacent to the plant. Nuisance drift and fog should be considered in locating equipment. If visual screening is necessary, adequate clearances must be maintained to ensure that the equipment performs as desired. Note that even though heat rejection equipment can be concealed within the building, such installations usually result in higher usage of power due to air movement against added static pressure.

Maintenance and service   Maintenance and service must be considered early in space planning. Essential issues include the following:

- Equipment should be arranged so as to provide for the removal and replacement of those elements that may require major maintenance. In most cases, boilers, chillers, engines, generators, and other equipment in this category will not be replaced in their entirety. It will be necessary, however, to remove major components of these items, such as tube bundles, compressors, impellers, and motors. Adequate aisle space and paths of egress should be provided. Consideration must be given not only to the mechanical spaces themselves, but also to adjoining corridors, vestibules, elevators, and areaways through which the equipment must be removed.
- Storage space for tools, materials, and spare parts should be considered. In accordance with the preference of the owner, shelves and cabinets can be included to allow for organized storage.
- Shop space should be provided for minor repairs to plant equipment. Minimum provisions include a workbench, tool cabinets, machine tools, etc.
- Record-keeping space should be provided for maintaining plant records and for housing such items as building automation equipment, monitoring, and annunciator panels. The space should be constructed to limit the penetration of noise and to provide security.
- Toilet and washroom facilities with locker space are necessary in plants that require extensive monitoring of their equipment.

4.6.7 Case Study 1

A large building complex requires three 1500-ton chillers. Figures 4–26 and 4–27 illustrate the layout and actual installation of these chillers. The room is actually 24 ft high, twice the height of a normal build-

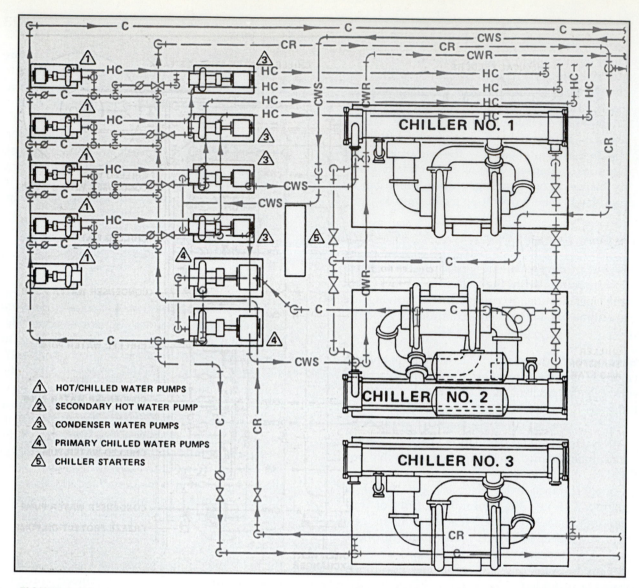

■ FIGURE 4–26

Floor plan of a large chilled water-plant with three chillers and their associated pumps.

■ FIGURE 4–27

Large chilled water plant.

124

ing floor. The room appears spacious with adequate service access as a result of good planning.

4.6.8 Case Study 2

A large computer center demands high cooling capacity due to its extremely high heat gain generated by the computers. Load analysis indicates that the high heat gain can be economically utilized to heat other buildings without using additional heating energy in the winter. Thus, the plant includes a heat reclaiming chiller (Figure 4–28, chiller No. 4). The floor plan also illustrates the large amount of space needed for the chiller's transformers and starters, which require nearly the same amount of floor space as the chillers. Figure 4–29 shows the arrangement of pumps and conduits.

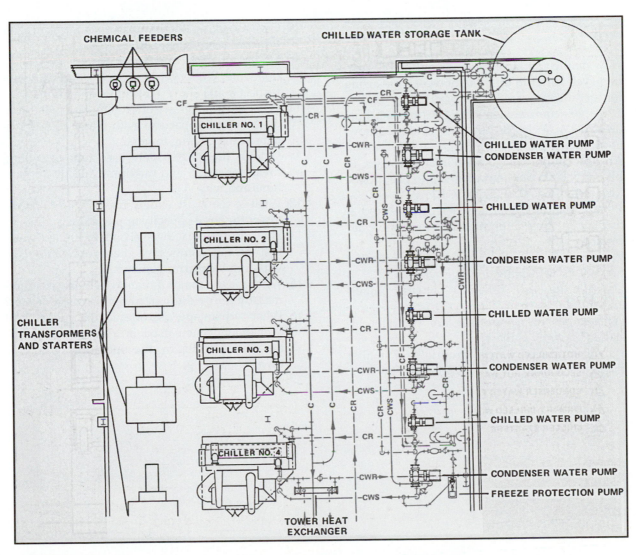

■ FIGURE 4–28
Floor plan of a large chilled-water plant with heat reclaiming feature.

■ **FIGURE 4–29**
Lineup of pumps and major piping headers and conduits over the equipment access aisle. This arrangement enhances the visibility of pipe and conduit routing, which is important for better plant maintenance.

QUESTIONS

4.1 How do the vapor compression and the absorption cycle differ in their methods of condensing refrigerant?

4.2 A vapor compression refrigeration machine uses 20 kW of electric power to produce 26 tons of cooling. What is its COP?

4.3 An absorption refrigeration machine uses 15 kW and 800 lb per hour of steam to produce 60 tons of cooling. What is its COP?

4.4 What is the difference between direct evaporative air cooling and indirect evaporative air cooling?

4.5 What are the limitations of DX equipment that prevent its application to large systems?

4.6 Describe the basic difference between unitary and split DX systems.

4.7 What types of compressor are typically installed on water chillers?

4.8 Which compressors are appropriate for smaller machines? Which for large machines?

4.9 Give a ranking of compressors in terms of their typical energy efficiency.

4.10 Under what circumstances would an absorption water chiller be an economical choice with respect to the energy cost of its operation?

4.11 What advantages are offered by air-cooled condensers in comparison with water-cooled systems using cooling towers?

4.12 What is the most widely used type of cooling tower for small- and medium-capacity applications and why?

4.13 What is the most widely used type of cooling tower for large-capacity applications and why?

4.14 How do primary and secondary chilled-water loop arrangements save energy in comparison with unit loop arrangements?

4.15 What are the major advantages of ice as a thermal storage medium in comparison with water?

4.16 What advantage is offered by plate heat exchangers when used in a system to generate chilled water by the operation of cooling towers?

5

HEATING PRODUCTION EQUIPMENT AND SYSTEMS

AN UNRELIABLE HEATING SYSTEM CAN RENDER A building uninhabitable and result in a loss of productivity far greater in value than any savings that might be realized by accepting a marginally designed plant. More than any other portion of the HVAC system design, heating plant designs need a proper consideration regarding the location, space, reliability, and operating efficiencies of the heating system. This chapter describes available heating equipment and systems, with an emphasis on the most commonly used systems—air, water, and steam. Other systems, such as infrared and heat pump systems, are only briefly mentioned.

5.1 TYPES OF HEATING SYSTEMS

Heating systems may be classified according to the following criteria:

1. *Energy source.* Coal, electric, gas, oil, biomass, energy from air or water via a heat pump cycle, or a combination of these sources.
2. *Energy transport medium.*
   - *Steam.* Low pressure (below 15 psig), medium pressure (15 to 100 psig), and high pressure (100 to 350 psig).
   - *Water.* Low-temperature (below 250°F), medium-temperature (250° to 350°F), and high-temperature (over 350°F).
   - *Air.* Gas, oil, or electric furnaces, in-space heaters, etc.
   - *Direct radiation.* Gas or electric infrared heaters, etc.

The basic components of a heating system are as follows:

1. *Steam systems.* Steam boilers, heat transfer equipment (exchangers, coils), combustion air supply and preheating, makeup air supply and preheating, flue gas venting (breaching, flue stack or chimney), condensate (return, deaeration), water (makeup, deaeration, treatment, and pumping), fuel (supply, pumping, heating, burners, and storage), control and safety devices, etc.
2. *Water systems.* Hot water boilers (generators) and circulating pumps. Other components are the same as in the steam system, except that condensate return and deaeration equipment are not required.
3. *Air systems.* Furnaces, in-space air heaters, ductwork, fuel, combustion air, and flue gas components are similar to those in steam or hot water systems. Electric furnaces do not require flue gas removal or fuel systems.
4. *Infrared systems.* Heaters (electric, gas), flue, gas venting.
5. *Heat pump systems.* Air-to-air, air-to-water, water-to-water and air-to-refrigerant systems, pumps, and compressors.

5.2 HEATING ENERGY SOURCES

Heat for HVAC systems is commonly produced by combustion of fuels or electric resistance. Heat pumps (reverse refrigeration) and heat reclaiming from processes are alternative methods of producing heat that can conserve fuels and electricity.

Fuels vary in cost and convenience of utilization. Coal was once a common fuel even for small residential heating systems. Now, due to the inconvenience of mining and transporting it, and because of the impact of its products of combustion on the environment, coal is appropriate only for large-volume users who can afford the high equipment cost to take advantage of the low energy cost. Similarly, heavy oils are economical only for large-volume users. Most residential, commercial, and institutional HVAC systems use light oil or gas as fuel for combustion.

Gas and light oil are burned to produce heat in several types of HVAC equipment. Furnaces and space heaters heat air, which is distributed either by natural convection or forced air, the latter of which employs a fan. Fuels can also be burned to heat water or produce steam used in convectors or forced-air devices.

Electric resistance elements can similarly be used to produce heat directly in convectors, in coils to heat air, or in boilers to distribute heat by hot water or steam.

Several factors must be considered before a fuel can be selected:

1. *Availability.* Adequate supplies must be maintained for normal operating conditions and during emergencies in critical facilities such as hospitals.

2. *Fuel Storage.* Natural gas is normally supplied by a utility company through piping to the site; therefore, no storage is required. All other fuels, including liquid petroleum gas, oil, and coal, must be stored on the site.

3. *Code Restrictions.* Building and boiler codes limit the use of certain types of fuels for reasons of safety and their impact on the environment. Heavy oils, coal, and other solid fuels must be used in a manner that will meet the requirements of the Environmental Protection Agency and local ordinances.

4. *Cost.* Cost is a major element of the fuel selection process, both in terms of capital investment and annual energy costs. Cost is also affected by the degree of maintenance required. In general, electric heating will result in a low initial cost and low maintenance; however, the cost of energy is high. Gas and light oil heating are higher in initial cost, but will be less expensive overall. Heavy oil and coal heating are most expensive to install and prone to high maintenance, but are significantly less expensive than light oil, gas, or electric heating.

The fact that coal is lower in cost than either gas or oil compensates somewhat for the higher cost of the plant. On the other hand, because of its simplicity and favorable regional and time-of-day rates, electric heating may be economical in some cases. There is no need for combustion air, chimneys, or fuel storage, which can be expensive, especially in high-rise buildings.

Some properties of energy sources, including gas, oil, coal, and electricity, are shown in Table 5–1.

TABLE 5–1
Energy value of common sources

Type of Energy	Unit of Measure	Conventional Unit (Btu)	SI Unit (KCal)
Electricity	Kilowatt-Hour (kWH)	3413	860
Gasoline	Gallon (gal)	128,000	32,000
Fuel Oil (No. 2)	Gallon (gal)	140,000	35,000
Residual Oil (No. 6)	Gallon (gal)	147,000	37,000
Natural Gas	Cubic Feet (CF)	1000	250
	Therms (TM)	100,000	25,000
Liquid Petroleum Gas (Butane)	Gallon (gal)	102,600	26,600
Liquid Petroleum Gas (Propane)	Gallon (gal)	93,000	23,300
	Cubic Feet (CF)	2500	625
Steam (at 14.7 psia)	Pound (lb)	1150	290
Steam (at 200 psia)	Pound (lb)	1200	300
Coal (Average)	Pound (lb)	10,000	2500
Wood (12% Moisture)	Pound (lb)	8000	2000

Note: Btu = British thermal unit; KCal = Kilocalorie. Energy values shown in the table may vary with the source.

5.3 FURNACES AND AIR HEATERS

Furnaces are packaged air heaters that are generally used in small buildings, in residences, or in small decentralized systems for large buildings. There are, however, large furnaces that produce up to millions of Btu/hr (Btuh) for industrial plants or warehouses, where the hot air can be distributed into open spaces. They are often installed in combination with an air conditioner. Natural gas, oil, and electricity are the most common energy sources. Other air-heating devices differ from furnaces with regard to their configuration and installation.

It is not practical to transfer heat by air for long distances, due to the physical space required for ductwork and the cost of electricity to run fans. In applications requiring extensive distribution of heat from a central plant, boilers are used to generate steam or hot water, which can then be distributed in a compact piping system.

5.3.1 Type of Furnaces

Furnaces are classified and described according to the following categories:

- *Type of fuel or energy.* Gas, oil, gas-oil, or electricity.
- *Process of combustion.* Open combustion chamber (atmospheric or power blower) or a sealed chamber (impulse).
- *Design and construction.* Cabinet configuration (vertical or horizontal), air flow (up or down flow), air delivery (ducted or free delivery), construction (indoor or outdoor, pad mounted or roof top mounted).
- *Services.* Heating only or heating and cooling combined.

5.3.2 Combustion Efficiencies

The open-chamber furnace usually has an annual fuel utilization efficiency (A.F.U.E.) between 75 and 80

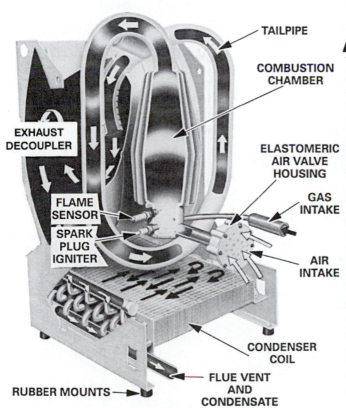

TAILPIPE

COMBUSTION CHAMBER

EXHAUST DECOUPLER

ELASTOMERIC AIR VALVE HOUSING

GAS INTAKE

FLAME SENSOR

SPARK PLUG IGNITER

AIR INTAKE

CONDENSER COIL

RUBBER MOUNTS

FLUE VENT AND CONDENSATE DRAIN

PROCESS OF COMBUSTION

The process of combustion begins as gas and air are introduced into the sealed combustion chamber with the spark plug igniter. Spark from the plug ignites the gas/air mixture, which in turn causes a positive pressure build-up that closes the gas and air inlets. This pressure relieves itself by forcing the products of combustion out of the combustion chamber through the tailpipe into the heat exchanger exhaust decoupler and on into the heat exchanger coil. As the combustion chamber empties, its pressure becomes negative, drawing in air and gas for the next pulse of combustion. At the same instant, part of the pressure pulse is reflected back from the tailpipe at the top of the combustion chamber. The flame remnants of the previous pulse of combustion ignites the new gas/air mixture in the chamber, continuing the cycle. Once combustion is started, it feeds upon itself allowing the purge blower and spark plug igniter to be turned off. Each pulse of gas/air mixture is ignited at a rate of 60 to 70 times per second, producing from one-fourth to one-half of a Btu per pulse of combustion. Almost complete combustion occurs with each pulse. The force of these series of ignitions creates great turbulence which forces the products of combustion through the entire heat exchanger assembly resulting in maximum heat transfer.

■ **FIGURE 5–1**
A cutaway section of the sealed combustion chamber of an impulse type of furnace. (Courtesy: Lennox Industries, Inc., Richardson, TX.)

percent. It uses either gas or oil. The impulse type of furnace is designed for gas only; it has a high A.F.U.E. of from 92 to 95 percent. The principle of operation of the impulse furnace is shown in Figure 5–1.

5.3.3 Typical Furnace Arrangements

Figure 5–2 illustrates a typical open-chamber furnace and a number of its design arrangements.

5.4 BOILERS

5.4.1 Types of Boilers

In terms of their design and construction, boilers can be classified and described as follows:

- *Source of energy.* Gas, oil, gas-oil, electric, coal, biomass, etc.
- *Heat transfer surface.* Combustion type, including water tube or fire tube boilers, and electric type, including resistance or electrode boilers.
- *Design of combustion chamber.* High-fire box or low-fire box (Scotch marine) type.
- *Construction.* Cast iron sectional or packaged, steel or copper tubes, etc.

5.4.2 Fire Tube Boilers

Fire tube boilers are so named because the combustion gases travel through the fire box and through the insides of steel tubes, usually in two or more passes, before being exhausted to the stack. In such boilers, the water is contained in the boiler shell, which com-

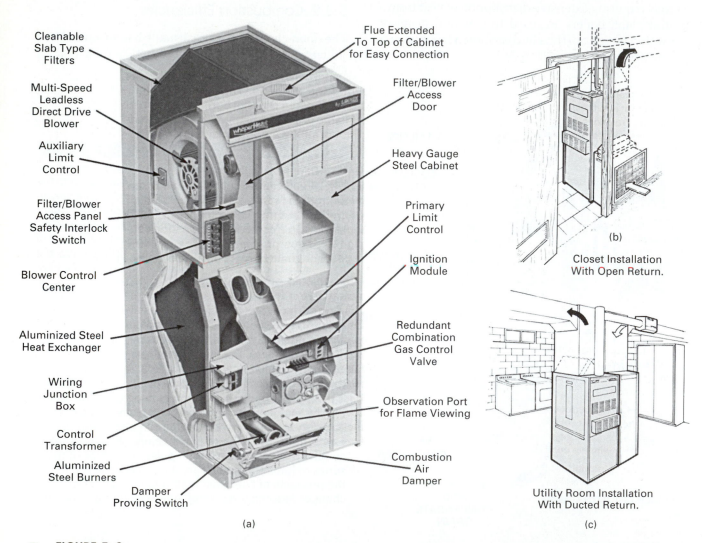

■ FIGURE 5–2
(a) A cut-away view of a typical open-chamber type gas fired furnace; (b) typical closet installation; and (c) typical utility room installation. (Courtesy: Lennox Industries, Inc., Richardson, TX.)

pletely circumvents and submerges the tubes. In hot-water boilers, the shell is completely filled with water. In steam boilers, the water level is lowered to allow a space for evaporation.

Scotch marine-type boilers are fire tube units that have low contours. They were initially designed for use on board ships, where head room is limited. Low height is an advantage in building spaces that are restricted in height. They are suitable for stable

loads similar to the HVAC loads of commercial buildings. Figure 5–3 illustrates a typical Scotch marine boiler.

5.4.3 Water Tube Boilers

Water tube boilers are so named because water is on the inside of the tubes and the combustion gases pass around the outside of the tubes. Water tube boilers

(a)

(b)

■ **FIGURE 5–3**
(a) Typical configuration of a low fire box Scotch marine-type water heater of boiler; (b) A cutaway view of the Scotch marine-type unit showing combustion chamber and the fire tubes. (Courtesy: Cleaver Brooks, Milwaukee, WI.)

generally have collection headers and upper drums. There are, however, other arrangements proprietary to various manufacturers. Convective water flow takes place within the tubes connecting the drums and headers.

Water tube boilers are available in all sizes, ranging from small residential units to very large units used for generating power. This type of boiler generally contains less water than is contained by a fire tube boiler of equivalent capacity. It therefore has a quick response to load fluctuations, such as those experienced in a hospital. Figure 5–4 illustrates a typical small, modular type of low-fire box boiler; Figure 5–5 shows a large high-fire box boiler. The latter is used in

(a)

(b)

(c)

■ FIGURE 5–4

(a) Typical removable tube-type of water tube boiler. Portion of the the boiler panel is cut out to show the interior tubes. The model shown is a hot-water unit. (b) Bend tube, water, and flue gas passages, in the boiler. (c) Manifolding of three identical boilers for modular design. (All illustrations courtesy: Bryan Steam Corporation, Peru, IN.)

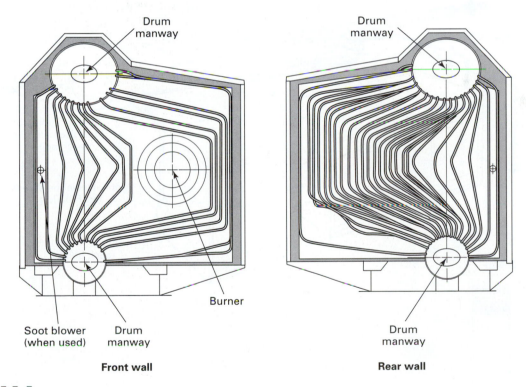

Drum
manway

Drum
manway

Burner

Soot blower
(when used)

Drum
manway

Drum
manway

Front wall

Rear wall

■ **FIGURE 5–5**
Typical large high-fire box type of water tube boiler. (Courtesy: Cleaver Brooks, Milwaukee, WI.)

central plants for very large structures, in industrial plants, and in campus types of applications.

5.4.4 Cast Iron Sectional Boilers

Cast iron sectional boilers are units consisting of a series of vertical cast iron sections filled with water. Each section is shaped like an inverted "U" with a bottom connecting link. When the sections are stacked, the void inside the "U" forms the furnace or firing volume. These boilers are used primarily in residential and small to medium-size commercial buildings. They can be assembled in the field, section by section, and, as such, are adaptable to many existing installations where it is not possible to get a completely assembled boiler through the entry to the building. Also, they can be expanded for additional capacity by adding sections. Cast iron boilers are not applicable to high-pressure systems, due to the limited strength of cast iron. They are generally rated for 15 psi for steam and a maximum of 100 psi for water.

See Figure 5–6 for typical construction and piping arrangements.

5.4.5 Electric Boilers

Electric boilers are made for either hot water or steam. The most popular type is the immersion resistance boiler, with capacities up to 1500 kW of input. Larger boilers may be of the electrode design, which utilizes water as the current conducting medium to generate steam by electrolysis. The immersion resistance elements are usually made of a high-temperature alloy and sheathed within Monel or Incoloy alloy tubes to separate the heating element from the water. The tube bundles are usually constructed in "U" configurations to facilitate wiring and replacement.

With the electric boiler, electrical energy is converted directly into heat energy with nearly 100 percent efficiency. The only losses are from radiation and convection. There is no need for breeching and a flue stack. This is a very important factor in designing

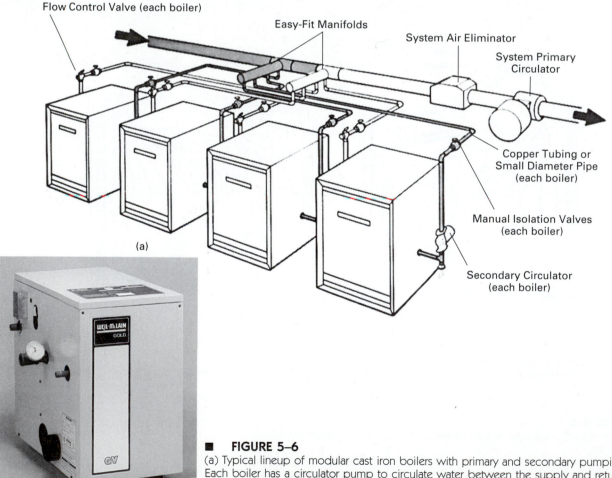

■ **FIGURE 5–6**

(a) Typical lineup of modular cast iron boilers with primary and secondary pumping. Each boiler has a circulator pump to circulate water between the supply and return headers. The system's primary circulator provides circulation to all loads. With the primary and secondary pumping system, operation of the boiler can be sequenced to conserve energy. (b) The simple modular design of a cast iron boiler. (Courtesy: Weil-McLain, Michigan City, IN.)

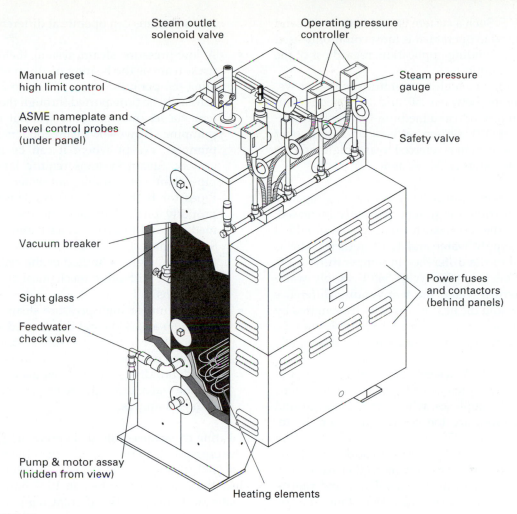

Steam outlet
solenoid valve

Operating pressure
controller

Manual reset
high limit control

Steam pressure
gauge

ASME nameplate and
level control probes
(under panel)

Safety valve

Vacuum breaker

Power fuses
and contactors
(behind panels)

Sight glass

Feedwater
check valve

Pump & motor assay
(hidden from view)

Heating elements

■   **FIGURE 5–7**
A typical low-resistance type electric boiler. (Courtesy: Chromalox Industrial Heating
Products, Pittsburgh, PA.)

high-rise buildings. From an energy-cost point of view, electrical energy is always more expensive than gas or oil. The selection of an electric boiler is based on an evaluation of life cycle costs, distributed between capital and operating costs.

Electric boilers are ideal to operate in conjunction with fuel-fired boilers for off-peak, off-season, and standby operations. Figure 5–7 illustrates the construction of a typical low-resistance type of electrical boiler.

5.5 SELECTION OF MEDIUM AND EQUIPMENT

5.5.1 Selection of Medium

Steam vs. hot-water generation   Several factors must be evaluated to decide whether a central heating plant should utilize hot-water or steam boilers. If the

system supported by the boiler plant is strictly a heating system, the most logical choice is hot water. With steam loads of this type, it may be desirable to generate steam for the specific requirements and use steam-to-water heat exchangers for heating hot water. Another alternative is to use hot-water generators for the building heating load and a separate steam boiler to serve the steam load. When the basic system has been selected, the subject of the operating temperature and pressure must be addressed. The following is a generalization of system characteristics; other uses, however, such as kitchen uses, laundry uses, sterilizers, industrial processes, or absorption cooling, may require steam.

Hot-water systems can operate at different temperatures:

■   A low-temperature hot-water system (below 250°F) consists of one or more hot-water generators (boilers) normally operating between 180°

and 240°F. Such a system is simple to design and economical to operate. It is most commonly used for single-building application, regardless of the size of the building.

- A medium- or a high-temperature hot-water system operates between 250° and 350°F and is commonly referred to as a medium-temperature hot-water (MTW) system. When operating over 350°F, the system is referred to as a high-temperature hot water (HTW) system.

At the higher temperature level, the heating capacity of a given distribution system is greatly increased. For example, the heat transfer that can be provided with 200°F supply water and 120°F return water is proportional to the difference in temperature at the two conditions, i.e., 80°F. With 360°F supply water and 120°F return water, the temperature difference will be 240°F, and the heating capacity transmitted by the distribution system is tripled, i.e., the size of pipe required to carry the same load can be substantially reduced.

MTW and HTW systems are applicable to campus-type facilities such as colleges, research centers, and industrial complexes, where large underground distribution pipes are too costly. However, equipment for MTW and HTW systems must be rated for higher temperature and pressure. In addition, piping must be designed to allow for more thermal expansion; all pipes must be heavily insulated, and a nitrogen pressurization system must be installed to prevent the hot water from flashing into steam. All these extra features result in a higher capital cost, which must be evaluated against the cost savings from the reductions in the size of the pipes.

Steam systems can operate at different pressures:

- A low-pressure steam system (below 15 psig) uses steam as the heating medium with an operating temperature range of 220° to 250°F. Steam can easily be transported through the piping system via its own pressure without the need of pumping. However, steam condensate must be pumped, except where it can be returned by gravity. Steam systems require larger sizes of pipe than comparable hot-water systems do. Pipes for both steam and condensate must be properly pitched to facilitate the return of the medium and to avoid water hammer and noise. Steam may not be suitable for buildings in which the floor height is limited or the ceiling space is congested with other mechanical or electrical installations.

- A medium- or high-pressure steam system (between 60 and 350 psig) is preferred when steam loads exist. Medium-pressure systems up to 100 psig are commonly used for hospitals and research buildings. The heating load can be served by hot water through the use of steam-to-water heat exchangers.

Heating capacities of fluids in pipes are governed by the pressure drop per unit length of the pipe (usually psi/100 ft) and by the velocities of flow. Table 5–2 lists the heating capacities, in 1000 Btuh (MBH), for water and steam with different temperature drops. For long-distance transmission, the pressure drop per unit length of pipe becomes the governing factor for the size of the pipe, particularly with low-pressure steam. The reader should notice that the design veloc-

TABLE 5–2

Heating capacity of heat media in pipes

| Heat Medium | Heating Capacity (MBH) @ Pipe Size | | | |
	1"	2"	4"	8"
Low-Temperature Water (from 160° to 200°F)	120 @ 2.5 fps	700 @ 4.5 fps	5000 @ 6 fps	30,000 @ 10 fps
Medium-Temperature Water (from 200° to 300°F)	300 @ 2.5 fps	1750 @ 4.5 fps	12,500 @ 6 fps	75,000 @ 10 fps
High-Temperature Water (from 200° to 400°F)	600 @ 2.5 fps	3500 @ 4.5 fps	25,000 @ 6 fps	150,000 @ 10 fps
Low-Pressure Steam (from 5 psig to 180°F water)	50 @ 50 fps	300 @ 100 fps	2000 @ 150 fps	11,500 @ 200 fps
Medium-Pressure Steam (from 50 psig to 180°F water)	200 @ 60 fps	1400 @ 100 fps	6000 @ 150 fps	25,000 @ 130 fps
High-Pressure Steam (from 150 psig to 180°F water)	800 @ 100 fps	4000 @ 130 fps	15,000 @ 130 fps	60,000 @ 130 fps

Notes: MBH = 1000 Btu/hr; MTW and HTW require extra-heavy pipe material; fps = velocity, in feet per second; fpm = 60 fps.

ity for steam is about 20 to 40 times faster than that for water. This is because steam, as a vapor, does not have the mass to erode the piping material. The limitation on velocity is governed by the noise level that can be tolerated.

5.5.2 Selection of Equipment

Once the system and the type of fuel have been determined, the next design step is the selection of boilers and auxiliary equipment, including burners, feedwater heating, water treatment, combustion air heating, safety devices, and other equipment. Good practice dictates that some redundancy be provided for critical equipment that is subject to routine maintenance or mechanical failure. The following represent general criteria for boiler plant design:

1. *Boilers.* Boilers are selected on the basis of their load demand and load profile. Depending on these characteristics, the plant can utilize two or more boilers to carry the major portion of the load when one boiler is out of service. In general, with a two-boiler configuration, each boiler can be sized for 65 percent to 75 percent of the load; with a three-boiler configuration, each boiler can be sized for 50 percent of the load.

 Efficiency varies according to the type of boiler, the type of burner, and the control method. In general, most boilers with power burners can maintain a fuel-to–heating medium efficiency between 80 and 85 percent under varying load conditions. Atmospheric burners use more excess air and therefore have lower overall efficiencies. With electric boilers, 100 percent of the electrical energy is converted to the heating medium. The only losses are the result of radiation and convection associated with the boiler shell and appurtenances.

2. *Burners.* There are a number of gas, oil, and combination burners that will work with a variety of boilers. Safety provisions for the control of gas and oil burners are subject to building codes and insurance regulations.

3. *Feedwater Systems.* Feedwater systems provide treated water to steam boilers. The water quality, the capacity of the system, and adequate pressure are essential.

4. *Fuel Supply Systems.* When liquid fuels are used, equipment for the transfer of the fuels from storage tanks to the point of usage at the boilers is necessary.

5. *Combustion Air Supply.* A large quantity of air is required to support proper fuel combustion. The air should be heated to improve the efficiency of the boiler.

6. *Flue Gas Discharge.* Flue gas must be vented through one or more stacks, which can influence the selection of the fuel and location of plant. The stack must extend vertically above the highest part of the building or any adjacent taller buildings. In the case of high-rise construction, the cost of the stack and the net space occupied by the stack passing through each floor can become prohibitively expensive when the boiler is located at the base of the building.

7. *Water Treatment Systems.* Water treatment systems consist of water softeners and chemical feeders to control the corrosive and scale-forming characteristics of water.

5.6 AUXILIARY SYSTEMS

5.6.1 Burners

Burners are designed to mix fuel and air and ignite the mixture for combustion within the boiler. Burners may be classified into atmospheric (natural draft) and power types.

Atmospheric burners are used for small and medium-size gas boilers. Such burners depend on the natural draft provided by the stack for the introduction of combustion air. Where adequate stack height cannot be provided, and where the breeching must be abnormally long, induced-draft fans can be used to provide the necessary draft.

Atmospheric burners are generally of the ribbon type, with one or more pilots strategically located to provide a smooth shut-off. These burners are the simplest available and operate on low gas pressure (3 in. to 5 in. water gauge [WG]).

Power burners incorporate a blower that supplies combustion air at a pressure adequate to compensate for the resistance to airflow offered by the boiler. The blower also provides a means of including prepurge and postpurge cycles. This adds an element of safety in that the boiler is scavenged of any possible combustibles prior to lighting off and subsequent to the burning cycle.

Burners used for firing oil are equipped with a means of atomizing the oil for proper combustion. With small burners, pressure atomization is used. With larger, modulating types of burners, atomization is accomplished with the use of compressed air or steam. Steam atomizers are commonly used on large institutional and industrial types of installations firing heavy oil. Air atomizers are used when firing light oil.

Power burners allow the use of smaller diameter breechings and stacks, which must be gastight because they are under positive pressure. Power burners also provide a means for controlling the percentage of excess air for efficient combustion.

Combination gas and oil burners are a hedge against the uncertainty of the availability of fuel, cost fluctuations, and restrictions imposed by utilities or gas consumption during certain high-demand periods. Such burners incorporate the features of both gas and oil burners and can be easily switched from one fuel to the other.

5.6.2 Boiler Blowdown

Blowdown is the process of draining water from the boiler to remove sludge from the bottom of the boiler shell and to skim off suspended materials at the surface of the boiler water. Sludge is formed as a precipitate composed of waterborne solids combined with water treatment chemicals.

Before any effluent is discharged to the floor drain of a building, the effluent must be cooled to a temperature below 140°F. This is a code requirement and is generally accomplished by using a blowdown separator and cooler designed to separate flashing steam from the effluent and to cool the effluent to an acceptable temperature. The flashing steam is vented, and cooling is accomplished with a heat exchanger using once-through city water or boiler feedwater.

Intermittent blowdown is implemented via a quick-opening valve at the boiler blowdown outlet, piped in series with a slow-opening valve. The quick-opening valve is opened first, followed by the slow-opening valve. When the slow-opening valve is fully opened, it is immediately closed. The quick-opening valve is then closed as well. The frequency of blowdown is a function of the concentration of solids and degree of water treatment required for the boiler water.

Continuous blowdown is common on large systems and serves to remove suspended materials from the surface of the boiler water. The quantity of blowdown is manually or automatically controlled.

5.6.3 Feedwater System

Steam is condensed into water after releasing its latent heat. The water is called *condensate*, which should be returned to the boiler for reuse. A feedwater system is used for this purpose. In addition, fresh water will have to be added to the boiler to make up for losses due to boiler blowdown and any other drains and leakages. A feedwater system may consist simply of a feedwater pump, for small systems, or an assembly of complex equipment, for large systems. A complete feedwater system may serve any or all of the following functions:

- To return the condensate.
- To make up the shortage due to leaks and blowdown.
- To preheat the water to the boiler operating temperature, preventing thermal shock.
- To remove the undissolved air and uncondensed gases in the system.
- To accumulate the condensate in order to compensate for the time delay between the output of steam and return of condensate.

Figure 5–8 illustrates a typical feedwater unit consisting of a storage tank, steam heater feedwater pumps, and vents.

5.6.4 Fuel Handling

Gaseous fuels, such as natural gas or liquid petroleum gas, are usually supplied with pressure and thus can flow without pumping. Liquid fuels such as oil are usually stored in tanks prior to usage and must be pumped, unless the oil tank is elevated, permitting a gravity flow. Each heating plant must be individually analyzed to determine its pumping requirements. Factors to evaluate include the height of equipment above the storage level and the friction loss of piping and fittings.

There are two basic oil pump systems:

- *Simple suction system.* This type of system utilizes the pump that is normally furnished with the boiler to pump oil from the storage tank to the boiler/burner. Such a system works reasonably well, as long as the suction lift to the pump is not exceeded and the tanks are reasonably close to the boiler room (within say, approximately 50 ft). The maximum practical lift for priming a gear-driven pump is about 12 in. of mercury vacuum.

   When priming is lost (usually due to a leaky check valve or foot valve or an air leak in the piping), it becomes necessary for the pump to purge the line of air and to draw oil from the tank into the line. This can require several minutes and create a start-up problem.
- *Pressurized loop system.* A pressurized loop system is used whenever the pumping head and suction lift of the burner pump are exceeded. This system will ensure that oil is always available at adequate pressure at the point of usage. It

common. The most acceptable method of storage is the use of buried fuel oil tanks located outside the building in an accessible location for servicing from a delivery truck.

- *Direct-burial tanks.* Double-walled glass fiber tanks are used in direct-burial applications. These tanks are corrosion resistant, and the double-wall feature traps leaking oil to prevent the surrounding soil from becoming contaminated. Monitoring of the cavity for leaks is required. Steel tanks are more fire resistant and durable than glass fiber is and can generally withstand more internal pressure. Steel tanks, however, are more susceptible to corrosion, and resulting leaks can cause fire and groundwater contamination. When buried, steel tanks must be properly coated and should be provided with impressed-current cathodic or sacrificing anodic (aluminum) protection. Buried tanks must be weighted down by heavy (concrete) pads to avoid buoyancy due to underground water. Figure 5–9 gives some details of buried tanks.
- *Above-ground tanks.* Steel is preferred for above-ground applications.
- *Inside tanks.* These tanks are subject to stringent requirements dictated by the National Fire Protection Association and local codes. In general, inside tanks must be housed in three-hour fire-rated enclosures, properly vented and constructed to retain oil within the enclosure in the event of a rupture or leak.

An oil heating system is required for boilers utilizing low-cost heavy oil (generally No. 5 or 6). Heating must be provided at several locations to ensure satisfactory circulation:

- *In the tank.* Electric or steam heaters are needed in the oil tank to reduce the viscosity of oil for easy pumping. The tank temperature should be maintained at approximately 110°F.
- *At the pipes.* Underground or outside fuel oil piping should be electrically or steam traced and insulated. If these lines are not traced, and no circulation occurs for a period during severe weather, oil flow may not be possible.
- *Burner preheater.* This device should be provided at the boiler to raise the oil temperature to between 150° and 180°F for immediate firing. Heaters can be of the combination electric or steam type. The electric heater is deactivated once steam pressure has developed to operate a steam-to-oil heat exchanger.

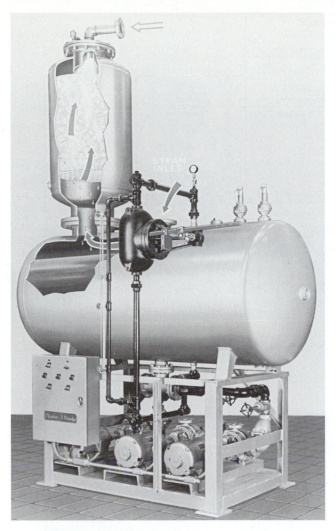

■ FIGURE 5–8
Typical feedwater heater, deaerator, and pumping assembly. (Courtesy: Cleaver Brooks, Milwaukee, WI.)

also eliminates loss of priming at the burner pump, as is often experienced with systems utilizing the burner pump only.

A pressurized loop system circulates oil from the tank to the point of usage and returns unused oil to the tank. A pressure-relief valve in the loop maintains positive pressure in the segment of the loop supplying the boiler burners. The burner pumps become secondary pumps, deriving oil from the pressurized portion of the loop and returning unused oil to the return side of the loop.

The amount of oil storage is dependent on the desired frequency of delivery of oil, the availability of the oil, and the rate of fuel consumption. A storage capacity equal to three or four weeks of peak consumption is

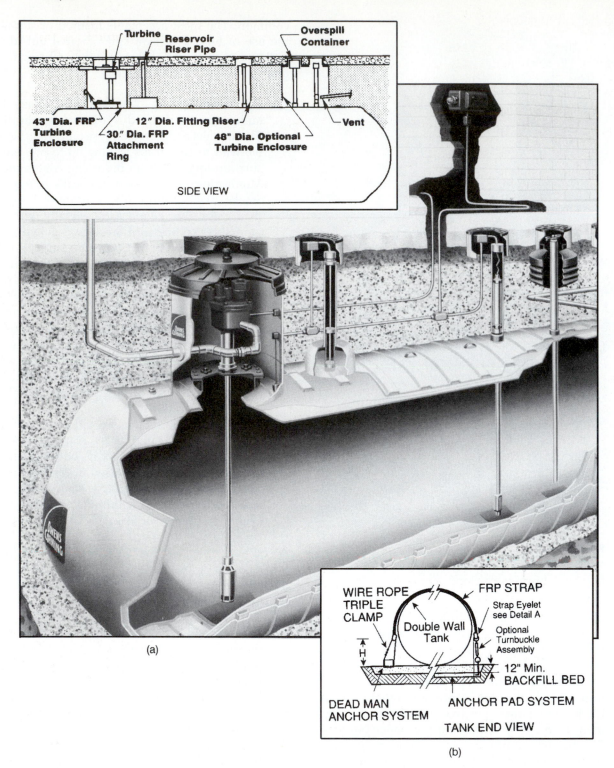

Turbine   Reservoir
Riser Pipe

Overspill
Container

43" Dia. FRP
Turbine
Enclosure

12" Dia. Fitting Riser

30" Dia. FRP
Attachment
Ring

48" Dia. Optional
Turbine Enclosure

Vent

SIDE VIEW

(a)

WIRE ROPE
TRIPLE
CLAMP

FRP STRAP

Strap Eyelet
see Detail A

Double Wall
Tank

Optional
Turnbuckle
Assembly

H

12" Min.
BACKFILL BED

DEAD MAN
ANCHOR SYSTEM

ANCHOR PAD SYSTEM

TANK END VIEW

(b)

■   **FIGURE 5–9**
(a) Double-wall buried fiberglass fuel oil tank with manholes with fuel level measurement
risers. (b) A typical mounting detail showing a concrete anchor pad with dead man
anchors to prevent the tank from floating during high ground water conditions. (Courtesy:
Owens Corning Company, Corning, IL.)

5.6.5 Combustion Air

Air is a vital part of the combustion process when gas, oil, or coal is the fuel. Basically, all fuels contain varying amounts of combustibles, primarily carbon (C), hydrogen (H), and sulfur (S). Combustion is the result of the following chemical reaction between oxygen (O) and the combustibles:

$$C + O_2 = CO_2 \qquad (5\text{--}1)$$

$$2H_2 + O_2 = 2H_2O \qquad (5\text{--}2)$$

$$S + O_2 = SO_2 \qquad (5\text{--}3)$$

Air contains approximately 21 percent oxygen by volume. Thus, for each cubic foot of oxygen required, 4.79 cubic feet of air are required. The density of air at sea level and 70°F is about 13.5 cubic feet per pound of dry air. Accordingly, the amount of air required for the stoichiometric combustion of various fuels can be calculated as shown in Table 5–3. In stoichiometric combustion of a hydrocarbon fuel, fuel is reacted with the exact amount of oxygen required to oxidize all combustibles.

Air must be supplied to the combustion chamber in excess of the theoretical quantity required to assure proper combustion. In fact, there are varying degrees of incomplete combustion, even with excess air. If there is too much air, the combustion efficiency is also lowered due to the heat absorbed by the excess air. The amount of air required for combustion is proportional to the firing rate and thus should be modulated according to this rate or the heating load. Practically speaking, between 10 and 15 percent of excess air is necessary to assure proper mixing of fuel with air. Some newer burners can operate with as little as 5 percent of excess air with turndown ratios of 5 to 1 and still achieve a high quality of combustion.

Example: A gas-fired boiler is rated for 3,000,000 Btu/hr with an efficiency of 80 percent. The gas is natural gas having a heating value of 1000 Btu/cf. Calculate the amount of natural gas the boiler requires and the amount of combustion air with 25 percent excess air the boiler would require.

(1) The amount of natural gas required is

$$\text{Gas} = 3{,}000{,}000 \,/\, (1000 \times 80\%) = 3750 \text{ cu ft/hr}$$

(2) The amount of combustion air with 25% excess air is

$$\text{Combustion air} = 3750 \times 12 = 45{,}000 \text{ cu ft/hr}$$
$$= 750 \text{ cu ft/min (cfm)}$$

When large amounts of combustion air are introduced into the boiler room, the space may become extremely cold during the winter in the northern hemisphere, even to the degree that it causes localized freezing of water piping and other components. Thus, combustion air must be heated to a reasonable temperature to prevent portions of the system from freezing. In fact, many designers mistakenly avoid heating combustion air, believing that they will save energy. This is not true; air must be heated to several thousand degrees in the combustion chamber (fire box), regardless of whether it has been preheated in the boiler room. If it is not preheated, then additional energy will be needed, and thus, more energy will be expended anyway.

Note that combustion air must be provided in a manner that will not result in the boiler room's being under a negative pressure, which can affect the performance of the burner and be hazardous.

Ventilation is required in boiler rooms to prevent the room from overheating during hot weather. The boiler room should be maintained at positive pressure relative to atmospheric pressure. A well-designed boiler plant should utilize the combustion air as part of the ventilation air, although the requirements of the two do not always coincide. Properly co-

TABLE 5–3

Approximate amounts of air required for the stoichiometric combustion of fuels

Type of Fuel	Basic Content	Theoretical Air for Combustion	Air Required with 25% Excess
Coal (Anthracite)	C, S	9.6 lb/lb	160 cf/lb
Coal (Bituminous)	C, S	10.3 lb/lb	175 cf/lb
Oil (No. 2)	C, H, S	106 lb/gal	1800 cf/gal
Oil (No. 5)	C, H, S	112 lb/gal	1900 cf/gal
Oil (No. 6)	C, H, S	114 lb/gal	1900 cf/gal
Natural Gas	CH_4, C_2H_6	9.6 cf/cf	12 cf/cf
Propane (Liquid Petroleum)	C_3H_8	24 cf/cf	30 cf/cf

ordinated, combustion and ventilation air can be consolidated into a common system with the necessary redundancy and safety features to ensure the satisfaction of both requirements. Ventilation of the boiler room may be either by exhaust with exhaust fans or by pressurization with outside air fans.

In some boiler rooms, combustion air can be adequately provided by an intake louver communicating directly with the outside. In larger plants, a more active method may be desirable. This can be accomplished by installing a makeup air unit arranged to heat and deliver tempered outside air to the plant. Preheating the air with a tempered air unit will eliminate problems with freezing. Return and outside air dampers can be installed at the makeup air unit, allowing control of both ventilation and combustion air. With this type of system, a stationary louver should serve as a means of relief for any quantity of air delivered by the fan unit that is not used for combustion. The louver will also serve as a secondary combustion air inlet if the fan unit is out of service. Such a system maintains the boiler room at a neutral or slightly positive pressure.

5.6.6 Fuel Train

Fuel trains consist of regulators, gauges, valves, test cocks, and pressure switches to convey fuel to burners safely. Standardized fuel trains for both oil and gas are designed for optimum safety.

Fuel trains are available from the factory on all commercial and industrial types of boilers. Two separate motor-operated valves are installed, in addition to modulating valves. In the case of gas trains, a motor-operated or solenoid vent valve is provided between the two gas valves. A typical insurer-approved gas train is illustrated in Figure 5–10. This type of train provides a high degree of safety. The two motor-operated, two-position gas valves in the main gas piping are slow-opening, fast-closing valves that close in less than 1 sec. These valves respond in unison to boiler start-up and shutdown commands and always close upon a signal that the boiler's malfunctioning, such as a condition of high pressure, high temperature, low water, failure of the flame, or high or low fuel pressure.

A main gas pressure regulator should be provided upstream of the gas train. Also, a gas strainer should be considered and may or may not be required, depending on the gas distribution system.

5.6.7 Flue Gas Handling

The products of combustion from boilers and furnaces are called *flue gas*. Flue gas is toxic and must be ducted and vented to the outdoors through flue stacks or chimneys. The horizontal duct from the flue gas outlet on the boiler or furnace to the vertical uptake is called the *breeching*. The vertical uptake is called the *(flue) stack* or chimney. The terms *flue stack* and *chimney* are used interchangeably, although *chimney* is more correct to use when this component is constructed as part of a building or on a freestanding structure.

Flue gas temperature   For most combustion processes, the temperature of the flue gas is over several thousand degrees—e.g., 1500 to 3000°F for incin-

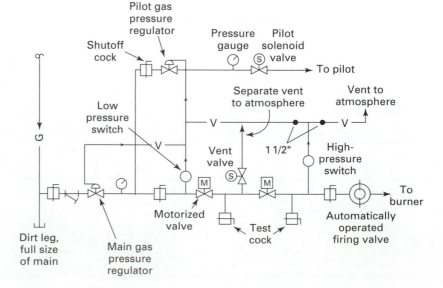

■ **FIGURE 5–10**
A typical gas train of a gas-fired boiler.

erators. The actual flue gas temperature at the outlet of heating appliances, which has been cooled by the heat transfer medium (air, water, or steam) is much cooler—e.g., 350°–400°F for gas appliances and 400°–550°F for oil-fired equipment. The flue gas temperature continues to drop after the gas travels through the breeching and flue stack. A mean temperature below 300°F may cause condensation of moisture (containing sulfuric acid) inside the flue stack. Too high a discharge temperature is an indication of energy loss. A proper design should avoid these extreme conditions.

Flue gas velocities   In breeching and stacks, flue gas velocities vary from 300 to 3000 fpm. High velocity causes a pressure drop and noise. A natural draft system is usually designed for 300 to 500 fpm, whereas a forced draft system is used between 1000 and 2000 fpm. Dispersal of the effluent to improve the ambient air quality may occasionally require a minimum upward chimney outlet velocity of 3000 fpm. A tapered exit cone best meets these requirements.

Draft   Draft is negative static pressure, measured relative to atmospheric pressure. It is normally calculated in inches of water. The draft needed to overcome chimney flow resistance is

$$\Delta p = D_t - D_a \qquad (5\text{--}4)$$

where Δp = draft, inches of water column (in. w.c.);
$\quad D_a$ = available draft—i.e., the natural draft produced by the buoyancy of the hot gas in the chimney relative to the cooler gases in the atmosphere;
$\quad D_t$ = theoretical draft—i.e., the draft needed at the equipment outlet.

The approximate theoretical draft at sea level (14.7 psig, or 30″ Hg) may be calculated by using Equation 5–4. The theoretical draft at a given elevation is proportional to the atmospheric pressure at that elevation; that is,

$$D_t = 7.7H(T_m - T_o) / (T_m \times T_o) \qquad (5\text{--}5)$$

where H = height of chimney above equipment outlet, ft;
$\quad T_m$ = mean chimney temperature, °R (Rankine);
$\quad T_o$ = ambient temperature, °R.

The available draft D_a needed at the appliance outlet should be provided by the appliance manufac-

turer. If the theoretical draft D_t created by the chimney is insufficient, then an induced or forced draft fan will be required. If the draft is too high, then a pressure balancing damper should be installed. The complete procedure for calculating the chimney size and height of a combustion system is given in Chapter 31 of the systems and equipment volume of the *ASHRAE Handbook* (1992).

Mechanical draft   When natural draft is insufficient to produce the required draft, mechanical draft-producing equipment is used. There are two basic types of such equipment:

- *Forced-draft type.* This is normally incorporated into the fuel burner installed at the front of the boiler or furnace to provide a positive pressure (or negative static pressure) in the combustion chamber.
- *Induced-draft type.* This is a fan assembly installed with the breeching to make up for the shortage of draft.

Flue stack height   The height of the flue stack is determined by the draft required, but in no case is lower than the building. Building codes also provide guidelines regarding stack heights in relation to the surrounding building, roof lines, or other obstructions.

Construction   Breeching is commonly fabricated of 10-gauge steel and covered with high-temperature insulation. Manufactured breechings are also available.

Breechings are subject to operating temperature variations, ranging from ambient temperature to the temperature of the flue gases. Because of the wide temperature range, provisions to compensate for expansion must be considered. Manufactured expansion joints are available for this purpose and should be utilized whenever the breeching design cannot otherwise accommodate the calculated changes in length.

Steel, masonry, and manufactured flue stacks are commonly used. Manufactured stacks are generally fabricated of double-wall steel with an air gap separating the two walls or of a steel casing with a refractory lining. For boilers with power burners, the stack and breeching are under positive pressure and must be gastight.

In the case of atmospheric burners, the draft developed by the stack must satisfy the requirements of the boiler-burner unit or be supplemented by an induced-draft fan. The height of the stack and temperature of the flue gas determine the theoretical available draft.

Figure 5–11 illustrates the construction and installation of a typical double-wall metal stack used for furnaces or low-temperature hot-water heaters, as well as a heavy-duty factory-insulated metal stack for boiler and high-temperature equipment, such as an incinerator.

5.6.8 Water Treatment

All heating (and cooling) systems involve a temperature change that affects the solubility of dissolved solids (minerals) and oxygen in water. The precipitation of dissolved solids and the released oxygen from wa-

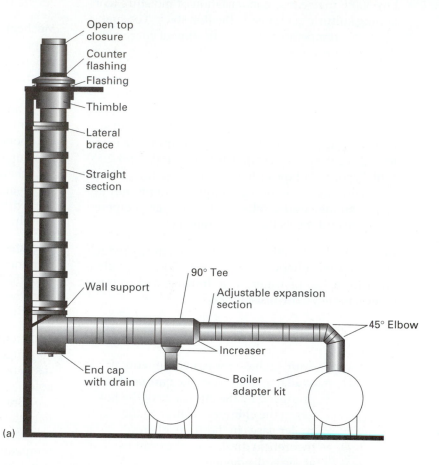

■ **FIGURE 5–11**
(a) Double-wall sheet metal flue components and installation of breeching and stack. (Note: Horizontal portion of flue is called the breeching and the vertical portion of the flue is called the stack.) (b), (c), (d): Step-by-step construction of an indoor stack. The view in (c) shows the acid-resistant refractory lining, which varies from 2 to 6 inches. (Courtesy: Van Packer Company, Beach Haven, NJ.)

Labels in figure (a): Open top closure, Counter flashing, Flashing, Thimble, Lateral brace, Straight section, Wall support, End cap with drain, 90° Tee, Adjustable expansion section, Increaser, Boiler adapter kit, 45° Elbow

(b)

(c)

(d)

ter can cause three symptoms: a reduction in the heat transfer rate, reduced water or steam flow, and corrosion or destruction of the equipment. Good water treatment is important for maintaining the capacity of the equipment, avoiding breakdown of the equipment, and prolonging the life of the boiler. Water treatment chemicals are used to control corrosion, scaling, embrittlement, and foaming.

Principles of treatment

- *Removal of solids in water.* Scale is a hard substance deposited on the interior surface of pipes and heat-exchanging equipment. It is caused primarily by the precipitation of calcium (Ca) salts in water. One method of removing the scale is to replace calcium ions with sodium (Na) ions through the Zeolite ion exchange process. Blowdown is performed to reduce the concentration of total dissolved solids, which build up when water is lost from the system through evaporation.
- *Removal of oxygen.* Oxygen is the primary cause of corrosion, particularly in the steam condensate system. Deaeration is an effective way to remove oxygen.
- *Reduction of acidity in water.* Acidity is another cause of corrosion. Adjustment of the water to a pH of 7.0 (neutral) is not sufficient to stop corrosion. A pH of 10.5 (alkalinity) is desired.
- *Inhibition.* Chromate has been a very effective agent for inhibiting corrosion. However, it is believed to be a carcinogen (cancer-causing substance) and hence has been banned in most countries. Sodium nitrite has been used as a substitute.

Planning   Water treatment is a must for steam systems and is desirable for closed-circuit hot-water systems. Automatic makeup provisions should be metered and monitored. Often, leaks occur through pump packings, gaskets, vents, etc. The lost water is automatically made up with fresh, untreated water that can cause scaling and oxygen corrosion.

Because of the varying characteristics and quality of water throughout the country, a specialist should be consulted to assure that the proper types and quantities of chemicals are utilized. Overtreatment can be just as damaging to a system as undertreatment.

5.7 OPERATING AND SAFETY CONTROLS

Whether a furnace or a boiler, any heating equipment involves the combustion of fuels. A malfunction of any equipment involved in combustion fuel can cause fire, destruction of property, or loss of life. Building codes and insurance regulations must be strictly followed. Boilers are required to have pressure, temperature, water level, fuel, and flame controllers to assure their safe operation. The following controls are applicable for commercial and industrial installations:

1. Operating temperature and pressure controls are two-position or modulating types of controls that are arranged so as to start and stop or modulate the burner when necessary.
2. Operating limit controls serve to shut down the burner if the temperature or pressure of the system continues to increase with the boiler in the low-fire or minimum-fire mode.
3. High-limit temperature and pressure controls have manual reset devices that are arranged so as to shut down the burner upon the occurrence of abnormal conditions.
4. Low-water control is used for steam boilers to cycle a boiler feedwater pump at a given water level and to shut down the burner upon further lowering of the water level.
5. High-/low- gas/oil pressure controls are arranged to shut down the burner when abnormal pressures occur. These are manual reset devices. A time delay may be required on the low-oil pressure switch to allow time for pressure to build up.
6. Flame failure controls are coupled with a programmer and arranged to shut down the burner upon failure of either the pilot or the main flame. Insurers and underwriters' laboratories require a manual reset for this control function. When applied to a forced-draft burner, the programmer also serves the function of programming the firing cycle.
7. Pressure or pressure/temperature relief valves are mechanical devices without any electrical connection and are designed to open at the rated boiler design pressure or temperature. These valves should be ASME stamped and sized to relieve the full boiler capacity.
8. With power burners, the combustion air damper can be controlled by the position of the fuel valves through a linkage and cam arrangement. As the fuel valves modulate in response to the firing demand, the air damper modulates proportionally. This type of control provides reasonably good results. To improve the efficiency of the burner and to control the quality of the flue gas emissions even more, independent control of the combustion air damper is desirable.

9. Flue gas control is used to continuously measure the products of combustion in terms of O_2, CO_2, CO, H_2, etc. The controller can be designed to adjust the combustion air damper and fuel valve for maximum efficiency of combustion.

10. When two or more boilers are installed in parallel, it may be desirable to sequence the boilers on line automatically as required by the load demand. This can be accomplished by utilizing a steam flow meter or water temperature sensor to energize an additional boiler when the load cannot be met by the boilers that are already on line. This portion of the control system would serve only to activate the controls of an additional boiler.

11. Additional pressure or temperature sensors located in the common steam or hot-water header are used as a controller to modulate all boilers on line in parallel. With this type of control, all of the boilers will fire at approximately the same rate and will feed the common header in a uniform manner.

5.8 HEATING PLANT DESIGN

The rules for good heating plant design are similar to those set forth in Chapter 4 for cooling production equipment and systems.

5.8.1 Location of the Plant

The heating plant may be located at the lower levels (ground level or basement) or upper levels (top floor or penthouse) of the building, or a separate or attached building. In general, heating and cooling plants are integrated and thus should be in the same location. (See Chapter 1 for a discussion of the location of the plant for high-rise buildings.)

5.8.2 Factors to Consider in Selecting Heating Plant Locations

Chimneys for combustion-type boilers must be extended above the roof of the building and must be clear of surrounding buildings. Locating the boiler plant at the base of a high-rise building involves considerable cost and loss of floor space, compared with locating it near the top of the building or in an ancillary building attached to or separate from the main building or complex.

While the boiler plant enclosure can be attractively designed in harmony with the main building itself, many projections from the building, such as chimneys, exhaust fans, vents, etc., are difficult to conceal and may limit the desirability of the plant's being close to the main building or complex. For a campus-type design, the boiler plant itself can be located in a totally separate building remote from the load and utilizing underground piping that is directly buried or that runs through tunnels to serve all building loads.

The proximity of the heating plant to the fuel supply and to the combustion air must also be considered in selecting a location for the heating plant, as must the effect of the location on the equipment rating. For example, a boiler located at the base of a high-rise building is subject to the high-static pressure of the piping system and therefore increases the required pressure rating of the equipment.

Finally, because the boiler plant generates noise and products of combustion, both of these should be considered in locating the plant. The plant should not be located where noise will be transmitted to other spaces. Also, the stack discharge should be coordinated to avoid the reentry of combustion gases through the outside air intakes.

5.8.3 Heating Piping Circuits

Piping distribution Hot-water systems normally consist of a closed loop, and the arrangement is relatively simple. However, care must be exercised regarding the proper selection of expansion tanks, air elimination devices, and pumping arrangements. Steam return systems are more complicated and thus require more attention to detail.

Heating-only steam plants In heating-only steam plants, essentially all the steam generated is converted to condensate and returned to the boiler. The only losses are those that occur from leaks in the system, such as from pump seals, evaporation through vents, blowdown, etc. Since very little water is lost from the system, only a nominal quantity of makeup water is required, and preheating of the condensate is not necessary.

Steam condensate The condensate is collected in receivers or surge tanks and pumped back to the boiler. Calculations must be made to determine the necessary capacity of the receiver tanks to accommodate the amount of condensate required to fill the distribution system. There will be a period between the start-up of the system and the return of condensate to the receiver tank. During this time, there should be enough reserve capacity in the receiver to satisfy the makeup water requirements of the boiler. This approach will minimize the required amount of city makeup water and prevent any waste of water due to the tank's overflowing.

Deaerating feedwater heaters These heaters are used for boiler plants operating at 75 psi or over and plants using 25 percent or more makeup water. The devices are designed to operate at slightly above atmospheric pressure by injecting steam to the deaerating tank. A deaerating heater, as the name implies, both heats and eliminates oxygen from the feedwater prior to its introduction into the boiler.

QUESTIONS

5.1 If gas is available at $0.50 per therm and #2 oil is available at $1.00 per gallon, which will be the more economical fuel source on the basis of cost per Btu? (Assume equal efficiencies.)

5.2 If gas is available at $0.50 per therm and the efficiency of utilization is 80%, what is the net cost of heating energy in dollars per million Btu?

5.3 If electric is available at $0.1 per kWh, what is the cost of electric for heating in dollars per million Btu?

5.4 What is the chief limitation preventing air furnaces from being used in heating systems for large buildings?

5.5 When might steam boilers be preferable to hot-water boilers for a large building system?

5.6 Name several factors that should be considered in locating a central boiler plant for a large building.

5.7 What characteristics of a Scotch marine boiler should be considered in selection for a building application?

5.8 What characteristics of a water tube boiler should be considered in selection for a building application?

5.9 What characteristics of a cast iron sectional boiler should be considered in selection for a building application?

5.10 Describe some advantages and disadvantages of power burners in comparison with atmospheric burners.

5.11 When will a pressurized-loop oil supply system be preferable to a simple suction system?

5.12 What special provisions must be made for a heavy oil storage and distribution system?

5.13 What special provisions must be made for underground oil tanks to comply with environmental regulations?

5.14 What is the purpose of boiler blowdown?

5.15 Name the chief safety features applicable to commercial and industrial boilers.

5.16 What is the nominal A.F.U.E. of impulse-type furnaces?

5.17 Which of the fire tube and water tube boilers are more responsive to load changes and why?

5.18 Why is nitrogen pressurization required for MTW and HTW systems? Why not air?

5.19 A heating plant requires 5000 MBH of heating capacity. Select the approximate size of pipe for the main if the plant is:
(a) a 200°F hot-water system
(b) a 5-psig low-pressure steam system
(c) a 150-psig high-pressure steam system

5.20 For the same heating plant as in Question 5.19, but using natural gas with a heating content of 1050 Btu/cf, what should be the gas flow rate (cf/hr)? How much combustion air is required with 25% excess air?

5.21 What are the major chemical components of natural gas?

5.22 What is the natural draft (in. w.c.) created by a chimney 100 feet tall with a mean temperature of 400°F and the ambient temperature of 60°F? (*Note:* Equation (5–4) is in degrees Rankine, which is (°F + 460°).)

5.23 What are the four basic principles for water treatment?

5.24 Name the factors considered in selecting the location of a heating plant.

6

AIR-HANDLING EQUIPMENT AND SYSTEMS

As MENTIONED IN CHAPTERS 3 AND 4, AIR IS THE most common medium used to deliver a heating or cooling effect to a space. In such a case, the delivery system is called an *air-handling system*. The air-handling system consists of air-handling units, duct-work, and air devices.

6.1 AIR-HANDLING EQUIPMENT

6.1.1 General Construction

The air-handling system consists of one or more fan sections, heat exchange sections for heating and/or cooling, an air filtration section, a section for mixing return air with outside air, and a discharge air plenum. A small air-handling unit may simply contain a supply air fan, coils, and an air filter, such as a fan coil unit. A large air-handling unit up to 100,000 to 300,000 cfm may be custom built to contain an array of components. The following components can be found in air-handling units:

- Fan sections, for supply air and return air/relief air fans
- A cooling section, for chilled-water or refrigerant cooling coils
- A heating section, for hot-water or steam coils, a gas heat exchanger, or an electrical coil
- A humidification section for extra humidity, if required

- Filter sections, for prefiltering filtering, and post-filtering
- An air-mixing section, for outdoor air to mix with recirculated air
- A discharge air plenum
- Other components, for electrical power, controls, operating a motor, drainage, etc.

Depending on their size, air-handling units can be delivered as a single package or assembled from modular components. Custom applications may even be built on the job site from panels or sheet metal applied to a metal frame.

Materials of construction must be suitable for the environment. Outdoor units generally are galvanized and painted for corrosion resistance or are constructed of aluminum. Indoor units may be simply painted or constructed of galvanized steel. Casings are insulated to prevent thermal losses and possible condensation on the exterior surface of cooling section. If the air-handling units are large enough for walk-in maintenance, their insulation may be protected by woven wire fabric or a sheet metal liner.

Casings are subject to positive and negative pressures with respect to the surrounding air. High-pressure units (up to 10 inches of water column) require adequate bracing to resist pressure and need to be well sealed to prevent excessive leakage.

149

6.1.2 Air-Handling Unit Configurations

The components of air-handling units can be arranged in many configurations, regardless of whether the fan blows through or draws through the coils or whether the housing is in a horizontal or vertical arrangement. Figure 6–1 illustrates the various arrangements.

6.2 HEAT TRANSFER

Heat transfer is the heart of a cooling and heating system. In the air-handling system, it occurs at the heat exchange section of the unit. Depending on the medium, heat transfer is accomplished through one or a combination of components.

6.2.1 Water Coils

Water coils are the most common component for transferring heat between the circulating air and the medium, such as chilled water, refrigerant, hot water, or steam. The coils are normally constructed of copper tubes with aluminum fins. A special coating and special materials such as stainless steel may be appropriate for corrosive environments, including salt spray and industrial pollution. Drain pans, required under cooling coils to collect condensate, are coated with a corrosion-resistant material or made of stainless steel. Figure 6–2 illustrates the construction of a typical water coil.

The performance of a heating or cooling coil depends on the design of its tubes and fins, as well as the size of the coil, including the depth and face area. The depth of the coils is expressed in rows, which represent layers of tubes that conduct the heating or cooling fluid. In general, deeper coils have more capacity because air is in contact with them for longer time and the temperature of the air can more closely approach the temperature of the heating or cooling medium.

The number of rows not only is significant for the performance of the coil, but also affects the location of inlet and outlet piping. A one-row coil would consist of a single layer of tubes conducting the heating or cooling from one side of the coil to the other. A two-row coil would have two layers of tubes with supply and return on the same side. Odd numbers of rows result in opposite-side piping, even rows in same-side piping. Coils constructed with even numbers of rows are preferred for simplicity of piping.

Heating coils can be designed with a high temperature difference between the heating fluid and the heated air. Heating coils are generally one to four rows deep. The temperature difference between the cooling medium and the cooled air is generally small, so cooling coils require four to eight rows of depth.

The height and width of a coil establish its face area. The area must be sufficient to handle the required airflow with acceptable velocities across the coil. If the velocity is too high, the pressure drop on the airside will be excessive, and condensate moisture will carry over into the airstream from the cooling coils.

Heating coil face velocities of up to 1200 feet per minute can be accommodated with a reasonable pressure drop, since heating coils are generally shallow, rarely exceeding four rows. For cooling coils, however, face velocities should be kept well below 600 feet per minute to avoid carry-over. Often, heating coils in series with cooling coils have similar face areas for convenience of construction.

6.2.2 Steam Coils

In steam coils, the tubes are designed for easy drainage of the condensate. Figure 6–3 illustrates two types of steam coil: the general-purpose or the conventional type, with supply (steam) and return (condensate) at different ends of the coil, and the steam-distributing type, in which steam is distributed evenly from an inner orifice tube within the outer heat transfer tube. The latter design assures an even temperature across the face of the coil and is most effective in preventing freezing of the coil in below-freezing temperatures.

6.2.3 Electrical Coils

Electrical coils may be designed as a part of the air-handling unit or may be installed on the ductwork exterior to the air-handling unit. The heating elements are usually made of a nickel-chromium alloy. Open electrical heating elements must be provided with safety features, such as a flow switch and thermal cutouts. They may be circuited for single-step on-off heating, heating with two or more steps, or modulated heating. Figure 6–4 illustrates a typical electrical duct heater.

6.2.4 Direct Expansion (DX) Coils

When the cooling medium is a refrigerant (e.g., freon), the cooling coil is designed to allow the refrigerant to vaporize in the coil, thus absorbing heat from the air. A typical DX coil consists of a refrigerant heater and many distribution tubes. The tubes may

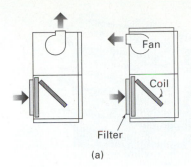

(a)

■ **FIGURE 6–1**

(a) A basic fan coil unit. The unit may be in a vertical or horizontal configuration for an exposed or concealed mounting. (b) A packaged horizontal air-handling unit having a single supply air fan. The unit shown has one return air inlet and one air supply outlet. (c) A built-up horizontal air-handling unit showing one return air fan with mixing dampers, combination filters, and cooling and heating coils, and one supply air fan with a sound-attenuating section and a discharge plenum. (Courtesy: York Applied Systems, York, PA.)

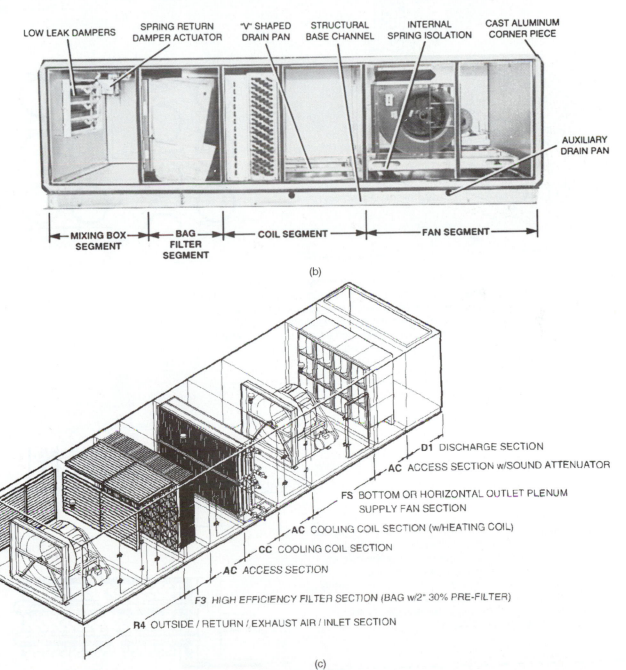

(b)

(c)

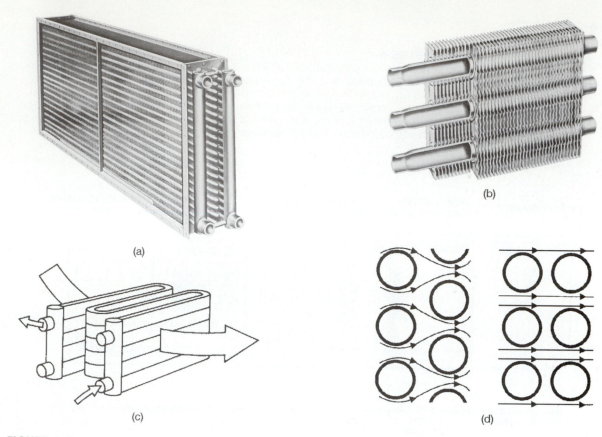

(a)

(b)

(c)

(d)

■ FIGURE 6–2

(a) Typical serpentine, finned tube, multirow water coil assembly. (b) A cutaway view of the finned tube construction. The fins are rippled and corrugated to improve the rate of heat transfer. (c) Coils are normally circuited for counterflow—i.e., the inlet air and inlet water flow in opposite directions to have maximum overall heat transfer. (d) Tubes of adjacent rows are staggered to gain more contact between air and coil (left); when tubes are lined up (right), the surface between tubes and air is reduced. (Courtesy: McQuay International, Minneapolis, MN.)

■ FIGURE 6–3

(a) Conventional steam coil with a single-tube design. Coil temperature is likely to be lower at the far end of the coil, as steam is condensed inside the finned tubes. (b) Steam distribution-type coil with tube-in-tube design. Steam is evenly distributed. This type of steam coil is ideal for low-temperature preheating and special process applications. (Courtesy: Trane Company, LaCrosse, WI.)

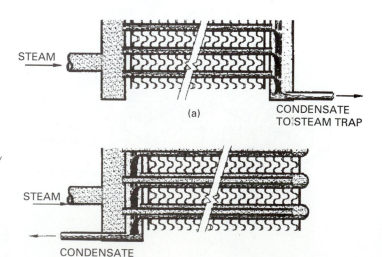

STEAM

(a)

CONDENSATE TO STEAM TRAP

STEAM

CONDENSATE TO STEAM TRAP   (b)

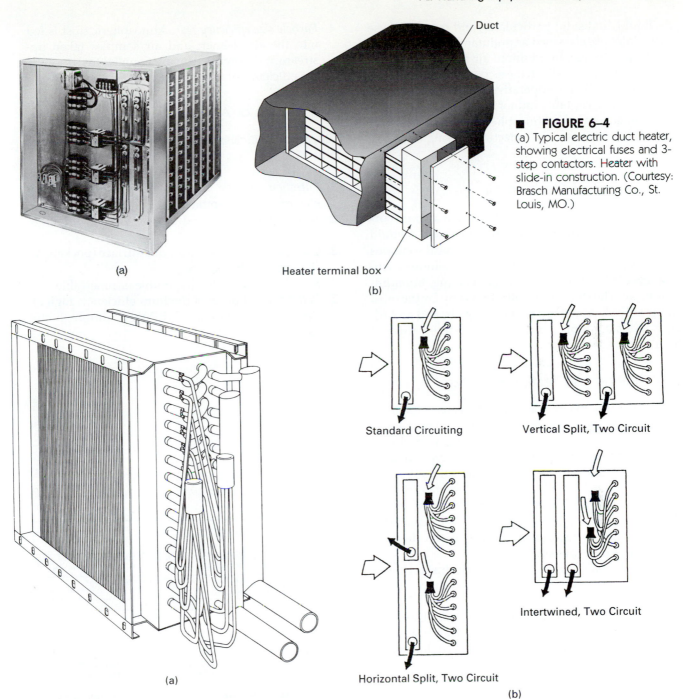

■ **FIGURE 6–4**
(a) Typical electric duct heater, showing electrical fuses and 3-step contactors. Heater with slide-in construction. (Courtesy: Brasch Manufacturing Co., St. Louis, MO.)

Duct

Heater terminal box

(a)

(b)

Standard Circuiting

Vertical Split, Two Circuit

Horizontal Split, Two Circuit

Intertwined, Two Circuit

(a)

(b)

■ **FIGURE 6–5**
(a) Typical DX coil construction and (b) the circuiting arrangements for effective coil temperature controls. (Courtesy: Trane Company, LaCrosse, WI.)

be single circuited for small coils or horizontally and vertically split for even distribution of the refrigerant to the coil, depending on the performance required. Figure 6–5 illustrates a typical DX coil and circuiting.

6.3 AIR CLEANING

Mountain air, although apparently clean and pure, actually contains impurities. Air in urban environments contains even more impurities, in the form of gas, liquid, and solid particulates, and many of these particulates are classified as pollutants, such as smog, smoke, pollen, etc. In addition, air may contain bacteria and viruses. All of these pollutants are detrimental to health. It is imperative that air be cleaned to maintain an acceptable indoor air quality. Table 6–1 lists some common atmospheric particulates that should be evaluated in selecting among filtration media.

6.3.1 Means of Cleaning Air

Air can be cleaned by passing it through a liquid curtain or spray or through a dry filter medium. A liquid curtain or spray may use water or chemical solutions to remove the air particulates but these solutions usually serve other functions, such as cooling, humidification, etc. The dry type of filtration is by far the most commonly used method for cleaning air, and accordingly, we shall limit our discussion to this variety of filter.

6.3.2 Rating and Testing Air Filters

There are three major operating characteristics of air filters: efficiency, resistance to airflow, and dust-holding capacity. No single test can adequately specify all three characteristics, so, in general, four types of tests are used for rating the efficiency of air filters:

1. *Dust weight arrestance test.* Particles of synthetic dust of various sizes are fed into the air cleaner (filter), and the fraction by weight of the dust removed is determined.
2. *Dust spot efficiency test.* Atmospheric dust is passed into the air cleaner, and the discoloration is observed.
3. *Fractional efficiency or penetration test.* Uniform-sized particles are fed into the air cleaner, and the percentage removed by the cleaner is determined.

4. *Particle size efficiency test.* Atmospheric dust is fed into the air cleaner, and air samples taken upstream and downstream are counted to determine the efficiency of removal of each size of particle.

6.3.3 Types of Air Filters

Filters may be classified according to the following criteria:

1. *Filtration principle.* Filtration by the medium or by electrostatic precipitation.
2. *Impingement.* Dry medium or viscous impingement.
3. *Configuration.* Flat or extended surface (pockets, V-shaped or radial pleats).
4. *Service life.* One-time disposable or renewable.
5. *Performance.* Low and medium efficiency, high efficiency particulate air (HEPA), or ultrahigh efficiency (UEPA).

TABLE 6–2

Comparative performances of viscous impingement and dry filters (reprinted with permission from *ASHRAE Handbook*)

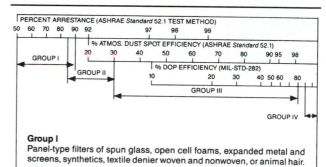

Group I
Panel-type filters of spun glass, open cell foams, expanded metal and screens, synthetics, textile denier woven and nonwoven, or animal hair.
Group II
Pleated panel-type filters of fine denier nonwoven synthetic and synthetic-natural fiber blends, or all natural fiber.
Group III
Extended surface supported and nonsupported filters of fine glass fibers, fine electret synthetic fibers, or wet-laid paper of cellulose-glass or all-glass fibers.
Group IV
Extended-area pleated HEPA-type filters of wet-laid ultra-fine glass fiber paper. Biological grade air filters are generally 95% DOP efficiency; HEPA filters are 99.97 and 99.99%; and ULPA filters are 99.999%.

Notes: 1. Group numbers have no significance other than their use in this table.
2. Correlations among the test methods shown are approximations for general guidance only.
3. DOP = di-octylphthalate, an oily substance used to test the penetration property of a filter medium.

TABLE 6–1

Common suspended particulates in urban air

Particulate	Normal Size (μm)*	Particulate	Normal Size (μm)
Fumes	0.001–1	Tobacco Smoke	0.01–1
Smog	0.001–2	Bacteria	0.3–30
Dust	0.001–20	Pollen	10–100
Viruses	0.003–0.05	Human Hair	40–200

Source: *ASHRAE Handbook of Fundamentals*, 1993.
*μm = micron or micrometer (10^{-6} meter).

6. *Special features.* Odor absorption, disposal of radio-active material, etc.

Table 6–2 compares the performance of the various types of filter, based on the ASHRAE standard 52.1 test methods. For convenience, the filters are placed into four groups.

6.3.4 Typical Air Filters

Filters are available in a wide variety of designs to suit particular applications. Figure 6–6 shows a range of designs for general applications. Figure 6–7 shows special filters for applications requiring odor control or high degrees of air cleanliness and electrostatic filters.

6.3.5 Application of Air Filters

Filters are selected on the basis of their efficiency, but no less important are the resistance to airflow and the life cycle cost of the system. High-resistance filters require more power and thus more energy. The resistance can be reduced when the design airflow rate is lowered. In general, the filter manufacturer provides a range of face velocities and initial and final resistances for each type of filter. Table 6–3 gives an overview of filter applications for various kinds of buildings. In general:

1. *For residential and commercial buildings,* low-efficiency filters are adequate. Typical low-efficiency filters are made of fiberglass installed in a rigid frame of cardboard. Bag-type or pleated filters are

(a)

(b)   (c)

■ **FIGURE 6–6**
(a) Low-efficiency filter with replaceable medium; low-efficiency washable aluminum filter; low-efficiency disposable fiberglass filter; low- to medium-efficiency pleat-type panel filter. (b) High-efficiency HEPA filters in a rack. (c) Medium-efficiency extended-surface bag filter.

(a)

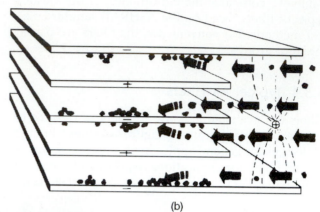

(b)

(c)

■ **FIGURE 6–7**

(a) Odor-removal carbon filters in a rack. (b) Principle of operation of an electrostatic filter. Dust and fumes first are positively charged at 14,000 volts and then enter a second electric field where they are attracted to the collector plates. (c) The assembly of an electrostatic filter. (Courtesy: AAF International, Louisville, KY.)

used for higher efficiency. These designs provide more surface area for airflow, resulting in a lower velocity through the filter medium and, accordingly, a lower pressure drop. Excessive pressure drops should be avoided to conserve energy.

2. *For health care facilities and laboratories,* the filtration requirements are often dictated by codes and regulations. Other special applications include clean rooms and special laboratories. HEPA (high efficiency particulate air) filters are used in these applications for very clean environments or removal of hazardous particles. HEPA filters are expensive and exhibit a high pressure drop, so their use is limited to these special applications.

TABLE 6–3

Application of air filters in air-handing systems[a]
(reprinted with permission from *ASHRAE Handbook of Fundamentals*)

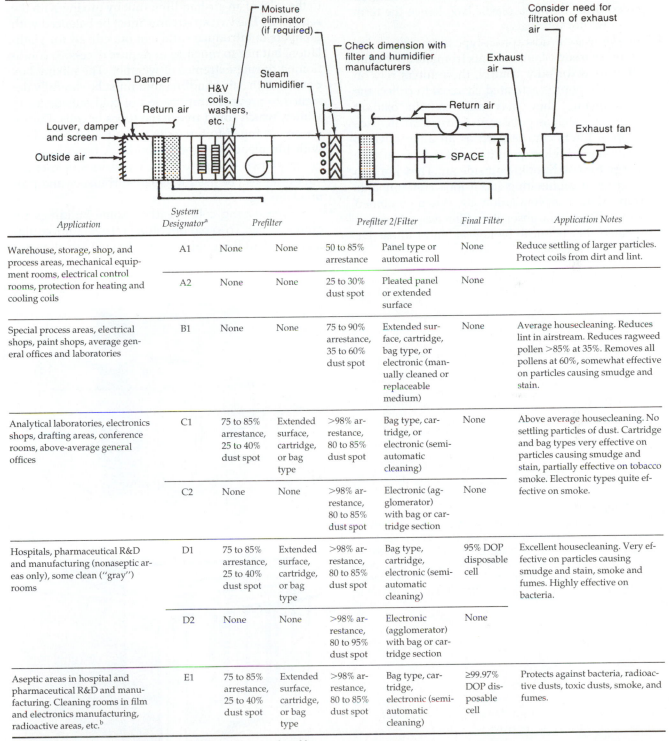

Application	System Designator[a]	Prefilter		Prefilter 2/Filter		Final Filter	Application Notes
Warehouse, storage, shop, and process areas, mechanical equipment rooms, electrical control rooms, protection for heating and cooling coils	A1	None	None	50 to 85% arrestance	Panel type or automatic roll	None	Reduce settling of larger particles. Protect coils from dirt and lint.
	A2	None	None	25 to 30% dust spot	Pleated panel or extended surface	None	
Special process areas, electrical shops, paint shops, average general offices and laboratories	B1	None	None	75 to 90% arrestance, 35 to 60% dust spot	Extended surface, cartridge, bag type, or electronic (manually cleaned or replaceable medium)	None	Average housecleaning. Reduces lint in airstream. Reduces ragweed pollen >85% at 35%. Removes all pollens at 60%, somewhat effective on particles causing smudge and stain.
Analytical laboratories, electronics shops, drafting areas, conference rooms, above-average general offices	C1	75 to 85% arrestance, 25 to 40% dust spot	Extended surface, cartridge, or bag type	>98% arrestance, 80 to 85% dust spot	Bag type, cartridge, or electronic (semiautomatic cleaning)	None	Above average housecleaning. No settling particles of dust. Cartridge and bag types very effective on particles causing smudge and stain, partially effective on tobacco smoke. Electronic types quite effective on smoke.
	C2	None	None	>98% arrestance, 80 to 85% dust spot	Electronic (agglomerator) with bag or cartridge section	None	
Hospitals, pharmaceutical R&D and manufacturing (nonaseptic areas only), some clean ("gray") rooms	D1	75 to 85% arrestance, 25 to 40% dust spot	Extended surface, cartridge, or bag type	>98% arrestance, 80 to 85% dust spot	Bag type, cartridge, electronic (semiautomatic cleaning)	95% DOP disposable cell	Excellent housecleaning. Very effective on particles causing smudge and stain, smoke and fumes. Highly effective on bacteria.
	D2	None	None	>98% arrestance, 80 to 95% dust spot	Electronic (agglomerator) with bag or cartridge section	None	
Aseptic areas in hospital and pharmaceutical R&D and manufacturing. Cleaning rooms in film and electronics manufacturing, radioactive areas, etc.[b]	E1	75 to 85% arrestance, 25 to 40% dust spot	Extended surface, cartridge, or bag type	>98% arrestance, 80 to 85% dust spot	Bag type, cartridge, electronic (semiautomatic cleaning)	≥99.97% DOP disposable cell	Protects against bacteria, radioactive dusts, toxic dusts, smoke, and fumes.

[a]System designators have no significance other than their use in this table.

[b]Electronic agglomerators and air cleaners are not usually recommended for clean room applications.

3. *For hazardous materials,* the filter housing may be constructed so that the filter can be replaced without exposing it to the environment. The housing is designed so that the contaminated filter can be pulled directly into a plastic bag; hence the term ''bag out'' for this feature.

4. *For odor removal,* adsorption-type filters are used to remove gaseous contaminants from the airstream. Often used for odor control, these filters rely on extremely porous activated charcoal to collect the contaminant. They may also contain oxidant chemicals, such as potassium permanganate. Adsorption filters can be used as an alternative to maintaining higher quantities of fresh air to save energy for conditioning outside air. They are also effective at eliminating hazardous contaminants from exhaust or ventilation air. When combined with HEPA filters, absorption filter assemblies can be rated for chemical, biological, and radioactive (CBR) applications.

6.4 AIR MIXING

Outside air required for a building is usually ducted to the inlet of an air-handling unit by mixing with the return air. The two airstreams must be balanced with dampers to introduce sufficient outside air for ventilation, but not so much as to require excessive conditioning during extremes of weather. The mixing box portion of the air-handling unit must be carefully designed to prevent stratification of cold outside air in winter, which could freeze the tubes of coils. Dampers are arranged to force the two airflows to collide, with turbulence for good mixing. A typical mixing box is shown in Figure 6–8. Accessories or rotating vanes can also be used to create turbulence and promote mixing.

Even during cold weather, some buildings require cooling. To avoid operating refrigeration equipment, air-handling units can be designed to allow in-

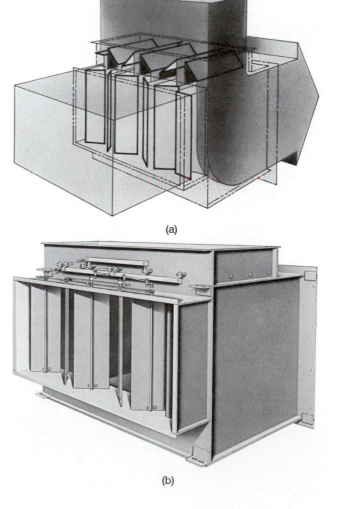

(a)

(b)

■ **FIGURE 6–8**
(a) Typical air-mixing box. In this design, the opposed blade dampers between the two airstreams are staggered so as to provide proper mixing. Other blade configurations, such as parallel blades, are also applied, depending on the duct inlet conditions. (b) Same mixing box as in (a), with the dampers shown for both airstreams. (Courtesy: Trane Company, LaCrosse, WI.)

creased amounts of outside air for free cooling. This feature, called an *economizer cycle,* involves controlling the position of the outside and return air dampers so as to result in a mixed air temperature that will supply cooling for the building—generally, 55° to 60°F.

Care must be taken in the design of mixing sections for air-handling units used in variable air volume applications. Low airflow during cold weather results in low velocity through the mixing sections, stratification of cold air, and the risk of freezing coils.

Introducing large quantities of outside air for ventilation or free cooling will tend to over-pressurize a building, unless provisions are made for relieving air. Gravity or motor-operated dampers can be used to relieve air from small buildings. Large buildings require more positive control by using relief fans. Relief fans can be located on the roof or at pe-

rimeter walls. They can also be designed as part of the air-handling unit.

If the pressure drop in the return ductwork is significant, a return air fan may be required. This fan can also be used to relieve the excess air in the building during the operation of the economizer cycle. Figure 6–9 shows three schemes for air relief from a building. The choice between using a return air fan or a relief air fan depends on the relative resistance of the air paths. A long relief air path that causes a high pressure drop would favor the use of a relief air fan. Otherwise, a return air fan is sufficient.

6.5 FANS

A fan is a mechanical device that moves air or other gases to deliver energy. Although heating energy can

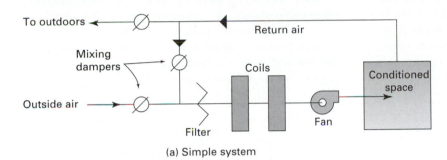

(a) Simple system

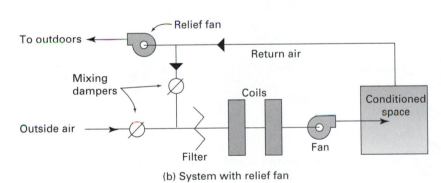

(b) System with relief fan

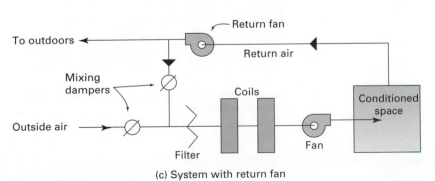

(c) System with return fan

■ **FIGURE 6–9**
Three relief air schemes are commonly used to utilize outside air to cool a space when the outside air temperature is below the set-point temperature of the space.
(a) Simple damper control scheme without using fans. This scheme is applicable to small air-handling systems within a large building where return air can easily be expelled from the building without overly pressurizing the building. This scheme is simple and low in cost.
(b) Relief air fan scheme. A relief air fan is used when the relief air path to the outdoors after the mixing dampers is long and restrictive.
(c) Return air fan scheme. A return air fan is used when the return air path to the mixing dampers is long and restrictive. In large systems, both relief air and return air fans are often used.

also be delivered by means of radiation, conduction, or convection without a force airflow, there is little choice but to use air as the practical means to deliver cooling energy. Without air movement created by the fan, cold air tends to be stagnant and may result in condensation of moisture on room surfaces.

All fans have a rotating impeller with blades that increases the kinetic energy of air by changing its velocity. The increased velocity is then converted to pressure. There are two basic fan designs—centrifugal and axial—each with many variations, but the same operating principle. Figure 6–10 illustrates the general configuration of these two types of fans.

6.5.1 General Classification of Fans

Both centrifugal and axial fans are used in HVAC systems. Centrifugal fans draw air into the center of a rotating wheel. Vanes on the wheel project the air radially and tangentially toward the outside of the fan housing, which directs the air to the fan discharge. Axial fans move air through their housings in a direction parallel to the wheel axle. The following commonly used fans are illustrated in Figure 6–11.

Centrifugal fans General-purpose centrifugal fans can be freestanding or housed within cabinets or air-handling units. Double-inlet fans are most common, but the geometry of the system may require the use of a single-inlet fan. In-line centrifugal fans and plug fans are used when saving space and simplicity of installation are important considerations.

The design of the blade affects its performance. Small centrifugal fans most often use a forwardly curved configuration. Backwardly inclined blades are common on larger fans, and blades can be constructed with an efficient airfoil shape rather than uniform in thickness.

Roof ventilators are centrifugal exhaust fans that are suitable for outdoor installation. For general exhaust applications, the discharge opening of the fans normally face downward toward the roof so the fan housing will not accumulate rainwater. For kitchen or other exhausts which contains grease or odor, upblast configurations are normally used. The upblast configuration prevents grease from depositing locally near the fan and helps to disperse the exhaust air into a higher elevation, above the roof level.

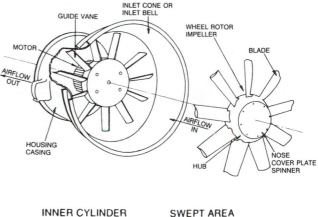

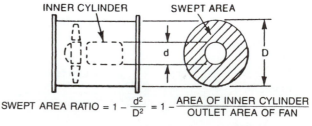

$$\text{SWEPT AREA RATIO} = 1 - \frac{d^2}{D^2} = 1 - \frac{\text{AREA OF INNER CYLINDER}}{\text{OUTLET AREA OF FAN}}$$

Note: The swept area ratio in axial fans is equivalent to the blast area ratio in centrifugal fans.

(b)

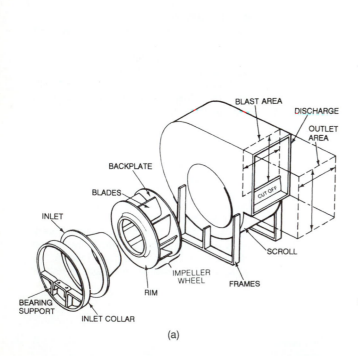

(a)

■ **FIGURE 6–10**

Typical construction and components of fans (reprinted with permission of ASHRAE). (a) Centrifugal fan components. (b) Axial fan components.

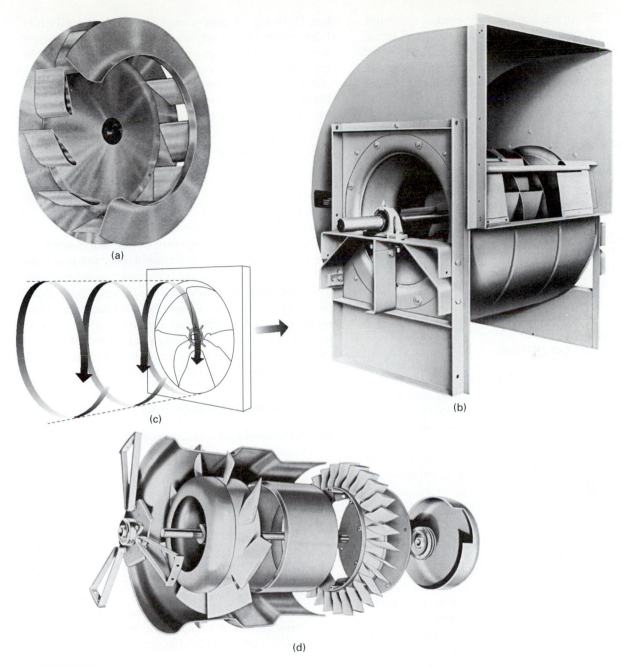

Types of Fans.
(a) Illustrates a typical backward inclined centrifugal fan wheel. The blades may be either straight radial, backward or forward inclined. (b) Illustrates the assembly of a centrifugal fan with horizontal discharge which may also be vertical or down discharge. The scroll housing may have either single or double inlet. (Courtesy: Trane Company, LaCrosse, WI.)
(c) Illustrates the air movement of discharge air of a typical propeller fan.
(d) Illustrates the components of a vane axial fan including the rotor assembly which consists of a nosepiece, hub, and blades. The rotor assembly is directly connected to the motor. (Courtesy: Joy Manufacturing Company, Philadelphia, OH.)

Utility sets   Utility sets are weatherproof, single-inlet, generally roof-mounted centrifugal fans used in exhaust applications. Horizontal discharge is preferred for general exhaust, upward discharge for kitchen or fume exhaust. Special materials or coatings will be required for corrosive exhaust, as might be discharged from laboratories. If the exhaust is potentially toxic, a duct should be installed at the discharge with its opening at least seven feet above the roof level.

Axial and propeller fans   These types of fans can be used for general-purpose applications and have the advantage of being compact when installed in line with ductwork. Propeller fans are generally installed at building walls and are suitable for large-volume, nonducted applications, such as in greenhouses and for ventilation of manufacturing space. Propeller fans can be installed at the roof if proper attention is paid to cover the discharge and prevent rain from entering the space.

6.5.2 Application of Fans

Air-handling unit fans   Air-handling fans are generally of centrifugal design, although axial fans are occasionally used. The supply fan section of the air-handling unit can be placed in the draw-through or blow-through position with respect to the coils. Many factors should be considered in choosing between draw-through and blow-through configurations.

Draw-through units are generally more compact and less expensive to construct than blow through units. However, since the fans are closer to the supply duct system, they are more prone to duct noise. In addition, the heating effect caused by the fan is introduced downstream of the cooling coil, requiring lower discharge temperatures for a desired supply temperature. This can result in slightly higher power requirements for refrigeration equipment.

Blow-through units have the advantage of promoting mixing of air upstream of the coils, minimizing stratification and the risk of freezing during cold weather. However, they require provisions for even distribution across the coils.

Return air fans   These fans may be part of the air-handling unit or may be installed separately in ductwork upstream of the unit. Return fans can be centrifugal or axial and need not be the same type as the supply fans.

6.5.3 Volume (Flow Rate) Controls

Fans may be required to deliver a constant or a variable volume of air. If variable flow is required, there are several options for controlling the output of the fan. Discharge dampers are common for small units. Inlet guide vane dampers are more energy efficient and are used on larger air-handling units. Variable-frequency motor speed controllers are most energy efficient and have become common in recent years. Figure 6–12 compares the energy input of the three methods of volume control.

6.5.4 Fan Drives

Motors can be placed within the fan cabinet or air-handling unit or can be mounted externally. Fan wheels can be coupled directly to the motor or can be driven by belts and pulleys or by sheaves.

Direct-drive fans operate at the same speed as the motor and are not adjustable. Sheaves and belts allow fans to be designed for slower, quieter operation, and the performance of the fans can be tailored by proper selection of the sheave diameter. Some sheaves are adjustable, and sheaves can be replaced to slow down or speed up the operation of the fan in the event that higher or lower performance is needed. Belt guards are needed if the belt drive is exposed.

6.5.5 Fan Performance and Fan Laws

The performance of a fan is measured by the following characteristics:

- Volume of air delivered per unit time (airflow rate), normally in cfm
- Pressure created (static, velocity, and total pressure), normally in inches of water column (w.c.), inches of mercury column (Hg), or pounds per square inch (lb/sq in., or psi). (See appendix for conversion to SI units.)
- Power input, horsepower (hp).
- Mechanical and static efficiency, percentage.
- Other factors, such as sound level, in noise criteria (NC) or decibels (dB), etc.

Performance data are provided by the fan or equipment manufacturer either in tabulated form or in charts. A typical fan performance chart is shown in Figure 6–13. Plotted are the air-handling system curves with resistance or static pressure drop versus the flow rate. In general, for the same system (equipment and ductwork), the resistance or the pressure drop varies with the square of the flow rate.

The fan laws most frequently used are as follows:

- The system resistance increases with the square of the airflow rate.
- For the same fan size and system, the airflow rate varies directly with the speed, the pressure var-

(a)

(b)

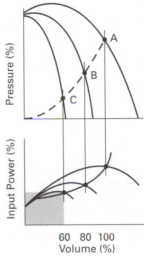

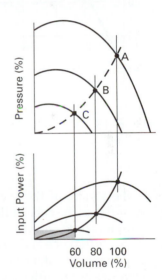

Outlet dampers are metal plates positioned in the airstream on the outlet side of the fan. The fan operates at a constant speed. Volume is reduced by closing the damper. However, closing the damper increases pressure in the system, resulting in little energy savings. *At 60% volume, input power is 88%.*

Inlet guide vanes mounted in the fan inlet impart a whirl to the air entering the fan wheel, reducing the pressure. Volume is reduced by altering the vane position. Fan speed remains constant, so lowering the volume produces only a small reduction in input power. *At 60% volume, input power is 62%.*

IGBT drives control volume by lowering fan speed. Because input power varies with the cube of fan speed, as speed is reduced, input power is substantially reduced. *At 60% volume, input power is 28%.*

(c)

■ **FIGURE 6–12**
(a) An inlet vane assembly of a centrifugal fan. (Courtesy: York Applied Systems, York, PA.) (b) Interior view of a variable-frequency drive employing insulated-gate bipolar transistor (IGBT) technology. (c) Fan performance curves, showing power savings of various volume control methods—outlet dampers, inlet guide vanes, and GPD drives. (Courtesy: MagneTek, Clawson, MI.)

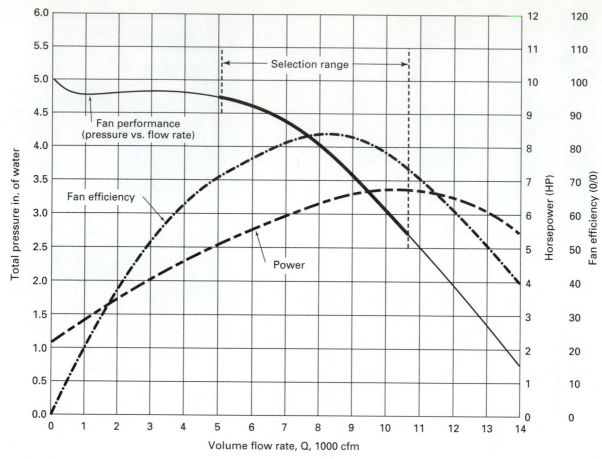

■ **FIGURE 6–13**
Typical fan performance curves.

ies as the square of the speed, and the horse-power varies as the cube of the speed.

- For constant speed and varying fan size, the flow rate varies as the cube of the fan size, the pressure varies as the square of the fan size, and the horse-power varies as the fifth power of the fan size.

For purposes of estimation, the following formulae can be used to determine the airflow rate, pressure required, and fan power:

- *Relation between Pressures*

$$TP = SP + VP \qquad (6\text{--}1)$$

where TP = total pressure developed or required by the fan, in inches of water
VP = velocity pressure, that is, the pressure created by the air velocity at a point in the air system selected for calculations, in inches of water

SP = static pressure, that is the pressure exerted on the sides of the duct at a point in the air system selected for calculations, in inches of water

- *Velocity of Flow to Velocity Pressure*

$$V = 4005 \times \sqrt{VP} \qquad (6\text{--}2)$$

where V = velocity of the airflow at a point in the air system

- *Power Required at the Rated Flow and Pressure (at sea level)*

$$AHP = \frac{CFM \times TP}{6356} \qquad (6\text{--}3)$$

$$BHP = \frac{CFM \times TP}{(6356 \times FE)} \qquad (6\text{--}4)$$

$$EHP = \frac{CFM \times TP \times 0.746}{6356 \times FE \times ME}$$ (6–5)

$$= \frac{0.000117 \; CFM \times TP}{FE \times ME}$$

where AHP = air horsepower at 100% fan
efficiency, in HP

BHP = brake horsepower at the selected
fan efficiency, in HP

EHP = electrical power at the selected
motor efficiency, in kW

FE = mechanical efficiency of the fan,
per unit (Pu)

ME = motor efficiency of the selected
motor, per unit (Pu)

6.5.6 Examples of Fan Performance

An air-handling system is designed to circulate
15,000 cfm at a system pressure of 2.7 inches w.c. A
fan is selected on the basis of the graph in Figure 6–14.

Plotted on this graph is also the air-handling system
curve, which is constructed according to the fan laws.

a. With this particular choice of fan, what should be
the fan speed?

b. What is the BHP operating at this condition?

c. If the same fan is operating at 825 rpm, what will
be the fan delivery, the air-handling system static
pressure, and the BHP required?

(*Note:* Instead of using the graph, you may use the fan
laws to obtain the same results.)

Answers:

a. From graph, the intersection of 15,000 cfm with
the system curve demands that the fan run at 1125
rpm.

b. The fan should be driven by a motor having a minimum 8.55 BHP.

c. At 825 rpm, the fan can deliver 1100 cfm, operating at 1.46 in. w.c. and drawing 3.39 BHP.

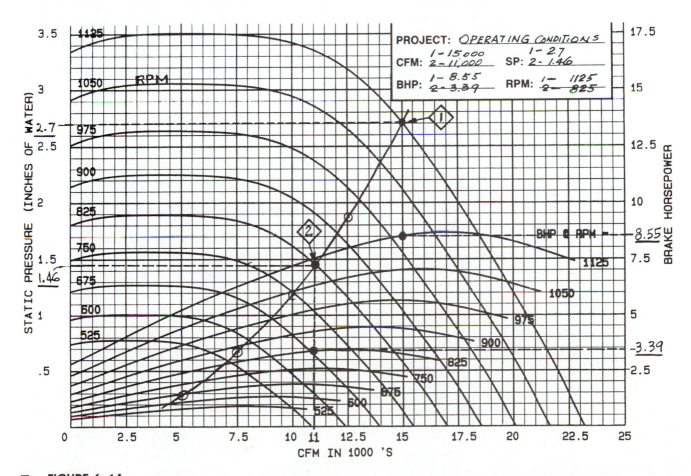

■ **FIGURE 6–14**
Fan performance curves for the fan in example 6.5.6. (Fan graphs are based on Barry
Blower Model Versacon.)

6.6 DUCT SYSTEMS

Ductwork is part of the air-handling system and includes the supply, return, outside air, relief air, and exhaust air ducts. While the supply and return air ducts must be connected to the air-handling units, the other ducts may be run independently. Ducts are usually fabricated from sheet metal, such as galvanized steel, aluminum, or stainless steel, thus duct work is also called sheet metal work although some ducts are made with non-metals, such as plastics.

6.6.1 General Classifications for Ductwork

Duct systems for supplying air may be classified as low pressure/velocity, medium pressure/velocity and high pressure/velocity. For a given airflow, lower velocities reduce friction in the duct, and power for distribution. In addition, lower velocities reduce air noise.

Low-velocity ductwork is used for small airflow requirements, generally at final branches of the system. It can also be used for large air quantities where space is available for larger ductwork and the initial cost for the extra duct material is warranted. High airflow requirements generally dictate higher velocities to conserve space and duct materials. Table 6–4 shows typical ranges of acceptable velocities for various airflows through ductwork. According to the Sheet Metal and Air Conditioning Contractors' National Association, Inc. (SMACNA), there are seven velocity/pressure classifications for duct systems. There are also three types of sealing requirements that govern the type of joints and sealant.

6.6.2 Symbols for Sheet Metal Work

There is a variety of standard symbols for sheet metal work to be used in drawings. One of the commonly accepted standards is the SMACNA Symbols for Ventilation and Air Conditioning, which includes sheet metal work and component devices related to ventilation and air conditioning systems. These standard symbols are found in most engineering handbooks (see Figure 6–15).

6.6.3 Duct Shapes and Insulation Methods

Low-velocity ductwork is most often rectangular in cross section. Insulation is often applied to the interior of the duct and provides acoustical absorption as well as thermal benefits. When applied internally, insulation is called *duct liner*. Liner, which is applied in the fabrication shop, reduces heating or cooling loss from the duct. In cooling applications, liner also prevents condensation on the outer surface of the duct.

Round or flat oval ducts can be used for low-velocity applications, but these shapes are generally reserved for medium- and high-velocity ductwork. If insulation is required, it is usually applied externally in the form of a fiberglass blanket wrap with an external vapor barrier for cooling applications. The vapor barrier is intended to prevent the migration of humid air through the insulation, which could result in condensation on the surface of a cold duct. External insulation is more expensive than liner, due to the field labor required in installing it.

Internal insulation is also available for round or flat oval ductwork. Internal insulation is preferred if ductwork is to be exposed to view. The material is scored and formed to the interior contour of round and flat oval ducts. Fitting this type of insulation is difficult, especially at elbows and fittings.

6.6.4 Materials of Construction

Galvanized steel is the most widely used material for general-purpose ductwork. Aluminum, stainless steel, or plastic might be used for ducts installed in

TABLE 6–4

Pressure/velocity classification of ductwork

Class Type	Class Pressure	Operating Pressure, in. w.c.	Maximum Velocity, Ft per Min (fpm)	Design Velocity, Ft per Min (fpm)	
				Main	Branch
1	½" w.c.	−½" to +½"	2000	Up to 2000	Up to 800
2	1" w.c.	−1" to +1"	2500	Up to 2500	Up to 1200
3	2" w.c.	−2" to +2"	2500	Up to 2500	Up to 1500
4	3" w.c.	−3" to +3"	4000	Up to 4000	Up to 2000
5	4" w.c.	−3" to +4"	4000	Up to 4000	Up to 3000
6	5" w.c.	0" to +6"	5000	Up to 5000	Up to 4000
7	6" w.c.	0" to +10"	5000	Up to 5000	Up to 4000

SYMBOL MEANING	SYMBOL	SYMBOL MEANING	SYMBOL
POINT OF CHANGE IN DUCT CONSTRUCTION (BY STATIC PRESSURE CLASS)		SUPPLY GRILLE (SG)	20 × 12 SG / 700 CFM
DUCT (1ST FIGURE, SIDE SHOWN 2ND FIGURE, SIDE NOT SHOWN)	20 × 12	RETURN (RG) OR EXHAUST (EG) GRILLE (NOTE AT FLR OR GLG)	20 × 12 RG / 700 CFM
ACOUSTICAL LINING DUCT DIMENSIONS FOR NET FREE AREA		SUPPLY REGISTER (SR) (A GRILLE + INTEGRAL VOL. CONTROL)	20 × 12 SR / 700 CFM
DIRECTION OF FLOW		EXHAUST OR RETURN AIR INLET CEILING (INDICATE TYPE)	20 × 12 GR / 700 CFM
DUCT SECTION (SUPPLY)	S 30 × 12	SUPPLY OUTLET, CEILING, ROUND (TYPE AS SPECIFIED) INDICATE FLOW DIRECTION	20 / 700 CFM
DUCT SECTION (EXHAUST OR RETURN)	E OR R 20 × 12	SUPPLY OUTLET, CEILING, SQUARE (TYPE AS SPECIFIED) INDICATE FLOW DIRECTION	12 × 12 / 700 CFM
INCLINED RISE (R) OR DROP (D) ARROW IN DIRECTION OF AIR FLOW	R	TERMINAL UNIT. (GIVE TYPE AND/OR SCHEDULE)	T.U.
TRANSITIONS: GIVE SIZES. NOTE F.O.T. FLAT ON TOP OR F.O.B. FLAT ON BOTTOM IF APPLICABLE		COMBINATION DIFFUSER AND LIGHT FIXTURE	
STANDARD BRANCH FOR SUPPLY & RETURN (NO SPLITTER)	S R	DOOR GRILLE	DG 12 × 6
SPLITTER DAMPER		SOUND TRAP	ST
VOLUME DAMPER MANUAL OPERATION	VD	FAN & MOTOR WITH BELT GUARD & FLEXIBLE CONNECTIONS	
AUTOMATIC DAMPERS MOTOR OPERATED	SEC. MOD	VENTILATING UNIT (TYPE AS SPECIFIED)	
ACCESS DOOR (AD) ACCESS PANEL (AP)	OR AD	UNIT HEATER (DOWNBLAST)	
FIRE DAMPER: SHOW VERTICAL POS. SHOW HORIZ. POS.	FD AD	UNIT HEATER (HORIZONTAL)	
SMOKE DAMPER	SD AD	UNIT HEATER (CENTRIFUGAL FAN) PLAN	
CEILING DAMPER OR ALTERNATE PROTECTION FOR FIRE RATED CLG	C	THERMOSTAT	T
TURNING VANES		POWER OR GRAVITY ROOF VENTILATOR-EXHAUST (ERV)	
FLEXIBLE DUCT FLEXIBLE CONNECTION		POWER OR GRAVITY ROOF VENTILATOR-INTAKE (SRV)	
GOOSENECK HOOD (COWL)		POWER OR GRAVITY ROOF VENTILATOR-LOUVERED	
BACK DRAFT DAMPER	BDD	LOUVERS & SCREEN	36 × 24L

SYMBOLS FOR VENTILATION & AIR CONDITIONING   SMACNA

■ FIGURE 6–15
Standard symbols for ventilation and air conditioning. (Courtesy: SMACNA.)

(a)

(b)

■   **FIGURE 6–16**

(a) An assembly of high pressure/velocity round ducts with low pressure rectangular ducts. (b) Connecting round ducts. (Courtesy: United McGill Corporation, Groveport, OH.)

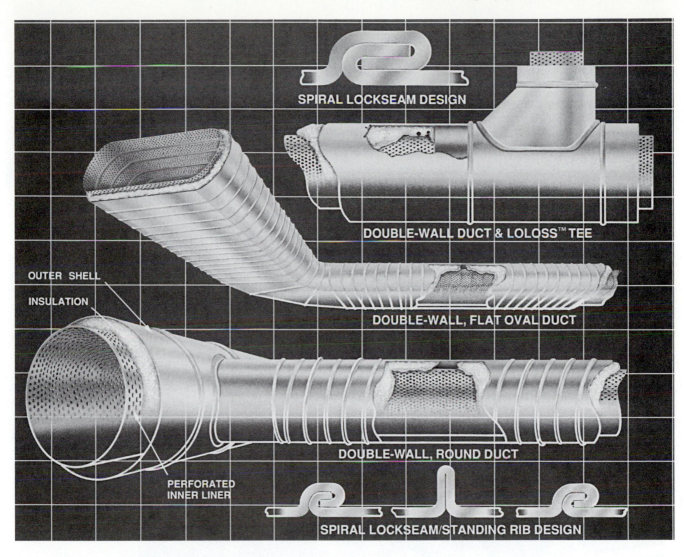

(c)

■ **FIGURE 6–16**
(c) Typical construction and configuration of double-wall flat and round ducts. The perforated inside duct is separated from the outside duct by acoustic insulation material. (Courtesy: United McGill Corporation, Groveport, OH.)

humid environments or ducts that carry moist air, such as a dishwasher exhaust. Heavyweight, fire-resistant steel ductwork is used for ducts exhausting kitchen hoods over ranges and fryers. Laboratory fume hoods are often called upon to carry corrosive materials. Stainless steel, plastic, or regular steel duct-work with corrosion-resistant coatings are used in fume exhaust applications. Figure 6–16 illustrates typical low- and high-velocity duct construction and fittings.

All but the smallest rectangular ductwork requires bracing for rigidity. Bracing consists of angles or channels attached at intervals transversely to the duct. Bracing prevents the duct walls from deflecting excessively under pressure. Requirements for bracing depend on the cross-sectional dimension and internal pressure of the duct.

6.6.5 Duct Assembly

The photos shown in Figure 6–16 show the duct assemblies in an industrial plant, the method of connecting round ducts, and the variation of duct configurations. Round ducts have the most area per unit area of sheet metal, and thus are most economical. When ceiling space is limited, oval ducts fit the need.

Round or oval ducts are available in double-wall construction, which incorporates perforated interior duct and insulation material to reduce the air transmitted and duct-radiated noises.

There are many standard fittings designed to create smooth air flow, change of direction, or to accommodate the change of duct dimensions. All fittings cause air turbulence, resulting in pressure drop and power loss. These problems can be minimized by using low-loss fittings, such as long-radius or segmented elbows, or special accessories, such as turning vanes or extractors.

6.6.5 Coordination of Ductwork with Other Building Elements

In most commercial buildings, ductwork shares space above the ceiling with other elements, including structural supports, fireproofing, electrical conduit, sprinkler piping, and light fixtures. All of these elements must be coordinated to fit together and allow flexibility for future changes in their layout and function.

Clearances must be provided between ductwork and the lighting fixtures below them to allow the fixtures to be removed. Generally, 3 in. is adequate to raise and remove a fixture from a standard tee-bar grid. The clearance envelope around ductwork must allow for external insulation and bracing.

Bracing may extend 1 to 3 in. from the surface of the duct, depending on the duct's size and pressure classification.

Detailed planning should recognize that buildings are not constructed to tight tolerances. Also, a structural element will deflect under a load. A minimum of 2 in. of extra clearance should be planned. Figure 6–17 illustrates clearances and coordination with other elements in the ceiling.

6.6.6 Materials and Fittings for Sound Control

Sound attenuators, also called sound traps, are special duct fittings containing sound-absorbing material faced with perforated metal. They are designed to provide a large surface area to be in contact with the airstream. Sound traps are installed at the discharge of air-handling units to prevent fan noise from being transmitted into the distribution system. Sound traps are also used at return air openings or ductwork between ceiling plenums and noisy equipment rooms.

Double-walled ductwork (Figure 16(c)) is also used for sound attenuation. The inner wall is constructed of perforated metal, the outer wall of regular sheet metal. The cavity between the walls is packed with sound-absorbing material. Double-wall duct installed for 20 to 50 feet downstream from the air-handling unit will help absorb fan noise.

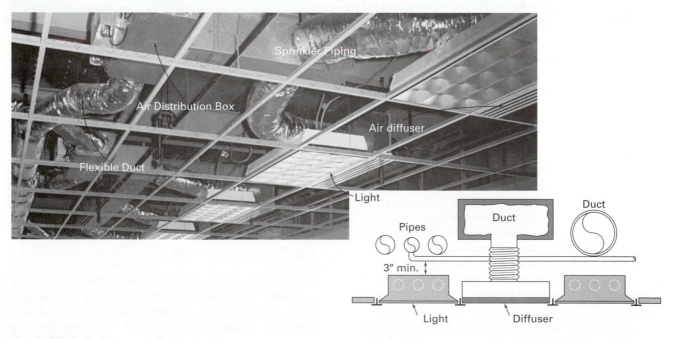

■ FIGURE 6–17
Congested ceiling space and the importance of coordination between systems.

6.7 AIR DEVICES

6.7.1 General Classifications

Air devices, used for supplying air and removing air from spaces, are among the few parts of the HVAC system that are visible to the occupants of a building. The physical appearance of the devices and their organization with other building elements are important considerations, as are their capacity, performance, and maintainability. Air devices include diffusers, grilles and registers, flow control devices, and other accessories. Figures 6–18 and 6–19 illustrate a few of these devices.

6.7.2 Grilles and Registers

Grilles and registers are air devices equipped with vanes for directing airflow. They can be located in ceilings, walls, or floors. Vanes may be fixed or adjustable. Grilles simply contain vanes; registers contain a control damper behind the vanes.

Grille faces may be constructed of blades, bars, or egg crate of various finishes and materials. Heavy-duty grilles are available for floor locations and other applications that are subject to physical abuse.

6.7.3 Ceiling Diffusers for Air Conditioning

Diffusers are ceiling-mounted air supply devices with louvers, slots, or vanes. Their function is to mix or diffuse supply air with room air without undue draft or localized hot or cold areas. Diffuser accessories and options include volume dampers, pattern control (direction of discharge) mechanisms, and means for removal of face elements to facilitate cleaning.

Ceiling diffusers rely on a phenomenon called the *Coanda effect*, which involves cold air clinging to the ceiling upon being discharged from the diffuser. The Coanda effect allows the airstream to fall gradually and mix with the room air over a large area. If the effect is not established, the chilled air will fall near the diffuser and create a cold spot beneath. This is called *dumping* and results in an uncomfortable cold draft. Figure 6–20 illustrates the effect of air velocity and the distribution pattern of diffusers.

If there is a discontinuity in the surface of the ceiling, dumping can result if diffusers are placed adjacent to light fixtures with parabolic lenses or on the bottom side of soffits.

Common diffuser types include those with a louvered face, perforated face, linear slot, and troffers

placed over lighting fixtures, all illustrated in Figures 6–18 and 6–19. Louvered-face diffusers are economical and perform well over a wide range of airflow. They are available in round, half-round, square, rectangular, and other shapes, with mounting frames for virtually all types of ceiling construction. In modular lay-in ceilings, they may be installed in the face of a ceiling tile or be provided with a flat extension to fill the module.

Perforated-face diffusers are designed with small louvers or deflectors behind a flush layer of perforated sheet metal. Perforated-face diffusers are less obtrusive than louvered-face diffusers. Performance at low airflow is less effective, and perforated-face diffusers are prone to dumping at low airflows, which occur in variable-volume applications.

For a given size of diffuser, the horizontal velocity will depend on the quantity of air supplied. Low airflow will result in low velocity, high airflow in high velocity. For a given airflow requirement, the horizontal velocity will depend on the size of the diffuser. A small diffuser will result in high velocity, a large diffuser in low velocity.

The size of the diffuser must be matched with the airflow to develop an appropriate velocity and avoid dumping. This can be a problem with VAV systems, for which a diffuser is required to perform over a range of airflows. Some diffusers cope well with varying airflow, while some do not. Louvered-face and linear slot diffusers work well for VAV systems. Perforated-face diffusers tend to dump at low airflow and should be avoided in spaces where large load variations are expected.

Different types of diffusers can be used in the same space. For example, a large office might effectively use perforated-face diffusers at the interior, which exhibits a relatively constant load, and linear slot diffusers at the perimeter wall, where load variations are significant.

6.7.4 Special Concerns for Warm Air Supply

Areas close to cold surfaces, such as large windows, are subject to uncomfortable downdrafts. If air is supplied at the floor or sill under the cold surface, the downdraft can be offset. If ceiling devices are used to deliver warm air, the velocity must be sufficient to create turbulence and mixing with cold air at the offending surface.

Using the same ceiling diffuser to deliver heating and air conditioning can be tricky. Warm air has a tendency to stay at the ceiling, due to its low density. If the warm air simply floats to the nearest return opening, the space will be deprived of any heating ef-

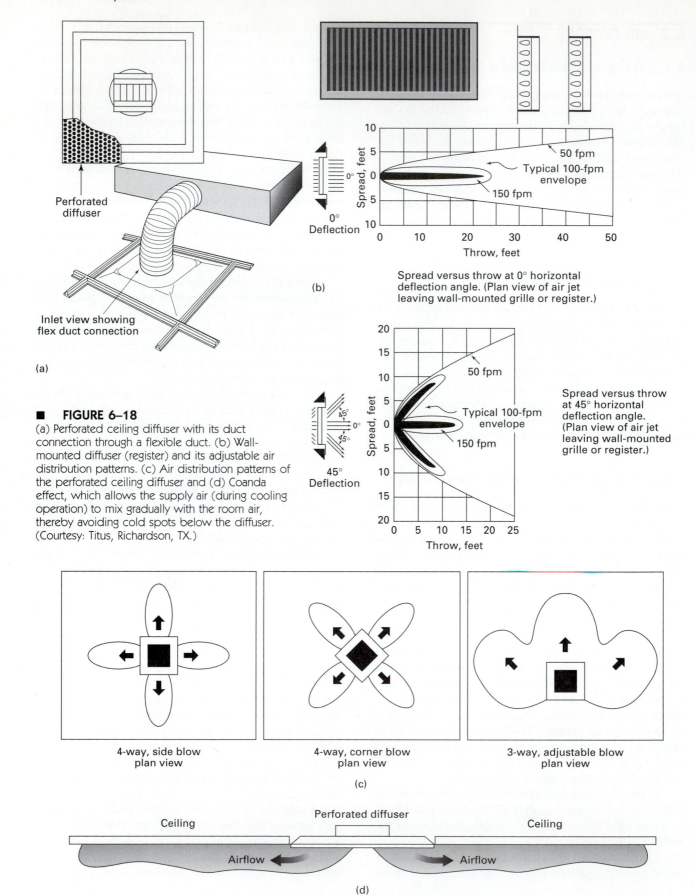

Perforated diffuser

Inlet view showing flex duct connection

(a)

(b)

Spread versus throw at 0° horizontal deflection angle. (Plan view of air jet leaving wall-mounted grille or register.)

0° Deflection

45° Deflection

Spread versus throw at 45° horizontal deflection angle. (Plan view of air jet leaving wall-mounted grille or register.)

■ **FIGURE 6–18**
(a) Perforated ceiling diffuser with its duct connection through a flexible duct. (b) Wall-mounted diffuser (register) and its adjustable air distribution patterns. (c) Air distribution patterns of the perforated ceiling diffuser and (d) Coanda effect, which allows the supply air (during cooling operation) to mix gradually with the room air, thereby avoiding cold spots below the diffuser. (Courtesy: Titus, Richardson, TX.)

4-way, side blow plan view

4-way, corner blow plan view

3-way, adjustable blow plan view

(c)

Ceiling

Perforated diffuser

Ceiling

Airflow

Airflow

(d)

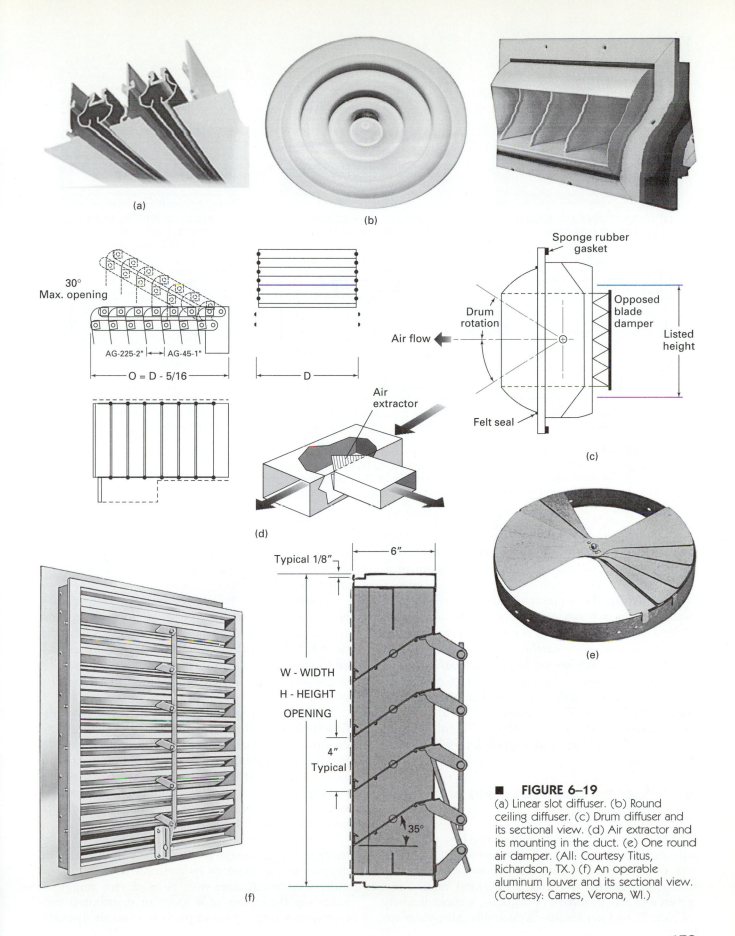

(a)

(b)

30°
Max. opening

AG-225-2" ← → AG-45-1"

O = D - 5/16

D

Sponge rubber
gasket

Drum
rotation

Air flow ←

Opposed
blade
damper

Listed
height

Felt seal

(c)

Air
extractor

(d)

Typical 1/8"

6"

W - WIDTH

H - HEIGHT

OPENING

4"
Typical

35°

(e)

(f)

■ FIGURE 6–19
(a) Linear slot diffuser. (b) Round
ceiling diffuser. (c) Drum diffuser and
its sectional view. (d) Air extractor and
its mounting in the duct. (e) One round
air damper. (All: Courtesy Titus,
Richardson, TX.) (f) An operable
aluminum louver and its sectional view.
(Courtesy: Carnes, Verona, WI.)

173

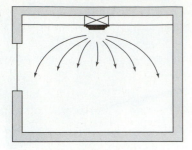

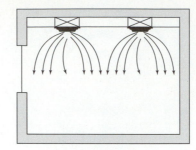

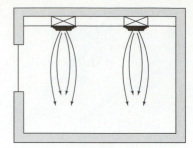

(a) GOOD COVERAGE BASED
     ON PROPER THROW

(b) GOOD COVERAGE WITH
     PROPERLY SELECTED
     MULTIPLE DIFFUSERS

(c) POOR COVERAGE. DUMPING
     OF COLD AIR. THROW
     IS NOT ESTABLISHED

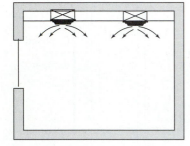

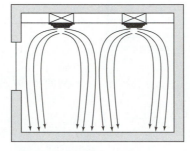

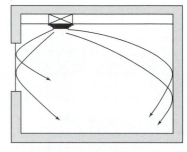

(d) INSUFFICIENT THROW.
     STAGNANT AIR

(e) COLD DOWNDRAFTS
     FROM EXCESSIVE THROW

(f) EXCESSIVE THROW.
     AIR TURBULENCE

■   **FIGURE 6–20**

Air distribution pattern from ceiling air diffusers. The distribution shown is in cooling mode. During heating mode, a different throw may be desired. The diffusers in (b) will be good for both cooling and heating, whereas those in (d) will be worse during heating mode.

fect. Again, adequate velocity is essential because it will promote turbulence and mixing with room air to distribute heat.

Generally, the amount of warm air required to heat a perimeter space is less than the amount of chilled air required to cool the same space. If the diffuser is selected solely on the basis of good chilled air distribution, it may be too large to impart adequate velocity to the smaller quantity of warm air. In this case, the warm air might simply float at the ceiling and be ineffective. The problem can be avoided by compromising on a diffuser that is slightly smaller than ideal for cooling and slightly larger than ideal for heating.

6.7.5 Spacing, Distribution, and Area of Coverage

The area that can be effectively served by a diffuser is affected by a parameter called *throw*, which is a measure of the distance from the diffuser at which the room air velocity is within a specified range for a given airflow. In most applications, a room velocity of 25 to 75 feet per minute is desirable. Minimum air-

flow is required to avoid complaints of stuffiness. If the airflow is too high, complaints of drafts will result. (See Figure 6–20.)

Diffusers must be selected and spaced to result in good coverage based on throw. If they are too far apart, areas of dead airflow will result; they are too close together, airflows might impinge at the ceiling and create downdrafts. Uncomfortable downdrafts might also occur if air is distributed too close to a wall.

Diffusers are placed and selected to deliver airflow in the right directions for good distribution. The direction is adjustable, as in the case of linear slot diffusers, which contain blades to vary the amount of air delivered from either side. Directionality can also be accomplished by the selection and adjustment of louvers or pattern control devices, as in the case of louvered-face or perforated-face diffusers.

The direction of blow should be considered in selecting the right design based on the placement of the diffuser in the room or its position with respect to other devices. A four-way blow device might be placed in the center of a room or distributed uniformly in a large space to provide even air distribu-

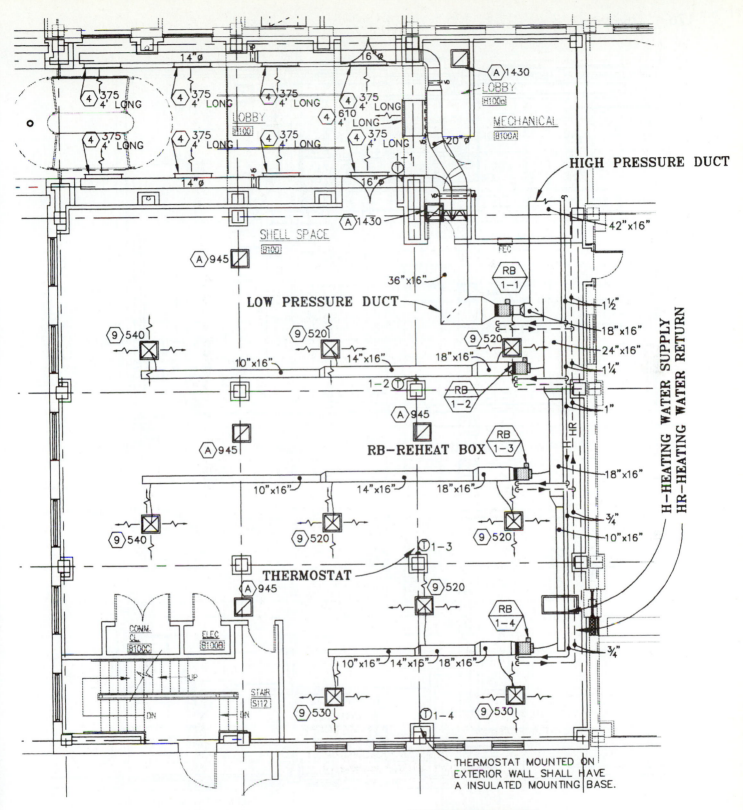

■ **FIGURE 6–21**

Ductwork plan of a simple HVAC system showing supply air ducts and heating water piping. Design criteria notes are added for reference only.

DESIGN CRITERIA NOTES
Total airflow, 9450 cfm
High pressure duct velocity, 2000 fpm
Low pressure duct velocity, 1200 fpm
Low pressure duct loss, 0.15"w.c./100 ft.
Diffuser neck velocity, 400 fpm

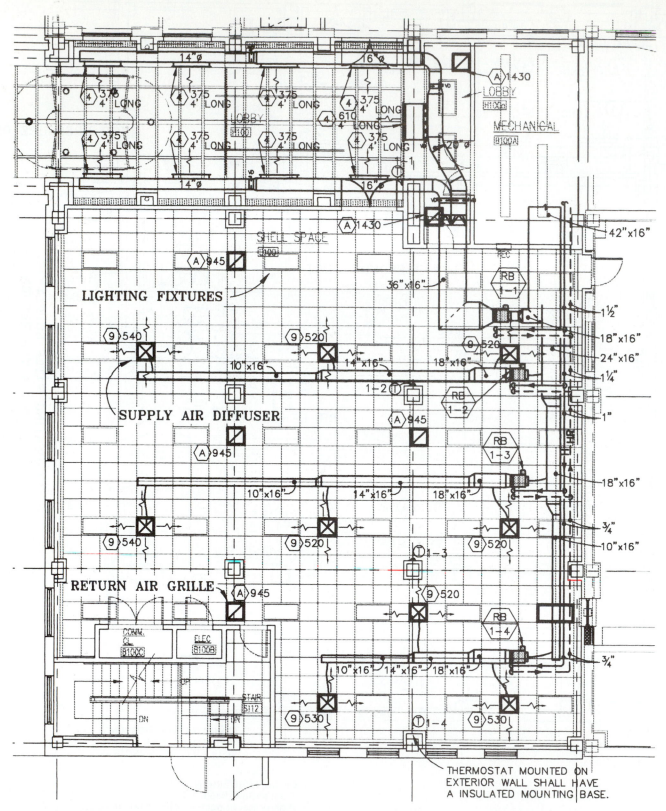

■ **FIGURE 6–22**
Ductwork plan of the same HVAC system shown in Figure 6–21. Ceiling grid and lighting
fixtures are added for coordination between different space elements.

tion. Other options for the geometry of placement and distribution are shown in Figure 6–20.

Linear slot diffusers perform well in variable air volume applications. Due to their shape, they are strong visual elements, requiring attention to their organization with other ceiling devices, including light fixtures and sprinkler heads.

Troffers placed over light fixtures deliver air from slots designed integrally with the light fixtures. This method of air distribution is least obtrusive visually. Distribution characteristics are satisfactory for variable air volume applications, but good workmanship is essential to avoid leaks at the joint between the troffer and the light fixture.

6.8 GENERAL GUIDELINES FOR DUCT SYSTEM DESIGN

A well-designed duct system will result in the lowest installation cost and most energy-efficient operation of the air-handling system. Duct sizes are selected that will produce the highest possible velocity consistent with a reasonable friction loss. Extensive duct systems in multistory buildings will have higher duct velocities in order to minimize the size and associated cost of the ductwork. A penalty is paid in a higher operating cost of the system because the fans operate against a higher pressure. Smaller duct systems typical of one-story buildings will have lower duct velocities, lower system pressures, lower installation costs, and lowest operating costs.

Duct installations are best when maximum use is made of straight duct runs and the number of fittings is minimized. Elbow fittings should have turning vanes. Butt-head tee fittings should be avoided. Branch ducts in higher velocity duct systems should be connected to the main duct using factory-made aerodynamically designed fittings.

Duct systems that are not properly designed will create sounds that are objectionable to the occupants of a building. Insulation placed inside the ductwork will attenuate noise radiating from the system. Lined ductwork is particularly cost effective for cold air ducts, in that the lined duct is an insulator as well as a sound attenuator. Return air and exhaust air ducts are at a neutral temperature and are lined only when acoustical considerations dictate doing so.

Doors should be provided in the duct walls to allow access for cleaning inside the duct and for maintaining components that are sometimes installed in the ductwork. Figure 6–21 is a simple supply air duct drawing, with design criteria and nomenclature notes added to identify the ductwork, airflow rate, air ve-

locity in ducts, friction loss, and fittings. Figure 6–22 illustrates the same simple ductwork coordinated with the ceiling grid and lighting fixtures. Coordination of the ductwork with the lighting layout is an important criterion in the architectural design process.

QUESTIONS AND EXERCISES

6.1 What is the difference between an air-handling unit and a fan coil?

6.2 Why are the casings of air-handling equipment insulated?

6.3 What materials of construction are appropriate for indoor air-handling unit casings? Outdoor?

6.4 Why can heating coils be designed for higher face velocity than cooling coils can?

6.5 What is the advantage of bag-type filters over flat filters?

6.6 What type of air filters would be appropriate for hazardous particulate materials?

6.7 What is the chief consideration in designing air-handling unit sections for mixing outside air and return air?

6.8 What are the advantages of blow-through air-handling units in comparison with draw-through units?

6.9 When might a plug fan be considered instead of a centrifugal fan?

6.10 What is the advantage of belt-drive fans compared with direct-drive fans?

6.11 What is the limitation of low-velocity ductwork in large distribution system applications?

6.12 When might aluminum or stainless steel be preferred over galvanized steel duct construction?

6.13 What are the benefits and limitations of flexible ductwork?

6.14 What is the function of a diffuser in an HVAC distribution system?

6.15 Name several conditions that could cause uncomfortable dumping of cold air from a diffuser.

6.16 Briefly compare the relative benefits of commonly used diffuser types.

6.17 What potential problems do variable air volume systems pose in the selection of diffusers?

6.18 What parameter is important in the selection of a diffuser for good room coverage?

6.19 What is the significance of NC levels listed in catalogues? Will NC levels be higher or lower as the airflow increases for a given diffuser? Why?

6.20 What are the benefits of using light fixture slots for air return from the room to the ceiling?

7
PIPING EQUIPMENT AND SYSTEMS

IN ADDITION TO THE AIR-HANDLING SYSTEMS, PIPING systems are another means of conveying (transporting) heating or cooling energy. In fact, for most applications, air-handling and piping systems must complement each other. For example, without piping to carry the chilled water or refrigerants to coils, air cannot be cooled by the coils and, consequently, cannot cool the building.

Hot and chilled water, refrigerants, steam and condensate, and gas and oil are fluids transported through piping in HVAC systems. This chapter addresses the basic components, accessories, and materials of construction for piping used in the systems described in other chapters.

The flow of fluid in a piping system depends on a number of factors, such as properties relating to fluidity (viscosity and specific gravity), phase (liquid, vapor, or gas), and physical status (temperature, pressure, and velocity).

Similarly to ductwork, piping should be designed large enough to prevent a high pressure drop (friction) and high noise generation. There are special accessories designed for piping systems, such as valves, couplings and air controls in water piping, and oil separators in refrigerant piping systems. Other special considerations in piping design include corrosion control, pressure control, flow control, and control of pipe expansion and contraction.

7.1 PIPING SYSTEMS AND COMPONENTS

Piping systems are classified as follows:

1. *Service provided.* Heating, cooling, oil supply, condensate return, etc.
2. *Energy medium conveyed.* Steam, hot water, chilled water, refrigerant, oil, gas, condensate, chemicals, etc.
3. *Pressure class.* Low, medium, or high pressure classes. The pressure ranges vary with the types of media to be conveyed, such as steam, water, etc.
4. *Temperature class.* Low, medium, or high temperature classes. The temperature ranges vary with types of media to be conveyed.
5. *Piping arrangement.* One-pipe (monoflow) or two-pipe system, direct or reverse return, series or parallel flow, etc.
6. *Hydrology.* Gravity or forced flow, open or closed.
7. *Piping materials.* Steel, copper, plastic, nonmetallic, etc.

7.1.1 Forced or Gravity Flow

Most piping systems utilize pumps or compressors to maintain the flow of the fluid. This is called a *forced*

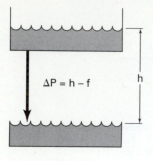

$$\Delta P = h - f$$

(a)

*h – height, ft.
f – friction, ft.
P – pump head, ft.
ΔP – Net pressure, ft.

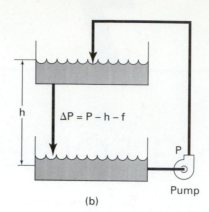

$$\Delta P = P - h - f$$

Pump

(b)

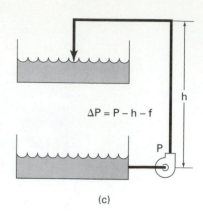

$$\Delta P = P - h - f$$

P

(c)

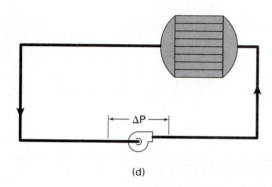

$$\Delta P$$

(d)

■   **FIGURE 7–1**

Hydraulic circuit of water systems. (a) Open system with gravity flow and no return. (b) Open system with forced flow. (c) Open system with forced flow and no return. (d) Simple closed system. (e) Closed system with reverse return. (f) Closed system with direct return. (g) Closed system with series pumping. (h) Closed system with parallel pumping.

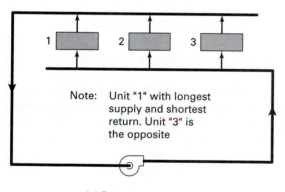

Note: Unit "1" with longest supply and shortest return. Unit "3" is the opposite

(e) Reverse return

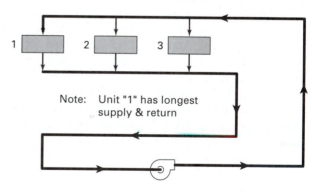

Note: Unit "1" has longest supply & return

(f) Direct return

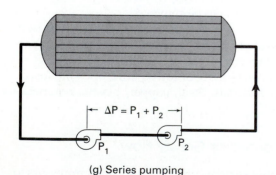

$$\Delta P = P_1 + P_2$$

P_1 P_2

(g) Series pumping

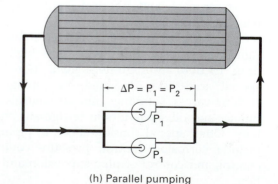

$$\Delta P = P_1 = P_2$$

P_1

P_1

(h) Parallel pumping

system. However, pumps are not always required. Steam flows in a piping system due to the pressure differential between the steam source and the devices that use the steam. Steam condensate flows by gravity or, if adequately cooled, can be pumped. Refrigerant is generally forced through a piping system by the action of a compressor if the refrigerant is in the vapor phase or a pump if it is in the liquid phase.

7.1.2 Open or Closed System

In an open system, the fluid surface is exposed to the atmosphere. (See Figures 7–1(a) to 7–1(c).) The flow may be either by gravity or by pumping. In a closed system, fluid is not in contact with the atmosphere. The flow must be by pumping. Most HVAC piping systems are closed systems. Figures 7–1(d) to 7–1(h) illustrate various closed-system piping arrangements.

In an open system, the net pressure for the pump to overcome is frictional pressure plus the pressure equivalent of the height of the fluid being lifted. In a closed system, the net pressure for the pump to overcome is only the frictional pressure. It makes no difference whether various pieces of equipment is at different elevations, since the elevation is balanced at both sides of the pump. Closed systems are always preferred when designing a pumping system.

In the open system (Figure 7–1f), equipment closest to the pump has the least pressure loss, and thus may have more or inbalanced water flow. The reverse return piping arrangement (Figure 7–1e) is preferred, since the water flows through multiple units of HVAC equipment and is inherently balanced. Reverse return piping systems usually cost more because more pipe is needed.

For equal-size pumps, a series arrangement doubles the head (Figure 7–1g), and a parallel arrangement doubles the flow. Either arrangement can be used to provide continuity of service and energy savings if one of the two pumps is shut off. (See Section 7.2.)

In an open, pumped system, the pressure that causes the fluid to flow is the pump pressure minus the static height and the frictional loss. This relation is expressed as

$$\Delta P = P - h - f \qquad (7\text{--}1)$$

Where ΔP = net pressure available at the discharge, ft

P = pump pressure at the flow rate, ft

h = static head or difference in elevation, ft

f = frictional loss at the rated flow, ft

Pressure expressed in feet (of water column) can be converted to other units by the following relations:

- 1 psi = 34 ft/14.7 psi = 2.31 feet w.c.
- 1 psi = 29.9 in./14.7 psi = 2.03 in. of Hg (mercury)
- 1 psi = 2.31 ft × 12 in./ft = 27.7 in. of w.c.
- 1 in. w.c. = 0.036 psi

In a closed system, static height is not a factor, because it is balanced on both the suction and the discharge sides of the pump. This relation is expressed as

$$\Delta P = P - f \qquad (7\text{--}2)$$

By comparing equations (7–1) and (7–2), it should be evident that a closed system is the preferred piping system when fluid must be circulated at different elevations.

7.1.3 Water as the Basic Medium

Water is the most popular fluid used in HVAC systems, because water is chemically stable, nontoxic, and economical. Water has a density of 62.4 pounds per cubic foot and is considered to have a specific gravity of 1.0 compared with other fluids.

7.1.4 Basic HVAC Piping Systems

Common piping systems for HVAC applications are water, steam (water in the vapor phase), refrigerant (freon, ammonia, water), fuel (oil, gas), and other fluids. Pipes are used to transport the fluids to and from system components, which may be pumped to provide power for transport or heat-exchanging equipment to transfer the fluid energy. Valves that govern the flow of fluid in the pipes are important components in a piping system and will be discussed later in the chapter. Another important aspect in piping system is the instrumentation used. These include temperature, pressure and flow gauges. See Figure 7–2.

■ **FIGURE 7–2**

Some basic instruments. (a) A pressure gauge. (b) A thermometer. Without proper instrumentation, the performance of a fluid system cannot be confirmed, tested, balanced, or adjusted. Instruments are usually installed in (c) thermal wells to separate the instruments from direct contact with the fluid. (Courtesy: Palmer Instruments, Inc. Cincinnati, OH.)

7.2 PUMPS

7.2.1 Classification

Pumps are basic equipment in a hydronic system. Depending on their operating principles, construction, and services performed, they may be classified and used as follows:

1. *Operating principles.* Positive displacement or centrifugal type.

   - For positive displacement type—reciprocating piston, reciprocating plunger, rotary blades, rotary roots, screw types, etc.
   - For centrifugal type—radial flow, axial flow, or mixed flow, etc.
   - For turbine type—axial flow, mixed flow, etc.

2. *Casing design.* Horizontal split case, vertical split case, submerged, etc.
3. *Mounting.* Base mounted, in-line mounted, etc.
4. *Connection with driver (motor or engine).* Flexible coupled, close coupled, belt driven, etc.
5. *Construction material.* Cast iron, stainless steel, bronze, concrete, plastic, fiberglass, etc.
6. *Service functions.* Hot-water supply, condensate return, vacuum, oil supply, boiler feedwater, etc.

Centrifugal pumps are used in most HVAC applications. A variety of commonly used pumps is shown in Figure 7–3. Small pumps can be mounted in line with the piping system; larger pumps are mounted on the floor. Virtually all pumps are flexibly coupled with the motor that drives them.

Turbine pumps can be used effectively for applications requiring water to be pumped from a sump into a piping system. The most common HVAC system application is pumping condensing water from a cooling tower sump to water chillers.

Positive displacement pumps use pistons or rotary gears to pump oil for boilers and engine generators. Because of their self-priming feature, positive displacement pumps can be used to draw oil up from underground tanks.

Pumps are selected to meet flow criteria and to develop appropriate pressure to move fluid through the system. Water flow is measured in gpm; pressure, often called "head," is measured in psig or, more commonly, feet of water. Pumps are usually driven by motors rated at standard induction speeds. For 60-hertz power, the common induction motor speeds are 1150, 1750, and 3500 rpm. These speeds correspond to the synchronous speed at 1200, 1800, and 3600 rpm. The difference between the induction speed and the synchronous speed is due to the slip of the rotor in induction motors.

7.2.2 Pump Performance

Performance curves A specific size of pump operating at a given speed can perform a range of flows and pressures, depending on the diameter of the impeller

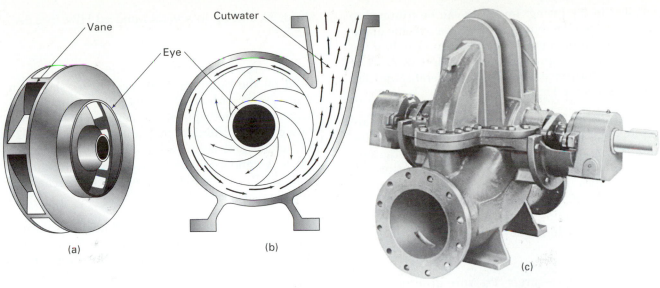

(a)   (b)   (c)

(d)

(f)

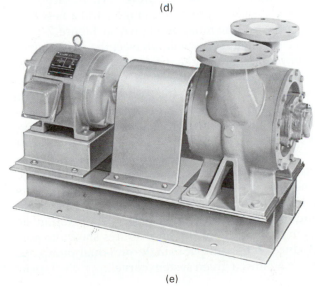

(e)

■ **FIGURE 7–3**

Centrifugal and turbine pumps.
(a) Basic impeller. (b) The directional arrows indicate the flow pattern and illustrate why the impeller direction must be toward the discharge opening for proper operation. Reversing the direction of rotation causes turbulence and reduces pump capacity, at the same time increasing motor loading. (c) A horizontal split-case centrifugal pump. (d) An end suction, direct-coupled centrifugal pump. (e) A pump-motor assembly with mechanical coupling. (The coupling is covered by the protection guard and is not visible.) (f) A multi-stage turbine pump. The impeller of each stage may be removed individually. (Courtesy: ITT Bell & Gossett, Morton Grove, IL.)

installed in the casing. Impellers can be trimmed to match the precise requirements for a specific application. When the impeller becomes too small or too large, a new pump casing size is selected. A set of pump performance curves for different impeller trims is shown in Figure 7–4(a).

System curve   The system curve is the pressure requirement of the piping system at various flow rates. For a given flow rate, the pressure is the sum of the pressure drop of the system components, including pipes, fittings, equipment, control devices, and static head. If the flow rate is changed, the pressure will change in proportion to the square of the flow. The following equations show the basic relations:

System closed pressure = Pressure drop of
(piping + fittings + equipment + controls)   (7–3)

$$(q_2/q_1)^2 = p_2/p_1$$
$$\text{or } p_2 = p_1 \times (q_2/q_1)^2 \qquad (7\text{–}4)$$
$$\text{or } (q_2)^2 = (q_1)^2 \times (p_2/p_1)$$

where p_1 = system pressure required at flow rate q_1

p_2 = system pressure required at flow rate q_2

q_1 = system flow rate at condition 1

q_2 = system flow rate at condition 2

A system curve can be plotted as long as one of the operation conditions (e.g., q_1 and p_1) is known. All other points on the system curve can be plotted using Equation (7–4) and the known q_1 and p_1.

An example of system flow characteristics is plotted on the pump curves shown in Figure 7–4(b). The pump operating point is the point where the pump curve intersects the system curve. At this point, the head and flow characteristics of the pump match those of the system.

Power requirement   Another set of curves on the pump performance curve (Figure 7–4(a)) indicates horsepower requirements. This information is used to select the proper size of motor. Often, a motor will be selected that is capable of the maximum requirement indicated by the pump curve, regardless of the normal system operating conditions. If the motor is grossly oversized (or underloaded), it will operate at a low power factor. This situation should be avoided by selecting a smaller motor. Standard integral HP motor sizes are 1, 2, 3, 5, 10, 15, 20, 25,

30, 40, 50, and up to several hundred HPs. (See Section 7.3.)

Example—see Figure 7–4(b)
Based on the pump performance curves shown in Figure 7–4(a) with a 12" impeller, the pump should deliver about 1800 gpm at about 55 feet of head. Operating at this condition, the pump will require a 32–BHP motor and is operating about 80 percent efficiency. A standard 40 BHP rated motor will be used. (*Note:* BHP = brake horsepower, the mechanical equivalent of electrical power.)

Net positive suction head   When fluid enters the impeller "eye" of a centrifugal pump, fluid pressure is reduced due to the greatly increased velocity at the impeller. The reduced pressure may cause the fluid to "flash" or "vaporize," inside the pump. This phenomenon causes cavitation and damage to the pump. *Net positive suction head* (NPSH) is the characteristic of a centrifugal pump that which must be considered during the design of the piping system. Each and every centrifugal pump has a specific NPSH pressure rating that must be assured by the piping system design. For a given pump, the required NPSH increases with the flow rate, varying from a few feet up to 30 or 40 feet. NPSH is not normally critical in closed systems. However, in open systems, such as for cooling tower water, the placement of the pumps in the piping system is critical.

Capacity control   Capacity control for variable-flow water systems can be accomplished by throttling (restricting) valves, staging of multiple pumps, variable-speed motors, or combinations of these methods. In any case, the pump or pumps are selected for the nonthrottling peak design flow. When throttling occurs, the system pressure changes, and a new system curve is generated as shown in Figure 7–5(a). The new system curve will intersect the pump curve at a lower flow and higher pressure. Another strategy for variable-flow systems is to use variable-speed pumping. In this case, the system curve remains constant, and the pump curve expands or contracts according to the change in speed, as shown in Figure 7–5(b).

Arranging two or more pumps in parallel or in series is another method of varying the performance of the pumps. Individual pump heads are additive, as shown in Figure 7–5(c) for series operation, and the pump flow rate is additive for parallel operation, as shown in Figure 7–5(d). These figures show the performance of the pump for single- and dual-pump operation against a given system curve.

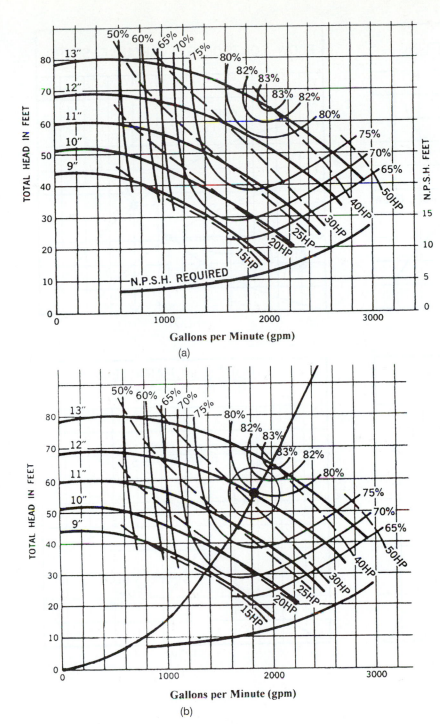

(a)

■ FIGURE 7–4

(a) Typical pump performance curves indicating head, flow rate, brake horsepower and efficiency for different size impellers. (Note: N.P.S.H = net positive suction head.) (b) Same pump performance curves with system curve plotted.

(b)

7.2.3 Basic Hydronic Formulae

The flow of fluid in a piping system is similar to that of airflow in a duct system. Two major differences exist, however. First, fluid in the liquid phase is incompressible and thus follows closely the first law of thermodynamics. In the case of a pumped circuit, the fluid power (flow × pressure) is increased by the addition of the pump power. The second difference is that fluids have different viscosities and specific grav-

ities that require different power in transporting the fluids.

The basic formulae for determining power are as follows:

$$WHP = (q \times h \times SG)/3960 \qquad (7\text{–}5)$$

$$BHP = WHP/Ep \qquad (7\text{–}6)$$

$$Ep = WHP/BHP \qquad (7\text{–}7)$$

$$Q = 500 \times TD \times q \qquad (7\text{–}8)$$

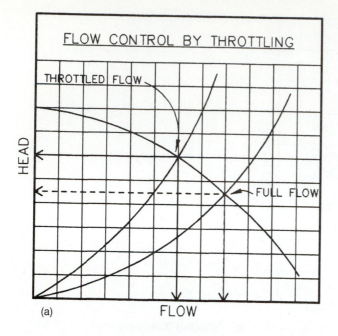

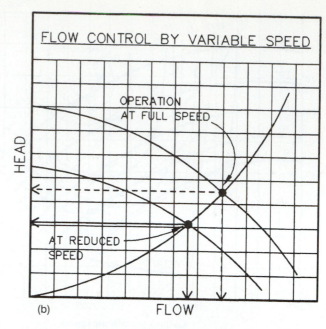

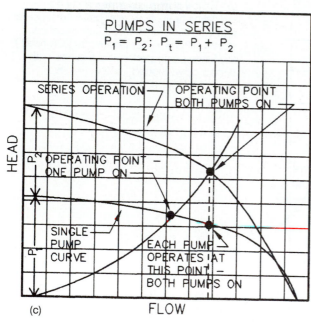

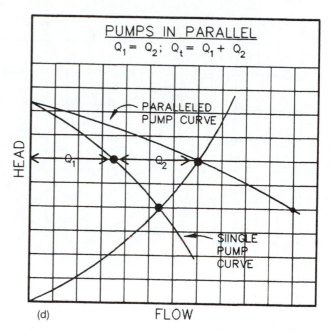

■   FIGURE 7–5
Method of capacity control for pumping operations.

where q = flow rate, GPM.

h = system head, ft w.c.

SG = specific gravity. For water = 1.0; for other fluids, SG is determined by the ratio of the specific weight of water to the specific weight of the other fluids

Ep = pump efficiency, per unit

WHP = water horsepower (the theoretical power required)

TD = temperature difference, °F

Q = heat, Btu/hr (Btuh)

BHP = brake horsepower (the actual power required, including the pump's efficiency)

Affinity laws provide methods for approximating of the effects of changed impeller diameters and rotational speed on the pump curve. These laws are as follows:

▪   Pump capacity (GPM) varies directly as the speed (RPM) or the ratio of impeller diameters.

TABLE 7–1
Effect of change of impeller diameter and speed

Change of	Capacity	Head	BHP
Impeller Diameter	$q_2 = \dfrac{D_2}{D_1} q_1$	$H_2 = \left(\dfrac{D_2}{D_1}\right)^2 H_1$	$P_2 = \left(\dfrac{D_2}{D_1}\right)^3 P_1$
Speed	$q_2 = \dfrac{RPM_2}{RPM_1} q_1$	$H_2 = \left(\dfrac{RPM_2}{RPM_1}\right)^2 H_1$	$P_2 = \left(\dfrac{RPM_2}{RPM_1}\right)^3 P_1$

where Q = Flow rate, GPM; D = Diameter, in.

- Pump head varies directly as the square of the speed (RPM) or the ratio of impeller diameters.
- BHP varies directly as the cube of the speed (RPM) or the ratio of impeller diameters.

Some useful formulae are given in Table 7–1.

7.2.4 Examples

This example covers flow rate calculations (Section 7.4.4), pipe sizing (7.4.4), and pump selection (7.2.2 and 7.2.3).

A hot-water system consists of a boiler (hot-water generator) and several air-handling units (AHUs), shown in Figure 7–6(a). Copper pipes and fittings are used. The total heating load of all the AHUs is 1,500,000 btu/hr.

The maximum pressure required to pump water through the AHU with highest pressure drop is 40 feet in total head. The inlet and outlet temperatures at the AHUs are designed to be 180°F and 150°F, respectively.

1. What is the required total flow rate (GPM)?
2. Allowing 30% extra capacity for load pickup, what should be the boiler capacity?
3. What should be the main pipe size, based on a flow velocity of 4.5 feet per second?
4. Select a pump from the manufacturer's pump performance curves that will fit the need. (*Note:* The pump selected is running at 1750 rpm.) Plot the system curve on the pump curves. What size impeller is required for this pump operating at this condition?
5. If the piping system is redesigned to lower the total system pressure down to 30 feet, what pump impeller should be used?
6. What is the brake horsepower required at both operating conditions?

Answers

1. From Equation (7–8), 100 GPM is required.
2. The boiler should be rated for 1,950,000 Btu/hr.

3. From Figure 7–11(b) for copper tubing, a 3" pipe may be used.
4. From Figure 7–6(b), the impeller size should be slightly larger than 6½" diameter.
5. The flow still should be 100 GPM. At 30 feet total head, a new system curve is established in Figure 7–6(c).
6. From Figure 7–6(b), the horsepower required is 1½ BHP, when the pump operates at 40 feet head and is reduced to about 1¼ BHP when it operates at 30 feet head.

7.3 HEAT EXCHANGERS

Heat exchangers are used to transfer heat from one medium to another, such as from steam to hot water, or from water at a higher temperature to water at a lower temperature. There are two basic types of heat exchangers: The shell-and-tube type and the plate type.

7.3.1 Shell-and-tube type

The shell-and-tube type of heat exchanger consists of an externally insulated steel sheet and a bundle of tubes in the shell, as shown in Figure 7–7(a) and (b). For HVAC applications, the primary medium is either steam or water, which flows in the shell. The secondary medium is always water, which flows through the tubes. The tubes are partitioned to allow single or multiple passes in order to increase the temperature and the heat transfer.

7.3.2 Plate-Type Heat Exchangers

Plate-type heat exchangers consist of thin metal plates forming alternate compartments to allow the primary water to flow in one compartment and the secondary water in the other. Figure 7–8(a) illustrates the typical construction of this type of heat exchanger. Such a heat exchanger is ideal for heat transfer when the temperature difference between two wa-

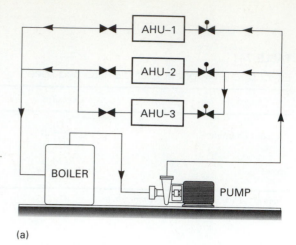

■ FIGURE 7–6

Examples

(a) Piping diagram of the examples. (b) Pump performance and system curves. (c) Pump performance and system curves under revised condition.

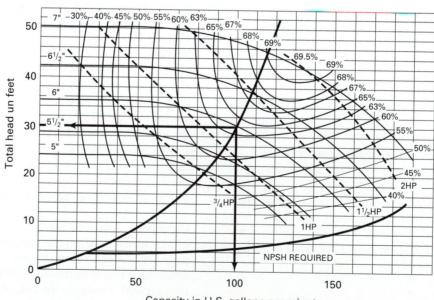

(a)

(b)

(c)

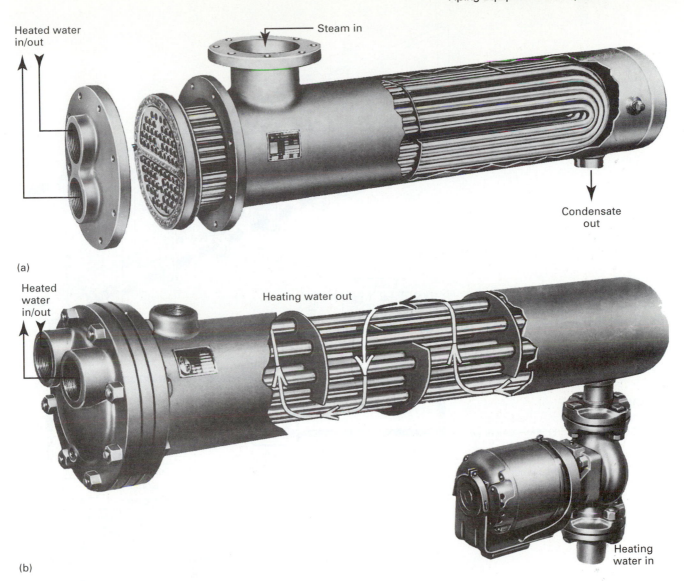

Heated water in/out

Steam in

Condensate out

(a)

Heated water in/out

Heating water out

Heating water in

(b)

■ FIGURE 7–7
(a) A typical steam-to-water U-tube heat exchanger. (b) A typical water-to-water U-tube heat exchanger with water circulating pump shown. (All illustrations courtesy of ITT Bell & Gossett, Morton Grove, IL.)

ter media is very close. A typical application of a plate-type heat exchanger is the economizer cycle of a chilled-water system during winter, as shown in Figure 7–8(b). In this application, the condenser water is cooled through a heat exchanger by the cold water in the cooling tower. The cooled condenser water is then used to cool the evaporator water, thereby conserving electrical energy without using the chiller compressor.

7.4 PIPING

Piping for fluid systems is analogous to ducts for air systems. Both serve to transport energy from equipment to equipment. Piping includes pipes, fittings,

and accessories, which differ with the type of fluid, services, and application.

7.4.1 Materials for Pipes

General-purpose piping for HVAC systems can be constructed of plain (black) or galvanized steel, of copper, or of cast iron. Special applications requiring corrosion resistance may call for nonmetallic or more expensive materials, such as stainless steel, aluminum, or even glass.

Fiberglass and plastic pipe can be an economical alternative to metal. However, these materials are fragile and require extra design attention to prevent damage from water hammer. Thermal expansion also

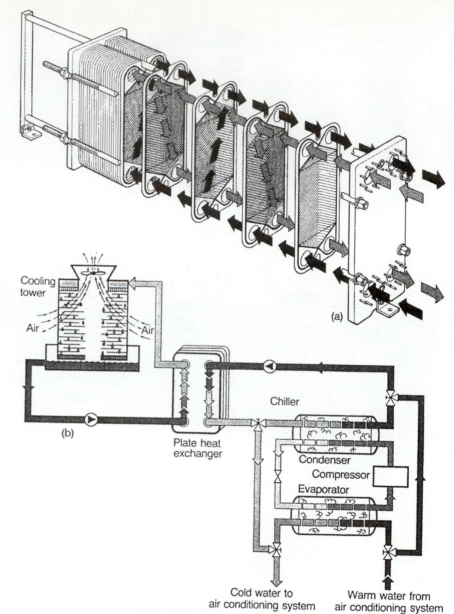

(a)

(c)

■   FIGURE 7–8
(a) Operating principle of a plate-type
heat exchanger. (b) Piping arrangement of
an economizer chiller cycle (free cooling
cycle) during cold weather (winter in
northern hemisphere and summer in
southern hemisphere). (c) Typical
installation of a plate-type heat
exchanger. (Courtesy: Alfa-Laval Company,
Richmond, VA.)

needs to be considered carefully; plastics expand at a much higher rate than metals. Commonly used plastic or fiberglass pipes include:

- Polyvinyl chloride (PVC)
- Polybutylene (PB)
- Polypropylene (PP)
- Fiberglass-reinforced plastic (FRP)
- Chlorinated PVC (CPVC)
- Polyethylene (PE)
- Reinforced thermosetting resin (RTR)

Table 7–2 lists preferred materials for various HVAC piping services, along with the range of pressures and temperatures over which they operate.

Actual pipe diameters are slightly different from their nominal dimensions and will vary depending on the material of construction. Plastic piping is manufactured in the same dimensions as steel.

7.4.2 Method for Joining Pipes

Joints can be soldered, brazed, welded, screwed, flanged, solvent welded, or mechanically coupled.

Table 7–3 shows joining methods applicable to piping of various sizes and materials. Figure 7–9 shows several pipe joints, including a popular mechanical coupling system that is less expensive than welding and has the added advantage of inherent flexibility to relieve stress and thermal expansion.

7.4.3 Valves and Accessories

Valves are used in piping systems to adjust and control the flow of the medium. They may be on-off valves or adjustable either manually or automatically. Figure 7–10 shows a variety of valves commonly used in HVAC piping systems. Table 7–4 lists typical selections for the functions described next.

Functions of valves

- *Shutoff valves* are installed to isolate components and portions of systems. These valves avoid the inconvenience of draining and refilling entire systems or portions of systems during maintenance or repairs. If properly planned, they also allow continued service to portions of the system not affected by the work.

TABLE 7–2

Preferred materials for HVAC piping and their operating ranges

Services	Material	Temperature (°F)	Pressure (Psi)
Water	black steel, copper, plastic	40° to 250°	to 400 for steel
Steam	black steel, copper		
Steam Condensate	steel, copper, black steel		
Condensing Water	galvanized, black steel	40° to 250°	to 150 for copper
Refrigerant	copper		
Fuel Oil	black steel	40° to 200°	to 125 for FRP, PVC
Gas	black steel, copper		

TABLE 7–3

Piping material and normal methods of joining

Piping material	Welded	Threaded	Flanged	Soldering	Bracing	Solvent	Flared/Compression	Grooved Pipe
Black steel, up to 2"		●	●					●
Black steel, 2½" and up	●		●					●
Galvanized steel		●	●					●
Type M or L copper			●	●	●			●
Type K copper			●		●			●
Stainless steel	●		●					
Rigid aluminum	●	●	●					
Rigid plastic			●			●		
Fiberglass			●			●		

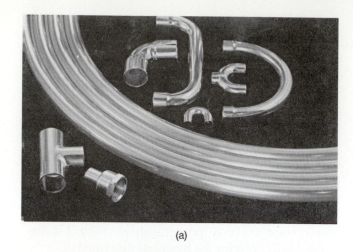

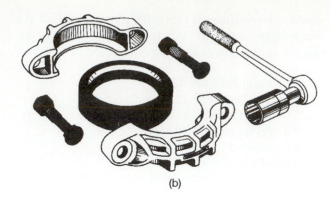

(a)

(b)

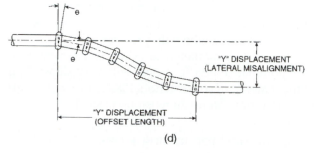

(d)

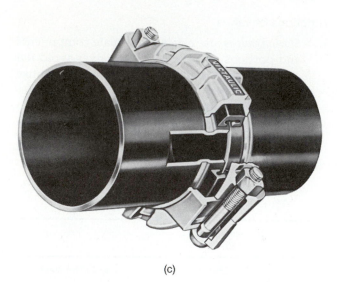

(c)

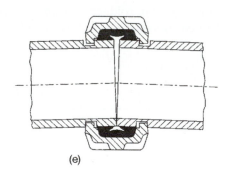

(e)

■ FIGURE 7–9

Two of many methods of joining pipes.
(a) Soldering fittings for copper, stainless steel, and aluminum pipes. (Courtesy: Mueller Brass Co., Port Huron, MI.) (b) Mechanical joints (couplings) on grooved steel, iron or other heavy wall pipes. (c) A mechanical coupling typically is constructed of a split ring malleable iron housing, a synthetic rubber gasket, and nuts, bolts, locking toggle or lugs to secure the unit together. (d) An unique characteristic of mechanical joints is their flexibility, which allows the pipes to deflect without causing leakage. (e) A cutaway view of the pipe and coupling showing a deflected or misaligned pipe joint. (Courtesy, (b) through (e): Victaulic Company of America, Easton, PA.)

TABLE 7–4

Typical and maximum recommended pressure drops and velocities in water systems

Applications	Velocity ft/sec		Pressure drop, ft/100 ft	
	Typical	Maximum	Typical	Maximum
Hot/chilled mains	1.5–4	6	6	10
Hot/chilled branches	1–3	4	4	6
Heat exchangers				10' total
Coils				10' total

- *Drain valves* are installed at equipment and at the low points of piping systems. Hose end fittings are normally used, and fluid is conducted to the nearest floor drain. Floor drains are required in mechanical rooms and should be placed close to condensate discharge from cooling coils to avoid piping over the floor. This is generally a convenient location for draining hot- and chilled-water coils.

- *Balance valves* are installed to adjust the flow through the system in proper amounts to serve loads. These valves need to be adjusted when a

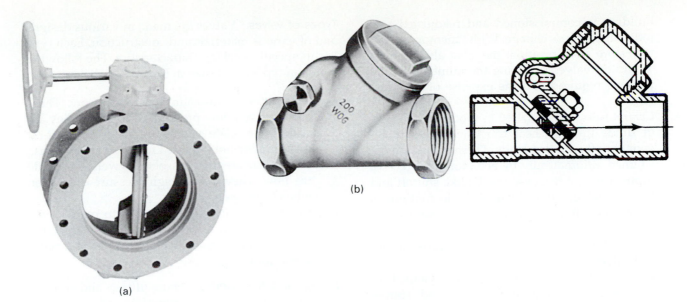

(a)

(b)

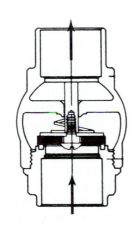

(c)

■ **FIGURE 7–10**
Commonly used valves.
(a) Flanged butterfly valve. (b) Swing check valve. (c) Spring-
loaded center post check valve. (d) Globe valve. (e) Gate valve.
(Courtesy: Grinnell Corporation, Exeter, NH.)

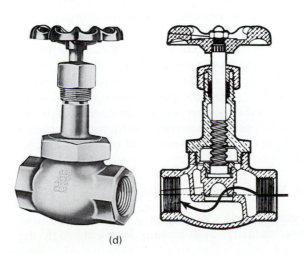

(d)

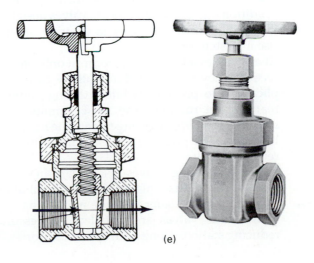

(e)

building is commissioned and readjusted periodically as needs change. With "memory stops" to restore a set position, they can also be used as shutoff valves, reducing the number of components required and the expense of construction.

- *Control valves* are installed to modulate the flow of fluids through heat transfer devices in response to fluctuating loads. These valves are positioned automatically by actuators that receive input from control systems. Unlike shutoff and balance valves, control valves are in continuous operation. They are designed to wear well in movement and be accurate over a wide range of operating conditions. There are two types of control valves:

  - A *two-way* control valve is installed in series with the load device and simply throttles the flow in response to the load. This control method results in a variation of flow.
  - A *three-way* control valve is arranged to force fluid through the load device or to bypass fluid around the load device, resulting in a constant system flow.

Two-way valve arrangements are simpler and less expensive to install, but under light load conditions, choking effects on the overall system can result in high pressures and poor control. (See the discussion of pump and system curves on p. 182.) Often, a combination of two-way and three-way valves is designed to blend economical installation with acceptable control. The three-way valves maintain enough flow during light load conditions to prevent an undue rise in system pressure by the choking effect of the two-way valves.

- *Back-pressure regulators* can also be used to avoid unacceptably high system pressure at light load conditions. These regulators are valves designed to modulate toward the open position upon a rise in system pressure.
- Check valves allow flow to occur in only one direction. They are installed in multiple pump applications to prevent backflow through pumps that are not running. The most common construction is a flapper arrangement. The flapper is held closed by pressure opposite the desired direction of flow.
- Pressure-reducing valves (PRVs) are in steam systems to reduce high-pressure distribution steam to a lower, safer pressure in building systems. These valves are installed in pressure-reducing stations along with other accessories.

Types of valves   Valves are made in various designs and of various materials and construction. Each type has its specific working characteristics. The following common types are shown in Figure 7–10 below:

- *Gate valves* are used to isolate the flow. These valves have a low pressure drop when they are wide open. Gate valves are used for shut-off duty and are not intended for flood regulations.
- *Globe valves* are used to isolate as well as regulate the flow. They have a high pressure drop even when wide open.
- *Ball and butterfly valves* are used to regulate the flow. They can also be used for isolating the flows although gate or globe valves are better suited for this purpose.

Fittings and Accessories   Many fittings and accessories are required to make a complete and operational piping system. Some are used in all systems, while others are applicable only to a particular system. Following is a list of some common fittings and accessories:

- *Pipe fittings* include elbows, tees, flanges, couplings, unions, etc. A union is used on threaded pipe joints to facilitate disconnecting the pipes for maintenance or repair.
- *Strainers* are installed to prevent foreign materials from damaging sensitive components that have moving parts or close clearances, mainly pumps and control valves. Strainers contain a perforated basket to entrain objects large enough to cause damage. They are designed with provisions for removal and inspection of the baskets.
- *Pressure and temperature taps* are installed to allow the insertion of gauges to check the performance of systems and components. The flow through a piece of equipment can be determined if the pressure drop across the equipment can be measured. Measurements of inlet and outlet pressures are also necessary to determine whether pumps are operating effectively and to diagnose possible system blockages. Knowing inlet and outlet temperatures of heat transfer equipment is essential to verify the performance of the equipment and diagnose possible problems. Taps can be equipped with permanent thermometers and gauges for the convenience of maintenance personnel.

7.4.4 Water-Piping Systems

Pipe sizing criteria   Water piping is sized on the basis of flow, velocity, and friction losses through the

TABLE 7–5

Equivalent length of straight pipes for various fittings and valves for turbulent flow only. (Courtesy: Hydraulic Institute.)

Fittings			Pipe size										
			$\frac{1}{2}$	$\frac{3}{4}$	1	$1\frac{1}{4}$	$1\frac{1}{2}$	2	$2\frac{1}{2}$	3	4	5	6
Regular 90° elbow	Screwed	Steel	3.6	4.4	5.2	6.6	7.4	8.5	9.3	11.0	13.0		
		C.I.								9.0	11.0		
	Flanged	Steel	.92	1.2	1.6	2.1	2.4	3.1	3.6	4.4	5.9	7.3	8.9
		C.I.								3.6	4.8		7.2
Long Radius elbow	Screwed	Steel	2.2	2.3	2.7	3.2	3.4	3.6	3.6	4.0	4.6		
		C.I.								3.8	8.7		
	Flanged	Steel	1.1	1.3	1.6	2.0	2.3	2.7	2.9	3.4	4.2	5.0	5.7
		C.I.								2.8	3.4		4.7
45° elbow	Screwed	Steel	.71	.92	1.3	1.7	2.1	2.7	3.2	4.0	5.5		
		C.I.								3.3	4.5		
	Flanged	Steel	.45	.59	.81	1.1	1.3	1.7	2.0	2.6	3.5	4.5	5.6
		C.I.								2.1	2.9		4.5
Tee inline flow	Screwed	Steel	1.7	2.4	3.2	4.6	5.6	7.7	9.3	12.0	17.0		
		C.I.								9.9	14.0		
	Flanged	Steel	.69	.82	1.0	1.3	1.5	1.8	1.9	2.2	2.8	3.3	3.8
		C.I.								1.9	2.2		3.1
Tee branch flow	Screwed	Steel	4.2	5.3	6.6	8.7	9.9	12.0	13.0	17.0	21.0		
		C.I.								14.0	17.0		
	Flanged	Steel	2.0	2.6	3.3	4.4	5.2	6.6	7.5	9.4	12.0	15.0	18.0
		C.I.								7.7	10.0		15.0
180° Return bend	Screwed	Steel	3.6	4.4	5.2	6.6	7.4	8.5	9.3	11.0	13.0		
		C.I.								9.0	11.0		
	Reg. Flanged	Steel	.92	1.2	1.6	2.1	2.4	3.1	3.6	4.4	5.9	7.3	8.9
		C.I.								3.6	4.8		7.2
	Long Rad. Flanged	Steel	1.1	1.3	1.6	2.0	2.3	2.7	2.9	3.4	4.2	5.0	5.7
		C.I.								2.8	3.4		4.7

Fittings			Pipe size										
			$\frac{1}{2}$	$\frac{3}{4}$	1	$1\frac{1}{4}$	$1\frac{1}{2}$	2	$2\frac{1}{2}$	3	4	5	6
Globe valve	Screwed	Steel	22.0	24.0	29.0	37.0	42.0	54.0	62.0	79.0	110.0		
		C.I.								65.0	86.0		
	Flanged	Steel	38.0	40.0	45.0	54.0	59.0	70.0	77.0	94.0	120.0	150.0	190.0
		C.I.								77.	99.0		150.0
Gate valve	Screwed	Steel	.56	.67	.84	1.1	1.2	1.5	1.7	1.9	2.5		
		C.I.								1.6	2.0		
	Flanged	Steel						2.6	2.7	2.8	2.9	3.1	3.2
		C.I.								2.3	2.4		2.6
Angle valve	Screwed	Steel	15.0	15.0	17.0	18.0	18.0	18.0	18.0	18.0	18.0		
		C.I.								15.0	15.0		
	Flanged	Steel	15.0	15.0	17.0	18.0	18.0	21.0	22.0	28.0	38.0	50.0	63.0
		C.I.								23.0	31.0		52.0
Swing check	Screwed	Steel	6.5	6.3	7.0	8.4	9.3	12.0	13.0	16.0	21.0		
		C.I.								18.0	17.0		
	Flanged	Steel	.38	.53	.72	1.0	1.2	1.7	2.1	2.7	3.8	5.0	6.3
		C.I.								2.2	3.1		5.2
Coupling or union	Steel		.21	.24	.29	.36	.39	.45	.47	.53	.65		
	C.I.									.44	.52		
Bell mouth inlet	Steel		.10	.13	.18	.26	.31	.43	.52	.67	.95	1.3	1.6
	C.I.									.55	.77		1.8
Square mouth inlet	Steel		.96	1.3	1.8	2.6	3.1	4.3	5.2	6.7	9.5	13.0	16.0
	C.I.									5.5	7.7		13.0
Sudden enlargement			$h = \dfrac{(V_1 - V_2)^2}{2g}$ feet of fluid; if $V_2 = 0$ $\qquad h = \dfrac{V_1^2}{2g}$ feet of fluid										

pipes, fittings, and equipment. Friction in pipes causes a pressure drop, which is normally expressed in feet of head per 100 feet of straight pipe. The flow velocity in the pipe should be fast enough to move water with its entrained air through the system, yet slow enough to prevent undue noise and erosion. Table 7–5 shows typical and maximum recommended pressure drops and velocities based on the type of service. These guidelines can be used to select reasonable pipe sizes for a given application.

Pipe sizing data The flow of water in pipes is affected by the material of the pipe, the smoothness of the interior surface, the types of fittings in the pipe, and the velocity of the water. Figure 7–11 gives fundamental data relating to pipes. The pressure drop of a fitting or a valve may be as much as many linear feet of straight pipe. Table 7–6 gives data on pressure drop for valves and fittings. Additional data can be found in the *ASHRAE Handbook of Fundamentals*.

Flow rate calculations Water is heated or cooled to transport the energy for heating and cooling systems. The amount of water that needs to be circulated depends on the designed temperature difference of the flow entering and leaving the equipment, such as a coil or a heat exchanger. The basic formula for water flow calculations is

$$Q = 500 \times FL \times (t_2 - t_1) \qquad (7\text{–}8)$$

or

$$FL = Q/500(t_2 - t_1)$$

where Q = heat being transferred, Btu/hr
  FL = flow rate, GPM
  $t_2 - t_1$ = temperature change (rise or drop), °F.

For example, the water flow rate required to transfer 25,000 Btu/hr of heat through a heating unit with hot water entering at 180°F and leaving at 140°F is as follows:

$$FL = 25,000/500 \times (180 - 140) = 1.25 \text{ gpm}$$

Air management A primary concern in hydronic system design is dissolved air in water. The solubility of air in water decreases as the water temperature rises; the air then migrates to high points of the piping system and may curtail the water flow to portions of the system. Air in heat transfer equipment can also reduce the effective heat transmission. Therefore, undissolved air must be eliminated from the system.

There are two approaches to air management—air control and air elimination. In the air control approach, all air separated from water (undissolved, or free, air) must be returned to a tank where air is accumulated at the top of the tank with water at the bottom of the tank to maintain a consistent pressure head. The tank is known as an air/water interface compression tank. In the air elimination approach, a diaphragm type tank is used to maintain the pressure head. Excessive free air is vented to the atmosphere through separate air vents. Figure 7–12 (a) shows a centrifugal action type air separator and (b) shows a typical piping diagram.

TABLE 7–6

Properties of saturated steam
(condensed from Keenan and Keyes, *Thermodynamic Properties of Steam*, by permission of John Wiley & Sons, Inc.)

Gauge Pressure PSIG	Absolute Pressure, PSIA	Steam Temperature, °F	Heat of Saturated Liquid, Btu/lb	Latent Heat, Btu/lb	Total Heat of Steam, Btu/lb	Specific Volume, cu ft/lb
0.0	14.696	212.00	180.07	970.3	1150.4	26.80
1.3	16.0	216.32	184.42	967.6	1152.0	24.75
2.3	17.0	219.44	187.56	965.5	1153.1	23.39
5.3	20.0	227.96	196.16	960.1	1156.3	20.09
10.3	25.0	240.07	208.42	952.1	1160.6	16.30
15.3	30.0	250.33	218.82	945.3	1164.1	13.75
20.3	35.0	259.28	227.91	939.2	1167.1	11.90
25.3	40.0	267.25	236.03	933.7	1169.7	10.50
30.3	45.0	274.44	243.36	928.6	1172.0	9.40
40.3	55.0	287.07	256.30	919.6	1175.9	7.79
50.3	65.0	297.97	267.50	911.6	1179.1	6.66
60.3	75.0	307.60	277.43	904.5	1181.9	5.82
70.3	85.0	316.25	286.39	897.8	1184.2	5.17
80.3	95.0	324.12	294.56	891.7	1186.2	4.65
90.3	105.0	331.36	302.10	886.0	1188.1	4.23
100.0	114.7	337.90	308.80	880.0	1188.8	3.88

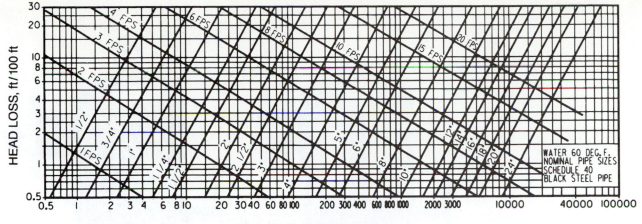

(a)

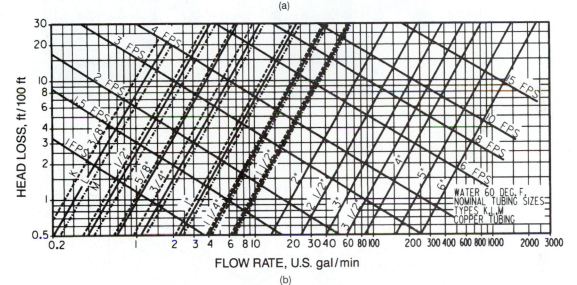

(b)

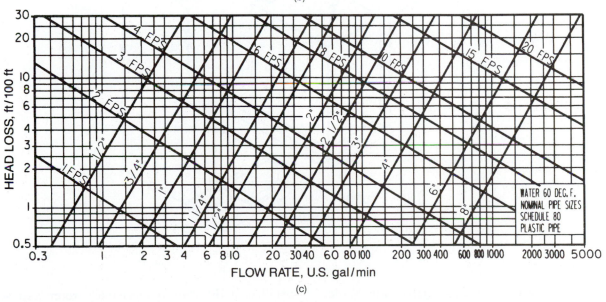

(c)

■ **FIGURE 7–11**

Friction loss for water in straight pipes (reprinted from *ASHRAE Handbook* with permission from ASHRAE). (a) For steel pipes (Schedule 40). (b) For copper tubing (Types K, L, M). (c) For plastic pipes (Schedule 80).

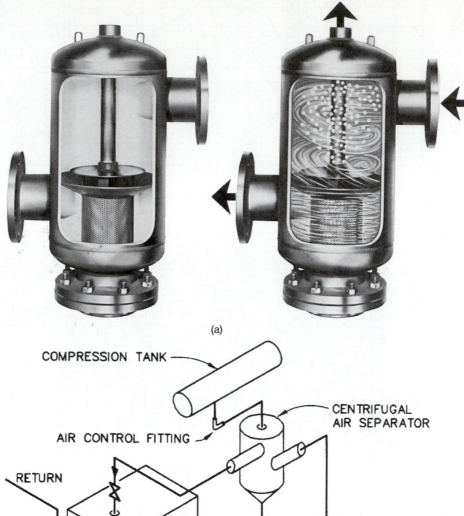

(a)

FIGURE 7-12
(a) A centrifugal action air-control device utilizing the whirlpool action to separate air and water. The lighter air vents from the top of the tank. (Courtesy: ITT Bell & Gossett, Morton Grove, IL.) (b) An isometric piping diagram showing the relative location of the air separator and the compression tank in a heating water system.

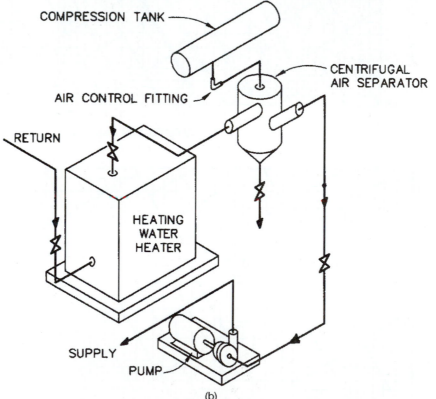

(b)

Air vent may be of the manual or automatic type. Manual vents require periodic maintenance to release air. Automatic vents reduce maintenance, but can be prone to nuisance water leakage. An air separator with an automatic air vent is normally installed at the central mechanical room. Air separators are designed to slow the water flow locally and allow entrained air to float upward for removal. Air separators are installed in the warmest portion of the piping system. This is the return side of chilled-water systems and the supply side of hot-water systems.

Specialty devices for water system In addition to valves and pipe fittings, there are a number of specialty devices unique to water systems. Some of these, illustrated in Figure 7–13, are as follows:

- Because water expands when heated, an *expansion tank* must be installed as a reservoir for the expanded water. The expansion tank is usually maintained by air pressure. In an open tank the air is on the top half of the tank, and in a diaphragm type of tank the air and water are separated from each other.
- A *compression tank* adds to the system pressure for positive water flow. Again, air is used as the pressurization medium.
- *Suction diffusers* are devices that allow a low-pressure drop entering the pump section.
- *Airtrol fittings* are devices that allow the air in a tank or equipment to escape, without allowing the water to leak out of the system.

Typical piping details Figure 7–14 illustrates several piping details of water system equipment. (Note the flexible hose connections at the pump inlet and outlet, as well as the air vents at top of the coils.) Manual air vents (MAVs) are shown. All coil and equipment connections are boiler piping, such as thermometers (M), pressure gauges (P) and flow control valves (M).

7.4.5 Steam-Piping Systems

The use of steam as a source of heat energy has gradually diminished, except for large buildings, campuses, hospitals, or industrial plants where steam plants are already in place. When used, steam is usually converted to hot water for final distribution. One of the limitations of steam systems is that the piping must maintain a correct slope so that the condensate can drain properly. Therefore, such systems require higher floor-to-floor space. One popular source of steam is the central or district supply, where steam is piped underground from remote steam-generating plants. The steam may be a by-product of an industrial process, nuclear power plant, or waste treatment plant. By providing district steam in the central distribution of a major city, the need for individual heating plants in each building is eliminated. This could result in considerable savings in construction costs and energy for the entire municipality.

Heating value of steam and condensate Steam is water in the vapor phase. Its heating value includes the heat of steam vapor (latent heat of vaporization) and the heat of liquid (sensible heat), with 32°F and 212°F the freezing and boiling points, respectively, of water. The total heat of vapor and liquid is called the total heat of steam. Table 7–6 lists the properties of steam and condensate at various pressures. Complete listings of properties of steam and liquid can be found in numerous engineering handbooks and the Keenan and Keyes steam tables. For preliminary design purpose, 1000 Btu/lb can be used as the total heat of one pound of steam at pressures from 0 to 15 psig and a condensate temperature of 200°F. Thus,

$$H = 1000 \times W \tag{7–9}$$

$$h = 1000 \times w \tag{7–10}$$

where H = total heating value of steam and condensate at 200°F Btu
W = mass of condensed steam, lb
h = total heat rate of steam and condensate, Btu/hr
w = steam rate, lb/hr.

As shown in the table, the specific volume of steam is about 20 cu ft/lb at 5 psig and decreases rapidly as pressure increases. The specific volume at 30 psig is 9.4 and at 100 psig is 3.88. It should be clear that a considerable reduction in the size of piping can be realized by using high-pressure steam. On the other hand, equipment and accessories required for high-pressure steam are considerably more costly, and the piping design is more complex. For HVAC applications, steam systems and their distribution are usually limited to 30 psig. When district steam is the source, it is usually reduced to lower pressure before distribution.

Pipe sizing

- *Steam supply* Piping is generally sized on the basis of flow requirements and maximum velocity, to prevent noise and erosion. For building heating systems, the velocity of steam should be between 8000 and 12,000 fpm for quiet operation. Data on the sizing of pipes are available in piping handbooks. If the allowable pressure drop is limited, larger pipe sizes might be required. Pipes should slope downward 1" in 40' in the direction of flow.
- *Condensate return* Piping from heating devices may flow by gravity in a system that is open to atmospheric pressure. A downward slope is required for piping toward the condensate tank. Closed, pumped condensate piping need not be

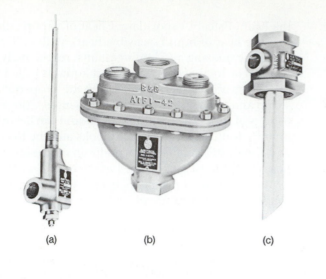

(a)   (b)   (c)

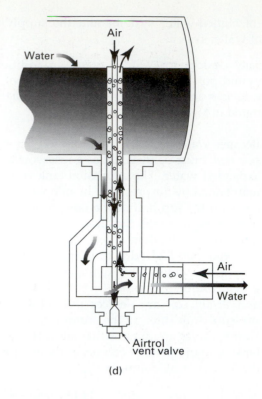

Water

Air

Air

Water

Airtrol
vent valve

(d)

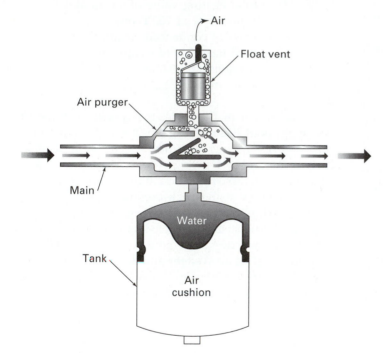

Air

Float vent

Air purger

Main

Water

Tank

Air
cushion

(e)

■ **FIGURE 7–13**

(a) thru (d) Illustrate several air control fittings and their operating principles. (a) Tank fitting with built-in vent tube. (b) Tank fitting with separate vent tube. (c) Boiler fitting. (d) Cutaway section of a tank fitting as in (a). (e) The diaphragm type tank assembly consists of an air purger to separate air from water and a float vent to bleed off air. (f) The diaphragm which separates air from water flexes to balance the air and water pressure. Courtesy: (a)–(d), ITT Bell & Gossett, Morton Grove, IL; (e) and (f), AMTROL, Inc., West Warwick, RI.

Air
cushion

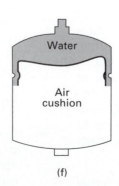

Water

Air
cushion

Water

Air
cushion

(f)

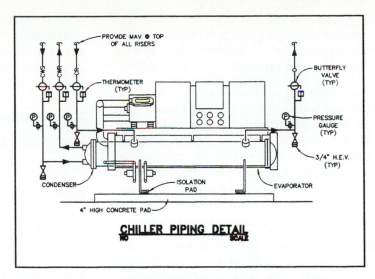

CHILLER PIPING DETAIL
NO SCALE

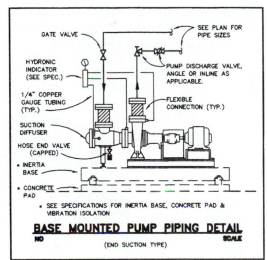

BASE MOUNTED PUMP PIPING DETAIL
NO SCALE
(END SUCTION TYPE)

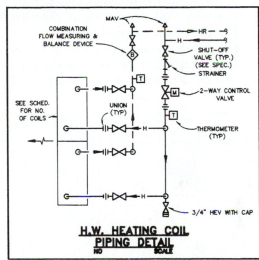

H.W. HEATING COIL PIPING DETAIL
NO SCALE

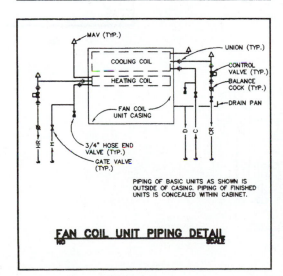

PIPING OF BASIC UNITS AS SHOWN IS OUTSIDE OF CASING. PIPING OF FINISHED UNITS IS CONCEALED WITHIN CABINET.

FAN COIL UNIT PIPING DETAIL
NO SCALE

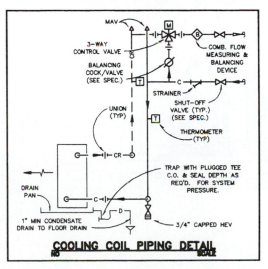

COOLING COIL PIPING DETAIL
NO SCALE

■ **FIGURE 7–14**

Typical piping details of equipment and coils.

sloped. Generally, for low-pressure steam systems, condensate piping will be half the diameter of the steam service upstream. Most designers attempt to maintain a pipe slope of 1" per 40'.

Steam traps   Steam traps are used to separate and remove condensate from steam piping and heat transfer devices. There are many designs, such as the inverted bucket, float, and thermostatic thermodynamic. Steam traps are selected on the basis of the condensate temperature, steam pressure, flow rate, and other properties. It is advisable to get recommendations from manufacturers for different applications. Figure 7–15 illustrates several steam traps.

Accessories   Steam-piping systems require considerably more accessories than water systems do. Piping must be properly pitched to avoid noise, vibration (steam hammer), and reduction in capacity. Illustrated in Figure 7–16 are typical connection details at coils, steam traps, pressure-reducing stations, and heaters. Note in particular the steam heating coil piping. The coil is divided into two sections, each of which consists of two traps to assure a uniform steam distribution. Pressure-reducing valves are used to reduce the steam pressure from up to 150 psig down to 50 psig for medium pressure-rated equipment and 5 to 10 psig for HVAC heating equipment.

7.4.6 Refrigerant Piping Systems

Refrigerants flow in three forms through refrigeration systems: cold suction gas, compressed hot gas, and liquid. Most refrigeration piping is copper or brass, with the notable exception of piping carrying ammonia, which is not often used in HVAC systems. Ammonia is low in cost and very efficient: however, it is toxic and has a strong smell. It is prohibited for use in building HVAC systems by many municipalities. However, recent concerns about depleting the ozone layer with CFC-type refrigerants have revived ammonia applications in central plant designs. Ammonia pipe is normally black steel.

Refrigerants   With the concern about ozone depletion and global warming, several of the popular

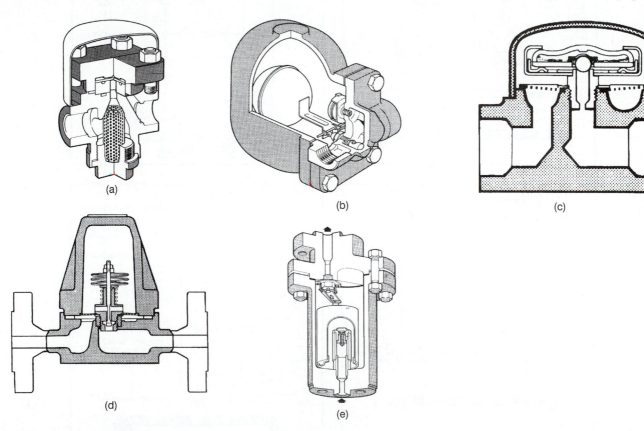

(a)

(b)

(c)

(d)

(e)

■ **FIGURE 7–15**
Several steam traps.
(a) Thermodynamic type. (b) Float and thermostatic types. (c) Balanced pressure type.
(d) Bimetallic type. (e) Inverted bucket type. (Courtesy: Spirax Sarco, Inc., Allentown, PA.)

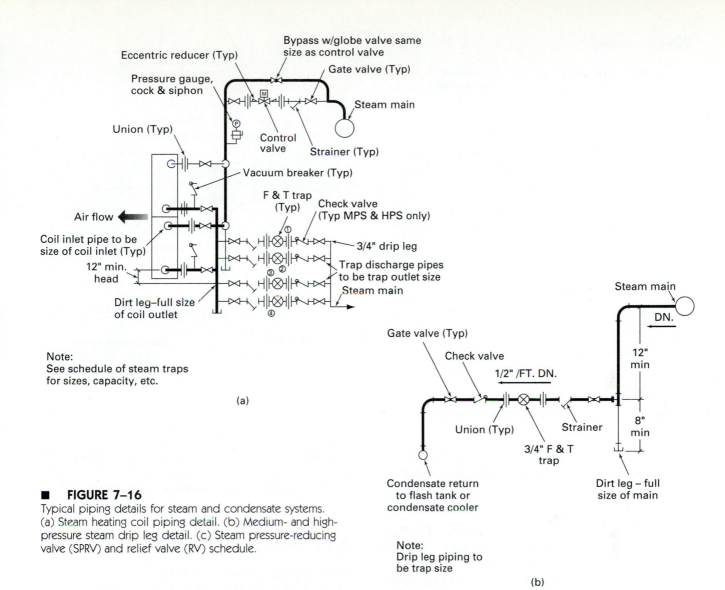

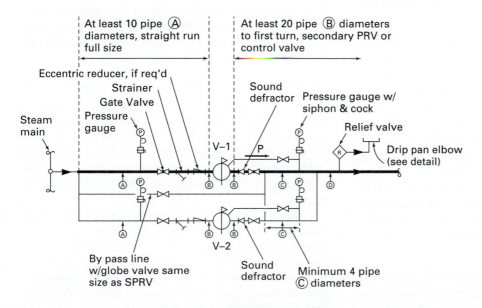

■ FIGURE 7–16

Typical piping details for steam and condensate systems.
(a) Steam heating coil piping detail. (b) Medium- and high-pressure steam drip leg detail. (c) Steam pressure-reducing valve (SPRV) and relief valve (RV) schedule.

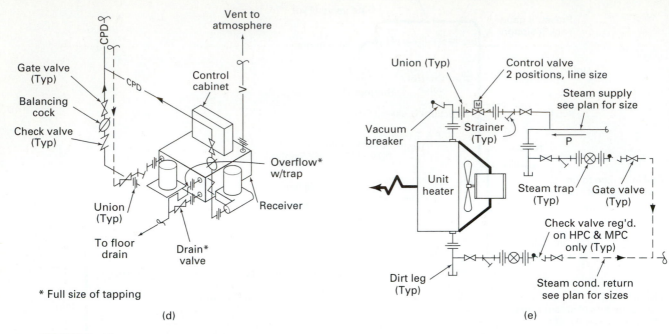

■ **FIGURE 7–16**
(d) Duplex condensate pump piping detail. (e) Steam unit heater piping detail.

refrigerants, such as freon nos. 11, 12, and 22 are being phased out. Newer types of refrigerants, such as HCFC-123 and HFC-134A, have gained acceptance. Table 7–7 is a comparison of their physical properties.

Thermodynamic properties of refrigerants   The thermodynamic properties of refrigerants vary widely. In general, it takes about 2.9 and 3.3 lb/min. of refrigerant to produce 1 ton (12,000 btu/hr) of refrigeration effect for the refrigerants CFC-22 and SUVA-123, respectively.

Pipe sizing criteria   Most refrigerant piping is contained within packaged equipment. Field-installed piping between system components is generally limited to the suction and liquid lines from the condensing unit to the coil or the hot gas and liquid lines from a unitary device to an outdoor condenser. Sizing is best left to manufacturers' recommendations. Losses due to friction are a critical criterion for preventing undue loss of compressor capacity. The velocity must be high enough for the flowing refrigerant gas to draw oil back to the compressor. When the flow is upward, a specially designed oil trap detail is generally used. Figure 7–17 is a DX coil piping riser connections.

Joints   In refrigerant piping, joints are generally soldered or brazed, although special screwed connectors are often used for connection to small residential and light commercial air-conditioning equipment.

Refrigeration piping diagram and accessories   The basic DX refrigeration system illustrated in Figure 7–17 shows the arrangement of components and accessories.

7.4.7 Other Piping Systems

Other piping systems in buildings include gas and oil piping for fuel-fired equipment such as water heaters, boilers, furnaces, etc. Gas and oil are combustible fuels. Leaks could be very hazardous. Building codes and utility regulations have strict rules governing the installation of these piping systems.

Gas Piping   Gas is usually measured by volume. The normal heating value for natural gas is approximately 1000 btu/cu ft. Heating values can also be in therms, which equal 100,000 Btu. The heating value may vary with the gas composition. The exact heating value should be obtained from the supplier. For preliminary design purposes, 1000 btu/cu ft can be assumed.

Black steel or type K copper pipes are used. Pipes should be exposed for ready inspection and repair. When installed above ceilings, pipes should be enclosed in a duct that is vented to the atmosphere, or they should have all welded joints. Liquid petroleum gas has been used for residences when natural gas was not available. Liquid petroleum gas is heavier than air and is generally odorless. Leaks will vapor-

TABLE 7–7

Physical properties of Du Pont SUVA™ and FREON® refrigerants (reprinted with permission from Du Pont Company).

Refrigerant	SUVA Centri-LP (HCFC-123)	FREON 11 (CFC-11)	SUVA Cold-MP (HFC-134a)	FREON 12 (CFC-12)
Chemical Name	2,2-dichloro, 1,1,1,trifluoro-ethane	trichloro-fluoro-methane	1,1,1,2-tetrafluoro-ethane	dichloro-difluoro-methane
Chemical Formula	$CHCl_2CF_3$	CCl_3F	CH_2FCF_3	CCl_2F_2
Vapor Pressure (psia) @ 25°C (77°F)	14	16	96	95
Boiling Point °C @ 1 atm	27.9	23.8	−26.5	−29.8
°F	82.2	74.9	−15.7	−21.6
Freezing Point °C	−107.0	−111.0	−101.0	−158.0
°F	−161.0	−168.0	−149.8	−252.0
Flammability Limits in Air vol%	None	None	None	None
Ozone Depletion Potential	0.02	1.0	0	1.0
Halocarbon Global-Warming Potential	0.02	1.0	0.26	2.8
Toxicity	100 (AEL)*	1000 (TLV)**	1000 (AEL)*	1000 (TLV)**

*AEL (allowable exposure limit) is a preliminary toxicity assessment established by Du Pont; additional studies may be required.

**TLV (threshold limit value) is a registered trademark of the American Conference of Government Industrial Hygienists.

ize, but the vapor tends to stay near the floor or migrates to the lower part of a building. It is extremely hazardous. Proper inspection and ventilation of the space where liquid petroleum gas pipes are located is essential. Figure 7–18 illustrates some equipment connection details, including many safety provisions.

Fuel oil piping Similar to gas piping, fuel oil piping is made of black steel or copper. The heating value of oil varies with the classification of the oil. No. 1 or No. 2 (diesel) are the lightest. No. 5 and No. 6 (bucket fuel) are the heaviest and require heating before they can be pumped in piping systems.

7.4.8 Thermal Expansion

Once filled, an HVAC piping system will be subject to changing temperatures and accompanying expansion and contraction of the water volume. Expansion must be compensated for to prevent damaging pressures; contraction must be compensated for to avoid negative pressures, which might draw air into the system. Expansion tanks are installed to provide high and low limits on the pressure.

Expansion tanks Tanks can be of the open gravity variety or closed and pressurized. Gravity tanks need to be placed at the very top of the system and are open to the atmosphere. As water in the system expands and contracts, the level in the expansion tank rises and falls. The surface exposed to the atmosphere is a source of dissolved air, which can cause corrosion and air binding.

Closed tanks need not be placed at the high point of the system. They can be constructed with the water surface exposed to a charge of air, or they can be separated from water by a rubber bellows. Bellows-type expansion tanks are preferred, to avoid problems with dissolved air.

Piping expansion Heating system piping must operate over a wide variety of temperatures, and thermal expansion must be accommodated. Good design employs anchorage to firm structures at regular intervals and provisions for expansion between anchors. Expansion compensation devices include bellows fittings and concentric slip fittings. Slip fittings require seals and packing, which need maintenance to prevent leakage.

Elbows, Z-shaped and U-shaped expansion loops can also be designed to use the inherent flexibility of the piping materials to absorb expansion. Elbows and expansion loops are generally the least expensive method for accommodating expansion, and they require the least maintenance, but they also require supports designed for lateral movement of piping and clearance for this movement. Care is also required to size expansion legs adequately to prevent undue flexure stress.

7.4.9 Insulation

Purpose of insulation Insulation is desirable to prevent thermal losses, reduce the hazard of touching hot piping surfaces, and avoid condensation on cold

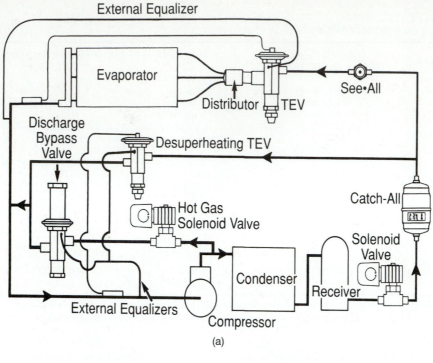

(a)

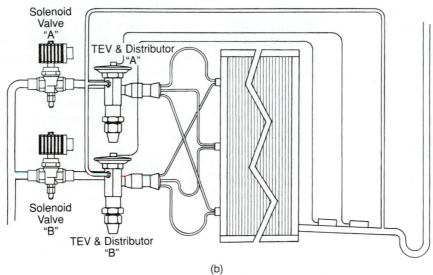

(b)

■ **FIGURE 7–17**

(a) Typical DX system diagram showing major components. (b) One method of capacity control of DX system is to unload the compressor cylinders and to install the corresponding number of solenoid valves (SV) and thermal expansion valves (TEV) to control the evaporator coils. This diagram shows a single evaporator coil with two SVs and two TEVs which can be controlled in sequence resulting in capacity reduction. (Courtesy: Sporlan Valve Company, St. Louis, MO.)

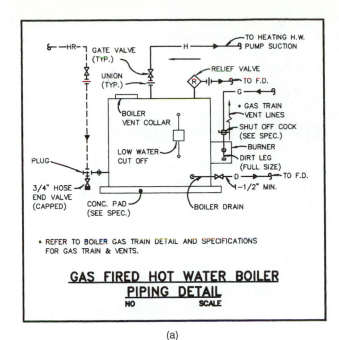

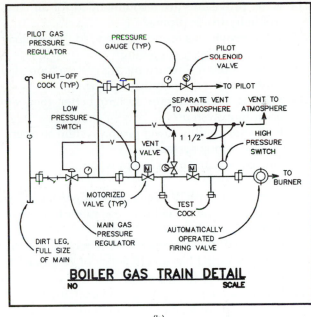

(a)   (b)

■ **FIGURE 7–18**
(a) Gas piping connection of a gas-fired boiler and (b) gas train details.

piping. Commonly used materials for piping insulation include calcium silicate, foam glass, fiberglass, and foam rubber. Calcium silicate and foam glass are durable, rigid materials that are suitable for hot or cold piping applications. Fiberglass is less durable, but equally suitable for the range of temperatures encountered in HVAC systems.

Insulating materials Most insulating materials require a protective jacket. Materials for the jacket include fabric wrap applied with wet plaster, thin plastic casing with taped joints, and sheet aluminum. Plastic and aluminum function as barriers to vapor, to prevent condensation on the surface of cold piping. For small-diameter piping, foam rubber insulation can be used without a jacket, but it should be avoided in high temperatures and in locations subject to mechanical abuse. Refer to the manufacturer's data on the properties of insulation material.

QUESTIONS

7.1 What factors are considered in selecting the size of pipe for fluid flow applications?

7.2 What feature most differentiates turbine pumps from centrifugal pumps? How are turbine pumps applied in HVAC systems?

7.3 What feature most differentiates positive displacement pumps from centrifugal pumps?

How are positive displacement pumps applied in HVAC systems?

7.4 For a given water piping system, a pressure of 50 ft of head is required to produce 20 GPM water flow. How much pressure will be required to flow 40 GPM?

7.5 Compare the differences in flow developed for two identical pumps operating in parallel vs. operating in series.

7.6 What are the main advantages and disadvantages of plastic piping in comparison with steel?

7.7 What factors limit maximum velocities in piping systems? Minimum velocities?

7.8 What are the main differences in the characteristics of calcium silicate, fiberglass, and foam rubber insulation that would affect their application in piping systems.

7.9 What is the functional difference between a control valve and a balance valve?

7.10 Under what circumstances would three-way control valves be used rather than two-way control valves, despite the higher expense of three-way valves?

7.11 What is the purpose of a check valve?

7.12 What are some limitations and disadvantages of open expansion tanks for hot-water systems?

7.13 What are the advantages and disadvantages of expansion loops in comparison with slip joints compensating for thermal expansion?

7.14 What provisions are recommended for air control in hydronic piping?

7.15 What provisions are recommended for condensate control in steam piping?

7.16 Why is a reverse return piping system more desirable than a direct return system?

7.17 Why is a closed-loop system better than an open-loop system?

7.18 Is a water-cooling tower piping system an open- or closed-loop system?

7.19 One psi is equal to 34 feet of water. (True) (False)

7.20 Calculate the WHP (water horsepower) of a water-piping system with a flow rate of 50 GPM and 20 feet of head.

7.21 What BHP brake horsepower is required if the pump selected has an efficiency of 70% operating as specified in question 7.20?

7.22 What are the two types of fluid exchangers?

7.23 Which type of heat exchanger is ideal for a low temperature differential?

7.24 Grooved-pipe mechanical joints can be used for steel pipe only. (True) (False)

7.25 The advantage of grooved-pipe mechanical joints is their extremely strong coupling. (True) (False)

7.26 Soldered joints are used for type K copper. (True) (False)

7.27 What is the equivalent straight pipe length of a 2″ screwed globe valve? A 2″ gate valve?

7.28 Why is it necessary to have air venting in a hot-water system?

7.29 What is the latent heat of steam at 10.3 psig?

7.30 The total heat of steam at 50 psig (including the sensible heat of condensate at 200°F and the latent heat of vapor) is _____.

7.31 What is the absolute pressure (psia) at 70.3 psig?

7.32 What is the temperature of saturated steam at 85 psia?

7.33 Name one or more refrigerant that have low potential for depleting ozone?

7.34 What refrigerant has the least potential for contributing to global warming?

7.35 How many pounds of refrigeration per minute are needed to produce 1 ton of refrigeration?

8

PLUMBING EQUIPMENT AND SYSTEMS

BROADLY DEFINED, PLUMBING SYSTEMS COVER ALL aspects of water and drainage systems exterior to and within buildings. The scope of plumbing systems includes the following:

- Water supply, distribution, treatment, quality, and temperature conditioning
- Selection and installation of plumbing fixtures and drainage devices
- Waste and soil collection, and treatment and disposal of drainage
- Collection, retention, and disposal of storm water
- Building equipment provisions, including water and waste for HVAC, food service, pools, fountains, processing, etc.

The design of each system or subsystem requires a voluminous data base. The data can be found in appropriate handbooks published by technical and engineering organizations. Thus, this chapter will not duplicate the available data, but rather, will present sample data sufficient for illustrating the design methodology. Some leading organizations that maintain a plumbing data base are:

- American Society of Plumbing Engineers (ASPE)
- American National Standards Institute (ANSI)
- American Society of Heating, Refrigerating and Air Conditioning Engineers (ASHRAE)

- American Society of Sanitary Engineers (ASSE)
- American Society for Testing Materials (ASTM)
- American Water Works Association (AWWA)
- Building Officials and Code Administration (BOCA)
- Basic National Plumbing Code (BOCA/NPE)
- Cast Iron Soil Pipe Institute (CISPI)
- National Standard Plumbing Code (NSPC)
- Southeast Building Code Conference (SBCC)
- Uniform Plumbing Code (UPC)

Since plumbing systems and installations directly affect personal hygiene and public health, plumbing codes are quite specific regarding the design and installation of sanitary facilities and their piping systems. As a rule, all applicable codes specify the minimum number of plumbing fixtures that must be provided in a building and the specifics of piping connections. These specific code requirements have greatly simplified the design of a sanitary piping system, to the extent that detailed engineering calculations are seldom necessary. However, the simplified process cannot be applied to the design of water systems, which depend strictly on the design engineer's decisions. The method of dealing with water-piping systems will be addressed in this chapter.

Graphic symbols are used extensively in illustrations of plumbing systems to explain their operational principles and construction details. Commonly used graphic symbols are shown in Figure 8–1.

209

Commonly used plumbing and piping symbols

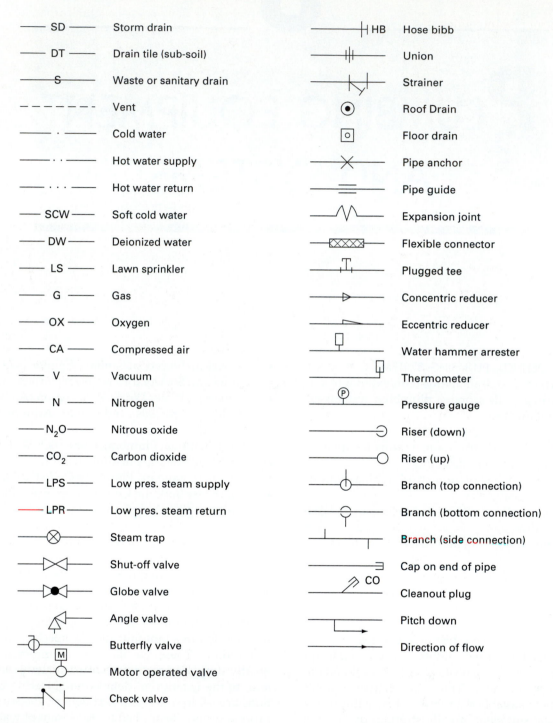

—— SD ——	Storm drain
—— DT ——	Drain tile (sub-soil)
—— S ——	Waste or sanitary drain
– – – – – – –	Vent
—— · ——	Cold water
—— · · ——	Hot water supply
—— · · · ——	Hot water return
—— SCW ——	Soft cold water
—— DW ——	Deionized water
—— LS ——	Lawn sprinkler
—— G ——	Gas
—— OX ——	Oxygen
—— CA ——	Compressed air
—— V ——	Vacuum
—— N ——	Nitrogen
—— N₂O ——	Nitrous oxide
—— CO₂ ——	Carbon dioxide
—— LPS ——	Low pres. steam supply
—— LPR ——	Low pres. steam return
⊗	Steam trap
⊲⊳	Shut-off valve
Globe valve	Globe valve
Angle valve	Angle valve
Butterfly valve	Butterfly valve
Motor operated valve	Motor operated valve
Check valve	Check valve

HB	Hose bibb
Union	Union
Strainer	Strainer
⊙	Roof Drain
▫	Floor drain
✕	Pipe anchor
=	Pipe guide
⋀	Expansion joint
▨	Flexible connector
Plugged tee	Plugged tee
▷	Concentric reducer
Eccentric reducer	Eccentric reducer
Water hammer arrester	Water hammer arrester
Thermometer	Thermometer
Ⓟ	Pressure gauge
Riser (down)	Riser (down)
Riser (up)	Riser (up)
Branch (top connection)	Branch (top connection)
Branch (bottom connection)	Branch (bottom connection)
Branch (side connection)	Branch (side connection)
Cap on end of pipe	Cap on end of pipe
CO	Cleanout plug
Pitch down	Pitch down
→	Direction of flow

■ **FIGURE 8–1**
Graphic symbols commonly used in illustrating plumbing and piping installation.

8.1 WATER SUPPLY AND TREATMENT

Providing water is among the most critical services in a modern building. Without water, a building is not suitable for human occupation or most industrial operations. Water is also needed for fire protection.

Generally, water of potable quality is supplied from a public water system. For buildings remote from the public water system, an alternative source of water must be found, such as a well or a lake. Water supplied to a building for potable use must meet the quality standards prescribed by the governing public health agencies. If the quality of the water does not meet these standards, then physical and chemical treatment of the source is needed.

Units of measure frequently used to describe properties relating to water and its quality are as follows:

- 1 liter = 1000 cubic centimeters (cc)
- 1 pound (lb) = 454 grams (gm)
- 1 kilogram (kg) = 1000 grams (gms) = 1,000,000 milligrams (mg)
- 1 lb = 7000 grains (gr)
- 1 grain = 0.0648 gram (gms)
- 1 gallon (gal) of water weighs 8.33 pounds (lb)
- 1 cubic centimeter (cc) of water weighs 1 gram (gm)
- 1 cubic foot (cu ft) of water = 28.3 liters (L)
- 1 cu ft of water ≈ 7.5 gal
- 1 cu ft of water at room temperature weighs 62.4 lb
- 1 liter of water at room temperature weighs 1000 gm (1 kg)
- 1 part per million (ppm) of chemical contents in water = 1 mg/L = 1 mg/kg
- 1 grain per gallon (gpg) = 17.1 ppm

8.1.1 Water Quality

For use in a building, the water supply must meet a minimum level of quality based on several major characteristics:

1. *Physical characteristics.* The water supply must contain only a limited amount of suspended material in terms of turbidity, clarity, color, acceptable taste, odor, and temperature. To qualify for drinking, the water supply should be lower than 5 turbidity units (TU).
2. *Chemical characteristics.* The water supply must contain no more than the maximum content prescribed by health standards pertaining to hardness

and dissolved matters, such as minerals and metals. The preferred hardness of a water supply is lower than 200 ppm of calcium carbonates.
3. *Biological and radiological characteristics.* The water supply should be practically free of bacteria, viruses, and radioactive materials.

Upon request, the public water utility will provide a typical analysis of a building's water supply. The utility is responsible for treating the water to meet the quality standards of the local health department. Typical public water supply characteristics in the United States are shown in Table 8–1. Plainly, the chemical and physical properties of water vary considerably among geographical regions. Nonetheless, to qualify as suitable for drinking from the tap without further treatment or filtration, water must be better in quality than the minimum standards set forth by the Environmental Protection Agency and local health departments. The minimum standards shown in Table 8–1 are consolidated values in the United States. Standards may differ in other countries due to the unique properties of their water sources and local life-styles.

8.1.2 Processes Commonly Used to Improve Water Quality

Depending on the initial quality of water and the purpose for which the water will be used in a building, one or more of the following processes improve the quality of the water:

1. *Sedimentation.* Allows suspended matter to settle out of water by precipitation. Improves the turbidity of the water.
2. *Coagulation.* Removes suspended matter from water by means of chemicals, such as alum (hydrated aluminum sulfate), to improve the turbidity, color, and taste of the water.
3. *Aeration.* Introduces air into water to improve its taste and color.
4. *Filtration.* Removes suspended material and bacteria from water to improve its overall quality, including its turbidity, potability, color, and taste.
5. *Disinfection.* Uses chemicals, such as chlorine gas, hyperchloride solids, and ultraviolet light, to disinfect whatever bacteria remain in water after previous treatment.
6. *Fluoridation.* Adds a fluoride chemical in water, primarily to prevent tooth decay.
7. *Softening.* Replaces the calcium and magnesium ions with sodium ions to reduce scale formation in

TABLE 8–1

Analysis of typical water supplies and minimum quality standards

| Substance | Chemical Symbol | Water Samples* | | | | | | Typical Quality Standard |
		(1)	(2)	(3)	(4)	(5)	(6)	(7)
Silica	SiO_2	2	12	37	10	22	—	—
Iron	Fe_2	0	0	1	0	0	—	0.3
Calcium	Ca	6	36	62	92	3	400	—
Magnesium	Mg	1	8	18	34	2	1300	—
Sodium	Na	2	7	44	8	215	11,000	—
Potassium	K	1	1		1	10	400	—
Bicarbonate	HCO_3	14	119	202	339	549	150	—
Sulfate	SO_4	10	22	135	84	11	2700	250
Chloride	Cl	2	13	13	10	22	19,000	250
Nitrate	NO_3	1	0	2	13	1	—	10
Dissolved Solids	—	31	165	426	434	564	35,000	500
Carbonate Hardness	$CaCO_3$	12	98	165	287	8	125	—
Noncarbonate Hardness	$CaSO_4$	5	18	40	58	0	5900	—

Numbers indicate location or area, as follows:
(1) Catskill supply, New York City
(2) Niagara River (filtered), Niagara Falls, NY
(3) Missouri River (untreated), average
(4) Well waters, public supply, Dayton, OH, 30 to 60 ft.
(5) Well water, Smithfield, VA, 330 ft.
(6) Ocean water, average
(7) Minimum quality standards for drinking water based on the composite values of Environmental Protection Agency (EPA) and health departments in the U.S.A.

*Condensed from Table 3, Chapter 43, Application Volume, *ASHRAE Handbook*, 1991.

**All units are in ppm or mg/L, rounded to the nearest whole number.

---

the hot-water piping system and to improve sudsing for doing laundry.

8.1.3 Water Conditioning

When the quality of water does not meet the minimum standards for which the water is to be used, the water must be conditioned by the various treatment processes on the site, either within or outside of the building. The treatment may be applied to the entire water source or only to a portion of the water supply. For example, even if the water supply source to a hospital is good enough for general use, a portion of the water may need to be softened for laundry use, filtered for food preparation, deionized for laboratory use, and distilled for use in research. Figure 8–2 is a schematic diagram of water treatment systems for a research hospital.

Water softening system

Water from rain or snow usually contains no minerals. However, after flowing through mountains or rivers, it picks up dissolved minerals such as calcium or magnesium salts in the form of carbonates or sulfates. When the water is heated, carbonates easily drop from the solution, forming hard scale. This scale can reduce the heat transfer efficiency and the net cross-sectional area of the pipe, as well as corrode the equipment. Carbonates in water are termed temporary hardness. When they are in excess of 200 ppm, they should be removed from the water.

The most popular water-softening process is the Zeolite system. In this system, water is softened by an ion exchange process in which calcium or magnesium ions are replaced by sodium ions. The softener consists of vertical tanks filled with small beads of exchange resin saturated with sodium cations (positive ions). The resin preferentially removes calcium and magnesium cations from the water while releasing sodium cations to the water. Thus, the water becomes softened. When the resin beads are saturated with calcium or magnesium cations, they are replenished with sodium cations of brine through the regenerative cycle. Figure 8–3 illustrates the construction of typical water softener tanks. While the Zeolite system effectively reduces the hardness of the water by replacing calcium or magnesium with sodium, it does not reduce the total dissolved minerals in the water.

Water purification systems

A *deionizing system* is similar to the Zeolite softener system, except that the tank or tanks contain a mixed bed of both cation- and anion-absorbing resins made of porous polymer minerals. The deionizing system

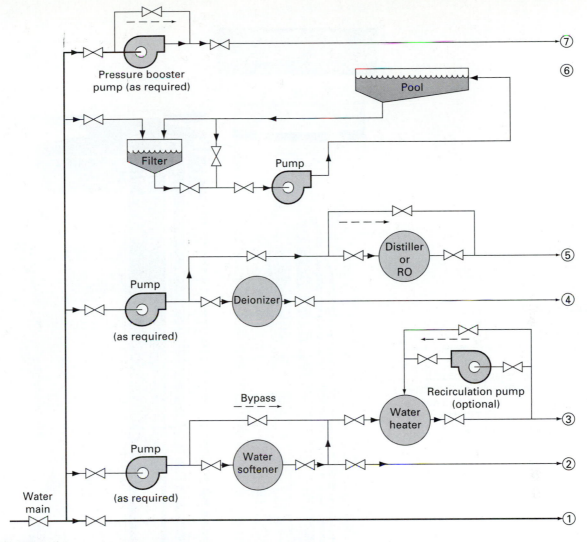

■ FIGURE 8–2

Schematic diagram of a water distribution system for a large hospital.

① Cold-water supply to lower floors for general-purpose use
② Soft cold-water supply for building heating system
③ Soft domestic hot water for laundry and plumbing fixtures
④ Deionized cold water to laboratory
⑤ Distilled or reverse osmosis (RO) water for laboratory testing services
⑥ Filtered water for swimming pool, therapeutic pool, etc.
⑦ Cold-water supply to upper floors, with pressure booster pump as required to overcome static height and frictional losses

can produce water that approaches the theoretical limits of purity. The system reduces both the mineral content and the hardness of water. Figure 8–4 illustrates the operating principle of deionizers.

A *reverse osmosis* (RO) system is another effective system used to purify water and reduce its hardness. The system operates on the principle of diffusion rather than ion exchange. Osmosis is a natural phe-

nomenon that occurs when water solutions of different concentrations are separated by a semipermeable membrane. Water tends to flow from lower concentrations to higher concentrations. RO works on the principle of reversing water flow by applying high external pressure (200 to 400 psi) to the side with the higher concentration (the more impure side) to force pure water to flow into the side with the lower con-

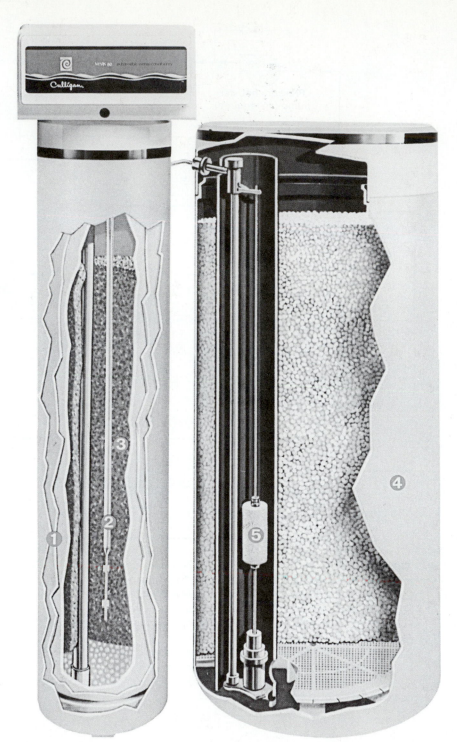

■ **FIGURE 8–3**

Typical construction of a Zeolite water softener showing: (1) a plastic lined steel tank, (2) immersed sensor to monitor the Zeolite solution and to signal the controller when recharging is needed, (3) Zeolite resin in the tank, (4) brine tank to regenerate the Zeolite and (5) a refill system with float shut-off. (Courtesy: Culligan Company, Northbrook, IL.)

centration (the more pure side). As a result, water downstream of the membrane is highly purified. The RO system is used to obtain highly purified water that is normally associated with research or manufacturing and is generally considered too costly for general building use. Figure 8–5 illustrates the principles of the system.

A *distillation system* is a traditional method of obtaining highly purified water. The water is heated to water vapor and condensed into highly purified water that is practically free of impurities. Distillation systems are used in research, hospitals, and manufacturing. However, they are less energy effective than the other two types of purification systems.

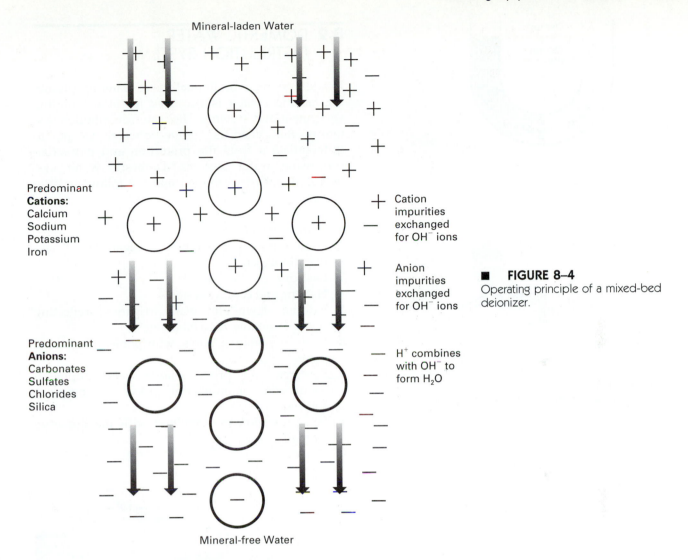

Mineral-laden Water

Predominant
Cations:
Calcium
Sodium
Potassium
Iron

Predominant
Anions:
Carbonates
Sulfates
Chlorides
Silica

Cation
impurities
exchanged
for OH^- ions

Anion
impurities
exchanged
for OH^- ions

H^+ combines
with OH^- to
form H_2O

Mineral-free Water

■ **FIGURE 8–4**
Operating principle of a mixed-bed
deionizer.

Water filtration systems

Water filtration systems differ from water purification systems in that they only remove (filter out) the undissolved matter, such as dirt, suspended matter, and debris, and will not alter the dissolved matters in water, such as chlorine, calcium carbonates, and salt. Filtration systems are used primarily for large bodies of water, such as swimming pools, reflection pools, and fountains. They are also used for drinking water systems if the water turbidity is unacceptable.

There are two basic filtration systems for large bodies of water:

1. *Sand filters* can be of either the pressure or gravity variety. The main equipment consists of one or more large tanks containing graduated layers of sand and circulating pumps to circulate the water through the layers. The water passes through the layers, but the solids are trapped in them. When the layers are clogged with trapped solids, the pump reverses its direction of circulation and washes out the solids. The reversed flow process is called the backwash cycle. Figure 8–6 shows the construction of a typical pressurized-sand filtering system. Figure 8–7 illustrates the operation of the filtration and backwashing cycles.

2. *Diatomaceous earth* (silica, or SiO_4) is a fine mineral powder found in nature. In the filtering process, water is again pumped through the filter bed to trap the solids, and the cycle is reversed when backwash is needed. The diatomaceous earth filtering system is more compact than sand filters and requires less physical space, but is more costly to operate.

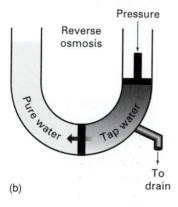

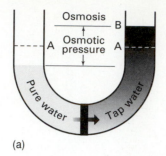

(a)

(b)

■ FIGURE 8–5
Operating principles of (a) osmosis and (b) reverse osmosis. (Courtesy: American Society of Plumbing Engineers.)

8.2 DOMESTIC WATER DISTRIBUTION SYSTEMS

In addition to the traditional use of water for plumbing facilities, water is also used for other building equipment and systems. The water distribution system is often called the "domestic water system," to differentiate it from fire protection and industrial-processing water systems. Domestic water system loads may be grouped into the following categories:

1. Plumbing facilities
2. Food service—preparation, refrigeration, washing, dining, etc.
3. Laundry
4. Heating and cooling systems
5. Exterior—lawn and plant irrigation, reflecting pools, fountains, hose bibbs, etc.
6. Pools—swimming pools, whirlpools, therapeutic pools
7. Research and process—laboratory equipment, commercial or industrial processes, computer equipment
8. Fire protection (if combined with the domestic system)
9. Others

■ FIGURE 8–6
Construction of typical water softener tanks. The cutaway sections illustrate the following:
1. Layers of minerals with lightweight chips at top to retain large, flat debris. Second layer collects coarse particles, third layer removes finer particles, and bottom layer polishes water down to micron-size insolubles.
2. Timer controller assembly.
3. Automatic valve system to accomplish service, backwash, and rinse cycles.
4. Pressure-rated steel tanks.
5. Conical distributor system for small systems.
6. Hub-radial distributor for large systems.
(Courtesy of Culligan Co., Northbrook, IL.)

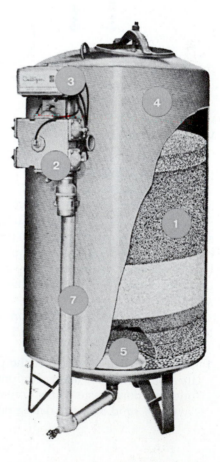

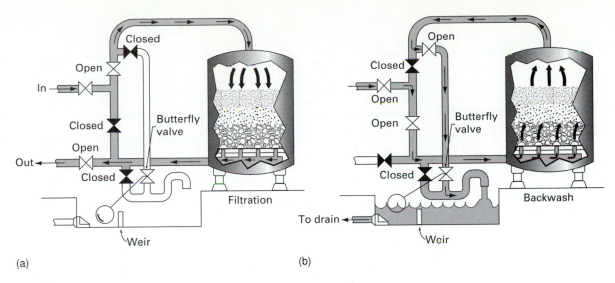

Water required for all of these categories must be accounted for in the design of a domestic water system. In general, water demand for plumbing usage is the highest in most building occupancies except some industrial plants where water is used in the production process. Accordingly, this chapter will concentrate on the design of plumbing facilities.

Units of measure commonly associated with water distribution systems are:

- 1 gallon per minute (GPM) = 0.063 liters per second (L/s)
- 1 liter per second (L/s) = 15.87 GPM
- 1 foot per minute (fpm) = 5.08 millimeters per second (mm/s)
- 1 foot per second (fps) = 0.305 meter per second (m/s)
- 1 cubic foot per minute (CFM) = 0.472 (L/s)
- 1 Liter per second (L/s) = 2.12 CFM
- 1 pound per square inch (psi) = 6.9 kilopascal (kPa)
- 1 psi = 2.04 inches of mercury column (in. Hg)
- 1 psi = 2.31 feet of water column (w.c.)
- 1 atmosphere (atm) = 14.7 psi = 101.4 (kPa)
- 1 atmosphere (atm) = 30 in. Hg
- 1 water supply fixture unit (wsfu) ≈ a numerical weighing factor to account for the water demand of various plumbing fixtures, using the lavatory as 1 wsfu. (1 to 1.5 GPM water flow rate)
- 1 drainage fixture unit (dfu) ≈ 0.5 GPM of drainage flow rate)

Most water distribution system materials, fittings, control devices, and pumps are similar to those used in other mechanical systems. Basic information on the selection of pipes, valves, controls, and pumps is covered in detail in Chapter 7 and will not be repeated in the current chapter.

8.2.1 Determination of Domestic Water System Load

The required water capacity of a building depends on the coincidental peak load demand (CPLD) of all load categories, based on an assumed time of day in the heavy-demand season. For example, the highest CPLD for an office building would be noontime in the summer, when the building is fully occupied, plumbing facilities are in heavy use, and air conditioning is near its peak. The highest CPLD for an apartment building would be around dinner time in the summer, when most people are home taking showers, washing, and preparing meals. Table 8–2 provides a convenient form for tabulating the estimated loads of various categories and the method used to determine the gross system capacity required for the building.

Plumbing facilities

Water demand for plumbing facilities depends on the number and type of fixtures actually installed. Section 8.3 provides data on the various loads of different types of plumbing fixtures, and Section 8.4 provides data on the minimum number of fixtures required by codes. In practice, the actual number of plumbing fixtures in modern buildings is usually in excess of the minimum required by codes, particularly in public assembly facilities and high-rise office buildings.

TABLE 8–2

Determination of demand load for domestic cold-water systems[a]

Connected Load			Net Load Demand of Each Category[b]		Coincidental Peak Demand[c] (ep)	Remarks
No.	Load Category	Description	in (wsfu)	in (ep)		
1.	Plumbing Facilities					
2.	Food Service					
3.	Laundry					
4.	HVAC Systems					
5.	Exteriors					
6.	Pools					
7.	Data Processing					
8.	Research/Process					
9.	Fire Protection					
10.	Other ( )					

Coincidental peak demand --- _____ ep
Spare capacity anticipated --- _____ ep
Gross system capacity required -------------------------------------- _____ ep
Net system capacity installed [d] ----------------------------------- _____ ep

[a]Domestic cold-water demand of each category includes the cold water required to generate hot water.

[b]Net load demand for plumbing facilities is based on Tables 8–3A and 8–3B; demand for other categories is determined from code and design references or the inputs of user/consultants.

[c]The coincidental peak demand is based on the selected season and time of day when peak water demand is likely to occur.

[d]Explain if different than gross system capacity.

Each plumbing fixture is assigned a wsfu rating, representing the relative water demand for its intended operating functions. For example, a lavatory that does not demand a heavy flow of water is given a wsfu of 1, and a flush valve–operated water closet that demands a heavy flow of water (even for only a few seconds) is given a wsfu of 10. The wsfu values of all other fixtures are shown in Table 8–6. For special plumbing equipment that does not carry a wsfu rating, a wsfu of 1 may be used for each GPM of flow rate.

It is recognized that in most installations, only a fraction of the total number of connected plumbing fixtures is expected to be in use concurrently. The portion of these fixtures that could be concurrently active is a function of the occupancy of the building, the pattern of use of the fixtures and other demographic parameters.

An early study by Roy B. Hunter, of the National Institute of Standards and Technology (formerly the National Bureau of Standards), established a statistical base for estimating the peak demand flow rate of plumbing systems having mostly intermittent (noncontinuous) flows. These peak demand values are shown in Table 8–3(a). The values are conservative, but are useful for sizing risers or branches of a water distribution system. Recent studies have resulted in a considerably lower set of *maximum probable flow values* for residential, light commercial, motel, and similar services. These values are shown in Table 8–3(b).

Example 8–1   If an office building is installed with 10 water closets, 4 urinals, and 8 lavatories, what is the total wsfu installed?

Based on Table 8–6, we have the following calculations:

		wsfu
Water closets (flushometer)	10 × 10	100
Urinals, blowout (flushometer)	4 × 5	20
Lavatories	8 × 1	8
Total plumbing fixtures installed		**128 wsfu**

Note: According to most building codes, flushometer valves must be used in public buildings, such as offices.

Example 8–2   For the office building in Example 8–1, what is the estimated demand for cold water, in GPM?

Based on Table 8–3(a) (office building), the estimated demand for cold water for these installed

TABLES 8–3

Estimation of probable demand for water

(a) For building occupancies other than residential, light commercial, motel and similar services, and for sizing individual risers of a multi-riser water distribution system

For predominantly flush tanks		For predominantly flush valves		For either tank or valve systems	
wsfu	Demand GPM	wsfu	Demand GPM	wsfu	Demand GPM
1 to 6	5			1000	208
10	8	10	27	1250	240
15	11	15	31	1500	270
20	14	20	35	1750	300
25	17	25	38	2000	320
30	20	30	41	2250	350
40	25	40	47	2500	375
50	29	50	52	2750	400
80	38	80	62	3000	430
100	44	100	68	4000	525
160	57	160	83	5000	600
200	65	200	92	6000	645
250	75	250	101	7000	685
300	85	300	110	8000	720
400	105	400	126	9000	745
500	125	500	142	10,000	770
750	170	750	178		

(b) For residential, light commercial, motel and similar services

For predominantly flush tanks		For predominantly flushometer		For either tank or flushometer systems[a]	
wsfu	Demand GPM	wsfu	Demand GPM	wsfu	Demand GPM
up to 6	8	up to 6	30	1000	126
10	12	10	35	1250	152
15	13	15	40	1500	175
20	13	20	44	1750	200
25	14	25	48	2000	220
30	15	30	50	2250	240
40	16	40	55	2500	260
50	17	50	59	2750	282
80	21	80	68	3000	300
100	24	100	72	4000	352
150	31	150	79	5000	395
200	35	200	85	6000	425
250	42	250	91	7000	445
300	48	300	96	8000	456
400	60	400	102	9000	461
500	71	500	107	10,000	462
750	100	750	118		

[a]For large public assemblies, such as a concert hall, sports arena, or stadium, the demand for water would be greater due to the pattern of use of the facilities. In such applications, Table 8–3(a) should be used.

plumbing facilities is between 68 GPM for 100 wsfu and 83 GPM for 160 wsfu. By interpolation, the demand for cold water for 128 wsfu is about 75 GPM.

Food services

Water demand for food services varies considerably between residential and commercial equipment. In general, food preparation and cooking do not require much water. The major demand for water is for washing in sinks or dishwashers. The use of water for washing in sinks has been accounted for in the plumbing fixture units. Thus, the only load needed to be added is for the dishwashers, which require from 10 to 15 gallons per wash for residential units. In estimating the demand load, 2 to 3 GPM per machine may be used. The demand load for commercial machines could be 5 to 10 times greater. Data on water demand should be obtained from the product manufacturer.

Laundry services

Water demand for laundry again varies between residential and commercial equipment. Residential clothes washers require 20 to 40 gallons of water per wash, depending on the size of the machine and design of the wash cycle. In estimating water demand, 4 to 6 GPM per machine may be used. The demand load for commercial laundry should be provided by the laundry manufacturer or a consultant.

Heating and cooling systems

Normally, heating and cooling systems are closed-circuit systems that do not require constant water makeup. One exception to this rule is water-cooling towers for condenser water. The amount of water used to make up the evaporation and drift losses in cooling towers is huge, since water is in continuous demand as long as the tower is in operation. In general, 3 to 4 GPM of condensing water is required for every refrigeration ton of cooling load. A 3-percent makeup rate will require about 0.1 GPM per ton, or about 6 gallons per ton-hour. This flow rate may not seem significant, but the consumption on an annual basis is astonishing.

Example 8–3 If the office building in Example 8–1 has a gross floor area of 25,000 sq ft and is installed with a 100-ton chiller, what are the water demands for the cooling system and the annual water consumption if the system operates 12 hours a day for 200 days?

Assuming that the circulating rate of the condensing water is 3 GPM/ton with a 4-percent water makeup, the demand (continuous) and annual water consumption will be as follows:

- Water demand: 100 tons × (3 GPM/ton × 4%) = 12 GPM
- Annual consumption: 12 GPM × 60 min/hr × 12 hr/day × 200 days/year = 1,728,000 gallons

Exteriors

Water usage for building exteriors depends on the size of the lot and the portion that is landscaped. In some luxury residences, campus-type institutions, or country clubs, water demand for irrigation could be far greater than the usage within buildings. No generalization can be made for a load of this nature. The demand load must be determined on a project-by-project basis:

- For manual watering of plants and lawns with 1/2" to 5/8" hoses, water demand is between 5 and 15 GPM per hose outlet and may be ne-

glected from the overall calculations if watering occurs at off-peak hours.
- For landscape sprinkler systems, the demand water flow rate of sprinkler heads ranges from 1 to 10 GPM for 1/2"-diameter pipe inlet to 30 GPM for 1"-diameter pipe inlet utilizing 20 to 60 psi of water pressure. Usually, a landscape sprinkler system is timer controlled to operate on off-peak hours.
- For fountains, each nozzle may have a flow rate from several GPM for a fine spray pattern to several thousand GPM for 3" or larger pipe geysers. Fountains are usually designed for recirculation. A 10-percent makeup capacity should be provided.

Swimming pools

Swimming pools vary widely in size, from residential pools containing a few thousand gallons of water to Olympic size containing several hundred thousand gallons. Normally, the flow rate of the circulating pump is designed to turn over (circulate) the entire volume of water in the pool in 6 to 8 hours, or 3 to 4 times in 24 hours. About 1 or 2 percent of the pumped circulation rate should be provided as continuous makeup water demand to overcome evaporation, bleed-off, and spillage losses. To fill the pool initially, a separate quick-fill line should be provided to do the job in 8 to 16 hours. However, filling is usually done at off-peak hours. Thus, the demand flow rate need not be considered in the system demand calculations, unless it outweighs the demand of all other demands even during the off-peak hours.

Example 8–4 If a swimming pool contains 100,000 gallons of water, and the circulating pump is designed to change the water in 6 hours, what is the capacity of the pump, and what is the water demand load for makeup?

If the turnover rate of the filtering system pump is 6 hours, then:

- Pump circulation rate = 100,000 gallons/6 hr = 16,600 gph; 16,600/60 = 277 GPM
- Demand load for makeup water = 277 × 2% = 5.5 (use 6 GPM)

Data-processing equipment

Most data-processing equipment in buildings is air cooled. However, in large industrial, institutional, and government data-processing centers, water-cooled central processing units are also used. The demand water load is obtained from the computer system manufacturers.

Research and processing

The use of water for research and processing in special buildings could be very high. The demand water load should be obtained from those involved.

Fire protection

Normally, the water supply for fire protection is not included in the domestic water system. However, there are two fire protection system components that may be combined with the domestic water system.

When a *standpipe system* is connected to a domestic water system, the domestic system must be capable of supplying a minimum of 100 GPM of additional demand for small buildings, using 1½" hose to 500 GPM or more for large buildings. Design criteria will depend on the building code requirements.

If a limited area within a building is to be protected with automatic sprinklers (with fewer than 20 heads) even though the rest of the building lacks sprinklers, these *limited area sprinklers* may be connected to the domestic water supply system.

The demand water flow for a sprinkler head depends on the pressure available (residual pressure) at the head and the size of the orifices in the head, expressed by a K factor (varying between 2.7 and 8). For estimating demand loads, the flow rate of each sprinkler can be based on 30 GPM (at 30 psi with $K = 5.5$).

If the building is installed with a standpipe (and hose) system, the limited-area sprinklers are connected to the standpipe system risers. Such a riser is normally rated for 500 GPM for the first standpipe and 250 GPM for each additional standpipe. For determining the water demand load, only the greater of the two loads (standpipe and sprinklers) needs to be considered.

It should be mentioned that when a single water service (main) is used to serve both the fire protection and the domestic system, the combined water main must be installed with a special bypass meter which consists of two components—one small meter to measure the normal domestic water flow and an oversized detector check valve to allow large volume of water flow during a fire. The detector check valve assembly usually consists of a (normally) closed check valve that opens automatically when the pressure drop across the small domestic meter reaches a preset value, say 1.5 to 4.5 psi.

Example 8–5 If the office building in Example 8–1 is required to have one standpipe (500 GPM) and two rooms each with 15 sprinklers, what demand load should be included in the domestic water system?

The demand water load for limited-area fire protection as part of the domestic water system is as follows:

- For the standpipe = 500 GPM
- For water demand in one of the two areas with sprinklers = 15 sprinklers at 30 GPM = 450 GPM

Since the sprinkler demand is lower than that of the standpipe system, only the standpipe system load is added to the overall demand load (500 gpm) of the domestic water system.

Example 8–6 For the same office building in Example 8–5, estimate the gross and net system demand flow rate by considering the following:

- For plumbing facilities: 78 GPM (Example 8–1)
- Food service: There will be a future small dining and kitchen area with two kitchen sinks (3 wsfu) and one dishwasher (10 wsfu)
- Exterior: Two hose bibbs at 4 wsfu (5 GPM) each, to be used during office hours
- Air conditioning (HVAC): 12 GPM (Example 8–3)
- Fire protection: 500 GPM (Example 8–5)

The demand load is determined by using Table 8–2 as a basis (see next page). It is assumed that the coincidental peak demand for this office building occurs during a summer weekday and includes water for air conditioning.

8.2.2 Determination of Water Pressure

A water system must be maintained with positive pressure in order to establish a flow in the distribution system and through the plumbing fixtures or equipment. Furthermore, positive water pressure prevents water from being contaminated by external sources, since at a positive pressure, water tends to leak out of the pipe, thus preventing it from being infiltrated by foul water external to the pipe, such as groundwater in a buried piping system.

Water pressure should be sufficient to overcome any pressure loss due to friction, differences in elevation, and flow pressure at outlets or equipment.

Fixture or equipment Every plumbing fixture or connection that uses water must have the proper pressure to maintain the required flow. The minimum flow pressures required at standard plumbing fixtures are given in Table 8–4. Flow pressure is defined as the pressure at the plumbing fixture or equipment while water is flowing at the required

No.	Connected Load — Load Category	Description	Net Load Demand of Each Category in (wsfu)	in (ep)	Coincidental Peak Demand (ep)	Remarks
1.	Plumbing Facilities	Plumbing Fixtures	128	75	75	TABLE 8–3(b)
2.	Food Service	Kitchen Equipment	16	40	0	Future
3.	Laundry	—	—	—	—	
4.	HVAC Systems	100 Tons Cooling	—	12	12	
5.	Exteriors	Two Hose Bibbs	8	10	5	one in use
6.	Pools	—	—	—	—	
7.	Data Processing	—	—	—	—	
8.	Research/Process	—	—	—	—	
9.	Fire Protection	Stand. P. + Sprink.	500		500	Stand pipe only
10.	Other ( )					

Coincidental peak demand -   __592__   ep

Spare capacity anticipated -   __40__   ep   NOTE (2)

Gross system capacity required -   __632__   ep   (__500__ used)

Net system capacity installed -   __600__   ep

Notes: (1) The standpipe water supply is connected ahead of the domestic water meter through a detector check valve.

(2) The water main need be sized only for 500 GPM, since the domestic water load will not be active during a fire. The domestic water supply from the water meter to the building is sized for 132 GPM.

Remarks: 600 ep is adequate for stand pipe (500 ep) and other loads (totalling 100 ep) during a fire.

---

flow rate. It is different from static pressure, which is pressure at no-flow conditions only. The flow pressure required for equipment varies widely, and the requirements must be obtained from manufacturers.

When the pressure required for some equipment exceeds the system pressure that can be economically provided, the better solution is to install a booster pump for that equipment, rather than raising the overall system pressure to meet the special need. In

general, pressure higher than 80 psig should be reduced by pressure-reducing devices connected to standard plumbing fixtures.

Water meter and backflow preventer Depending on the design and size, a pressure drop between 5 and 10 psi (or 10 psi average) can be expected for a water meter, and similarly for the backflow preventer.

Piping Depending on the piping material (e.g., copper, steel, PVC), the nominal internal diameter, and the flow rate selected for design, the pressure drop for water flow in the pipes is given in terms of psi per unit length, usually psi/100 ft. The data is readily available in most code and engineering handbooks. The pressure drop chart for fairly smooth (Type L copper) pipes is shown in Figure 8–8. The chart shows the pressure loss due to friction, in psi/100 ft, corresponding to the flow rate, in GPM, and the resulting flow velocity in the pipe. Good practice should limit the flow velocity to below 10 fps to avoid noisy flow conditions.

TABLE 8–4

Minimum flow pressure required for typical plumbing fixtures and equipment

Fixture or Equipment	Minimum Flow Pressure (psi)
Lavatory, sink, bathtub, shower, bidet, drinking fountain, water closet (tank)	8–10
Water closet and urinal (flushometer)	20–25
Garden hose, lawn sprinkler, dishwasher, clothes washer	15–20
Commercial dishwasher (self-contained pump)	30–50
Fire protection sprinkler	25–30
Fire hose (1½″)	65
Fire hose (2½″)	65

Example 8–7 If the coincidental peak demand of a water system in a commercial building is 120 GPM,

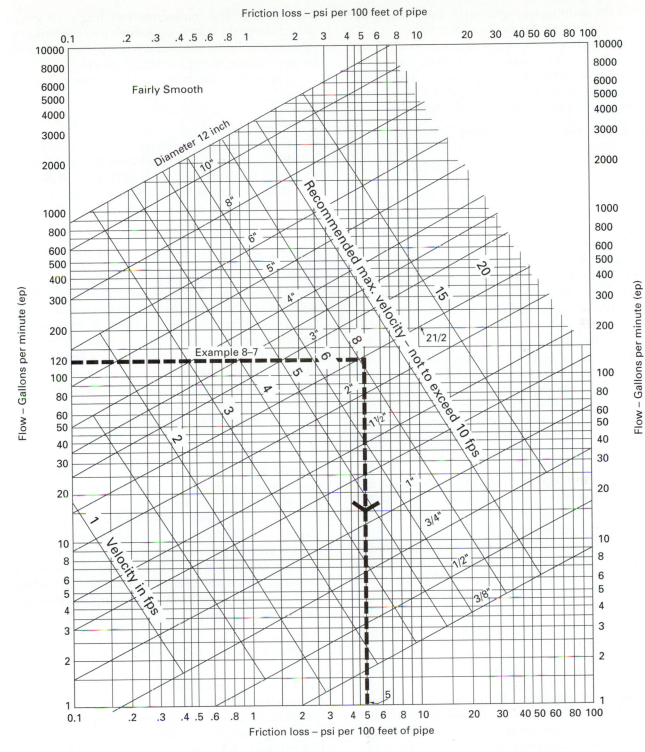

Friction loss – psi per 100 feet of pipe

■ **FIGURE 8–8**

Pressure loss (psi/100 ft) vs. flow rate (GPM) for fairly smooth pipe (e.g., copper, PVC).

what is the main water service size, based on copper pipe and a flow velocity not to exceed 8 fps?

From Figure 8–7, interpolating between 100 and 200 GPM and the 8-fps velocity line, the service should be a 2½" line for a pressure loss of 5 psi/100 ft. If the velocity is limited to 6 fps, then a 3" line should be selected for a pressure loss of 2.5 psi/100 ft.

Pipe fittings and control devices Pressure loss in pipe fittings, such as elbows, tees, valves, and controls, is a significant part of the total piping system. In fact, an improperly selected control valve would impose a higher pressure loss than all of the straight pipes in the entire system. Pressure loss by fittings and devices is significantly higher than straight pipes. The pressure loss is usually given in the unit of "equivalent length of straight pipe," "equivalent length" (EL) for short. For example, a 1½" gate valve has a pressure drop of 1 EL, whereas the same size globe valve has a pressure drop of 40 EL. A 45° elbow has about 2 EL and a 90° elbow 4 EL. Values for standard pipe fittings can be found in most codes and handbooks. In engineering design, the number and character of all fittings and control devices are identified, and their pressure loss in EL is accounted for. For preliminary estimating purposes, generally 50 percent to 100 percent is added to the linear pipe length as the total equivalent length of the piping system.

Example 8–8 If the most remote part of the plumbing system for the commercial building is about 200 feet from the service entrance, what is the pressure loss due to the piping system?

Since the piping system for a commercial building is fairly simple, in that no unusual control device is required, it can be safely assumed that the EL for pipe fittings will be 50 percent of the straight pipes, or EL = 100 feet. Thus, the total piping length is 300 EL, and the pressure loss to the piping system at a nominal 5 psi/100 ft is

$$300 \text{ ft EL} \times 5 \text{ psi/100 ft} = 15 \text{ psi}$$

Differences in elevation Water required at floors higher than the water service entrance to the building must overcome the force of gravity. This is known as *static height* due to a difference in elevation. From simple units of conversion, 1 psi is equivalent to 2.31 feet of water column. Thus, in a building 231 feet in height, the plumbing fixtures on the top floor would require 100 psi of static water pressure to reach that elevation. This static pressure must be added to the pressure required to cover friction losses and flow pressure.

Example 8–9 If a commercial building is 70 feet in height above the water service entry to the building, what should be the minimum water pressure in order to serve the top-floor plumbing fixtures?

The water pressure required at the entry is as follows:

- Water meter–15 psi
- Backflow preventer–15 psi
- Piping pressure loss (see Example 8–8)–15 psi
- Static height (elevation) = 70 ft/2.31–30 psi
- Flow pressure at fixtures (water closets)–25 psi
- Total pressure required at the service: 15 + 15 + 15 + 30 + 25 = 100 psi

Available water pressure and pressure boosting

Normally, the flow pressure of underground water mains from the utility company is between 50 and 100 psi. However, in municipalities where the infrastructure lags behind the development, water pressure may be lower than that which is required, particularly during peak hours. Other cases of low pressure are often found downtown in older cities where the water mains are clogged with scale or are too fragile to stand higher internal pressure. In such a case, an interior pressure-boosting system may need to be installed.

Example 8–10 If the water pressure available at the water main is 60 psi, how can the commercial building in Example 8–7 be served?

The water pressure required for this building is 100 psi. (See Example 8–9.) Since the pressure at the main is only 60 psi, water pressure must be boosted, or the requirement must be reduced. To boost the pressure, pumps are inserted in the system at the base or at the upper floor level to make up the shortage. The added pressure is

Booster Pressure Required = 100 − 60 = 40 psi

Life cycle cost analysis will determine that the proper solution is to reduce the pressure required by installing oversize pressure drop components, or to insert a booster pump system, or to have a combination of both techniques.

Water pressure may be boosted by installing one or more pumps at strategic locations. These pumps may be designed to run either constantly, intermittently with storage tanks, or constantly with variable speed. Figure 8–9 illustrates a water distribution riser diagram and the pumping schemes of a high-rise building.

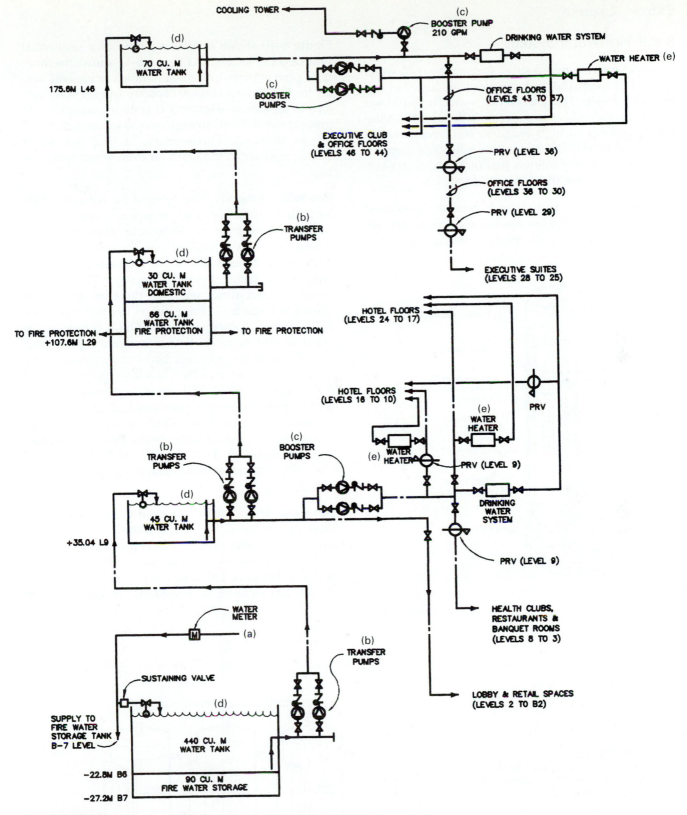

■ FIGURE 8–9

Schematic and riser diagram of cold- and hot-water system of a high-rise building, showing (a) water main, (b) three sets of transfer pumps, (c) three sets of water pressure booster pumps, (d) house water tanks at three upper floor levels. Lower portion of the tanks at the B6 and L29 levels are reserved for fire protection use. (e) The water heaters at three upper levels.

8.2.3 Hot-Water System

A hot-water system is a subsystem of the cold-water system. The demand for hot water is included in that for cold water. The use of hot water in buildings varies considerably, from very little use in office-type buildings to high usage in residences, restaurants, and hotels. The design of a hot-water distribution system closely follows that of a cold-water system with several additional considerations.

Source of energy   Hot water is normally generated in the building by the installation of water heaters using oil, gas, steam, or electricity as an energy source.

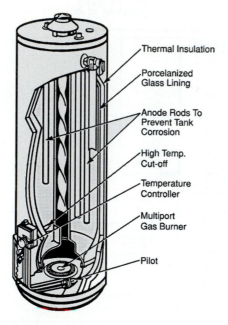

Thermal Insulation

Porcelanized Glass Lining

Anode Rods To Prevent Tank Corrosion

High Temp. Cut-off

Temperature Controller

Multiport Gas Burner

Pilot

■ **FIGURE 8–10**
Cutaway view of a typical gas-fired residential water heater. (Courtesy: State Industries, Inc. Ashland, TN.)

Figure 8–10 shows a cutaway view of a residential type of water heater, and Figure 8–11 shows the manifold pipe of three heaters in parallel. In general, water heaters have a storage capacity from several gallons to hundreds of gallons. It is more economical to preheat the water in storage than to generate hot water on demand. Furthermore, storing hot water makes it easier to maintain an even water temperature.

Hot-water demand   The demand for hot water varies with the user. For example, a person may take a 3-minute shower or a 15-minute shower. While the demand flow rate (GPM) of hot water remains the same in both cases, the latter requires 5 times the amount of water than the former. The average amount of hot water usage in the United States is shown in Table 8–5. Usage differs in other countries. For purposes of energy conservation, a conscientious effort should be made to discourage the prolonged use of hot water, through better management and proper selection of water heating equipment and flow control devices. The use of solar energy panels for hot-water heating is both practical and economical in most regions of the world. Figure 8–12 is a typical diagram of a solar water-heating scheme. In practice, an auxiliary heat source, such as electrical elements, is also installed in the storage tank to serve as a standby.

Hot-water temperature   Desired temperatures vary with the intended use of the water. For residential use, the hot-water storage tank is usually kept at 130° to 140°F, and the mixed water (between cold and hot water) is between 100° and 110°F. Water for washing dishes and clothing in a residence is satisfactory at

■ **FIGURE 8–11**
Manifold piping connection of three water heaters in parallel. (Courtesy: State Industries, Inc. Ashland, TN.)

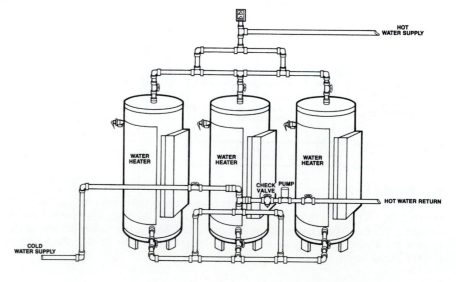

TABLE 8–5

Hot-water consumption

Usage	Demand, GPM	Total Consumption, gal
Shower, 5 min	2–6	10–30
Bath, in tub	5–15	25–30
Dishwashing, residential	1–2	10–15
Dishwashing, commercial	10–30	50–200
Clothes washer, residential	2–4	10–50

140°F, whereas 180°F is required for commercial applications. This higher water temperature is usually accomplished by local booster heaters attached to the machines.

Maintenance of water temperature Since the hot-water temperature of 140°F is about 60° to 70°F higher than ambient room temperature, hot water in the piping system tends to cool off when not in constant use, even if the pipes are insulated. For better maintenance of desired water temperatures at the point of use (plumbing fixtures), a second pipe known as the *hot-water circulation line* (or *re-circ. line*) is installed. The re-circ. line can be designed for gravity circulation based on the principle that hot water, when cooled off, tends to drop down the second pipe, thus creating a natural circulation, or it can be a pump to force a slow, but constant, circulation. Figure 8–13 is a schematic diagram of a hot-water system.

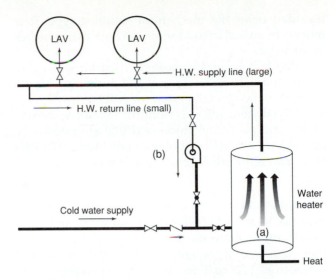

■ FIGURE 8–13

Schematic diagram of a hot-water system, showing the circulation pattern of water from the water heater to typical plumbing fixtures. (a) Hot water rises within the water heater, because it is lighter than cold water. (b) An electrical pump may be used (optional).

Hot-water capacities Applicable building or plumbing codes frequently require a minimum hot-water system capacity in residential buildings. These requirements vary from 20-gallon storage capacity with a minimum recovery rate for a single bedroom to 80-gallon storage capacity with a correspondingly higher recovery rate for three or more bedrooms. The appropriate hot-water capacity for residences will have to be analyzed on an individual basis.

8.2.4 Design Considerations for Water Distribution Systems

Piping material Copper, plastic, galvanized steel, and stainless steel are approved materials for potable water services.

Copper is the most commonly used water-piping material due to its physical strength, durability, and resistance to corrosion. It is made in three different classifications: Type K (heavy wall gauge), Type L (medium gauge), and Type M (light gauge). Type L is the normal choice for interior distribution systems. It is jointed with lead-free soldering compounds.

Stainless steel is sometimes used in lieu of copper when the sulfur content in the water or air is high, such as in the area of hot springs. The joints are either welded or threaded. Stainless steel pipes and fittings are more expensive than copper pipes and fittings.

Hot-dipped galvanized steel is standard low-carbon steel dipped into molten zinc. Joints must be

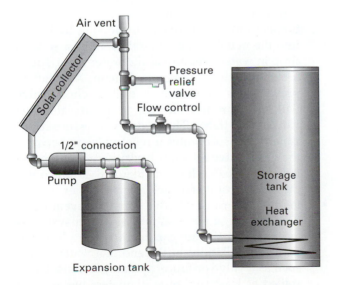

■ FIGURE 8–12

Piping diagram of a solar-energy-heated water-heating system. (All illustrations courtesy of State Industries, Inc., Ashland, TN.)

threaded or of the mechanical coupling type. It is more economical to use for pipes larger than 6 inches in diameter.

Plastic is being increasingly used for water distribution due to its lower cost and durability. Plastic pipes must be specially rated for potable water service. Polyethylene (PE), acrylonitrile butadiene styrene (ABS), and polyvinyl chloride (PVC) are rated only for cold-water use, and polyvinyl dichloride (PVDC) is approved for use for either cold- or hot-water services.

Thermal insulation   Pipes are insulated with thermal material, such as fiberglass, mineral wool, and foam plastic, to maintain the temperature of water for either chilled, cold, or hot water. Insulation material varies from 1/2" to 2" in thickness, depending on the desired heat transfer resistance. Cold-water pipe may need to be insulated in spaces with high relative humidity, such as an industrial plant with no air conditioning, or where usage is very frequent, such as in a public toilet.

Acoustic isolation   When the noise of water flowing in a pipe is annoying or disturbing in a quiet space, such as a conference room or residence, both cold- and hot-water piping are insulated for thermal and acoustical purposes.

Expansion or contraction   When the ambient temperature to which the piping system is exposed or the water temperature in the pipe is changed, or when the relative coefficient of expansion between the building and the piping material is different, there will be a differential movement created between the piping and the building. Flexibility must be built into the piping system to allow for such movement. The commonly used methods are to install expansion loops or joints to compensate for the physical expansion (or contraction) of the pipes. Figure 8–14 illustrates a typical bellows type of expansion compensator.

Backflow prevention   The water distribution system must safeguard against being contaminated. This provision is governed strictly by code. Some of the techniques used are as follows:

- A check valve allows water to flow in one direction only. Figure 8–15 illustrates several check valves.
- A vacuum breaker is installed at the branch connection to an equipment or plumbing fixture, such as a sink, dishwasher, boiler, water closet,

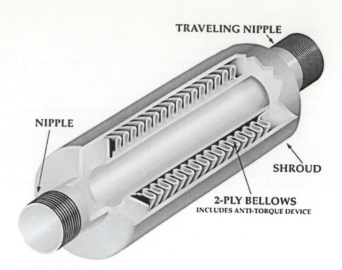

■   **FIGURE 8–14**
Cutaway view of a metal bellows type of expansion compensator showing the externally pressurized accordion-type bellows and threaded pipe ends. (Courtesy of Flexonics, New Braunfels, TX.)

urinal, etc. A vacuum breaker will automatically open the piping system to atmospheric pressure (14.0 to 14.7 psig) when pressure in the piping drops to below the atmospheric pressure level. This will prevent foreign material or foul water in the equipment or fixture from being siphoned into the piping system. Figure 8–16 illustrates the construction and operating principle of a vacuum breaker.

- A backflow preventer (BFP) is used at the entrance to the main water supply and at critical branches to stop water from flowing backward into the main system should there be a sudden drop in water pressure at the water main due to a break in the underground main. A backflow preventer is essentially a double check valve with pressure-test fittings to assure that the valves are tight. The reduced-pressure-relief-zone BFP incorporates a relief valve in between the check valve for a more positive action against failure of the check valve. Figure 8–17 illustrates the construction of a backflow preventer.
- Perhaps the most positive way to prevent backflow of foreign material into the water system is by means of the air gap required on plumbing fixture installations at the end of a water outlet, such as a valve or faucet. An air gap must be installed at least two pipe diameters higher than the receptor (bowl or basin), so that in case of water overflow above the receptor rim foreign ma-

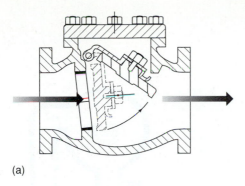

■ FIGURE 8–15
(a) Wing-type check valve showing the check in closed and open positions. (b) A center-pivoted design with reduced pressure drop. (Courtesy of Watts Regulator Co., N. Andover, MA.)

(b)

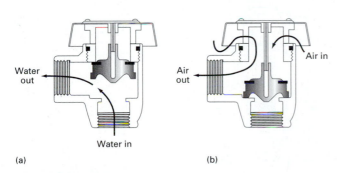

(a)   (b)

■ FIGURE 8–16
(a) Water pressure pushes the spring-loaded plunger up. The air port is closed and the water port is open. (b) When water pressure is below atmospheric pressure, the plunger is down, the water port is closed and the air port is open. (Courtesy of Sloan Valve Company, Franklin Park, IL.)

terial will not get into the piping system. Figure 8–18 illustrates air gap requirements.

Shock absorption   When the flow of water in a pipe is abruptly stopped, such as by the closing of a faucet or a flushometer, the dynamic (kinetic) energy in the water must be absorbed. If it is not the energy will be converted into a loud noise and vibration known as *water hammer*. To avoid the disturbing water hammer, pipes at the end of a branch or next to a quick-closing fixture must be installed with an air chamber in which the air acts as a cushion. If space is limited, mechanical shock absorbers can be used instead, as they have a large neoprene chamber that operates on the same principle. Figure 8–18(c) illustrates the construction of an air and shock absorber, also known as a water hammer arrestor.

8.2.5 Water-Pumping System

Water pressure required for water distribution is determined from a number of design parameters: water demand (flow rate), elevation, pipe sizes, material, routing, type of fittings, accessories, etc. All these factors contribute to pressure loss. If the water supply does not have sufficient pressure to overcome the total pressure loss, then the pressure must be boosted by a pump. However, a pump consumes energy and requires maintenance; thus, it should be avoided whenever possible. (See Chapter 7 for a discussion of the various types of pumps and a guide to their selection.)

A special need for pumping is manifested in the hot-water recirculation system, where a low-flow in-line pump is installed to assure the immediate delivery of hot water at the far end of a hot-water piping system. Figure 8–19 illustrates a typical in-line centrifugal pump and its pressure-flow characteristics.

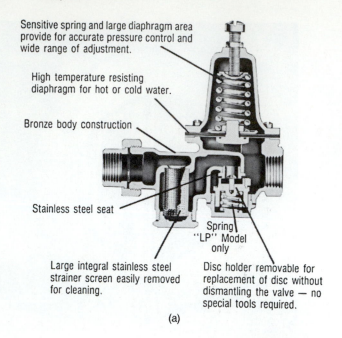

Sensitive spring and large diaphragm area provide for accurate pressure control and wide range of adjustment.

High temperature resisting diaphragm for hot or cold water.

Bronze body construction

Stainless steel seat

Large integral stainless steel strainer screen easily removed for cleaning.

Spring "LP" Model only

Disc holder removable for replacement of disc without dismantling the valve — no special tools required.

(a)

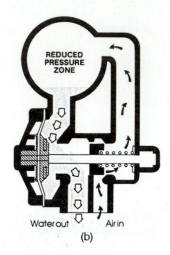

REDUCED PRESSURE ZONE

Water out   Air in

(b)

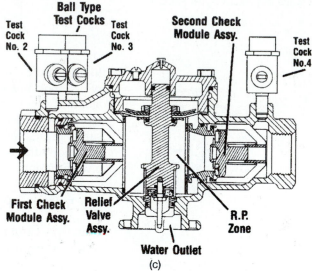

Ball Type Test Cocks

Test Cock No. 2

Test Cock No. 3

Second Check Module Assy.

Test Cock No.4

First Check Module Assy.

Relief Valve Assy.

R.P. Zone

Water Outlet

(c)

■ FIGURE 8–17

(a) Cutaway section of a typical spring-loaded water pressure reducing valve.

(b) Schematic of the operating principle of the reduced-pressure-zone relief valve.

(c) Cutaway section showing the internal construction of a reduced pressure zone backflow preventer. When the supply pressure is reduced, the first check valve will close and the relief valve will open to drain off any water in the assembly. Should both check valves fail, the relief valve will allow air to enter the valve to relieve the negative pressure in the water supply side and to allow backflow (potentially foul water) to circulate to the drain.

(Courtesy of Watts Regulator Co., N. Andover, MA.)

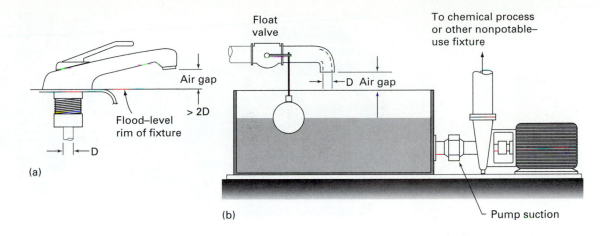

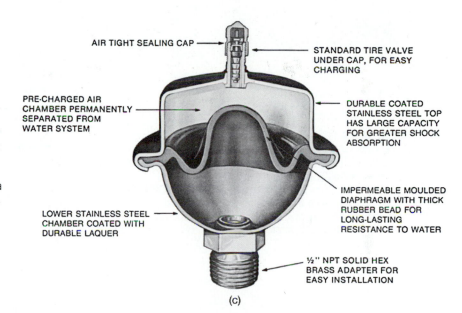

■ **FIGURE 8–18**
(a) and (b) An air gap is required for any connection to a fixture or equipment that may backflow into the potable-water system. (Source: National Standard Plumbing Code.) (c) Cutaway section of a water hammer arrestor or shock absorber. (Courtesy of Watts Regulator Co., Lawrence, MA.)

8.3 PLUMBING FIXTURES AND COMPONENTS

Plumbing fixtures are installed receptacles, devices, or appliances that are supplied with water or that receive liquid-borne wastes and then discharge wastes into the drainage system. The following are some commonly used plumbing fixtures for building services, together with the symbols that abbreviate them in diagrams:

- Water closets (WC)
- Sinks (SK)
- Bidets (BD)
- Urinals (UR)
- Bathtubs (BT)
- Service sinks (SS)
- Kitchen sinks (KS)
- Lavatories (LAV) (LV)
- Showers (SH)
- Drinking fountains (DF)

8.3.1 Plumbing Fixtures

As illustrated in Figures 8–20 and 8–21, plumbing fixtures are normally made of dense, impervious materials, such as vitreous china, enameled cast iron, stainless steel, or some other acid-resistant material. Vitreous china is most suitable for water closets, urinals, bidets, and lavatories, whereas sinks are more often made of stainless steel for durability and abrasion resistance.

Water closets   Water closets are normally made of vitreous china with hollow interior walls to direct the

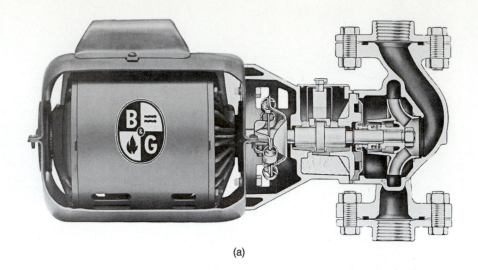

(a)

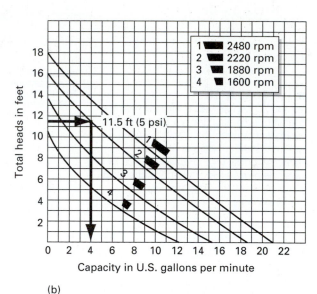

(b)

■ FIGURE 8–19
(a) Cutaway view of an in-line centrifugal pump.
(b) Pressure-flow chart of a typical multiple-speed small in-line pump operating at 1600, 1880, 2220, and 2480 rpm.
Example: Select the pump speed that will provide a 4-GPM flow rate at 5 psi. *Answer:* From the chart, at 11.5 feet (5 psi) and 4 GPM, the two lines meet at speed "2." The pump should accordingly be selected to operate at 2220 rpm. (Courtesy: ITT Bell & Gossett, Morton Grove, IL.)

passage of water and an integral water seal trap to separate the fixture from the drainage system. Water closets are usually the most prevalent plumbing fixture in a building, both in number and in water demand; thus, they have the most impact on the capacity of water and drainage systems. For this reason, the designer should thoroughly understand the various types of water closets available, their similarities and differences, and their operating principles. Water closets may be classified in terms of the following features:

1. *Method of mounting.* There are floor-mounted and wall-mounted (-hung) models. The wall-hung variety is more costly to install, but easier to maintain. Wall-hung fixtures must be mounted on concealed carriers.

2. *Cleansing action of the bowl.* Siphon-jet and washdown varieties are the quietest in operation and are universally used in private and residential applications. The blowout type used in public facilities is noisier and more positive in cleansing action. Figure 8–22 illustrates the flushing action of siphon-jet and blowout-type bowls. There are also air-water and air-vacuum bowls.

3. *Method of water control.* In the tank type of control, the water closets have 2- to 4-gallon water storage tanks. The water supply is connected to the tank through a vacuum breaker and air gaps. Water is discharged into the bowl by gravity. Figure 8–23 illustrates the operating principle of the tank type of water closet.

In the flushometer type of control, the water closets are equipped with a flush valve that admits

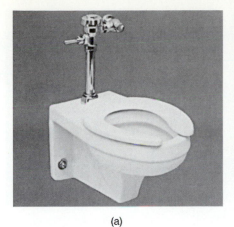

(a)

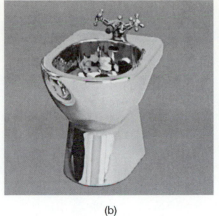

(b)

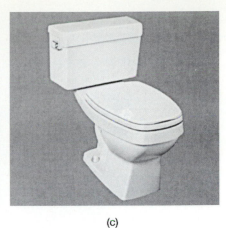

(c)

(d)

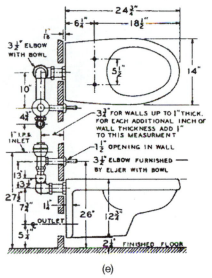

(e)

■ **FIGURE 8–20**

Photographs and dimensions of typical plumbing fixtures.
(a) Water closet with elongated bowl and flushometer valve.
(b) Bidet with hot-cold water-mixing valve.
(c) Water closet with standard flush tank.
(d) Water closet with low-profile flush tank.
(e) Top and side views of water closet in (a).
(f) Top and side views of a floor mounted water closet with flush tank.
(g) Top, side, and front views of a bidet.
(All fixtures Courtesy of Eljer Plumbing Products, Plano, TX.)

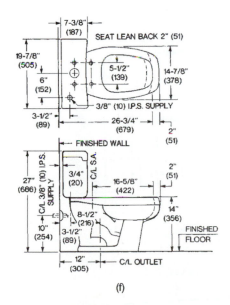

(f)

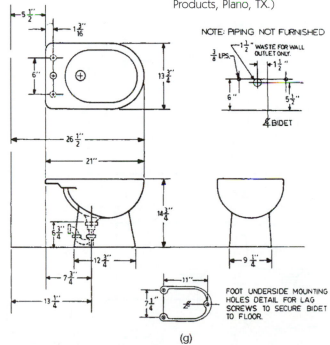

(g)

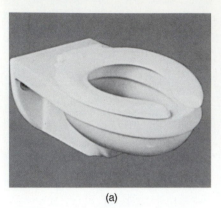
(a)

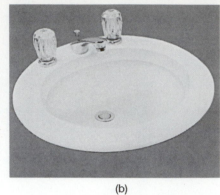
(b)

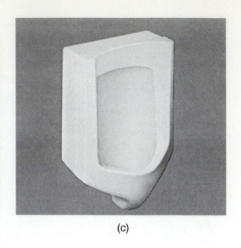
(c)

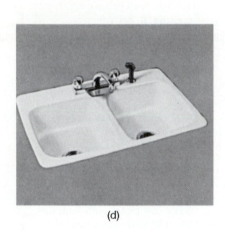
(d)

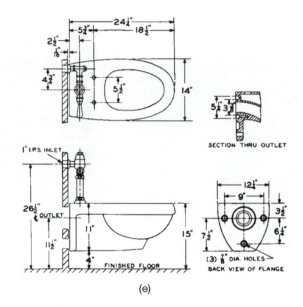
(e)

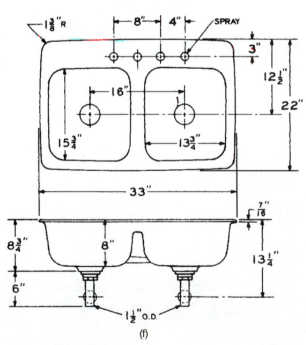
(f)

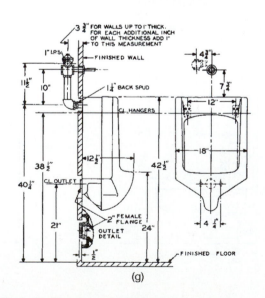
(g)

■ **FIGURE 8–21**
Photo and dimensions of typical plumbing fixtures.
(a) Water closet with elongated bowl.
(b) Counter-top-mounted lavatory.
(c) Wall hung urinal.

(d) A double compartment service sink.
(e) Top and side views of water closet in (a).
(f) Top and side views of a service sink.
(g) Side and front views of a wall-hung urinal.
(All fixtures Courtesy of Eljer Plumbing Products, Plano, TX.)

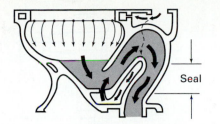
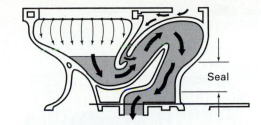

■ **FIGURE 8–22**

Flushing action of water closets. (a) Blowout design: Water enters through the rim for cleansing and a strong jet at the base to blow out the contents. (b) Siphon-jet design: Water enters through the rim for cleansing and jets in up the leg of the trapway to create a siphon.

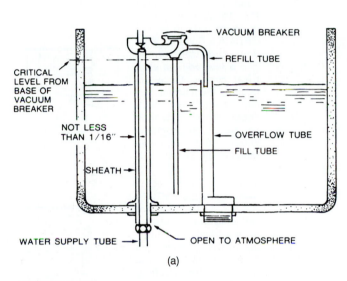

(a)

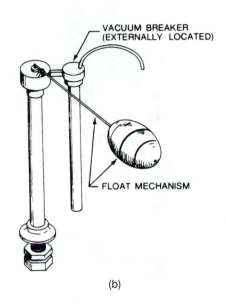

(b)

■ **FIGURE 8–23**

Operating principle of a flush tank. (a) Water feeds through the supply tube and down to the fill tube through a vacuum breaker. A portion of the water bleeds into the overflow tube (or compartment) to refill the bowl. (b) The raising of a float valve stops the water supply until the next flushing. (Courtesy of National Standard Plumbing Code.)

a time-measured (adjustable between 5 and 10 seconds) amount of water into the bowl under the water pressure. The amount of water admitted is between 2 and 4 gallons, comparable to that of the tank type of control; however, the instantaneous water demand rate is considerably higher (around 20 to 30 GPM). The flushometer type of water closet can be ready for use immediately after flushing and therefore is required in all public occupancies. Figure 8–24 illustrates the operating principle of a diaphragm type of flushometer. It has become increasingly popular to use automatic sensing devices to operate the flushometer in public facilities, such as airports, restaurants, hotels, theaters, arenas, and sport stadiums. Automatic flushing design and construction with solid-state

components have become very reliable and cost effective, and are a more positive way to ensure sanitation conditions. Figures 8–24 (a)–(d) illustrate several of the flushometer designs and their operations.

In the pressure tank type of control, the water closets are equipped with a pressurized tank within a conventional gravity tank. The pressurized tank is charged with air and water under 25-psi water pressure. When the plunger at the base of the tank is released, the air-water mixture is forced into the bowl to blow out its contents. Because the blowout action is under pressure, 1.5 gallons of water are sufficient; thus, a considerable amount of water is conserved. With water shortages in many parts of the world, water conserva-

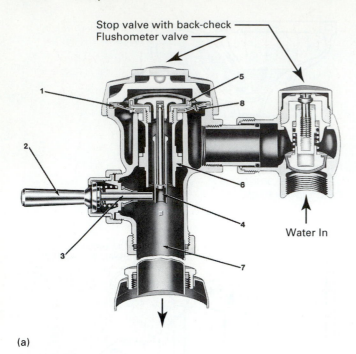

Stop valve with back-check
Flushometer valve

1
2
3
5
8
6
4
7

Water In

(a)

(c)

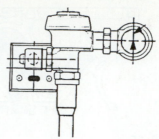
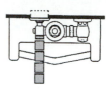
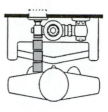
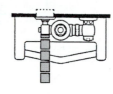

1.
A continuous, invisible light beam is emitted from the OPTIMA Sensor.

2.
As the user enters the beam's effective range (15" to 30") the beam is reflected into the OPTIMA's Scanner Window and transformed into a low voltage electrical circuit. Once activated, the Output Circuit continues in a "hold" mode for as long as the user remains within the effective range of the Sensor.

3.
When the user steps away from the OPTIMA Sensor, the loss of reflected light initiates an electrical "one-time" signal that operates the Solenoid (24V AC) and initiates the flushing cycle to flush the fixture. The Circuit then automatically resets and is ready for the next user.

(b)

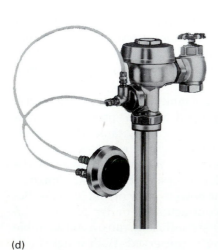

(d)

■ **FIGURE 8–24**

(a) Principle of operation of a diaphragm type of flushometer. (Courtesy of Sloan Valve Co., Franklin Park, IL.) When the flush valve is in the closed position, the segment diaphragm (1) divides the valve into an upper and lower chamber with equal water pressure on both sides of the diaphragm. The greater pressure on top of the diaphragm holds it closed on its seat. Movement of the handle (2) in any direction pushes the plunger (3), which tilts the relief valve (4) and allows water to escape from the upper chamber. The water pressure in the lower chamber (below the segment diaphragm (1)) now being greater, raises the working parts (1), (4), (5), and (6) as a unit, allowing water to flow down through the valve outlet (7) to flush the fixture. While the valve is operating, a small amount of water flows through the bypass (8) of the diaphragm, gradually refilling the upper chamber and equalizing the pressure once more. As the upper chamber fills, the diaphragm (1) returns to its seat to close the valve.

(b) Electronic sensor (Optima model) and its principle of operation.

(c) Battery-operated model operates on 4-AA 1.5-volt batteries.

(d) Hydraulic-operated model with remotely mounted push button. (Courtesy: Sloan Valve Company, Franklin Park, IL.)

tion should be of primary concern to all building owners and designers. The pressure tank type of water closet will become increasingly more popular. Figure 8–25 illustrates the construction and operating principle of the air-water type of water closet.

The vacuum type of water closet operates on a central vacuum piping system. When the valve below the bowl is opened, the contents of the bowl are sucked into the drainage piping system under a vacuum using only 0.3 gallon of water per flushing. This type of water closet is commonly used on airplanes and oceangoing ships. Figure 8–26 illustrates its construction and operating principle.

Example 8–11   Based on the office building in Example 8–1, if the water closets were of the pressure-tank type with a wsfu rating of 2, what size should the domestic water riser be to serve the plumbing fixtures?

- If the water closet is rated for 2 wsfu, and the other fixtures remain the same, then the total wsfu installed will be 48 in lieu of 128, as calculated in Example 8–1.
- From Table 8–3B, the demand flow will be about 17 GPM (taken from the "predominantly flush tank" column), instead of 59 GPM (taken from the "predominantly flushometer column").
- From Figure 8–8, the pipe size for 17 GPM (at 8 fps velocity) is between 3/4" and 1"-diameter pipe, whereas that for 59 GPM (at 8 fpm) is a 1½"-diameter pipe.

Urinals   Similar to water closets in construction and operating principle, urinals come in siphon-jet, blow-out and washdown varieties. They are either wall or floor mounted. Wall-mounted units are most commonly used, because they require less space and are easier to clean. Urinals should have a visible water seal with no strainer above the seal. Water-flushing action should thoroughly clean the entire interior wall surface. Construction deviating from these principles is prohibited by many governing codes.

Lavatories   Lavatories are designed in a variety of sizes and shapes. Fittings, such as faucets, drains, and other accessories, are unlimited in design and material.

Sinks   General-purpose sinks and kitchen sinks are available in single, double and triple compartment models. Stainless steel sinks are preferred due to their durability and ease of cleaning. As a special application, service sinks are installed in public buildings for cleaning crews. They may be wall, floor, or recess floor mounted and made of stainless steel, enameled cast iron, precast terrazzo, or reinforced glass fiber.

Drinking fountains   Drinking fountains can be wall or floor mounted and are usually nonrefrigerated. However, they may be piped from a central chilled-water supply with a separate filtering system.

Electric water coolers   Electric water coolers are individual drinking fountains with self-contained water-chilling units. The individual electric water cooler is preferred over the central system due to its flexibility and low operating costs.

Bathtubs   The most popular bathtub is 5 feet long with a net water basin dimension of 4'–6". Recent trends show an increased use of models 6 feet or longer, as well as wider, with built-in whirlpool nozzles and pump. Bathtubs are made of enameled cast iron, porcelain enamel on pressed steel, or fiberglass-

■ **FIGURE 8–25**
Pressure tank system operating on 25 psi or higher water supply system. The tank contains a 1.5-gallon pressurized air-water reservoir. When the flush lever is activated, it allows the mixture of air and water to rush into the bowl and thus expels (blows out) the contents of the bowl under pressure. After flushing, a charge cycle takes place to refill the reservoir until the 25-psi (or higher) pressure is reached. (Courtesy of Kohler Co., Kohler, WI.)

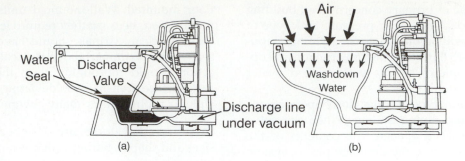

■ FIGURE 8–26

Cross-sectional views of a vacuum type of water closet. (a) Water seal in the bowl when the water closet is not in use. (b) When an electrical or pneumatic push button is pushed, it will open the discharge pipe, and air at atmospheric pressure will force the contents of the bowl through the discharge valve (open for 4 seconds) and into the discharge line under the vacuum. Concurrently, a small amount of water (0.4 gal) will open (for 7 seconds) to provide washdown and water seal in the bowl. (Courtesy of Environvac, Inc., Rockford, IL.)

reinforced plastics. Incorporated with a bathtub is usually a shower head with combined water control fittings. The base of a tub should be designed with a nonslip surface.

Showers Showers can be integrated with a bathtub or be independently constructed into shower stalls or a battery of showers. One major concern is the control of the water temperature by a mixing valve between cold and hot water. Thermostatic mixing valves mix the water depending on the desired outlet water temperature. Better designed control valves also have a pressure balance feature so that the water temperature will not fluctuate when there is a sudden change in the relative water pressure between the cold- and hot-water systems.

Bidets Bidets are small baths, the size and shape of a water closet, used primarily for personal hygiene. Hot or cold water enters around the flushing rim, with a spray rinse as an optional feature. The rinse is also referred to as a douche. Bidets are popular in European countries.

8.3.2 Components of Fixtures

Traps To prevent the backup of sewer gas into a building through the drainage connection of plumbing fixtures, every plumbing fixture must be connected by means of a trap seal. As a rule, all water closets, bidets, and urinals are manufactured to have an integral trap, whereas other plumbing fixtures use external traps.

The trap is a portion of piping in a U shape and filled with a water seal. It operates on the principle that two columns of water balanced within the two legs of the trap prevent sewer gas, noxious fumes, or vermin from passing from the sewer side to the building. The water column in a trap is normally between 2 and 4 inches in depth; it thus requires a minimum of 2 inches of w.c. pressure differential to break the water seal. There are various trap designs, each with its own strict applications. P-traps are normally used for standard plumbing fixtures.

A water seal is always filled with water if the fixture is periodically used. However, if a building is unoccupied for long periods of time—say, several months—water in the trap of any fixture may dry out due to normal evaporation. In such case, the trap should be filled with some lighter-than-water but nonevaporative fluid, such as glycerine, to cover the water seal surface thus retarding the rate of water evaporation.

Air gap To avoid the possibility of soil or waste contaminating the water supply system, connections between water supply components and the plumbing fixtures must be separated vertically through an air gap. The gap is normally designed to have a minimum separation between the bottom opening of the water supply and the fixture rim, which is the highest point of the fixture. With an air gap, any backflow of the waste will overflow at the fixture rim, and the potential for contaminating water is eliminated. (See Figure 8–18(a).)

Prevention of backflow Water flowing in the pipe is normally under positive pressure. When the water pressure within the piping system drops suddenly, it may be possible for soil or waste, to which the piping is connected, to be sucked or siphoned into the water system. Figure 8–16 illustrates the operation of pipe-mounted vacuum breakers. Figure 8–21 shows the vacuum breaker required on a tank-type water closet. Figure 8–25 shows the combination stop valve and backflow preventer ahead of a flushometer. Prevention of backflow is required on any equipment that is connected to water piping without an air break (gap). Components of fixtures, such as faucets with "hose-end" connections, dishwashers, etc., are required to be installed through vacuum breakers.

8.3.3 Fixture Units

The demand for water and the load imposed on the drainage system of plumbing fixtures differ considerably. In fact, a fixture with low water demand may impose a high drainage load, depending on the mode of operation of the fixture. For example, a bathtub usually does not demand a high water flow, but it could impose a high drainage load. It normally takes over 10 minutes to fill a tub, but only 2 to 3 minutes to drain it. For this reason, a fixture may have a different water supply rating than its drainage rating.

TABLE 8–6

Drainage and water fixture unit values[1] and minimum pressure and pipe size requirements

| | Drainage | | Water | | | | |
| | Load Value (dfu)[2] | Minimum[3] Pipe Diameter | Load values (wsfu)[2] | | | Minimum at Fixture[4] | |
Fixture			Cold	Hot	Total	psig	Pipe Diameter
PUBLIC OR GENERAL							
Water closet, flushometer	6	4	10	—	10	20–25	1
Water closet, gravity tank	4	4	5	—	5	8–10	½
Water closet, pressure tank[5]	2	4	2	—	2	25–30	½
Urinal, flushometer	4	2–3	5	—	5	15–20	¾
Urinal, tank	2	2	3	—	3	8–10	½
Lavatory	1	2	1.5	1.5	2	8–10	⅜
Bathtub	2	2	3	3	4	8–10	½
Shower	2	2	3	3	4	8–10	¾
Service sink	3	3	2.25	2.25	3	8–10	¾
Kitchen sink	3	2	3	3	4	8–10	¾
Electric water cooler	0.5	2	0.25	—	0.25	8–10	⅜
PRIVATE							
Water closet, flushometer	6	4	6	—	6	20–25	1
Water closet, gravity tank	4	4	3	—	3	8–10	½
Water closet, pressure tank[5]	2	4	2	—	2	25–30	½
Lavatory	1	2	0.75	0.75	1	8–10	⅜
Bathtub	2	2	1.5	1.5	2	8–10	½
Shower	3	2	1.5	1.5	2	8–10	½
Bathroom group, flushometer[6]	8	4	6	6	8	20–25	1
Bathroom group, tank[6]	6	4	4	4	6	15–20	1
Kitchen sink	2	2	1.5	1.5	2	8–10	½
Combination fixtures	3	4	2.25	2.25	3	10–15	½
Dishwasher	2	2	—	1	1	10–15	½
Laundry, clothes washer (8 lb)	3	2	1.5	1.5	2	10–15	½

		Drainage				Water		
		Load Value (dfu)	Min Pipe Diam				Load values, wsfu	
						Pipe Size	Private	Public
Fixtures not listed above, but with trap size given	1¼ in.	1	2		Fixtures not listed above, but with pipe size given			
	1½ in.	2	2					
	2 in.	3	2					
	2½ in.	4	2½			⅜ in.	1	2
	3 in.	5	3			½ in.	2	4
	4 in.	6	4			¾ in.	3	6
						1 in.	6	10

[1]For fixtures with continuous water flow, the load is added without a demand factor.

[2]Dfu = drainage fixture unit; wsfu = water supply fixture unit.

[3]Minimum-drainage pipe diameter, in inches, for fixture branches. Fixture outlet tailpieces may be smaller in diameter, as per manufacturer's specifications.

[4]Minimum pressure and pipe size should be verified with fixture manufacturer.

[5]Dfu and wsfu are estimated for the pressure-tank type.

[6]A bathroom group is defined as consisting of not more than (1) WC, (1) LV, (1) BT and (1) SH, or not more than (1) WC, (2) LV, and (1) BT or (1) separate SH.

The *water supply fixture unit* (wsfu) is a measure of the probable hydraulic demand on the water supply by various types of plumbing fixtures. The value of the wsfu for a particular fixture depends on the rate of supply, the duration of a single operation, and the frequency of operation of the fixture. A 1/2" residential-type lavatory faucet is rated for one wsfu (about 1 to 1.5 GPM flow rate). A 1" flushometer (about 35 GPM) is assigned with 10 wsfu, although the instantaneous demand for water is more than 10 times that of the lavatory faucet. This is due to the effect of duration and frequency of operation. For determining the wsfu of nonstandard plumbing fixtures, one wsfu rating may be used for each 1.5 GPM of flow rate. The wsfu ratings of standard plumbing fixtures are given in Table 8–6.

The *drainage fixture unit* (dfu) is a measure of the probable discharge into the drainage system by various types of plumbing fixtures. The value of the dfu for a particular fixture depends on the rate of drainage discharge, the duration of a single operation, and the frequency of operation of the fixture. Laboratory tests show that the rate of discharge of an ordinary lavatory with a nominal 1.25-in. outlet, trap, and waste is about 7.5 GPM (0.5 L/s). It is assigned as dfu = 1. However, the dfu equivalent for continuous-flow (or semicontinuous-flow) equipment is drastically different, which should be 1 dfu for each 0.5 GPM of continuous drain. The dfu ratings of typical plumbing fixtures are given in Table 8–6.

8.4 PLANNING PLUMBING FACILITIES

Plumbing facilities for buildings are primarily toilets, bathrooms, washrooms, and lockers for private or public use. Because plumbing facilities usually handle the major loads for water supply and of drainage systems, the requirements for such facilities are normally analyzed and planned by the architect to fulfill the needs of the building's occupants. However, planned facilities must equal or exceed the minimum

TABLE 8–7

Minimum plumbing fixture requirements in a building
(Values shown are consolidated from different codes. Follow the applicable code requirements in actual design.)

		Recommended Maximum Number of Occupants per Fixture					
Building Use Group		Water closets (WC) (Urinal see note a)		Lavatories (LV)	Bathtubs (BT) Showers (SH)	Drinking Fountains (DF)	Service Sinks (SS)
		Male	Female				
Theaters, Churches	First Fixture	50	50	½ of WC	—	1000	1/floor
	Additional	150	150				
Stadiums/Arena	First 3 Fixtures	50	50	½ of WC	—	200	1/floor
Ball Parks, Terminals	Additional	300	150				
Restaurants	First 3 Fixtures	50	50	½ of WC	—	200	1/floor
Convention Halls	Additional	200	100				
Recreational	First Fixture	40	40	½ of WC	—	75	1/floor
Night Clubs	Additional	40	20				
Business/Stores, Banks	First Fixture	15		½ of WC	—	100	1/floor
Offices, Shopping Centers	Additional	25					
Educational	Preschools	15		½ of WC	—	30	1/floor
	Elementary	25				40	
	Secondary	30				50	
Hospitals	Ward	1:8 patients		½ of WC	1:20 patients	100	1/floor
	Private Rooms	1/room		1/room	—		
Nursing Home		15		15	1/units	100	1/floor
Institutional, Prisons		1/cell		1/cell	1:6 inmates	—	1/floor
Hotels/Motels		1/unit		1/unit	1/units	—	1/floor
Dwelling Units		1/unit		1/unit	1/units	—	(c)
Dormitories		20		½ of WC	8	100	1/floor
Industrial		30		½ of WC	(b)	75	1/floor

(a) One urinal may be installed in lieu of one water closet in male toilets for up to 50 percent of the required water closets.

(b) See applicable code for detail requirements.

(c) 1 kitchen sink and 1 clothes washer connection per dwelling unit.

code requirements. In practice, the plumbing facilities in high-quality buildings have always been more than are required by the code, for the following reasons:

- *To improve convenience.* Facilities should be dispersed and thus frequently duplicated, so that occupants need not travel too far on one floor or to other floors to use them.
- *To accommodate the fluctuation in a building's occupants.* The proportion of a building's occupants of each sex is constantly in a state of flux; thus, extra facilities must be built in to accommodate this fluctuation. This is particularly true for assembly and sports facilities.
- *To avoid congestion.* The use of a building's plumbing facilities is usually concentrated at certain times of the day, e.g., at lunch time, at the end of an event, and during intermissions of events of one kind or another. To avoid congestion, extra plumbing fixtures beyond the number specified as a minimum by the code should be provided. It is quite common for modern sporting arenas and stadiums to have more than twice the minimum number of facilities required by code to provide satisfactory services.

8.4.1 Minimum Requirements by the Code

Nationally recognized plumbing codes normally include the minimum numbers of plumbing fixtures that must be installed in a building based on the building's population (occupants) and use. Table 8–7 lists such requirements. For determining the number of occupants within the building, the architect or engineer shall make the estimate based on the following:

1. The maximum number of occupants intended, but not less than that specified in the governing building code. Table 8–8 is one method commonly used to determine the number of occupants of a building. For special or unlisted occupancies, the occupant load shall be established by the architect or engineer, subject to the approval of the building official.
2. The plumbing fixtures shall be distributed equally between the sexes, based on the percentage of each sex anticipated in the occupant load, and shall be composed of 60 percent of occupants for each sex if the estimated occupants are evenly divided.
3. In commercial buildings and institutions, facilities for employees or workers may need to be calculated separately from those for customers or inmates.

TABLE 8–8

Method for calculating number of occupants of a building
(excerpt from BOCA National Building Code; see Section 10 of BOCA/NBC for a complete listing of occupancies and exceptions)

Occupancy	Occupant/Seat or Floor Area/Occupant
Assembly, with fixed chairs	1 occupant/seat
with fixed seating benches	1 occupant/18"
with fixed seating booths	1 occupant/24"
with movable chairs	7 NFA[a]
with movable chairs and tables	15 NFA
with standing space only	3 NFA
Business area in a building	100 GFA[b]
Educational, classroom area	20 NFA
Shops and other vocational areas	50 NFA
Library reading rooms	50 NFA
Library stack areas	100 GFA
Institutional, inpatient areas	240 GFA
outpatient area	100 GFA
sleeping area	120 GFA
Mercantile, basement and grade floor area	30 GFA
Area on other floors	60 GFA
Storage, stock, shipping areas	300 GFA
Parking garages	200 GFA
Residential	200 GFA
Miscellaneous (storage, mechanical equipment, etc.)	300 GFA

[a]NFA (net floor area) is defined as the actual occupied area of the space and shall not include unoccupied accessory areas or the thickness of walls.
[b]GFA (gross floor area) is defined as the floor area within the perimeter of the outside walls of the building under consideration.

8.4.2 Substitution for Water Closets

Urinals may be substituted for water closets in male toilets on a one-for-one basis, but the substitution should not be more than 50 percent of the required water closets.

Example 8–12 If a 100,000-GFA office building has 80 percent of the GFA for office occupancy and 20 percent for miscellaneous spaces (storage and mechanical rooms), determine the number of building occupants and the minimum number of plumbing fixtures required.

- According to Table 8–8, the number of building occupants shall be:
  Business (office) area:
80,000 sq ft/100	800 occupants
Miscellaneous spaces:
20,000 sq ft/300	67 occupants
Total occupants	**867**
Assumed male occupants, 50%
(use 60%)	520

Assumed female occupants,
50% (use 60%)   520

- From Table 8–7, the minimum required plumbing fixtures are as follows:

	WC	UR	LV	BT	DF	SK
Men	11	11	11	0	—	—
Women	22		11	0	—	—
For building	—		—	0	10	1/floor

8.4.3 Planning Guidelines

Plumbing fixtures shall be installed so as to afford easy access for cleaning and maintenance. More space is required for the physically disabled. For example, a 42″ clearance on one side of a WC stall is necessary to make room for a wheelchair.

Figure 8–27 shows a typical layout of residential bathroom groups with piping in one or more walls. Toilets and washrooms should be located where they are easy to access or likely to be expected, such as near building entries, waiting areas, elevator lobbys, stairways, telephone stations, etc.

Men's and women's toilets should be located in the same general area. They may be placed back to back or with some separation between them. When they are placed back to back, the cost of piping is substantially reduced for small and low-rise buildings.

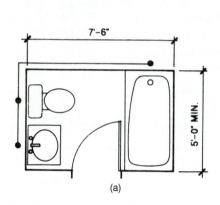

(a)

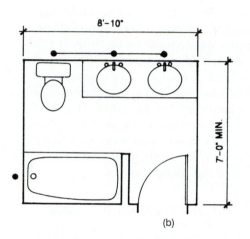

(b)

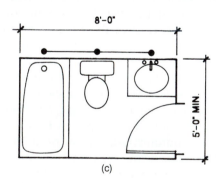

(c)

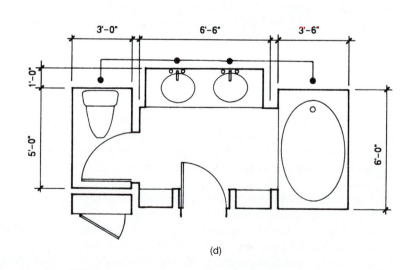

(d)

■ **FIGURE 8–27**
Typical residential bathroom plan.
(a) A basic 3-fixture plan with piping on two walls.
(b) A more spacious plan with piping on two walls.
(c) A basic 3-fixture plan with piping on one wall (most economical).
(d) A deluxe plan including whirlpool and private water closet compartment.
(Courtesy: Tao and Lee Associates, Inc., Architects, St. Louis, MO.)

The cost of piping for large and high-rise buildings may not have the same degree of impact, since the large number of fixtures to be installed is likely to require multiple stacks and risers in any case.

When men's and women's toilets are placed back to back or near to each other, the noise and cross-talk transmitted between the common stacks, pipes, partitions, louvers, and ducts will be avoided. Figure 8–28 shows a group of public toilets for a sports arena in which men's and women's toilets are separated by a corridor.

The designer must be careful to provide a visual barrier between the public passageway and the interior of the toilets when the door to the room is opened and there is a possibility of a reflected image being seen through a mirror. Allow ample distance between fixtures and room surfaces. Refer to codes and standards for the minimum space requirements.

8.5 SANITARY DRAINAGE SYSTEMS

Drainage in buildings consists of three major components: sanitary waste, storm water, and specialty waste, such as toxic, radioactive, chemical, or other processing wastes. Sanitary waste and storm water may be piped separately or combined, depending on the public sewer system to which the drainage is connected. Combined storm and sanitary sewer systems still exist in some major cities, carried over from the early practice of discharging untreated sewage into rivers. This practice is no longer permitted in the United States and most other developed, as well as developing, countries. Because of their pollutants, specialty wastes must be piped and treated separately.

Following are definitions of several terms commonly used in drainage systems:

- *Waste* (liquid). The liquid discharged from water-consuming equipment.
- *Sanitary waste*. The liquid discharged from plumbing fixtures.
- *Soil* (waste). The liquid discharged from plumbing fixtures that contains or potentially contains fecal matter, such as liquid from a floor drain within a toilet.
- *Sanitary drain*. The main drain of the sanitary drainage system within a building.
- *Sanitary sewer*. The extension of the sanitary drain at the exterior of a building for connection to the public sewer or to a sewage disposal system.

- *Storm water*. The rainwater collected from building roofs (roof drains) and from exterior area (area drains). Depending on the purity of the water collected, rainwater may be classified as waste or may be stored for reuse, even as potable water.
- *Storm drain*. The main drain of the storm water drainage system within a building.
- *Storm sewer*. A sewer exterior to a building and that contains storm water only.
- *Combination sewer*. A sewer that contains sanitary waste and storm water.

A sanitary drainage system is thus a drainage system designed to carry away sanitary wastes (including soil wastes) from within a building to a public sewer or to a sewage disposal plant. The system may be designed to flow by gravity (gravitational force) without resorting to the use of mechanically or electrically powered equipment, or it may be designed to flow under pressure by pumping. The gravity system is, of course, more reliable and more economical to operate and thus should be used whenever feasible. The discussions in this section will deal with gravity flow only.

8.5.1 Drainage-Waste-Venting

When the sanitary drainage system is connected to a public sewer, it is conceivable that sewer gas, insects, or rodents may enter the building through the system. To overcome this problem, all drainage equipment (including all plumbing fixtures) connected to the sanitary drainage system must be separated by a liquid seal trap that acts to separate the building from the sewer. Each trap must be adequately vented to the atmosphere to prevent the liquid seal from being siphoned or sucked dry should there be a pressure differential created between the building and the sewer due to the flow of the drainage. Figure 8–29 illustrates the operating principle of a liquid seal trap. Because of the importance of venting, a sanitary drainage system is often referred to as a drainage-waste-venting (DWV) system. Some of the important terms associated with a DWV system and their definitions are as follows:

Stack the vertical portion of a DWV-piping system.

Waste stack the vertical portion of a waste-piping system.

Soil stack the vertical portion of a soil-piping system.

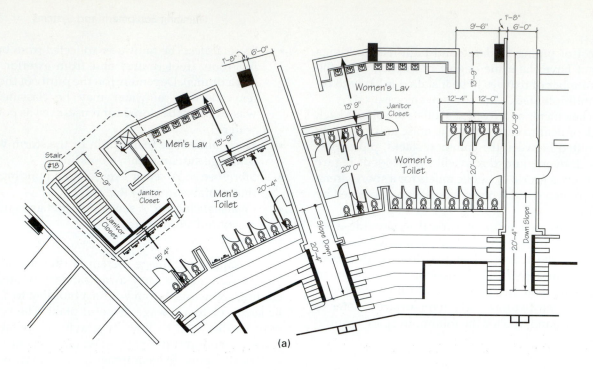

(a)

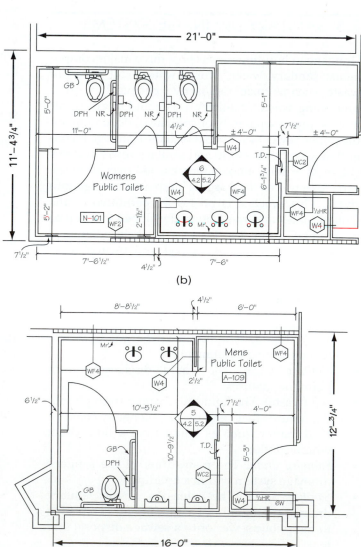

(b)

■ **FIGURE 8–28**

(a) Public toilets in a large sports arena. Note the separate entry and exit for each toilet, the toilets for the handicapped, and the janitor's closets. The men's and women's toilets are separated by a vomitory for noise control. (b) Typical public toilet layouts. Note the drinking fountain at the entry to the women's toilet, the screen walls for privacy, and the water closets for the handicapped.

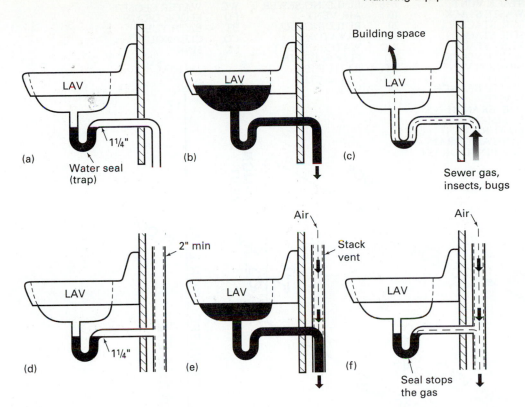

■ FIGURE 8–29
Illustrations and operating principle of liquid seal traps.
(a) With no vent, a trap may be started with a seal at no flow.
(b) But during heavy drainage flow, the momentum of the flow could siphon out the liquid from the seal.
(c) As a result, the trap will be only partially filled and will allow sewer gas and other substances to enter the building.
(d) With proper venting, a trap is filled with liquid from the previous drainage flow.
(e) Even under heavy flow, air enters the vent pipe to break up the siphoning effect.
(f) As a result, the trap will continue to maintain a full depth of liquid.

Stack vent the open-ended extension of a waste or soil stack above the highest horizontal drain connected to the stack.

Branch interval a section of a soil or waste stack corresponding to one story in height (but in no case less than 8 feet).

Vent a pipe open to the atmosphere.

Vent stack a stack that does not carry waste of any kind and that is installed primarily for providing circulation of air to and from any part of the DWV system.

Branch vent a branch of the venting system.

Common vent a vent connected at the common connection of two fixtures.

Circuit vent a branch vent that serves two or more traps and that extends from the downstream side of the highest fixture connection of a horizontal branch to the vent stack.

Crown vent a vent connected to the crown of a trap.

Developed length the total length of a pipe, measured along the centerline of the pipe.

Figure 8–30 is an isometric diagram of a DWV system identifying the various components of the system.

The conventional universally accepted DWV system is essentially a two-pipe system consisting of a waste-piping network that drains the waste and a separate venting network to equalize the air pressure in the system. A specially engineered single-pipe system known as a *sovent system* is also available. It is an all-copper DWV system consisting of standard copper drainpipes from fixtures to a single stack, a specially designed aerator fitting at each floor, and a deaerator fitting at the base of each stack. The single-pipe system has been used satisfactorily. Its applica-

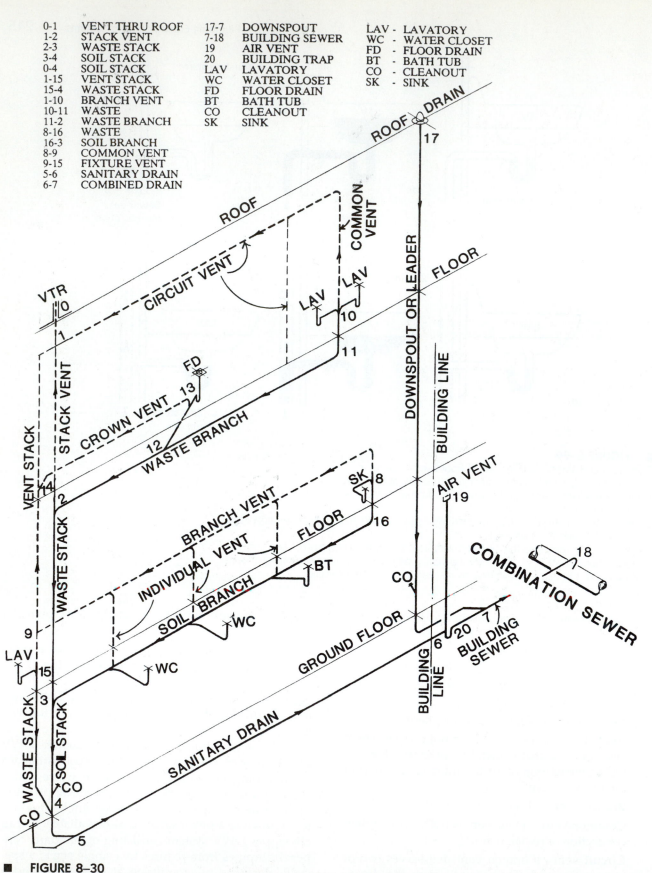

0-1	VENT THRU ROOF	17-7	DOWNSPOUT	LAV - LAVATORY
1-2	STACK VENT	7-18	BUILDING SEWER	WC - WATER CLOSET
2-3	WASTE STACK	19	AIR VENT	FD - FLOOR DRAIN
3-4	SOIL STACK	20	BUILDING TRAP	BT - BATH TUB
0-4	SOIL STACK	LAV	LAVATORY	CO - CLEANOUT
1-15	VENT STACK	WC	WATER CLOSET	SK - SINK
15-4	WASTE STACK	FD	FLOOR DRAIN	
1-10	BRANCH VENT	BT	BATH TUB	
10-11	WASTE	CO	CLEANOUT	
11-2	WASTE BRANCH	SK	SINK	
8-16	WASTE			
16-3	SOIL BRANCH			
8-9	COMMON VENT			
9-15	FIXTURE VENT			
5-6	SANITARY DRAIN			
6-7	COMBINED DRAIN			

■ **FIGURE 8–30**

Isometric diagram of a DWV drainage system identifying the various components of the system. (Note: when the public sewer is separated as sanitary and storm sewers, building sewers should also be separated.)

tion depends on the relative costs of labor and materials in the localities in question.

8.5.2 Piping Materials and Fittings

The materials used for sanitary drainage systems are as follows, in the general order of their preference or popularity:

1. Aboveground—cast iron (C.I.), ABS and DWV plastics, DWV copper, and stainless steel
2. Underfloor (drains)—C.I., DWV plastics
3. Underground (sewer)—vitrified clay pipe (VCP), PVC sewer plastics with pipe stiffness (PS) 35 psi or heavier.

All pipes and fittings must have the joining methods appropriate for the material, such as the following:

1. Caulked joints, for hub-and-spigot types of C.I. pipes only
2. Threaded joints, for steel or heavy wall plastic pipes
3. Soldered joints, for copper or DWV pipes
4. Brazed joints, for heavy copper pipes only
5. Compression joints, for no-hub C.I., VCP, or plastic pipes
6. Gasket joints, for hub-and-spigot C.I. pipes
7. Flexible compression joints, for VCP pipes
8. Solvent joints, for plastic pipes
9. Transition joints, specially made for joining of different materials

See Figure 8–31 for illustrations of the various types of joints.

8.5.3 Sizing of Drain Lines

The capacity of (waste or soil) drainpipes of the DWV system depends on two major factors: the slope of the pipes and the dfu they serve. All horizontal drain lines have a uniform downward slope in the direction of flow. If the slope is too small, solid contents in the liquid may drop out. If the slope is too great, turbulence flow and erosion of the pipes may occur. In general, the slope shall not be less than ¼" per foot (approximately a 2-percent slope) for 3"-diameter and smaller horizontal lines and ⅛" per foot (approximately a 1-percent slope) for 4"-diameter and larger horizontal lines. To determine the drainage capacity and velocity of horizontal main and branches:

1. If the flow GPM is known, use Table 8–9 to find the flow velocity in fps for various slope and pipe sizes.
2. If the diameter is known, use Table 8–10 to determine the maximum number of dfu that can be connected for various size horizontal branches and stacks.
3. If the diameter of the building drain or sewer is known, use Table 8–11 to determine the maximum number of dfu that may be connected.

8.5.4 Sizing of Vent Pipes

The capacity of vent pipes depends on three major factors: the size of the stack, the number of dfu connected on the stack, and the developed length of the vent pipe. In addition, when routing of vent pipes is not direct, auxiliary vents, such as relief vent, a loop vent, or crown vent, will be required. Requirements may vary with applicable codes. In general, vents should comply with the following rules:

TABLE 8–9

Approximate discharge rate (GPM) and velocities (fps) in half-full[1] sloping drains
(data: condensed from National Plumbing Code)

Actual Inside Pipe Diameter	1/16 in./ft		1/8 in./ft		1/4 in./ft		1/2 in./ft	
Inches	GPM	fps	GPM	fps	GPM	fps	GPM	fps
1½					3.9	1.42	5.5	2.01
2					8.4	1.72	11.9	2.43
3			17.6	1.59	24.8	2.25	35.1	3.19
4	26.7	1.36	37.8	1.93	53.4	2.73	75.5	3.86
6	78.5	1.78	111.	2.52	157.	3.57	222.	5.04
8	170.	2.17	240.	3.07	340.	4.34	480.	6.13
10	308.	2.52	436.	3.56	616.	5.04	872.	7.12
12	500.	2.83	707.	4.01	999.	5.67	1413.	8.02

[1]*Half full* means filled to a depth equal to one-half of the inside diameter.

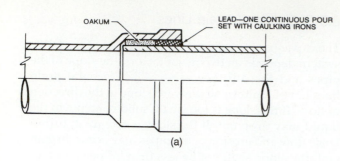

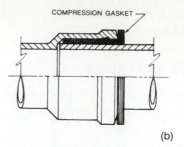

FIGURE 4.2.11.2b
TYPICAL HUBBED JOINT WITH
A COMPRESSION GASKET

(a)

(b)

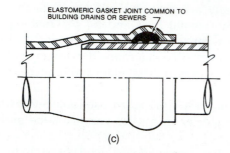

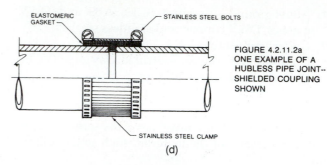

FIGURE 4.2.11.2a
ONE EXAMPLE OF A
HUBLESS PIPE JOINT--
SHIELDED COUPLING
SHOWN

(c)

(d)

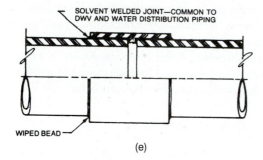

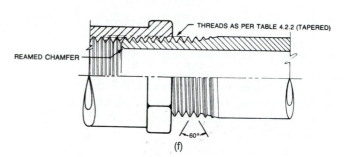

(e)

(f)

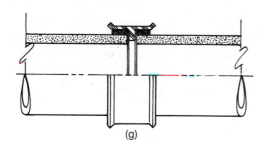

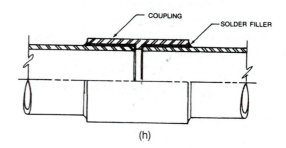

(g)

(h)

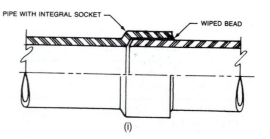

(i)

■ FIGURE 8–31

Typical methods of jointing pipes. (a) Cast iron (C.I.), caulked; (b) C.I., compression gasket; (c) DWV, plastic elastomeric gasket; (d) C.I./plastic, joint-shielded coupling; (e) Plastic, solvent; (f) Steel/plastic, threaded; (g) Clay (VCP), plastic bell; (h) Copper, soldered; (i) Plastic, solvent with socket. (Courtesy of National Standard Plumbing Code.)

TABLE 8–10

Maximum number of fixture units (dfu) that may be connected to branches, branch intervals (BI), and stacks (data: condensed from National Plumbing Code)

| Diameter of Pipe, Inches | Maximum Number of dfu to Be Connected | | | |
| | Horizontal Branches[1] | Stacks with < or = 3BI | Stacks with > 3BI | |
			Total	Each BI
1½	3	4	8	2
2	6	10	24	6
4	160	240	500	90
6	620	960	1900	350
8	1400	2200	3600	600
10	2500	3800	5600	1000
12	3900	6000	8400	1500

[1]Does not include branches of the building drain.

TABLE 8–11

Maximum number of fixture units (dfu) that may be connected to building drain or sewer (data: condensed from National Plumbing Code)

Pipe Diameter, Inches	1/16 in./ft	1/8 in./ft	1/4 in./ft	1/2 in./ft
2			21	26
3			42	50
4		180	216	250
6		700	840	1000
8	1400	1600	1920	2300
10	2500	2900	3500	4200
12	2900	4600	5600	6700

1. Individual vents shall not be smaller than one-half the diameter of the required drainpipe served.
2. No vent shall be smaller than 1¼″ in diameter.
3. Vents exceeding 40 feet in developed length shall be increased by one nominal pipe size. The nominal pipe sizes are 1¼″, 1½″, 2″, 2½″, 3″, 4″, 5″, 6″, 8″, 10″, 12″, and 15″.
4. Vents exceeding 100 feet in developed length shall be increased by another nominal pipe size.

Example 8–13 Based on the architectural floor plan of two office building toilets (Figure 8–32(a)) determine and show the drainage and vent piping sizes of the DWV system in the isometric diagram. For design purposes, the vent stack is assumed to be 30 feet total developed length (DL), and the floor drains are 2″ outlet size. The plumbing floor plan is shown in Figure 8–32(b).

First, from Table 8–6, identify the dfu load for each section of the DWV system it serves. The dfu loads are marked in Figure 8–32(c). (*Note:* The floor drains are not normally used, except during washdowns or during fixture overflows; thus, they are not counted in the dfu loads.)

Next, select the drain and stack sizes from Table 8–10. The answers are shown in Figure 8–32(c).

Finally, determine and show the individual vent, circuit vent, and vent stack sizes in accordance with the rules given in this section. These answers are also shown in Figure 8–32(d). (*Note:* 2″ minimum vent size

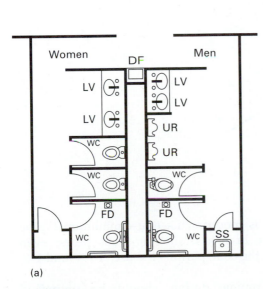

(a)

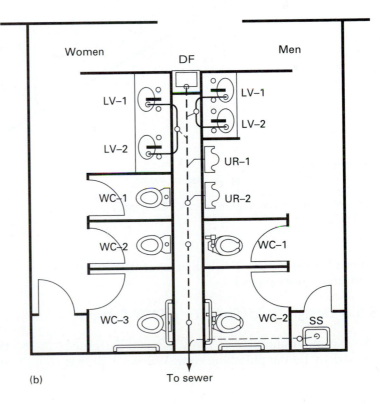

(b)

■ **FIGURE 8–32**

Architectural floor plan of two toilets and plumbing drawings showing a DWV piping system. (a) Architectural floor plan. (b) Plumbing floor plan.

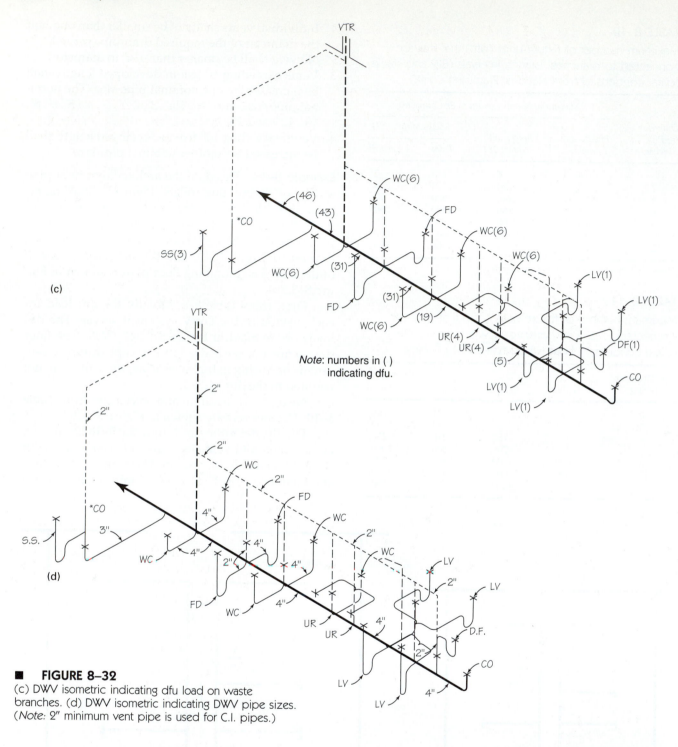

■ FIGURE 8–32
(c) DWV isometric indicating dfu load on waste
branches. (d) DWV isometric indicating DWV pipe sizes.
(*Note:* 2″ minimum vent pipe is used for C.I. pipes.)

is used throughout. A 2″ C.I. pipe is actually cheaper
than smaller size pipes.)

8.5.5 Design Guidelines for a DWV System

1. The minimum size of DWV pipes is governed
   strictly by the plumbing code. When C.I. pipe is

used, it is good practice to use minimum 2″ pipe
size for branch drain and vent pipes, even
though the code permits the use of 1¼″ and 1½″
pipes.

2. DWV piping design should be optimized
   through the proper coordination of pipe chase
   sizes and locations.

3. The layout of the toilet room is usually initiated by the architectural plan. However, it is the responsibility of the plumbing engineer to determine the need for floor drains and to coordinate with the architect on the location of cleanouts, etc.

4. Indirect waste shall be provided for all equipment that contains toxic or harmful chemicals. The drainage shall be piped to a separate receptor for sedimentation, neutralization, or filtration before being discharged into the public sewer system.

5. Other than intermittent discharges into the drainage system from dishwashing and laundry equipment with water at a temperature of 140°F or above, no high temperature waste or steam pipe shall discharge into the drainage system without subcooling the effluent prior to connecting to the sanitary sewer.

6. All plumbing fixtures or drainage equipment without a built-in trap must be connected through an external trap.

7. All traps must be vented.

8. All horizontal drainage piping shall be installed in alignment at a uniform slope not less than ¼″ per foot for a diameter of 3″ and less, and not less than ⅛″ per foot for a diameter of 4″ or more. In rare cases, ¹⁄₁₆″ per foot is permitted by agreement with the code-enforcing authority. A slope greater than 1″ per foot should be avoided when the waste contains solid matter.

9. Cleanouts shall be installed at the base of drainage stacks and at the beginning of main horizontal branches so that the entire DWV system can be cleaned and cleared from clogging.

10. Grease-laden waste from kitchens should be piped directly to the building drain or stack whenever practical. A grease trap shall be installed for commercial kitchens, prior to connection to the waste pipe.

11. Waste containing high volumes of insoluble matter, such as sand, plaster, etc., shall be intercepted by sediment basins or catch basins prior to discharging into the sewer.

12. Waste containing oil, such as drains from a commercial garage, shall be connected through an oil interceptor.

8.6 SEWAGE TREATMENT AND DISPOSAL

To protect water resources and the greater environment, all waste from buildings and industrial processes must be treated to meet certain standards of quality. Domestic sewage from dwellings and DWV systems in buildings are permitted to be discharged into the public sewer system, which provides the necessary treatment prior to its discharge into nature. When public sewers are not accessible, or when there is no public sewer system in the vicinity of a building or buildings, a private sewage treatment system will have to be constructed.

Following are the definitions of some commonly used terms related to the subject of sewage treatment methods and disposal processes:

1. *Digestion.* That portion of the sewage treatment process in which biochemical decomposition of organic matter takes place, resulting in the formation of simple organic and mineral substances. Also known as *aerobic (bacterial) digestion.*

2. *Influent.* Untreated sewage flowing into a treatment system.

3. *Effluent.* Treated or partially treated sewage flowing out of a treatment system.

4. *Septic tank.* A watertight tank that receives influent from a DWV system designed to separate solids from the liquid, to digest organic matter through a period of detention, and to discharge the effluent to an approved method of disposal.

5. *Sedimentation.* Formation of layers of heavy particulates in the influent.

6. *Aerobic (bacterial) digestion.* Digestion of the waste through the natural bacterial digestive action in a septic tank or digestion chamber.

7. *Active sludge.* The sewage sediment, rich in destructive bacteria, that can be used to break down fresh sewage more quickly.

8. *Dosing tank or chamber.* A component of the septic tank system that periodically discharges effluent to an approved method of disposal.

9. *Filtration.* A means of filtering out any solid matter from the effluent.

10. *Disinfection.* A process to disinfect the effluent with chemicals.

11. *Percolation.* The flow or trickling of a liquid downward through a filtering medium.

12. *Drain field* or *leaching field.* A set of trenches containing open-ended, or perforated, pipes designed to allow the treated effluents to percolate into the ground.

8.6.1 The Sewage Treatment Process

The sewage treatment process may be divided into three major steps:

1. *Primary Treatment,* which is subdivided into:
   - *Sedimentation and retention:* Raw sewage is retained for the preliminary separation of indigestible solids and the start of aerobic action
   - *Aeration:* Introduction of air through natural convection or mechanical blowers to accelerate the decomposition of organic matters
   - *Skimming:* Removal of scum that floats on top of the partially treated sewage
   - *Sludge removal:* Disposal of heavy sludge at the bottom of treated sewage
2. *Secondary Treatment,* namely, the removal of fine suspended matter from the effluent through a filtration process, such as the use of sand filters, drain fields, seepage pits, etc.
3. *Tertiary Treatment,* that is, the disinfection of effluent by the addition of chemicals, such as chlorine.

8.6.2 Sewage Treatment Plants

The design of sewage treatment plants for large buildings, building complexes, and municipalities follows precisely the same processes described above. However, modern treatment plants do require considerable mechanized equipment and controls in order to be efficient and reliable. The design of these treatment plants is done by engineers who specialize in the subject.

8.6.3 Septic Tank System

Rather than a sewage treatment plant, the septic tank system is most commonly used in rural areas for small capacity applications. It consists of a septic tank serving as primary treatment and a method of filtering the effluent as secondary treatment. The septic tank is usually constructed of concrete and has a retention volume from 1000 to 10,000 gallons. A typical

design is shown in Figure 8–33(a). A septic tank system is normally designed to flow by gravity, although ejector pumps are used when the topography does not permit gravity flow.

Filtering of effluent may be by means of a filter pit, sand filters, or drain fields. A drain field, which consists of multiple runs of underground trenches, is popular because it requires less maintenance. Typical construction of a drainage trench is shown in Figure 8–33(b) and that for a drain field is shown in Figure 8–33(c). Figure 8–33(d) illustrates the addition of a tertiary treatment for a septic tank system.

To design a septic tank system, one begins with an analysis of the sewage load based on the number of occupants in and type of occupancy of the building on a 24-hour basis. Table 8–12 provides the estimated load for various occupancies. The data are applicable to most regions in the United States and should be modified for other cultures and countries.

Next, a suitable location is selected. Normally, septic tanks and their disposal fields are located away from heavy traffic and maintain at least a minimum distance from the building, wells, water services, etc., as prescribed in the local health or building codes. In general, septic tanks and disposal fields are a minimum of 10 feet from building water lines and property lines, 50 feet from deep wells, and 100 feet from shallow wells.

The required size of a septic tank for use in residences is determined from either the number of fixture units (dfu) served or the number of bedrooms in the residence. Table 8–13 provides a sizing guide.

The size of the septic tank for nonresidential building) is given by the following general rule: Use a multiplier of 1.5 applied to the calculated daily sewage rate (gpd), with gradual reductions of the multiplier for systems larger than 1500 gpd.

TABLE 8–12

Average daily sewage load for occupancies in the United States
(in gallons/day/person unless otherwise indicated)

Occupancy	Load	Occupancy	Load
Schools (without cafeteria and showers)	15	Hospital (general)	100
Schools (with cafeteria and showers)	35	Public institutions	100
Schools (boarding)	100	Restaurants (per serving)	25
Day camps	25	Motels	60
Trailer parks, etc.	50	Stores (per toilet)	400
Swimming pools and beaches	10	Airport (per passenger)	5
Luxury residences/estates	150	Assembly hall (per seat)	2
Hotels (2 persons/room)	100	Churches (small)	3–5
Factories	50	Churches (large)	5–7
Nursing homes	75	Subdivision or homes	75

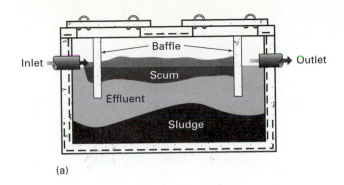

(a)

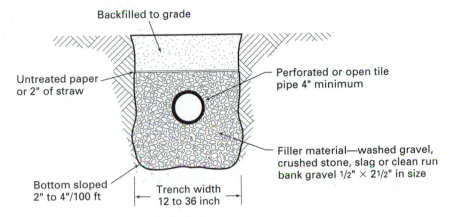

Backfilled to grade

Untreated paper or 2" of straw

Perforated or open tile pipe 4" minimum

Filler material—washed gravel, crushed stone, slag or clean run bank gravel 1/2" × 21/2" in size

Bottom sloped 2" to 4"/100 ft

Trench width 12 to 36 inch

(b)

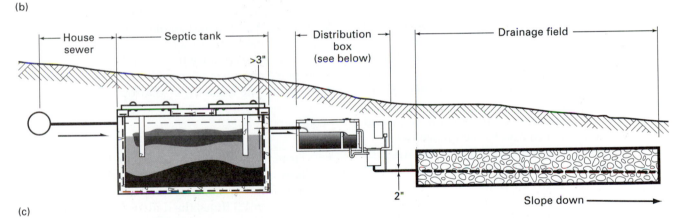

House sewer

Septic tank

>3"

Distribution box (see below)

Drainage field

2"

Slope down

(c)

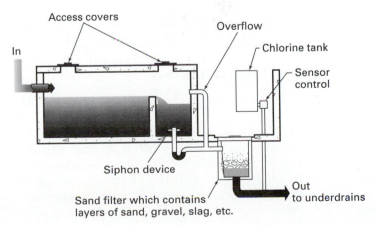

Access covers

Overflow

Chlorine tank

Sensor control

In

Siphon device

Sand filter which contains layers of sand, gravel, slag, etc.

Out to underdrains

(d)

■ FIGURE 8–33

(a) Longitudinal section of a typical septic tank. The tank is normally constructed of precast or reinforced concrete. Watertight access covers are required for cleanup. (b) Cross section of a typical underground drainage trench. (c) General layout of a septic tank sewage treatment system showing the primary treatment (septic tank) and the secondary treatment (drain field). (d) Addition of tertiary treatment (chlorine) and a dosing chamber to provide an intermittent flow of the effluent will improve distribution. (Figures a, b, and d courtesy of National Standard Plumbing Code.)

TABLE 8–13

Required capacity of septic tanks for residential units

Single Family Dwellings (Number of Bedrooms)	Multiple Dwelling Units or Apartments (One Bedroom Each)	Other Uses: Maximum Fixture Units Served	Minimum Septic Tank Capacity, Gallons*
1–3		20	1000
4	2	25	1200
5 or 6	3	33	1500
7 or 8	4	45	2000
	5	55	2250
	6	60	2500
	7	70	2750
	8	80	3000
	9	90	3250
	10	100	3500

Extra bedroom	150 gal. ea.
Extra dwelling units over 10	250 gal. ea.
Extra fixture units over 100	25 gal/dfu

*Septic tank sizes in this table include sludge storage capacity and the connection of domestic food waste disposal units without further volume increase.

The drain field includes one or more rows of drainage trenches complying with the following guidelines:

- *Trench construction.* Perforated or open-jointed tile surrounded with washed gravel or crushed stone and covered with backfill. (See Figure 8–33(b).)
- *Drain tile size.* Minimum 4-inch-diameter VCP or plastic.
- *Trench slope.* Bottom of trench shall be sloped downward between 2 and 4 inches per 100 ft.
- *Trench depth.* Below the frost line of the area, but in no case less than 18 inches below grade.
- *Trench width.* Between 12 and 36 inches. Wider trenches have a higher percolation capacity. (See Table 8–14 for capacity ratings.)
- *Trench spacing.* Minimum 6 ft between adjacent trenches for 12- to 24-in.-wide trenches and up to 9 ft for a 36-in.-wide trench.
- *Length of trench.* Not to exceed 100 feet.

The minimum size of the drain field depends on the width of the drain trench, makeup of the soil, absorption capacity of the soil, and total length of the drain tiles. The absorption capacity of the soil is determined by percolation test to determine the time taken for water in the drain field trenches to drop 1 inch. The total linear length of drain pipe required can be found from Table 8–14. When the water table of the field is too close to the bottom of the drain field trenches, al-

TABLE 8–14

Drain tile required
(in linear foot for each 100 gallons of sewage/day)

Time in Minutes for 1-inch Drop in Percolation	Tile Length for Trench Widths of		
	1 foot	2 feet	3 feet
1	25	13	9
2	30	15	10
3	35	18	12
5	42	21	14
10	59	30	20
15	74	37	25
20	91	46	31
25	105	53	35
30	125	63	42

source: National Standard Plumbing Code.

ternative filtering methods, such as mounted drain fields and sand filters at higher elevations, should be used.

Example 8–14   Design a septic tank system for an elementary school based on the following criteria:

1. The school has a cafeteria and showers.
2. The total student and faculty population is 200.
3. A drain field system with 2-ft wide trenches shall be used as secondary treatment.
4. A percolation test of the soil indicates that the absorption rate is 2 min/in. of drop.

Answers

- From Table 8–12, the estimated daily sewage flow is 35 gallons per person, and the total daily flow for 200 persons is 35 × 200, or 7000 gpd.
- Based on 1.5 times the daily flow rate, the septic tank volume should be 1.5 × 7000, or 10,500 gallons.
- Based on Table 8–14, the number of linear feet of drain field should be 15 feet per 100 gpd; thus

$$\text{Total length of trenches} = \frac{10{,}500 \times 15}{100}$$
$$= 1575 \text{ ft (use 1600 ft)}$$

- The plan would favor the use of two (2) 5000-gallon tanks, each with 800 ft of drain tiles divided among eight (8) 100-ft-long trenches on 6-ft centers. Figure 8–34 is a preliminary layout of the system prior to a consideration of other concerns.

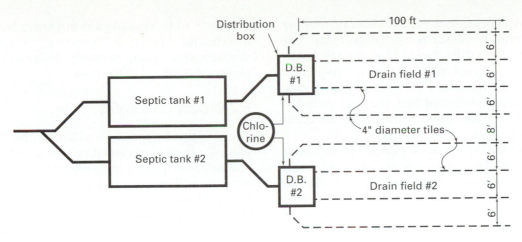

■ FIGURE 8–34
Flow diagram of a typical septic tank sewerage treatment and disposal system.

8.7 STORM DRAINAGE SYSTEM

A storm drainage system conveys rainwater or melting snow from a building or site to the points of disposal. Among these locations to be drained are roofs, patios, areaways, etc., of buildings and parking lots, and roadways, lawn, gardens, etc., on the site. In general, other than small or incidental areas, all exterior storm drainage should be connected externally to the building storm drainage system.

Following are some of the commonly used terms related to storm drainage systems:

1. *Roof drain.* Device installed on a roof to receive water collected on the surface of the roof and to discharge it to the point of disposal through a conductor (or downspout).
2. *Area drain.* Similar to a roof drain, except that it is installed in an exterior area, not over a building space.
3. *Conductor* or *downspout* Vertical portion of a storm drainage system installed in the interior of a building.
4. *Gutter.* Exterior trough, installed below a roof, which collects rainwater from the roof and discharges it to the point of disposal through a leader.
5. *Leader.* Vertical portion of a storm drainage system installed at the exterior of a building.
6. *Subsoil drain.* Drainpipe that collects subsurface water and conveys it to a place of disposal.
7. *Controlled storm drainage system.* Storm drainage system that collects storm water on a roof and releases the flow slowly to the drainage system to allow the load to drain within a longer time frame.
8. *Primary drainage system.* Basic storm water drainage system designed for normal use.
9. *Secondary drainage system.* Additional storm drainage system that will handle any storm water overflow that may occur when heavier storms occur.
10. *Sump.* Receiver or pit that receives liquid waste, storm water, or groundwater located below the elevation of a gravity system.
11. *Sump pump.* Pump that removes liquid from a sump.
12. *Projected roof area* or *horizontal projected roof area.* Horizontal component of a sloping or pitched roof. The projected roof area is in effect equal to a flat roof at the base of the pitched roof. When rain falls at an oblique angle, one side of the pitched roof will intercept more rain than the other side. However, the average rainfall on both sides of the roof is still no more than that of a flat roof. (See Section 8.7.3 for a discussion of the sizing of storm drainage pipes.)

8.7.1 Design Principle

The fundamental principle behind the design of storm water systems is to install a piping or conductor system to lead the storm water away from the building and site in a reasonable length of time to keep the surfaces dry. The size of the system depends on the rate of rainfall, rather than the total rainfall in a day or in a year. The rate of rainfall varies with intensity and frequency of occurrence—for example, inches of rainfall per hour during a 15-minute period as it may happen once per 100 years (15 min per 100-year return). It should be understood that "per 100-year return" actually means "1-percent probability," and "per 10-year return" means "10-percent probability," that the stated intensity of rainfall may occur. As with any statistical data, the same probability may occur in consecutive years or may not occur at all for many years.

Although the design of an exterior storm drainage system is similar in principle to that of an interior system, the circumstances surrounding each are considerably different. Exterior drainage must be closely coordinated with the capability of the public sewer or public drainage system. It will not be cost effective to drain a large area, such as a parking lot or a park, on the same basis as a building area. In many instances, the instantaneous drainage of storm water on a large exterior area may overtax the public sewer system. If this occurs, a method of delaying the runoffs may be necessary or may be directed by the public sewer district. The delaying methods include the undersizing of drainpipes, by letting the ground gradually absorb the water, or by creating a storm water holding area, a technique known as "ponding." Exterior storm drainage systems are normally designed by civil engineers specialized in the topic.

8.7.2 Data on Rainfall

The selection of a design value for the intensity of rainfall is usually governed by code and the economic and safety factors of the project. Overdesigning for intensity will assure a faster drainage of storm water from the building or the site, whereas underdesigning will cause inconvenience, safety hazards, or possible structural failures. Careful coordination between the architect and structural and plumbing engineers will result in an optimum design. Most codes specify the design rainfall rate to be based on 1-hour and 15-min duration rates in a 100-year return period for the primary and secondary drainage systems, respectively.

The recognized rainfall rates of selected cities in each of the 50 states are shown in Table 8–15.

TABLE 8–15

Rate of rainfall for U.S. cities
(rainfall rates, in inches per hour (in./hr), based on a storm of 1-hr duration and a 100-year return period)

State	City	in./hr	State	City	in./hr
Alabama	Mobile	4.6	Montana	Ekalaka	2.5
Alaska	Fairbanks	1.0	Nebraska	Omaha	3.8
Arizona	Phoenix	2.5	Nevada	Las Vegas	1.4
Arkansas	Little Rock	3.7	New Hampshire	Concord	2.5
California	Los Angeles	2.1	New Jersey	Newark	3.1
	San Fernando	2.3	New Mexico	Albuquerque	2.0
Colorado	Denver	2.4	New York	New York	3.0
Connecticut	Hartford	2.7	North Carolina	Charlotte	3.7
Delaware	Wilmington	3.1	North Dakota	Bismarck	2.8
District of Columbia	Washington	3.2	Ohio	Cincinnati	2.9
Florida	Miami	4.7	Oklahoma	Oklahoma City	3.8
Georgia	Atlanta	3.7	Oregon	Eugene	1.3
Hawaii	Hilo	6.2	Pennsylvania	Philadelphia	3.1
	Honolulu	3.0	Rhode Island	Providence	2.6
Idaho	Boise	0.9	South Carolina	Charleston	4.3
Illinois	Chicago	3.0	South Dakota	Rapid City	2.9
Indiana	Indianapolis	3.1	Tennessee	Memphis	3.7
Iowa	Des Moines	3.4	Texas	Dallas	4.0
Kansas	Wichita	3.7		Houston	4.6
Kentucky	Lexington	3.1	Utah	Salt Lake City	1.3
Louisiana	New Orleans	4.8	Vermont	Burlington	2.1
Maine	Portland	2.4	Virginia	Norfolk	3.4
Maryland	Baltimore	3.2	Washington	Seattle	1.4
Massachusetts	Boston	2.5	West Virginia	Charleston	2.8
Michigan	Detroit	2.7	Wisconsin	Milwaukee	3.0
Minnesota	Minneapolis	3.1	Wyoming	Cheyenne	2.2
Mississippi	Biloxi	4.7			
Missouri	Kansas City	3.6			
	St. Louis	3.2			

Source: U.S. National Weather Service.

8.7.3 Design of Storm Drainage Systems

To design a storm drainage system, first determine where drainage is required; other than roofs, area-ways, driveways, and walkways toward spaces of lower elevation, entrances into and exits from buildings are areas that should be considered. Whenever possible, all exterior storm drains should be connected to a storm sewer outside of the building. Drains lower than the elevation of the sewer may have to be collected in a sump and be pumped out automatically.

After this, determine the roof drain locations. Roof drains should be located at lower spots or in depressed areas of the roof. The locations should be determined by the architect and structural engineer. A 0.5-percent minimum slope toward each roof drain is necessary to assure positive drainage without standing water or ponding. Where controlled discharge is desired, in order to use the roof as a reservoir, the roof structure must be designed to carry the extra weight of water (at 62.4 lb/cu ft). In general, a minimum of two roof drains should be installed for a building, and a minimum of four for buildings over 10,000 sq ft.

Next, determine the roof drain design criteria, based on or exceeding the code requirements for the rate of rainfall for primary and secondary roof drain systems. Secondary drainage is a safety provision during an extremely heavy rainfall or in the event that drains become clogged. A secondary system may be piped separately or may be designed to spill over the roof, as is permitted by the code.

Then select the appropriate drainage fittings. Numerous designs of roof and area drains are available to suit the construction methods. Typical roof and area drains are shown in Figure 8–35. Drainage receptors and trenches are shown in Figure 8–36.

Finally, select the piping materials and method of installation. Different piping materials are used for above- and underground locations. Commonly used piping materials are:

- *Above ground (interior).* C.I. with caulked or mechanical joints, galvanized steel with mechanical joints (welded joints are not permissible), copper pipe (types K or L), copper tube (DWV), aluminum, plastic (ABS or PVC).
- *Above ground (exterior).* C.I., galvanized steel, copper leader (round or rectangular), aluminum leader (round or rectangular).
- *Below ground (within building).* C.I., copper (type K), reinforced concrete, plastic (ABS or PVC).

- *Below ground (exterior to building).* Extra-strength VCP, reinforced concrete, plastic (ABS or PVC).
- *Subsoil drainage.* Open-jointed or perforated VCP, plastic, bituminized fiber, concrete pipes embedded in crushed stone, rock, or gravel. [See details in Figure 8–36(b).]

When plastic pipes are used, consideration should be given to contraction and expansion of the material, which is more severe than in metal. Thermal insulation may be needed for metal pipes to avoid external condensation of moisture on the surface of the pipes, which may cause staining on ceilings or walls.

In all of the preceding, the size of the drainage pipe is determined by three factors: the equivalent horizontal projected area (EHPA), the rate of rainfall, and the slope of the drainpipe. The EHPA is determined as follows:

- For a horizontal roof, use the actual roof area.
- For a two-sided pitched roof, use the horizontal projected roof area.
- For a single-sided pitched roof, use 1.4 times the horizontal projected roof area.
- For a flat roof adjacent to a vertical wall, use the flat roof area plus 50 percent of the vertical wall area above the roof.
- For a flat roof adjacent to two or more vertical walls, use the flat roof area plus 25 percent of the average of the two largest walls.

Size of storm drain pipes are determined from the following:

- For horizontal drain pipes (interior), use Table 8–16.
- For gutters (exterior), use Table 8–17.
- For conductors (interior) or leaders (exterior), use Table 8–18.

Example 8–15 Determine the size of horizontal drainpipes and the main vertical conductor (downspout) for a building having two roof drains. Each roof drain serves a flat roof area of 2000 sq ft. The building is located in St. Louis, Missouri.

- From Table 8–15, the design rainfall rate for St. Louis, Missouri, is 3.2 in./hr, based on a 1-hr duration and a 100-year return period.
- Based on a ¼-inch-per-foot slope, each roof drain shall be connected with a 4-inch horizontal drainpipe. (*Note:* A 4-inch drain line at a ¼″

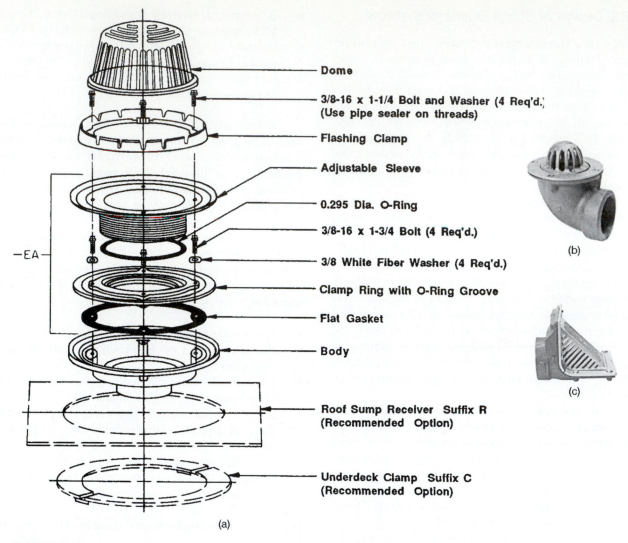

Dome

3/8-16 x 1-1/4 Bolt and Washer (4 Req'd.)
(Use pipe sealer on threads)

Flashing Clamp

Adjustable Sleeve

0.295 Dia. O-Ring

3/8-16 x 1-3/4 Bolt (4 Req'd.)

3/8 White Fiber Washer (4 Req'd.)

Clamp Ring with O-Ring Groove

Flat Gasket

Body

Roof Sump Receiver Suffix R
(Recommended Option)

Underdeck Clamp Suffix C
(Recommended Option)

(a)

(b)

(c)

■   **FIGURE 8–35**
(a) Exploded view of a roof drain showing the drain body (sump), clamp ring,
adjustable roof desk sleeve, flashing clamp/gravel guard, and dome strainer. (b) Cornice
drain. (c) Oblique scupper drain. (All illustrations courtesy: Zurn Industries, Inc., Erie,
PA.)

slope can serve a 5300- and 2650-sq-ft EPRA for 2″ and 4″ rainfall rates, respectively. Interpolating for the 3.2-inch rainfall rate, a 4-inch pipe should be selected for the horizontal drainpipes.)

- The main horizontal drainpipe should be a 5-inch pipe.
- From Table 8–18, interpolating between 2 inches and 4 inches for the 3.2-inch rainfall rate, we find that a 4-inch conductor can serve up to 4844 sq ft. However, a 5-inch downspout still should be used, since reducing the size of pipe downstream of the flow is prohibited.

8.7.4 Subsoil Drainage System

When the lower level of a building (e.g., the basement) is below ground, there are two potential problems: Surface water may flow into the building through cracks or seepage of the basement walls, and groundwater may push into the building due to hydrostatic pressure if the normal groundwater level (known as the *water table*) is higher than the lowest floor level.

To avoid water seepage, all walls below ground must first be waterproofed, and the grade next to the building is often sloped away from the building. If

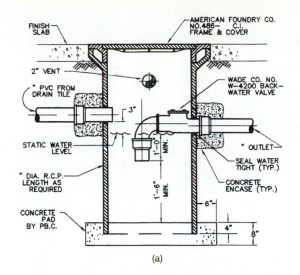

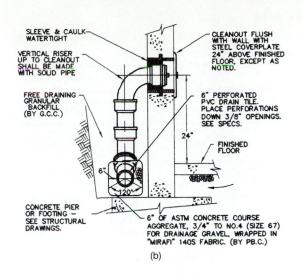

(a)   (b)

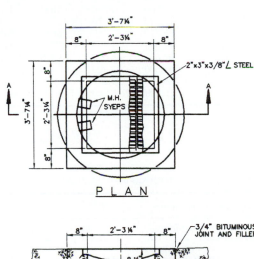

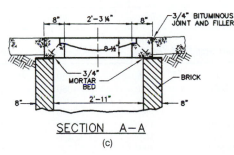

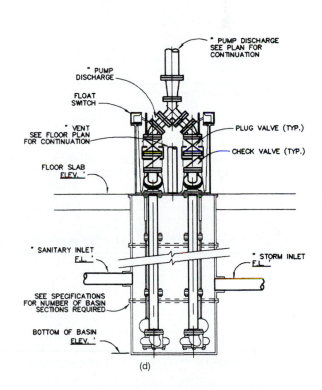

(c)   (d)

■   **FIGURE 8–36**

Typical storm drainage details.

(a) A sand interceptor where sand entrained in the storm water is settled at the bottom of the basin with only clear effluent to flow out.

(b) A subsoil drainpipe installation detail with drain pipe embedded in 6″ of granular fill. Wall cleanouts are provided at frequent intervals.

(c) A storm water drainage manhole or catch basin with cast iron grating.

(d) A combination storm and sanitary basin with duplex submersible sump pumps.

TABLE 8–16

Allowable flow rate (GPM) and projected roof or surface area (sq ft) for horizontal storm drainpipes (inches) at various slopes and design rainfall rates

Dia. of Drain (Inches)	Design Flow in Pipe (GPM)	Allowable Projected Roof Area at Various Rates of Rainfall per Hour (Sq Ft)				Dia. of Drain (Inches)	Design Flow in Pipe (GPM)	Allowable Projected Roof Area at Various Rates of Rainfall per Hour (Sq Ft)			
		1"	2"	4"	6"			1"	2"	4"	6"
Slope 1/16 in./ft						*Slope 1/4 in./ft*					
2						2	17	1632	816	408	272
3						3	48	4640	2320	1160	775
4						4	110	10,600	5300	2650	1770
5	100	9600	4800	2400	1600	5	196	18,880	9440	4720	3150
6	160	15,440	7720	3860	2575	6	314	30,200	15,100	7550	5035
8	340	32,720	16,360	8180	5450	8	677	65,200	32,600	16,300	10,870
10	620	59,680	29,840	14,920	9950	10	1214	116,800	58,400	29,200	19,470
12	1000	96,000	48,000	24,000	16,000	12	1953	188,000	94,000	47,000	31,335
						15	3491	336,000	168,000	84,000	56,000
Size	GPM	1"	2"	4"	6"		GPM	1"	2"	4"	6"
Slope 1/8 in./ft						*Slope 1/2 in./ft*					
2						2	24	2304	1152	576	384
3	34	3290	1645	822	550	3	68	6580	3290	1644	1100
4	78	7520	3760	1880	1250	4	156	15,040	7520	3760	2510
5	139	13,360	6680	3340	2230	5	278	26,720	13,360	6680	4450
6	222	21,400	10,700	5350	3570	6	445	42,800	21,400	10,700	7130
8	478	46,000	23,000	11,500	7670	8	956	92,000	46,000	23,000	15,330
10	860	82,800	41,400	20,700	13,800	10	1721	165,600	82,800	41,400	27,600
12	1384	133,200	66,600	33,300	22,200	12	2768	266,400	133,200	66,600	44,400
15	2473	238,000	119,000	59,500	39,670	15	4946	476,000	238,000	119,000	79,330

Source: National Standard Plumbing Code.

TABLE 8–17

Allowable flow rate (GPM) and projected roof or surface area (sq ft) for semicircular roof gutters (inches)

Diameter of Gutter (Inches)	1/16-inch-per-foot Slope		1/8-inch-per-foot Slope		1/4-inch-per-foot Slope		1/2-inch-per-foot Slope	
	Square Feet	Gallons per Minute	Square Feet	Gallons per Minute	Square Feet	Gallons per Minute	Square Feet	Gallons per Minute
3	680	7	960	10	1360	14	1920	20
4	1440	15	2040	21	2880	30	1080	42
5	2500	26	3520	37	5000	52	7080	74
6	3840	40	5440	57	7680	80	11,080	115
7	5520	57	7800	81	11,040	115	15,600	162
8	7960	83	11,200	116	14,400	165	22,400	233
10	14,400	150	20,400	212	28,800	299	40,000	416

Source: BOCA/National Plumbing Code. Data are based on design rainfall rate of 1 inch per hour and shall be multiplied by the appropriate rainfall rate of the locality.

the anticipated water table is close to or higher than the building floor level, a subsoil drainage system consisting of 4- to 6-inch-diameter drain tiles should be installed.

Note also the following:

- Minimum 4-inch-diameter VCP or plastic pipe laid end to end with about a ½-inch gap between sections or perforated pipes should be used.

- The drainpipes should be located on the exterior of foundation wall if the water table is higher than the footing. Figure 8–36(b) shows the construction of a subsoil drain outside of the foundation wall.

- Pipes shall be surrounded by graded, crushed rocks, or gravel and be screened on the outer area by a mesh material to prevent dirt, sand, or backfill material from washing into the tiles.

TABLE 8–18
Allowable flow rate (GPM) and projected roof or surface area (sq ft)
for various conductors or leaders (inches) at various design rainfall rates

Dia. of Pipe or Leader (Inches)	Design Flow in Pipe (GPM)	Allowable Projected Roof Area at Various Rates of Rainfall per Hour (Sq Ft)			
		1"	2"	4"	6"
2	23	2176	1088	544	363
2½	41	3948	1974	987	658
3	67	6440	3220	1610	1073
4	144	13,840	6920	3460	2307
5	261	25,120	12,560	6280	4187
6	424	40,800	20,400	10,200	6800
8	913	88,000	44,000	22,000	14,667

- Tiles shall be sloped to drain, if achievable, or to be pumped after being collected in a sump. A sand interceptor (trap) may be necessary at the end of drain tile to allow solid material to drop out prior to disposal of the liquid. Figure 8–36 (a) is a typical layout of a sand interceptor.

8.8 PLUMBING SERVICES FOR OTHER BUILDING EQUIPMENT

Plumbing services for other building equipment are part of the plumbing work that must be designed and/or coordinated by the plumbing engineer. These services include water supply, treatment, drainage, and waste disposal for the following:

- *HVAC equipment.* Boilers, heaters, chillers, cooling towers, etc.
- *Fire protection equipment* (for buildings without an independent fire service). Fire hose cabinet, limited-area automatic sprinklers, etc.
- *Food service equipment.* Kitchen sinks, dishwashers, disposals, grease interceptors, etc.
- *Laundry equipment.* Clothes washers, oil and fluid separators, etc.
- *Landscape.* Irrigation sprinklers, reflecting pools, fountains, etc.
- *Athletic and health equipment.* Health club whirlpools, baths, steam baths, swimming pools, etc.
- *Data-processing equipment.* Water-cooled computers, etc.
- *Research-processing equipment.* High-quality water, speciality gases such as oxygen and nitrogen, compressed air, and a vacuum-piping system in hospitals and research laboratories.

Design criteria for the foregoing equipment are usually provided by the owner/architect, manufacturers, or specialty consultants and evaluated for implementation by the plumbing engineer.

QUESTIONS

8.1 Do National Plumbing Codes provide sufficient data and guidelines for designing a sanitary drainage system? If so, explain.
8.2 Do National Plumbing Codes provide sufficient data and guidelines for designing a domestic water distribution system? Explain.
8.3 What are the equivalents between pounds and grains, gallons and cubic feet, ppm and mg/L?
8.4 Name the major influences on the quality of water.
8.5 Name seven processes used to improve the quality of water.
8.6 Name the five substances (impurities) that are limited in minimum quality standards for drinking (potable) water systems in the USA.
8.7 What mineral content is used to determine the hardness in water?
8.8 Describe the Zeolite water softening process.
8.9 What is osmosis? Reverse osmosis?
8.10 What are the two commonly used water filtration systems?
8.11 What are the equivalents of gpm vs. L/s, cfm vs. L/s, psi vs. kPa?
8.12 A light commercial building is installed with four flush-type tanks, water closets, two urinals with flushometers, and four lavatories. What is the total water supply fixture (units wsfu)? What is the probable water demand?
8.13 A building is installed with 200 tons of air conditioning. The air conditioning unit is water cooled through a cooling tower. How much makeup water should be provided?

8.14 What is the pressure drop due to friction loss for 200 feet of one-inch PVC pipe with a flow of 10 gpm?

8.15 What is usually assumed as the equivalent length (EL) of pipe fittings in a plumbing piping system? (Note: This procedure applies to preliminary design only.)

8.16 What is the pressure at the base of a water pipe feeding a plumbing facility 500 feet above the base due to static height alone?

8.17 What is the hot water demand for a residential dishwasher in gpm and total consumption?

8.18 What type copper pipe is normally used for domestic water system?

8.19 What is the function of a vacuum breaker?

8.20 Why is backflow prevention important in a water distribution system? What is the most effective method used to prevent backflow?

8.21 What is the function of a shock absorber? Where should it be installed?

8.22 Based on the pump shown in Figure 8–18, what pump speed should be selected for a flow of 12 gpm requiring a total head of 7 feet of water column (w.c.)? If the same pump is used requiring a flow of 21 gpm, what total head can the pump develop?

8.23 Why is a flushometer required for water closets in lieu of flush tanks in public facilities?

8.24 What is the water demand for a pressure tank type water closet per flushing? For standard gravity tanks?

8.25 What is the purpose of a drainage trap? Do water closets require traps?

8.26 What is the nominal rating in drainage fixture units (dfu) of a water closet with flushometer? With gravity tank?

8.27 What is the difference between waste pipe and soil pipe?

8.28 What is a bathroom group? What is the wfu of a bathroom group with flush tank? What should be the minimum soil pipe diameter?

8.29 A school is planned for 2000 students, teachers and staff. Assuming the population is 50% male and 50% female, determine the minimum number of plumbing fixtures required. In your judgment, is this adequate?

8.30 Calculate the minimum number of plumbing fixtures for a 500-seat movie theater. In your judgment, is this adequate?

8.31 What is a stack? A vent? A stack vent? A vent stack?

8.32 What is the purpose of a stack vent?

8.33 What are the piping materials commonly used in drainage piping systems?

8.34 What are the joining methods for cast iron pipes? copper? plastic?

8.35 Normally, municipalities in the USA have separate sanitary and storm sewer systems. There are, however, several major cities which still have combination storm and sanitary sewer systems. In such cases, can the sanitary and storm piping systems within the building be combined?

8.36 What is the maximum number of dfu that can be connected to a 4-inch horizontal branch?

8.37 What is the maximum number of dfu that can be connected to a 4-inch building drain with 1/8" per foot fall? 1/2" per foot fall?

8.38 Give the major steps in a sewage primary treatment system, and the secondary treatment process.

8.39 What is the purpose of a dosing chamber in a septic tank sewage treatment system? Why?

8.40 A day camp has 100 occupants. Determine the septic tank size. Determine the linear feet of drain tile required based on a 10 minute/inch drop from percolation test and 2-foot wide trenches.

8.41 What is project roof area?

8.42 What is the rainfall rate of St. Louis, Missouri, based on a storm of 1-hour duration and a 100-year return period?

8.43 If the roof area is 1000 sq. ft., and the roof is adjacent to three vertical walls of 100, 200, and 500 sq. ft. each. What shall the roof drain or drains be designed for?

8.44 What is the minimum number of roof drains for any size building? For a building over 10,000 sq. ft. of roof area?

8.45 What size horizontal storm drain pipe should be installed for a building with 16,000 sq. ft. of projected roof area, in a region having 2-inch rainfall rate, and the drain pipe laid with 1/16"/ft. slope? With 1/4"/ft. slope?

8.46 What size roof drain leader should be installed for a 600 sq. ft. roof area in a region with 6"/hour rainfall rate?

8.47 In addition to the sanitary facilities, what other water drainage should be considered in plumbing system design for a large building complex?

REFERENCES

1. *National Standard Plumbing Code/1984*. Falls Church, Virginia: National Association of Plumbing/Heating/Cooling Contractors, 1992.

2. *Basic Plumbing Code/1993.* BOCA International, Inc., Chicago: Building Officials and Code Administrators International, Inc.

3. National Bureau of Standards, Report *BMS 79,* 1970 (now the National Institute of Building Sciences).

4. *ASHRAE Handbook and Product Directory, Systems, 1992.* Atlanta: American Society of Heating, Refrigerating and Air-Conditioning Engineers, Inc.

5. Building Officials and Code Administrators International, Inc., *National Building Code,* 1993.

6. *Fundamentals of Plumbing Design, ASPE Data Book/1992,* vol. 1. Sherman Oaks, California: American Society of Plumbing Engineers, 1992.

7. Tao, William, and Karl Norman, "Plumbing System Design." *Heating, Piping and Air Conditioning,* March 1989.

FIRE PROTECTION EQUIPMENT AND SYSTEMS

A FIRE PROTECTION SYSTEM IS A SYSTEM THAT includes devices, wiring, piping, equipment, and controls to detect a fire or smoke, to actuate a signal, and to suppress the fire or smoke. The two primary objectives of fire protection are to save lives and protect property. A secondary objective is to minimize interruptions of service due to a fire. Imagine that a fire is detected at the ground floor of a high-rise hotel. If the fire can be quickly isolated and extinguished, then there will be no need to disturb the guests on the upper floors. Otherwise, the entire building must be evacuated. Statistics indicate that to evacuate a 40-story building with 100 persons per floor through a single stairway will require one full hour. With two stairways, it will still take more than one-half hour. The chaos and financial consequences can hardly be overstated.

The cost of a fire protection system depends on a number of factors, such as the fire resistivity of the building, type of occupancy, number of floors below the ground level, height of the building, adequacy of egress, and degree of protection desired. The initial cost of a fire protection system may be as low as 1 percent, and is seldom more than 5 percent, of the total building construction cost. In many applications, the additional investment for a better fire protection system can have a positive payback in that the building's fire insurance premium will be reduced, in addition to achieving better protection of lives and property.

Current trends in building design and modern life-styles make for serious fire hazards. Some of the contributing factors are the following:

1. *High-rise buildings.* Most building codes define a high-rise building as a building more than 7 stories or more than 75 feet in height. The figure is more or less determined by the capability of fire department ladders in most cities. With rapid population growth in the congested urban areas, buildings become taller and more densely situated. Super high-rise buildings and skyscrapers from 50 to over 100 stories in height are a common sight in all countries. Thus, the design of fire protection systems must be closely coordinated with the local fire department's capabilities.

2. *Architectural design trend.* For better communication and productivity, modern buildings are also designed with larger areas and open spaces. Quite often, a single floor exceeds 50,000 sq ft without a separation. In such a case, building codes may require fire walls or partitions with separate zoning of the fire protection system.

3. *Controlled interior environment.* The exterior fenestration of modern buildings is usually constructed of fixed glass panels in lieu of operable windows in order to provide security and a controlled interior environment in terms of temperature, humidity, and air quality. Furthermore, the opening of a window on the upper floor of a high-

265

rise building may create a drastic updraft within the building in the winter, with air rushing out of the upper floors due to a stack effect. Conversely, outside air will rush in on the lower floors. The building fire code usually specifies the installation of operable, but locked, windows for access to fight fires and for smoke removal purposes. These windows should not be operated when there is no fire.

4. *Increased use of combustible materials.* While a building's construction materials are well controlled, the use of furnishings, equipment, and decorative finishes are not easy to manage. Materials such as plastic, synthetics, etc., are a source of toxic gas and smoke during a fire. Serious considerations should be given to selecting these materials.

Most building codes specify the type of construction methods and materials that can be used for building occupancies. In addition, building codes usually refer to the National Fire Codes as the basis for criteria and technical specifications. The National Fire Codes are published by the National Fire Protection Association (NFPA), which includes more than 100 separate codes, standards, and recommended practices dealing with diversified subjects related to fire protection. Some of the codes and standards related to building design and construction are the following:

- *NFPA 10* Standards for portable fire extinguishers
- *NFPA 11* Standards for foam extinguishing systems
- *NFPA 12* Standards on carbon dioxide extinguishing systems
- *NFPA 13* Standards for the installation of sprinkler systems
- *NFPA 14* Standards for the installation of standpipe and hose systems
- *NFPA 70* National Electrical Code
- *NFPA 78* Lightning Protection Code
- *NFPA 101* Life Safety Code
- *BOCA* National Fire Prevention Code

This chapter covers only the principles of fire protection. The reader should refer to the above codes and standards for more specific requirements.

Following are some commonly used terms pertaining to fire protection and their definitions:

- *Automatic fire suppression system.* An engineered system using an automatic sprinkler system (wet or dry), gas (carbon dioxide or inert gases), or chemicals (wet or dry) to detect and suppress a fire through fixed piping and nozzles. (*Note:* A standpipe-and-hose system is manually operated; thus, it is not an automatic fire suppression system.)
- *Fire area.* The aggregate floor area enclosed and bounded by fire walls.
- *Fire partition.* A vertical assembly of wall material designed to restrict the spread of fire. The overall assembly shall have a fire resistance rating (in hours) equal to or greater than that specified in the governing code. Depending on the use of the building and whether it has a sprinkler system, a fire partition shall have a rating up to two hours.
- *Fire door.* A door assembly that provides protection against the passage of fire.
- *Fire wall.* A fire-resistant wall that extends continuously from the foundation of a building up to or through the roof. Fire walls are rated in hours. Depending on the use of the building and the fire separation distance, required fire resistance ratings may vary from one hour to four hours.
- *Fire separation distance.* The distance in feet or meters measured from the face of a building to the adjacent building or the nearest public way.
- *Fire protection rating.* The time in hours or fractions thereof that an opening protective assembly, such as a wall or a partition, will resist exposure to fire.
- *Fire separation assembly.* A horizontal or vertical fire resistance–rated assembly of materials designed to restrict the spread of fire.
- *Flame spread.* The propagation of flame over a surface of construction materials for building materials and interior finishes.
- *Flame spread rating (FSR).* A rating describing how quickly fire will spread in certain materials. The values (Class I, 0–25; Class II, 26–75; Class III, 76–200) are relative values based on the testing procedure developed by the American Society for Testing Materials (ASTM). Materials with a rating of over 200 shall not be used in any building with human occupancy.
- *Means of egress.* A continuous and unobstructed path of horizontal and vertical travel from any point in a building to a public way, such as a street, an alley, or open land.
- *Use group.* The classification of building occupancy in accordance with the intended use of the building.
- *Fire loading.* The base used by building codes to estimate the nominal amount of combustibles

for determining the required fire resistivity of construction for various building use groups, expressed in lb per sq ft. Heat liberated from the total fire loading can, therefore, be calculated on the basis of the calorific value of the materials and the weight of the combustible materials.

- *Smoke-developed rating (SDR).* A numerical rating developed by ASTM to express the smoke developed by interior finishes and materials. With this method, the densities of smoke developed by various materials under identical testing (burning) conditions are compared. The method assigns SDR = 0 for asbestos cement board and SDR = 100 for red oak. A SDR rating of 50 is considered safe for most building occupancies, and an SDR rating over 450 is prohibited for use in any buildings.

9.1 CLASS OF FIRE HAZARDS AND CONSTRUCTION

Fire is the rapid combustion of materials in the presence of oxygen. Without oxygen, a fire cannot be sustained. The products of fire have two components: the thermal elements, which produce flame and heat, and the nonthermal elements, which produce smoke and toxic gases. Smoke is always associated with a fire. Indeed, statistics indicate that smoke is the primary cause of fire-related deaths. Thus, a fire protection system is used not just to extinguish a fire, but to remove the smoke as well. For simplification of discussion, the term "fire" shall mean "fire and smoke" to the extent applicable in this text.

9.1.1 Classification of Fires

According to the NFPA, there are four classes of fire:

1. *Class A.* Fires of ordinary combustible materials, such as wood, cloth, paper, rubber, and many plastics.
2. *Class B.* Fires in flammable liquids, oils, greases, tars, oil base paints, lacquers, and flammable gases.
3. *Class C.* Fires that involve energized electrical equipment. In such fires, it is important that the extinguishing medium not be a conductor of electricity.
4. *Class D.* Fires in combustible metals, such as magnesium, titanium, zirconium, sodium, lithium, and potassium.

9.1.2 Classification of Hazards

According to the NFPA, fire hazards may be grouped into three classes:

1. *Light (low) hazard.* Locations (buildings or rooms) where the total amount of Class A combustible materials, including furnishings, decorations, and other contents, is of minor quantity. Among these locations are offices, classrooms, churches, assembly halls, etc.
2. *Ordinary (moderate) hazard.* Locations where the total amount of Class A combustibles and Class B flammables are present in greater amounts than expected under light hazard occupancies. Ordinary hazards are divided into three groups: Group 1, stockpiles lower than 8 ft; Group 2, stockpiles lower than 12 ft; and Group 3, stockpiles exceeding 12 ft. These locations generally consist of offices, classrooms, mercantile shops and allied storage, light manufacturing, research operations, auto showrooms, garages, etc.
3. *Extra (high) hazard.* Locations where the total amount of Classes A and B materials are in storage, production or other use, etc.

Fire protection systems must be designed specifically to be most effective for each class of fire and hazard.

9.1.3 Type of Construction

According to BOCA, building construction is divided into types 1 through 5, with type 1 being the most fire resistant. The construction material of walls, partitions, ceilings, floors, roofs, structural system and exit envelopes shall all be constructed of noncombustible material having the minimum fire resistance rating specified in the code.

9.1.4 Use or Occupancy

The requirements for fire protection in a building also are governed by how the building is being used or occupied. Most building codes, including BOCA, UBC and SBC (see Appendix A), classify building occupancy by use groups and subgroups. When a building houses more than one occupancy, each portion of the building shall conform to the requirements for the occupancy housed therein. For example, an office building which is primarily used for administration work, typing, and data processing is classified as Use Group B. If the building also contains an auditorium the auditorium portion shall be classified as a place of assembly to comply with Use Group A. The following is the general classification of buildings by use group or occupancy:

Group A: Assembly Occupied by more than 1000 people (A–1), less than 1000 people (A–2), and other situations (A–3, A–4 and A–5).

Group B: Business Used for offices, professions or service-type transactions.

Group E: Educational Elementary schools (E–1 E–2), daycare (E–3).

Group F: Factory Moderate hazard (F–1), low hazard (F–2).

Group H: Hazard Groups H–1 through H–7, depending on the hazardous material being handled or stored.

Group I: Institutional Nurseries, hospitals, nursing homes (I–1), others (I–2 I–3).

Group M: Mercantile Used for display, storage and sale of merchandise.

Group R: Residential Hotels, motels, or boarding houses, (R–1); multi-family dwelling (R–2); one or two family dwellings (R–3); child care (R–4).

Group S: Storage Moderate hazard (S–1), low hazard (S–2), repair garage (S–3), open parking garage (S–4), aircraft (S–5).

Group U: Utility Buildings not covered by the above groups.

In addition, there are requirements for special building types such as malls, high-rise buildings, atriums, underground structures, etc. Refer to the applicable codes for specific requirements.

9.2 PLANNING FOR FIRE PROTECTION

Planning for fire protection starts with architectural and engineering design in all disciplines. Figure 9–1 illustrates the decision steps. Once the plan is formalized and implemented, a well-planned fire protection system usually operates sequentially as follows:

- *Step 1, detection.* The presence of a fire is detected manually or automatically.
- *Step 2, signaling.* The building's management, its occupants, and the fire department are notified of the presence of the fire. The occupants are advised of the actions to take.
- *Step 3, suppression.* Manual or automatic fire suppression equipment and systems are used to extinguish the fire and remove the smoke.
  - *3A, initial effort.* Portable and manual fire-fighting equipment, such as fire extinguishers, fans, and a first-aid fire hose, are used to extinguish the fire and to remove smoke by dilution or exhaustion.
  - *3B, main effort.* Fire suppression systems, such as automatic sprinklers, fire hoses, and

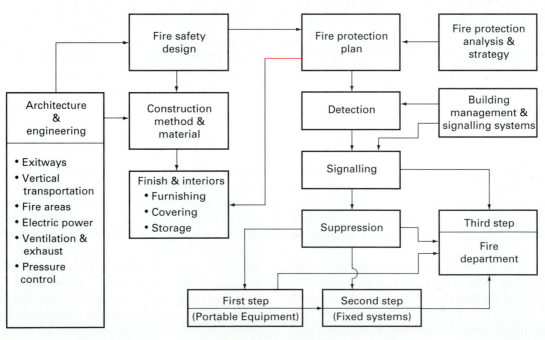

■ **FIGURE 9–1**
The fire protection decision tree.

other systems, are used to extinguish the fire, and smoke control systems are activated to remove or contain the spread of smoke.

- *3C, last effort.* The fire department takes over the fire-fighting effort when all previous efforts are ineffective.

Note that a fire may spread rapidly within minutes, and all three steps of fire protection may have to be initiated or activated concurrently to prevent the spread of the fire.

9.3 FIRE SAFETY DESIGN

Fire safety starts with the design and construction of a fire-safe building. Most building codes are specific on basic fire safety design requirements. Some of the fundamental criteria are as follows:

1. *Fire-resistant construction.* The construction of walls, partitions, ceilings, and floors shall meet or exceed the fire resistance ratings specified in the governing codes. The required ratings vary with building occupancy, size, and height. A 3- to 4-hour rated building assembly is considered a highly resistive design. For example, a 4-inch-thick reinforced concrete slab without additional ceiling material below it has only a 1-hour fire resistance rating.

2. *Smoke controls.* In addition to being made of fire-resistant construction, a building of any size must have proper smoke control by removal, dilution, and/or confinement of the smoke. Such control could be as simple as opening windows or as complicated as pressurizing the building.

3. *Length of travel.* All exits shall be located so that the maximum length of travel to access the exit, measured from the most remote point to an approved exit along the natural and unobstructed line of travel, shall not exceed the distances given in Table 9–1.

4. *Means of egress.* There shall be two separate means of egress from any space, except where a space is so small and arranged so that a second exit would not provide an appreciable increase in safety.

5. *Exit enclosures.* Exit enclosures such as stairways shall be used solely for exit purposes, and penetration by ducts, conduits, boxes, and pipes shall be limited and protected.

6. *Adequate lighting.* Egress passages should be illuminated to a minimum of 1 footcandle (fc), and preferably 3 fc, with clearly identified and

TABLE 9–1

Typical building fire protection requirements according to the building occupancy

	Required Protection		Max. Distance to Exit, ft	
Typical Occupancies	*Fire Alarm*	*Fire Suppression*	*Not Sprinkled*	*Sprinkled*
Theaters, TV studios	—	Yes	200	250
Amusement, entertainment	—	Yes	200	250
Churches and religious services	Yes	Yes	200	250
Business (2 or more stories)	Yes[a]	—	200	250
Education	Yes	Yes[b]	200	250
Factories (low hazard)	—	Yes[c]	200	250
Factories (moderate hazard)	—	—	300	400
Penal or correction institute	Yes	Yes	150	200
Hospital, child care	Yes	Yes[d]	150	200
Prisons, detention center	Yes	Yes	—	—
Mercantile	—	Yes[e]	200	250
Hotels, motels	Yes	Yes[c]	200	250
Apartments	Yes	—	200	250
One- or two-family dwellings	—	—	200	250
Storage (low hazard)	—	—	300	400
Miscellaneous	—	—	—	—

(Note: The requirements are based on the interpretation of several prevailing building codes, which vary slightly. In actual building design, the requirements of the prevailing code shall govern.)

[a]Manual boxes are not required for buildings below 75 feet.

[b]Except if the fire area is less than 20,000 sq. ft.

[c]Except less than 3 stories.

[d]Except child care facilities with 100 or less children.

[e]Except less than 12,000 sq. ft. for each fire area or less than 3 stories.

illuminated signs. (*Note:* Signs are normally mounted at a level of 7 ft or higher. Additional signs at a 2-ft level above the floor are effective ways to guide occupants to safety in a smoke-filled corridor.)

7. *Vertical openings* (other than elevator shafts). Vertical openings shall be sealed to limit fires to a single floor.

8. *Vertical transportation.* Elevator shafts shall be vented or pressurized, depending on the HVAC system. Elevators are not recognized as exits. Escalator floor openings shall be protected with fire shutters, unless they are protected by water curtains as part of the sprinkler system.

9. *Coordination with mechanical and electrical systems.* Mechanical and electrical systems shall be designed to meet the applicable codes, such as the National Electrical Code (NFPA 70), BOCA National Mechanical Code (BNMC), BOCA National Fire Protection Code (BNFPC), etc.

10. *Comply with code requirements for specific use groups.* The classification is generally consistent with that of other building codes. Table 9–1 lists the building occupancies and the requirements of fire protection systems.

11. *Coordination with fire department.* The fire marshal must be consulted in regard to the required access to the building and the locations of fire hoses, fire hydrants, electrical power panels, and alarm systems.

9.4 FIRE DETECTION AND SIGNALING DEVICES

When fire or smoke is detected either by the occupants of a building or by automatic detection devices, the building's management needs to evaluate immediately the severity of the fire and take appropriate action, such as activating the building alarm system, announcing the total or partial evacuation of the building, and notifying the fire department. For larger buildings, the fire code often requires the installation of automatic sensing (detecting) devices, as well as automatic alarm and signaling systems. Some basic devices and systems are briefly discussed in this section; Figures 9–2 and 9–3 illustrate several detection and signaling devices.

Detection and signaling devices may be addressable or nonaddressable. The addressable type can be individually addressed and identified so that the system can immediately identify the type and location of the initiating devices associated with a given address.

9.4.1 Manual Alarm Station

A manual alarm station is an electrical switch specially designed for fire protection service that activates an alarm system, such as bells, gongs, and flashing lights. To avoid accidental operation of the switch, the station is usually designed so that a person must break a glass panel or glass rod or must perform other preliminary actions before the alarm can be operated.

9.4.2 Thermal Detectors

Thermal detectors are temperature-activated sensors that initiate an alarm when the temperature in their immediate vicinity reaches a predetermined setting. These detectors are designed to meet various conditions. Thermal detectors are used for property (building) protection only. They are not intended for life safety. Some commonly used sensing devices are the following:

1. *Fixed-temperature type.* These detectors are either of self-restoring or nonrestoring design. The self-restoring type consists of normally open contact held by bimetallic elements that will close the contacts when the ambient temperature reaches a fixed setting. The setting is generally designed for operation at 135°F (57°C), 190°F (88°C) or 200°F (94°C). The contact will return to the normally open position when the ambient temperature drops back to normal. The nonrestoring type usually contains a fusible element which holds open the electrical contacts, and will melt to close the electrical signal, initiating a signal. The entire detector must be replaced.

2. *Rate-of-rise type.* This sensor reacts to the rate at which the temperature rises. A sealed but slightly vented air chamber within the device expands quickly when the temperature in the vicinity of the device rises quickly. When the air chamber expands faster than it can be vented, electrical contacts attached to the chamber begin to close and thus initiate an alarm. A rate-of-rise device reacts to fire much earlier than a fixed-temperature device; however, it may cause unnecessary false alarms if installed in spaces that are subject to normal rapid temperature fluctuations.

3. *Combination type.* This device reacts to both a fixed temperature and a rate of rise. It may be designed on the principle of differential expansion between two different metals or on the rate of expansion of the air chamber. The combination type of sensing device is desirable for most applications.

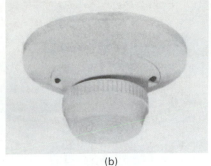 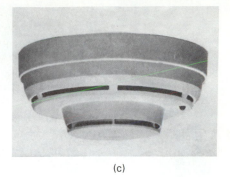

(a)   (b)   (c)

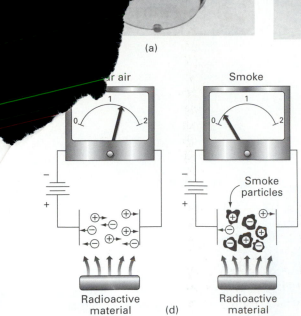

(d)

■ **FIGURE 9–2**

Fire and smoke detection devices.
(a) Ionization smoke detector (Source: Pyrotronics). (b) Ionization smoke detector (Source: System Sensor/Division BRK). (c) Ionization detector (Source: Fire Control Instruments, Inc.). (d) Principle of operation for ionization smoke detector. (e) Cross-section view of an ionization smoke detector (Source: Pyrotronics Reprinted with permission from *The Fire Protection Handbook,* 17th ed. copyright © 1991, National Fire Protection Association, Quincy, MA., 02269.)

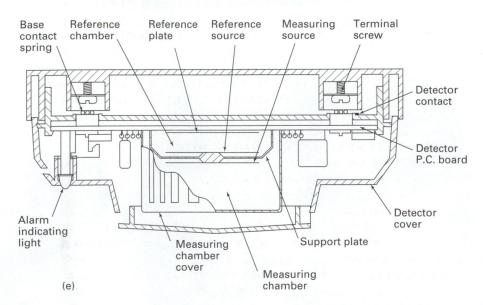

(e)

9.4.3 Smoke Detectors

Smoke detectors are more and more required by codes in most building occupancy groups including Group R and Group I, except when the building is fully protected by an automatic sprinkler system. Smoke detectors are more sensitive than thermal de-

tectors as long as the smoke generated by the fire is within the limits of detectability of the smoke detectors. In most situations, smoke detectors will sound an early warning. Two of the more commonly used smoke detectors, shown in Figure 9–2, are the following:

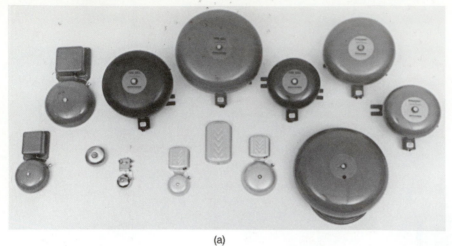

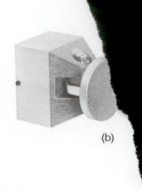

(a)

(b)

(c)

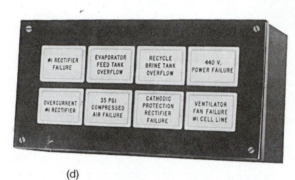

| #1 RECTIFIER FAILURE | EVAPORATOR FEED TANK OVERFLOW | RECYCLE BRINE TANK OVERFLOW | 440 V. POWER FAILURE |
| OVERCURRENT #1 RECTIFIER | 35 PSI COMPRESSED AIR FAILURE | CATHODIC PROTECTION RECTIFIER FAILURE | VENTILATOR FAN FAILURE #1 CELL LINE |

(d)

■ **FIGURE 9–3**
(a) Typical signal gongs and bells.
(b) Typical fire alarm with manual pull station.
(c) Typical magnetic door holder.
(d) Typical visual annunciation panel.
(e) Typical signal horns, general distribution and narrow projecting types (from 60 dB to 120 dB at a 10-ft range). (Courtesy of Edwards/General Signal, Farmington, CT.)

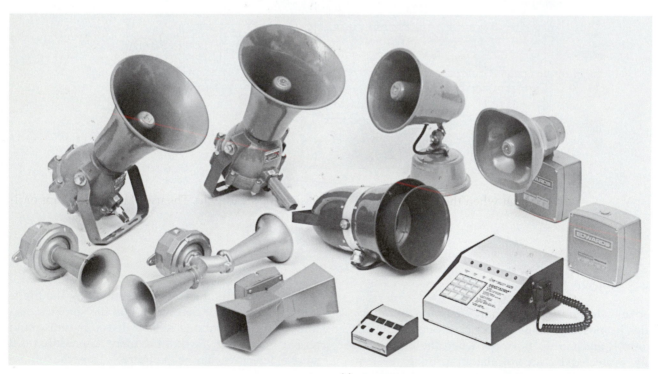

(e)

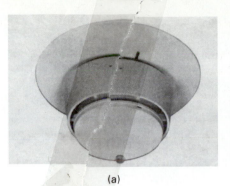

(a)

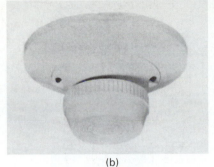

(b)

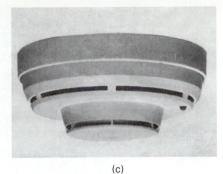

(c)

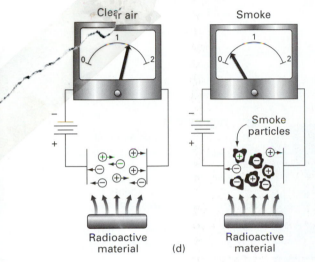

Clear air   Smoke

Smoke particles

Radioactive material   (d)   Radioactive material

■ **FIGURE 9–2**

Fire and smoke detection devices.
(a) Ionization smoke detector (Source: Pyrotronics). (b) Ionization smoke detector (Source: System Sensor/Division BRK). (c) Ionization detector (Source: Fire Control Instruments, Inc.). (d) Principle of operation for ionization smoke detector. (e) Cross-section view of an ionization smoke detector (Source: Pyrotronics Reprinted with permission from *The Fire Protection Handbook,* 17th ed. copyright © 1991, National Fire Protection Association, Quincy, MA., 02269.)

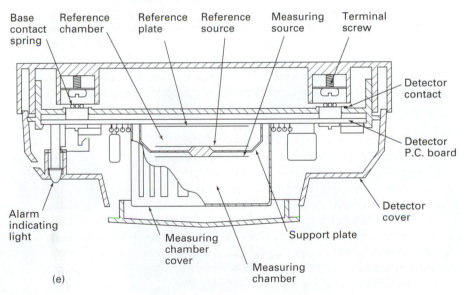

Base contact spring   Reference chamber   Reference plate   Reference source   Measuring source   Terminal screw

Detector contact

Detector P.C. board

Detector cover

Alarm indicating light

Measuring chamber cover

Support plate

Measuring chamber

(e)

9.4.3 Smoke Detectors

Smoke detectors are more and more required by codes in most building occupancy groups including Group R and Group I, except when the building is fully protected by an automatic sprinkler system. Smoke detectors are more sensitive than thermal de-tectors as long as the smoke generated by the fire is within the limits of detectability of the smoke detectors. In most situations, smoke detectors will sound an early warning. Two of the more commonly used smoke detectors, shown in Figure 9–2, are the following:

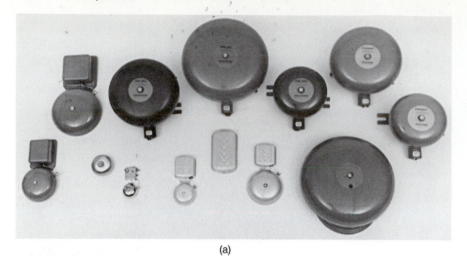

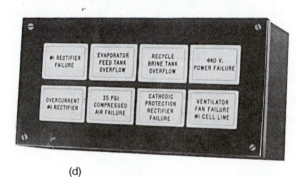

(a)

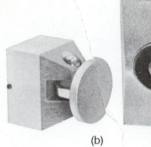

(b)

(c)

(d)

■ **FIGURE 9–3**
(a) Typical signal gongs and bells.
(b) Typical fire alarm with manual pull station.
(c) Typical magnetic door holder.
(d) Typical visual annunciation panel.
(e) Typical signal horns, general distribution and narrow projecting types (from 60 dB to 120 dB at a 10-ft range). (Courtesy of Edwards/General Signal, Farmington, CT.)

(e)

1. *Photoelectric type.* This variety operates on the principle of the scattering of light. Under normal conditions, the light beam generated from a light-emitting diode (LED) is reflected by a mirror within the detector chamber away from a photosensitive diode. When smoke is present in the detector chamber, the light beam is scattered, causing the diode to sense the scattered light and activate the electric circuit to sound an alarm.

2. *Ionization type.* This variety operates on the principle of changing conductivity of air within the detector chamber. The detector contains a minute source of radioactive material that emits alpha particles which ionize the nitrogen and oxygen molecules in the air. A voltage applied across the ionization chamber causes a very small electrical current to flow to an electrode of opposite polarity. When invisible or visible combustion particles enter the ionization chamber, the conductivity of the air is reduced, and the voltage on the electrodes increases, which in turn initiates an alarm when the voltage reaches a preset level. Ionization-type detectors are the most sensitive of smoke detectors, although the presence of radioactive material in them has been a concern in terms of health risks as well as reliability. However, improved technology has eliminated both problems. See Figure 9–2 for illustrations.

9.4.4 Flame Detectors

Flame detectors are used to detect the direct radiation of a flame in the visible, infrared, and ultraviolet ranges of the spectrum. There are four basic types: infrared, ultraviolet, photoelectric, and flame flicker detectors. Flame detectors are used chiefly in industrial processes, in mining, and for the protection of combustion equipment.

9.4.5 Magnetic Door Release

A magnetic door release is an electromagnetic device that holds a fire door open when the device is energized and releases the door to a closed position in case of fire.

9.4.6 Signal Devices

Signal devices are audio and video devices that satisfy the intended function. Among the devices commonly used are the following:

1. *Single-stroke bell.* A device containing a bell that is struck once each time electrical power is applied.

2. *Vibrating bell.* A device containing a bell that rings continuously as long as electrical power is applied.

3. *Buzzer.* A device consisting of an electromagnetic coil that, when electrical power is applied to it, will cause a thin metal piece (reed) to vibrate.

4. *Chime.* A device that produces a pleasing or musical tone each time electrical power is applied to it.

5. *Horn.* A device consisting of an electromagnetic coil that causes a metal diaphragm to vibrate and produce a sound that is amplified by a horn.

6. *Siren.* A device consisting of an electric motor that produces a continuous, high-pitched sound (up to 100 dB). A siren can be used only in a place with normally high ambient noise, such as a steel mill, auto assembly plant, etc.

7. *Light signal.* A flashing strobe or steady illuminated light with wording such as "Fire, leave the building," "Go to 20th floor," etc., for hearing-impaired persons. Strobe lights are required by most applicable fire codes.

9.4.7 Flow Detectors

Flow detectors are devices that indicate or initiate an alarm when water is flowing in the fire suppression systems. They are either sail-type switches that deflect under a flow of water or pressure switches that sense the pressure differential caused by a flow of water. Their operation depends on prior actuation of a sprinkler head or the operation of a fire hose. Flow detectors provide a positive means of monitoring the fire suppression system even when a building is unoccupied.

9.4.8 Visual Annunciation Devices

Visual annunciation devices are displays that may consist of single or multiple lights with marked messages such as "fire," "fire escape," "go to Area B," etc. The lights may be of different colors and may flash or be steadily lit. The use of long-life lamps or fiber optic light tubes (usually lasting over 30,000 hours) embedded in floors or walls to lead the occupants to safety is an effective means of visual signalling. This type of design is increasingly used in places of public assembly, such as theaters, auditoriums, sports arenas, airplanes, etc.

The selection and spacing of detectors and signal devices depend on the performance characteristics of the products and are governed by the specifications of manufacturers. Figure 9–3 shows signalling and activating devices.

9.5 FIRE ALARM SYSTEMS

Fire alarm systems are an integral part of a fire protection plan. They are basically electrical systems that are specially designed to announce the presence of fire or smoke. They are not intended to suppress or extinguish a fire. Fire alarm systems are described in Chapter 13.

9.6 FIRE SUPPRESSION SYSTEMS

Fire suppression is achieved by cooling the combustible material to below its ignition temperature or by preventing oxygen from reacting with the combustible material. Depending on the class of fire and the type of building occupancy, some fire suppression systems are more effective than others. For example, while water is an effective cooling agent, it is detrimental in an electrical fire because water is an excellent electrical conductor. Fire suppression systems may be classified in several ways:

- According to the the medium—water, foam, chemicals, halogenated gas, etc.
- According to the action of the device—a portable extinguisher, standpipe and hose, sprinkler system, etc.
- According to the method of operation of the device—manual or automatic.

Depending on the building use group, both manual and automatic suppression systems may be required. Requirements vary with the applicable building codes; the designer should check with the codes for specifics and exceptions. In general, a fire suppression system is required in the building use groups given in Table 9–1.

9.6.1 Water Supply

Water is the universal fire-fighting medium. It is readily available in large quantities and, in general, is more economical than any other fire-fighting medium. Public water is always preferred whenever available.

For fire protection purposes, the water supply should be separated from a building's domestic water system, even though the two are connected to the same public water main. Usually, the fire main is separated from the domestic main ahead of the water meter. A check valve is often required by the water department to detect the flow of water, but not to measure the quantity.

The pressure at various parts of the system should be higher than the following values when the system is at its rated flow:

- At the water main (residual pressure at maximum flow) 10–20 psig
- At the end of a hose (nozzle) 65 psig
- At each sprinkler head 15–30 psig
- At the top of the roof 15–30 psig

When the public water supply is inadequate to serve fire-fighting needs, alternative water sources such as lakes, ponds, wells, and storage tanks must be provided. The amount of storage required depends on the code. In general, a 30-minute storage capacity at the calculated flow rate is the norm. Figure 9–4 illustrates several alternative water supply systems.

9.6.2 Portable Fire Extinguishers

Portable fire extinguishers are used as the first line of fire protection. Often, they avert a major disaster. They are normally precharged with water or chemicals and are hand operated. Fire extinguisher requirements are governed by applicable building codes and NFPA Standard No. 10. In general, a portable fire extinguisher shall be installed in the following locations:

- In all occupancies in use groups A–1,2,3, E, I–2, R–1 and H, and in staff locations of I–3.
- In all special areas containing commercial kitchens, wherever fuels are dispensed or combustible liquids handled, and in laboratories and shops.

Extinguishers shall be selected for a Class A fire in all areas and for Classes B, C, and D fires in special areas.

Extinguishers containing various media are labeled as follows for different classes of fires:

- Water-stored pressure or cartridge operated   A
- Aqueous film-forming foam (AFFF)   A, B
- Film-forming fluoroprotein (FFFP)   A, B
- Carbon dioxide (CO_2)   B, C
- Halogenated (non-ozone-depleting variety)   B, C
- Dry chemical (potassium based)   B, C, D
- Dry chemical (ammonium phosphate based)   A, B, C, D

Typical sizes and effective ranges for various classes of fire extinguishers are summarized in Table 9–2.

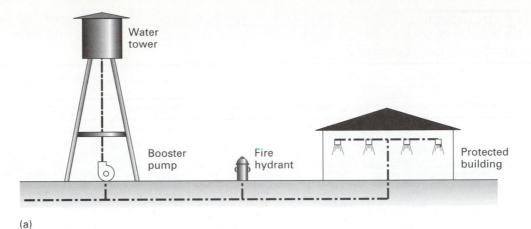

(a)

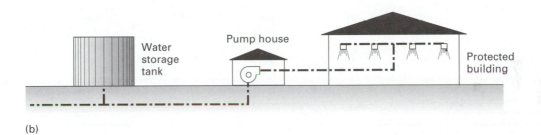

(b)

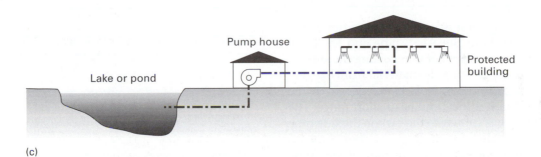

(c)

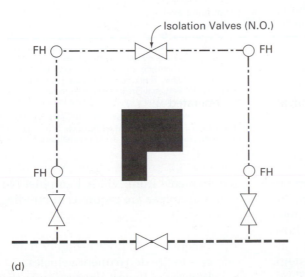

(d)

■ **FIGURE 9–4**

Schematic diagrams of alternative water supply systems.

(a) Elevated water tower: At night, the public water supply pressure is adequate to fill the elevated water tower. If the water pressure is insufficient to feed the tower, a booster pump will be required.

(b) Groundwater tank: A pump is used to transfer water from the tank.

(c) Lake or pond: A pump is used to transfer water from the lake or pond.

(d) A fire loop surrounds a building with fire hydrants (FH) and isolation valves. Water may be shut off from either direction if needed due to breaks or leaks.

TABLE 9–2

Summary of fire extinguisher applications
(Courtesy: Elkhard Manufacturing Company.)

Types	Water		Soda Acid	Foam	Gas	Loaded Stream	Dry Chemical
	Air Pressurized	Cartridge Operated			Halogenated/CO₂	Air Pressurized	
Typical Sizesᵃ (gal)	2.5–5	2.5–60	2.5–33	2.5–33	2.5–100	2.5–33	1–100
Subject to Freezing	Water—yes Antifreeze—no	Water—yes Antifreeze—no	Yes	Yes	No	No	No
Operating Method	Pull pin, squeeze handle	Invert and bump	Invert	Invert	Pull pin, squeeze handle	Pull pin, squeeze handle	Pull pin, squeeze handle
Effective Range	40–50'	40–50'	40–50'	30–40'	4–10'	40–50'	10–25'
Conductor of Electricity	Yes	Yes	Yes	Yes	No	Yes	No
Extinguishing Effect	Cooling	Cooling	Cooling	Blanketing and cooling	Blanketing, some cooling	Cooling	Blanketing
Class A Fires, Wood, Paper, Textiles, etc.	Yes	Yes	Yes	Yes	No—surface only	Yes	No—surface only
Class B Fires, Flammable Liquids and Gases	No	No	No	Yes	Yes	Yes	Yes
Class C Fires, Electrical	No	No	No	No	Yes	No	Yes
Maintenance	Visual inspection semiannually. Hydrostatic test every 5 years	Weigh cartridge yearly. Hydrostatic test every 5 years	Recharge yearly Hydrostatic test every 5 years	Recharge yearly Hydrostatic test every 5 years	Weigh semiannually Hydrostatic test every 5 years	Visual inspection semiannually Hydrostatic test every 5 years	Visual inspection semiannually

ᵃWall mounted and on carts.

The advantage of a water type fire extinguisher is its lower cost. Both the AFFF type and the FFFP type work on the principle of film coating the surface material and thus isolating the material from oxygen. The carbon dioxide type displaces oxygen from the electrical fire, but it is toxic and has a short range. The halogenated type is similar to the carbon dioxide type and may be inhaled for a short duration (several minutes). The major advantage of the gaseous type (carbon dioxide and other non-ozone-depleting gases) is that the gas does not deposit any film on the protected surface. This feature is very important for electrical and electronic equipment. Dry chemicals are effective for chemical or burning metals, but they are usually corrosive and difficult to clean.

The size and placement of extinguishers shall comply with the criteria given in Table 9–3.

9.6.3 Standpipe-and-Hose Systems

Standpipe-and-hose systems consist of piping, valves, hose connections, and allied equipment to provide streams or sprays of water for fire suppression purposes. For simplicity, standpipe-and-hose systems are abbreviated as standpipe systems, although a hose is always used in a standpipe system. Typical components and system diagrams are illustrated in Figure 9–5. Most building codes adapt their

TABLE 9–3

Fire extinguisher size and placement for Class A hazard

Item	Occupancy Hazard		
	Light	Ordinary	Extra
Minimum-rate single extinguisher	2–A	2–A	4–A
Maximum floor areaᵃ	3000	1500	1000
Maximum floor area for large extinguisherᵇ	11,250 ft²	11,250 ft²	11,250 ft²
Maximum travel distance to extinguisher, ft	75	75	75

Portions of this table reprinted with permission from NFPA 10, *Portable Fire Extinguishers*, copyright © 1994, National Fire Protection Association, Quincy, MA 02269. This reprinted material is not the complete and official position of the National Fire Protection Association on the referenced subject which is represented only by the standard in its entirety.

ᵃOne "1–A" unit is about 1½ gal. capacity; 2–A = 2½ gal.; 4–A = 5 gal.

ᵇThe area of a square inscribed within a circle having a radius of 75 ft as the maximum area of open space with the extinguisher located at the middle.

requirements from NFPA Pamphlet No. 14. In general, standpipes are required in building use groups or areas:

- Where the top floor is higher than 30 ft from access to fire department vehicles.
- Where the lowest floor is lower than 30 ft from access to fire department vehicles.

- Where any portion of the building is more than 400 ft from access to fire department vehicles, except if the building is equipped with an automatic sprinkler system or if the building is less than 10,000 sq ft or if the building is in any of the use groups A–4, A–5, F–2, R–3, S–2, and U, according to BOCA.
- That are malls.
- That are stages of a theater, auditorium, etc., where props are prepared and used.

According to NFPA, there are three classes of standpipe systems:

- Class I   2½″ hose stations for use by the fire department
- Class II   1½″ hose stations for use by the occupants of the building or the fire department
- Class III   both 1½″ and 2½″ hoses

Standpipe systems are classified as follows:

- *Wet systems.* Systems in which the standpipes are filled with water under pressure. Whenever the system is activated, water will charge into the connected hose immediately. As a rule, wet systems are used for most building occupancies.
- *Dry systems.* Systems in which the standpipes are not filled with water. Each standpipe is normally filled with compressed air. Upon opening the dry pipe valve, it will automatically admit water into the system. Dry systems may be of the automatic or semiautomatic (deluge) type. The semiautomatic type usually includes a deluge valve that is remotely controlled to allow the water to enter the standpipe. A dry system is used when the building space is subject to freezing conditions or when the use of water may cause other hazards or extreme property damage—for example, in a computer center or a power-generating station.
- *Manual systems.* These systems are used by fire departments exclusively. They may be of the dry or wet variety, depending on the application. A manual dry-type standpipe system is directly connected to a fire department connection (also called a Siamese connection) at the exterior of the building, where water is pumped into the standpipe through the use of a fire pump or via a connection to fire hydrants.

According to BOCA, standpipe systems installed in all buildings shall be of the wet variety, except when the highest floor of the building is:

- lower than 75 feet, or

- lower than 150 feet and the building is protected by an automatic sprinkler system, or
- lower than 150 feet and the building is an open parking structure

The required number of standpipe risers depends on the horizontal distance covered by the hoses. In general, a hose connection shall be provided at each intermediate landing between floor levels in every required stairway. The stairway is normally enclosed by fire walls. Thus, the standpipe riser is ideally located within the stairway enclosure. There shall be fire hose connections on each side of fire enclosures, so that hose connections can be made inside or outside of the stairway enclosure. Standpipe risers at some stairways may be omitted if:

- all parts of the building can be reached by a 30-foot "hose stream" (water stream) at the end of a 100-foot hose from other risers. (*Note:* The distance shall be the actual developed length measured along a path of travel originating from the hose connection), or
- all parts of the building are within 400 feet from access to fire department vehicles and not more than 200 feet from another hose connection.

For design purposes, the water supply, pressure, and flow rate shall be generally based on the following criteria:

- Water from a public water system or from an alternative water source shall be adequate for a minimum duration of 30 minutes.
- The minimum pressure shall not be less than 65 psig residual pressure at the topmost hose outlet. (*Residual pressure* is the remaining pressure at the outlet when water is discharging at the rated flow. *Static pressure* is the pressure available at the outlet when there is no flow. The difference between the two is the pressure drop.)
- If the calculated residual pressure is more than 100 psig, a pressure-regulating device (valve) shall be provided. If no fire hose is provided at the standpipe outlet, as in the manual standpipe system, the maximum pressure may be raised to 175 psig. If the water pressure is too high, the extremely high reaction force due to a high water flow will be too difficult to handle.
- In Class I and III systems—i.e., 2½″ systems, for calculation purposes, piping and risers shall be sized for 500 GPM for the first standpipe and 250 GPM for each additional standpipe, but normally not more than 1250 GPM for the total system, including the sprinkler water demand for buildings that have a limited-area sprinkler system.

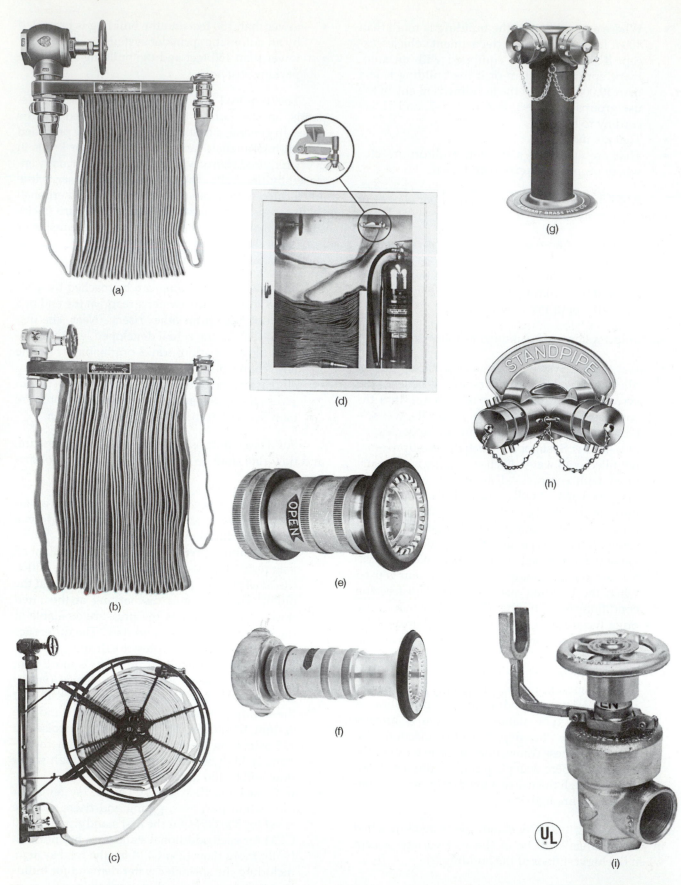

(a)

(b)

(c)

(d)

(e)

(f)

(g)

(h)

(i)

- In Class II systems—i.e., 1½″ systems, pipe sizes shall be based on 100 GPM for each riser and not more than 500 GPM for the total system.
- For buildings in use Groups B, I, R–1 or R–2, and with an automatic sprinkler system, the flow rate may be sized for 250 GPM for each standpipe and not more than 750 GPM for all standpipes. The demand flow rate for the automatic sprinkler system shall be calculated independently.

The sizes of standpipes shall not be less than those shown in Table 9–4.

Piping material in standpipe-and-hose systems is usually black steel with welded or mechanical fittings, galvanized steel with mechanical fittings, or copper with solder fittings. The use of mechanical fittings, which allows for limited pipe movement due to thermal expansion, contraction, or vibration, is desirable for most high-rise buildings.

A fire department connection, also known as a Siamese connection, shall be provided outside of the building. A minimum of two connections shall be installed for high-rise buildings. Hoses may be installed on racks, on reels, or in cabinets containing portable fire extinguishers and tools. The cabinet door will be latched.

A water flow alarm shall be provided for automatic and semiautomatic standpipe systems. Also, valves shall be provided for isolation of piping, prevention of back pressure, and shutoff purposes. All valves shall indicate at a distance whether they are open or closed. A standing stem valve is one of the approved varieties and shall be capable of being closed within 5 seconds. Isolation valves shall be locked in the open position.

Each standpipe shall be provided with a means of draining. The drainpipe shall be ¾ inch for a 2-inch standpipe and up to 2 inches for a 4-inch or larger standpipe. The drain line shall be 3 inches for standpipe equipped with pressure-regulating devices.

The standpipe system shall be limited to a vertical height of 275 feet. For high-rise buildings, separate standpipe systems shall be provided for each 275 feet of vertical height. (See Figure 9–6.)

TABLE 9–4

Pipe schedule, standpipes and supply piping (inches)

Total Accumulated Flow (GPM)	Total Distance of Piping from Furthest Outlet		
	<50 ft	50–100 ft	>100 ft
100	2	2½	3
101–500	4	4	6
501–750	5	5	6
751–1250	6	6	6
1251 and over	8	8	8

Portions of this table reprinted with permission from NFPA 14, *Installation of Standpipe and Hose Systems*, copyright © 1993, National Fire Protection Association, Quincy, MA 02269. This reprinted material is not the complete and official position of the National Fire Protection Association on the referenced subject which is represented only by the standard in its entirety.

9.6.4 Other Fire Suppression Systems

In addition to automatic sprinkler systems, to be discussed in Section 9.7, there are a number of specialty systems that do not use water.

Foam systems Foam systems are most effective for Class B fires (fires in which flammable liquid, oil, grease, paint, etc., is a factor). The foam is made by generators, which mix water with detergent or other chemicals to produce as much as 1000 gallons of foam for each gallon of water. Sprayed from large nozzles, the foam covers the fire, insulating it from oxygen, and cools down the temperature when water is evaporated into steam. Figure 9–7 (a) illustrates the operation of a foam system. Typical applications of foam systems are in printing plants and aircraft manufacturing plants, among others.

Gaseous fire suppression systems Gaseous systems are most effective for Class C fires (fires caused by electrical equipment). All of these gases are stored in liquid state under high pressure. There are three varieties:

The *carbon dioxide system* stores CO_2 in a liquid state in pressurized tanks. When discharged through nozzles, the liquid vaporizes and smothers the fire by displacing oxygen. The evaporation of the liquid to a gas also absorbs about 120 Btu of heat per pound of CO_2. To be effective, carbon dioxide systems should

■ **FIGURE 9–5**
Standpipe-and-hose system components.
(a) Hose rack with 2½″ valve and 1½″ hose (up to 125 ft).
(b) Hose rack with 1½″ valve and 1½″ hose (up to 100 ft).
(c) Hose reel with 1½″ valve and 1½″ hose and fire extinguisher, and enlarged detail of the hose clamp.
(d) Hose cabinet with 1½″ valve, 1½″ hose and fire extinguisher; enlarged detail of the hose clamp.
(e) Fog-type nozzle for use on an electrical fire (30° fog pattern).

(f) High-capacity (up to 250 GPM) direct connection nozzle.
(g) Sidewalk-type, freestanding two-way Siamese connection.
(h) Wall-mounted surface two-way Siamese connection.
(i) Automatic pressure-reducing valve to limit the water pressure at the nozzle under both flow and no-flow conditions.
(Courtesy: Elkhart Mfg. Co., Inc., Elkhart, IN.)

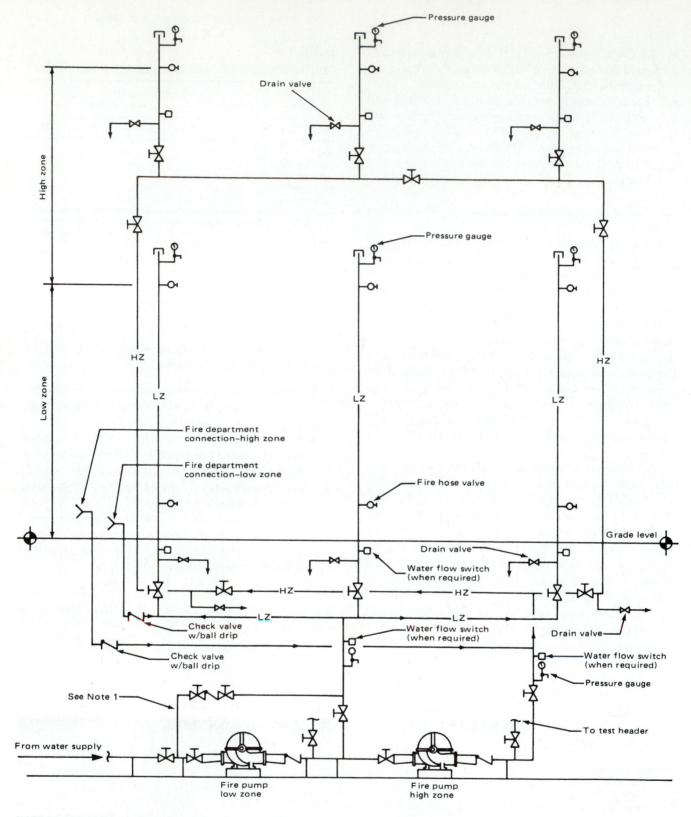

NOTE 1: Bypass subject to NFPA 20.
NOTE 2: High zone pump may be arranged to take suction directly from source of supply.

■ **FIGURE 9–6**

Riser diagram of a three-zone standpipe-and-hose system showing a fire pump for each zone and a storage tank for the high zone. (Reprinted with permission from NFPA 14, *Installation of Standpipe and Hose Systems,* Copyright © 1993, National Fire Protection Association, Quincy, MA 02269. This reprinted material is not the complete and official position of the National Fire Protection Association on the referenced subject which is represented only by the standard in its entirety.)

(a)

FIGURE 9–7
(a) Foam fire suppression system. (Courtesy: Kidde-Fenwal, Inc., Ashland, MA.) (b) Carbon dioxide fire suppression system in an industrial plant. (Courtesy: Ansul Fire Protection, Marinette, WI.)

(b)

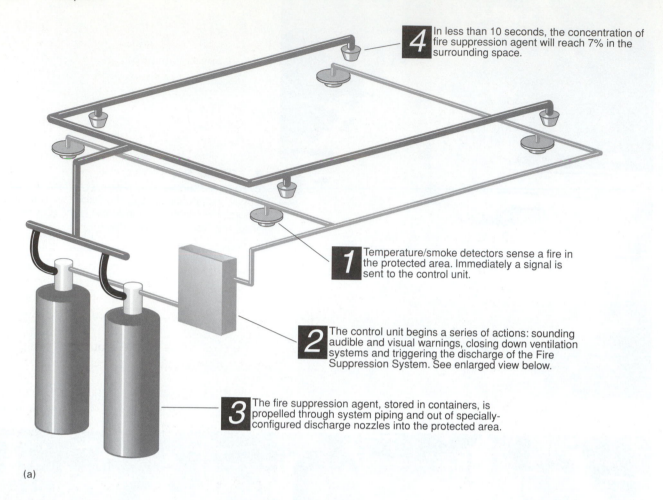

4 In less than 10 seconds, the concentration of fire suppression agent will reach 7% in the surrounding space.

1 Temperature/smoke detectors sense a fire in the protected area. Immediately a signal is sent to the control unit.

2 The control unit begins a series of actions: sounding audible and visual warnings, closing down ventilation systems and triggering the discharge of the Fire Suppression System. See enlarged view below.

3 The fire suppression agent, stored in containers, is propelled through system piping and out of specially-configured discharge nozzles into the protected area.

(a)

■ **FIGURE 9–8**

(a) A gaseous fire suppression system for a building space. System may be divided into a number of zones. (b) An enlarged view of the control unit. It can be designed to monitor diversity of functions or equipment, such as temperature, pressure, generator, entry, HVAC system, fuel, voltage surge, humidity, etc. The system may be interfaced with the building automation system. (Courtesy: Kidde-Fenwal, Inc., Ashland, MA.)

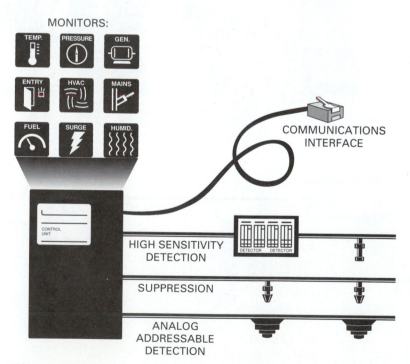

(b)

be used in confined and normally nonventilated spaces. The CO_2 is toxic: A few minutes of exposure in an atmosphere containing over 10 percent CO_2 could render a person unconscious. A carbon dioxide fire suppression system is shown in Figure 9–7(b).

Halogenated gas is a generic name for gases containing the elements of fluorine, chlorine, bromine or iodine. The most commonly used gases for fire extinguishing purposes are Halon 1301 and 1211. However, because of their ozone-depleting property, these gases have been banned and are being replaced with a number of alternatives. These include FE-25 (pentafluoroethane), FE-232 (dichlorotrifluoroethane), FE-13 (trifluoroethane), and FM-100 (bromodifluoromethane), FM-200 (heptafluoropropane), developed by a number of leading producers. These agents have zero or negligible ozone depletion potential (ODP). Their effectiveness as fire extinguishing agents is still being evaluated.

Atmospheric gas is a mixture of argon, carbon dioxide and nitrogen. A typical composition is 40 percent argon, 8 percent carbon dioxide and 52 percent nitrogen. The gas mixture is non-toxic, with zero ozone depleting potential (ODP) and zero global warming potential (GWP). This type gas system is effective, but requires more gas per volume of space than the halogenated gases. Figure 9–8 shows a typical application in a data processing space.

Dry chemicals Dry chemicals are used especially for Class D fires (fires where combustible metals or flammable liquids are a factor). Most of the dry chemicals contain bicarbonates, chlorides, phosphates, and other proprietary compounds. The use of water should be avoided on burning metals, since hot metal extracts oxygen from water, promotes combustion, and at the same time liberates hydrogen, which ignites readily. When water must be used to fight a Class A fire simultaneously, water must be of sufficient quantity which could help to cool off the burning metal and thus extinguish the fire.

9.7 AUTOMATIC SPRINKLER SYSTEMS

Automatic sprinkler systems are integrated fire suppression systems consisting of a water supply, a network of pipes, sprinkler heads, and other components to provide automatic fire suppression in areas of a building where the temperature or smoke has reached a predetermined level. Automatic sprinkler systems are the most effective fire suppression system for a Class A fire in buildings containing ordinary combustible materials, such as wood, paper, plastics, etc. The design and installation of the system

are strictly regulated by insurance companies in accordance with NFPA Standard 13. Specific requirements for various system applications are too numerous to be included in this section; described are only principles and practices to provide an overview of requirements, the scope of such systems, their features and design approaches. Construction drawings showing all dimensions of piping, sprinklers, etc., as well as detailed hydraulic calculations, are normally prepared by the installation contractors.

9.7.1 Required Locations

According to BOCA and similar building codes, the following building use groups or areas within the building are required to be protected by an automatic sprinkler system. When the use of water is detrimental or ineffective, other fire suppression systems, such as, dry chemical, foam, or halogenated gas systems, should be used.

The building areas or use groups for which automatic sprinkler systems are required are as follows:

1. Assemblies (use groups A–1, A–3, and A–4): All fire areas exceeding 12,000 sq ft, except for auditoriums (A–1 or A–3), naves or chancels (A–4), or sport areas (A–3) where the main floor is at the exit discharge level of the main entrance.
2. Assemblies (use group A–2): All fire areas exceeding 5000 sq ft or any portion of an assembly that is located either above or below the exit discharge level.
3. Business occupancies (use group B): This use group is a catch-all occupancy that is not included in the other use groups. Use group B includes offices, banks, professional services, and miscellaneous businesses, such as car wash, print shop, etc. (There is an overlap with use group M). In general, when a building is more than 12,000 sq ft on one floor or more than 24,000 sq ft on all floors, an automatic sprinkler system is required. Furthermore, sprinkler requirements in individual spaces or areas, such as an auditorium, a computer room, or a laboratory, shall comply with the provisions of those other use group requirements.
4. High-rise buildings (over 75 ft in height above the fire truck access): All buildings and structures shall be installed with an automatic sprinkler system.
5. Educational (use group E): All fire areas exceeding 20,000 sq ft.
6. High hazard (use group H): All fire areas, except magazines (explosive storage areas) if specially constructed.

7. Institutions (use group I): All fire areas, except child care facilities with 100 or fewer children located at the level of the exit discharge and with exit doors in all child care rooms.
8. Mercantile, storage, and factory (use groups M, S–1, and F–1): All fire areas exceeding 12,000 sq ft on one floor, or 24,000 sq ft on all floors, or more than three stories above grade.
9. Residential (use group R1, hotel, motel, etc.): All fire areas, except when all guest rooms are no more than three stories above grade and each guest room is directly open to the exterior exit.
10. Residential (use group R2, multifamily dwellings): All fire areas, except when the building is only two stories high with no more than 12 dwelling units per fire area.

9.7.2 Sprinklers

The major component of an automatic sprinkler system is the sprinkler, which discharges water in a specific pattern for extinguishing or controlling a fire. A sprinkler head consists of three major components: a nozzle, a heat detector, and a water spray pattern deflector. The fusible link type of heat detector is constructed of a "eutectic alloy," which has a single melting point rather than gradual softening. When the link temperature reaches its melting point, the link is pulled apart by the water pressure and opens the nozzle. The frangible bulb type of detector contains a glass bulb partially filled with a liquid that expands with temperature. At the rated temperature, the liquid will shatter the bulb and open the nozzle.

Since sprinklers must be exposed in the space which they protect, there are a variety of sprinkler styles to satisfy construction methods and aesthetics. Figure 9–9 illustrates several of these styles. Following is a brief summary of the various types of sprinkler and their performance data.

The *spray pattern* of a sprinkler may be a symmetrical or asymmetrical spray, a fine mist, or water droplets. Sprinklers may be *mounted* pendant, upright, flush with the ceiling, recessed into the ceiling, concealed in the ceiling, or mounted in a sidewall.

Response may be of the quick-response, quick-response and extended-coverage, quick-response and early-suppression, or early-suppression, and fast-response variety.

The *heat-sensing elements* of a sprinkler may be fusible links or frangible bulbs. Or the sprinkler may be the open type (with no heat-sensing element) or the dry type (with a built-in seal at the end of the inlet to prevent water from entering the nipple until the sprinkler operates).

The *temperature rating* of fusible links is divided into seven groups, starting with ordinary (135° to 170°F) and proceeding to intermediate (175° to 225°F), high (250° to 300°F), extra high (325° to 445°F) and on up. Sprinklers are color coded for ease of identification. Sprinklers with ordinary ratings are used in general occupied spaces, those with intermediate ratings are installed in boiler or mechanical rooms, and those with higher ratings operate in spaces where the normal ambient temperature is close to the range of the fusible links, such as industrial spaces.

Ordinary temperature-rated sprinklers shall be used throughout a space, except in locations near a heater, directly exposed to the sun's rays under a skylight, inside of unventilated show windows, or near commercial cooking equipment. In these locations, a higher temperature sprinkler shall be used.

The *flow rate* of a sprinkler depends on the size of its orifice and the residual pressure of the water supply. The flow rate is calculated from the fundamental fluid flow formula,

$$Q = K \times \sqrt{p} \qquad (9\text{--}1)$$

where Q = flow rate, GPM
K = flow constant, per unit; varies from 1.3–1.5 for a $\frac{1}{4}''$ orifice, 5.3–5.8 for a $\frac{1}{2}''$ orifice, and 13.5–14.5 for a $\frac{3}{4}''$ orifice
p = Pressure, psi (residual pressure at rated flow)

The most commonly used sprinklers are for $\frac{1}{2}''$ pipe thread with a $\frac{1}{2}''$ orifice (K factor of 5.5). The flow rate at 15 psi residual pressure is about 20 GPM (21.3 GPM, as calculated by Equation 9–1). With 15 psi as the base, the flow rate will be 28 GPM when the pressure is doubled to 30 psi and 40 GPM when the pressure is quadrupled. Sprinklers with other orifices or K factors have different flow rates varying from a $\frac{1}{4}''$ orifice ($K = 1.4$) to a $\frac{3}{4}''$ orifice ($K = 8.0$). Sprinklers with smaller orifices are ideally suited for residential occupancies, since most residential spaces are small, relatively speaking.

Based on a $\frac{1}{2}''$ standard orifice as 100 percent, the flow rate of a sprinkler with a $\frac{1}{4}''$ orifice is as low as 25 percent, and that of a sprinkler with a $\frac{3}{4}''$ orifice is as high as 250 percent.

Example 9–1 What is the flow rate of a sprinkler with a standard $\frac{1}{2}''$ orifice ($K = 5.5$) with 20 psi residual pressure?
Answer:

The flow rate $Q = 5.5 \times \sqrt{20} = 5.5 \times 4.47$
$$= 24.6 \text{ GPM}$$

(a)

Flow Rate (gpm)	Pressure (psi)	Enclosure Area Width x Length	
		(ft)	(ft)
25	9.3	14 x	18
30	13.4	14 x	20
30	13.4	16 x	18
35	18.2	16 x	20

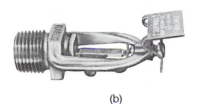

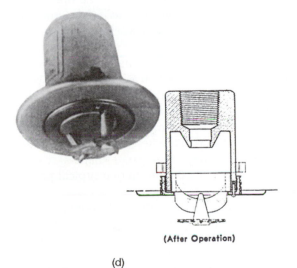

(b)

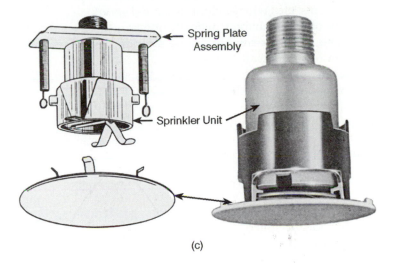

Spring Plate Assembly

Sprinkler Unit

(c)

(After Operation)

(d)

(e)

(f)

■ **FIGURE 9–9**

(a) Standard pendant sprinkler showing water deflected by a deflector plate. (Courtesy: Figgie (ASCOA) Fire Protection Systems, Charlottesville, VA.) (b) Sidewall type of sprinkler with flow rate and enclosure area (Reliable Sprinkler Co.). (c) Typical ceiling-concealed sprinkler (Star Sprinkler Corp., Milwaukee, WI). (d) Typical ceiling-recessed sprinkler (Star Sprinkler Corp.). (e) Fragile bulb type of sprinklers, upright and pendant (Grinnell Co. St. Louis, MO). (f) Standard fusible link type of upright sprinkler (Reliable Sprinkler Co. Mount Vernon, NY).

Example 9–2   What is the flow rate of a sprinkler with a ¾" orifice ($K = 14$) with 65 psi?

Answer:

The flow rate $Q = 14 \times \sqrt{65} = 14 \times 8.1$
$$= 112.9 \text{ GPM}$$

Compared with the ½" sprinkler at 24.6 GPM, the ¾" orifice can discharge 4.59 (112.9/24.6) times as much water.

Normally, the flow rates of individual sprinklers need not be calculated in the design of sprinkler systems. (See section 9.7.7.)

There is a number of specially designed sprinklers. Sidewall sprinklers are designed to discharge most of the water toward one side, in a pattern somewhat resembling one-quarter of a sphere. The forward throw may be as far as 20 to 25 ft, with a sideward throw from 7 to 8 ft. Sidewall sprinklers are ideal for hotel guest rooms, lobbies, executive offices, dining rooms, etc., where ceiling-mounted sprinklers would be objectionable aesthetically or where there is no ceiling space to run the pipes through. Figure 9–9(b) shows a sidewall sprinkler.

Residential sprinklers are designed especially for residential spaces, which are usually small, narrow, and with lower ceilings. The orifice factor K is as low as 1.5. The combination use of sidewall and residential sprinklers can accommodate most unusual architectural features in decorative spaces.

9.7.3 Types of Automatic Sprinkler Systems

There are numerous kinds of automatic sprinkler system, each ideally suited for certain spaces. Two major varieties are:

1. *Wet systems.* Wet pipe, antifreeze, and circulating closed loop.
2. *Dry systems.* Dry pipe, preaction, deluge, and combined dry-preaction.

A *wet-pipe system* is a piping system containing sprinklers under water pressure so that water discharges immediately from the sprinklers when they are opened by heat from a fire. Wet pipe is the basic automatic sprinkler system used for all building use groups. (See Figure 9–10 for a typical piping perspective and Figure 9–11 for the operation of some typical automatic sprinkler systems.)

An *antifreeze system* is a wet-pipe system containing antifreeze solutions used in areas subject to freezing. If the system is connected to the potable water system, then only pure glycerin or propylene glycol solutions are permitted. Suitable glycerin-water solutions and their freezing temperatures are given

in NFPA Standard 10. Antifreeze systems are similar to dry-pipe systems, except they have a faster response.

A *circulating closed-loop system* is a wet-pipe sprinkler system having non-fire-protection connections in a closed-loop piping arrangement for the purpose of utilizing sprinkler piping to circulate water for heating or cooling. Water is not removed or used from the system. A typical utilization of this system is the unitary hydronic heat pump system, where the sprinkler piping is used for circulating water through either the evaporator or the condensing coils, depending on the mode of operation. To qualify for fire suppression purposes, the operation of the HVAC system should not interfere with the water flow from the sprinkler when the automatic sprinkler function is activated. (See Chapter 7 for a discussion of hydronic heat pump systems.)

A *dry-pipe system* is a piping system filled with compressed air (or nitrogen). The air pressure prevents water from entering the pipes beyond a control valve known as a dry-pipe valve. When the air pressure is released at the opening of a sprinkler, the water flows into the piping system and out of the opened sprinkler. The water supply and the size of the dry-pipe valve shall be designed so that water is able to reach the far end of the system within 60 seconds. Dry-pipe systems are installed in areas that are subject to freezing, such as in a nonheated warehouse. (See Figure 9–10 for a typical piping perspective.)

A *preaction system* is a dry-pipe sprinkler system filled with air and having a supplemental detection system installed in the same area. Actuation of the detection system opens a valve that permits water to enter into the sprinkler piping system and to be discharged from any sprinklers that may be open. The detection system may be sensitive to either temperature or the density of smoke. Preaction systems are used in areas where water damage may be extremely detrimental to the property or may interfere with business operations. Yet, when a fire is out of control, water is readily available when any sprinkler fusible link is opened. A typical application of a preaction system is in a computer center of a large business or a central bank, where even a short interruption of services may cause huge financial losses. (See Figure 9–11 for the operation of a typical preaction system.)

A *deluge system* is a dry sprinkler system equipped with open-type sprinklers (no fusible links). The control valve is operated by a supplemental detection system in the same area. Actuation of the detection system opens the control valve, which allows water to flow through all the sprinklers at

Wet-pipe sprinkler systems employ automatic sprinklers attached to a piping system containing water and connected to a water supply so that water discharges immediately from sprinklers opened by a fire. Wet-pipe systems are the most reliable and simple of all sprinkler systems since no equipment other than the sprinklers themselves need to operate. Only those sprinklers which have been operated by heat over the fire will discharge water.

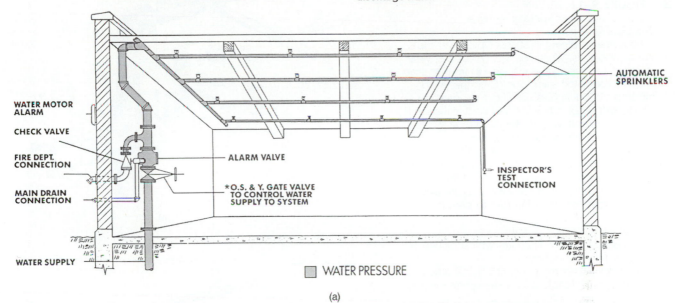

WATER MOTOR ALARM

CHECK VALVE

FIRE DEPT. CONNECTION

MAIN DRAIN CONNECTION

WATER SUPPLY

ALARM VALVE

*O.S. & Y. GATE VALVE TO CONTROL WATER SUPPLY TO SYSTEM

AUTOMATIC SPRINKLERS

INSPECTOR'S TEST CONNECTION

▢ WATER PRESSURE

(a)

Dry-pipe sprinkler systems employ automatic sprinklers attached to a piping system containing air or nitrogen under pressure, the release of which, as from the opening of a sprinkler, permits water pressure to open the dry-pipe valve. The water then flows into the piping system and discharges only from those sprinklers which have been operated by heat over the fire. Dry-pipe systems are installed in lieu of wet-pipe systems where piping is subject to freezing.

* O. S. & Y. – Outside screw and yoke gate valve.

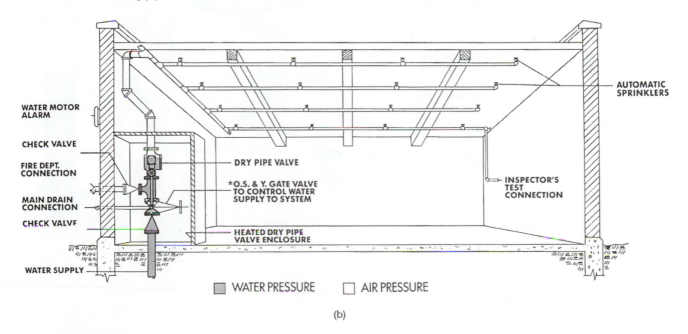

WATER MOTOR ALARM

CHECK VALVE

FIRE DEPT. CONNECTION

MAIN DRAIN CONNECTION

CHECK VALVE

WATER SUPPLY

DRY PIPE VALVE

*O.S. & Y. GATE VALVE TO CONTROL WATER SUPPLY TO SYSTEM

HEATED DRY PIPE VALVE ENCLOSURE

AUTOMATIC SPRINKLERS

INSPECTOR'S TEST CONNECTION

▢ WATER PRESSURE   ▢ AIR PRESSURE

(b)

■ **FIGURE 9–10**
Typical piping perspective of automatic sprinkler systems.
(a) Basic wet-pipe system.
(b) Basic dry-pipe system.
(Courtesy: Grinnell Fire Protection Systems Company, Exeter, NH.)

■ **FIGURE 9–11**

Operation of automatic sprinkler systems.

(a) Wet pipe system

The isometric piping illustrates typical wet pipe system components. When sprinkler (A) opens, water pressure lifts the alarm valve clapper (B) from its seat and flows through the alarm port (C) to the retard chamber (D), building pressure under the pressure switch (E) and sending the optional electrical alarm (F). The optional water-flow indicator (G) also activates an alarm and shows the location of the fire.

Water flows to the water motor alarm (H), sounding a mechanical signal. During surges or pressure fluctuations, excess pressure is trapped in the system, allowing only small amounts of water into the alarm port and retarding chamber in order to prevent a false alarm.

(b) Deluge and preaction system

The isometric piping illustrates typical deluge or preaction system components.

In the deluge system, the sprinklers are always open; there are no fusible links at the sprinkler heads. The upper chamber of the deluge valve (A) is pressurized in the ratio of 2:1 to keep the clapper seal on the inlet (B). When the release (C) detects a fire, the pneumatic actuator (D) opens by an air pressure drop, venting the upper valve chamber and permitting the valve to open. The pressure-operated relief valve (E) continues venting. Alarm (F) sounds, and water flows to the sprinklers (G). Since the sprinklers are open, water is applied to the fire immediately. Hydraulic and electric release systems may also be used.

In the preaction system, the sprinklers contain fusible links. In other words, the sprinkler heads are closed. When the preaction valve (A) is activated by one or more thermal or smoke detectors, water simply fills the pipes, and there is no water being discharged from the sprinklers until the fusible links in one or more sprinklers are melted due to heat. The piping air-pressure integrity is supervised by a monitoring system. Preaction systems may be interlocked singly, so that broken piping or broken sprinklers will not trip the system. It is also possible to double interlock the system to require the detection system to trip and a sprinkler to open before the valve opens.

(Courtesy: The Viking Corporation, Hastings, MI.)

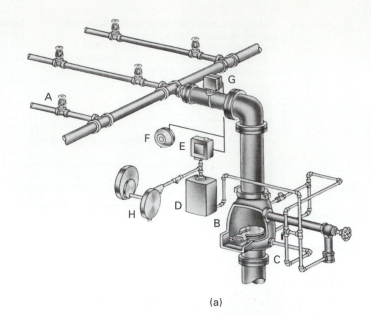

(a)

(b)

once. Sprinklers are closely spaced so as to create a "water wall" to separate different fire areas. This system is frequently used in a multilevel atrium, where the construction of fire walls to separate areas is undesirable for operational or aesthetic reasons. (See Figure 9–11 for the typical operation of a deluge system.)

9.7.4 Piping Material and Components

Piping material is usually black steel with welded or mechanical fittings. Galvanized steel is also used, but should not be welded, as welding will destroy the galvanized (zinc) coating. Copper may be used, but is too costly. Mechanical fittings of the groove-joining method are ideal for high-rise buildings when the differential expansion between the structure of the building and the piping system is substantial.

The fire department connection, water alarm, valves, and drain lines in an automatic sprinkler system are the same as in the standpipe-and-hose system.

The alarm check valve is the main control valve of every automatic sprinkler system. The valve assembly consists of single or double check valves, pressure gauges, water flow alarm test and drain valves, one or more check and control valves.

The water motor alarm is a water power–driven gong that produces a loud sound at the exterior of the building when water is flowing in the water-operated motor immediately inside the building. The water motor alarm does not require any electrical power, which is fortunate because such power may be cut off during a fire.

Figure 9–12 shows some components of an automatic sprinkler system.

9.7.5 Fire Pumps

Fire pumps are used when the residual water pressure at the most remote sprinkler heads of an automatic sprinkler system or at the hose outlets of a standpipe system cannot be met by the water supply system. A fire pump serves to boost the pressure to the necessary level and is selected on the basis of its flow rate (GPM), the pressure differential required, control features, and the driving equipment—an electrical motor or engine. The fire pump is usually of the split-case centrifugal variety for either horizontal or vertical mounting. A vertical pump assembly requires less floor space, but more floor-to-ceiling height. The control system is designed to start the pump whenever the system pressure drops below the required operating pressure. To avoid frequent start-

up of the fire pump, a small pressure maintenance pump known as a jockey pump is installed in parallel with the fire pump. When the system pressure drops below the preset level due to leakage or drainage, the pressure switch will first start the jockey pump to maintain the system pressure. If the system pressure drops rapidly, as in the case of the opening of sprinkler heads or the discharge of fire hoses, the jockey pump will not be able to maintain the required operating pressure, and the fire pump will start automatically. Figures 9–12 (f), (g), and (h) illustrate, respectively, a fire pump piping assembly, a diesel engine–driven fire pump, and a horizontal fire pump installation.

9.7.6 Planning Guidelines

Although the final layout of the automatic sprinkler system is prepared by the fire protection contractor, the architect/engineer should be knowledgeable regarding basic requirements of the system, so the layout is acceptable aesthetically while maintaining a reasonable construction cost. The following is a summary of planning guidelines for sprinkler systems:

1. The maximum floor area of a building that can be protected by a single system shall not exceed 52,000 sq ft for hazards classified as light or ordinary. The maximum area is reduced for extra-hazard classified buildings.
2. The maximum floor area that can be covered by a sprinkler shall not exceed that given in Table 9–5.
3. In addition to the maximum floor area limitation, sprinklers shall meet the following dimensional limitations:

- The maximum distance between sprinklers shall be no more than 15 ft for light and ordinary hazards and 12 ft for extra hazards and above, except that the spacing of sidewall-type sprinklers may be according to the approved area of coverage.
- When sprinklers are spaced less than 6 feet on centers, baffles shall be located between the sprinklers to prevent nonactivated sprinklers from being cooled off by the water discharge from adjacent sprinklers.
- The distance of a sprinkler from a wall shall be no more than half of the allowed distance between sprinklers—not more than 9 feet or less than 4 inches from a wall.
- The distance between vertical obstructions and the sprinkler shall not be less than that given in Table 9–6.

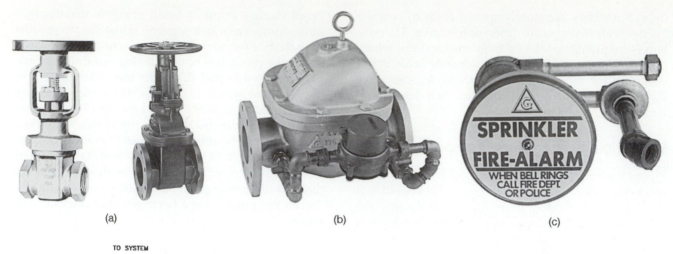

(a)

(b)

(c)

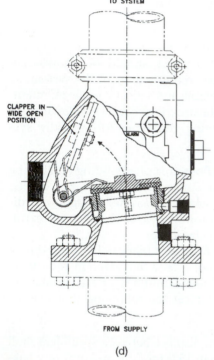

TO SYSTEM

CLAPPER IN
WIDE OPEN
POSITION

ALARM

FROM SUPPLY

(d)

(e)

(f)

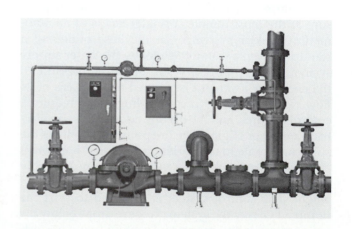

(g)

9.7.7 System Design Approaches

Piping for automatic sprinkler systems can be designed by either the pipe schedule or the hydraulic method.

Pipe schedule The pipe schedule is a traditional method permitted for new light- and ordinary-hazard occupancies of 5000 sq ft or less, and for the modernization of existing systems designed by the pipe schedule. The sizes of branches and risers can be determined from a pipe schedule, assuming that other requirements, such as residual pressure and flow rates are all in compliance with code. In general, the pipe schedule yields more conservative pipe sizes; it is very useful for preliminary planning and cost estimation. Table 9–7 gives the pipe schedule.

Example 9–3 A two-story office building totaling 4000 sq ft is interpreted as a light hazard (200 sq ft/sprinkler from Table 9–5). After adjusting for the room configurations, it was determined that there shall be 14 sprinklers installed at each floor. If black steel pipe is used, determine the riser size.

Answer The main riser is for 28 sprinklers. From Table 9–7, a 2½" riser that is allowed to serve up to 30 sprinklers will be used. If, instead, the building is interpreted as an ordinary hazard by the building code official, then a 3" riser will be used (good for 40 sprinklers).

Hydraulic method The hydraulic method actually calculates the pipe size of the entire piping system, based on the distribution of sprinklers, developed length, fitting losses, size and location of areas within the building, and the water density and pressure required. The engineering for hydraulic calculations is based on the following formulas:

TABLE 9–6

Minimum horizontal distance of sprinklers from any vertical obstructions (in ft)

Maximum Dimension of Obstructions	Minimum Distance
Less than ½ inch	no limit
Between ½ and 1 inch	6 inches
Between 1 and 4 inches	12 inches
More than 4 inches	24 inches

TABLE 9–5

Maximum sprinkler protection area (sq ft)

Construction	Classification of Hazard		
	Light	Ordinary	Extra
Unobstructed Construction[a]	200–225[b]	130	90–130[c]
Obstructed Construction:			
Noncombustible	200–225[b]	130	90–130[c]
Combustible	130–225[a]	130	90–130[c]

[a]For light combustible framing members spaced less than 3 feet on centers, the maximum area is 130 sq ft, and for heavy combustible framing members spaced 3 feet or more on centers, the maximum area is 225 sq ft.

[b]If pipe sizing is based on pipe schedule, the maximum protected area shall not exceed 200 sq ft.

[c]If pipe sizing is based on pipe schedule, the maximum area shall not exceed 90 sq ft. If based on hydraulic method with density less than 0.25 GPM/sq ft, the maximum area shall not exceed 130 sq ft.

TABLE 9–7

Pipe schedule for number of sprinklers allowed in a sprinkler system

Pipe Size (inch)	Hazard Classification			
	Light		Ordinary	
	Steel	Copper	Steel	Copper
1	2	2	2	2
1¼	3	3	3	3
1½	5	5	5	5
2	10	12	10	12
2½	30	40	20	25
3	60	65	40	45
3½	100	115	65	75
4	a	a	100	115
5	—	—	160	180
6	—	—	275	300
8	—	—	b	b

[a]One 4-inch system may serve up to 52,000 sq ft of floor area.

[b]One 8-inch system may serve up to 52,000 sq ft of floor area.

■ **FIGURE 9–12**

(a) Gate valves. (b) Detector check valve. (c) Water motor alarm. (Courtesy: Grinnell Fire Protection Systems Company, Exeter, NH.) (d) Cutaway section of an alarm check valve showing clapper in the closed position. When water is flowing, the clapper will be in the wide open position. (Courtesy: Reliable Automatic Sprinkler Co.,) Mount Vernon, NY.) (e) Wet-pipe sprinkler alarm check valve assembly showing auxiliary piping, flow control devices, and pressure gauges. (Courtesy: Reliable Automatic Sprinkler Co., Mount Vernon, NY.) (f) Side view of a fire pump assembly. (g) Diesel engine–drive fire pump, which has the advantage of being available during an electrical power failure and downsizing the emergency generator capacity. Engine-driven fire pump systems have a higher installation cost than those that are electrically driven. (Courtesy: General Signal Pump Group, North Aurora, IL.)

- Friction loss, $P_f = \dfrac{4.52\, Q^{1.85}}{C^{1.85}\, d^{4.87}} = F\, Q^{1.85}$ (9–2)

  where P_f = frictional resistance, psi/ft of pipe equivalent
  Q = water flow rate, GPM
  d = actual internal pipe diameter, inches
  C = friction loss coefficient of piping material
  F = combined value of $(4.52/c^{1.85}\, d^{4.87})$; See Table 9–9 for P_f values for pipes with C = 100, 120 and 150.

- Velocity pressure, $P_v = 0.001123\, \dfrac{Q^2}{d^4}$ (9–3)

  where P_v = velocity pressure, psi
  Q = water flow rate, GPM
  d = inside diameter of pipe, inches

- Normal pressure (acting perpendicularly to pipe wall)

  $$p_n = p_t - P_v \qquad (9\text{–}4)$$

  where P_n = normal pressure, psi
  P_t = total pressure, psi
  P_v = velocity pressure, psi

- Elevation pressure, $p_e = 0.433h$ (9–5)

  where p_e = elevation pressure, psi
  h = height of elevation, feet

- Velocity, $V = \dfrac{0.485\, Q}{d^2}$ (9–6)

  where V = velocity, fps
  Q = flow rate, GPM
  d = inside diameter of pipe, inches

For convenience, the equivalent lengths of pipe fittings and controls are given in Table 9–8, and the values of friction loss (P_f), velocity (V), and velocity pressure (P_v) for steel and copper pipes are given in Table 9–9.

Water demand Statistics indicate that, during the early stage of a fire, only a small portion of the sprinklers needs to be operated. Thus, fire protection codes normally require that only a small area be calculated for simultaneous flow demand. This area is known as the "area of sprinkler operations," or, more simply, the "design area." The area selected for calculations should be the most hydraulically remote from the water supply source. It varies from a minimum of 1500 sq ft to a maximum of about 5000 sq ft, depending on the hazard classification of the building, as interpreted by the authoritative code or the insurance company. The code also calls for a minimum water flow density per unit area (GPM/sq ft). The minimum water supply for the sprinkler system is determined independently of the actual size of the building. For example, the water demand for a 50,000-sq-ft and a 750,000-sq-ft office building may be the same as far as sprinkler systems are concerned. This is because a fire can be expected to start in only one area, unless it is deliberately set at multiple locations. If the sprinkler system also supplies inside and outside hoses, the water demand of the hoses should be included as well. The water demand for a combined sprinkler and standpipe system can be expressed by the formula

$$TWD = (AOP \times DD \times OVF) + HSD \qquad (9\text{–}7)$$

where TWD = total water demand, GPM
AOP = area of operations (see Figure 9–13)
DD = Density demand, GPM/sq ft (see Figure 9–13)
HSD = Hose stream demand, GPM (see Table 9–10)
OVF = Overage factor (usually 1.1)

TABLE 9–8

Equivalent pipe length chart. Applicable to pipes with the listed C values: cast iron (100), black steel (100 and 120), galvanized steel (120), plastic (120), cement-lined cast iron (140), and copper or stainless steel (150) (data from NFPA Standard 13.)

| | Fittings and Valves Expressed in Equivalent Feet of Pipe | | | | | | | | | | | | |
Fittings and Valves	¾ in.	1 in.	1¼ in.	1½ in.	2 in.	2½ in.	3 in.	3½ in.	4 in.	5 in.	6 in.	8 in.	10 in.
45° Elbow	1	1	1	2	2	3	3	3	4	5	7	9	11
90° Standard Elbow	2	2	3	4	5	6	7	8	10	12	14	18	22
90° Long Turn Elbow	1	2	2	2	3	4	5	5	6	8	9	13	16
Tee or Cross (Flow Turned 90°)	3	5	6	8	10	12	15	17	20	25	30	35	50
Butterfly Valve	—	—	—	—	6	7	10	—	12	9	10	12	19
Gate Valve	—	—	—	—	1	1	1	1	2	2	3	4	5
Swing Check	—	5	7	9	11	14	16	19	22	27	32	45	55

Note: Value in table is for C = 120; multiply table value by 0.713 for C = 100, 0.713 for C = 140, and 1.51 for C = 150.

TABLE 9–9

Friction loss, flow velocity, and velocity pressure for steel and copper pipes at the calculated flow rate. (Courtesy: Viking Corporation, Hastings, MI.)

Size (in.)	O.D. (in.)	Wall Thickness (in.)	I.D. (in.)	P_f(psi/lin. ft) Multiply $Q^{1.85}$ by		V (fps) Multiply Q by	$P_{v(psi)}$ Multiply Q^2 by
Steel Pipe—½–6 Sch 40							
8–12 Sch 30				For C = 120	For C = 100		
½	0.840	0.109	0.622	6.50×10^{-3}	9.11×10^{-3}	1.056	7.50×10^{-3}
¾	1.050	0.113	0.824	1.65×10^{-3}	2.32×10^{-3}	0.602	2.44×10^{-3}
1	1.315	0.133	1.049	5.10×10^{-4}	7.14×10^{-4}	0.371	9.27×10^{-4}
1¼	1.660	0.140	1.380	1.34×10^{-4}	1.88×10^{-4}	0.215	3.10×10^{-4}
1½	1.90	0.145	1.610	6.33×10^{-5}	8.87×10^{-5}	0.158	1.67×10^{-4}
2	2.375	0.154	2.067	1.87×10^{-5}	2.63×10^{-5}	0.096	6.15×10^{-5}
2½	2.875	0.203	2.469	7.89×10^{-6}	1.11×10^{-5}	0.067	3.02×10^{-5}
3	3.500	0.216	3.068	2.74×10^{-6}	3.84×10^{-6}	0.0434	1.27×10^{-5}
3½	4.000	0.226	3.548	1.35×10^{-6}	1.89×10^{-6}	0.0325	7.09×10^{-6}
4	4.500	0.237	4.026	7.29×10^{-7}	1.02×10^{-6}	0.0252	4.27×10^{-6}
5	5.563	0.258	5.047	2.43×10^{-7}	3.40×10^{-7}	0.0160	1.73×10^{-6}
6	6.625	0.280	6.065	9.91×10^{-8}	1.39×10^{-7}	0.0111	8.30×10^{-7}
8	8.625	0.277	8.071	2.47×10^{-8}	3.45×10^{-8}	0.0063	2.65×10^{-7}
Steel Pipe—1–3½ Sch 10 S							
4–8 .188 Wt.				For C = 120	For C = 100		
1	1.315	0.109	1.097	4.10×10^{-4}	5.75×10^{-4}	0.339	7.75×10^{-4}
1¼	1.660	0.109	1.442	1.08×10^{-4}	1.52×10^{-4}	0.196	2.60×10^{-4}
1½	1.900	0.109	1.682	5.12×10^{-5}	7.17×10^{-5}	0.144	1.40×10^{-4}
2	2.375	0.109	2.157	1.52×10^{-5}	2.13×10^{-5}	0.088	5.19×10^{-5}
2½	2.875	0.120	2.635	5.75×10^{-6}	8.05×10^{-6}	0.059	2.33×10^{-5}
3	3.500	0.120	3.260	2.04×10^{-6}	2.86×10^{-6}	0.0384	9.94×10^{-6}
3½	4.000	0.120	3.760	1.02×10^{-6}	1.43×10^{-6}	0.0289	5.62×10^{-6}
4	4.500	0.188	4.124	6.49×10^{-7}	9.09×10^{-7}	0.0240	3.88×10^{-6}
5	5.563	0.188	5.187	2.12×10^{-7}	2.98×10^{-7}	0.0152	1.55×10^{-6}
6	6.625	0.188	6.249	8.57×10^{-8}	1.20×10^{-7}	0.0105	7.36×10^{-7}
8	8.625	0.188	8.249	2.22×10^{-8}	3.11×10^{-8}	0.0060	2.43×10^{-7}
Copper Tube—Type "K"				For C = 150			
¾	0.875	0.065	0.745	1.79×10^{-3}		0.736	3.65×10^{-3}
1	1.125	0.065	0.995	4.36×10^{-4}		0.413	1.15×10^{-3}
1¼	1.375	0.065	1.245	1.47×10^{-4}		0.264	4.67×10^{-4}
1½	1.625	0.072	1.481	6.29×10^{-5}		0.186	2.33×10^{-4}
2	2.125	0.083	1.959	1.61×10^{-5}		0.106	7.63×10^{-5}
2½	2.625	0.095	2.435	5.59×10^{-6}		0.069	3.19×10^{-5}
3	3.125	0.109	2.907	2.36×10^{-6}		0.0483	1.57×10^{-5}
3½	3.625	0.120	3.385	1.12×10^{-6}		0.0357	8.55×10^{-6}
4	4.125	0.134	3.857	5.95×10^{-7}		0.0275	5.07×10^{-6}
Copper Tube—Type "L"				For C = 150			
¾	0.875	0.045	0.785	1.38×10^{-3}		0.663	2.96×10^{-3}
1	1.125	0.050	1.025	3.78×10^{-4}		0.389	1.02×10^{-3}
1¼	1.375	0.055	1.265	1.36×10^{-4}		0.255	4.39×10^{-4}
1½	1.625	0.060	1.505	5.82×10^{-5}		0.180	2.19×10^{-4}
2	2.125	0.070	1.985	1.51×10^{-5}		0.1037	7.23×10^{-5}
2½	2.625	0.080	2.465	5.26×10^{-6}		0.0672	3.04×10^{-5}
3	3.125	0.090	2.945	2.21×10^{-6}		0.0471	1.49×10^{-5}
3½	3.625	0.100	3.425	1.06×10^{-6}		0.0348	8.16×10^{-6}
4	4.125	0.110	3.905	5.60×10^{-7}		0.0268	4.83×10^{-6}

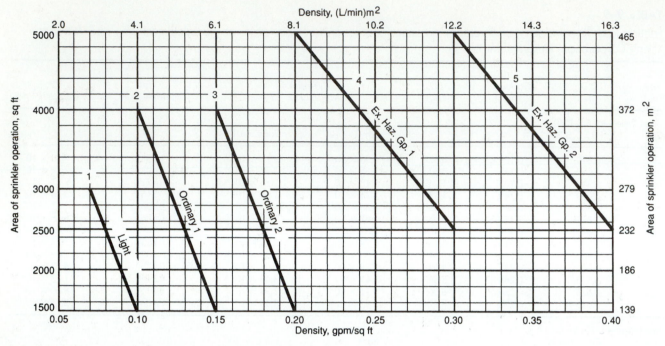

For SI Units: 1 sq ft = 0.0929 m²; 1 gpm/sq ft = 40.746 (L/min)/m².

■ **FIGURE 9–13**
Area of operations (AOP) vs. density demand.
(Reprinted with permission from NFPA 13, *Installation of Sprinkler Systems*, copyright © 1994, National Fire Protection Association, Quincy, MA 02269. This reprinted material is not the complete and official position of the National Fire Protection Association on the referenced subject which is represented only by the standard in its entirety).

Example 9–4 A school has 50,000 sq ft of floor area over which an automatic sprinkler system is to be installed. The building is four stories in height and is equipped with a standpipe system consisting of two fire hose cabinets per floor. The official code classifies the building as a light hazard. Determine the minimum water supply that must be provided or made available to the building.

Answer

The sprinkler water demand = AOP × DD × OVF

From Figure 9–13, select AOP = 1500 sq ft and DD = 0.10, so that

$$\text{Sprinkler water demand} = 1500 \times 0.10 \times 1.1$$
$$= 165 \text{ GPM}$$

From Table 9–10, for a light hazard, the HSD = 100 GPM. Thus,

$$\text{Total water demand (TWD)} = 150 + 100$$
$$= 265 \text{ GPM}$$

TABLE 9–10

Hose stream demand and water supply duration requirements

Hazard Classification	Inside Hose (GPM)	Combined Inside and Outside Hose (GPM)	Duration in Minutes
Light	0, 50, or 100	100	30
Ordinary	0, 50, or 100	250	60–90
Extra Hazard	0, 50, or 100	500	90–120

If the building were classified as an ordinary hazard (group 1), the sprinkler demand would be 1500 × 0.15 × 1.1 = 247 GPM, the HSD would be 250 GPM, and the TWD would be 497, or about 500, GPM.

If the water supply is an in-house storage tank, then the water storage should be of minimum 30-minute duration, or

$$\text{TWD} = 265 \text{ GPM} \times 30 \text{ min}$$
$$= 7950 \text{ (about 8000) gal}$$

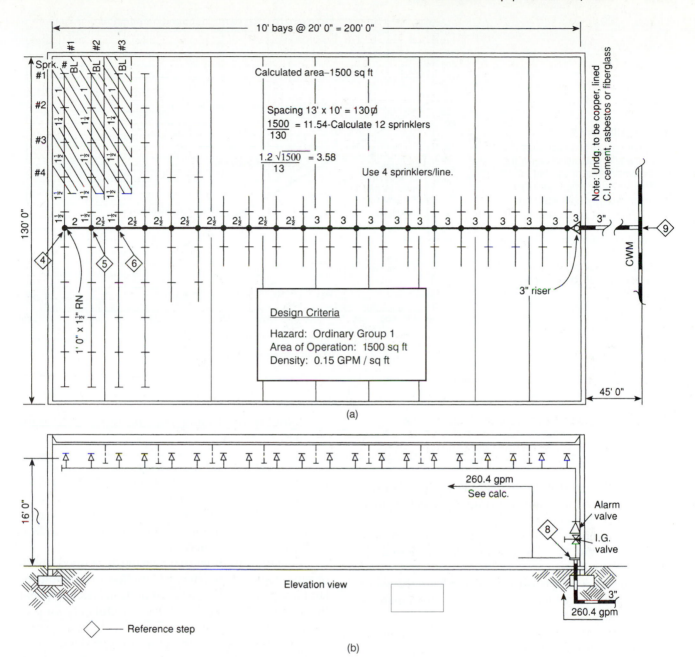

(a)

(b)

◇ —— Reference step

■ **FIGURE 9–14**

Sample hydraulic calculations. (a) Floor plan of a building. (b) Elevation. The building is an employee garage classified as an ordinary hazard, Group 1. From Curve 2 of Figure 9–13, the area of operation is selected to be 1500 sq ft with a water flow density of 0.15 GPM/sq ft. Based on Table 9–5, the maximum sprinkler protection area of unobstructed construction is 130 sq ft/sprinkler. Black steel pipe is used for interior pipes, with a coefficient of friction $C = 120$. The standard ½-inch orifice upright type of sprinkler ($K = 5.65$) is used throughout. From Table 9–10, the hose stream demand allowance for an ordinary hazard is 250 GPM. (Reprinted with permission from NFPA 13, *Installation of Sprinkler Systems*, copyright © 1994, National Fire Protection Association Quincy, MA 02269. This reprinted material is not the complete and official position of the National Fire Protection Association on the referenced subject which is represented only by the standard in its entirety.)

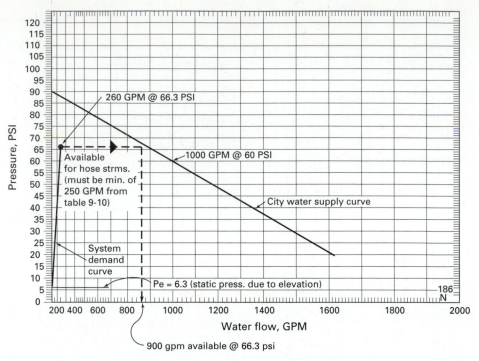

■ FIGURE 9–15

Hydraulic graph of water supply and hydraulic calculation form for the selected area of operation. (a) The hydraulic graph includes the water supply data resulting from a flow test at the nearest fire hydrant. The data are plotted on standard $N^{1.85}$ semilog graph paper, which provides a curve of residual pressure vs. linear flow on the graph for ease of interpolation.

(a)

9.7.8 Hydraulic Calculation Procedure

Hydraulic calculations require the following steps:

- Determine the building hazard classification and the allowed maximum sprinkler protection area, as given in Table 9–5.
- Determine the water demand of the sprinkler system and the hose stream system, based on Equation 9–7.
- Determine the pressure available from the water supply source by a flow test at the street main or storage system.
- Determine the required flow rate and pressure of the end head (last sprinkler).
- Calculate the residual pressure available from the end head to the water supply main, the required total pressure, and the average pressure loss (psi/ft), including the loss through pipe fittings.
- The sizes of branches, cross-mains, ceiling mains, and underground mains can be determined from the standard chart of friction loss vs. flow rate for the selected piping material (steel, copper, etc.), the calculated flow rate (total demand in GPM), the maximum water velocity (fps), and/or the selected friction loss (psi/100 ft). A typical chart for fairly smooth pipes (copper or plastic) is shown in Figure 8–7 in Chapter 8. Similar charts for other piping materials can be

found in most handbooks on water or hydraulics.

In contrast to the design of domestic or HVAC piping systems, which should limit the flow velocity to less than 10 fps, the flow velocity of water in a fire protection piping system is not constrained, since both noise and erosion caused by water flow are not of concern during a fire.

For large buildings in which the developed length of pipe main is extensive, the friction loss should be kept low to minimize the total friction loss. Note that the manual pipe-sizing method is useful only for preliminary design calculations. The final design, by computer program, will yield optimum pipe sizes.

Sample hydraulic calculations Hydraulic calculations for a complete automatic sprinkler system are somewhat tedious, although not complex. Computer software is available through various professional and trade organizations. A sample manual hydraulic calculation for a small single-story garage is shown in Figure 9–14 and 9–15. The sample is taken from NFPA Standard 13.

9.7.9 Combined Systems

In design applications, standpipe system and automatic sprinkler system are often combined into a single system to achieve economy. Figure 9–16 illustrates a combined system for a large building.

Step no.	Nozzle ident. and location	Flow in C.P.M.	Pipe size	Pipe fittings and devices	Equiv. pipe length	Friction loss P.S.I./foot	Pressure summary	Normal pressure	D = 0.15 GPM/sq ft Notes	Ref. step
Contract Name	*Group I: AOP=1500 sq ft*								D = 0.15 GPM/ sq ft	
①	1 BL-1	q	1		L 13.0	C = 120	Pt 11.9	Pt		
					F		Pe	Pv	q = 130 × .15=19.5	
		Q 19.5			T 13.0	.124	Pf 1.6	Pn		
②	2	q 20.7	1 1/4		L 13.0		Pt 13.5	Pt		
					F		Pe	Pv	q = 5.65 √13.5	
		Q 40.2			T 13.0	.125	Pf 1.6	Pn		
③	3	q 22	1 1/2		L 13.0		Pt 15.1	Pt		
					F		Pe	Pv	q = 5.65 √15.1	
		Q 62.2			T 13.0	.132	Pf 1.7	Pn		
④	4 DN RN	q 23.2	1 1/2	2T-16	L 20.5		Pt 16.8	Pt		4
					F 16.0		Pe	Pv	q = 5.65 √16.8	
		Q 85.4			T 36.5	.237	Pf 8.6	Pn		
⑤	CM TO BL-2	q	2		L 10.0		Pt 25.4	Pt	K = 85.4 / √25.4	5
					F		Pe	Pv		
		Q 85.4			T 10.0	0.7	Pf .7	Pn	K = 16.95	
⑥	BL-2 CM TO BL-3	q 86.6	2 1/2		L 10.0		Pt 26.1	Pt		6
					F		Pe	Pv	q = 16.95 √26.1	
		Q 172.0			T 10.0	.109	Pf 1.1	Pn		
⑦	BL-3 CM	q 88.4	2 1/2		L 70.0		Pt 27.2	Pt		
					F		Pe	Pv	q = 16.95 √27.2	
		Q 260.4			T 70.0	.233	Pf 16.3	Pn		
⑧	CM TO FIS	q	3	E5	L 119.0		Pt 43.5	Pt		8
				AV15	F		Pe 6.5	Pv	Pe = 15 × .433	
		Q 260.4		GV1	T 140.0	.081	Pf 11.3	Pn		
⑨	Thru Underground to City Main	q	3	E5	L 50.0	C = 150	Pt 61.3	Pt		9
				GV1	F 32.0	Type "M"	Pe	Pv	COPPER 21 X 1.51 = 32	
		Q 260.4		T15	T 82.2	.061	Pf 5.0	Pn		
		q			L		Pt 66.3	Pt		
					F		Pe	Pv		
		Q			T		Pf	Pn		
		q			L		Pt	Pt		
					F		Pe	Pv		
		Q			T		Pf	Pn		
							Pt			

(b)

■ FIGURE 9–15 *(continued)*

(b) The calculation, shown in nine steps starting from the first sprinkler to the water supply at the street main, indicates a total pressure (p_t) of 66.3 psi. The total pressure and water demand determine the end point on the system demand curve, indicating that the water supply is substantially greater than the sprinkler demand (260 GPM) and the hose stream demand (250 GPM). Interpolating from the graph yields the result that the water supply is capable of supplying about 900 GPM at 66.3 psi.

Legend:
AV—Alarm valve
BL—Branch line
RN—Riser nipple
CM—Ceiling main
L—Length of pipe
F—Fitting equivalent length
T—Total equivalent length
q—Individual sprinkler flow
Q—Accumulated flow
(Reprinted with permission from NFPA 13, *Installation of Sprinkler Systems*, copyright © 1994, National Fire Protection Association, Quincy, MA 02269. This reprinted material is not the complete and official position of the National Fire Protection Association on the referenced subject which is represented only by the standard in its entirety.)

9.8 SMOKE CONTROLS

Smoke is always present when there is a fire. The degree of smoke generated depends on the combustible material of the fire. Fire from wood and paper generates relatively light smoke, whereas fire from plastic or synthetic materials generates heavy, toxic smoke. It has been proven that loss of life due to smoke is considerably higher than from fire alone. Often, smoke spreads out to great distances from the origin of a fire. Smoke control, therefore, has become the emerging topic in building design. It requires the closest coordination between architectural planning and structural, HVAC, and fire protection systems. The practice of utilizing HVAC systems for smoke control is relatively new. The principles of effective smoke control are well documented in the *ASHRAE Handbook*. The following paragraph contains some excerpts of pertinent smoke control recommendations. For in-depth recommendations and the method of designing smoke control systems, see the *ASHRAE Handbook*.

9.8.1 Stack Effect

Smoke spreads in a building primarily because of two factors: hot air, which makes the smoke rise due to its

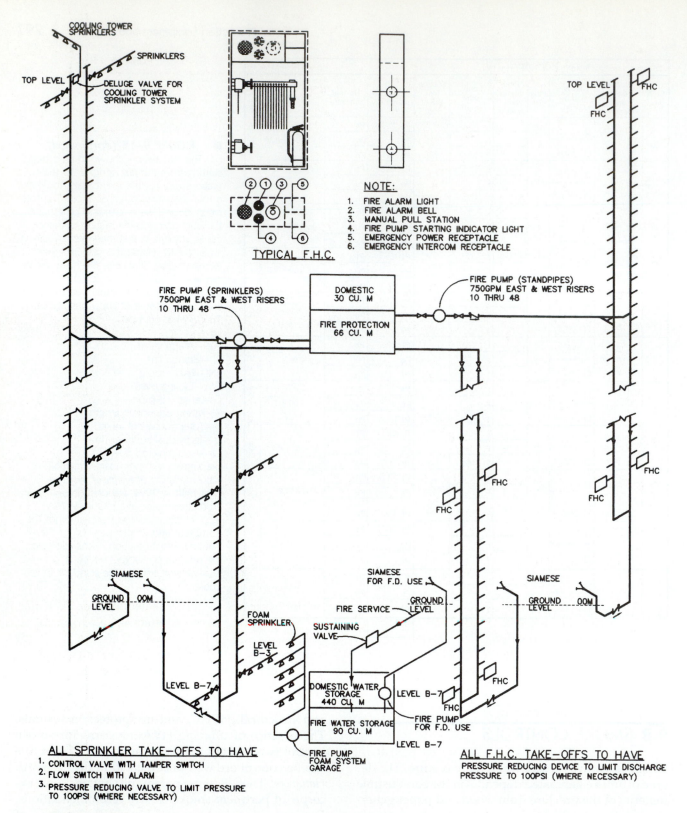

COOLING TOWER
SPRINKLERS

SPRINKLERS

TOP LEVEL

DELUGE VALVE FOR
COOLING TOWER
SPRINKLER SYSTEM

TYPICAL F.H.C.

NOTE:

1. FIRE ALARM LIGHT
2. FIRE ALARM BELL
3. MANUAL PULL STATION
4. FIRE PUMP STARTING INDICATOR LIGHT
5. EMERGENCY POWER RECEPTACLE
6. EMERGENCY INTERCOM RECEPTACLE

TOP LEVEL

FHC

FIRE PUMP (SPRINKLERS)
750GPM EAST & WEST RISERS
10 THRU 48

DOMESTIC
30 CU. M

FIRE PROTECTION
66 CU. M

FIRE PUMP (STANDPIPES)
750GPM EAST & WEST RISERS
10 THRU 48

FHC

FHC

SIAMESE

GROUND
LEVEL   00M

SIAMESE
FOR F.D. USE

FIRE SERVICE

GROUND
LEVEL   00M

SUSTAINING
VALVE

FOAM
SPRINKLER

LEVEL
B-3

LEVEL B-7

DOMESTIC WATER
STORAGE
440 CU. M

FIRE WATER STORAGE
90 CU. M

LEVEL B-7

FIRE PUMP
FOAM SYSTEM
GARAGE

FIRE PUMP
FOR F.D. USE

LEVEL B-7

SIAMESE

GROUND
LEVEL   00M

FHC

FHC

FHC

FHC

ALL SPRINKLER TAKE-OFFS TO HAVE

1. CONTROL VALVE WITH TAMPER SWITCH
2. FLOW SWITCH WITH ALARM
3. PRESSURE REDUCING VALVE TO LIMIT PRESSURE
   TO 100PSI (WHERE NECESSARY)

ALL F.H.C. TAKE-OFFS TO HAVE

PRESSURE REDUCING DEVICE TO LIMIT DISCHARGE
PRESSURE TO 100PSI (WHERE NECESSARY)

■ **FIGURE 9–16**

Riser diagram of a combination sprinkler and standpipe system. The system is for a 48-story high-rise building. The system is divided into two vertical zones. The domestic water distribution system shares the same storage tanks, except the lower part of each tank is exclusively reserved for fire protection purposes and cannot be drawn down by the domestic water pumps.

lower density, and pressure differences in the building, which cause air to migrate throughout the building. The following is a direct quotation from the *ASHRAE Handbook:*

> When it is cold outside, air often moves upward within building shafts, such as stairwells, elevator shafts, dumbwaiter shafts, mechanical shafts, or mail chutes. Referred to as normal stack effect, this phenomenon occurs because the air in the building is warmer and less dense than the outside air. Normal stack effect is great when outside temperatures are low, especially in tall buildings. However, normal stack effect can exist even in a one-story building.
>
> When the outside air is warmer than the building air, a downward air flow, or reverse stack effect, frequently exists in shafts. At standard atmospheric pressure, the pressure difference due to either normal or reverse stack effect, frequently exists in shafts. At standard atmospheric pressure, the pressure difference due to either normal or reverse stack effect is expressed as:

$$\Delta p = 7.64 \, (1/T_o - 1/T_i)h \qquad (9\text{–}8)$$

where Δp = pressure difference, inches of water

T_o = absolute temperature of outside air, °R

T_i = absolute temperature of air inside shaft, °R

h = distance above neutral plane, ft

For a building 200 ft tall with a neutral plane at the mid-height, an outside temperature of 0°F, and an inside temperature of 70°F, the maximum pressure difference due to stack effect would be 0.22 inch of water. This means that at the top of the building, a shaft would have a pressure of 0.22 inch of water greater than the outside pressure. At the bottom of the shaft, the shaft would have a pressure of 0.22 inch of water less than the outside pressure.

Stack effect is more serious in cold weather than in hot weather, since the temperature differential between the outdoors and the interior of the building is greater. For example, with the interior maintained at 75°F, the differential is 75°F when the outdoor temperature is 0°F in the winter, and only 20°F when the outdoor temperature is 95°F in the summer. Figure 9–17 illustrates pressure variations in a high-rise building during cold weather. Figure 9–18(a) shows air leaking into a shaft, such as an elevator shaft, when the shaft is vented at the top. Figure 9–18(b)

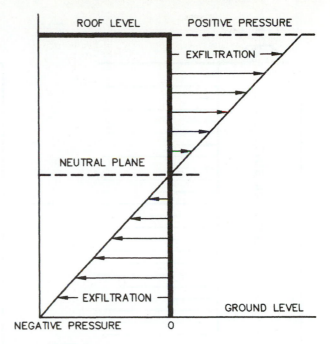

■ **FIGURE 9–17**
Stack effect. During cold weather, outdoor air is heavier than indoor air. Air begins to rise in the building, exfiltrating through the upper floor windows, cracks or openings. The cold outdoor air is drawn into the building through the lower floors. This is known as the stack effect. If a fire occurs on the lower floor, smoke can easily migrate into the upper floor.

shows an unvented shaft in which air infiltrates the lower floors with diminishing intensity (pressure) and leaks out (exfiltration) with increasing intensity (pressure) above the neutral zone (floor). The total amount of air leaking in must be equal to the total amount of air leaking out. The neutral zone is practically at the mid-level of the building, but is normally lower than that because lower floors are usually larger in area, and have more doors and windows at the ground floor. If a building is perfectly airtight and without exhausts, the air will not travel between floors even with stack effect. However, smoke still may rise to the upper floors due to a buoyancy force or wind pressure. For these reasons, current practice in smoke control depends heavily on the HVAC systems to counter the natural air movement in the building.

9.8.2 Pressure Control

Air flows only when there is a pressure difference between two areas. The flow is from the area of higher pressure to the area of lower pressure. If the fire area is maintained at a relatively low pressure by exhausting, then air containing smoke will not flow easily to

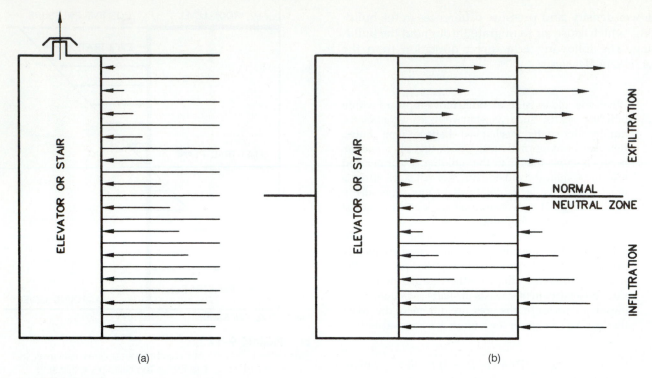

(a)   (b)

■   **FIGURE 9–18**

Typical building pressure profile. If a building shaft, such as an elevator hoistway, is vented at the top, (a) air will enter the shaft at all floor levels with diminishing amount at the upper levels, and smoke will be drawn into the shaft during a fire. For this reason, elevator lobbies should have positive air pressure in relation to surrounding spaces and elevators are deactivated after they are brought down to the ground level. (b) If a shaft is not vented, but not airtight, then air infiltration and exfiltration will behave as in Figure 9–17.

the other areas in the building. The principle can be applied to all types of buildings, low rise or high rise. Following are some common control practices.

Local exhaust   Exhaust by fans or relief by venting at the floor where fire is started will create low pressure in the fire zone, causing air in the other zones to rush in and thus confine the smoke.

Pressure sandwich   By the proper control of air supply, return, and exhaust, smoke at the fire zone will have less chance to migrate to the other zones. In general, if the pressure at the fire zone is between 0.1 and 0.15 inch w.c. lower than that at the other zones, smoke can be effectively contained. Figure 9–19 shows the pressure sandwich principle. This concept is increasingly used in high-rise buildings and is required by most building codes.

Compartmentation   The building is divided into two or three vertical compartments as if they were separate buildings stacked on top of each other. This practice is used only in skyscrapers or buildings taller than 50 stories. Figure 9–20 illustrates this principle. The drawback is the added cost and inconvenience caused by the required transfer floors between each zone.

Stair pressurization   The same concept of pressure differential used between zones applies to stairways, which are the major means of egress from a building. If positive pressure is maintained in stairways by a stair pressurization fan, and vestibules with elevators are maintained at a lesser positive pressure (with some exhaust relief) than the other areas of the building, smoke will not likely migrate into the stairways. Figure 9–21 illustrates a stair pressurization system. The design requires careful analysis and includes the following guidelines:

■   *Performance of the stair pressurization fan.* Overpressurization will make it difficult to open the stair doors. A 30-lb force at the door handle is considered the maximum that an average person can exert. In general, a minimum of 0.05 inch

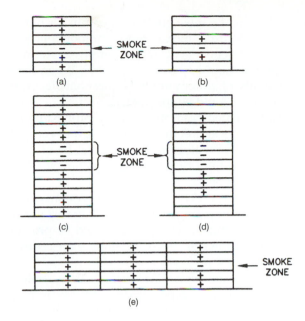

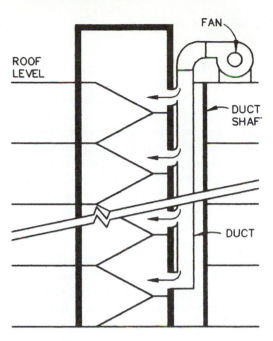

■ FIGURE 9–19

Pressure sandwich design. The smoke zone is indicated by a minus sign, and pressurized spaces are indicated by a plus sign. Each floor can be a smoke control zone as in (a) and (b), or a smoke zone can consist of more than one floor as in (c) and (d). A smoke zone can also be limited to a part of a floor, as in (e).

■ FIGURE 9–21

Stair pressurization. Rooftop-mounted fan injects air at every floor. A single point of injection is acceptable for buildings not over 10 stories. The fan may also be mounted at the base of the stairwell. Typically, a system will require about 5,500 L/s (12,000 cfm) at 60 Pa (0.25" wc) up to 20 stories, and proportionally increase for taller buildings.

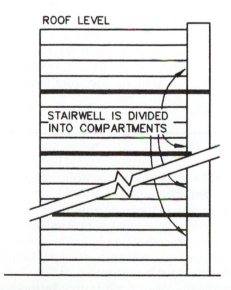

■ FIGURE 9–20

Compartmentation. Compartmentation can be an effective means of providing stairwell pressurization for very tall buildings when a staged evacuation plan is used. The drawback is the additional landing space required at the transfer floors.

w.c., and not more than 0.25 inch w.c., positive pressure shall be maintained.

- *Capacity of the stair pressurization fan.* This depends on the particular analysis and on the assumption as to how many doors may be opened at the same time. Obviously, the more doors that are opened at the same time, the more air capacity is required. As a rule, a minimum of three doors, or at least 20 percent of the doors, shall be considered fully open. The amount of air to be delivered for stairs is normally between 15,000 and 25,000 CFM, depending on the above assumptions and design of the stairway.

- *Points of air supply.* To assure even pressure at upper levels of the building, a multiple-injection system with air injecting into the stairway at every 5 to 10 floors is preferred. The division of stairs into compartments will further improve the pressure variation at different levels.

Sealing of all penetrations All openings for piping, ducts, or structural members in or out of the fire partitions, walls, floors, and shafts are paths of smoke. These openings shall be sealed and caulked. Ducts shall be equipped with smoke and fire dampers interlocked with the fire protection (signaling and suppression) systems.

Pressure control in the elevator shaft Normally, exhaust is required at the top of an elevator shaft to relieve any smoke that may leak into the elevator cabinet. This concept is, however, in conflict with the pressurization concept for the other areas of the building. Studies indicate that maintaining a positive pressure in the elevator shaft is preferred. This is particularly important for elevators designated for use by fire fighters and the physically disabled. Further studies are being conducted by ASHRAE and NFPA for new design guidelines. Present practice is to avoid the use of elevators during a fire. On the other hand, many occupants of a building, such as those in a high-rise hotel, may not be able to walk down many flights of steps due to their limited physical ability. In such case, safe refuge areas or floors may have to be provided until the fire is under control.

9.9 SUMMARY

Although this chapter covers only the fundamental principles and guidelines on a variety of fire protection topics, the information contained in the text can easily be expanded by using the reference codes listed at the end of this chapter. The following is a recap of the key issues in the planning of fire protection systems for buildings.

- The trends in architectural design and the increased use of combustible interior materials demand serious attention to fire protection planning and coordination.
- Most building and fire protection codes are similar, but not identical. While the information presented in this chapter is valid in principle, the designer must be aware of the specific variations which might exist with the governing codes for the building under consideration.
- Fire protection requirements differ widely for different building use groups. A list of the classifications is given in Table 9–1.
- An effective fire protection system starts with a carefully analyzed operational plan which involves multiple design disciplines, the building management and the fire department. The operational plan requires the input and implementation of all those mentioned.
- A safe building starts with good architectural layout as to the location and construction of the means of egress.

- Water is the most effective medium for fighting Class A fire, which is most common among all fires. Adequacy of water supply in terms of quantity and pressure is essential. When the reliability of water supply is questionable, on-site storage should be considered.
- Halogenated gas is most effective for extinguishing electrical fire, in stock trading and computer rooms. The halogenated gas must be of a type which will not cause ozone depletion at the upper atmosphere.
- Portable fire extinguishers are economical and effective in extinguishing a fire before it spreads beyond control. Thus, they should be provided and installed at more locations than codes require.
- An automatic sprinkler system can be designed using either the pipe schedule method or the hydraulic method. The former is applicable for small light hazard occupancies only. However, it is very useful for any occupancy as a preliminary design tool when coordination of piping and sprinkler heads with other building systems are critical. The engineering algorithm for hydraulic calculations is included in the text, but computer programs are readily available to avoid the tedious calculations.
- Statistics indicate that smoke generated during a fire is more hazardous than the fire itself. Smoke cannot be quenched by water and usually travels far beyond the spread of the fire. Smoke migration is difficult to contain, especially in high-rise buildings. A coordinated approach with HVAC system design is essential for effective smoke containment and removal. This includes smoke evacuation, pressure sandwich, stair pressurization, and compartment provisions.

QUESTIONS

9.1 What are the two primary objectives of building fire protection systems?

9.2 Name one of the most important secondary objectives for building fire protection.

9.3 What is the normal range of fire protection system costs as a percentage of total building construction cost?

9.4 Name the major contributing factors to serious fire hazard in modern buildings.

9.5 What are the classes of flame spread rating (FSR)?

9.6 What is the smoke developed rating (SDR)?

9.7 What are the governing codes for fire protection in the United States?

9.8 What are the four classes of fire according to NFPA?

9.9 What are the three classes of fire hazard? What is the fire hazard class for office buildings, laboratories, and production facilities?

9.10 What are the normal classifications of building occupancies?

9.11 Name the operating sequence of a well-planned fire protection system.

9.12 Name ten fundamental criteria for fire safety planning.

9.13 Does a single-family dwelling require a fire alarm system?

9.14 What is the maximum distance to exit for a hospital protected with sprinklers? Not protected with sprinklers?

9.15 Name as many detection devices in a fire protection system as you can.

9.16 What is the operating principle of an ionization-type smoke detector?

9.17 Name the four types of media used in fire suppression systems.

9.18 What is the pressure required for a sprinkler head during its rated flow?

9.19 What is the maximum floor area that can be served by a portable fire extinguisher with a 2–A rating for a Class A fire hazard?

9.20 What is a Class III standpipe and hose system?

9.21 What is the required minimum water flow for standpipe systems?

9.22 Describe the basic automatic sprinkler system.

9.23 What is a Siamese connection?

9.24 What is the building use group classification for an office building? If the building contains a 15,000 sq ft health club on the second floor, is an automatic sprinklers system required?

9.25 Determine the amount of water that can be delivered by a sprinkler head having a 1/2" orifice with a 5.5 K-factor, and installed on an automatic sprinkler system having 36 psi residual pressure.

9.26 What is the maximum sprinkler protection area for a building classified as light hazard?

9.27 If a room has a soffit 20 inches below the ceiling, what is the minimum distance that a sprinkler can be installed? A beam 4 inches below the ceiling?

9.28 What is the difference between pipe schedule method and hydraulic method in sprinkler system design?

9.29 What is the maximum number of sprinklers that can be served by a 3-inch steel pipe in a light hazard building? (Use the pipe schedule method).

9.30 Select an area of sprinkler operation (ASO) and the corresponding density demand (DD) for an automatic sprinkler system classified as "ordinary hazard-2". What is the water demand for the sprinkler system (with OVF = 1.1)? If the hose stream demand (HSD) for the standpipe system is 250 gpm, what is the total water demand (TWD) for the combined standpipe and sprinkler system?

9.31 Can a domestic water system and fire protection system be combined in an on-site water storage system? Which system has the priority, and how is this assured?

9.32 Should smoke control be considered in fire protection system planning?

9.33 What is stack effect in a building?

9.34 How does one overcome the stack effect of a high-rise building?

REFERENCES

1. *National Fire Codes.* Quincy, MA: National Fire Protection Association, 1993.
   - NFPA 10, Standards for portable fire extinguishers
   - NFPA 11, Standards for foam extinguishing systems
   - NFPA 12, Standards on carbon dioxide extinguishing systems
   - NFPA 13, Standards for the installation of sprinkler systems
   - NFPA 14, Standards for the installation of standpipe-and-hose systems
   - NFPA 70, National Electrical Code
   - NFPA 78, Lightning Protection Code
   - NFPA 90A, Installation of air-conditioning and ventilating systems
   - NFPA 101, Life Safety Code

2. *Fire Protection Handbook,* 17th ed. Quincy, MA: National Fire Protection Association, 1991.

3. *Building Codes.* Includes the current editions of:
   - BOCA Building Officials and Code Administrator's International, Inc.
   - SBC Standard Building Code
   - SBCCI, Southern Building Code Congress International, Inc.
   - UBC Uniform Building Code

4. *Fire Resistance Directory.* Underwriters Laboratories (UL), Inc.

5. ASHRAE, *Handbook of Fundamentals.* 1993, Chapter 47, "Smoke Control."

10

INTRODUCTION TO ELECTRICITY

ELECTRICITY IS STEADILY BECOMING THE PREFERRED source of power for lighting, heating, air conditioning, transportation, production equipment, and numerous appliances in homes. Statistics indicate that the nominal use of electrical power in offices increased from 1 to 3 watts per square foot in the 1940s, to 3 to 5 watts per square foot in the 1980s, to 5 to 10 watts per square foot in the 1990s, and use is still growing.

All matter is made of *atoms*. The atoms contain *electrons* which drift at random through the atomic structure of matter. If two dissimilar materials, such as nylon and fur, are rubbed together, the surface of one material will accumulate a surplus of electrons, resulting in a negative overall charge known as static electricity. When such a charge, or *electromotive force* (potential), is applied across the two ends of a wire, the electrons within the wire are induced to flow in a certain direction, establishing an electric current. This flow of electrons is known as *current electricity*.

One unique property of electricity is its ability to be transmitted through a relatively small space by means of wires and cables. The amount of power that can be transmitted by a particular wire depends on the voltage level at which the transmission occurs. A 1/2-in-diameter wire at high voltage—say, one million volts—can deliver enough power to serve the needs of a large municipality. The disadvantage of high-voltage power transmission is the inherent danger to human life and property. For efficient utiliza-

tion within a reasonable degree of safety, utility voltages in the United States are standardized at 120 volts for small appliances and up to 480 volts for larger equipment. In large buildings, voltages up to 34,000 volts have been used directly with equipment rated for such voltage class.

The flow of electrical current (and thus, power and energy) is proportional to the available electrical voltage. Electrical systems with voltages below some 30 volts are not easy to pass through the human body and thus are relatively safe to the touch. On the other hand, household electrical systems at 120 volts used in the United States and 240 volts used in Europe and some other countries may cause deadly shocks or death to persons in their path. For this reason, electrical wiring and equipment are strictly regulated. The primary code governing electrical systems and installations is the National Electrical Code (NEC).

The design of building electrical systems is normally done by electrical engineers with specialized knowledge of such systems. However, every member of the building design profession must have a fundamental knowledge of electricity and its operating principles, characteristics, and limitations. The electrical terms, equipment and units listed in Table 10–1 should be considered as the minimum knowledge for architects, engineers, contractors, and building managers.

The most effective means of illustrating an electrical design or electrical circuit is to use symbols and

TABLE 10–1

Electrical terms, equipment and units

ELECTRICAL FUNDAMENTALS

1. Alternating Current (AC)
2. Capacitance
3. Current (I)
4. Direct Current (DC)
5. Electricity
6. Electromagnetic
7. Energy
8. Frequency
9. Ground
10. Impedance
11. Kirchhoff's Laws
12. Magnetic Induction
13. Ohm's Law
14. Parallel Circuits
15. Power
16. Reactance
17. Resistance (R)
18. Series Circuits
19. Voltage (V or E)

ELECTRICAL SYSTEMS

1. Alternators
2. American Wire Gauge (AWG)
3. Ampacity
4. Auxiliary Systems
5. Boxes (Junction, Outlet)
6. Branch Circuit
7. Bus Bars
8. Bus Duct
9. Capacitors
10. Cellular Floor
11. Circuit Breakers
12. Circular Mill (CM)
13. Coax Cables
14. Conductors
15. Contactors

16. Contacts, Maintained
17. Contacts, Momentary
18. Controllers
19. Demand (Power)
20. Demand Change
21. Detectors, Smoke
22. Diagrams, Connection
23. Diagrams, One Line
24. Diagrams, Riser
25. Diagrams, Schematic
26. Disconnects
27. Diversity
28. Electric Metallic Tubing (EMT)
29. Energy Charge
30. Exit Signs
31. Fault Current
32. Feeders
33. Fiber Optics
34. Fuses
35. Generator
36. Ground Fault Interruptors (GFIs)
37. Ground
38. Grounding (System, Equipment)
39. Interrupting Capacity
40. Lighting Fixtures (Luminaires)
41. Load Factor
42. Loads, Connected
43. Loads, Demand
44. Loads, Inductive
45. Loads, Resistive
46. Meters, Ammeter
47. Meters, Volt
48. Meters, Watt-hour
49. Motor
50. Overcurrent
51. Overcurrent Protection
52. Panelboards

53. Power Factor
54. Power, Apparent
55. Power, Reactive
56. Power, Working
57. Public Address (PA)
58. Raceways
59. Resistors
60. Root Mean Square (RMS)
61. Short-Circuit Capacity
62. Single Phase
63. Switchboard
64. Switches, 3-way, 4-way
65. Switches, Double Pole
66. Switches, Double Throw
67. Three-Phase
68. Transformers
69. Twisted Pairs
70. Underfloor Ducts
71. Voltage Drop
72. Voltage Spread
73. Voltage, Equipment
74. Voltage, System

UNITS OF
MEASURE/QUANTITIES

1. Ampere (A)
2. British Thermal Unit (Btu)
3. Horsepower (hp)
4. Kilovolt-ampere (kVA)
5. Kilowatt (kW)
6. Volt-ampere (va)
7. Volt (V)
8. Watt (w)
9. Ohm
10. Megawatt (MW)

---

schematic diagrams. Table 10–2 shows some commonly used symbols for basic electrical circuitry. Additional symbols for electrical equipment and wiring are introduced in Chapters 11, 12, and 13.

10.1 DIRECT CURRENT

Direct current (or DC) is an electric current that flows through a circuit in only one direction, although the rate of current flow may vary.

10.1.1 Basic Properties of Electricity

The *electric charge* or *quantity of electricity (Q)* is the basis of electricity. It is expressed in coulombs (C). However, in applications, the term is not often used.

Current (I) is the flow of electricity in an electrical circuit. Conventionally, the current flow is from pos-

itive to negative voltage, even though the electrons that make up the current flow are from negative to positive voltage. The convention is still the standard of practice in applications, such as the marking of dry cell or automotive batteries. The positive terminal is considered the supply side, the negative terminal the return side.

The unit of current is the *ampere* (A). It is defined as flow of one coulomb per second. Smaller units are the *milliampere* (mA) or *microampere* (μA).

Voltage (V or E) is the electromotive force (EMF) or potential difference that causes an electric current to flow. The symbols E or V are used interchangeably, depending on the application. The unit of voltage is the *volt* (v).

Resistance (R) is an internal property of matter that resists the flow of electric current. A material with low resistance to electrical flow is called a *conductor*, and one with high resistance to electrical flow

TABLE 10-2
Basic symbols for electrical circuits[1]

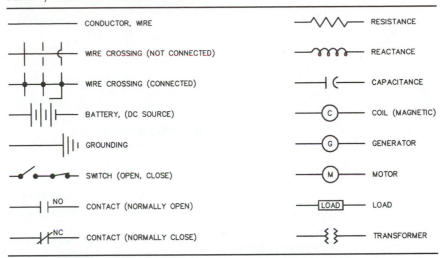

———————	CONDUCTOR, WIRE	—/\/\/—	RESISTANCE
	WIRE CROSSING (NOT CONNECTED)	—mmm—	REACTANCE
	WIRE CROSSING (CONNECTED)	—|(—	CAPACITANCE
	BATTERY, (DC SOURCE)	—(C)—	COIL (MAGNETIC)
	GROUNDING	—(G)—	GENERATOR
	SWITCH (OPEN, CLOSE)	—(M)—	MOTOR
	CONTACT (NORMALLY OPEN)	—[LOAD]—	LOAD
	CONTACT (NORMALLY CLOSE)	—§§—	TRANSFORMER

[1]Additional symbols for electrical equipment and wiring are included in later chapters.

is called an *insulator*. There is no rigid demarcation to separate conductors from insulators. Materials with limited conductivity are called *semiconductors*. The unit of resistance is the ohm (Ω), named for the scientist George Simon Ohm, who discovered the law also named for him.

Capacitance (C) is the property by virtue of which an electrical circuit stores energy in a field or a component. Capacitance provides an opposition force in a circuit to delay the change of voltage of the circuit due to its storage effect. Capacitance is present in every component of an electrical circuit when there is a dielectric material in the component, such as plastic, paper, rubber, liquids, and even air. Devices specially designed to store electrical charge are called *capacitors*.

The unit of capacitance is the farad (f), microfarad (μf) or picofarad (pf). One microfarad is one millionth of a farad, and one picofarad is one millionth of a microfarad.

10.1.2 Ohm's Law

In 1827, George Simon Ohm discovered a simple, but very important, relationship among the current, voltage, and resistance of a DC circuit. This relationship may best be visualized by comparing the electric circuit with a hydraulic circuit, as illustrated in Figure 10-1. In a hydraulic circuit, the rate of water flow is directly proportional to the pressure differential produced by the pump and inversely proportional to the resistance of the pipes and valves. Analogously, in an electric circuit, the electric current flowing through

the circuit is directly proportional to the voltage delivered by the battery and inversely proportional to the electrical resistance inherent in the wires and other components. This relationship, known as *Ohm's law*, may be expressed mathematically.

$$I = E/R, \text{ or } E = I \times R, \text{ or } R = E/I \quad (10-1)$$

10.1.3 Kirchhoff's Laws

Two other relationships important in the study of electrical circuits are Kirchhoff's laws. Kirchhoff's *first law* says that the algebraic sum of all electric currents into and out of any junction of an electric circuit is zero. In other words, the total current flowing into

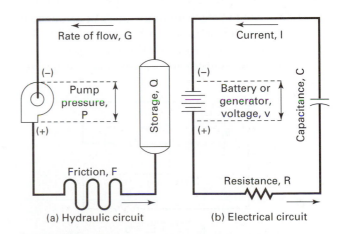

(a) Hydraulic circuit   (b) Electrical circuit

■ **FIGURE 10-1**
Comparison between hydraulic and electrical circuits.
(a) Hydraulic circuit. (b) Electric circuit.

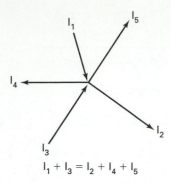

$$I_1 + I_3 = I_2 + I_4 + I_5$$

■ **FIGURE 10–2**
Kirchhoff's first law.

a junction must be equal to the total current flowing out of the junction. Using the hydraulic circuit as an analogy, if water flowing into a pipe flows out of two branch pipes, the total outflow from both pipes must be equal to the water flowing into the pipe. (See Figure 10–2.) Mathematically, Kirchhoff's first law may be expressed as

$$\Sigma I = 0 \qquad (10\text{–}2)$$

Kirchhoff's *second law* says that the algebraic sum of all voltages measured around a closed path in an electric circuit is zero. Again, using the hydraulic circuit as an analogy, in a closed pump and load circuit, the total rising pressure due to the pump must be equal to the total dropping pressure due to the loads. (See Figure 10–3.) Algebraically,

$$\Sigma E = 0 \qquad (10\text{–}3)$$

10.1.4 Resistance in Electric Circuits

Resistance used in electric circuits may be arranged in series, in parallel or in a combination of both. The mathematical relationship of the total circuit resistance to the individual resistances may be derived

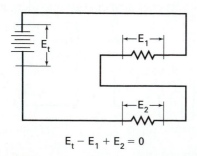

$$E_t - E_1 + E_2 = 0$$

■ **FIGURE 10–3**
Kirchhoff's second law.

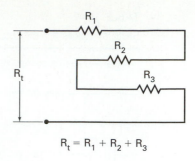

$$R_t = R_1 + R_2 + R_3$$

■ **FIGURE 10–4**
Resistances in series.

from Ohm's law and Kirchhoff's laws for either the series or the parallel arrangement.

Resistance in series Here, the resistances are arranged one after another, or end to end. (See Figure 10–4.)

$$R = R_1 + R_2 + R_3 \cdots + R_n \qquad (10\text{–}4)$$

Resistances in parallel The individual resistances are arranged side by side united at each end of the circuit. (See Figure 10–5.) Symbolically,

$$1/R = 1/R_1 + 1/R_2 + 1/R_3 \cdots + 1/R_n \qquad (10\text{–}5)$$

10.1.5 Power and Energy

Energy exists in many forms, such as mechanical, sound, light, electrical, nuclear, and chemical energy. All energy can be converted from one form to another, and all energy eventually degrades into heat energy. An overview of the energy used in buildings is given in Chapter 1.

Of all forms of energy, electrical energy is perhaps the most convenient to use. It is readily convertible to other forms, for example, to mechanical energy through a motor, to lighting energy through a lamp, to heating energy through a resistance heater, etc.

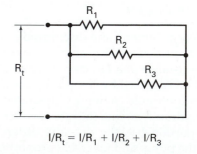

$$1/R_t = 1/R_1 + 1/R_2 + 1/R_3$$

■ **FIGURE 10–5**
Resistances in parallel.

Power is the *rate* of consuming energy. Thus, time is a crucial element in the consumption of energy: A large power load may consume less energy than a smaller power load if it is used over a shorter time. For example, a 1000-watt lamp operating for 1 hour consumes 1000 watt-hours of energy. However, a 40-watt lamp operating for 30 hours consumes 1200 watt-hours of energy. In this instance, the small 40 watt lamp actually consumes more energy than the large 1000-watt lamp.

Often, the concepts of power and energy are confused by the general public. Chapter 11 discusses the subject further.

Power is the product of voltage and current; that is,

$$P = \text{Power} = \text{Voltage} \times \text{Current} \quad (10\text{--}6)$$
$$P = E \times I$$

or

$$P = (I \times R) \times I = I^2 R \quad (10\text{--}7)$$

Power is expressed in units of watts (W).

Energy is the product of power and time; that is,

$$W = \text{Energy} = \text{Power} \times \text{time}$$
$$W = (E \times I) \times t \quad (10\text{--}8)$$
$$= I^2 \times R \times t$$

Energy is expressed in units of watt-hours (Wh) or kilowatt-hours (kWh).

Example

As a practical example of an electrical power application, suppose that one 100-watt lamp and one 200-watt lamp are plugged into a 120-volt circuit. For the purpose of this example, the power system may be either DC or AC. The two lamps are connected in parallel. Calculate the current flow through each lamp, the total current in the circuit, the resistance of each lamp, the total resistance of the circuit, the total energy consumed in a year, and the cost of electrical energy for the year.

Current flow through each lamp

From Eq. (10–6), $P = I \times E$, or $I = P/E$. So

$$I_1 = \text{current through lamp 1}$$
$$= 100 \text{ W} / 120 \text{ V} = 0.83 \text{ A}$$

$$I_2 = \text{current through lamp 2}$$
$$= 200 \text{ W} / 120 \text{ V} = 1.66 \text{ A}$$

Total current through the circuit

From Kirchhoff's first law,

$$I = I_1 + I_2 = 0.83 + 1.66 = 2.49 \text{ A} \ (\approx 2.5 \text{ A})$$

Resistance of lamps

From Ohm's law, $I = E/R$, or $R = E/I$
Thus,

$$R_1 \text{ (lamp 1)} = 20/0.83$$
$$= 44 \text{ ohms}$$

$$R_2 \text{ (lamp 2)} = 120/1.66$$
$$= 72 \text{ ohms}$$

Total resistance of the circuit

From Eq. (10–5), $1/R = 1/R_1 + 1/R_2$

$$1/R = 1/144 + 1/72 = 0.021$$
$$R = 1/0.021 = 47.6 \text{ ohms}$$

Total power of the circuit

$$P = I \times E = 2.5 \times 120 = 300 \text{ watts}$$

Total energy consumed in a year. Assuming that the lamps are used 10 hours per day and 200 days per year.

$$W = P \times t = (100 + 200) \times 10 \times 200$$
$$= 600{,}000 \text{ Wh} = 600 \text{ kWh}$$

Cost of electrical energy for the year. Assuming an energy rate of $0.10 per kWh, we get

$$\text{Cost} = \text{energy used} \times \text{energy rate}$$
$$= \text{kWh} \times \$/\text{kWh}$$
$$= 600 \times 0.10 = \$60 \text{ per year.}$$

10.2 DIRECT CURRENT GENERATION

One way to generate direct current is with batteries which convert chemical energy into electrical energy. Another method is with a generator that converts mechanical energy into electrical energy. This process involves the principle of *electromagnetism*. When electrical current flows in a wire, it induces a magnetic field around the wire, as shown in Figure 10–6a. Conversely, if a wire moves across a magnetic field, then an electrical voltage (and thus current) will be induced in the wire. Figure 10–6b illustrates the operating principle of a DC generator which contains a stator and a rotor. The stator consists of a permanent magnet or an electromagnet; the rotor consists of

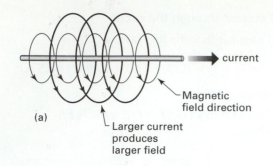

(a)

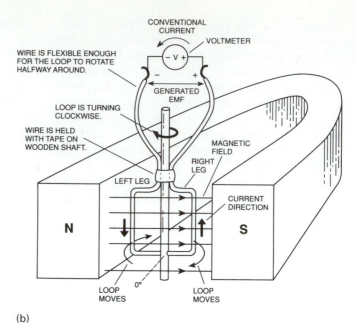

(b)

■ **FIGURE 10–6**
(a) Illustrates the phenomenon of electromagnetic induction that a magnetic field is induced surrounding the flow of an electrical current.
(b) Conversely, when an electrical conductor (coil) moves (rotates) across a magnetic field, EMF (voltage) is induced and the flow of electricity (current) is rectified through a commutator.
(c) The voltage and current are cyclic in nature complying with the sinusoidal relations. This cycle repeats itself every 180° rotation of the coil.
(Courtesy: Master Publishing, Inc., Ft. Worth, TX.)

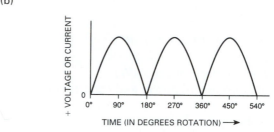

(c)

multiple coils of wire which rotate between the magnetic poles. (For simplicity, only a single coil is shown in the illustration.) When the coil rotates between the magnetic poles, it cuts the magnetic field at various angles, and thus generates a variable electrical voltage. The generated voltage follows the form of a sinusoidal curve. During the first 180° of the rotation, the voltage generated in the coil starts from zero to a peak value, and then drops to zero. During the second 180° turn (180° to 360°), another sinusoidal curve is generated, but with reverse polarity. When the ends of the rotating coil are connected to an external wiring system through segmented slip-rings known as the commutator where the voltage becomes rectified; i.e., the polarity remains unchanged. The resulting voltage (and current) characteristics are shown in Figure 10–6c.

10.3 ALTERNATING CURRENT (AC)

An alternating current (AC) system is an electrical system having its voltage and current reversing periodically or cyclically in the circuit; i.e., the voltage al-

ternates in polarity between the conductors and so does the current. Nearly all electrical power provided by the electrical utilities in the United States is through AC systems, in lieu of the DC systems which were the standard a long time ago. The advantages of AC over DC system will be discussed later.

The operating principle of an AC generator is the same as that of a DC generator except the connection between the generator and the external electrical system is through smooth slip rings, without the use of a segmented commutator. Figure 10–7 illustrates the operation of an AC generator (also known as an alternator).

10.3.1 Basic Properties of Alternating Current

Because current reverses its direction of flow rapidly in an alternating circuit, there are some properties unique to AC.

AC resistance   In an AC circuit, the reversing flow of electrons (current) tends to concentrate near the outer

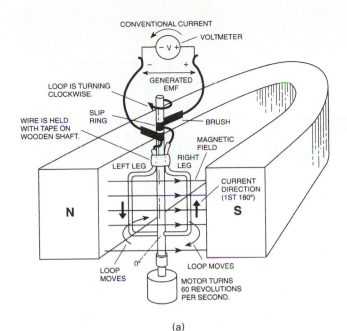

(a)

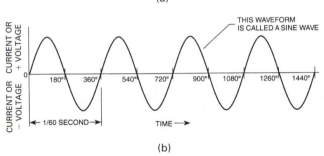

(b)

■ **FIGURE 10–7**

(a) The operating principle of an AC generator is identical to that of a DC generator except for the final connection between the generator and the electrical system. The alternator has smooth slip rings whereas a DC generator has segmented slip rings, or a commutator.

(b) The characteristics of the generated voltage are cyclic and reverse in polarity every 180° rotation of the coil. The flow of current is similar. For each revolution of the coil, it completes an electrical cycle. The frequency in number of cycles per second is called Hertz (Hz). The standard frequency in the USA is 60 Hertz; 50 Hertz is the standard in most European and Asian countries.
(Courtesy: Master Publishing, Inc. Ft. Worth, TX.)

diameter of the conductor resulting in increased resistance of the conducting material compared with a DC circuit. This phenomenon is known as the *skin effect*, and the effective resistance is called *AC resistance*. It is also expressed in ohms (Ω).

Figure 10–8 illustrates the voltage and current relationship of an electrical system containing pure resistance. In a pure resistive circuit, the current and voltage waves increase in unison or in phase.

Reactance   Reactance, which causes AC resistance, offers opposition to the reversing flow of current. There are two types of reactance; each is also expressed in ohms (Ω).

- *Inductive reactance* (X_L) appears through the presence of an electromagnetic field (via either an electromagnet or permanent magnet) in the equipment, such as in a motor, a relay, or a ballast for lights. Inductive reactance has a delaying effect on current in relation to the voltage. Thus, the current lags behind the voltage. Since most buildings contain lots of motors and electromagnetic devices, building electrical systems are usually more inductive than other electrical systems. Inductive reactance is expressed mathematically as

$$X_L = 2\pi f L \qquad (10\text{–}9)$$

where f = frequency, in Hertz
  and L = inductance, in henries.

Figure 10–9 (a) illustrates the voltage and current relationship of an electrical system containing pure inductive reactance with current lagging the voltage wave by 90°.

- *Capacitive reactance* (X_c) is created by the presence of capacitance in the equipment, such as insulation on cables and devices. Capacitive reactance has a storage effect on current in relation to voltage. Thus, the voltage lags behind the current, or the current leads the voltage. In general, the capacitive reactance of building components is quite small and can be neglected unless the system is equipped with capacitors to overcome the effect of inductive reactances in the system. Algebraically, we have

$$X_c = I \,/\, 2\pi \cdot f \cdot C \qquad (10\text{–}10)$$

where f = frequency in Hertz
  and C = capacitance in farads.

- Figure 10–9 (b) illustrates the voltage and current relationship of an electrical system containing pure capacitive reactance, with current leading the voltage wave by 90°.

Impedance

The combined effect of resistance and reactance is called impedance, also expressed in ohms. The rela-

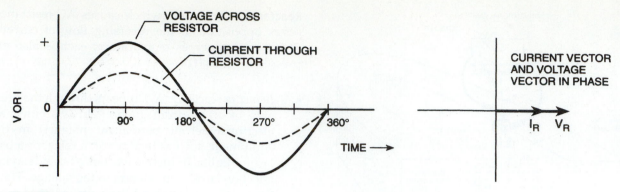

■ **FIGURE 10–8**
The current and voltage relationship in a resistive circuit. The vector angle between
current and voltage quantities is zero.

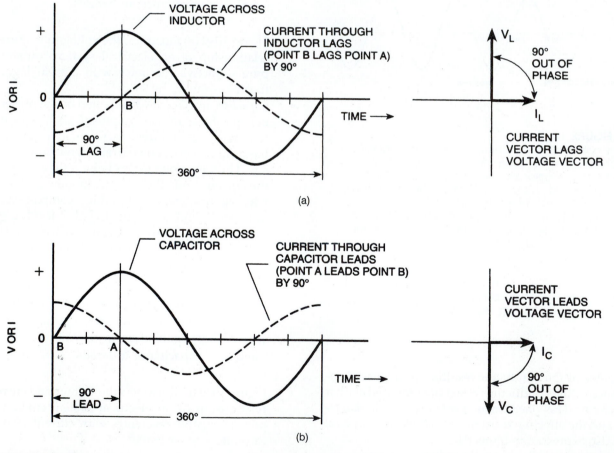

(a)

(b)

■ **FIGURE 10–9**
The voltage and current relationship in a reactive circuit. (a) Inductive circuit where
voltage leads current by 90°; (b) capacitive circuit where current leads voltage by 90°.
(Courtesy: Master Publishing, Inc., Fort Worth, TX.)

tionship between resistance, reactance and impedance is

$$Z^2 = R^2 + X^2 \qquad (10\text{--}11)$$

where Z = impedance in ohms.

10.3.2 Ohm's Law for AC

Since impedance is the AC equivalent of DC resistance, Ohm's law for AC can be written as

$$E = I \times Z \qquad (10\text{--}12)$$

10.3.3 Effective Value of Alternating Current and Voltage

The values of current and voltage in a typical alternating current system varies as the sine wave. Questions arise as to the representative value of the current or voltage in such a system. To compare alternating current with direct current, the *effective* values of current and voltage are used. These correspond to 0.707 (reciprocal of $\sqrt{2}$) times their maximum values. The effective values are frequently referred to as *root-mean-square*, (rms) values of current and voltage. Mathematically,

$$I = 0.707\, I_p \qquad (10\text{--}13)$$

$$E = 0.707\, E_p \qquad (10\text{--}14)$$

where I = effective rms current, in amperes (measured by an ammeter)

I_P = peak current of sine wave, in amperes (measured by an oscilloscope)

E = effective rms voltage, in volts (measured by a voltmeter)

E_P = peak voltage of sine wave, in volts (measured by an oscilloscope).

10.3.4 Frequency

The *frequency (f)*, or the number of cycles repeated per second, depends on the construction and rotation speed of the rotor of the generator. The standard adopted in the United States is 60 cycles per second (cps) or 60 Hertz (Hz), whereas in Europe and some other countries, the standard frequency is 50 Hz.

The speed of rotation of electrical apparatus, such as motors or clocks, depends on the frequency of the power system. A motor designed for 60-Hz

power will operate at $^{50}\!/_{60}$ of its designed speed when connected to a 50-Hz power source. The reverse ratio will be true for a 50-Hz apparatus operating on 60 Hz.

Higher frequencies of 400 to 3000 Hz have been used for specially designed fluorescent lighting systems with improved efficiency and economy. In fact, utility power in the United States may adopt a higher frequency—say, 120 Hz as a standard in the future. In aircraft, 400-Hz power is used as the standard to reduce the weight of electrical equipment on board.

10.4 ADVANTAGE OF AC OVER DC SYSTEMS

Because of their inherent advantages, alternating current systems have replaced direct current systems in almost all modern utility power systems. The two major advantages of AC over DC systems are as follows:

1. *Lower generating cost.* An AC alternator is simple to construct without the need for a complicated split ring commutator. Thus, the AC alternator permits a higher speed of rotation with lower maintenance cost and a lower cost of power generation.
2. *Easier voltage transformations.* The voltage of an AC system can be changed by the use of a simple piece of electromagnetic equipment known as a *transformer*. The transformer permits the transmission of large amounts of electrical power at higher voltages for long distances with relatively small transmission lines and thus a lower cost of transmission.

For example, utility power is normally generated between 16,000 to 25,000 volts at the generating plant and transmitted at 11,000 to 69,000 volts for distribution, but may be as high as 345,000 volts for long distance transmission. These transmission voltages are then stepped down to 4,160 volts at local substations for distribution to the users or buildings. The distribution voltage is then stepped down again at users' premises to utilization levels, such as 120, 208, 240, 277, 480 volts as appropriate for the connected loads.

Figure 10–10 is a one-line diagram of the power transmission scheme from a typical generating station through step-up and step-down substations and to a building. A substation typically consists of transformers and switches. The final step-down transformer(s) are usually located near the load or may be located within a building.

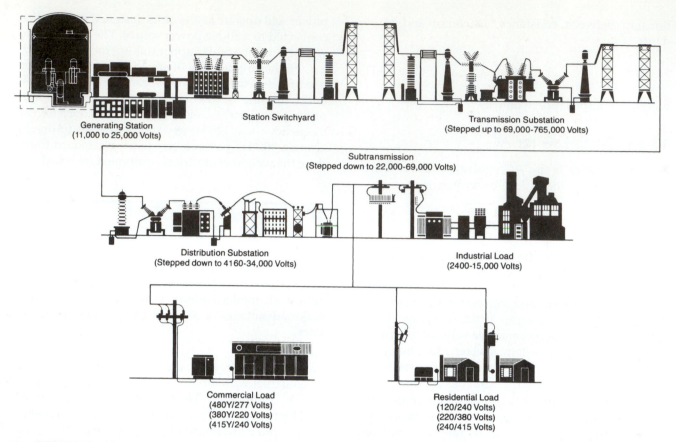

■ **FIGURE 10–10**
Power transmission of a utility system. (Courtesy: Cutler Hammer/Westinghouse, Pittsburgh, PA.)

10.5 AC-TO-DC CONVERSION

While AC is the dominant power system used for buildings, DC is needed for electronic equipment—for example, television receivers, computers, batteries, precision controls, and special power equipment, such as elevators and industrial processes. DC power for this type of equipment needs to be connected to an independent DC source or the AC-to-DC converter. The conversion from AC to DC is by means of a rectifier, and the conversion from DC to AC is by means of an inverter.

10.6 SINGLE-PHASE VS. THREE-PHASE ALTERNATOR

A single-phase generator is an alternator with a single set armature coil producing a single voltage waveform. A three-phase alternator has three sets of coils spaced 120 degrees apart and generates three sets of voltage waveforms. The three-phase alternator is the most commonly used alternator in the United States. Rarely in use are two-phase and six-phase alternators. When more than one coil is wound in the armature of the alternator, each coil will generate its own independent voltage. Figure 10–11 illustrates the phase relations of a three-phase power system generated by a three-phase alternator.

10.7 POWER AND POWER FACTOR

In a single-phase AC circuit that contains only resistance, the current and voltage will be *in phase* and the power consumed will simply be the product of voltage and current, similar to the DC circuit. In a three-phase circuit that contains only resistance, power is 1.73 times the line voltage and current because the separate powers of the three phases are staggered.

In an AC circuit that contains reactance in addition to resistance, the current and voltage will be out of phase by an angle, known as the power angle, and designated by the Greek letter (θ).

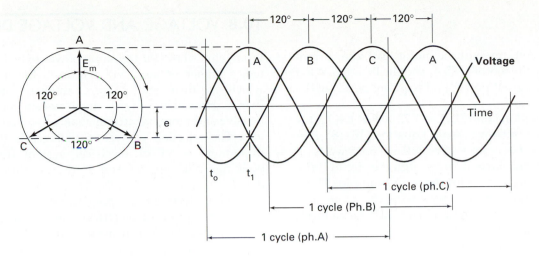

■ **FIGURE 10–11**
Phase relations of a three-phase power system.

For single and 3-phase AC systems, the power (P_a, P, P_r), line current (I), and voltage (E) relationships are as follows:

10.7.1 Apparent Power

The apparent power P_a, in volt-amperes (va) is the product of voltage and current, which may or may not be in phase with each other. For single phase power

$$P_a = E \times I \qquad (10\text{–}15)$$

For three phase power

$$P_a = 1.73 \times E \times I \qquad (10\text{–}16)$$

10.7.2 Reactive Power

The reactive power P_r, in volt-amperes reactive (var), is the component of AC power that does not perform useful work. For single-phase power

$$P_r = E \times I \times \sin \theta \qquad (10\text{–}17)$$
$$= P_a \times \sin \theta$$

For three phase power

$$P_r = 1.73 \times E \times I \times \sin \theta \qquad (10\text{–}18)$$
$$= 1.73 \times P_a \times \sin \theta$$

10.7.3 Working Power

The working power P, in watts (w), is the component of AC power that performs useful work. In practical applications, power always refers to working power. For single phase power

$$P = E \times I \times PF = E \times I \times \cos \theta \qquad (10\text{–}19)$$

Three phase power

$$P = 1.73 \times E \times I \times PF$$
$$= 1.73 \times E \times I \times \cos \theta \qquad (10\text{–}20)$$
$$= 1.73 \times P_a \times \cos \theta$$

where the voltage (E) is the *line-to-line* voltage (E), not the *line-to-neutral* voltage (E_n). The relationship between E and E_n is explained in Chapter 11.

10.7.4 Power Factor

The mathematical term $\cos \theta$ is also known as the power factor (PF). If the power angle of a load circuit is 30°, then $\cos \theta$ is 0.866, and the power factor is said to be 0.866 or 86.6 percent of the unit power.

As can be seen from the preceding equations, for the same values of voltage and current, a higher power factor will produce more useful power. Conversely, a low power factor load will require more current, compared with a high power factor load. Obviously a higher power factor is preferred by both the utility company and the user. For residential and commercial loads, the overall system power factors generally range from 0.8 to 0.9; for industrial plants, the overall power factor is generally lower due to the extensive use of motors and inductive equipment. In such case, some means to improve the power factor may be necessary.

Examples

1. In a 208-volt, three-phase motor load, the current drawn by the motor is 25 A, and the PF is 0.6. What are the apparent power and working power of the circuit?
   From Equation (10–15), $P_a = 1.73 \times E \times I = 1.73 \times 208 \times 25 = 8996$ va. From Equation (10–18), $P = 1.73 \times E \times I \times \cos\theta = 8996 \times 0.6 = 5398$ watts

2. What will the current be through the circuit if the PF is increased to 0.8 for the same motor load?
   The working power remains the same, and the current through the circuit will be reduced proportionately, or

$$I = 25 \times 0.6/0.8 = 18.75A$$

TABLE 10–3

Maximum allowable voltage drops (percent)

Portion of Distribution System	For Lighting and Power Loads	For Electric Heating	For Power Only
Service entrance to switchboard	1	1	2
Feeder to distribution centers	1	1	3
Branch circuits to connected load	3	1	3
Overall maximum voltage drop	5	3	8

10.8 VOLTAGE AND VOLTAGE DROP

From Ohm's law, for the same power to be transmitted, a higher voltage will require a lower current which in turn will require smaller conductors (wires). However, a high voltage system is more expensive and more hazardous than a low voltage system. The selection of system voltages is the most important aspect of electrical system design in buildings. Chapters 11 and 12 discuss this topic in detail.

Every electrical component, whether a wire, a switch or a piece of electrical equipment, contains more or less electrical resistance and reactance. The presence of these properties in an electrical circuit causes a voltage drop (loss) between the supply and the receiving ends of the circuit. The drop is unavoidable, but must be minimized.

In general, equipment and appliances are rated at about 5 percent lower voltage than the rated system voltage. For example, a lamp is normally rated at 115 volts for use on 120-volt systems. If the system voltage becomes lower than the rated equipment voltage, it may lead to a lower capacity, reduced speed, lower efficiency, a shutdown due to overheating, or some other malfunction. For critical equipment, such as TV broadcasting equipment, electron microscopes, research instruments, and computers, separate voltage regulators to maintain the voltage within 1 percent of the rated equipment voltage may be required.

TABLE 10–4

Summary of DC and AC properties (for single-phase circuits)

Quantity	Unit	Symbol	Formula
Voltage	Volts (V)	E (or V)	DC: $E = I \times R$
			AC: $E = I \times Z$
Current	Amperes (A)	I	DC: $I = E/R$
			AC: $I = E/Z$
Resistance	Ohms (Ω)	R	DC/AC: $R = E/I$
Inductive Reactance[1]	Ohms (Ω)	X_L	AC: $X_L = 2\pi \times f \times L$
Capacitive Reactance[2]	Ohms (Ω)	X_C	AC: $X_C = 1/2\pi \times f \times C$
Impedance	Ohms (Ω)	Z	AC: $Z^2 = X_L{}^2 + X_C{}^2$
Apparent Power	Volt-amperes (VA)	P_a	DC/AC: $P_a = E \times I$
Working Power	Watt (W)	P	AC: $P = E \times I \times \cos\theta$
		P	AC: $P = E \times I \times PF$
		P	DC/AC: $P = I^2R$
Reactive Power	VA reactive (VAR)	P_r	AC: $P_r = E \times I \times \sin\theta$
Energy	kilowatt-hours (kwh)	W	DC/AC: $W = P \times t$
Power Factor	Per-unit or percent	PF	AC: $PF = P/E \times I = \cos\theta$

[1]L—inductance, henries.
[2]C—capacitance, farads.

To avoid excessive voltage drop from wiring within a building, the National Electrical Code (NEC) limits the voltage drop for lighting and power loads within the building. Table 10–3 gives the recommended maximum voltage drop for different types of load.

10.9 SUMMARY

Table 10–4 summarizes the various properties of AC and DC circuits in relationship to their basic quantities, such as current, voltage, resistance, reactance, impedance, power and energy.

QUESTIONS

10.1 The flow of electrical current through a circuit is directly proportional to the EMF (voltage) for both AC and DC systems. (True) (False)

10.2 What is the relationship between current, voltage and impedance in an AC circuit?

10.3 What is the relationship between energy, power, current, and voltage in an AC circuit?

10.4 What is the standard voltage rating for convenience power (plug-in receptacles) in the USA? In Europe?

10.5 What is Kirchhoff's first law of electrical circuits? Is it possible to have all currents flow into a junction?

10.6 What is Kirchhoff's second law of electrical circuits? What will happen when a 120-volt appliance is connected to a 240-volt system?

10.7 Define power and energy.

10.8 What is the relation between peak current and effective (rms) current of an AC power system generated by commercial power companies in the USA?

10.9 What are the line-to-line and line-to-neutral voltages in a 120/240 volt, single-phase AC system? In a 120/208 volt, three-phase system? (See Chapter 11.)

10.10 What is power factor? What is the power factor of a DC system?

10.11 Is it desirable to maintain a high power factor in an electrical system? Can power factor be improved, and how?

10.12 What is the standard AC system frequency in the United States and in some other countries?

10.13 Name the two major advantages of an AC system over a DC system in building applications.

10.14 Can AC and DC systems be interconverted and if so, how?

10.15 What is the advantage of a three-phase system over a single-phase system?

10.16 Due to the resistance and/or reactance of wiring, the voltage at the end of a distribution system is always lower than its beginning. This is called voltage drop. A well-designed system should limit the voltage drop to a minimum. Answer the following:
(a) What is the recommended maximum overall voltage drop for a combination lighting and power system?
(b) For an electrical heating system?
(c) What is the nominal voltage rating of single-phase appliances in the United States?

10.17 An electrical load is rated for 1150 watts at 115 volts, single phase. What is the load current if the power factor is 1.0 or 100 percent?

10.18 If the current of the load in question 10.17 is 12 amps, what is the power factor?

10.19 If the same load as in question 10.17 (1150 watts) is 208 volt, three-phase, what is the current with 0.8 PF?

10.20 What is the apparent power in VA of the load in question 10.19?

10.21 What is the peak current of the load in question 10.19?

10.22 What is the peak voltage of a 120 volt, single-phase system? a 208 volt, three-phase system?

10.23 If the resistance and reactance of a circuit is 2 ohms and 4 ohms respectively, what is the impedance?

10.24 If three loads rated for 2, 3, and 4 kW are connected in parallel, what is the total load?

10.25 If the system voltage for the loads in question 10.24 is 480 volts, three-phase, what is the total current of the loads? (Note: since PF is not given in the question, then it must be assumed.)

11

ELECTRICAL EQUIPMENT AND SYSTEMS

11.1 ELECTRICAL DISTRIBUTION SYSTEMS

There are numerous power distribution systems used for buildings. The systems that are most appropriate for a specific building depend on the size of the building and the characteristics of the predominate loads, such as power ratings (HP or kW), voltages, and phases. Frequency is another important characteristic of a power system. However, the frequency is normally standardized in a country, such as 60 hertz (Hz) in the United States and 50 Hz in Europe, and the standard frequency should be complied with, except in unusual applications. Most large buildings contain loads with diversified characteristics, such as single-phase lighting and appliances and three-phase motors. Thus, it is quite common to have more than one power distribution system in the same building, although multiple systems are costly and difficult to maintain. Cost-benefit studies should be made to compare options. As a rule, if more than one system is necessary, the main system should be selected to satisfy the predominant loads, with one or more subsystems converted from the main system for the minor loads.

Electrical power systems used in buildings may be divided into four voltage classes:

- Extra-low voltage systems   50 volts and below (signal and communications systems)

- Low-voltage systems up to 600 volts (building utilization power)
- Medium-voltage systems up to 23,000 volts (building primaries and large power equipment)
- High-voltage systems up to 70,000 volts (exterior primaries)

There are variations within each voltage class. For power distribution within buildings, low-voltage systems up to 480 volts are most common. Medium- and high-voltage systems are used only in large buildings with several million square feet in floor area or in industrial plants. For the purpose of this book, we shall consider only the low- and medium-voltage systems.

The following sections give guidelines for selecting a system and knowing its limitations.

11.1.1 Loads Less Than 100 kVA

Residential and small buildings typically have a demand load less than 100 kVA. Loads are normally rated for 120 or 240 volts, single phase. Thus, a 120–240-volt, single-phase three-wire system is most appropriate (See Figure 11–1(a).

11.1.2 Loads Greater Than 100 kVA

Loads greater than 100 kVA are usually served by three-phase systems. An exception is the multiunit apartments or small shopping centers, where it is entirely appropriate to use a single-phase system, even if the total load far exceeds 100 kVA. As the load increases, systems at higher voltages are favored due to reduction of wire sizes.

Popular systems used in the United States are 120/208-volt, 240-volt, 277/480-volt, and 480-volt, three-phase systems. The 220/380-volt systems are quite common in countries where the utilization voltage for appliances is 220 volts rather than 120 volts. In such a system, 220-volt, single-phase lines will serve small appliances, and 380-volt, three-phase lines will serve larger loads. However, the 220/380-volt system will eventually be replaced by a 277/480-volt system, which is considerably more economical. For example, a 1000-kVA, 380-volt, three-phase system requires a 1600-amp main service, whereas a 480-volt, three-phase system requires only a 1200-amp main service. This 25-percent reduction in current affects the distribution equipment and feeders all the way down the

line and, thus, the overall system cost. (See Figures 11–1(b) and 11–1(c).

Medium-voltage systems such as 2400/4160 volts and high-voltage systems such as 7200/12,470 volts are also used for buildings in which the total load exceeds the service capacity of low-voltage systems.

11.1.3 Common Distribution Systems

Figure 11–2 provides a summary of power systems appropriate for various load capacities. For extremely large buildings with over one million square feet in gross floor area, or for campus-type facilities, power distribution may consist of many small systems, all interconnected to increase system flexibility and reliability. The figure shows five primary and secondary distribution systems, ranging from simple secondary radial to the more complex primary and secondary selectives. The term "primary" refers to

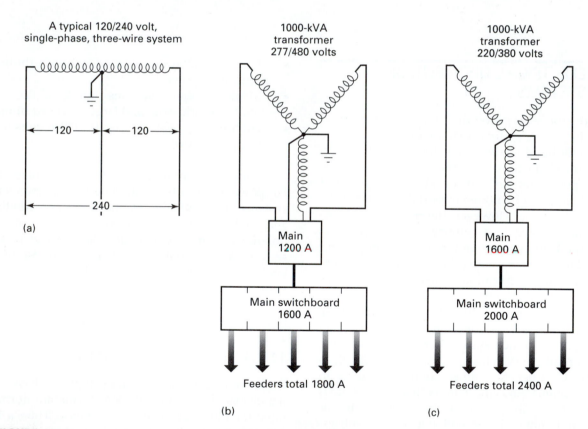

■ FIGURE 11–1
(a) One-line diagram of a 120/240-volt single-phase system. Similar systems of the same voltage class include 115/230 volts and 125/250 volts.
(b) A one-line diagram of a 1000-kVA, 277/480-volt, three-phase, four-wire power distribution system showing its size (the capacity of the system components).
(c) An identical 1000-kVA system as in (b), but at 220/380 volts requires about 25 percent higher ampacity of its distribution components and thus a higher capital investment.

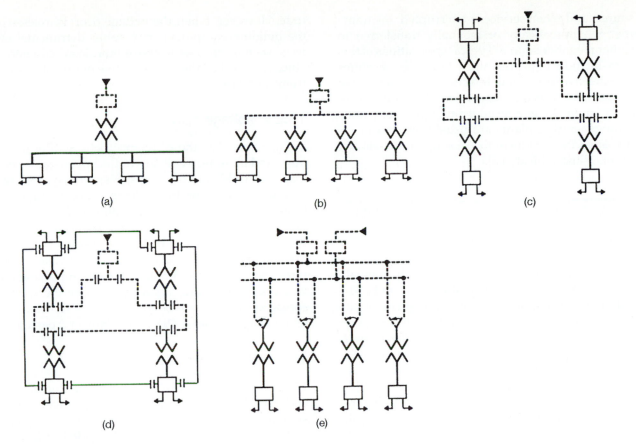

■ FIGURE 11–2

Typical power distribution systems. (a) Simple radial system. (b) Distributed radial system. (c) Secondary loop system. (d) Primary and secondary loop system. (e) Primary selective system.

the portion of a distribution system before a transformer, and "secondary" indicates the portion after a transformer.

Simple secondary radial   This system, shown in Figure 11–2(a), is the simplest of all. It usually contains a single step-down transformer that transforms the primary voltage of 4160 volts to a secondary voltage and 120–240 volts for distribution to the loads. Typical applications of the simple secondary radial system are small office buildings, stores, and large residences. The transformer may be external to the building, or shared with other homes, or may be within the building, depending on the needs.

Distributed radial   This system, shown in Figure 11–2(b), consists of multiples of simple secondary radial systems. Typical applications are shopping centers, apartment complexes, large department stores, schools, and institutional buildings.

Secondary loop   This system, presented in Figure 11–2(c), consists of several simple secondary radial

distribution systems with the secondary feeders connected in an open loop to enable feeders to be switched to a different transformer in case the normal transformer is out of service. Typical applications are high-rise office buildings, hospitals, computer centers, and industrial plants, where the reliability of electrical systems is critical.

Primary and secondary loops   This system (Figure 11–2(d)) is similar to the secondary loop. However, the primary side of the system is also tied to an open loop, so, upon interruption of the normal primary feeder to a transformer, the alternative primary feeder may be activated. Naturally, this will add to the reliability of the system, as well as the cost. Typical applications of this system include campus-type schools, offices, and industrial plants.

Primary selective   This type of system, shown in Figure 11–2(e), is similar to the distributed radial system, except it is served by two primary feeders or sources. The transformer is connected to one primary feeder or source, and when service is interrupted, the transformer is automatically or manually transferred to the other feeder or source. Typical applications of this system include high-rise office buildings, facilities with extensive uses, and hospitals. The alternative source may be a second feeder from the utility company or an on-site engine generating plant or plants. On-site generating plants are usually limited in capacity and are designed to back up only selected emergency and critical loads.

11.2 VOLTAGE SPREAD AND PROFILE

11.2.1 Voltage of the Utility

Most utility power systems automatically regulate their supply voltage at the generating plant and at the distribution substation to within 1 percent of the nominal voltage. However, when the utility power system is overloaded, its supply voltage may drop. A 5-percent fluctuation is quite common and is within the legal limits of utility regulations.

Under unusual circumstances, when a utility network is overloaded, the utility company may resort to the technique of lowering its distribution voltage by 5 to 10 percent of its normal voltage—for example, from 4160 volts to 3750 volts. This will correspondingly lower the secondary voltage in the building by a similar percentage and more or less reduce the building demand load by the same magni-

tude. However, when the voltage drop is excessive, the building equipment may suffer detrimental effects, such as an overheated motor, lowered energy efficiency, or the failure of sensitive electrical or electronic instruments.

11.2.2 Voltage Spread

Voltage spread is the difference between the maximum and minimum voltages of the system under no-load and full-load conditions, respectively. Voltage spread accounts for the voltage drop through the transformers, distribution equipment, feeders, and branch circuits. For example, if the voltage drop through the transformer of a 240-volt system is 6 volts between no load and full load, and the voltage drops through the feeder and branch circuits supplying the load are 2 and 4 volts, respectively, then the voltage spread at the load will be (6 + 2 + 4) = 12 volts. The total voltage spread is (100 × 12/240), or 5 percent.

11.2.3 Voltage Profile

The cumulative effect of utility voltage fluctuations and the voltage drop within the building distribution system may subject the load equipment to an overvoltage when the system is lightly loaded, such as at night or on weekends, and an undervoltage at full-load conditions. Excessive overvoltage or undervoltage may shorten the life of or even burn out the load equipment. In the case of incandescent lamps, a

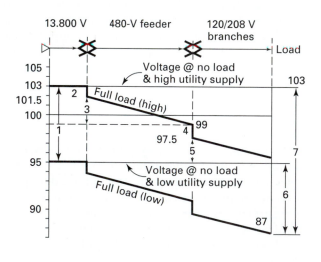

As Percent of Initial Voltage

1 Utility voltage spread
   = 103-95
   = 8%
2 Distribution transformer voltage drop
   = 103-101.5
   = 1.5%
3 Feeder voltage drop
   = 101.5-99
   = 2.5%
4 Building transformer voltage drop
   = 99-97.5
   = 1.5%
5 Branch circuit voltage drop
   = 97.5-96%
   = 1.5%
6 Utilization voltage spread (no load)
   = 96-87
   = 9%
7 Total system spread (full load)
   = 103-87
   = 16%

■ FIGURE 11–3
Voltage profile of a typical electrical distribution system. The profile shows extreme conditions. A high-quality utility service and well-designed distribution system should keep all drops and spreads below 50 percent of the figures shown.

5-percent voltage drop below their rated voltage will reduce the light output (lumens) by nearly 20 percent. For this reason, it is imperative to minimize the voltage spread within the building distribution system. Figure 11–3 gives a graphic illustration of voltages at various points of a distribution system under extreme conditions.

11.2.4 System and Equipment Voltage Ratings

Since, even with the best designed systems, the voltage of a power distribution system will unavoidably drop between its supplying end (source) and receiving end (load) due to the inherent impedances in the distribution components, there has to be a plan to overcome the variations. For this reason, equipment or appliance voltage ratings are usually about 5 percent lower than system voltage ratings. For example, single-phase electrical appliances are usually rated for 100–115 volts for use on the 120–125-volt system in the United States. Similarly, 460-volt mo-

tors are used on 480-volt, three-phase systems. Table 11–1 lists the differences between system and equipment voltages of common power distribution systems.

11.3 Grounding

11.3.1 Grounded System

When the electrical system is connected to the earth, either intentionally or accidentally, the system is said to be *grounded*. In general, all building electrical systems are intentionally grounded at the point where the voltage to ground is the lowest. There are several reasons for grounding:

1. Grounding protects the system and equipment from overvoltage due to accidental contact with higher voltage sources, such as the primary voltage side of the distribution system, which may exceed several hundred thousand volts.

TABLE 11–1

Voltage rating of common distribution systems showing the system vs. equipment voltages and the maximum equipment rating vs. the system capacity. In general, the maximum rating of a single piece of equipment should not exceed 15 to 20% of the system capacity if the system also serves other loads.

Distribution Systems	Nominal System Voltage (volts)	Nominal Equipment Voltage (volts)	Application Limits	
			Equipment Rating* (kva)	System Capacity (kva)
1 φ, 3 W (A, N, B)	120/240	115/230	20	100
3 φ, 4 W (A, B, N, C)	120/208	115/200	150	750
	220/380	210/365	500	3000
	277/480	265/460	500	3000
	2400/4160	2300/4000	5000	15,000
	7200/12,470	6900/12,000	10,000	50,000
	19,920/34,500	N/A	N/A	No maximum
3 φ, 3 W (A, B, C)	240	230	150	750
	480	460	500	3000
	600	575	500	3000
	4160	4000	5000	15,000
	13,800	13,200	10,000	50,000
	34,500	N/A	N/A	No maximum

*Refers to inductive load, such as a motor that can be connected on this system without causing excessive voltage dip.

2. When lightning strikes a building and its electrical system, the electrical wiring and insulation on equipment may break down, and lightning current will flash over to seek the ground. If the system is properly grounded, the lightning current will follow a direct path to the ground, bypassing the feeders and equipment.

3. Grounding protects people from heavy electrical shock. If the system and equipment are grounded and the electrical circuit is accidentally shorted to the equipment, the circuit protection device (fuses or circuit breakers) should trip and cut out the circuit. If the system is not grounded, then any accidental grounding of the wire through the equipment may pass through a person who happens to be touching the equipment.

4. A grounded system is more economical than an ungrounded one. The grounded side of a circuit need not be switched. Thus, on a single-phase circuit, only a single-pole switch is required, whereas a double-pole switch must be used on an ungrounded single-phase system.

11.3.2 Ungrounded Systems

While grounding is preferred for general building systems, it does have one major disadvantage: A grounded system can easily be "tripped," or put out of service, whenever the ungrounded side of the system accidentally touches ground. This is a problem especially with critical loads, such as lighting and power for hospital surgical rooms. For this reason, the electrical power supply to the surgical rooms is ungrounded. However, ungrounded power systems should be isolated from the main system through individual isolation transformers and equipped with a sensitive ground detection system to detect any accidental grounding. The detection system allows the power system to continue operation until the fault (accidental ground) is corrected.

11.3.3 Grounding Methods

System grounding   A common practice is to use an underground water main as the grounding electrode, because the water main is extensively and solidly buried in the ground. However, experience indicates that grounding an electrical system in this manner may cause corrosion of the water main. Because of this problem, water utilities discourage such grounding, by installing nonconductive couplings at the main or by using plastic piping materials. A less

effective, but still acceptable, alternative is to use copper rods or plates driven into the ground as the grounding electrode. A more reliable, but costlier, method is to install a grounding grid under the building or a grounding loop around the building as a system ground. There are two grounding requirements.

System grounding   The entire distribution system should be grounded at the neutral, or lowest voltage, point of the system. Figure 11–4 indicates the locations of grounded systems of some interior distribution systems.

Equipment grounding   All electrically conductive enclosures, such as the frame of a motor or the enclosure of an electrical appliance, should be grounded by the use of a separate green color-coded conductor. Although metal raceways can be used instead as the ground, the use of an equipment-grounding conductor is preferred. (See Figure 11–5(a).

11.3.4 Polarization

Since one side of all building interior distribution systems is grounded, the enclosure of any electrical equipment, such as the shell of a lamp socket, the metal case of an electrical tool, or the steel enclosure of a clothes-washing machine (see Figure 11–5(b)) should be connected to the grounded side of the electrical circuit to safeguard the person who operates the equipment. This requirement is satisfied by the installation of polarized receptacles with equipment-grounding terminals. The ungrounded side of the receptacle is smaller than the grounded sides; thus, it can receive only the smaller blade from the equipment plug. (See Figure 11–5(c).)

11.3.5 Ground Fault Circuit Interrupter

For added safety, certain receptacles and circuits that supply loads for swimming pools, kitchens, garages, bathrooms, laundries, or outdoor equipment should be of the type that can trip or cut out the circuit if an imbalance of current flow is sensed between the grounded and ungrounded sides of the circuit. Such an imbalance indicates that current is leaking through the ungrounded side of the circuit; thus, the situation should be corrected, by a ground fault circuit interrupter (GFCI). The requirements for using GFCI are given in Article 210 of the NEC. Figure 11–5 illustrates the configuration and operating principle of a GFCI receptacle.

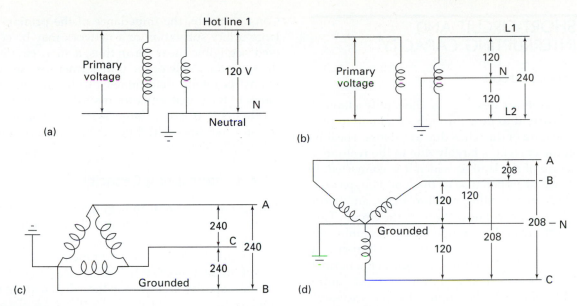

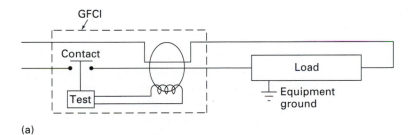

■ **FIGURE 11–4**

Typical grounded system.

(a) 120-volt, single-phase, two-wire system. Line to ground = 120 volts.

(b) 120/240-volt, single-phase, three-wire system. Line to ground = 120 volts.

(c) 240-volt, three-phase, three-wire system. Line to ground = 240 volts.

(d) 120/208-volt, three-phase, four-wire system. Line to ground = 120 volts.

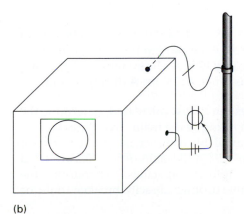

■ **FIGURE 11–5**

(a) Schematic diagram of the operating principle of a GFCI. Any unbalanced current flow between the two wires indicates a ground current leakage, which trips open the contact.

(b) The metal enclosure of a domestic washing machine or similar equipment is grounded either to the piping system or to the ground wire of the equipment-grounding system.

11.4 SHORT-CIRCUIT AND INTERRUPTING CAPACITY

11.4.1 Short Circuit

Electrical circuits are designed to allow only a limited amount of current to flow through the circuit to prevent overheating of the wires due to I^2R loss. The limitation on the current is largely due to the resistance or impedance of the load to which it is connected. If the resistance or impedance of the load is bypassed, or *shorted*, then, according to Ohm's law, an abnormally high current will flow through the circuit. This is called *short circuit*. Depending on the remaining resistance or impedance of the circuit, the short-circuit current could be 10 to 30 times as high as the normal current. At this abnormally high level, most equipment and wiring will be ruined by the excessive amount of heat generated.

11.4.2 Short-Circuit Calculations

To avoid the hazard mentioned in the previous section, all electrical circuits and equipment connected to the system must have an *interrupting rating*, or *interrupting capacity*, equal to or greater than the calculated *short-circuit capacity* of the system. The calculation of the short-circuit capacity of a power system is very involved and complex. Conservatively, it can be as high as 20 to 30 times the normal full-load current of the system. For example, if a power system is designed to carry a full-load current of 2000 amperes, then the short-circuit capacity could be in the neighborhood of 20 × 2000 (40,000) amperes, or even 30 × 2,000 (60,000) amperes. Computer programs are required to calculate a realistic level of short-circuit currents.

In its simplified form, the short-circuit current can be approximated by the equation

$$I_S = E / Z$$

or

$$I_S = 100 \times I/(Z_p + Z_t) \qquad (11\text{--}1)$$

where I_S = Short-circuit current, amperes
$\quad E$ = Voltage, volts
$\quad Z$ = Total impedance, ohms
$\quad I$ = Full-load current of transformer, amperes
$\quad Z_p$ = Impedance of the primary, %
$\quad Z_t$ = Impedance of the transformer, %

Conservatively, the impedance of the primary on a large utility substation or a network may be considered negligible (near zero); thus, a short circuit near the secondary side of the transformer (or the system mains) is totally determined by the internal impedance of the transformer (or mains). For example, the short-circuit (I_s) of a large system connected to a transformer rated with 5 percent impedance could be $100 \times I/5$, or $20I$.

11.4.3 Interrupting Capacity

To prevent the system and equipment from being destroyed due to a short circuit, all distribution wires and cables must be protected by devices such as fuses and circuit breakers. All equipment or devices installed on the power system must also carry an interrupting capacity greater than the calculated short-circuit current.

Examples

11.1 The full-load current of a building power distribution system is 1200 A. The building is served by a single transformer having 5-percent impedance. The utility power service supplying the transformer is from a nearby substation with practically unlimited power. Determine the available short-circuit current at the main switchboard.
From Equation (11–1),

$$I_s = 100 \times I/(Z_P + Z_t)$$

where I_s = 1200 A
$\quad Z_P$ = 0 (the impedance of the substation is considered to be near zero because the substation has an unlimited power supply)
$\quad Z_t$ = 5 (the transformer's impedance, as a percent of its own rating)

Thus, I_s = 100 × 1200/(0 + 5) = 24,000 A. In other words, even though the normal full-load current is only 1200 A, the system may instantly produce as much as 24,000 A during a fault.

11.2 For the system as just calculated, determine the minimum rating of the main circuit breaker.
Based on available commercial products, circuit breakers of 1200A/30,000A or 50,000A are rated for interrupting capacities. Therefore, the 30,000A interrupting capacity breaker could be selected.

11.5 EMERGENCY POWER SYSTEM

Emergency power systems are legally required by the building codes to assure the continuity of operation of the building when the loss of normal power may create a hazard to life, a fire hazard, or a loss of property or business.

11.5.1 Alternative Energy Power Supplies

Three types of emergency power supply are in common use:

1. *Tap ahead of the main switch.* When the main switch is disconnected, the emergency branch will still be active. This is the least reliable alternative emergency power system, since if the power is lost from utility, both the normal and emergency branches are out of commission. This alternative is good only for residences and small businesses. (See Figure 11–6.)

2. *On-site generator.* One or more electrical generators installed on-site or in the building automatically start up to provide power to the essential loads. This is the most reliable alternative, but is more costly to own and maintain.

3. *Separate power source.* The emergency load is automatically transferred to a separate power source either from the same utility or from another power source. This alternative is not always available in other than extremely large buildings.

11.5.2 Type of Loads

Emergency loads are electrical loads that must be maintained during a power failure or an emergency, such as a fire. These loads include emergency or exit lights, building control systems, a fire alarm system, elevators, etc. *Critical loads* are electrical loads that, when

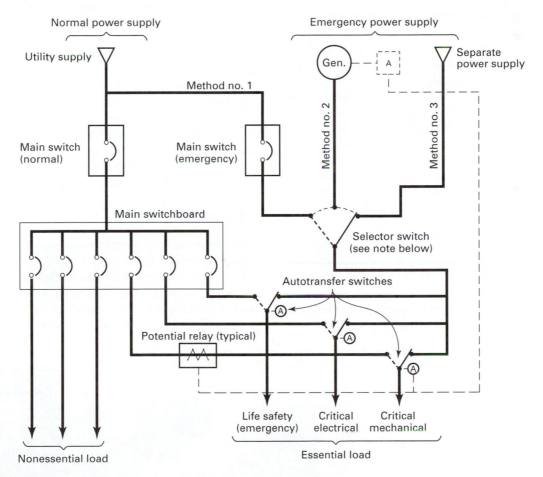

■ FIGURE 11–6
A one-line diagram of a power system showing three methods for emergency power supply. Method No. 1 is used for small systems only. (*Note:* The selector switch is not required if the alternative methods are mechanically and electrically interlocked.)

interrupted, may cause products, such as refrigerators or freezers, to become nonoperational. Sometimes, as in the case of a cash register, business is lost as well. Critical loads are specifically required in health care–related facilities.

11.5.3 Emergency Power Capacities

The capacity of essential loads (emergency and critical) varies with the design and occupancy of a building. Obviously, a single-story office building does not require anywhere near the essential load of a high-rise building, a hospital, a laboratory, or a special industrial plant. In general, the essential load for offices may range from 10 percent to 20 percent of the total connected load, and may even be as high as 40 percent, for super-high-rise buildings or hospitals.

11.5.4 Standby Power Generator Set

A *generator* is electrical equipment that generates electrical power. A *generator set* is a complete unit, including a prime mover that drives the generator and the necessary start-up, speed, and power controls. Automatic sensors initiate the start-up of the prime mover within seconds after the loss of normal power. The prime mover may be either a gasoline or diesel engine or a steam turbine. Standby power is required for many types of buildings, such as hospitals and hotels. It is also essential for high-rise buildings (over seven stories in height) regardless of what kinds of building they are—for example, high-rise offices or apartment buildings. Figure 11–7 is a photo of a generator set.

11.6 POWER EQUIPMENT

There are various levels of equipment in an electrical power system, starting with the power service, transformers, and distribution equipment and ending with the load-end protection devices.

11.6.1 Service Entrance

Utility power may enter a building through an overhead service drop for small systems or underground duct banks for large systems. The service entrance power may be at the building utilization voltage, such as 120/240 volts or 120/208 volts, or it may be at a higher voltage, which is then reduced to the utilization voltage through step-down transformers.

11.6.2 Switchboards and Panelboards

A *switchboard* is an assembly of switches and circuit protection devices from which power is distributed. The switchboard serves as the main distribution center of a small system or as a portion of the distribution center of a large system. (See Figure 11–8(a).)

A *panelboard* is an assembly of switches and circuit protection devices that serves as the final distribution point of the power distribution system. Figure 11–9(c) shows the interior of a small panelboard. Depending on the load to which it is connected, a panelboard may be identified as a lighting, power, heating, specialty, or combination panelboard. Figures 11–9(a) and (b) show several types of panelboards. There is no clear demarcation between a switchboard and a panelboard, although a switchboard is always referred to as the distribution equipment closest to the supply end of the system.

11.6.3 Transformers

Transformers are power transmission equipment primarily intended to convert a system's voltage from one level to another. All transformers operate on the principle of magnetic induction, wherein primary and secondary coils are wound on a common silicon steel core. The incoming side is the primary, the out-

■ FIGURE 11–7
Typical diesel engine–driven generator or generator set.

(a)

(b)

■ **FIGURE 11–8**

(a) Typical power substation for a large building complex. (b) Line-ups of motor control centers which contain motor starter modules in freestanding construction. Interlocking control wiring between motors can be conveniently located within each motor control center.

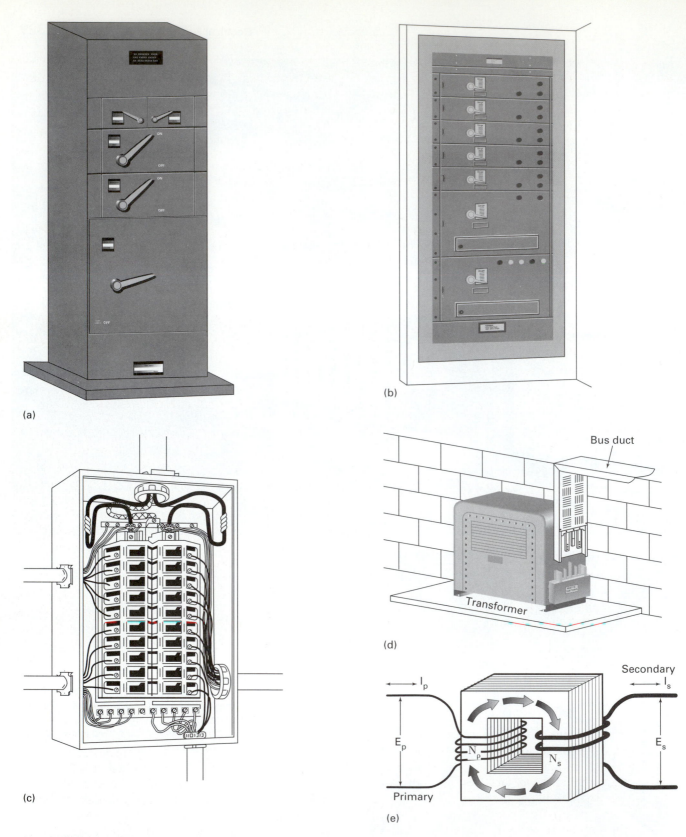

■ **FIGURE 11–9**

(a) A free-standing switchboard. (b) A wall-mounted power distribution panelboard. (c) The interior view of a lighting panelboard, showing 20 single-pole circuit breakers, the main feeder from the top of the panel, and branch circuit wires (black) from the circuit breakers to various loads through four feeder conduits. The white wires are neutral conductors for these branch circuits. (d) Small pad-mounted transformer with bus duct connection. (e) The operating principle of a transformer, where $E_s/E_p = N_s/N_p$. (Courtesy: Cutler-Hammer/Westinghouse, Pittsburgh, PA.)

going side the secondary. If the voltage is increased from primary to secondary, the transformer is a *step-up* transformer; if the voltage is decreased, the transformer is a *step-down* transformer. With magnetic induction, voltage is induced from the primary to the secondary. The resulting voltage is directly proportional to the ratio of the primary-to-secondary winding turns. That is, $E_s = (N_s/N_p) \times E_p$, i.e., $E_s = \frac{1}{2} E_p$, if N_p is twice N_s. (See Figure 11–10(e).)

Transformers are classified as liquid or dry types. The latter are used mostly in indoor applications. Liquid-type transformers have a lower impedance and are normally more efficient, but need to be installed in transformer vaults or enclosed spaces. Figure 11–8(a) shows an indoor liquid-type transformer in line with a large power distribution center. Figure 11–9(d) illustrates a small pad-mounted, dry-type transformer.

11.6.4 Motors and Motor Starters

Motors Virtually any equipment requiring motion, such as a pump, elevator, fan, air conditioner, or even equipment as small as an electric clock, requires a motor. Motors are classified according to the following characteristics:

- *Size.* Fractional or integral horsepower, etc.
- *Voltage.* 120, 208, 240, 277, 380, 480, 600, 2300, 4160 volts, etc.
- *Number of poles.* Two pole (3600 rpm), four pole (1800 rpm), six pole (1200 rpm), etc. (based on 60 Hertz)
- *Phase.* Single, double, triple phase, etc.
- *Operating principle.* Universal, split phase, induction (squirrel cage, wound rotor), synchronous, etc.
- *Construction.* Open drip proof, watertight, explosion proof, etc.
- *Starting characteristics.* High torque, low starting current, etc.

Most motors used in building equipment are of the squirrel-cage induction variety. Due to inductive reactance in the motor winding, induction motors always have a lagging power factor that can range from 70 to 80 percent at full load to as low as 10 to 20 percent during start-up. Consequently, the starting current of a motor may be as high as 10 times its full-load current.

The size of a motor is rated in horsepower (HP) which is equivalent to 746 watts, or 0.75 kW. The full-load current of a motor varies with the design of the motor. It may be calculated approximately from the basic power formula given in Chapter 10.

Operating on the principle of slippage, an induction motor has a normal speed that is slightly slower than its synchronous speed. For example, a two-pole motor normally has a synchronous speed of 3600 rpm (60 Hz × 60 sec/min), but a rated speed of 3450–3500 rpm running on a 60-Hz system.

Motor starters When a motor starts up, its current is many times higher than its normal full-load current for several seconds. The persistence of the starting current depends on how fast the equipment can be brought into full speed, which in turn depends on the inertia of the load. Ordinary-duty manual on-off switches are not capable of withstanding the momentary inflow of current. Thus, motor circuit-rated switches must be used. Automatic starters are required for large motors of one horsepower or higher. These starters allow a large inflow of current, as well as protection against continuous overload. Motor starters may be classified according to the following properties:

- *Operating principle.* Electromagnetic, solid state, etc.
- *Protection devices.* With or without a disconnect switch, with or without short-circuit protection.
- *Starting circuitry.* Across-the-line, reduced voltage (autotransformer type), reduced inrush (delta-wye, part-winding types).
- *Protection circuitry.* Overcurrent, overvoltage, undervoltage, reverse phasing, etc.
- *Construction.* General service, weathertight, waterproof, explosion proof, etc.

Figure 11–10 illustrates two of many motor starters which are used to reduce the inrush current during the startup of a motor load. Motor starters may be individually mounted or preassembled as a motor control center to facilitate in interlocking control wires in large systems.

11.7 CONDUCTORS

A conductor is an electrical component that conducts and confines the flow of electrical current within itself. Conductors are made of high-conductivity (low-resistivity) material to minimize the loss of power and drop in voltage. Normally, they are made in cylindrical form as wires, but they are also made in square or rectangular sections.

Depending on their construction, conductors are classified according to the following characteristics:

- Material Copper, aluminum, etc.
- Form Wire, cable, bus, bus-duct, etc.

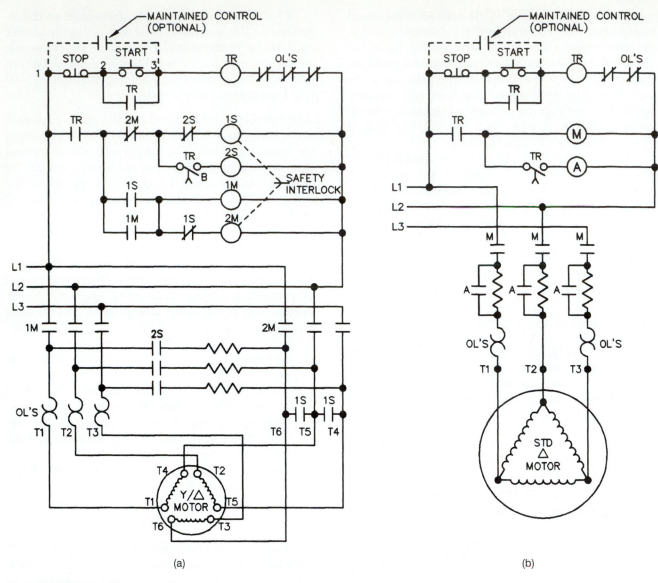

■ **FIGURE 11–10**

Reduced voltage and reduced in-rush motor starting methods. (a) Wye-Delta (Y/Δ) type starter where the motor is connected in Wye during startup (stator windings at 57.7% of line voltage), and connected in delta after an adjustable time delay (stator windings at 100% of line voltage). This method requires a special 6-leads motor. (b) Primary resistor type starter utilizes standard 3-leads induction motor, but requires large bank of resistors to reduce the line voltage during start-up.

- Composition   Solid, stranded, etc.
- Voltage class   100 volts, 250 volts, 600 volts, 5000 volts, etc.
- Insulation   Rubber, thermoplastic, asbestos, etc.
- Covering   Lead, aluminum, nonmetallic, cross-links polymer, etc.
- Temperature rating   60°C, 75°C, 90°C, 250°C, etc.

Figure 11–11 shows various kinds of wire and cable. Some of the more common 600-volt general building wires used in raceways are the following:

- *THHN*   Heat resistant, thermoplastic, 90°C for wet and dry locations; used mostly for branch circuits.
- THWN   Heat and moisture resistant, 75°C; used mostly for branch circuits.
- USE   Underground service entrance cable, 75°C, heat- and moisture-resistant insulation with nonmetallic covering.
- XHHW   Heat and moisture resistant, cross-linked synthetic polymer, 75°C for wet and dry locations; used mostly for larger feeders.

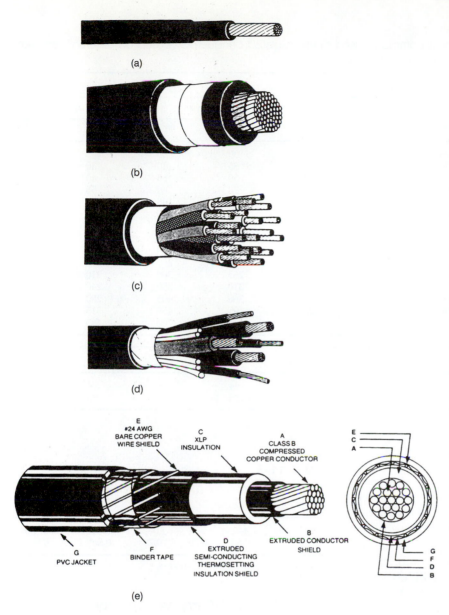

In addition to wires installed in raceways, certain wires and cables are permitted to be installed without raceways. The following such wires are used in buildings:

- *Mineral insulated (MI).* These are metal-sheathed wires with a temperature rating of 90° to 250°C. These MI cables may be directly buried in concrete, pavement, walls, or ceilings.
- *Nonmetallic (NM and NMC).* This sheathed cable is a factory assembly of two or more insulated conductors having an outer sheath of moisture-resistant, flame-retardant, nonmetallic material. It is used mostly in residential dwellings and other types of buildings not exceeding three stories in height. Such cable is commonly called Romex.
- *Armored cable (AC).* This cable contains two or more insulated conductors in a metallic enclosure. It can be used in exposed or concealed applications, mostly for small sizes. It is sometimes called BX.
- *Flat conductor cable (FCC).* This type of cable consists of three or more flat copper conductors placed edge to edge, separated, and enclosed within an insulating assembly. FCC can be installed under carpet tiles not larger than 3 ft square. It is not to be used in residences, schools, hospitals, or hazardous locations. FCC wire is commonly called flatwire.

TABLE 11-2

NEC-allowed maximum ampacities of most commonly used single insulated conductors (based on 30°C/86°F ambient temperature and not more than three conductors in raceway)

Size	In Raceways				In Free Air			
	Copper		Aluminum		Copper		Aluminum	
AWG & MCM	75°C	90°C	75°C	90°C	75°C	90°C	75°C	90°C
14	20	25			30	35		
12	25	30	20	25	35	40	30	35
10	35	40	30	35	50	55	40	40
8	50	55	40	45	70	80	55	60
6	65	75	50	60	95	105	75	80
4	85	95	65	75	125	140	100	110
3	100	110	75	85	145	165	115	130
2	115	130	90	100	170	190	135	150
1	130	150	100	115	195	220	155	175
1/0	150	170	120	135	230	260	180	205
2/0	175	195	135	150	265	300	210	235
3/0	200	225	155	175	310	350	240	275
4/0	230	260	180	205	360	405	280	315
250	255	290	205	230	405	455	315	355
500	380	430	310	350	620	700	485	545

Ambient	Ampacity Correction Factor							
88–95°F (31–35°C)	.94	.94	.94	.96	.94	.96	.94	.96
97–104°F (36–40°C)	.88	.91	.88	.91	.88	.91	.88	.91

11.7.1 Wire Sizes

American wire gauge Conductors are numbered according to the American wire gauge (AWG) from No. 36 to No. 0000 (#4/0). The numbers are retrogressive, i.e., a smaller number denotes a larger size. A #4/0 solid (nonstranded) wire should have a diameter of 0.5 inch. Each smaller size will have a reduction of diameter in the ratio of 1.123. In other words, the diameter of a solid #3/0 wire should be 0.5/1.123, or 0.405 inch. The actual diameter of stranded wire is, or course, larger than solid wire of the same AWG.

Circular mil and square mil Conductors larger than #4/0 are described in circular or square mils. A mil is defined as 1/1000 of an inch, and a circular mil is the area of a circle 1 mil in diameter. Similarly, a square mil is the area of a square having its sides 1/1000 in length. A thousand CM is also expressed as one MCM, where the first "m" stands for "1000." For example, a conductor of 300,000 CM is 300 MCM. Large conductors come in rectangular shapes or bars called busbars.

TABLE 11-3

Correction factors for more than three conductors in raceway or cable

Conductors	4–6	7–9	10–24	25–42	43 or more
Factor	0.80	0.70	0.70*	0.60*	0.50*

*Include the effects of a load diversity of 50%.

11.7.2 Current-Carrying Capacity

The current that can be safely carried by a conductor depends on the size of the conductor as well as the type of insulation, method of installation, number of wires within a raceway, and surrounding temperature. The allowable current carrying capacities, or ampacities of various types and sizes of wires, are given in the NEC. Table 11–1 is a condensed table of commonly used cables. (See the NEC for properties of other types of cable.) The allowable ampacity of conductors is reduced at ambient temperatures higher than 88°F. (See Table 11–2 for the relevant correction factors.) The allowable ampacity of conductors also is when more than three conductors are installed in the raceway. (Again, see Table 11–2 for the relevant correction factor.)

11.7.3 Dimension of Conductors

The NEC code provides data on bare and covered conductors for the purpose of sizing raceways. A condensed listing of the dimensions of rubber- and thermoplastic-covered conductors is given in Table 11–5. The use of the table will be demonstrated in Section 11.9.

Examples

11.3 What is the code-allowed maximum ampacity of a #8 AWG copper conductor in a conduit with covering rated for 75°C?

From Table 11–1, the answer is 50 A.

11.4 What is the allowed maximum ampacity if there are nine conductors in the same conduit?

From Table 11–2, the correction factor is 0.70; thus, $50 \times 70\% = 35$ A.

11.5 If the ambient temperature of the installation is 100°F, what is the allowed ampacity in Example 11.4.

From Table 11–2, the temperature correction factor is 0.88; thus, $35 \times 0.88 = 30.8$ A.

11.6 What are the diameter and area of a #1/0-type THHN conductor?

From Table 11–3, the diameter is 0.491 inch and the area is 0.1893 sq in.

11.8 WIRING METHOD

There are over 20 NEC-approved wiring methods. Those approved for use in buildings are generally wires and cables installed within raceways, with the exception of types NM (Romex), AC (BX), and FCC (flat wires), which can be installed without raceways. The following are commonly used wiring methods:

1. *Electrical metallic tubing* (EMT) is commonly referred to as thin-wall conduit. It is made in ½" to 4" sizes, using a crimp type of coupling for quick installation. EMT is the most commonly used wiring method for all building applications, except in locations where wiring must be watertight.

2. *Rigid conduit* is similar to EMT, except that it uses threaded couplings. It may be used for all applications, including explosive and wet locations. It is more expensive than EMT; therefore, it is used only for 4" and larger sizes or where EMT is not permitted.

3. *Wireways* are used to enclose a large number of wires. They are usually 3" to 8" in size, contain tens or hundreds of wires, and should be installed where they are accessible. Surface metal raceways, such as "wiremold" and "plugmold," are examples of smaller size wireways. (See Figure 11–12.)

4. *Bus ducts* are used for feeding large power distribution systems. They come with a feeder or plug-in type of design. (See Figure 11–13.)

5. *Underfloor ducts* are raceways cast into a poured floor slab for the purpose of supplying electrical wiring to the center of large rooms. Single, double, and triple ducts may be allocated for different electrical systems. (See Figure 11–14.)

6. A *cellular floor* is a combined structural floor and electrical raceway system for a modern office or research building where electrical power is required throughout the floor area. It is a very flexible system. A cellular floor system may use one or more structural cells as electrical cells, depending on the electrical system used. The cost of having the system installed is usually higher than that for a raceway system. However, on a life cycle basis, a cellular floor is both economical and necessary for modern office buildings. (See Figure 11–15.)

7. A *raised floor* installed above the structural floor creates space between the two floors to run wiring with or without additional raceways. Raised floors are commonly used for computer mainframe space and spaces with a concentration of electrical equipment. (See Figure 11–16.)

With the increased popularity of telecommunications, data transmission, and local area networks, most modern office buildings have some form of underfloor raceways. A raised floor is the more flexible, but the more costly, of the two designs.

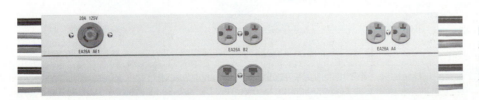

■ **FIGURE 11–12**
A section of surface wireway showing divided compartments, plug-in receptacles, and wires. (Courtesy: Wiremold Company, West Hartford, CT.)

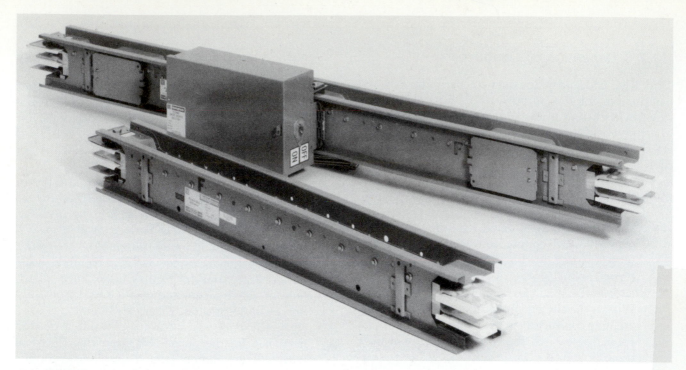

■ **FIGURE 11–13**

Sections of bus ducts. The upper one is the plug-in type ready to receive plug-in switches or circuit breakers. The lower one is a feeder bus duct that does not provide for a plug-in. (Courtesy: Cutler-Hammer/Westinghouse, Pittsburgh, PA.)

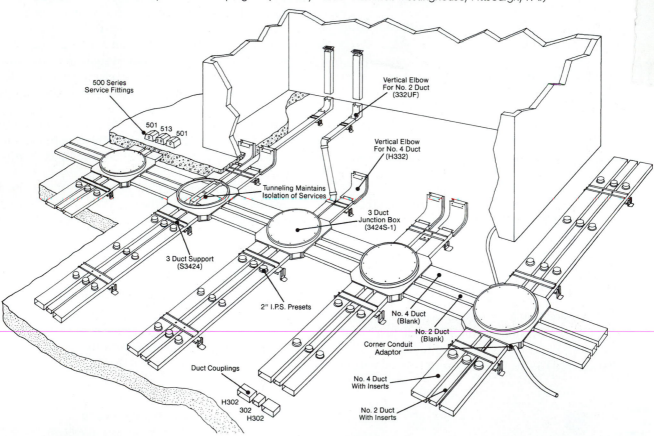

■ **FIGURE 11–14**

A single-level, three-cell underfloor duct system. Normally, the smaller cells are for power and data, and the larger cell is for communication wiring. For the proper class of wiring, data and communication wiring may be installed in the same cell. (Courtesy: General Electric Company.)

336

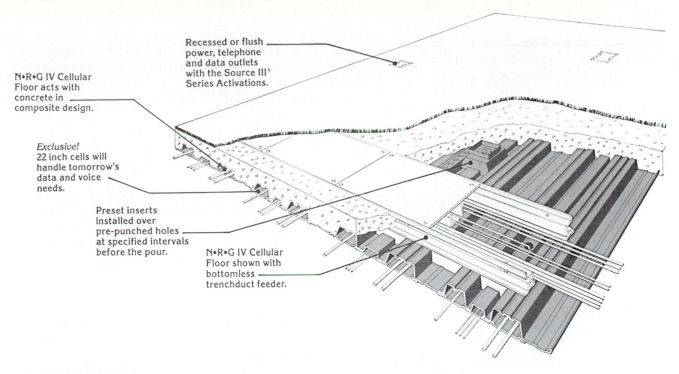

N·R·G IV Cellular Floor acts with concrete in composite design.

Recessed or flush power, telephone and data outlets with the Source III® Series Activations.

Exclusive! 22 inch cells will handle tomorrow's data and voice needs.

Preset inserts installed over pre-punched holes at specified intervals before the pour.

N·R·G IV Cellular Floor shown with bottomless trenchduct feeder.

■ **FIGURE 11–15**

A cutaway view of an underfloor electrical distribution system consisting of a trench duct above the cellular floor cells. The trench duct is for main distribution, and the cells are for wiring to floor outlets. (Courtesy: Walker Company, Parkersburg, WV.)

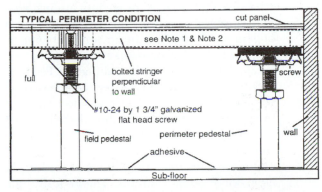

(a)

(b)

■ **FIGURE 11–16**

(a) A cutaway section of a raised floor with adjustable pedestal supports. The panels are normally 2-ft × 2-ft modules.
(b) Anchoring details of a pedestal. (Courtesy: Tate Access Floors, Inc., Jessup, MD.)

337

11.9 INSTALLATION OF WIRES IN RACEWAYS

The installment of wires (or cables) in raceways is strictly regulated. Generally, no more than 40 percent of the cross-sectional area of the raceway can be filled with wires or cables. The limitation is necessary for two key reasons:

1. *To prevent excessive heat buildup.* All wires have resistances and impedances. that create a power loss that turns into heat and, if unabated, may cause the breakdown of the insulation material or even a fire.
2. *To permit the physical installation of the wires.* Wires in conduits must be pulled into the conduits by special tools. A clear space must be provided for the wires to be pulled in easily, without damage.

Rules governing the number and size of wires that can be installed in a raceway are given in detail in the National Electrical Code. Tables 11–2, 11–3, 11–4, and 11–5 are condensed from the NEC tables for some cable types.

1. *Pull and junction boxes* When the raceway (conduit) is too long or contains too many bends, pull boxes must be installed at locations to facilitate the pulling of conductors into the raceway. When the conductors need to be spliced, a junction box must be installed. No conductor is allowed to have splices within a raceway other than at junction boxes or within equipment enclosures.

Examples

11.7   How many No. 12-type THWN wires can be installed in a 1-inch conduit?
From Table 11–6, the maximum number is 19.

TABLE 11–4

Dimensions and internal areas of electrical metallic tubing (EMT) and conduit. (See NEC for other conduit sizes and cable dimensions.)

Size, In.	Internal Diam., In.	Area, Sq In.
1½	1.610	2.04
2	2.067	3.36
2⅕	2.469	4.79
3	3.068	7.38
3½	3.548	9.90
4	4.026	12.72

11.8   What is the allowable ampacity of three 90°C-rated, 600-volt, No. 10 copper conductors in a conduit?
From Table 11–2, the maximum allowed is 40 A.

11.9   What is the allowable ampacity if 8 No. 10 copper conductors are installed in one conduit?
From Table 11–2, the maximum ampacity of 90°C-rated conductor is 40 amperes for not more than three conductors in the conduit.
From Table 11–3, the correction factor is 0.7. Thus, the allowable ampacity is 0.7 × 40, or 28 A.

11.10   For the same installation as in Example 11.9, but with the wires exposed to an ambient temperature of 100°F, what is the allowable ampacity?
From Table 11–2, the correction factor is 0.91; thus, the allowable ampacity is further reduced to 0.91 × 28, or 25.5 A.

11.11   Two sets of 120/208-volt, three-phase, four-wire distribution system feeders are installed in a common conduit that passes through a boiler room with a maximum ambient temperature of 102°F. The demand current of Feeder 1 is calculated to be 100 A, and that for Feeder 2 is 50 A. Determine the feeder sizes, based on 90°C copper wires (cables), and select the common conduit size. Assume the selected feeders are type THHN copper.
Answers to question 11.11
There are four wires in each set of feeders, or eight for Feeders 1 and 2. Theoretically, the neutral conductors may not carry any current if the load is balanced between Phases A, B, and C. However, recent design practices have been to treat the neutral conductor as a current-carrying conductor, due to the third harmonics of inductive loads such as PCs and electronic appliances. From Table 11–3, a correction (derating) factor of 0.70 must be applied.
The ambient temperature in the boiler room is 102°F; thus, a correction (derating) factor of 0.91 must be applied for the 90°C-rated wires (cables).
The overall derating factor for ampacity is 0.70 × 0.91 = 0.637; thus, Feeder 1 must be selected for 100 A/0.637 = 157 A, and Feeder 2 must be

TABLE 11–5

Dimensions of several rubber- and thermoplastic-covered conductors. (Refer to NEC for a complete listing of conductors as well as other properties.)

Size, AWG MCM	Types RFH-2, RH, RHH		Types TF, THW, TW		Types TFN, THHN, THWN	
	Approximate Diameter, In.	Approximate Area, Sq In.	Approximate Diameter, In.	Approximate Area, Sq In.	Approximate Diameter, In.	Approximate Area, Sq In.
14	.204*	.0327	.131	.0135	.105	.0087
12	.221*	.0384	.148	.0172	.122	.0117
10	.242	.0460	.168	.0222	.153	.0184
8	.328	.0845	.245	.0471	.218	.0373
6	.397	.1238	.323	.0819	.257	.0519
4	.452	.1605	.372	.1087	.328	.0845
3	.481	.1817	.401	.1263	.356	.0995
2	.513	.2067	.433	.1473	.388	.1182
1	.588	.2715	.508	.2027	.450	.1590
1/0	.629	.3107	.549	.2367	.491	.1893
2/0	.675	.3578	.595	.2781	.537	.2265
3/0	.727	.4151	.647	.3288	.588	.2715
4/0	.785	.4840	.705	.3904	.646	.3278

*The dimensions of RHH and RHW.

TABLE 11–6

Maximum number of conductors in conduits or tubing for most building wire types, including TW, XHHW, RHW, RHH, TW, THW. (See Chapter 9 of NEC for more specific conductors.)

AWG and MCM	Conduit or Tubing (inch)									
	½	¾	1	1¼	1½	2	2½	3	3½	4
14	9	15	25	44	60	99	142			
12	7	12	19	35	47	78	111	171		
10	5	9	15	26	36	60	85	131	176	
8	2	4	7	12	17	28	40	62	84	108
6	1	3	5	9	13	21	30	47	63	81
4	1	2	4	7	9	16	22	35	47	60
3	1	1	3	6	8	13	19	29	39	51
2	1	1	3	5	7	11	16	25	33	43
1		1	1	3	5	8	12	18	25	32
1/0		1	1	3	4	7	10	15	21	27
2/0		1	1	2	3	6	8	13	17	22
3/0		1	1	1	3	5	7	11	14	18
4/0			1	1	2	4	6	9	12	15
250			1	1	1	3	4	7	10	12
300			1	1	1	3	4	6	8	11
350			1	1	1	2	3	5	7	9
500				1	1	1	3	5	6	8

selected for 50A/0.637 = 78.5 A.

From Table 11–2, Feeder 1 must be a minimum size of 1/0 AWG, which is rated for 170 A under normal conditions, and Feeder 2 must be a minimum of size of No. 4 AWG.

From Table 11–5, No. 1/0 THHN cable has 0.1893 sq in. of cross-sectional area, and that for No. 4 cable is 0.0845 sq in. The total cross-sectional area of all the cables is

$$(4) \times 0.1893 + (4) \times 0.0845 = 1.160 \text{ sq in.}$$

Based on the maximum 40-percent fill rule, the conduit must have a minimum cross-sectional area of 1.160/40 percent, or 2.9 sq in. From Table 11–4, a 1-½ inch conduit has a cross-sectional area of 2.04 sq in., which is smaller than the required 2.9 sq in. Thus, the next larger size 2-inch conduit having a cross-sectional area of 3.36 sq in. must be used.

11.10 WIRING DEVICES

A variety of wiring devices—from switches, receptacles, and overcurrent protection devices to contactors and dimmers—are used in electrical systems. All wiring devices must be installed in code-approved boxes, regardless of the wiring system. The most commonly used devices are described in the rest of this section; some are shown in Figure 11–17.

11.10.1 Switches

A switch is a device for making, breaking, or changing the connections in an electrical circuit. Switches are classified, according to the following criteria:

1. *NEC rating.* General service, isolating, motor duty, etc.
2. *Method of engaging contact.* Sliding, snap, liquid (mercury), etc.
3. *Voltage rating.* 250, 600, 5000 volts, etc.
4. *Number of breaks.* Single break, double break, etc.
5. *Number of poles.* 1, 2, 3, 4 poles, etc.
6. *Number of closed positions.* Single throw, double throw, etc.
7. *Method of operation.* Manual, magnetic, motor operated, etc.
8. *Speed of operation.* Slow make/slow break, quick make/quick break, etc.
9. *Enclosure.* Open, enclosed, weathertight, watertight, explosion proof, etc.
10. *Control function.* Single acting, three way, four way, etc.
11. *Method of protection.* Nonfused, fused, circuit breaker, combination, etc.
12. *Performance of contacts.* Maintained contact, momentary contact, etc.
13. *Duty.* Light duty, heavy duty, load interrupting duty, etc.
14. *Other functions.* Dimming or voltage control, photoelectric, time clock, electrically or mechanically held, auxiliary controlled, such as pressure, temperature, flow, infrared, motion, proximity sensitive, etc.

Light switches are normally single-pole, single-throw switches. When lights need to be switched from more than one location, three-way and four-way switches are used. The operating principles of three-way and four-way switches are illustrated in Figure 11–18. As a rule, the first and last switches must be three-way switches, and the in between switches must be four-way switches.

11.10.2 Remote Control, Low-Voltage Switching

When the switching of lights and appliance loads is desired at multiple locations, a remote control, low-voltage switching system will provide flexibility and economy. With this system, all control wires are reduced to 24 volts or less and thus need not be installed in raceways (conduits). The loads, whether 120, 240, or 277 volts, are operated by one or more momentary contact-type electromagnetic relays. A typical remote control, low-voltage wiring system is shown in Figure 11–19.

11.10.3 Receptacles

A receptacle is a wiring device installed within an outlet box for the connection of electrical apparatus through an attachment plug. (See Figure 11–17.) Receptacles may be classified according to the following characteristics:

1. *Number of receptacles in the assembly.* Single, double (duplex), triple (triplex), etc.
2. *Current rating.* 5, 15, 20, 30, 50, 60, 100 A, etc.
3. *Voltage rating.* 125, 250, 277, 480, 600 volts, etc.
4. *Number of poles and wires.* Two-pole, two-wire; two-pole, three-wire grounding; three-pole, three-wire; three-pole, four-wire grounding; four-pole, five-wire grounding; etc.
5. *Shape of the blades.* Straight blades, locking blades (which can lock the plug in place by twisting).
6. *Configuration of the blades.* Parallel, angled, flanged, polarized, etc.

Commonly used receptacles for plug-in types of apparatus such as typewriters, portable lights, televisions, etc. are also known as *convenience outlets* and are normally rated for 125 volts and 15 amperes. They may be two wire with nonpolarized parallel blades, or three wire with polarized blades. Since all 120-volt electrical systems in a building are grounded, a two-wire parallel-blade nonpolarized receptacle is not able to distinguish the polarity of the wires and, hence, which side of the plug is grounded. Logically, there is a 50–50 chance that the equipment is plugged in on the wrong side and will subject a person to the

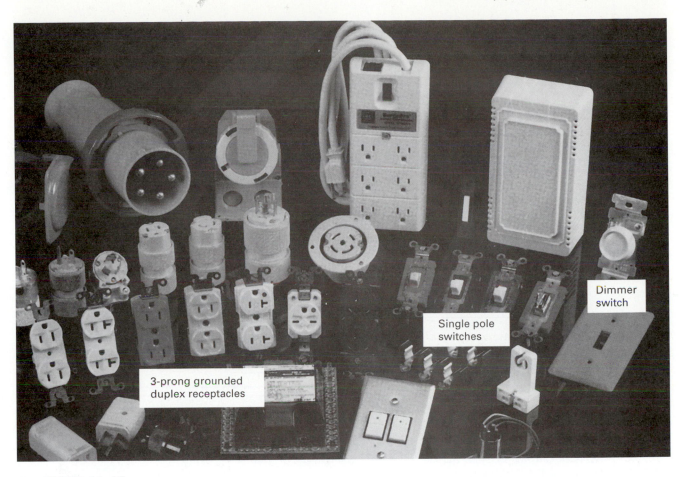

3-prong grounded
duplex receptacles

Single pole
switches

Dimmer
switch

■ **FIGURE 11–17**
Typical receptacles and switches. (Courtesy: General Electric Company.)

line voltage if the ungrounded side of the wire touches the equipment enclosure. Fortunately, the 125-volt system is not deadly when accidentally touched by a person, although an electrical shock will definitely be felt. For this reason, the NEC no longer approves the use of nonpolarized and ungrounded receptacles in new installations. All two-wire ungrounded receptacles will be replaced in time.

The National Electrical Manufacturers' Association (NEMA) has issued standard configurations of various wiring devices, including the plug-in type receptacles based on ampacity rating, blade configuration, and number of poles. Some NEMA standards and receptacles are shown in Figure 11–20.

11.10.4 Contactors and Relays

Contactors and relays are remote-controlled power-transmitting devices. The former is normally used to carry line voltage power, and the latter is normally used to carry line or low-voltage power. The con-

struction of these devices is the same and includes an electromagnetic coil and a set of electrical contacts. When the coil is energized, the magnetic force that is created causes the contacts, and thus the electrical circuit, to open or to close. Figure 11–20 illustrates the basic wiring diagrams of a contactor and a relay. The wiring diagram of magnetic motor starters shown in Figure 11–19 is, in effect, a combination magnetic contactor with overload relays. Contactors may be classified according to the following properties:

1. *Voltage.* 24, 48, 125, 250, 600, 5000 volts, etc.
2. *Number of contacts (poles).* 2, 3, 4, 6, 12 poles, etc.
3. *Type of contacts.* Button, mercury, knife blade, etc.
4. *Type of operations.* Maintained, momentary contact, etc.
5. *Current Rating.* 15, 30, 50, 60, 100, 200, 400, 800, 1200 A, etc.
6. *Duty.* Inductive, noninductive loads, etc.

Figure 11–20 shows a control scheme using momentary-contact stop-start push buttons. The start

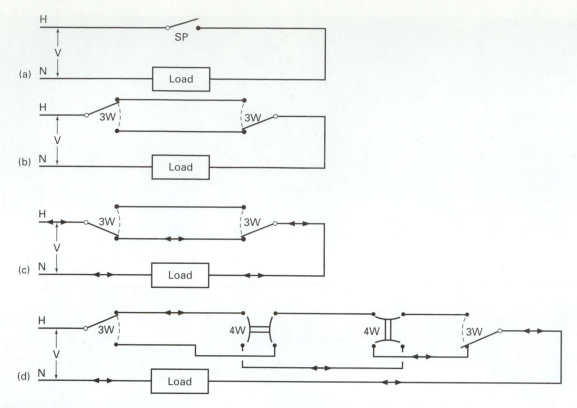

■ **FIGURE 11–18**
Schematic wiring diagrams.
(a) A single-pole (two-way) switched circuit.
(b) A load switched from two locations with two three-way switches. The position of the switches shows that the load is off.
(c) Same circuit as (b) with the load on. Change the position of any one of the three-way switches, and the load will be off.
(d) The load is controlled at four locations with two three-way and two four-way switches. The position of the switch shows that the load is on. Change the position of any switch, and the load will be off.

push button will close the control circuit through the coil (M), which closes the normally open (NO) main contact, allowing power to pass through L1 and L2. The auxiliary contact (M) wired across the start push button is closed to keep the control circuit energized even when the push button is released. Figure 11–20(b) is similar to Figure 11–20(a) except that the control switch is a maintained-contact switch, and the main contact is normally closed (NC). This means that when the control circuit is energized, the power circuit will be de-energized.

11.10.5 Dimmers

Dimmers are operating devices that reduce the input to, and thus the output from, an apparatus. Although most dimmers are used to control the intensity of light output, they are also used to control the speed of a fan, drill, etc. Dimmers usually operate on the prin-

ciple of adjustable resistance, adjustable voltage with autotransformer, or solid-state elements to control the shape of the wave or duration of the current through a load.

11.11 PROTECTIVE DEVICES

Electrical circuits which include feeders, distribution equipment, branch circuits, and the load equipment must be protected from exceeding their rated capacity which may occur as a result of many circumstances. Some examples are:

- *Overcurrent*, due to mechanical overload or internal or external electrical faults.
- *Overvoltage*, due to short circuiting between primary and secondary wiring or due to a lightning strike.

(a)

(b)

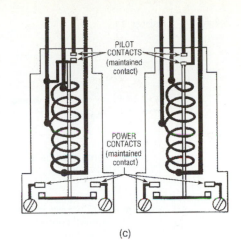

PILOT
CONTACTS
(maintained
contact)

POWER
CONTACTS
(maintained
contact)

(c)

(d)

■ **FIGURE 11–19**

Remote control, low-voltage switching.
(a) A low-voltage switch.
(b) A low-voltage relay.
(c) Sectional view of a low-voltage relay, illustrating the pilot and maintained contacts and the double-acting electromagnetic coils.
(d) The ganged master control for low-voltage switches.
(e) Typical wiring diagram of a remote control, low-voltage switching system.
(All illustrations courtesy of General Electric Co.)

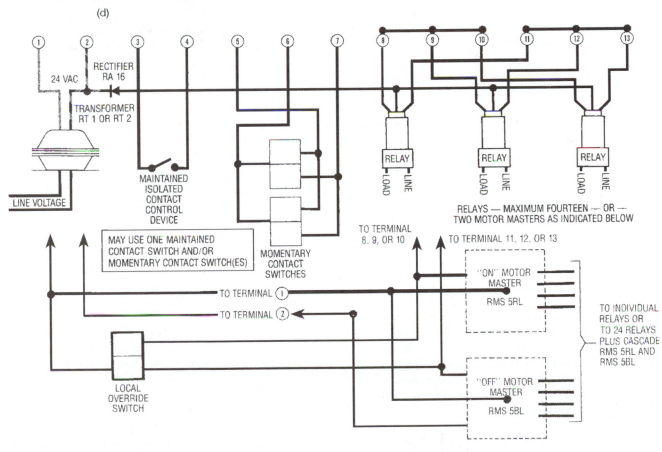

(e)

343

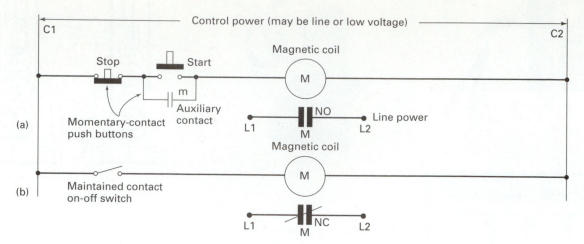

■ **FIGURE 11–20**
(a) The elementary diagram shows a control scheme using momentary-contact stop-start buttons (switches). The start push buttom will close the control circuit through the coil (M), which closed the normally open (NO) main contact, allowing power to pass through L1 and L2. The auxiliary contact (m) wired across the start push button is closed to keep the control circuit energized even when the push button is released.
(b) The elementary diagram shows a control scheme using a maintained-contact switch, and the main contact is normally closed (NC). This means that when the control circuit is closed, the main contact (M) of the power circuit will be opened (de-energized).

■ *Reversing polarity of a 3-phase system*, due to a change of power service.

The most common method used to prevent the damage caused by overloading conditions is the installation of protective devices at strategic locations, e.g., on switchboards, panelboards, at the beginning of a feeder, in a branch circuit, or at the equipment. These devices are divided into three general types: relays, circuit breakers, and fuses. Relays are normally used by utility companies to protect their primary distribution system or large primary equipment within a network. For building systems and equipment, circuit breakers and fuses are usually used.

11.11.1 Circuit Breakers

Classification of circuit breakers (CB) A circuit breaker is defined in NEMA standards as a device designed to open and close a circuit by nonautomatic means, and to open the circuit automatically on a predetermined overcurrent, without damage when properly applied within its rating. There are three types of circuit breakers.

1. *Molded case circuit breakers (MCCB).* The current carrying parts, mechanisms, and trip devices are completely contained within a molded case of insulating material. MCCBs are available in small and medium frame sizes from 30 A to 800 A, and with trip sizes from 15 A to 800 A.
2. *Power circuit breakers (LVPCB).* These CBs are also known as air breakers. They are primarily used in drawout switchgear construction. LVPCs have replaceable contacts, and are designed to be maintained in the field. LVPCBs are available in medium and large frame sizes from 600 A to 4000 A. LVPCBs are rated from 600 volts, whereas MVPCBs are rated for up to 72.5 kV, and HVPCBs for over 72.5 kV.
3. *Insulated case circuit breakers (ICCB).* These have the construction characteristics of both MCCB and LVPCB, and are used primarily in fixed mounted switchboards but are also available in drawout configurations. The frame sizes range from 600 A to 4000 A.

Construction and features of circuit breakers Circuit breakers are also classified by other construction and operating features according to the:

1. Arc quenching (extinguishing) media. Air or oil.
2. Operating principle. Thermal, magnetic, thermal magnetic, solid state, (electronic), etc.

3. Voltage class. 125, 250, 600 volts, 5, 12, 15, 35 kilovolts, etc.
4. Frame size. 30, 50, 100, 225, 400, 600, 800, 1200, 2000, 4000 A, etc.
5. Trip ratings. 15, 20, 30, 50, 90, 100, and up to the frame size ratings.
6. Interrupting capacity. 5000, 10,000, 15,000, 20,000, 30,000 A, and up.
7. Operating methods. Manual, remote operating, etc.
8. Other features. Overvoltage, undervoltage, auxiliary contacts, reverse current, reverse phase, etc.

Operating principles of CB   There are two types of circuit breaking (tripping) components within a CB.

1. *The bimetal/electromagnetic type.* Consists of a bimetal element which responds to temperature buildup inside the CB and an electromagnet which responds to the magnetic force caused by an abnormally high current flow. The bimetal provides thermal protection and the magnet provides short-circuit protection. The operating principles of these elements are illustrated in Figure 11–21(a), (b) and (c).
2. *The solid-state (electronic) type.* Consists of analog or digital devices to sense the electrical characteristics or circumstances of the circuit and process the data through a central processing unit (CPU) with pre-programmed actions. The analog type senses the peak current of a current, whereas the digital type senses the rms current, a more realistic representation of the AC current. The operating principle of a digital type solid-state CB is shown in Figure 11–21(d).

Advantages of circuit breakers   The advantages of circuit breakers over the other type protection devices, namely fuses, are:

- Easily resettable when the system is tripped
- More compact
- Adaptable for remote control and electrical interlock with other equipment
- Can serve as a disconnect switch (although should not be used as an operating switch)

Figure 11–21(e) and (f) shows several circuit breaker types—a single-pole molded case branch circuit breaker (MCCB) and a high-interrupting capacity (LVPCB) breaker with solid-state sensing and controls.

11.11.2 Fuses

A fuse is an electrical protective device that melts upon sensing an abnormal current and opens the circuit in which it is installed. It is a self-destructive device.

Classification of fuses   There are many types of fuses, classified into the following categories (see Figure 11–22):

1. *Voltage class.* 12, 24, 125, 250, 600, 5000, and higher voltages.
2. *Current rating.* From a fraction of an ampere to 6000 amperes.
3. *Construction.* Nonrenewable, renewable, single or dual elements, etc.
4. *Principle of operations.* Fast clearing, time delay, current limiting, etc.
5. *Short circuit interrupting capacity.* 5000 A to 200,000 A.
6. *Fusible material.* Lead, tin, copper, silver, etc.

Operating principle of fuses   The operating principle of a single-element fuse is very simple. The fusible link is made of a lead-tin-antimony eutectic alloy that has a single-temperature melting point without softening prior to reaching its melting point. The fusible element is precisely made, having bottlenecks (narrow sections) that melt upon overheating due to a higher current flow. See Figure 11–23(a).

The dual-element fuses as illustrated in Figure 11–23(b) are time-delay fuses. Each consists of two fusible elements. Under normal operating conditions, the time-delay element will break loose when the fusible material holding the "S" connector is melted. Because a heat sink is attached to the "S" connector, the melting is intentionally delayed to avoid nuisance tripping of the connected load, such as a motor load. A motor load has a high inrushing during start-up. If there is no time delay to compensate for the high inrush current, then the motor load will be cleared before it can be started. Figure 11–23(b) demonstrates the action of a dual element fuse under various conditions. The second fusible element, like the single-element fuse, melts when it senses an abnormally high short-circuit current.

Advantages of fuses   Two major advantages of fuses are that they are fast acting and self-coordinating in a properly designed distribution system. If the same class of fuse is used at all levels of protection, the lower level fuses (naturally, smaller in ampere rat-

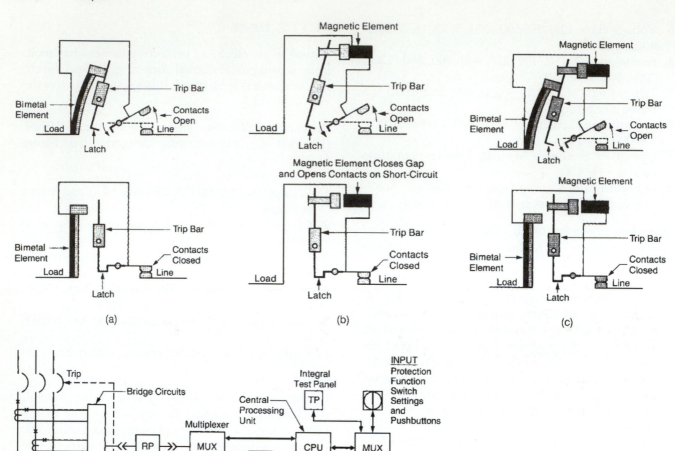

■ **FIGURE 11–21**

Operating principle of circuit breakers.

(a) The action of thermal element of the tripping mechanism is achieved through the use of bimetal heated by the load current. On a sustained overload, the bimetal will deflect, causing the operating mechanism to trip (open). The speed of action of the thermal element is inversely proportional to the magnitude of the current.

(b) The magnetic trip action is achieved through the use of an electromagnet whose winding is in series with the load current. When short circuit (fault) occurs, the magnetic field caused by the abnormally high current flow will trip (open) the contacts. The action of the magnetic trip is nearly instantaneous in order quickly to isolate electrical system and equipment from the fault.

(c) Most circuit breakers are designed to include both thermal and magnetic protection devices.

(d) An electronic type circuit breaker. A digital trip unit normally starts with analog inputs, multiplexer and analog/digital converter feeding into a microprocessor or central processing unit (CPU). Input data is continually scanned and updated with appropriate action taken. The microprocessor can be reprogrammed to fit the load changes.

(Courtesy: Cutler-Hammer/Westinghouse, Pittsburgh, PA.)

Flexible Copper Shunt Isolates
Load Terminal From Bimetal
Assembly Assuring Calibration
Accuracy Free From Load
Torque Effects

Permanently Set Steel Frame
Calibration With Bimetal
Welded Onto The Steel Frame

All Internal Ferrous Parts
Plated To Prevent Corrosion

Steel Arc Chute In Every Pole
Of Every Circuit Breaker

(e)   (f)

■ **FIGURE 11–21** (Cont'd)
Typical construction of circuit breakers.
(e) Illustrates the internal mechanism of a single-pole, molded case, thermal-magnetic
circuit breaker. Nominal rating from 15 A to 100 A, with an interrupting capacity not to
exceed 10,000 A.
(f) Illustrates a large draw-out type low voltage Power Circuit Breaker (LVPCB) Nominal rating
from 600 A to 4000 A, 480 V to 5000 V, with interrupting capacity up to 200,000 A.
(Courtesy: (e) Challenger Electrical Equipment Corp., (f) Cutler-Hammer/Westinghouse,
both Pittsburgh, PA.)

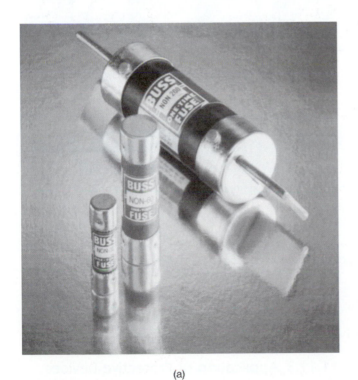

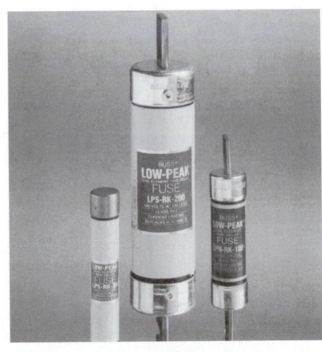

(a)   (b)

■ **FIGURE 11–22**
Several cartridge-type fuses. (a) General purpose, single or dual element type. (b) High
current limiting dual element type.

1. Cutaway view of a typical single-element fuse.

2. Under a sustained overload, a section of the link melts and an arc is established.

3. The open single-element fuse after opening a circuit overload.

4. When the fuse is subjected to a short-circuit current, several sections of the fuse link melt almost instantly.

5. The open single-element fuse after opening a shorted circuit.

(a)

Overload section consists of spring loaded "S" connector held in place by fusing alloy   Short-circuit sections are connected through the heat absorber

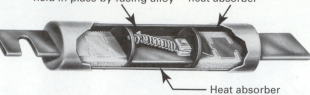

Heat absorber

1. The dual-element fuse has distinct and separate overload and short-circuit elements.

2. Under a sustained overload, the trigger spring fractures the calibrated fusing alloy and releases the connector.

3. The open dual-element fuse after opening a circuit overload.

4. A short-circuit current causes the restricted portions of the short-circuit elements to melt and causes arcing to burn back the resulting gaps until the arcs are suppressed by the arc-quenching material and increased arc resistance.

5. The open dual-element fuse after opening a short circuit.

(b)

■ **FIGURE 11–23**

(a) Sequence of operation of a single-element fuse. (b) Sequence of operation of a dual-element (time-delay) fuse. (Courtesy: Bussmann Division, Cooper Industries, St. Louis, MO.)

ings) will clear first and prevent interruption of the upper-level fuses. Figure 11–24(a) illustrates the normal trip time of a molded-case circuit breaker, which is over one cycle. Figure 11–24(b) is the clearing time of a fuse, which is normally less than half a cycle. Figure 11–24(c) illustrates the natural coordination of fuses in a distribution system in which the feeder fuse has a shorter clearing time than the main fuses.

11.11.3 Application of Protective Devices

Through improved mass production technology, small low-voltage circuit breakers for single and dou-

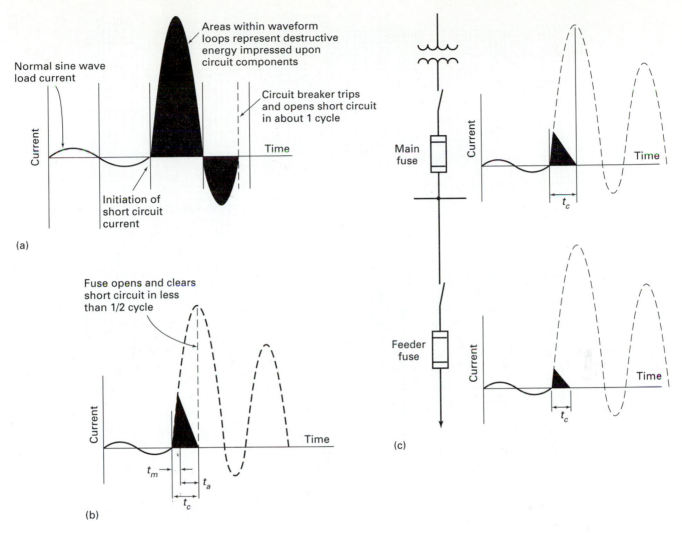

■ **FIGURE 11–24**
(a) Typical operating character of a molded-case circuit breaker, which opens a short circuit in about 1 cycle (1/60 of a second on 60 Hertz power system).
(b) Typical operating characteristic of a current-limiting fuse, which opens a short circuit in less than one-half of a cycle (t_m = melting time, t_a = arcing time, and $t + c$ = clearing time).
(c) Smaller fuses have a faster clearing time than larger fuses of the same class and thus will open first to selectively protect the distribution system from nuisance tripping.
(Courtesy: Bussmann Division, Cooper Industries, St. Louis, MO.)

ble-pole breakers up to 200 A are now as economical in cost as the corresponding fused switches. Thus, circuit breakers are universally used for lighting and receptacle panels. However, for large-ampacity loads—and especially systems with 40,000 amperes or higher of available short-circuit current—fused switches are definitely more economical. A combination of fuses and circuit breakers is often used for systems having an available short circuit in excess of 50,000 amperes.

Coordination is extremely important among different levels of protective devices, so that upon overload or fault, only the protective device next to the load or fault will open, while the others remain in service. Without proper coordination, more than one device may open, unnecessarily interrupting the services of other loads. Figure 11–25 illustrates the coordination between circuit breakers and relays. The same time-current graphs are used to coordinate fuses or fuses and circuit breakers.

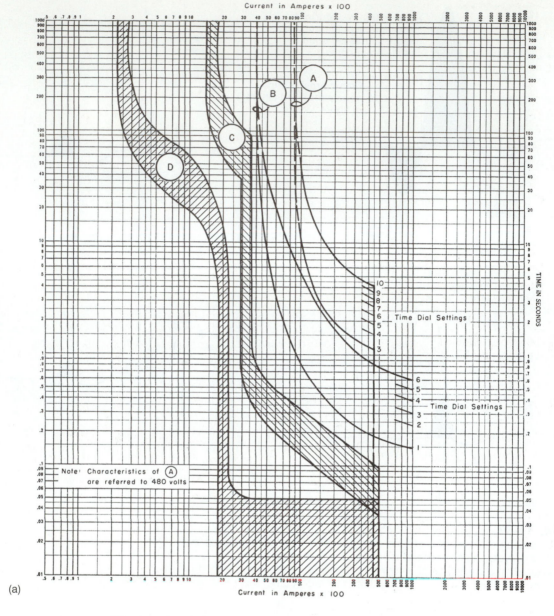

(a)

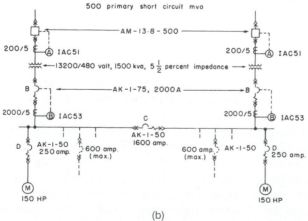

(b)

■ **FIGURE 11–25**

Coordination between protective devices. The time-current curves of Relay A and circuit breakers (B), (C) and (D) are selected so that (D) will trip first, followed by (C), (B) and (A). With proper clearance between the tripping time, system integrity and reliability are maintained and there is less chance of nuisance tripping or system interruption. (Courtesy: General Electric Company, Plainville, CT.)

QUESTIONS

11.1 What electrical distribution systems are normally used to serve residences? Small business buildings with load about 100 kVA?

11.2 What electrical distribution systems are normally selected to serve a combination of single phase and three-phase loads?

11.3 What is meant by voltage spread of any electrical distribution system? Voltage drop?

11.4 What is the recommended maximum kVA rating of a single motor-driven equipment on a 100 kVA, 120/240 volt, single-phase, three-wire system? Why is the recommended maximum considerably lower than the system rating?

11.5 What is the grounding voltage of a 120/208 volt, three-phase, four-wire system?

11.6 What is the grounding voltage of a 480 volt, three-phase, three-wire system?

11.7 Why should the interior electrical distribution system be grounded?

11.8 Must all electrical systems be grounded?

11.9 Should the interrupting capacity (IC) of the switchboard be rated higher or lower than the available short circuit current (I_s)?

11.10 Name a few NEC classified wires for building interior wiring systems and temperature ratings.

11.11 What on-floor power system would you recommend for a high-rise office building designed for flexibility and capability of using PC computers and electronic equipment?

11.12 If you wish to control a bank of lights in the room from two different locations, how many three-way and four-way switches should be used?

11.13 Name several advantages of fuses. Advantages of circuit breakers?

11.14 A 120/240 volt, single-phase system is a system with its center tap as the neutral point. (See drawing below.) If the load of circuit No. 1 connected between Line 1 and neutral is 1500 watts and that of circuit No. 2 between Line 2 and neutral is 1000 watts, what is the current in amperes carried by each circuit and the combined neutral?

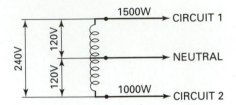

11.15 If the full load current of a transformer having 5% impedance is 500 amperes, what could be the expected short circuit current at the main switchboard?

(Note: For simplicity, a conservative approach is to assume that the utility power network serving this transformer is sufficiently large to have practically zero impedance during a fault.)

11.16 According to NEC what is the allowable (maximum) current that can be carried by a single #3/0, 90°C rated conductor in conduit? Six conductors in conduit, and what conduit size shall be used?

11.17 A 20,000 sq ft. school building has the following estimated demand loads:

- Lighting 20,000 watts 120 volt, 1-phase
- Mechanical equipment 100 HP* 480 volt, 3-phase
- Convenience power 1 W/sq ft. 120 volt, 1-phase

What power distribution system or systems would you choose?

(*1 HP = 746 watts, 0.75 kW, or approximately 1kVA @ 0.75 PF.)

12

ELECTRICAL DESIGN AND WIRING CONSIDERATIONS

ELECTRICAL SYSTEM DESIGN IS AN INTEGRAL PART OF the overall building design process. Nearly all mechanical equipment, such as air conditioning, pumps, and fans, as well as building equipment, such as elevators and appliances, is electrically powered. Even gas- or oil-fired equipment, such as boilers and heaters, require electric power. The selection of an electrical-powered system is often influenced by what mechanical system is chosen. For example, if an apartment building containing 20 or so apartments uses individual room air conditioners, the electrical distribution system would likely be a 120/240-V, single-phase system, whereas if each apartment is served by a central chiller and air-handling system, the electrical distribution system would likely be a 120/208-V, three-phase, four-wire system.

With the escalation of communication systems in buildings, the demand for electrical power has increased drastically in recent years. For example, the demand for convenience power (plug-in types of equipment) in an office building has increased severalfold in recent years. What had been acceptable standards a few years ago are now inadequate for use with data-processing and communication equipment in all types of buildings, including residences.

Generally speaking, electrical systems do not require much building space, compared with mechanical systems. However, most electrical operating devices are normally exposed in occupied spaces; therefore, their location, configuration, and aesthetics

must be precisely coordinated with architectural and interior design. More often than not, these electrical operating devices, such as switches, receptacles, controls, and alarms, are installed indisciminantly, without regard to their location, size, shape, and color.

12.1 ELECTRICAL DESIGN PROCEDURE

There are five basic steps in electrical system design:

- Analyze building needs
- Determine electrical loads
- Select electrical systems
- Coordinate with other design decisions
- Prepare electrical plans and specifications

In practice some of these steps may be temporarily bypassed in order to keep pace with the project's progress. For example, the power needs for a building are determined by its lighting, mechanical and building equipment loads from which the electrical power system is selected. However, before these steps can be completed sequentially, the utility company may wish to know the types of services in advance; the architect would like to know the electrical equipment room sizes and locations; the structural engineer would like to know the weight of all major equipment; and the HVAC engineer would like to know any system options for evaluating the cost of

HVAC equipment. Under these circumstances, reasonable assumptions must be made based on experiences or statistical data, some of which are given in this book. The assumptions will have to be assessed, modified and finalized at each phase of the design process.

12.2 ANALYSIS OF BUILDING NEEDS

The first step in electrical system design is to identify the needs for the building established by the architectural program. Following are the major factors affecting electrical systems:

1. *Occupation factors.* Type of building occupancy, number of occupants, present and future electrical appliances to be installed or anticipated in the building, etc.
2. *Cost factors.* Whether the building has an austere budget, is of average quality, or is a high-image building, etc.
3. *Architectural factors.* Size of the building, number of floors, floor-to-floor height, building footprints, elevations, etc.
4. *Building environments.* Whether the building is heated, whether it is air conditioned, whether the systems are central or unitary, etc.
5. *Illumination criteria.* The lighting level and the predominant type of light sources to be used.
6. *Other mechanical systems.* Need for electricity for cold water, hot water, sewage disposal, fire protection systems, etc.
7. *Building equipment.* Vertical transportation systems, food preparation, recreational equipment, processing equipment such as computers, other production equipment requiring electrical power, etc.
8. *Auxiliary systems.* Systems such as building management, time clock, fire alarm, telecommunication, radio and TV antenna, public address, etc. Specialty systems may be needed in hospitals, hotels, and industrial buildings. (The impact of an auxiliary system on a building power distribution system is negligible.)

12.3 DETERMINATION OF ELECTRICAL LOADS

The use of electrical power in buildings—particularly office buildings—has risen drastically. For example, the demand for convenience power for portable or plug-in equipment in office buildings has increased from an average of 2 to 3 watts/sq ft (on net floor area basis) in the 1980s to an average of 3 to 5 watts/sq ft since 1990 for convenience power in intelligent-type office buildings, which have a higher concentration of personal computers, telecommunication equipment, and fax and copy machines. Similarly, the number of electrical appliances at home has also increased steadily. Personal computers, microwave ovens, food processors, multiple units of television sets, and videotape players are just a few appliances that are becoming a standard of living rather than a luxury.

The increase in the electrical load of a building also has a direct effect on the building's air-conditioning (cooling) system. Heat generated from electrical loads in an environmentally controlled space must be removed through ventilation or the air-conditioning system in order for the space to be maintained at a comfortable level. In general, all electrical power input (watts) will become internal heat gain (Btuh) for the building. Following the fundamental law of energy conversion, 1 watt of electrical power used for 1 hour will generate 3.4 Btuh of heat. It must be understood, however, that the total heat gain of all equipment should be based on the net *demand load* rather than the gross *connected load* of all loads. The difference between demand and connected loads is explained in later sections.

Electrical loads for buildings can be analyzed according to a number of different categories.

12.3.1 Lighting

Lighting design is usually the coordinated effort of the architect, interior designer, lighting designer, and electrical engineer. Lighting accounts for one of the largest electrical loads in most buildings. In general, lighting fixtures are designed for 120-volt, single-phase power. However, they may also be designed for use on 208-, 240-, and 277-volt, single-phase power systems.

A continuous improvement in light source technology has increased the efficiency of converting electrical energy to lighting energy, which in turn has reduced the need for electrical power for lighting in buildings. While it was common to require 3 to 5 watts/sq ft for lighting in office buildings in the 1980s, only about 2 watts/sq ft is needed in the early 1990s, and the downward trend is continuing.

For electrical power–planning purposes, the average power required by various types of indoor areas or activities is given in Table 12–1. The figures shown are only approximations and should not be used for lighting design purposes. (See Chapter 17 for lighting design calculations.)

TABLE 12–1

Unit (Lighting) Power Density (UPD)—W/ft^2 (W/m^2) for common and specific areas

Common Spaces for All Building Occupancies	UPD	Building Occupancy or Specific Space Functions	UPD
Office-enclosed	1.5 (16.1)	Exhibit hall	1.0 (10.8)
Office-open plan	1.3 (14.0)	Religious building	2.0 (21.6)
Conference/multipurpose	1.0 (10.8)	Museum (general exhibition)	1.0 (10.8)
Classroom	1.5 (16.1)	Professional sports (court area only)	5.3 (57.0)
Library		Amateur sports (court area only)	3.9 (42.0)
■ file/cataloging	1.4 (15.1)	Professional sports televised (court + perimeter	3.8 (40.9)
■ stacks	1.8 (19.4)	playing area)	
■ reading area	1.7 (18.3)	Gymnasium (playing area)	1.9 (20.4)
Lobby (except theaters)	1.8 (19.4)	Auditorium (seating area)	1.6 (17.2)
Lobby (auditorium, theater)	0.8 (8.6)	Airport (concourse)	0.7 (7.5)
Atrium (first 3 floors)	1.4 (15.1)	Airport (ticket counter)	1.5 (16.1)
Atrium (each addl.' floor)	0.2 (2.2)	Hospital	
Food preparation	1.2 (12.9)	■ surgery/emergency	3.2 (34.0)
Restroom	0.8 (8.6)	■ nurse station	2.1 (22.6)
Corridor/transition spaces	0.6 (6.5)	■ corridor	1.0 (10.8)
Stairs-active	0.8 (8.6)	■ examination/pharmacy	2.7 (29.1)
Stairs-inactive	0.6 (6.5)	Hotel/motel/multi-family/dorm	1.2 (12.9)
Storage-active	1.2 (12.9)	Bank (banking area)	2.3 (24.8)
Storage-inactive	0.3 (3.2)	Restaurant	1.4 (15.1)
Warehouse (fine material)	1.7 (18.3)	Retail (sales area)	
Warehouse (medium/bulky)	1.2 (12.9)	■ Department store	3.3 (35.5)
Garage (parking area)	0.3 (3.2)	■ Fine merchandise	5.1 (54.9)
Elec./mech. space	0.8 (8.6)	■ Mall concourse	1.7 (18.3)
		■ Mass merchandising	1.8 (19.4)
		■ Personal services	2.0 (21.6)
		■ Supermarket	2.3 (24.8)

Note: The UPD values given in this table are for electrical power load planning purpose only and should not be used as a basis for lighting design. Refer to Chapter 16 for lighting calculation procedures. Data is condensed from the 1996 revision of ASHRAE/IESNA Standard 90.1–1989R.

12.3.2 Mechanical Equipment

Electrical power required for mechanical equipment varies widely with the building, type of climate, architectural design, size of the building, type of mechanical systems, and intended method of operation of these systems. For planning purposes, preliminary data on power are obtained from the design mechanical engineers and are updated periodically as the design progresses.

The mechanical system includes the HVAC, plumbing, and fire protection systems. Equipment, such as chillers, boilers, pumps, and fans, usually requires a large power capacity and is more economically designed for higher voltages, such as 208-, 240-, and 480-volt, three-phase power. However, residential equipment is normally designed for 120- or 240-volt, single-phase power.

12.3.3 Building Equipment

Building equipment includes vertical transportation equipment (elevators and escalators), food service equipment, and household, recreational, and miscel-laneous building operational equipment. The power required for this type of equipment varies widely in capacity and operating characteristics. The electrical power required for building equipment is usually gathered by the architect, the user, or the various specialty consultants.

12.3.4 Auxiliary Systems

Normally, auxiliary systems do not require a large power capacity; thus, they are usually designed for 120- or 240-volt, single phase power. Depending on the type of occupancy—residential, commercial, institutional, or industrial—each building may require one of more of the following systems:

- Building management systems
- Security system
- Time clock systems
- Fire alarm systems
- Telecommunication systems
- Radio and TV antenna system
- Specialty systems
- Public address systems

For power-planning purposes, usually one or two 20-A, single-phase circuits are sufficient for each auxiliary system in most buildings.

12.3.5 Convenience Power

Convenience power is power provided for plug-in types of equipment such as household appliances, personal computers, office equipment, laboratory instruments, service equipment, portable lights, and audio and video equipment. With the proliferation of electrical appliances used in homes, offices, and schools, the demand for on-floor power for plug-in equipment has increased drastically.

The total appliance load for typical offices and homes has already exceeded that required for lighting and becomes a dominant load of the building air-conditioning systems. The design of convenience power circuits must, therefore, be carefully analyzed with regard to capacity and appropriate location. Table 12–2 provides typical power rating of household appliances. Table 12-3 provides the suggested convenience power to be allowed for various interior spaces or building occupancies.

12.3.6 Connected and Demand Loads

Connected load   The connected load of an electrical power system is the algebraic sum of all electrical loads connected to the system. It does not take into account how and when these loads are being used.

Demand load   The demand load of an electrical power system indicates the net load that would likely

TABLE 12–2

Typical power rating of household appliances

Appliances	Range of Power (VA)
Window Air Conditioner	
■ ½ ton, 115 V	700–800
■ ¾ ton, 125V	1100–1200
■ 1 ton, 230V	1400–1600
■ 2 ton, 230V	2500–3000
Refrigerators, 115V	300–500
Freezers, 115V	300–800
Washing Machines, 115V	800–1200
Dryers, 115/230V	3000–5000
Gas Dryers, 115V	200–300
Water Heaters, 115/230V	3000–6000
Toaster, 115V	500–1000
Oven, 115/230V	3000–5000
Microwave Oven, 115V	500–1000
Combination Oven, 115/230V	1000–5000
Television, 115V	300–1000

TABLE 12–3

Typical power allowance for convenience (portable or fixed loads)

Type of Space	Power Allowance*
Offices	3 to 5 w/sf
Classrooms—general (no PCs)	1 to 2 w/sf
Classrooms—specific (with PCs)	Per load
Meeting Rooms	3 to 5 w/sf
Residential Spaces:	
■ Kitchen	(2) to (6) 20A circuit
■ Dining and Family	(2) 20A circuits
■ Bedrooms, not Air Conditioned	(1) 20A/room
■ Bedrooms, Central Conditioned	(1) 20A/2 rooms
■ Laundry	(2) 20A circuits
■ Exterior	(1) 20A/exposure

*In net occupied spaces.

be used at the same time of each load group. When all connected loads are used at the same time, the demand load is equal to the connected load. However, in most buildings, the demand load is always lower than the connected load. For example, to estimate the load of plug-in receptacles (convenience outlets), NEC recommends using 1.5A or 180VA for each duplex outlet. If the building contains 1000 duplex outlets, then the connected load is 180,000VA, but in reality, the demand may be at most 20 percent of the connected load or 54,000VA. In this example, the demand factor is 30 percent. The simple relationship between connected load (CL), demand load (DL), demand factor (DF), and gross demand load (GDL) is expressed in equation 12–1 and 12–2.

$$DL = \frac{CL}{DF} \qquad (12\text{–}1)$$

$$\text{and } GDL = \Sigma\, DL \qquad (12\text{–}2)$$

Diversity coefficient   The diversity coefficient accounts for the diversity of demand between different groups of loads. For example, when there is a very high appliance and lighting load in a space, the heat generated from these loads may cause the heating system to be turned off and reduce the demand load of the system. Another typical example is the diversification between heating and cooling systems when only one system is in use at a time.

The diversity coefficient is also called the diversity factor (DF). However, to avoid possible confusion with demand factor (also DF) in nomenclature, the author prefers to use the term diversity coefficient (DC), and it is used as a divisor rather than a multiplier. The diversity coefficient is time-consuming to calculate. For the purpose of this book, DC = 1.0 is used for systems which do not have obvious load diversification, to DC = 1.2 for large systems or systems with diversified load groups. When the diversification of loads is greater than the suggested range of diversity coefficients, a detailed analysis of connected loads under all operating conditions will have to be made on an hour-by-hour or minute-by-minute basis. Often, diversity coefficient is neglected by designers, resulting in oversized systems. The relationship between the net demand load (NDL), gross demand load (GDL), and the diversity coefficient (DC) is expressed in equation 12–3.

$$NDL = \frac{GDL}{DC} \qquad (12\text{–}3)$$

Example   A building electrical system is calculated to have the following connected loads by load groups. The estimated demand factors are shown in the table below. What is the net demand load of the system, assuming the diversity coefficient is 1.1?

Load Group	Connected Load (kW)	DF	Demand load (kW)
Lighting	125	0.9	112.5
Receptacles	85	0.2	42.5
Mechanical equipment	200	0.8	160.0
Building equipment	150	0.6	90.0
Connected load (CL)	560		
Gross demand load (GDL)			405 kW
Net demand load (NDL), (405 / 1.1)			368 kW
			(use 370)

12.4 SELECTION OF ELECTRICAL SYSTEMS

12.4.1 Three-phase vs. Single-phase Systems

Based on the load analysis, electrical power systems are selected to complement the load characteristics. For example, if the loads in the building are predominantly 120 volt, single phase, then the system should be either a 120/240-volt, single-phase, three-phase, three-wire or a 120/208-volt, three-phase, four-wire system. One of the advantages of a three-phase, four-wire system not mentioned earlier is its inherent economy. It allows one common neutral wire to serve up to three single-phase loads. However, with the increased use of inductive loads, such as computers in offices, there is a tendency to overheat the common

neutral wire due to the third harmonics. In such a case, individual neutral wires for each circuit are recommended.

12.4.2 Common Voltage Ratings

The electrical power system for a large building could be a combination of single-phase and three-phase, low-voltage systems supplied by one or more high-voltage systems. Figure 12–1 illustrates a simplified one-line diagram of a larger power distribution system in which the primary power, 13,800-volt, three-phase three-wire, passes through step-down transformers to three different low-voltage systems, namely, 120/208-volt, three-phase, four-wire, 277/480-volt, three-phase, four-wire, and 480-volt, three-phase, three-wire systems.

Naturally, for smaller buildings, a single power system will suffice. However, in some applications, it is desirable to separate the large power load from the lighting and appliance loads to minimize voltage fluctuation of the system due to the on-off nature of the larger loads. The design electrical engineer must select the system early and inform the architect and the mechanical engineers of any options, so that equipment selected by them will match properly. In general, the following equipment greatly affects the selection of the system during the preliminary design period.

Elevators   Elevators usually range from 15 to 200 horsepower, depending on their speed and load capacity. They are usually equipped with three-phase motors whose voltage is compatible with that of the building power system.

Food service equipment

Food service equipment of 30 kW or less could be either single phase or three phase; those of higher capacity are usually three phase. In all cases, the equipment voltage must match that of the building system. For example, if the building power system is 208 volts and the equipment is rated for 240 volts, then the equipment output will likely be reduced by 15 percent in capacity unless a local transformer is installed. The reverse is true when the system voltage is higher than the equipment voltage.

Mechanical equipment

While large pieces of mechanical equipment are always designed for three-phase power, smaller equipment, such as fans, coils, and unitary equipment, may be made for either single-phase or three-phase

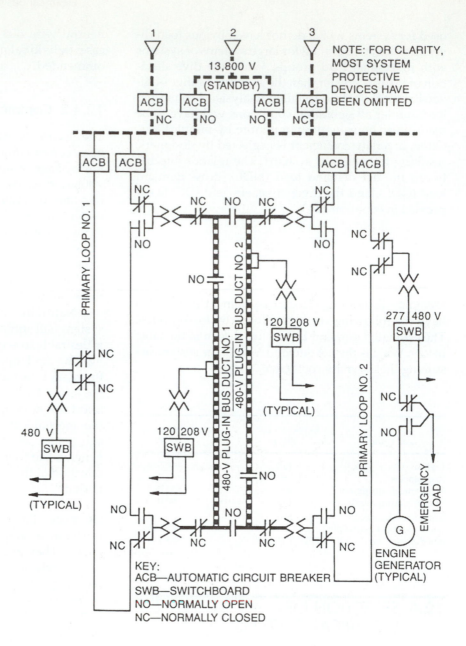

NOTE: FOR CLARITY, MOST SYSTEM PROTECTIVE DEVICES HAVE BEEN OMITTED

■ FIGURE 12–1
Simplified one-line diagram of a large power distribution system.

KEY:
ACB—AUTOMATIC CIRCUIT BREAKER
SWB—SWITCHBOARD
NO—NORMALLY OPEN
NC—NORMALLY CLOSED

power. The selection of such equipment must be a joint decision of the mechanical and electrical engineers during the preliminary design phase.

Household appliances

Although there are other options, most household appliances are made for a 120/240-volt, single-phase system. If a 120/208-volt, three-phase, four-wire system is selected for a residence, then either specially designed appliances must be provided or local load-size transformers must be installed at the equipment to boost the system voltage to match that of the equipment.

12.5 COORDINATION WITH OTHER DESIGN DECISIONS

12.5.1 Interfacing of Building Systems

The simplified one-line diagram shown in Fig. 12–1 clearly illustrates the complex interfacing required of all building systems. For example, in a high-rise apartment building consisting of 100 apartments served by a central cooling and heating system, the main electrical power system should be a 480-volt, three-phase, three-wire system for the central equipment and a 120/208-volt, three-phase, four-wire system for lighting and appliance loads. On the other hand, if each apartment has its own electrical meter for all connected loads—including individual unitary heating and cooling units—then the electrical power system would likely be multiples of small 120/240-volt systems, one for each apartment—even though the building is large. Usually, the selection of the mechanical system affects the selection of the electrical system.

12.5.2 Space Planning

Space planning is not an exact science, yet it is so vital to the orderly progression of the building design process. Once the electrical or mechanical space is allocated within the building, it is difficult to change its size or location, as doing so will undoubtedly affect the work in progress of many other disciplines or systems.

Unlike the design of a bathroom, where the size of plumbing fixtures is fairly standardized, the size of electrical or mechanical equipment varies with the capacity, features, and manufacturers of the equipment. The educated guesses and experience of the designers play important roles in this connection. A more detailed discussion and guidelines for electrical and mechanical equipment space planning will be presented in the next volume of this book. The following are general guidelines.

Accessibility

All equipment and devices must be accessible for inspection, service, and replacement.

Safety

Electrical equipment is hazardous. Sufficient clearance must be provided on all sides that require access. As a rule, the NEC requires the following minimum clearance in front of all accessible sides of the equipment. The larger values are for equipment containing exposed live components:

- Up to 150 volts: 3 ft
- 151 to 600 volts: 3 to 4 ft
- 601 to 2500 volts: 3 to 5 ft
- 2501 to 9000 volts: 4 to 6 ft

Common access spaces

To minimize the loss of useful building space due to access requirements, double-loaded, center-aisle design is preferred. The common aisle space between two lineups of motor control centers shown in Figure 11–8(b) of Chapter 11 is a typical example. As pas-

sageways, corridors are frequently utilized for wall-mounted switchboards and panelboards. The NEC describes other safety requirements that are referred to in design of building electrical systems.

Integration of electrical and structural elements

The electrification of structural floors is a popular method for distributing on-floor power. The system may be an underfloor duct system (Figure 11–17) or a cellular floor system (Figure 11–18). It is most applicable for offices, especially with an open landscape design. Although the initial cost of installation of underfloor distribution systems is greater than that of other systems, statistics have proven their overall economy with short payback periods.

12.6 DRAWING UP OF ELECTRICAL PLANS AND SPECIFICATIONS

12.6.1 Graphic Symbols

Graphic symbols are used to illustrate the various aspects of electrical design, including the indication of equipment, devices, wiring, and raceways. Without these symbols, electrical design would be difficult to illustrate. Standardized symbols are intended for use as a common language for communication. Unfortunately, current recognized standards, such as those published by the Institute of Electrical and Electronic Engineers (IEEE) and the American National Standards Institute (ANSI), have not kept up with the growth of electric systems in building applications. As a result, many nonstandard or custom symbols have been developed. A symbol schedule must, therefore, be included in each set of plans. A schedule of the most commonly used graphic symbols is included in Figure 12–2. These symbols are used in this text.

12.6.2 Electrical Plans

Electrical plans usually consist of the following:

1. *Floor plans.* In a floor plan, electrical devices and equipment are superimposed on an architectural background. For clarity, the architectural background may be screened or have a lighter line weight so that the electrical features will stand out. Electrical plans for major buildings are usually further divided into lighting, power, and auxiliary system plans to show the wiring design as clearly as possible. Figure 12–3(a) is a perspective of a

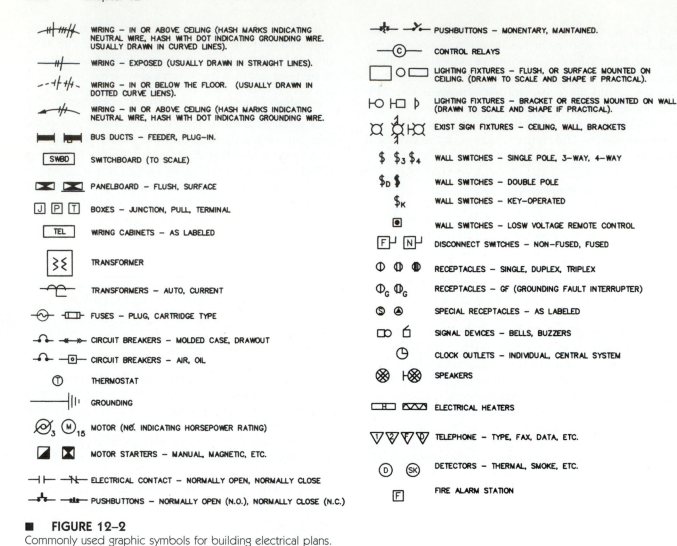

■ FIGURE 12–2

Commonly used graphic symbols for building electrical plans.

room with a light controlled by one switch, and Figure 12–3(b) is the electrical floor plan of the room.

2. *Schematic diagram.* Also called an elementary diagram, the schematic diagram illustrates the circuitry of a system and is the basis for understanding the functions of an electrical system. (Figure 12–3(c) is the elementary diagram of the lighting installation and its controls.

3. *Connection diagram.* Also called a wiring diagram, the connection diagram, illustrated in Figure 12–3(d), provides instructions regarding the connections between the wiring terminals of various devices and equipment. The connection diagram is not intended to illustrate the operating principles of the circuitry; rather, it is used for installation by electricians.

4. *One-line diagram.* This is a simplified system diagram that shows the principal relationships among major equipment. Figure 12–1 is a typical one-line diagram of a power distribution system.

5. *Riser Diagram.* This diagram expresses the physical relationship between different pieces of equipment or devices and is frequently used to show the vertical relationship between floors. (See Figure 12–4.)

12.6.3 Specifications

Specifications are the written portion of a design document. They are used to supplement the drawings in the document. Drawings and specifications are used jointly, and what is called for in one is considered to be called for in both.

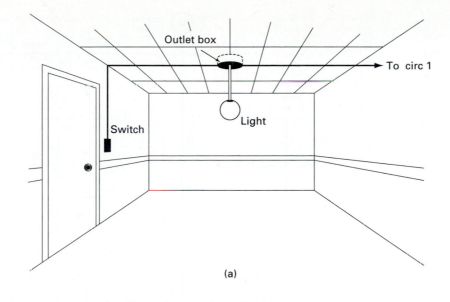

Outlet box

To circ 1

Switch

Light

(a)

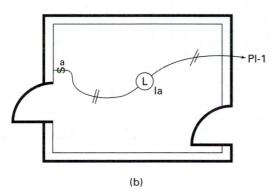

a

L la

PI-1

(b)

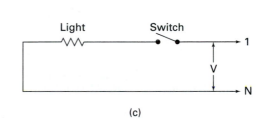

Light   Switch

V

1

N

(c)

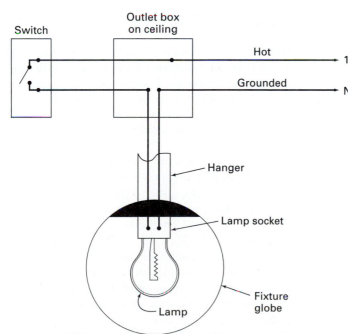

Switch

Outlet box
on ceiling

Hot

Grounded

Hanger

Lamp socket

Lamp

Fixture
globe

(d)

■ **FIGURE 12–3**

Various forms of electrical plans and diagrams of a simple lighting installation. (a) A perspective of a room with one light controlled by one switch. (b) Electrical floor plan of the room in (a). (c) Schematic diagram of the circuitry in the room. (d) Connection diagram of the circuitry.

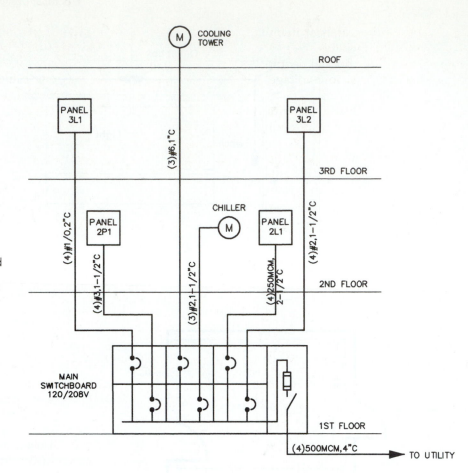

■ FIGURE 12–4
Typical riser diagram of a power distribution system, showing feeders between main switchboard, panels, and major motor loads.

12.7 NATIONAL ELECTRICAL CODE

The NEC is the governing code for electrical work in buildings and facilities in the United States. It is also widely used in many developed and developing countries throughout the world. Although there are variations in the electrical codes of different countries due to local customs, manufactured products, and differences in standards of measurements, the underlying principle of all electrical codes is the same. They provide a minimum standard for safety.

The design, installation, and maintenance of an electrical system in compliance with the NEC will result in an installation that is essentially efficient, convenient, and safe. Building officials usually inspect electrical systems prior to the issuance of occupancy permits. Thus, the NEC has been used as a design reference, as well as a resource book, by all disciplines involved in the design, construction, and operation of electrical systems.

Hazards often occur because the wiring system is overloaded beyond the NEC recommended values, commonly in older systems designed when the use of electricity was only a fraction of what it is today. Elec-

trical system design should allow for growth; 20 to 30 percent spare capacity is the norm.

12.7.1 Scope and Arrangement of NEC

The NEC covers methods of connection to electrical power and the installation of electrical conductors and equipment in public and private facilities. The code is divided into nine chapters, as follows:

- *Chapter 1.* General guidelines for compliance with the code.
- *Chapter 2.* Wiring and protection: methods for sizing feeders and branch circuits, overcurrent protection devices, and grounding requirements.
- *Chapter 3.* Wiring methods and materials: types of conductors and methods of installing conductors.
- *Chapter 4.* Equipment for general use: construction requirements for lighting fixtures, appliances, electrical heaters, motors, generators, and other equipment using or producing electrical power.
- *Chapter 5.* Special occupancies: requirements for hazardous locations, places of assembly (such as

auditoriums, arenas), manufacturing, health care and other facilities.

- *Chapter 6.* Special equipment: requirements for elevators, low-voltage systems, swimming pools, etc.
- *Chapter 7.* Special conditions: requirements of emergency systems, standby power, and optical fiber cables.
- *Chapter 8.* Communication systems: requirements for radio and TV antennas and equipment.
- *Chapter 9.* Tables and examples. Data pertaining to conductors and raceways, and examples for calculating loads and capacities for dwellings.

12.7.2 Essentials of NEC

The NEC contains hundreds of pages of rules and technical data. For each rule, there are numerous exceptions to cover various applications. Accordingly, any attempt to condense the code for simplification purposes could lead to misunderstanding and even violations of the code rules. Nonetheless, it is of great value to designers and users alike to understand the essentials of the NEC in order to achieve more useful and safer electrical systems. In general, the NEC rules may be encapsulated as follows:

1. Electrical power systems shall have a minimum capacity for various types of building occupancies. Naturally, this requirement varies for different countries, depending on local needs. For example, the electrical service for a single-family dwelling in the United States should not be less than 60A for a 120/240-volt, three-wire service and not less than 100 A for initial loads of 10 kVA or more. This requirement is less stringent in most other countries.

2. Alternating current systems of 50 to 600 volts shall be grounded wherever possible such that the maximum voltage to ground on the ungrounded conductors does not exceed 150 volts. Exceptions to this general rule include special systems in health care and industrial facilities.

3. Only wires listed in the code shall be used in buildings. Application of the wires should not exceed the temperature and voltage ratings of their insulation for approved dry or wet locations.

4. The feeder capacity and the number of branch circuits shall not be lower than either the actual demand loads or the calculated demand loads given under Articles 200, 210, 215, and 220, whichever is greater.

5. There are numerous code-approved insulated wires and cables. Most of these shall be installed in protective raceways, such as conduits, wireways, or cable trays. For practical purposes, all commercial, institutional, and industrial buildings are designed for using wires in raceways, primarily in EMT or rigid conduits (heavy walls).

6. The allowable ampacities of insulated conductors are given in NEC Article 310, including the derating factors for ambient temperatures and when there are more than three conductors in the raceway. Unless otherwise stated, all references are based on copper conductors.

7. Nonmetallic sheathed cables, types NM and NMC, may be used without raceways in one-family or multiple-family dwellings and other structures, except when any dwelling or structure exceeds three floors above grade.

8. All connections or splices of wires and cables must be made within code-approved boxes that are accessible for inspection and service.

9. All wires and equipment should be protected from abnormal situations by the proper installation of overcurrent, overvoltage, and overload protection devices, such as relays, fuses, or circuit breakers.

10. Feeders, branch circuits, and equipment shall be provided with proper means of disconnection or isolation switches to disengage them from the electrical system for service and repair. The means of disconnection shall be readily accessible and in sight of the equipment being disconnected.

11. Motors and electrical equipment shall be provided with means of disconnection, branch circuit protection, controllers, and overload protection. The means of disconnection shall be in sight of the equipment or shall be capable of being locked in an open position for the safety of service personnel.

12. To avoid the buildup of heat and to facilitate pulling wires into the conduit, wires shall occupy only a small portion of the conduit. In general, the overall cross-sectional area of all conductors shall not exceed 40 percent of the cross-sectional area of the conduit when there are three or more conductors in the conduit.

12.8 BRANCH CIRCUITS

According to the NEC, a branch circuit is that portion of a wiring system extending beyond the final overcurrent device protecting the circuit and the outlet or the load. A branch circuit may be classified as an appliance or as a general-purpose, individual, multiwire, or motor circuit. Every branch circuit shall be protected from overcurrent by overcurrent devices, which shall be located at the power supply end of the circuit.

There is no limit on the size of motor branch circuits, provided that the conductors are properly protected in accordance with NEC Article 430. In general, overcurrent devices with higher ampacity ratings are allowed in motor branch circuits than are allowed in general-purpose branch circuits, so as to accommodate the extremely high current that builds up during the start-up of a motor due to its inductive reactance or lower power factor. The ampacity ratings of general-purpose branch circuits are limited to 20A for lighting and 120V appliances, 30A for lighting with heavy-duty lamp holders and high wattage appliances, and 50A for cooking and laundry equipment in dwellings.

12.8.1 Branch Circuiting Design

Branch circuiting is the basis for the wiring of individual loads, and in the final design of the power distribution system, branch circuiting work is mostly tedious and time consuming. On the other hand, such work can be greatly simplified if the following rules are observed:

1. No wire smaller than no. 14 AWG shall be used for dwelling applications.
2. No wire smaller than no. 12 shall be used for commercial, industrial, and institutional applications.
3. For two-wire circuits, the continuous load per circuit should be limited to 1200 watts for 15A circuits and 1500 watts for 20A circuits.
4. For heavy-duty lamp circuits, the load per circuit shall not exceed 2000 watts for no. 10 wires, 2500 watts for no. 8, and 3000 watts for no. 6.
5. As a rule, NEC code requires that the rating of any one portable appliance shall not exceed 80 percent of the branch circuit rating; the connected load of inductive lighting load, such as fluorescent or HID fixtures, shall not exceed 70 percent of the branch circuit rating; and the total load of fixed wired appliances shall not exceed 50 percent of the branch circuit rating if lighting or portable appliances are also connected to the same circuit.
6. Where the run from a panelboard to the first outlet of a lighting branch circuit exceeds 75 ft, the size of the wire shall be at least one size larger

than that determined by any of the preceding considerations.

7. No run longer than 100 ft between a panelboard and the first outlet of a lighting branch circuit shall be made, unless the intended load is so small that the voltage drop can be restricted to two percent between the panelboard and any outlet on that circuit.
8. Where the run from a panelboard to the first outlet of a convenience outlet circuit exceeds 100 ft, no. 10 wires shall be used.
9. No convenience outlet shall be supplied by the same branch circuit that supplies ceiling or show-window lighting outlets.
10. The maximum number of convenience outlets included in one circuit shall be based on the following:
    - Outlets supplying specific appliances and other loads based on the ampere rating of appliances
    - Outlets supplying heavy-duty lamp holders—use 5 A receptacle
    - Other outlets (general purpose)—use $1\frac{1}{2}$ A/ receptacle
11. Ten to 15 percent of spare circuits shall be provided in each general-purpose panel, in addition to spare circuits for known future loads.
12. Motor branch circuits shall comply with NEC Article 430. In general, branch circuit conductors shall be rated as follows:
    - Single-motor circuit: at least 125 percent of full-load rating of motor
    - Multiple-motor circuit: at least 125 percent of the highest rated motor, plus the full-load rating of all other motors

12.8.2 Branch Circuiting Layout

Once the branch circuit design is formulated, it is shown on the electrical floor plans indicating the circuiting number, wire size (other than the basic no. 12 or no. 14 AWG), home runs, interconnection between outlets or boxes, points of control, and special notation.

Branch circuiting with NM cables

NM cables are used mostly for dwellings and building interior finishes. NM cables come with two conductors (black and white) or three conductors (black, white, and red). Each cable contains a bare grounding conductor within the nonmetallic protective jacket. Since NM cables are limited in their variety, the indication of complete branch circuit wiring on floor plans is quite straightforward and, in fact, could be eliminated in plans of simple dwellings. A typical electrical floor plan with NM wiring is presented in the Questions at the end of the chapter. In general:

- Use minimum no. 14 AWG for 15-A circuits and no. 12 for 20-A circuits.
- Use two-wire cable for single-circuit, single-switch, and 120-volt receptacles.
- Use three-wire cable for two circuits, three-way switches, and 240-volt appliances.
- Use the white wire for the grounded side of the power system only.
- Use the bare wire for grounding the conductive housing of the lighting fixture, wiring devices, equipment, and boxes.

Branch circuiting with wiring in conduit   Laying out branch circuits in conduits requires careful planning. When properly done, it may reduce complications in the field with regard to rerouting or enlarging some incorrectly sized conduits, and the cost of construction. The goal in laying out conduit is to minimize the length and size of the conduits. Following are some simple rules:

- EMT shall be used primarily for dry locations, although it may be used in wet locations with corrosion-resistant supporting components.
- EMT is limited to 4", and no conduit smaller than ½" shall be used.
- All wiring splices shall be made at outlet boxes, gutters of panelboards, or other junction boxes. All boxes must be accessible.
- Conductors for different voltage systems may be installed in the same conduit, provided that all conductors are rated for the highest voltage enclosed.
- In general, conductors for signals and for radio and TV systems may not occupy the same conduit or enclosure that lighting and power systems occupy, except when otherwise permitted under NEC Articles 800 through 820.
- Telephone wires shall use separate raceways.
- Conductors of a three-phase circuit shall be installed in the same conduit to reduce the induction effect (induced heat). The neutral conductor of a circuit shall also be installed in the same conduit whenever possible.
- No conduit shall have more than four equivalent 90-degree bends between boxes, including the bends at the boxes.
- It is good practice to feed the branch circuit first at the lighting fixture outlet box and then at the switch boxes. This will help in future alterations and in troubleshooting.
- Separate convenience outlet and appliance circuits from lighting circuits, except for incidental loads.
- Avoid using conduits larger than 1" for general-purpose branch circuits.
- Avoid having more than nine conductors in a single conduit, other than with low-voltage communication-type systems.

A typical wiring plan for branch circuiting in conduit is illustrated in Figures 12–5 and 12–6.

12.9 TABLES AND SCHEDULES

In the design of any building systems or subsystems, not all pertinent information can be expressed by drawings. Tabulated data and combination of graphics with tabulated data are effective ways to provide the design information of a project. In electrical system design, the following schedules are usually required:

Panelboard schedules   A panelboard schedule is the most important schedule of power system design. It serves to summarize the individual branch circuit loads from which the overall load system is calculated. A complete panelboard schedule used for large projects is shown in Figure 12–7. It includes four subschedules:

- Branch circuit connections. Including the circuit number, load (VA, watts, kW or kVA), phase to which the load is to be connected (x or y for single phase system or A, B, C for three-phase system), type of overcurrent protection devices (OCP) and current rating (A), number of poles (P).
- Demand load calculations. Based on the load types, the estimated demand factor (DF) and the estimated diversity coefficient (DC).
- Power supply data. Describing how the power is fed.
- Branch data number, poles and rating of poles.
- Specifications. The capacity rating, short-circuit rating, OCP ratings, physical dimensions and construction features.

The panelboard schedule can also be used as a switchboard schedule or a combined schedule when switchboard and panelboard are combined into one for small projects.

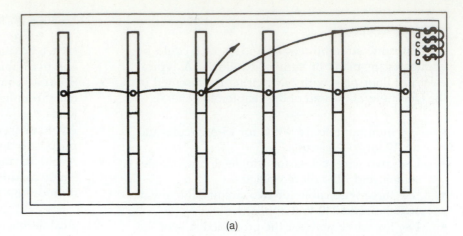

(a)

■   FIGURE 12–5

(a) The room is an architectural design studio with wall-to-wall windows on the south exposure. It is desired that these lights be controlled by four switches and the first two fixtures on each row be switched separately from the remaining fixtures in that row. This is to utilize the available daylight near the south side of the room.

(b) Wiring in conduits is shown in the wiring plan with the function of each wire.

(c) Room A has four fluorescent fixtures. These fixtures shall be connected to circuit #1. The upper two fixtures shall be controlled by switch *a* and the lower two by switch *b*. Room B has four incandescent fixtures. These fixtures shall be connected to circuit #3 and controlled by switch *c*. One switch shall be located at each door location.

(d) Branch circuit #1 and 3 shall share the same neutral on a 120/208 volt, 3-phase, 4-wire system. The number of wires in a conduit is shown by hash marks across the conduit. The long hashes represent the neutral wire and the short hashes represent the circuit or switch wires. For learning purposes, the function of each wire is identified alongside the marks. In practice, these identifications are omitted.

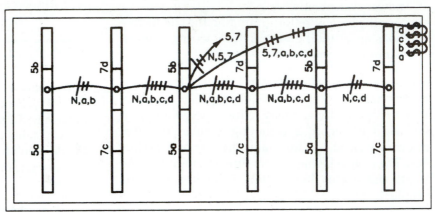

(b)

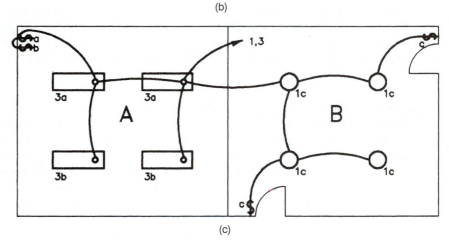

(c)

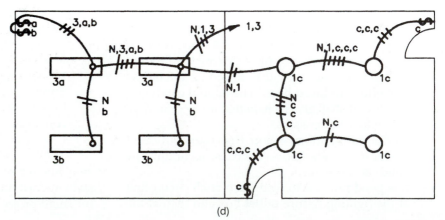

(d)

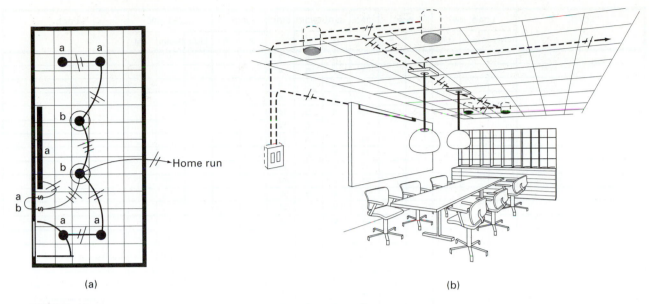

(a)   (b)

■   **FIGURE 12–6**
(a) An electrical branch circuit wiring plan showing the conduit layout, switching, and the number of wires in conduits.
(b) The conduit and wiring of the branch circuit plan are superimposed on an architectural perspective view of the room. The perspective clearly illustrates how conduits are running from outlet to outlet.

Feeder schedule   A feeder schedule usually contains the feeder designation, type of cable insulation, size in AWG or MCM, method of installation (with or without raceways), overcurrent protection, connected load in amperes, and demand load in amperes.

Mechanical equipment data schedule   Most modern buildings contain electrically operated equipment, such as HVAC equipment, plumbing and fire protection equipment, elevators, food preparation equipment, etc. This equipment is usually motor driven which constitutes inductive loads. Wiring design for inductive loads differ from the lighting loads in that they have large in-rush (starting) current and thereby require different design criteria and protection. Figure 12–8 shows a typical mechanical equipment data schedule. It consists of two parts: basic description of the load and the wiring data.

Lighting fixture schedule   Lighting could consist of up to 50 percent of the total connected load in a building. Figure 12–9 list a typical lighting fixture schedule.

Control and automation schedules   A control and automation schedule usually contains the name of equipment to be controlled and the required control functions in a matrix format. The schedule is to be used in conjunction with the technical specifications and control diagrams. The schedule normally contains the following:

- Operating mode: manual, programmed or automatic start or off for equipment.
- Control functions and indications: temperatures, pressure or flow for various media, such as air, water, or fluid, and for equipment, such as fans, pumps, and valves.
- Control point adjustments of the media and equipment.
- Interfacing of building systems between HVAC, fire protection, plumbing, security, transportation, lighting, and standby power systems.

12.10 DESIGN PROBLEM

Design the electrical system for a small apartment with load data given on the next page and the equipment layout shown in Figure 12–10.

A wiring layout is a major part of an electrical system design. It provides the information for the electrician to rough in all concealed wires, boxes and raceways before the building's interior finishes are installed. It consists of instructions for lighting, receptacles, switches and any equipment which requires electrical power or controls. The procedure for wir-

LOAD	LOAD, KW			OCP		CIRC	SEQUENCE		CIRC	OCP		LOAD, KW			LOAD
	Light	Recept	O/M	A	P		1∅	3∅		P	A	Light	Recept	O/M	
						1	X	A	2						
						3	Y	B	4						
						5	X	C	6						
						7	Y	A	8						
						9	X	B	10						
						11	Y	C	12						
						13	X	A	14						
						15	Y	B	16						
						17	X	C	18						
						19	Y	A	20						
						21	X	B	22						
						23	Y	C	24						
						25	X	A	26						
						27	Y	B	28						
						29	X	C	30						
						31	Y	A	32						
						33	X	B	34						
						35	Y	C	36						
						37	X	A	38						
						39	Y	B	40						
						41	X	C	42						
Sub-Total															Sub-Total
Total															

DEMAND LOAD

LOAD	CL (KW)	DF (PU)	DL (KW)
Lighting		1.0	
Receptacle			
Other			
Motor			
a Present Total DL			
b Spare @ %			
c Gross Overall DL			
d Est. Div. Coef, DC			
e Net Overall DL			
f Estimated PF			

Current 1∅   (1000) $\dfrac{\text{KW}}{\text{PF} \underline{\hspace{1cm}} V}$ =   A

Demand 3∅   (580) $\dfrac{\text{KW}}{\text{PF} \underline{\hspace{1cm}} V}$ =   A

POWER SUPPLY DATA

From		
Feeder	Type & Size	Phase	
		Neutral	
		Ground	
	Length, Ft.	
	Voltage Drop %	
Raceway Quan/Size		
OCP	CB	Frame/Trip	
	FS	Size/Fuse	

BRANCHES

Quanity	Poles	Rating A.	Total Poles
( )			
( )			
( )			
( )			
( )			
( )			
Total Poles			

SPECIFICATION

Available Fault, Symmetrical Amp		
IC Rating Symmetrical Amp		
Branch Devices	Type (CB) (FS)	
	Model	
Special		
Ground Bus Contactor, Split Bus, Detector		
Dimension	Depth Width Height	

Main

Feed From		
Lugs Per ∅	Quan.	
	Size Range	
Size and Type		
System V-∅-W		

LEGEND:

OCP	—	Over Current Protection	CL	—	Connected Load
O/M	—	Other or Motor Loads	DF	—	Demand Factor
IC	—	Interruption Capacity	DL	—	Demand Load
V-∅-W	—	Voltage-Phase-Wires	DC	—	Diversity Coefficient

If Multiple Section, also see Panel	PANEL DESIGNATION

■ **FIGURE 12–7**

A typical switchboard and panelboard schedule. (Courtesy: William Tao & Associates, St. Louis, MO.)

L I N E	Nameplate Designation		Motor or Equipment Data			N O T E
	Equipment	Mark	HP (KW)	V/φ/Speed	Equipment Location	

L I N E	Controller Data				Branch Circuit Data						Remarks ◇
	By	Access-ories	Type	Size	Panel Design	Device		Conductors		Conduit Size	
						Type	Size	Phase	Ground		

■ **FIGURE 12–8**
Typical mechanical equipment data schedule.

Type	General Description and Application		Lamp and Power Data					Specification Description	Notes
	General Description	Typical Applications	Type	Quan. x w/Lamp	Lamp Code	Supply Volts	Watts Per Fix	Manufacturer and Catalog Number Series	

■ **FIGURE 12–9**
Typical lighting fixture schedule.

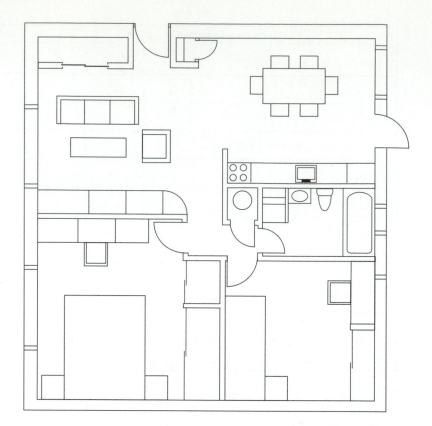

■ **FIGURE 12–10**
Architectural floor plan with furniture layout of a small apartment.

ing layout can best be illustrated with a practical example.

Figure 12–10 shows the architectural floor plan of a small apartment including the intended furniture layout, which is essential in electrical wiring planning. Should a furniture layout not be available, as is often the case for buildings during the planning phase, then reasonable assumptions and allowances have to be made by the engineer in consultation with the architect and the owner.

Based on the furniture layout, the lighting, receptacles and switch positions can be located. Methods for lighting design and layout will be covered in Chapters 14 through 17.

Convenience receptacles are located as needed to coordinate with the furniture layout. The NEC and other building codes have specific requirements for locating convenience receptacles in residential dwellings. In principle, duplex receptacles should be installed in the wall so that no point is more than 6 feet from a receptacle. An appliance or a lamp with a flexible cord attached may be placed anywhere along a wall and be within 6 feet of the receptacle. In certain instances, when a room is quite large, receptacles may also be installed in the floor.

Lighting fixtures selected for this apartment are as follows:

Type A   100 watt incandescent, wall mounted.

Type B   60 watt incandescent, ceiling recess mounted.

Type C   50 watt fluorescent including ballast, 4-ft/fixture, ceiling mounted.

Type D   100 watt 2-lamp fluorescent, 4-ft/fixture.

Type E   150 watt total of 9 incandescent lamps, pendent mounted.

Type F   150 watt total of 3 incandescent lamps, ceiling mounted.

The mechanical and household equipment load as determined by the owner, architect and the mechanical engineer are as follows:

Cooling: Window air conditioners, two 1 kW each, 120 volt, single phase.

Heating: Electrical heaters, two 1 kW each, 120/240 volts, single phase in the living room only. No heat is required in other spaces.

Water: Hot water heater, 3 kW, 240 volt, single phase.

Cooking equipment: Range, 8 kW, 240 volt, single phase. Oven, 3 kW, 240 volt, single phase. Microwave oven, 1 kW, 120 volt, single phase.

Kitchen appliances: Dishwasher, refrigerator, etc., 2 kW total, 120 volt, single phase.

Other appliances: To be commercial type, with electrical power rating within the nominal rating of standard receptacles (5 amperes/outlet).

Auxiliary system: Low-voltage systems, such as doorbell, security, TV antenna, etc., are not to be considered for this practice problem, although these are always installed in all homes in the US.

Based on the above information, work out the following problems:

1. Determine the total connected loads by load groups and the estimated demand loads of the apartment for summer and for winter.
2. What is the calculated full load (net demand load) current?
3. Determine the minimum electrical service capacity if the system is a 120/240-volt single-phase system.
4. What should be the size of the service entrance conductor? (Use type XHHW).
5. Complete the electrical plan, using NM cables without raceways.
6. Complete the electrical plan, using wires in conduit.

Answer to design problem

1. The total connected load by load group is determined as follows:

Load Group	Summer (watts)	Winter (watts)
Lighting (approximate)	1200	1200
Mechanical equipment	5000	5000
Kitchen equipment	12,000	12,000
Convenience power	3000	3000
Spare (approximately 10%)	2000	2000
Total connected load	23,200	23,200

In this example, the connected load for summer and winter happens to be the same. This is not the case for most buildings.

2. The total demand load for the building is obtained by applying an appropriate demand factor to the connected load of each load group. Because the building is very small, code does not allow the use of a demand factor, except that the spare load may be ignored at this time. For the same reason, the diversity coefficient is considered to be DC = 1. Thus, the total net demand load is:

$$\text{Total net demand load} = 23,200 - 2000$$
$$= 21,200 \text{ watts}$$

3. Assuming the overall power factor (PF) is 90%, then the full load demand current is:

$$I = P / (E \times PF) = 21,200 / (240 \times 90\%)$$
$$= 98A \text{ (use 100A)}$$

4. From Table 11–2 for conductors in conduit, the main service feeder in conduit shall be three #3 AWG single conductor cables. The NEC allowed maximum ampacity of a cable varies with type of insulation and number of cables in the raceway (conduit). The reader should refer to the various NEC rules and data for specific cable types.

5. Electrical wiring plan (using NM cables without conduit).
   (a) The preparation of electrical wiring plan starts with a clean architectural floor plan as shown in Figure 12–11. The furniture should be in lighter or finer line weight. If furniture plan is not available, then the electrical floor plan could be as shown in Figure 12–12.
   (b) The next step is to prepare an electrical floor plan showing all electrical load and devices, such as lighting fixtures, switchgear, receptacles, and electrical equipment. This is shown in Figure 12–13.
   (c) A completed electrical wiring plan is shown in Figure 12–14.

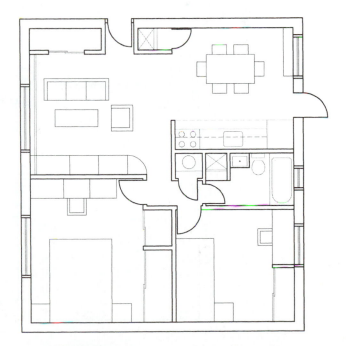

■ **FIGURE 12–11**
Electrical background floor plan with furniture layout.

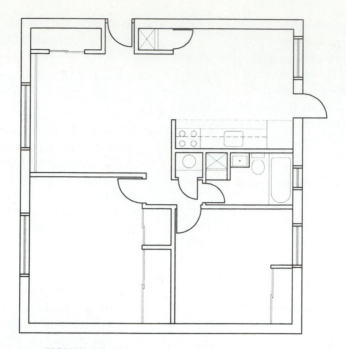

■ FIGURE 12–12
Electrical background floor plan without furniture layout.

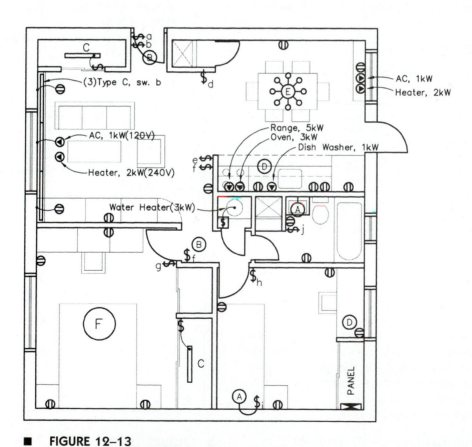

■ FIGURE 12–13
Electrical floor plan including lighting fixtures and electrical devices prior to wiring.

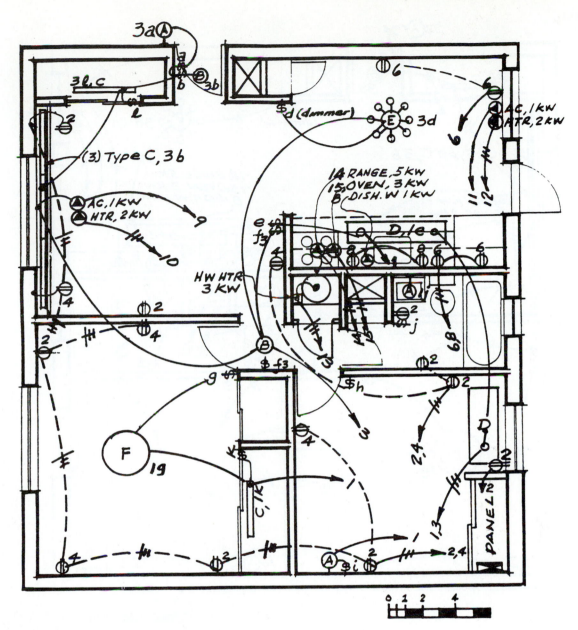

■ **FIGURE 12–14**
Completed wiring plan.

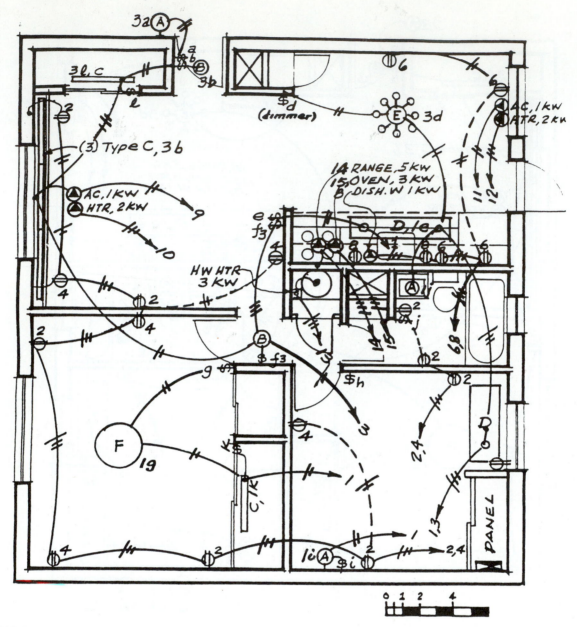

FIGURE 12–15
Completed wiring plan.

QUESTIONS

6. Electrical wiring plan (using wires in conduit). The procedure is same as in problem 5 except the wiring layout is based on single conductors in conduit. The wiring plan is shown in Figure 12–15.

12.1 Although electrical equipment does not require as much space as most mechanical equipment, the design of electrical systems and the layout of electrical components demand even closer interfacing with the architectural elements in space. Why?

12.2 What are the sequential steps in electrical design? Must they be followed in the same order?

12.3 Unit power density (UPD) is a suggested allowance or budget of electrical power for lighting in a building. It is given in watts per square foot of the floor area. The UPD values may be conveniently used for estimating the amount of electrical power required for spaces or buildings during the preliminary planning phase. It, however, is not intended for final lighting design which varies widely in layout and in the selection of light sources. As practice, estimate the electrical power required for lighting of an office building having the following spaces.

Area or activities	Description	Area, sq ft
General office	Large, with no partitions	1000
Private offices	Small, enclosed	500
Accounting offices	Small, enclosed	200
Storage	Active and fine visual task	300
Transient spaces	Corridors, stairs, etc.	400
	Total gross floor area	2400

12.4 Determine the gross and net demand loads of a power distribution for a large building complex with the following design data:

Lighting	2000 kW connected, with 0.9 DF
Convenience power	1500 kW connected, with 0.1 DF
Mechanical systems	3000 kW connected, with 0.5 DF
Elevators	1000 kW connected, with 0.5 DF
Building systems	1000 kW connected, with 0.8 DF
Estimated diversity coefficient	1.2

12.5 For safety reasons, adequate clearance must be maintained around the front and any side with access to the interior of all electrical equipment. What is the minimum clearance for a 120 volt wall-mounted panelboard? A free-standing 480 volt switchboard with rear access? And a 4160 volt controller which is not totally enclosed?

12.6 Name several methods for providing convenience electrical power to the middle of a space not accessible from wall outlets?

12.7 What is the difference between a schematic or elementary diagram and a connection or wiring diagram?

12.8 A room is installed with eight fluorescent fixtures. Each fixture uses three fluorescent lamps totaling 150 watts including the ballast. It is desired to switch on and off the center lamp of each fixture for low level of illumination and the other two lamps by another switch for medium level of illumination. Switching on all three lamps will provide the highest level. The switching shall be accessible from either of the two entries to the room. The room configuration, lighting layout and switching locations are shown in the floor plan below. Determine the number of branch circuits required, and prepare the branch circuit wiring plan.

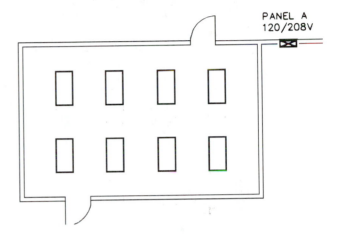

PANEL A
120/208V

12.9 Complete the electrical floor plan for the wiring of convenience outlets in a room as shown in the floor plan below. Section 12.8.1 provides the suggested loads for general purpose outlets which is 1.5A per outlet, or 3A per duplex outlet. If an outlet is identified to serve equipment, then the load for that outlet shall be based on the nameplate rating of the equipment, for example, a TV, a computer, etc. For purpose of this problem, the load at each receptacle is identified as follows:

C—general purpose receptacle   3A
D—special equipment   5A
E—special equipment   10A

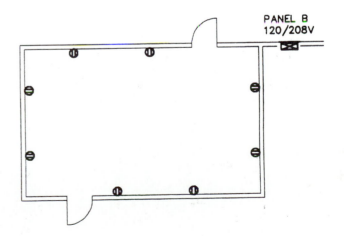

PANEL B
120/208V

13

Auxiliary Electrical Systems

Auxiliary electrical systems include systems that use electrical power to generate, process, store, or transmit information. Thus, auxiliary electrical systems are also considered information systems.

The scope of auxiliary electrical systems in buildings is growing at a very rapid pace. What was considered a luxury or state-of-the-art technology only a few years ago could easily be a minimum requirement in newer buildings. For example, a major office building built to operate into the 21st century should be equipped with data, communications, security, television, cable, closed circuit TV (CCTV) and/or satellite TV (SATV) systems, in addition to the traditional telephone, sound, signal, building automation, and fire alarm systems. Provisions should be made for raceways and/or wires if some of these systems will not be installed initially.

With the advances in existing technology and the introduction of totally new technology, it is unreasonable to expect the architect or the building systems design engineer to be fully cognizant of all available auxiliary systems, even though they may excel in one or more of these systems. This chapter introduces only the scope and operating principles of some basic auxiliary systems frequently required in modern buildings. Appendix A includes acronyms and abbreviations of basic terms with which a professional, regardless of his or her area of specialization, should be familiar. Some of these terms are discussed in the body of the chapter, and the others are explained in Appendix A.

13.1 FORM AND FUNCTION OF INFORMATION SYSTEMS

According to Davis and Davidson,[1] all information can be expressed by one or more forms and functions. There are four forms and four functions:

Forms

- Data   numbers, quantitative values
- Text   written language
- Sound   voice, signal or music
- Image   video, motion or still

Functions

- Generation   generating information into various forms
- Processing   editing, calculating, analyzing, synthesizing, expanding, interpolating, or extrapolating information
- Storage   storing, filing, or memorizing information for retrieval
- Transmission   sending or receiving information

[1]Stan Davis and Bill Davidson, *2020 Vision* (1991: Simon & Schuster, N.Y., N.Y.).

377

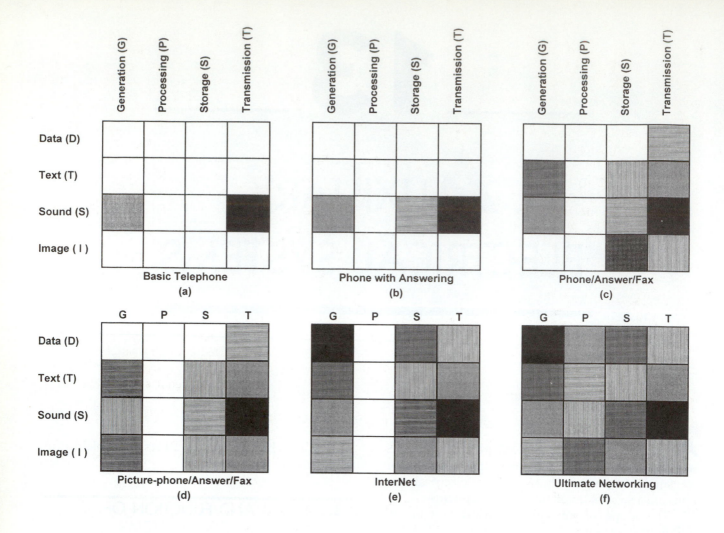

■ FIGURE 13–1

All information systems can be expressed by their forms and functions.

(a) A telephone system generates and transmits sound.

(b) A phone with an answering machine generates, transmits and stores sound.

(c) A phone with a fax also stores text and transmits data, text and images.

(d) A picture phone with an answer and a fax machine adds images to its form and function.

(e) A computer-phone interface adds processing capability.

(f) The networking of computer and telecommunications can perform all forms and functions.

(g) The ultimate in information systems will increase in intelligence until they may supersede that of human beings.

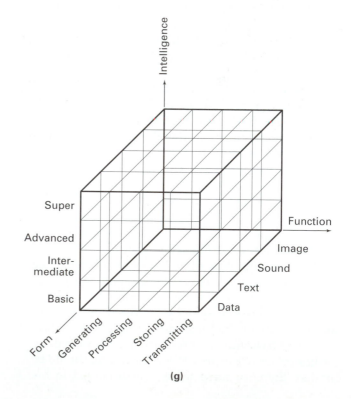

As illustrated in Figure 13–1, a basic telephone system is an information system that transmits sound, and only sound (one form and one function). A more sophisticated telephone system can encompass the transmission and storage of data, text, sound, and images (four forms and one function). The interfacing of a telephone system with computer technology adds processing capability (four forms and two functions). Finally, the networking of computer, telephone, and telecommunication systems completes all four forms and four functions. As science and technology continue to advance, the time will come when the intelligence level of information systems approaches or even surpasses that of humans (see Figure 13–1(g)).

13.2 COMMON CHARACTERISTICS OF AUXILIARY OR INFORMATION SYSTEMS

Several characteristics common to all systems are helpful in planning building systems:

1. *Auxiliary systems do not require heavy electrical power.* The power demand of most auxiliary systems is normally less than a few kW. In many, it is as low as a fraction of a watt.
2. *Most auxiliary systems operate on 30 volts or less, either AC or DC.* These systems are frequently described as low-voltage or extra-low-voltage systems. At this voltage level, wiring may be installed without raceways, although raceways or cable trays are often used to facilitate installation and to avoid tampering.
3. *Auxiliary systems do not require a large space for equipment.* The required space is easier to accommodate than that required for electrical power and HVAC systems. Other than operating consoles and signal devices, equipment may be located anywhere in a building with little effect on overall system performance. Whenever practical, it is desirable to centralize all equipment in head-end rooms to facilitate maintenance of the equipment.
4. *Manufactured products usually contain proprietary features.* It may thus be desirable to bid these systems as separate packages or on a performance basis.

13.3 CLASSIFICATION OF AUXILIARY SYSTEMS

There are two major groups of auxiliary systems:

1. *Communication Systems*
   - Audio — public address, intercom, music, radio, etc.
   - Video — TV, CCTV, SATV, etc.
   - Telephone — public private exchanges (PAX, PBX), etc.
   - Data — modem, local area network, wide area network, etc.
   - Signals — time, program, signals, etc.
   - Multimedia — combination of one or more systems
2. *Building Operational Systems*
   - Safety — fire alarm, sprinkler alarm, emergency evacuation, etc.
   - Security — access control, intrusion detection, etc.
   - Automation — BAS, BMS, BMAS, etc.
   - Specialty — sound masking and systems unique for special building occupancies such as hospitals, defense, retail stores, food services, theatrical, etc.

13.4 COMPONENTS AND WIRING

13.4.1 Basic Components

Most auxiliary systems operate on DC circuitry, because DC devices are more sensitive than AC. The current drawn could be in the level of milliamperes (1/1000 A) or microamperes (1/1,000,000 A). The following components are common to most systems.

Power supply unit The power supply unit consists of electromagnetic transformers to transform a utility AC system from 110/220 volts to less than 30 volts. The AC circuit is then rectified into DC by a solid-state device known as diode or silicon-controlled-rectifier (SCR). Figure 13–2(a) illustrates the voltage transformation and rectifying circuitry of most auxiliary systems.

Sensing and signaling devices Most auxiliary systems detect, control, or amplify variations in energy, such as sound, light, motion, temperature, color, infrared, ultraviolet, heat, microwave, and other forms of low-level energy, and convert these forms of energy into electrical energy to operate signaling devices, such as speakers, telephones, clocks, lights, etc. Converting and controlling this energy is increasingly done with semiconductors or solid-state materials.

(a) A typical full-wave rectifier circuit consists of a transformer and two diodes. The output is a pulsating DC varying voltage having the same frequency as the AC power supply. (Source: Radio Shack: *Building Power Supplies.*) (b) The upper diagram illustrates how a telephone voice signal is sampled at four equal intervals and converted into four eight-bit digital signals to represent the original analog input in millivolts. The lower diagram illustrates the conversion in reverse at the receiving end. (Source: Stephen Bigelow, *Understanding Telephone Electronics,* Prentice Hall Computer Publishing, Indianapolis, IN) (c) A typical unshielded twisted pair (UTP) cable. Generally, a 2-in. to 6-in. twist is used. For high-speed data transmission, a tighter twist less than 1 inch is required. (d) A typical shielded twisted pair (STP) cable with metallic shield and a separate ground wiring. (e) A typical coaxial cable. (f) Illustrated at the left is a typical four-pair UTP cable, and at the right is a typical four-pair STP cable. Multipair UTP and STP cables are made in sets of up to 24 or more pairs. In all of the illustrations of cable, the sizes are enlarged for clarification. In general, the conductors of UTP, STP, and coaxial cables are made with AWG no. 20 to no. 24 copper wires. The external diameter of the cable depends on the construction of the cable, as well as its shielding, insulation, and jacket thickness. A four-pair shielded cable may have an overall diameter of 0.25 inch. (Courtesy: The illustrations of cable are provided by Berk-Tek, Inc., New Holland, PA.)

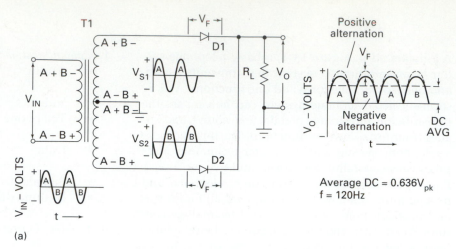

Average DC = 0.636V$_{pk}$
f = 120Hz

(a)

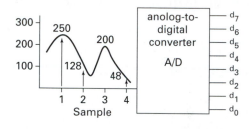

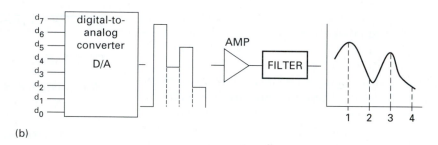

Sample	Code (8-bit)	Equivalent analog input
1	11111010	250 mV
2	10000000	128 mV
3	11001000	200 mV
4	00110000	48 mV

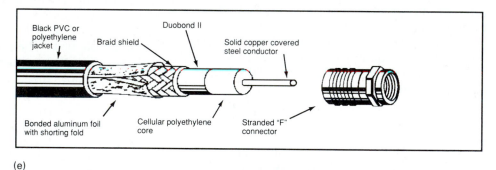

(b)

(c)

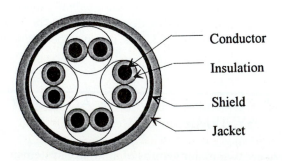

Black PVC or polyethylene jacket

Duobond II

Braid shield

Solid copper covered steel conductor

Bonded aluminum foil with shorting fold

Cellular polyethylene core

Stranded "F" connector

(d)

(e)

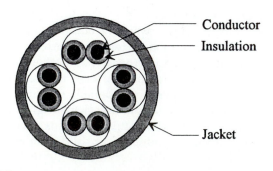

Conductor

Insulation

Jacket

Conductor

Insulation

Shield

Jacket

(f)

A semiconductor is usually defined as a material that has an electrical resistance between that of a conductor, such as metal, and an insulator, such as plastic. However, this simple definition does not take into account the impurities in a semiconductor content. Semiconductors are, in general, crystalline materials containing varying degrees of impurities. Silicon and germanium are the most popular semiconductors. Others include copper oxide, selenium, cadmium sulfide, etc. Semiconductors conduct electricity because of the presence of free electrons in them. Semiconductors may carry negative charges (*n* type), positive charges (*p* type), or both (*p-n* junction type). These types form the various semiconductor devices and circuits, such as thyristors (silicon unilateral switches) and rectifiers (SCRs).

Control devices Control devices are manufactured to provide on-off, variable-output (voltage or current), maintained or momentary contact, mechanical or electronic, direct or remote-controlled switching functions. Associated with switching are the necessary relays, circuit boards, signal and sensing devices.

13.4.2 Wiring for Auxiliary Systems

Line voltage (110 volts or higher) wiring for auxiliary systems follows the same guidelines as other wiring systems for power, lighting, and equipment discussed in Chapters 10 through 12. However, most auxiliary systems operate on low voltage and with limited power. Thus, the wiring system will not be a life or a fire hazard. Low-voltage wiring is, therefore, normally installed without raceways, except when these are desired for reasons of security or aesthetics.

13.4.3 Basic Wiring for Low-Voltage and High-Frequency Systems

One of the characteristics of low voltage is low signal strength. This is particularly true of telecommunication systems, such as voice (telephone), sound (music), radio, and video (TV) signals. The power levels of these systems are usually measured in milliwatts or microwatts, and thus, the systems are very sensitive to external disturbances, such as electromagnetic interference. The transmission wiring is evaluated according to the following characteristics.

Characteristic impedance Most telecommunication systems operate at much higher frequencies than do power systems. Power systems operate at 60 to 400

hertz, whereas telecommunication systems operate in the kilo- (1000), mega- (1,000,000) or giga- (1,000,000,000) hertz ranges. At these high frequencies, it is extremely important to match the impedance of the transmission wiring to that of the terminal equipment to minimize the loss of or attenuation of the signal. The term *characteristic impedance* is used to express the performance characteristics of the wires. It relates more to frequency than length. Characteristic impedances of typical low-voltage wires for high-frequency telecommunication systems are rated between 50 and 300 ohms.

Transmission capacity The transmission capacity of a medium is measured differently for different types of signals. *Analog signals* are continuous waves. Telephone and TV signals are typically of this variety. The transmission of an analog signal is measured by the bandwidth of the signal, in hertz. The *bandwidth* may be defined as the highest frequency that can be transmitted in a given medium without excessive attenuation. Typically, the upper limit is specified as the point at which the signal's strength has dropped by 3 dB. The wider the bandwidth, the higher the frequency a medium can serve.

The bandwidth of a wiring system is not only limited to the construction of the wires, but is also limited by mechanical components (connectors, jacks, patch panels, etc.), by imperfections in the wires (incorrect installation, damage occurring during installation, etc.), and by effects of the environment (heat, moisture, proximity to power and other signals, etc.).

Digital signals are separate on-off pulses. Data and computer signals are transmitted in digital format. The speed of transmission of digital signals is measured in bits, megabits, or gigabits per second. A 100 megabit-per-second (MB/S) wiring system can carry as much information as a 10-MB/S system. The one drawback of digital transmission is the increased bandwidth required to transmit digital pulses, compared with the bandwidth required for equivalent analog signals. A major advantage of digital transmission is the fidelity of the regenerated signal, which is relatively independent of the distance over which the signal is transmitted.

Analog and digital signals can be converted from one to the other. For example, the telephone network converts analog signals (the voice from the telephone handset) into digital signals for long-distance transmission and then converts the signal back to analog form on the other end. For a telephone system containing signals up to 4000 Hz (4 kHz), scanning (or sampling) is made at twice the frequency (8000 times/per second). With a seven-bit code that contains the values from 0 to 127, the result is a very realistic reproduction of the voice waveform. With 8000 samples per second and a seven-bit-per-second digital rate, the speed of data transmission is 7 × 8000, or 56,000 bits per second. For high-fidelity speech or music, the sampling rate is usually four times faster, or 32,000 samples per second and 224,000 bits per second. (See Figure 13–2(b) for a graphic illustration of analog and digital signals.)

There is no precise equivalence between the digital and the analog transmission capacity of a medium. An approximate equivalency exists between 10,000 bits per second of digital capacity and 10 MHz analog capacity.

Basic wiring systems   There are two choices for wiring low-voltage and telecommunication systems: twisted pair or coaxial cable.

Twisted pair (TP) wire is the most commonly used type of wire and is constructed as shown in Figure 13–2(c). TP consists of a pair of wires, generally copper, twisted around each other, with the length of the twist—the *lay*—selected to reduce interference from external electrical and magnetic fields. TP is normally constructed with an outer jacket for physical protection. Additional protection against external fields is afforded by installing an overall shield (a metallic tape or braid), as shown in Figure 13–2(d). These two types of TP wiring are referred to as unshielded twisted pair (UTP) and shielded twisted pair (STP), respectively.

UTP and STP have been used for many years in such applications as voice communications (telephone and public address systems), control systems (low-voltage lighting, building management, fire alarm, systems, etc.), low-speed digital signals in direct digital control (DDC) building management systems, and short-haul computer communications. The recent proliferation of computer use and the demand for high-speed digital transmission has resulted in significant improvement in UTP and STP performance (bandwidth), permitting their use in applications requiring rates as high as 100 megabits per second. According to the Telecommunication Industry Association (TIA)/Electronics Industry Association (EIA), UTP and STP are further divided into five classifications—categories 1 through 5, corresponding to increasing performance and permitting the selection of wiring appropriate to the application and the budget.

Coaxial cable is constructed as shown in Figure 13–2(e). It consists of two concentric conductors sharing the same axis (hence the name), separated by an insulating material and surrounded by an overall protective jacket. (Harsh environments may necessitate two or more layers of protective jackets.) The center conductor is almost always solid wire, while the outer conductor is braided or metallized tape. Although the outer conductor has an appearance similar to the shield in UTP, it is in fact a current carrier and affords no shielding properties. Consequently, coaxial cable is more susceptible to external interference and has a lower bandwidth capability than high-quality twisted pair (both UTP and STP).

13.4.4 Fiber Optics

Fiber optics is a technology that uses light as a medium to transmit information or data. This emerging technology utilizes the refractive property of a transparent material such as high-purity glass or plastic to transmit light with negligible loss. It promises to be the preferred means of transmission of information in the near future.

Light transmission within transparent fiber   As an electromagnetic wave, light travels in a straight line. It can be deflected, however, by reflection from a surface or by refraction at the boundary or interface between two transparent media, such as between air and glass or between glass and plastic. The degree of deflection or bending depends on the refractive indexes of the two media and the angle at which the light strikes the interface (boundary) between two media. (See Figure 13–3(a).) If light within the medium with the higher index hits the boundary with the medium with the lower index at a flat angle smaller than a certain critical angle, the light beam will be totally refracted within the medium. This is known as *total reflection* (no longer described as refraction). (See Figure 13–3(b).)

A fiber used in fiber optics is a hair-thin, all-glass filament. It consists of two solid portions—the core and the cladding, which are inseparable from each other. The light travels through the core, while the cladding keeps the light contained within the core because of its different refractive index.

Conversion of electromagnetic waves   Since light and electrical energy are all part of the electromagnetic spectrum, they differ only in wavelength and frequency. Light and electrical energy can, therefore, be converted from one form to another. This unique property becomes a great asset in modern communication technology when electrical signals in either analog or digital form are converted into light signals for low-loss transmission through a tiny optic fiber and then are converted back to electrical signals for

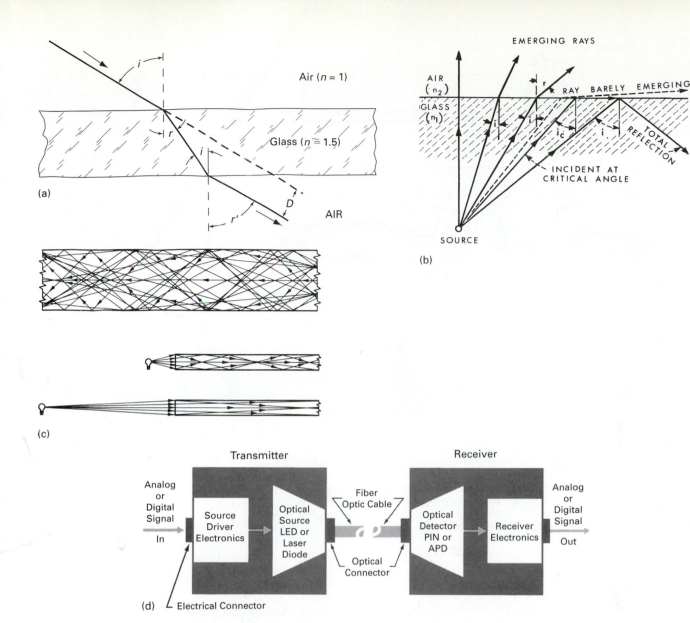

■ **FIGURE 13–3**

(a) Refraction of a light ray at a plane surface causes bending of the incident ray and displacement (D) of the emergent ray. Snell's law of refraction is expressed as

$$n_1 \sin i = n_2 \sin r,$$

where

n_1 = index of refraction of the first medium,
i = angle of incidence (angle the incident ray forms with the normal of the surface),
n_2 = index of refraction of the second medium,
r = angle of refraction.

(b) As the angle of incidence increases, the light ray bends more and more. The angle at which time the reflected ray is parallel to the surface is called the *critical angle* (i_c). A light ray incident at any angle greater than the critical angle is totally refracted (or reflected) back to the medium. (c) With angles of incidence greater than the critical angle between the glass core and its glass (or plastic) cladding, a light ray is totally reflected, but scattered. This phenomenon is called *pulse dispersion*. (Top illustration.) Light incident with a lower acceptance angle (bottom illustration) has a low pulse dispersion and thus a higher speed of transmission. (Source: Preceding illustrations reproduced with permission from *IES Lighting Handbook,* Illuminating Engineering Society of North America.)

(d) This simple schematic diagram shows an optical transmitter and receiver connected by a length of optical cable in a point-to-point link. The transmitter converts electronic signal voltage into optical power, which is launched into the wire by an LED, a laser diode, or a laser. At the photodetector point, either a positive-intrinsic-negative (PIN) or avalanche photodiode (APD) captures the light pulses for conversion back into electrical current. (Source: Belden Wire and Cable Co., Richmond, IN.)

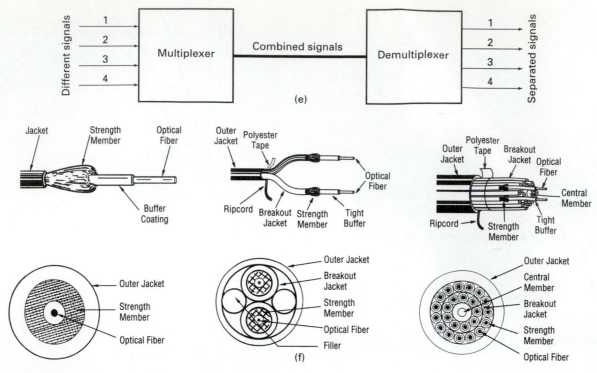

■ **FIGURE 13–3**

(e) Several signals can be combined or multiplexed into a single signal through a device called a multiplexer. The combined signal is then transmitted through a single transmission line, which greatly increases the capacity of transmission. A demultiplexer at the receiving end separates the signals into their original forms. (f) From left to right: Construction of fiber optic cables. Left: single fiber, tight buffer; center: two fibers, tight buffer; right: 24 fibers, tight buffer. The fiber size varies from 50 to 100 microns. (Source: Belden Wire and Cable Co., Richmond, IN.)

processing. (See Figure 13–3(d).) Furthermore, many signals may be transmitted simultaneously within the same fiber, a process known as *multiplexing*. (See Figure 13–3(e).) In practice, a complex fiber optic cable can transmit thousands of signals. With the proper light source (an LED), cable construction, and channels (pulse dispersion), a multimode fiber optic cable may transmit several millions of digital signals simultaneously.

Advantages and application of fiber optics The primary advantages of fiber optics technology are its high transmission capacity and low loss characteristics. In addition, the light signal is not subject to electromagnetic interference. Some typical applications of fiber optics are the following:

- Long-distance telephone lines, on land and at sea
- Local and wide area networks
- Cable television between microwave receivers and head-end equipment

- Transmission of signals where electromagnetic interference is a problem.
- High-security applications, such as financial, military, and intelligence systems

Fiber-optic systems Fiber optics is a technology in which light is used to transmit information from one point to another. The transmitting medium is constructed of thin filaments (strands) of glass wires through which light beams are transmitted. For illumination, light may be in any part of the visible spectrum. For data transmission applications, light must be of a single wavelength in order to be totally reflected (refracted) within the fiber.

A fiber-optic system consists of many components, including the light source, transmitter, receiver, repeater, regenerator, optical amplifier, and cable, as well as accessories. To transmit light at low loss and for flexibility, *optic fiber cables* are made of fine fibers about 100 to 200 microns in diameter. The cable consists of a glass fiber core coated with a thin layer of glass or plastic as cladding and covered with one or more layers of material for dielectric isolation and physical protection. (See Figure 13–3(f).)

Since optic fiber must be very thin (or small in diameter) in order to keep the light within the angle of total reflection, the *light source* must be very tiny. It has been found that semiconductor lasers or LEDs are the best sources. In addition, lasers at near-infrared wavelengths can keep loss at a minimum in long-distance transmissions. The popular laser and LED light sources are generated from gallium (Ga), aluminum (Al) and arsenic (As) compounds. A wavelength of 750 to 900 nm is used for low-cost, short-distance applications, in which the typical light power loss is about 1.5 to 2.5 dB/km. A 1300-nm wavelength is used for medium-distance applications, in which the typical light power loss is about 0.35 to 0.5 dB/km. And a wavelength of 1500 nm is used for long-distance applications, in which the typical light power loss is about 0.2 to 0.3 dB/km.

The *transmitter* is used to convert electrical signals into light signals. The electrical signals may be in digital format (e.g., computer data) or analog format (e.g., TV and radio). Light transmission can accommodate either of these types of signal. In general, digital signals can tolerate distortion, whereas analog signals are subject to disturbance. The operating speed of a fiber-optic transmitter is measured by its bandwidth for analog signals and data rate for digital signals.

The *receiver* does the opposite of a transmitter, converting light signals into electrical signals. *Repeaters*, *regenerators*, and *amplifiers* are used to reinforce a signal or to minimize attenuation in long-distance applications.

Finally, *connectors* and *splicers* are used to make connections and splices in fiber-optic cables. A secure and square connection with low dimensional offset is extremely important. The most accepted connector designs are the ST, SC, and the FDDI types. (See Figure 13–4.)

13.4.5 Selection of Wiring Systems

The selection of wires and cables among twisted pairs, coaxial cables, optic fiber, and combinations thereof depends on the application, the performance of the medium in terms of bandwidth, transmission speed, attenuation, impedance, shielding, etc., and, of course, the cost of installation of the wires and cables. Optic fiber is best as regards bandwidth and speed, but is the most costly of the three media. However, the cost gap is steadily closing. Optic fiber is now being installed in backbone wiring in buildings and in

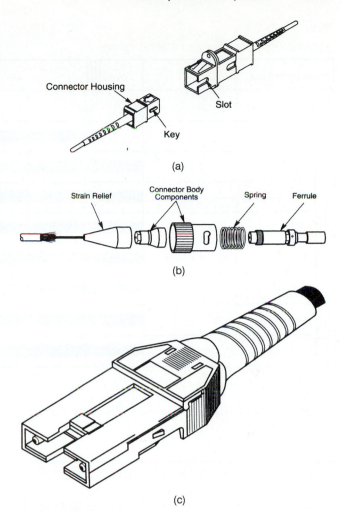

■ **FIGURE 13–4**
During earlier years of fiber cable development, each cable manufacturer often developed its own design for connecting or splicing cables. This leads to the problem of interfacing between equipment and systems. To eliminate the proliferation of connector designs, the International Standards Organization (ISO) and the Telecommunications Industry Association (TIA) have endorsed the square connector type SC, round connector type ST, and the Fiber Distributed Data Interface (FDDI) connector as preferred standards. Configurations of these connectors are shown in (a), (b) and (c). (Courtesy: Jeff Hetch, *Understanding Fiber Optics*, Sams Publishing, Indianapolis, IN.)

local and wide area networks. For design and installation guidelines, the EIA and TIA standards listed in the reference section of this chapter are the best resources.

Table 13–1 lists the basic applications of the popular transmission media and the ranges of speed at which they may be used. As with the design of any building system, the future needs of the building and the trends in technology must be carefully weighed in

Table 13–1

Applications of telecommunication cables

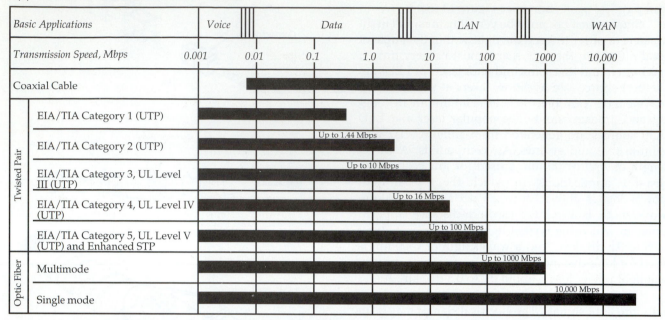

Basic Applications		Voice	Data		LAN		WAN
Transmission Speed, Mbps		0.001 0.01	0.1 1.0	10	100	1000	10,000
Coaxial Cable							
Twisted Pair	EIA/TIA Category 1 (UTP)						
	EIA/TIA Category 2 (UTP)		Up to 1.44 Mbps				
	EIA/TIA Category 3, UL Level III (UTP)		Up to 10 Mbps				
	EIA/TIA Category 4, UL Level IV (UTP)		Up to 16 Mbps				
	EIA/TIA Category 5, UL Level V (UTP) and Enhanced STP		Up to 100 Mbps				
Optic Fiber	Multimode		Up to 1000 Mbps				
	Single mode		10,000 Mbps				

the selection of wiring systems for all electrical auxiliary systems.

13.5 TELECOMMUNICATION SYSTEMS

A telecommunication, or telecom, system is defined as any electrical system that transmits, emits, or receives signals, images, sound, or information of any nature by wire, radio, video, or some other form of energy within the electromagnetic spectrum. Thus, telephone, radio, microwave, radar, intercom, public address, CCTV, broadcasting TV, and SATV are all telecom systems. Telecom technology is so dynamic that a new system may become obsolete within a few years. Thus, this section discusses only the general concept, without dealing with specific design data or guidelines.

13.5.1 The Electromagnetic Spectrum

The electromagnetic spectrum is a graphic representation of radiant energy in an orderly arrangement according to its wavelength or frequency. Radiant energy within the spectrum differs in wavelength and frequency, but its speed of transmission is constant. The speed of radiant energy in vacuum is 299,793 kilometers per second (km/s), or 186,282 miles per second (mi/s). The speed of radiant energy in air is slightly slower (299,724 km/s). In round numbers, the speed may be considered to be 300,000,000 m/s (3

$\times 10^8$ m/s). The following equation expresses the relationship among speed, frequency, and wavelength:

$$s = f \times G \qquad (13\text{–}1)$$

Where s = speed, 3×10^8 meters/second (m/s)
f = frequency, cycles per second (cps or hertz)
G = wavelength, meters (m)

Example 13–1   The standard frequency of electrical power in the United States is 60 Hertz. What is the wavelength of the electrical power?

Answer   The wavelength of 60-hertz power is $3 \times 10^8/60 = 5 \times 10^6$ meters.

Example 13–2   If the radio frequency of an FM station is 100 MHz, what is the wavelength?

Answer   The wavelength of the FM radio is $(3 \times 10^8)/(100 \times 10^6) = 3$ meters.

13.5.2 Standard Radio and Video Frequencies

The region of radio and video frequencies within the electromagnetic spectrum extends from 3 kHz to 300 GHz. (One GHz is 1 billion Hz.) The boundary of each subdivision of a region is gradual; there are no specific demarcation lines.

To minimize interference between transmitted signals, the applications of frequencies are strictly controlled by governments. The regulatory agencies in the United States are the National Telecommunications and Information Administration (NTIA) of the U.S. Department of Commerce and the Federal Communications Commission (FCC). Table 13–2 lists some typical applications of FCC/NTIA standards to power, radio, video, and radar signals. The governments of all developed and developing countries also control their transmission frequencies. In general, the frequencies allocated by other countries are comparable to those used in the United States, forming the basis for international standards.

13.5.3 Audio and Video Systems

Audio and video (A/V) systems refer to radio and TV systems that receive their electromagnetic signals "off air" (on the air), cable networks, or signals generated locally, such as tapes or compact disks. This section discusses systems that receive and transmit off air only.

Exterior antennas An antenna is a device for receiving or transmitting off-air A/V signals. Although most A/V equipment—i.e., radio or TV—has built-in antennas, heavy concrete or steel building walls and roofs attenuate these signals to the extent that they are often distorted or noisy. In most buildings, exterior antennas are necessary to assure good reception of the signal.

The presence of an antenna on or near a building has a major impact on the aesthetics of the building. The designer must be keenly aware of the size and location of the antenna during the design process, so

TABLE 13–2

Typical applications of FCC/NTIA frequency standards[1]

Frequency Bands[2]	Frequency	Applications
POWER BANDS (Extremely Low to Infralow Frequencies)		
Standard frequency, Hz	50–60	Public utilities
Intermediate frequency, Hz	400	Aircraft power
High frequency, Hz	3000	High-frequency lighting
RADIO BANDS (Medium to Ultrahigh Frequencies)		
Amplitude modulation (AM), kHz	530–1600	Off-air radio
Amateur, kHz	1800–1900	Ham radio
Telephone, Mhz	1.6–1.8	Wireless telephony (short range)
Amateur, MHz	3.5–4.0	Ham radio
High frequency, MHz	5–40	International radio
Frequency modulation (FM), MHz	88–108	Off-air radio
Telephone, MHz	400–500	Mobile telephone system
Telephone, MHz	824–894	Cellular telephone system
Products, MHz	902–928	Wireless, applications (short range)
TELEVISION BANDS (Very High to Ultrahigh Frequencies)		
VHF (low band), MHz	54–88	Channels 2–6
VHF (midband), MHz	121–169	Channels 14–22
VHF (high band), MHz	174–216	Channels 7–13
VHF (superband), MHz	217–295	Channels 23–36
UHF (hyperband), MHz	301–451	Channels 37–62
SATELLITE BANDS (Ultrahigh to Superhigh Frequencies)		
Ultrahigh frequency, MHz	950–1750	Satellite TV
Superhigh frequency, low band, GHz	3.7–4.2	C-Band satellite
Superhigh, midband, GHz	11.7–12.2	KU-Band satellite
Superhigh, midband, GHz	12.2–12.7	Direct broadcast system (DBS)
MICROWAVE BANDS (Ultrahigh to Tremendously High Frequencies)		
Superhigh frequency, GHz	2–10	Long-distance transmission
Extremely high frequency, GHz	10–50	Radar
Tremendously high frequency, GHz	50–500	Long-distance transmission

[1]Includes only those applications commonly related to audio and video reception and distribution systems in buildings.

[2]*Classification of Frequency Bands*

ELF	below 300Hz	Extremely low frequency
ILF	300–3000 Hz	Infralow frequency
VLF	3–30 kHz	Very low frequency
LF	30–300 kHz	Low frequency
MF	300–3000 kHz	Medium frequency
HF	3–30 MHz	High frequency
VHF	30–300 MHz	Very high frequency
UHF	300–3000 MHz	Ultrahigh frequency
SHF	3–30 GHz	Superhigh frequency
EHF	30–300 GHz	Extremely high frequency
THF	300–3000 GHz	Tremendously high frequency

that these elements may best be coordinated with the overall design, rather than be added as an afterthought. This is particularly important for buildings such as hotels and offices, which are increasingly dependent on satellite for data transmission.

The ideal location of an exterior antenna is on the roof of a building or in an open space on the ground. Antennas should be mounted in such a manner that there is no major obstruction between the antenna and the transmitted signals.

External antennas come in all shapes and sizes. Some popular models are the yagi type (straight bars) for VHF frequencies, loop type for FM/UHF frequencies, and dish type for satellite frequencies. Figure 13–5(a) illustrates a typical radio and VHF antenna. Higher frequency UHF signals are more directional. Their behavior is close to that of visible light, which travels in a straight line, but can be reflected by a reflector or focused by a parabolic-shaped dish. Figure 13–5(b) illustrates a typical dish-type antenna.

If outlets are limited in number and in distance, such as in a residence, the strength (gain) of a signal from an outdoor antenna should be sufficient without further amplification. On the other hand, larger buildings having tens or hundreds of outlets will require additional head-end equipment to filter, boost, convert, amplify, modulate, and combine the various frequencies into a total distribution system. By combining the signals, the distribution from the head-end equipment to the individual outlets can be accomplished by a single coaxial cable looping between outlets, rather than with multiconductor cables. The economic benefit is obvious.

Telecommunication satellites   Most telecommunication satellites are of the geostationary type, placed in an orbit 22,300 miles directly above the equator. The satellite maintains the same relative position and attitude with respect to the earth through small station-keeping jets. The electrical power required to operate these jets is obtained from batteries recharged by photovoltaic solar collectors.

The satellite itself is an assembly of many receivers and transmitters, with associated amplification. Each satellite is designed to cover certain areas on the earth. The signal is weaker at the edge of an area, and as the site of the receiving station moves further from the equator, the size of the receiving station is increased to compensate. In other words, the size of the satellite dish in Alaska is considerably larger than that in Florida in order to receive signals from the same transmitter. The angle of the receiving station is lower the further it is from the equator.

The speed of the signal is the speed of light. This means that there is a minimum known delay in the signal of at least the time it takes to travel the 22,300 miles between the receiver/transmitter and the satellite, or 0.24 second. The further from the equator the receiver/transmitter is, the longer is the distance to cover, and the greater the reduction in signal strength and the lower the angle of the receiver/transmitter. This lower angle makes it more probable that obstructions will be encountered. Processing time at the satellite (to turn the signal around) usually makes the transmission time approximately 0.4 second, or 400 ms or greater.

Satellite receiver/transmitter for building services   A satellite receiver/transmitter installed for building services is referred to as a *VSAT*, which stands for *very small aperture transceiver*. A transceiver is a combination transmitter and receiver. The dish portion of the device is a section of a parabola. The focal point concentrates the incoming signal at the receiver for amplification at the horn. The alignment of the dish is critical for reception and transmission of the signal to the satellite.

The angle and azimuth necessary to acquire the satellite signal are known from any location that can ''see'' the satellite. The viewing angle must be clear. Although it is possible to acquire a signal through deciduous trees. This situation should be avoided, since obstruction will attenuate the signals.

A VSAT ranges from 1 meter up to 10 meters in diameter. The required size depends on the altitude and longitude of the building and the satellites. With the advance of technology, antenna dishes have diminished drastically in size in recent years.

Smaller antenna terminals with more powerful amplifiers are being developed. The diameter ranges from about 1 ft for ultra SAT (USAT) to less than 1 ft for tiny SAT (TSAT). Associated with the reduction in size are an increase in frequency and transmission speed and a narrowing of the beam pattern.

Signal distribution   Each of the signals (frequencies) received is processed by a separate controller that is connected to the A/V and VSAT antenna via coaxial cable. The allowable distance between the antenna and the controller is limited and varies with the total configuration. Figure 13–5(c) is a schematic diagram of an A/V satellite signal distribution system in a large building.

13.6 DATA DISTRIBUTION SYSTEMS

Data distribution through networking is no longer limited to major facilities, such as computer centers, hospitals, hotels, department stores, and corporate

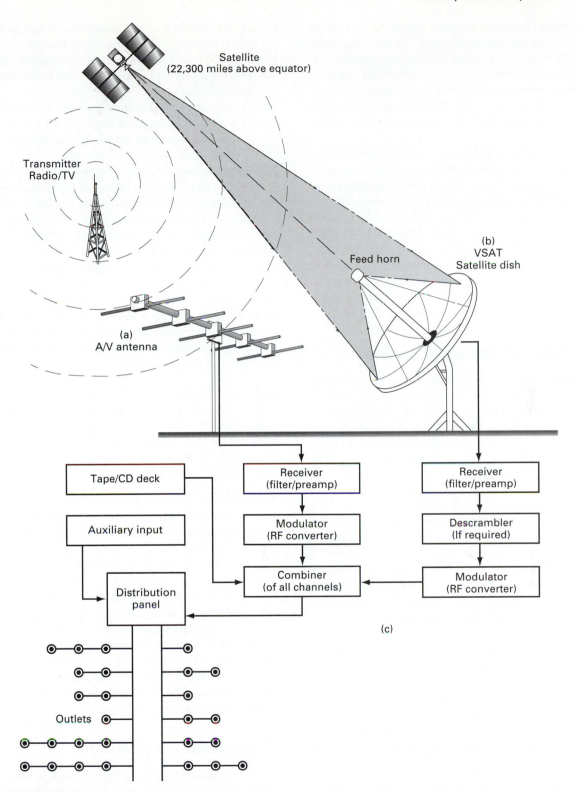

■ **FIGURE 13–5**
(a) A typical radio and TV antenna (54–451 MHz)
(b) A typical VSAT satellite dish (0.9 to 12.7 GHz)
(c) A typical A/V system distribution wiring diagram. The signals are filtered, amplified, modulated, and combined prior to distribution.

headquarters. Any building or group of buildings equipped with central computers, remote terminals, or multiple units of PCs should consider the installation of a data distribution system. With the continuing improvement in networking technology and with lower costs, it is conceivable that a data distribution system will be installed even in most homes in the near future.

The biggest advantages of networking are that it allows the sharing of files, programs, and equipment and it saves money. Networking also makes it easy to set up an electronic mail (E-mail) system between clients (users). With data sharing, data need be entered only once, thus eliminating the wasted effort of reentry of the data at each station. There are, of course, some disadvantages to a network system—for example, its complex procedures and system backups and its higher setup costs. However, the advantages often outweigh the disadvantages in most applications.

13.6.1 Types of Distribution Systems

When different pieces of equipment are connected to distribute data or files to each other, the equipment is said to be connected as a network. The data to be distributed are usually in digital format, although they may be in analog format as well. Networking may be confined to within a building or may be extended to thousands of miles away.

Local area network   A local area network (LAN) is a system that connects computers and peripheral equipment within a building or within several buildings in proximity to each other for the purpose of sharing resources. A LAN system is practical within several thousand feet. Thus, it is equally applicable to campus-type facilities, such as universities, corporate headquarters, and industrial complexes.

LANs are made up of servers and workstations, or PCs. The server needs a faster processor and a larger hard disk than the workstation does. One item of hardware needed to set up a LAN is a network adapter or a network interface card. The card is the interface between the server peripheral equipment, and the workstations on the network. There are many network adapter cards, such as Ethernet, ARCnet, Token Ring, and Zero-Slot LANs.

LAN topologies   The physical layout of a LAN is called its *topology*. The most common LAN topologies are as follows:

- *Point to point* is the simplest format, connecting one server and one workstation.

- A *star network* is an extension of the point-to-point topology to multiple workstations.
- A *bus network* has one cable connecting all the workstations and the server. This is the topology for Ethernet LANs.
- A *ring network* is similar to a bus network, except that the ends are tied together to form a ring. This is the topology for IBM Token Ring LANs.

The star topology consists of home-run cabling from each node back to a central point of connectivity, or hub. The hub maintains whichever logical topology is appropriate for the network protocol. The Star topology is the basis for the structured cabling system in buildings defined in the TIA/EIA, 568: "Commercial Building Telecommunications Wiring Standard."

In a ring network, the cabling between nodes forms a ring, and in an Ethernet network, the cabling between nodes forms a bus. Figures 13–6(a) through (c) illustrate the physical topologies of data distribution systems. Figure 13–6(d) is a block diagram of a combination data, telephone, and A/V system.

Wide area network   A wide area network (WAN) is defined as a data communication network that uses common carrier lines to extend a LAN beyond the building or campus it serves. WANs typically transmit over phone lines, microwave towers, and satellites without limitation.

Wireless LANs   Data and A/V signals may be transmitted through a wireless network, which usually operates on infrared or radio frequencies. This type of technology can transmit information at relatively high speeds, but needs a clear line of sight between the receiving and transmitting devices. Its major advantage is flexibility: Systems can be expanded without prewiring. It is usually more costly initially, but may be less expensive in the long run. Wireless LANs are limited to smaller systems. Another disadvantage of wireless LANs is the possibility of external interference with the signal. The infrared system is susceptible to interference from light sources, the radio frequency and from electromagnetism.

13.6.2 Design Considerations

Structured wiring is wiring that is used to coordinate the multimedia transmission of data, voice, and video signals. A plan for the installation of structured wiring should cover seven design elements:

- Entrance facilities, i.e., conduits, manholes, and cables

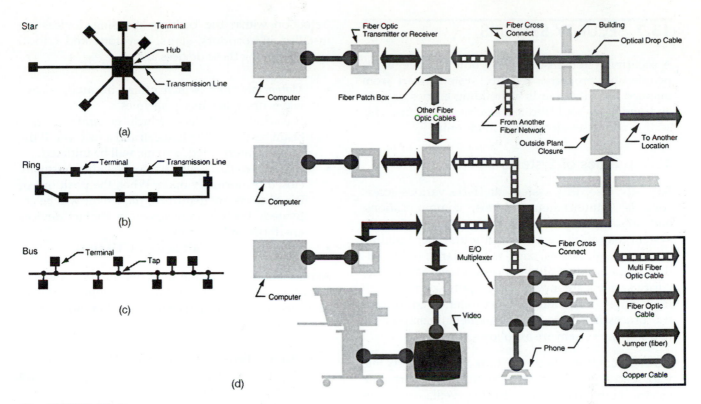

■ **FIGURE 13–6**

(a) Star LAN is arranged around a single hub. The hub can be simply a passive coupler or an active controller. This is the preferred arrangement for all LAN installations.

(b) Ring LAN is linked point to point. A bit pattern (token) is circulated to each node (workstation or peripheral equipment), to control the data transmission.

(c) Bus LAN uses carrier-sensing multiple access with collision detection to control data transmission.

(d) A LAN wiring diagram incorporating telephone, computer, and A/V links with multiple- and single-mode fibers as risers (backbone) and twisted pair or coaxial wires for equipment connections. (Courtesy: Belden Wire & Cable, Richmond, IN.)

- Equipment rooms, including construction and space requirements
- Telecommunication closets, particularly their locations and sizes
- Backbone wiring, including risers and the infrastructure
- Horizontal wiring, i.e., wiring between backbone and workstations
- The work area
- Administration

Specific recommendations for each design element are given in EIA/TIA, 568, standards. When the recommendations are followed faithfully, they will ease network segmentation and enhance computerized tracking and documentation, as well as minimize attenuation of the signal. The wiring recommended for backbone and horizontal distribution includes 100-ohm UTP, 150-ohm STP, 50-ohm coaxial cables and 62.5/125 micrometer optical fibers.

Wiring within a building for a LAN is usually made with twisted pair or fiber-optic cables. The selection depends on the type of data to be transmitted. For high-speed transmission and broad bandwidth, optic fiber is preferred at a higher cost. (See Table 13–1 for the applications of telecommunication cables.)

The connection and termination of cables are best made at one or more wiring closets distributed strategically throughout the building, preferably not more than 200 ft apart. The maximum length of LAN cable between the server and all its workstations should be less than 2 km, whether the network is confined to one building or stretches across several buildings.

Wiring for (digital) data distribution systems may be combined with (analog) A/V and telephone systems when the proper interface devices are provided. (See Figure 13–6(d) for a typical wiring diagram of a combined data, telephone, and A/V system.)

13.7 SECURITY SYSTEMS

A security system installed in a building safeguards people and property. Security systems start from manned security guards at building entries or exits and end with sophisticated electrical systems. This section addresses electrical systems only.

13.7.1 Types of Systems

Basically, a security system interfaces various access controls, annunciations, alarms, communications, and information-processing components. A security system may itself be interfaced with other building management systems, such as fire alarm, public address, and building automation systems. Modern security systems are inevitably computer systems with logic chips, programmable controllers, and/or central processing units. The fundamental components of a security system are as follows.

Intrusion controls   Security starts at the property line with fences or walls that may or may not incorporate electrical surveillance. Electrical surveillance is usually operating on infrared (active and passive), acoustical (audible and ultrasonic), microwave (beam pattern and field effect), or vibration detection principles.

Access controls   An access control system identifies the person seeking to enter or exit a building. The techniques include the following:

- *I.D. cards*, which may incorporate magnetic strips, proximity-tuned passive circuits, coded pattern capacitors, infrared optical marks, or mechanical (Hollerith) coded holes arranged in a pattern.
- *Biometric identifications*, which make use of several unique physiological characteristics of a person, such as fingerprints, a retinal scan, hand geometry, a signature, a voiceprint, etc. Of these characteristics, fingerprints and eye retina identification are the most reliable and are thus desirable for high security applications.
- Use of a *TV camera* in a CCTV system. With a camera, visual recognition is relatively simple and economical for small or medium-sized facilities. However, it would not be practical to use only a camera in large facilities, where hundreds or thousands of personnel must be recognized by the security personnel. Nonetheless, a camera works very effectively in conjunction with other methods of identification.

Detection within the building   Sensing devices are installed in corridors, stairs, elevators, and critical spaces. Among these devices are:

- *Motion detectors*, which utilize infrared, microwave, or capacitance principles.
- *Photoelectric detectors*, which generate a light beam between the photocell and a receiver. If the beam is interrupted, a signal will be initiated.
- *Electrical contacts*, which may be either normally closed or normally open. When the positions of the contacts are disturbed, the circuit will be activated. Typical applications of contact devices are doors and windows.
- *TV cameras*, as a part of the CCTV system.

Annunciation   When the detection devices are activated, the security system should announce the event at strategic locations in the building. Commonly used devices include:

- *Alarms*—bells, horns, buzzers, etc.
- *Annunciators*, i.e., lights and lighted panels with locations identified.
- *TV monitors*, as part of the CCTV system.
- *Speakers*, which interface with the building's public address or intercom system.
- *Digitized voice messages*.
- A *wireless radio system*, which interfaces with the building's radio communication system.

Information recording   All security-related activities should be recorded through a computer-based printer or file to register the sequence of events.

13.7.2 Closed Circuit Television

CCTV is a wired and self-contained television system. Although popularly used in security applications, it is also used in industrial process controls, business promotion, sports training, traffic control, experimentation, data filing, etc.

A basic CCTV system consists of an electronic camera that converts an optical image into analog signals, a transmission medium (usually coaxial cables), and a monitor to convert the signal back to an image. The signal is stored on tape or compact disk through a videorecorder or video printer. The system may be expanded into a multiple camera, monitor, and recorder system. A CCTV system may have one or more of the following features:

- *Mode.* Black and white or color; continuous, sequential switched, or time-lapsed monitoring.

(a)

Camera
WV-CP100

Color Monitor
WV-CM110A

Camera
WV-CP100

Record   Playback

Video

Camera
WV-CF20

Power

Time-Lapse   VCR

Camera
WV-CP100

Camera
WV-CP100

Camera
WV-CF20

Video

Power

Color Monitor
WV-CM110A

Time-Lapse VCR

Camera
WV-CP100

Camera Housing
WV-40

Auto Panning
Head
WV-35A

Camera
WV-CP100

Auto Panning
Head
WV-35A

Camera
WV-CP100

Camera
Extension Unit
WV-AD110A

Panning Controller
WV-32

Playback

Record

(b)

■ FIGURE 13–7

Closed circuit TV system
(a) Basic components: camera, monitor, VCR, switching units, etc.
(b) A typical CCTV system diagram.
(c) Illustration of a time-lapse mode recording system with automatic and field switching modes.
(Courtesy: Panasonic Broadcast & Television Systems Company, Secaucus, NJ.)

Camera-A

Camera-B

Camera-C

Digital Field Switcher
WJ-FS20

VCR

24H Time-Lapse Mode Recording Pattern

1sec   (Field Switching)

Video Tape	camera A	camera B	camera C	camera A	camera B	camera C	camera A	camera B	camera C
Recorded Pictures									

(c)

- *Cameras.* Fixed position or pan/tilt/zoom (PTZ); sensitivity to normal or infrared spectrum; freeze action; remote controls.
- *Monitors.* Single or multiple units; full or split screen; various sizes of screen; various resolutions (300 to 1200 lines per screen).
- *Recorders.* Videocamera recorders; time lapse recorders; sequential switches; time-date generators.

Figure 13–7(a) shows the various CCTV components, and Figure 13–7(b) shows the connection diagram of a system with a multicamera, time-lapse videotape recorder, sequential switching, and a date generator. In practice, CCTV systems are usually interfaced with other systems into a combined security system. (See Figure 13–7(c).)

13.8 TELEPHONE SYSTEMS

Since the invention of the telephone in 1878 by Alexander Graham Bell, telephony has become the primary means of communication in modern societies. The level of development of a nation or a region can often be judged by the number of telephones installed per 1000 capita.

Telephone services are usually provided by public utilities, which may be owned by private enterprise or the government. With the exponential development of new telephone technology, such as cellular telephone systems, private branch exchanges (PBX), wireless telephones, pagers, personal communication service (PCS), and facsimile (FAX), the planning of telephone systems in a building has become more complex. Often, the advice of the public utility or an independent telecommunication consultant is necessary if a new facility is to possess state-of-the-art capabilities.

The fundamental principle of a telephone system is very simple. The system starts with individual telephone sets, a central switching or exchange facility, a DC power supply, and distribution wiring in between. The traditional telephone set consists of a transmitter, a receiver, a ringing circuit, and switching and coding devices. The transmitter is in effect a microphone, and the receiver is a speaker. When a person talks into the transmitter, the acoustical (sound) energy compresses a carbon-granule-filled diaphragm, causing a change in its electrical resistance, which in turn varies the electrical current (at the milliampere level) through the loop circuit to the receiving party's telephone. The receiver then converts the electrical current back into acoustical energy. Figure 13–8(a) illustrates the construction of a traditional telephone transmitter. Nowadays, transmitter/receivers are increasingly the electrodynamic type, utilizing the principle of mutual inductance between electricity and magnetism (Figure 13–8(b)). The electret type utilizes the property of electrostatic charge on a coated dielectric diaphragm. Vibration of the diaphragm will change the electric voltage and current in the loop circuit. With amplification, the signal is transmitted to the receiving telephone (Figure 13–8(c)).

For long-distance transmission, the analog voice signals are converted into digital signals so that they may be integrated with other digital signals utilizing multiplexing technology. The quality of new telephone systems is improving steadily, to the point that long-distance conversation over thousands of miles away is just as clear as if it were next door.

Some mobile and cellular telephone systems operate on narrow radio frequency bands. (See Table 13–2 for their operating frequencies.) The signals are currently transmitted from closely spaced transmitter towers to cover a defined area. The technology will most likely migrate to satellites and eliminate the costly local transmission towers.

13.8.1 Types of Systems

Telephone services are directly provided by the telephone company. There are two types of dialing systems: pulse-tone (a rotary dial) and touch-tone (a push-button key pad). The latter is a newer technology that can interface with digital signals. It offers many more features, such as data transmission, voice mail, and fax, that are not easily achievable with the pulse-tone system. A touch-tone telephone has 12 push buttons (0 to 9 plus *, and #). When pushed, each button generates two tones, one low-frequency tone between 697 and 941 Hz and one high-frequency tone between 1209 and 1477 Hz. This feature, known as dual-tone multifrequency (DTMF), is designed to prevent false signals.

A *private branch exchange* (PBX) is a private telephone system serving one building or a group of buildings. A PBX system connects to the public telephone system through a switchboard or automatic switching equipment. The switchboard may or may not require an operator. The PBX abbreviation is used interchangeably with PABX, which stands for private automatic branch exchange, since current PBX systems are all of the automatic type. Most PBX systems have direct outward dialing, a feature wherein one dials ''9'' to connect to the public telephone trunk line. A PBX system can be interfaced with computer systems. Figure 13–8(d) illustrates a PC-based PBX system. Some typical features of a PBX system are the following:

- Ancillary device connection
- Automated attendant
- Automatic call distribution

- Automatic redial
- Automatic trunk-to-trunk traffic
- Call forwarding
- Call transfer
- Call holding
- Call privacy

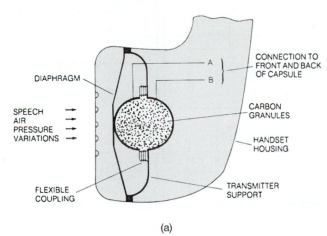

(a)

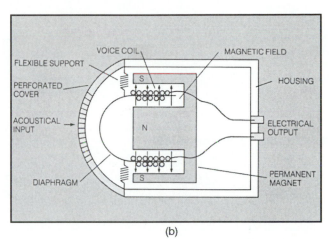

(b)

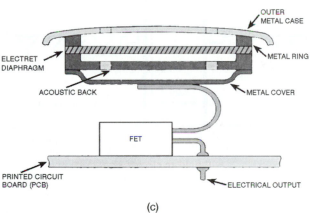

(c)

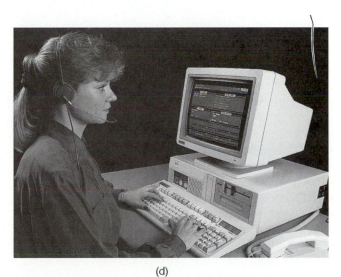

(d)

■   **FIGURE 13–8**

Basic construction and features of three types of transmitter (microphone) of a telephone set. The traditional carbon-granule type will soon be replaced by the newer electrodynamic and electret types. (a) Carbon-granule type. (b) Electrodynamic type. (c) Electret type. (Source: Stephen J. Bigelow, *Understanding Telephone Electronics*, Sams Publishing, Indianapolis, IN.) (d) A computer-based PBX system. The system can be fully automatic without anyone in attendance. When employed, the operator is usually assigned with other duties when not in demand. (Courtesy: Mitel Corporation, Kanata, Ontario, Canada.) (Figure continues page 398.)

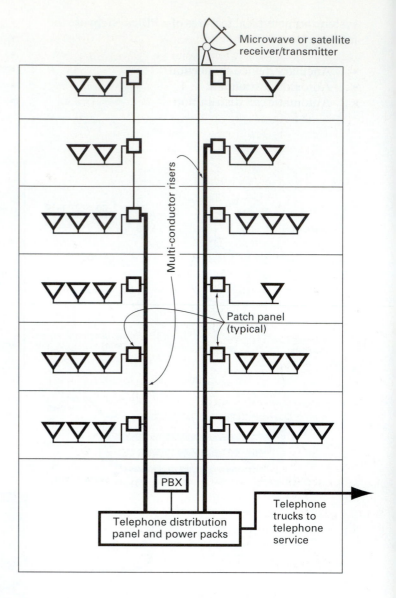

■ FIGURE 13–8 (continued)
(e) A typical telephone wiring distribution (riser) diagram.

- Conference calls
- Code restriction
- Customized messaging
- Delay announcement
- Direct inward dialing
- Direct outward dialing
- Speed dialing
- Intercom
- Fax interface
- Hot line
- Key/multifunction registration
- Night answer
- Off-hook voice announcement
- Off-premise extension
- PC programming
- Route advance block
- System/station speed dial
- System data up/down load

- Security code (password)
- Toll restrict
- Trunk-to-trunk transfer
- Voice mail integration
- Voice recording service
- Centrex compatibility
- All call

Centrex is a business telephone service offered by the local telephone company. It is basically single-line telephone service with special features, such as intercom, call forwarding, call transfer, least cost routing, etc. The advantages of Centrex over a PBX system are Centrex's lower initial investment, equipment space savings, continuous updating with new technology, and lower level of maintenance. However, the special features are limited, and the service may not be available in many areas.

Facsimile (FAX) is the technology that allows written material, data, and images to be transmitted through the switched telephone system and printed out at the receiving end. There are five internationally accepted equipment groups: 1, 2, 3, 3E (enhanced), and 4. Groups 3 and 3E are the most popular types, with transmission speeds between 9600 and 14,400 bps. Group 4 is designed for digital lines transmitting at speeds between 56,000 and 64,000 bps.

13.8.2 Design Considerations

Wiring for telephone systems need not be installed in raceways. However, for security and aesthetic reasons, all wiring within finished spaces should be concealed. Receptacles should be provided within the telephone closets for equipment. Telephone trunk lines should be terminated in strips in boxes or in telephone closets for splicing and distribution. Figure 13–8(e) illustrates the distribution diagram of a telephone system. For preliminary purposes, one linear foot of wall or closet space is allowed for each 3000 sq ft of finished floor space.

13.9 FIRE ALARM SYSTEMS

A fire alarm system is not limited to initiating an alarm during a fire; it may also serve to identify the location of the fire, transmit audio or visual messages, activate fire-extinguishing systems, and interface with building management systems. A more sophisticated fire alarm system is in effect a fire management system. This section introduces various fire alarm systems and their electrical wiring requirements.

13.9.1 Code Requirements

The requirements of a fire alarm system and the locations of devices are governed by the National Fire Protection Code (NFPA 72), Americans with Disabilities Act (ADA), and local building codes, which vary with the occupancy, location, size, and height of the building. In general, a fire alarm system is required in public assembly buildings, businesses, educational institutions, hotels, factories, and hospitals. Theaters and entertainment facilities are exempted, as a false alarm may cause a panic.

13.9.2 Types of Systems

A fire alarm system is usually a combination of several basic systems. Among these are the following:

Central station vs. local system   The alarm signals are transmitted to a central monitor remote from the building. With modern communication technology, the central station may be miles away from the building, at the fire department or even in another city to which transmission occurs through telephone or modem lines. Local systems are wired completely within the building. A key factor to a reliable fire alarm system is the reliability of the electrical power supply. On-site generators or battery backup is preferred.

Manual or automatic system   A fire alarm system may simply consist of manual alarm stations connected to sound bells or horns. However, for building use groups that involve the public and for residential buildings over three stories in height, automatic detection devices are usually mandatory. In automatic systems, signals are initiated by automatic detection devices, as well as from manual alarm stations.

Coded or noncoded system   A fire alarm system may provide the alarm signal continuously until the system is manually turned off. This is known as a noncoded system. By contrast, if the signal is intermittent in duration or frequency, then the system is said to be coded. A coded system is usually designed to produce three rounds of signals to identify the location of the fire or the initiating device. For example, a signal with one short and one long sound may mean that the fire is at the first-floor east wing, whereas a signal with two short and two long sounds may mean that the fire is at the second-floor west wing. A coded system is a necessity for multi-wing or multisectional buildings. The coding may have to be subzoned for large buildings when it becomes too complex.

Supervised or nonsupervised system   With a nonsupervised system, accidental grounding or breaking of the wiring or contacts will unknowingly disable the system until the problem is discovered. In the meantime, the system is inoperative and the building is unprotected. In a supervised system, the signal notifies the building management of the problem until it is fixed. Supervised or double-supervised systems are preferred in large buildings.

Single or zoned system   In a single-zone system, all alarms are activated at once; in contrast, a zoned system may divide the alarms into two or more zones according to the locations of the alarm devices. For example, a building with east, center, and west wings may be divided into 3 zones, 1 for each wing. If the building is five stories high, then it may be divided into 15 zones, 1 for each floor of each wing. Caution must be exercised not to overdivide the zones, as the signal may become too complicated to be recognized.

Single-stage or two-stage systems With a single-stage system, all alarms within each zone will be energized to signal that the building should be evacuated. Two-stage systems may provide a preliminary warning or alert to the occupants during the first stage and a signal to evacuate the building only when the second stage is energized.

General alarm or presignal system In buildings where a general alarm of a minor fire may cause a panic and the building is under full time supervision by qualified personnel, a presignal system may be desirable. With this system, detection devices or manual alarm stations will send the signal only to limited locations so that management can determine whether and when a general alarm is to be activated. Schools and stadiums are good candidates for such a system. Note that the use of a presignal system in hospitals is prohibited by the NFPA, unless specifically approved by code officials.

Voice communication system Modern fire-signaling systems can also include a public address system to provide instructions to the occupants by the building management and, in later stages, by the fire department. The public address system may also be supplemented by two-way communication devices, such as telephones, intercoms, radio frequency modules, etc. Speakers may be used to generate a tone in lieu of bells or horns. Voice communication systems are required in all high-rise buildings and public assembly use groups.

Addressable or nonaddressable system A system is addressable when each fire detection or signaling device is assigned a unique coded frequency (address). The address may be adjustable to allow reassignment in the field. With this capability, the fire protection plan (zoning and staging) may easily be changed to fit a building's operational program. Addressability is especially useful for a tenant-occupied building when the tenants are frequently changing. Nonaddressable systems do not have this flexibility, but are, or course, more economical. A system may combine addressable and nonaddressable devices through an interfacing device. Figure 13–9 illustrates a typical multizone fire alarm system.

Standalone or integrated system The fire alarm system may be a standalone system or may be interfaced with other building systems, such as an automatic sprinkler, building management, life safety, and vertical transportation systems. The advantages of interfacing are many, including space and cost savings and better coordination. The primary disadvantages

are the complexity of the system and the mutual dependence among its components.

Thus, a fire-alarm system combination may be a system such as the following:

- A manual, single-zone, single-stage system
- A manual and automatic, multizone, noncoded, single-stage system
- An automatic, multizone, coded, two-stage with presignal and voice communication system

Naturally, a more sophisticated system offers better protection and requires a greater capital investment and more supervision. The selection of a system is governed primarily by the code and determined partially by the evaluation of the specific needs of the building.

13.9.3 Design Considerations

Because of the critical nature of fire-signaling, and life safety, wiring for fire alarm systems should be in conduits or raceways separate from other, noncritical power or low-voltage systems. Electrical power should be supplied from the emergency section of the power system and, preferably, backed up by an independent power system, such as battery packs or on-site engine generators. Some fire alarm systems can also be provided with power packs.

Fire and smoke detection devices should be located where there is a likely hazard, such as in storage rooms, electrical-mechanical rooms, and public egress areas (e.g., in entrances, elevators, or lobbies, near stairways, etc.). The fire command center, where the control panel or the command console is located, should be in a secure locale, constructed of fire walls and, preferably, with a direct means of egress.

Audible fire annunciation devices, such as bells, horns, chimes, and tone generators, should have the proper sound power (dB), so that all occupants in the building can hear the signals. On the other hand, too loud a sound may create a deadening effect on the ears and possible panic. The sound level should be higher than 60 dB at any location, but not higher than 120 dB near the annunciation device. This range is usually the guideline for determining the selection and spacing of such devices.

Visual annunciating devices, such as flashing or strobe lights, should be provided for the hearing impaired, as per ADA requirements.

The fire alarm system should be coordinated with the building HVAC control system so that the HVAC system may operate in a manner that assists in the safe and speedy evacuation of the building's oc-

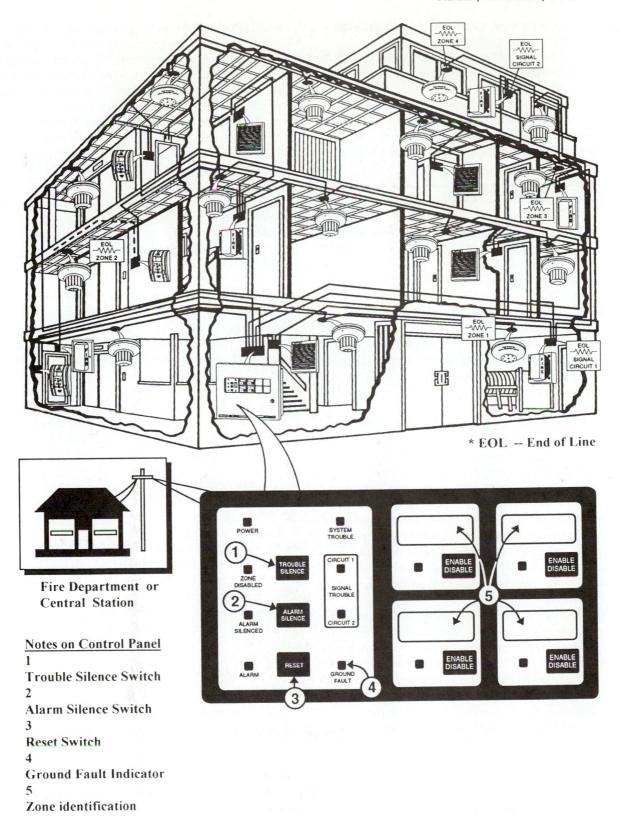

* EOL -- End of Line

Fire Department or Central Station

Notes on Control Panel

1
Trouble Silence Switch

2
Alarm Silence Switch

3
Reset Switch

4
Ground Fault Indicator

5
Zone identification

■ **FIGURE 13–9**
A small multizone fire alarm system showing the location of the control panel, signal and alarm devices. System shown is a four-zone system. Additional zones can be accommodated. (Courtesy: G. S. Edwards/A Unit of General Signal, Farmington, CT.)

cupants. This is usually achieved through the efforts of the HVAC design engineer, who attempts to sequence the quantities of supply, return, exhaust, and outside air to maintain a smoke-free environment at the building exitways.

In designing the fire alarm system, the designer should study the feasibility and economics of establishing an integrated fire alarm and building management system that is most appropriate for the building.

13.10 SOUND SYSTEM

Sound systems in buildings normally involve the distribution of voice or music from one part of the building to another. The system may be capable of one- or two-way communication and may be either wired or wireless.

A sound system consists of one or more input devices, such as a microphone, tape recorder, radio, or telephone. The input is converted to electric signals that are amplified and transmitted to output devices, such as speakers or telephone receivers. In broad terms, a sound system is analogous to an amplified telephone system.

Sound systems may be classified into two groups. A *public address system* is designed to broadcast a voice message, such as an announcement or speech, from one or more locations to other parts of the building. It also may carry or mix with other inputs, such as radio, tape, or telephone. It is usually a one-way system with output devices that have talkback capabilities. Public address systems are installed in schools, hotels, department stores, hospitals, industrial plants, and other buildings in which the administration or management can make announcements or transmit music.

An *intercom system* is a two-way communication system between two or more stations. It may consist of one or more master stations, which can communicate with each other, or slave units, which can communicate only with the master unit to which they are connected. Intercom systems are installed in the home, apartments, offices, and specific departments within a building, such as between the sales area and the warehouse of a store, between offices, and between the lobby and apartments in a multiunit apartment building.

13.10.1 Basic Components

Microphones   A microphone is an input device that reacts to and converts variable sound pressure (speech, voice, or music) into variable electrical cur-

rent. The most popular microphones are the condenser type, which operates through changes in capacitance caused by vibrations of its conductive diaphragm, and the dynamic type, which has an electrical coil attached to a diaphragm moving within a magnetic field. A low-cost microphone may have a frequency response range between 50 and 10,000 Hz (the normal frequency range for human hearing), whereas the high-end microphone could extend the range to between 20 and 20,000 Hz. The response to the sound pressure may be very narrow (directional), broad (cardioid) or nondirectional (omnidirectional). A microphone must be selected so as to match the impedance with other components of the system. Figure 13–10(a) illustrates the internal construction of a dynamic microphone.

Speakers   A speaker is an output device that converts electrical signals into sound signals through electromagnetic voice coils and a driver that vibrates a speaker cone. Speakers are designed to respond to either the full frequency range or tuned frequencies—that is, bass (a woofer, for low frequencies) or treble (a tweeter, for high frequencies). Speakers are hung from ceiling surfaces, are recessed into the ceiling, or are mounted on the wall. Most speakers are provided with matching transformer and volume controls. Figure 13–10(b) illustrates several types of speaker.

Amplifiers   An amplifier receives a sound-generated electrical signal and boosts the signal for transmission to output devices. Amplifiers for public address systems are from 15 watts to 300 watts of sound power with a frequency response from 20 to 30,000 Hz. They may be self-contained or may be designed to include switches, a radio tuner, tape or compact disks, and a telephone interface. Figure 13–10(c) illustrates a desktop model of a public address system and its block diagram.

13.10.2 Design Considerations

In designing a building sound system, one should determine the location of a control center where messages are normally initiated and the locations, sizes, and ambient noise levels of rooms to which the messages are to be transmitted. Figure 13–10(d) is a one-line diagram of a typical public address system.

Next, one must determine the approximate number of speakers required for each space. In general, ceiling speakers (usually 8″ in diameter) can cover 250 sq ft of floor area at an 8-ft ceiling height and up to 600 sq ft at a 12-ft ceiling height. A trumpet (horn) type of speaker may cover 8000 sq ft in large, quiet

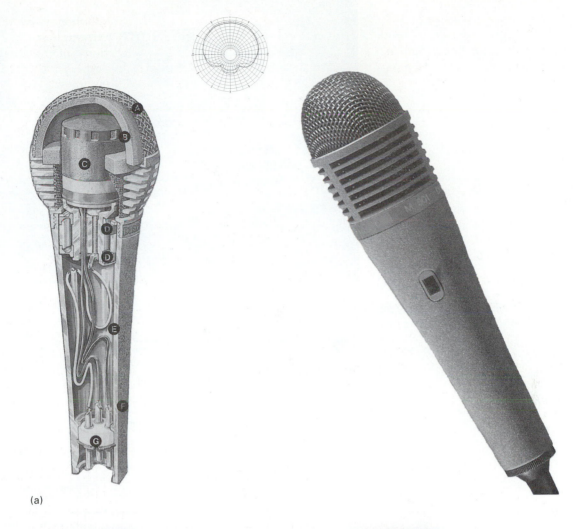

(a)

(b)

■ **FIGURE 13–10**
(a) Cutaway and exterior views of a dynamic microphone. (Courtesy: PASO, Inc., Pelham, NY.) (b) Typical ceiling speaker assembly including speaker with matching transformer and ceiling baffle (top); typical trumpet speaker (bottom). (Courtesy: Atlas/Soundolier, Fenton, MO.) (Figure continues page 402.)

FIGURE 13–10 (continued)
(c) Desktop model of PA system control panel with telephone interface, and a block diagram of system components. (Courtesy: Dukane Corporation, Saint Charles, IL.) (d) Line diagram of a typical PA system.

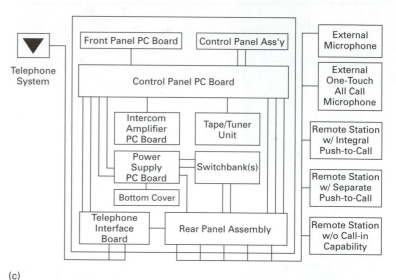

(c)

(d)

open spaces; the area of coverage is reduced in noisy spaces.

For planning purposes the power required for different speakers is as follows:

- 8" ceiling or single-projection wall speakers 1 watt
- 12" ceiling or double-projection wall speakers 2 watts
- Column speakers   10–30 watts
- Trumpet speakers   10–30 watts
- Signal (ring) generators   2–4 watts

The amplifier is sized for the total power of the speakers, plus an allowance for future expansion. To achieve good sound reproduction, it is good practice not to load the amplifier to its maximum rating.

Wiring of sound systems is primarily with twisted pair or coaxial cables. In conduits or wireways, shielded twisted pair (STP) is preferred.

13.11 TIME AND PROGRAM SYSTEMS

With watches becoming increasingly affordable, the need to indicate the time has diminished drastically in most buildings. However, time and program systems are still necessary in schools, sports facilities, and certain industries where synchronized time indication is important. Such systems consist of a control center (master clock, code generator, switches, etc.), individual clocks (secondary or slave clocks), program signaling devices (bells, buzzers, or annunciators), and wiring. Synchronization of the time may employ any of the following principles:

- *Synchronous motor driven type.* Secondary clocks and signaling devices are wired to the control center. The secondary clock runs by a synchronous motor, but its indication of time is monitored and adjusted by the master clock.
- *Impulse driven type.* Secondary clocks and signaling devices are wired directly to the control center. The secondary clock is driven by an impulse signal of the master clock and advances once every minute in unison with the master clock. During power failure, the control center will accumulate missed minute impulses for up to 24 hours by its standby battery power. Upon restoration of normal power, rapid correction pulses will be sent to advance the secondary clocks to the correct time.
- *Carrier frequency type.* Individual clocks are plugged into the standard 120-volt power without special wiring, but the indication of time is corrected by a high-frequency signal (between 3000 and 8000 Hz) carried (i.e., superimposed) through the 120-volt power wiring system. With the carrier frequency system, dedicated wiring between the control center and the individual clocks (and signaling devices) is eliminated.
- *Radio frequency type.* The individual clocks are plugged into the standard 120-volt power system without special wiring. The time is corrected by a radio frequency generator. The frequency may be tuned to the National Institute of Science and Technology (NIST) universal time for extreme accuracy in special research applications.

Figure 13–11 illustrates several components and the wiring diagram of a typical time and program system.

13.12 MISCELLANEOUS AND SPECIALTY SYSTEMS

Because of space considerations, many commonly used auxiliary systems have not been covered in this chapter. Nonetheless, they should not be overlooked in building design. For example, a doorbell system, although very simple, is necessary in many kinds of buildings. In addition, facilities such as offices, apartment complexes, hotels, hospitals, department stores, schools, and entertainment and sporting facilities often have many other unique systems that must be addressed during the initial planning phase of the project. Some of these systems are mentioned herewith without explanation. They serve as a reminder to the designer that auxiliary electrical systems are becoming more and more important in the building design process. The following list mentions some of these facilities and the systems often found in them:

- *Hospitals and health-care facilities.* Nurse's call; doctor's paging; patient-monitoring systems in intensive care units; bedside monitoring and control panel; telemedicine diagnoses; remote imaging readout.
- *Hotels and motels.* Do-not-disturb and room makeup systems; maid call; bedside environmental, lighting, and entertainment panel; CCTV; front-of-the-house and back-of-the-house management; central reservation and accounting point of sale.
- *Sporting and convention facilities.* Central lighting controls, scoreboards, CCTV, and videotapes;

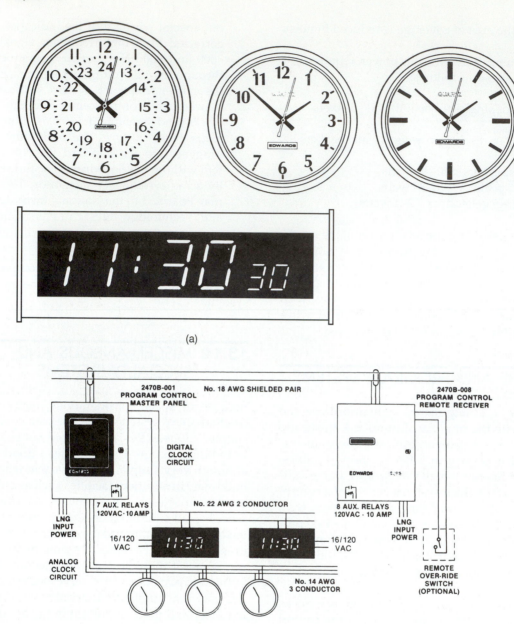

(a)

(b)

■ **FIGURE 13–11**

(a) Typical clocks include standard numerals with a 24-hour dial, a 12-hour dial, a graphic dial and digital readouts.

(b) A typical time and program system wiring diagram: program master control panel, programmable modules, back-up batteries, relays and zone switches. The system may also include signal devices, such as bells, buzzers, or horns.

(Courtesy: Edwards/General Signal, Farmingham, CT.)

two-way radios; special open-space fire and smoke (infrared beam).

- *Residences and apartments.* Door intercom; intrusion and security; wired sound; TV and satellite antenna.
- *Department and retail stores.* Shoplifting detection systems; general surveillance; point of sales; TV and satellite antennae distribution, staff-locating systems.

Table 13–3 is a checklist of auxiliary electrical systems that should be considered during the planning of the foregoing facilities.

TABLE 13–3

Checklist of auxiliary systems for buildings

Legend:
- ▨ Should be installed or required by Code
- ▥ (vertical lines) Should be considered to improve safety or operations
- ▤ (horizontal lines) May be considered in lieu of other options
- ☐ Not applicable

System to be considered	Residence	Office	School	College	Hotel	Hospital
Telecommunications System						
▪ Direct Telephone Lines	▨	▨	▨	▨	▨	▨
▪ Public Phones	☐	▥	▨	▨	▨	▨
▪ PBX System	☐	▤	▤	▤	▤	▤
▪ Fax/Modem/Data Lines	▤	▨	▨	▨	▨	▨
▪ Data Distribution Network	☐	▨	▤	▤	▨	▥
▪ LAN/WAN	☐	▥	▥	▥	☐	▤
▪ Microwave Dishes	☐	▤	▤	▤	▤	▤
Sound Systems						
▪ PA System	☐	☐	▥	▥	▨	▨
▪ Intercom	☐	▥	▥	▥	▨	▨
▪ Voice/Beep Paging	☐	☐	☐	☐	▨	▨
Radio and TV System						
▪ Radio & TV Antenna	▥	▥	▥	▥	▨	▥
▪ Satellite Dishes	▥	▤	▤	▥	▨	▥
▪ Cable TV	▥	▤	▤	▤	▨	▤
▪ Wired Music	☐	☐	☐	☐	▥	▤
Security Systems						
▪ Access Controls	▨	▥	▥	▥	☐	▥
▪ Intrusion Controls	▥	▥	▥	▥	▤	▤
▪ Closed Circuit TV	▤	▤	▥	▤	▨	▨
Door Bell and Intercom	▨	▨	▨	▨	☐	▨
Time and Program Systems	☐	▤	▨	▤	☐	▤
Special Systems						

QUESTIONS

13.1 What are the forms and functions of information systems?

13.2 What are the forms and functions that can be provided by radio? TV? A computer with a modem?

13.3 What is the difference between analog and digital signals? Can they be converted from one to the other?

13.4 Human hearing frequency extends beyond 10,000 Hz. What is the highest frequency that can normally be transmitted by a conventional telephone system?

13.5 What are commonly used wires for telecommunications systems?

13.6 What is fiber optics? What is an optic fiber?

13.7 Name two primary advantages of an optic fiber over twisted pair or coax cables.

13.8 What is meant by critical angle in an optic fiber?

13.9 The index of refraction of glass is 1.5 ($N_1 = 1.5$) and the index of refraction of air is 1.0 ($N_2 = 1.0$). What is the critical angle when light passes from glass to air?

13.10 What are the metal compounds commonly used to generate laser in fiber optic systems? What are the range of wavelengths in nanometers (nm)?

13.11 What is the applicable transmission speed range in Mbps for a coax cable? A single mode optic fiber?

13.12 What are the frequency bands assigned for FM radio in the USA? VHF/TV channels (2 to 6)? Cellular telephones? Satellite UHF?

13.13 What is meant by geostationary-type telecommunication satellite? What is its orbit (miles above the earth)?

13.14 What is a LAN? A WAN?

13.15 What are the disadvantages of wireless LAN?

13.16 What are the commonly employed principles in intrusion detection?

13.17 What are the techniques commonly used in access control?

13.18 What are the detection devices commonly used within a building?

13.19 What is PBX, DOD, and DID in a telephone system?

13.20 What are the two types of telephone dialing systems?

13.21 What is a coded, a supervised, a zoned, and an addressable fire alarm system?

13.22 Why are flashing or strobe lights required in fire alarm systems?

13.23 What is a public address (PA) system? What are its basic components?

13.24 What size ceiling or wall-mounted speaker is normally used in buildings and what is the approximate area of coverage?

13.25 What wire types are normally used for sound systems? Can optic fiber be used?

13.26 What is a time and program system? Where is it usually installed?

13.27 How are clocks synchronized in the time-clock system?

13.28 Name as many as you can of the auxiliary systems which should be considered in a large residence; a hospital; a hotel; and a department store.

REFERENCES

1. NFPA-70, National Electrical Code, National Fire Protection Association, Quincy, MA.

2. EIA/TIA 586, Commercial Building Wiring Standard.

3. Jeff Hecht, *Understanding Fiber-Optics.* Indianapolis: Prentice Hall Computer Publishing, 1993.

4. Harry Newton, *Newton's Telecom Dictionary,* 7th ed. New York: Flatiron Publishing, 1994.

5. Stephen Bigelow, *Understanding Telephone Electronics.* Indianapolis: Prentice Hall Computer Publishing, 1993.

6. Robert T. Paynter, *Introductory Electronic Devices and Circuits.* Englewood Cliffs, NJ: Prentice Hall, 1991.

7. *Lighting Handbook.* New York: Illuminating Engineering Society of North America, 1993.

14

LIGHT AND LIGHTING

LIGHTING IS THE UTILIZATION OF EITHER NATURAL OR artificially generated light to provide a desired visual environment for work and living. This chapter covers the properties of light, human vision, and the fundamentals of lighting applicable to the building's environment.

14.1 LIGHT AND VISION

14.1.1 Natural Light as Radiant Energy

Natural light is radiant energy that originates from the sun. Based on the wave theory, natural light is a portion of the electromagnetic energy spectrum sensitive to the human eye. This portion of radiant energy extends from 380 to 780 nanometers (nm) in wavelength. One nanometer is defined as one billionth of a meter.

All forms of radiant energy are transmitted at the same speed in vacuum: 186,282 miles per second, or 3×10^8 meters per second. In air, water, and other media, radiant energy travels somewhat slower. Each form of radiant energy differs in wavelength and thus, frequency. Equation (13–1) in Chapter 13 expresses the relation between the frequency and wavelength of radiant energy. The frequency of visible radiation energy is 800×10^{12} to 380×10^{12} Hz for violet to red, respectively. The overall spectrum of radiant energy is shown in color Plate 11; note the visible spectrum, or the portion that causes human visual sensation. Table 14–1 lists other properties of solar energy related to HVAC and lighting design.

14.1.2 Light Converted from Other Energy

Light may be converted from other forms of energy; for example, the burning of oil converts chemical energy into heat and light energy. The most efficient means of converting energy into light is through electrical energy, which is also part of the electromagnetic spectrum, but with a considerably lower frequency than that of light—for example, 60 cycles per second. With modern technology, the conversion efficiency of electrical energy to lighting energy has advanced steadily, and all modern light sources used in buildings are, without exception, electric light sources.

14.1.3 Vision

The human eye is so constructed that when light enters the cornea and lens, light energy is focused on the retina and transferred to the brain by optic nerve cells. The brain then translates the information back to the eyes forming the optical image. The impressions from both eyes integrate the information into three-dimensional images, a very complex process indeed. A cross-sectional view of the human eye is illustrated in Figure 14–1.

TABLE 14–1

Facts about solar energy (values approximate)

	Conventional Units	SI Units
Solar intensity (maximum)	330 Btuh/ft²	3800 KJ/m²
Solar heat gain through clear glass (vertical) (40° North Latitude)	250 Btuh/ft²	2800 KJ/m²
Daylight (sunlight + skylight)	10,000 footcandles	108,000 lux
Daylight (overcast sky)	1000 footcandles	10,800 lux
Sky luminance near horizon (clear day)	1500 candela/ft²	16,000 candela/m²
Sky liminance near horizon (overcast day)	500 candela/ft²	5400 candela/m²
Sky luminance overhead (clear day)	500 candela/ft²	5400 candela/m²
Sky luminance overhead (overcast day)	1500 candela/ft²	16,000 candela/m²

Human beings see in two ways: by color difference and by luminance (brightness) contrast. The details of color and luminance will be covered in later sections; here, we introduce some fundamental concepts relating to these properties. We can see red objects on a blue background, but not easily on a background of the same red color. There must be a *color difference* for our eyes to see. Similarly, we can read black print on white paper, but not easily on dark gray paper and not at all on black paper. Thus, there must also be a *luminance contrast* in order for our eyes to see.

Luminance and luminance contrast Simply stated, *luminance* is the amount of light sensed by our eyes from an object. Light may be generated from the object, such as the sun or a light source; may be reflected from the object, such as the tabletop or the paper of a book; or may be transmitted from the object, such as a stained glass window or a lamp shade. *Luminance contrast* can be defined in a number of ways. For a reading task, we generally use the formula

$$C = |L_t - L_b|/L_b \qquad (14\text{–}1)$$

where C = Contrast
L_t = Luminance of the task, candela per unit area
L_b = Luminance of the background, candela per unit area

(*Note:* The absolute value of $|L_t - L_b|$ is used when L_t is darker than L_b, and C is always less than unity.)

HOW THE EYE WORKS

Light waves enter the eye through the cornea, which roughly focuses the pattern of waves on the foveal pit in the retina. The waves are fine-focused as they pass through the lens. The iris acts as a diaphragm that expands or contracts the pupil (opening in the iris to the lens), controlling the amount of light that is permitted to enter the eye.

The rods and cones are the ultimate receivers for individual parts of the image. They transform the received optical image pattern from radiant energy into chemical energy, which energizes millions of nerve endings. The optical pattern then becomes a series of electrical impulses traveling within a very special group of nerves that connect to the optic nerve. The optic nerves (from both eyes) combine and transmit the selective impulses to the brain, where they are interpreted.

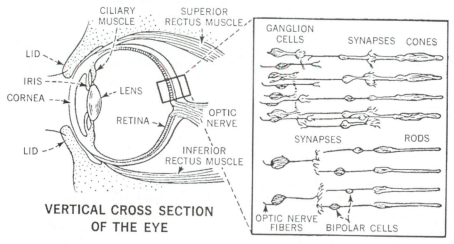

VERTICAL CROSS SECTION OF THE EYE

■ **FIGURE 14–1**

How the eye works. (Courtesy: General Electric.)

Plate 12a provides an excellent illustration of typed text over a gradually changing background. The difference in readability or visual acuity of the text is obvious. Contrast is even more important in the case of a fast-moving object, such as a baseball thrown at a speed exceeding 100 mi/hr. Plate 12b shows a baseball on a white background compared with one on a dark background. There are true stories about great professional players losing sight of routine fly balls in some indoor stadiums. The cause is very simple: There is no contrast between the white ball and the white stadium roof.

Color differences   When the foreground and background are different colors, visual acuity improves even at the same luminance level. Plate 13 illustrates this effect in its view of a tennis ball (normally yellow) against several colored backgrounds. As shown in the so-called color wheel (Plate 21), the further the color difference, the better the visual acuity becomes.

Visual comfort   While high luminance contrast improves visual acuity, too high a contrast may cause eyestrain with prolonged viewing. It is often more desirable to print black text on colored paper, such as tan, ivory, or light blue. Obviously, visual acuity is diminished when black text is printed on dark-colored paper.

Glare   When the luminance of a light source or an object is so high (so bright) that it begins to interfere with vision, it is called *glare*. When glare is so strong that it causes physiological discomfort, it is called *discomfort glare*. When glare actually affects the ability to see, it is called *disability glare*. The glare originated from light sources is called *direct glare*, and that reflected from a surface is called *reflected, or indirect, glare*. Direct glare can be avoided by relocating the light source away from the line of sight, if practical; indirect glare can be minimized by replacing the reflecting surface with nonglare (matte) or low reflectance (dark) surfaces.

Modeling   Modeling is the ability of the lighting system to reveal the three-dimensional image of an object. Without modeling, an object would appear to be flat. Modeling is extremely important in sports. Plate 18 illustrates the effect of lighting on the appearance of a three-dimensional tennis ball.

14.1.4 The Visible Spectrum

As illustrated in Plate 11, the visible spectrum extends from 380 to 780 nm in wavelength. Different wavelengths of light produce different sensations of color in the eyes. Wavelengths of colored light are not physically divided, but gradually shift from one color to the other.

Representative colors   The common recognizable colors of light are blue, blue-green (cyan), green, green-red (yellow), red, and red-purple (violet or magenta). Violet has the shortest wavelengths, from 380 to 430 nm, and red has the longest wavelengths, from 630 to 780 nm. The wavelengths of other colors are in between.

White light and the rainbow   White light is a mixture of all colors in the visible spectrum. A perfectly white light is the result of mixing all colors of equal energy level. Daylight at noon is close to being perfect, but daylight varies drastically throughout the day and is affected by the weather. (See Plate 19.) Thus, there really is no perfect white light, but rather, different white lights—bluish, greenish, yellowish, or reddish white lights. The fact that white light is a mixture of colored lights can be demonstrated by using a glass prism, as illustrated in Plate 14. When white light enters the prism, it bends by refraction. Since light of various wavelengths bends differently, the result is the splitting of white light into its component colors. This phenomenon occurs in nature when water vapor refracts sunlight to create a rainbow after heavy rain.

Ultraviolet and infrared   The shortest wavelength that is still easily visible is about 380 nm and is called violet. Wavelengths shorter than violet, in the region between 200 and 380 nm, are somewhat visible. This region is called ultraviolet, meaning "beyond violet." The longest wavelength that is only somewhat visible is in the region between 780 nm and beyond. This region is called infrared, meaning "below red."

Primary colors   Of all the colors, red, green, and blue are the most dominant. Because they are the basis of forming all the other colors, these three colors are called *primary colors* of light. This is contrary to pigments or paints, where magenta, yellow, and cyan are the primary colors. Mixing all three primary lights at equal energy levels produces white light. This is illustrated in Plate 15.

Secondary colors   When two of the three primary colors of light are mixed, they produce secondary colors of light: magenta (red + blue), cyan (green + blue), and yellow (green + red). Adding the primary color with its complementary secondary color will produce white light. For example, adding green light to magenta (red + blue) produces white.

Subtractive primaries   In pigments or paints (colorants), a primary color is defined as one that subtracts or absorbs a primary color of light and reflects or transmits the other two. So the primary color in pigments (also called subtractive primaries) are magenta, cyan and yellow, which correspond with the secondary colors of light. The subtractive nature of pigment is easily demonstrated by placing magenta, cyan, and yellow pigment filters over a white light source. Each of the filters absorbs or subtracts one of the primary colors from the light. Where two filters overlap, only the opposite primary color can pass through. For example, the yellow filter absorbs blue (transmitting red and green); the magenta filter absorbs green (transmitting red and blue). Together, the two filters only transmit red (see plates 15 and 16).

When a white light passes through all three subtractive primary filters, all colors are absorbed, and the result is no light, or black.

Reflection of light Light can be reflected from opaque or translucent materials. In fact, reflected light is how we see most objects in our visual field. As illustrated in Plate 16b, a red surface is a material that reflects red light and absorbs all other colors of light.

14.2 COLOR

14.2.1 Color Specifications

The human eye is sensitive to the visible part of the spectrum. (See Plate 11.) Color is a matter of visual perception, which varies with a person's subjective interpretation. A red apple may be described in common terms as red, deep red, bright red, rosy red, or even apple red. However, these terms are only a general description of a color. They are inadequate to define a color precisely. This section introduces some basic concepts used to quantify the color of a light or an object. In this regard, the references at the end of the chapter are valuable.

Three basic characteristics of color The color of a light or an object can be described by the following three characteristics:

- *Hue* is the basic color, such as red, yellow, green, blue, and the mixture of these colors, such as red yellow, blue green, red blue, etc. The basic color hues are illustrated in Plate 21a, commonly known as a color wheel.

- *Value* is the shade of color, such as light or dark red. In pigment or paint applications, the lighter color is the result of mixing the hue color with

white, and the darker color is the result of mixing the hue color with black. In lighting applications, the value of light is often disregarded, since black is simply the absence of light.

- *Chroma* is the intensity or degree of color saturation—that is, whether the color is vivid or dull. In lighting applications, it is the result of the dilution of spectrum (saturated) light with white lights.

Color specifications There are many systems for specifying color. Two of them commonly used for lighting are the Munsell system and the C.I.E. color system.

Munsell system A method of notation for describing the color of an object or a surface, such as a wall or carpet. The Munsell system uses two color charts. The first is a color wheel containing 20 basic hues of saturated colors. Each color is assigned an alphanumeric label, such as 5B for blue, 10G for the mixture of green and blue green, etc. (See Plate 21a.) The second chart is a set of color chips of the basic hues, but with different values and chromas. The value (vertical) scale is a lightness-darkness scale ranging from 1 for black to 10 for white. The value scale is useful in lighting calculations, and the value is approximately equal to the square root of the percentage reflectance of the color. For example: A surface with a Munsell value of 6 should have a reflectance value of approximately 36%. The chroma (horizontal) scale is a purity or saturation scale with the pure spectrum color as 10 and the fully diluted color as 1. (See Plate 21c.) The green coffee cup shown in Plate 21b can be described in Munsell notation as 5G5/10 green.

C.I.E color system The Commission Internationale de l'Éclairage (C.I.E.) adopted a color notation system for lighting applications. The system consists of a color diagram of the spectrum known as the *chromaticity diagram* (see Plate 22d). All colors can be found on this diagram, whether they are emitted, transmitted, or reflected. The horseshoe-shaped diagram is plotted on the x (red) and y (green) axes. The coordinate scales, from 0 to 1, are fractions of the red and green colors of the total primary colors, including blue. The z (blue) values can be determined by subtracting x and y from 1. Plate 17 shows a standard C.I.E. chromaticity diagram. All saturated colors are located on the perimeter of the diagram, which is known as the *spectrum locus*. All colors fade into white at the center of the diagram, which is called *equal-energy white* (0 percent saturated). Any color can be expressed in terms of its x, y-coordinates. For example, a saturated yellow is identified at $x =$

0.44 and $y = 0.56$, and a diluted yellow at $x = 0.38$ and $y = 0.46$.

14.2.2 Color Temperature

A theoretically black object, or *Planckian radiator*, which absorbs all radiant energy incident upon it, is called a *black body*. Such a body will become dull red when heated to 800°K, bright red at 2000°K, yellow at 3000°K, white at 5000°K, near equal-energy white at 6500°K, pale blue at 8000°K, and brilliant blue at 60,000°K. Thus, color temperature is frequently used to express the color of light sources. Representative values of light sources expressed in color temperatures are shown in Table 14–2.

The black body or Planckian radiator temperature locus can be found on the C.I.E. chromaticity diagram 17. Plotted on the same diagram are the color temperatures of several fluorescent and incandescent lamps, along with daylight sources of illumination. The color temperature rating of a lamp is a convenient way to describe the lamp's approximate color characteristics. It has nothing to do with the operating temperature or with the color-rendering quality of the lamp.

14.2.3 Color Rendering

Mixing of colors   White light may be obtained by mixing various amounts of the three primary colors. In fact, it can also be produced by the proper mixing of intermediate or unsaturated colors, since all unsaturated colors are mixtures of saturated colors. The color of an object is perceived by the composition of the light reflected or transmitted from the object. Thus, colored objects such as paints, textiles, etc., may appear to be different colors under different light sources—even different "white" light sources.

In general, standard white fluorescent lamps are rich in blue and lacking in yellow, whereas incandes-

cent lamps are rich in yellow and red, but are lacking in blue. Neither light source renders the true color of an object. The spectrum distribution characteristics of various light sources are given in Chapter 15.

The C.I.E. chromaticity diagram is a very useful tool in lighting design because it predicts color effects in merchandising or on stage when there are overlapping light sources. As an illustration (see Plate 17), a white light (W) may be produced by mixing equal power (lumens) of light source A ($x = 0.35$, $y = 0.52$) and light source B ($x = 0.26$, $y = 0.14$). The same white light may also be produced by mixing light source C ($x = 0.18$, $y = 0.45$) with light source D ($x = 0.46$, $y = 0.20$), or by any other combination of light sources in proper proportion.

Matching colors   When matching the color of two objects, one should view the objects under the light source that will be used. One may have an experience similar to that of matching wall paints or drapery materials in a store under one kind of light, only to find that they are not the same color under another light at home. A practical way to match colors is to observe the materials under two or more different light sources to see whether they match under all sources.

Psychology of color   Psychologically, people tend to associate the longer wavelength colors (red through orange) as warm, exciting, and dynamic and the shorter wavelength colors (blue through green) as cool, peaceful, and calm. Wavelengths in the middle (cyan, yellow, and tan) are neutral. In general, cooler colors are preferred in a hot climate or environment, and warmer colors are preferred in a cold climate or environment.

Color harmony and contrast   In general, colors adjacent in hue on the color wheel (see Plate 21) are similar. For example, red and orange (Munsell 5R and 10R) are next to each other. Colors separated by five or more positions, such as red and yellow (Munsell 5R and 5Y) or red and purple (Munsell 5R and 5P), are contrasting colors. Colors on the opposite side of the color wheel, such as red and blue green (Munsell 5R and 5BG), are complementary. Whether the color combination in a space should be similar, contrasting, or complementary depends largely on applications and individual preferences.

Color rendering index   One way to quantify the color rendering quality of light sources is the color rendering index (CRI) tested for the light sources. The CRI is a measure of the color shifts when standard color

TABLE 14–2

Color temperatures of various light sources

Light Source	Color Temperature (°K)
Candle Flame	2000
Gas-filled Incandescent Lamp	3000
Warm White Fluorescent Lamp	3500
Daylight Incandescent Lamp	4000
Cool White Fluorescent Lamp	4500
Daylight Photoflood Lamp	5000
Daylight Fluorescent Lamp	6500
Skylight (varies with time)	5500–28,000

samples are illuminated by the light source, as compared to a reference (standard) light source. This is explained in more detail in Chapter 15.

14.2.4 Color Selection

Color preference varies among individuals and is often affected by one's ethnic background, education, age, gender, etc. A color scheme for a building and its interior also depends on the building occupancy, spatial relations, geographical locations, climate, user's and designer's preferences, etc. Certain fundamentals, however, should be followed when selecting light sources to work with colored surfaces:

1. In general, warmer white light sources—e.g., incandescent, warm fluorescent, or high-pressure sodium lamps—should be used to enhance warm colors such as red, orange, etc. Cooler white light sources—e.g., cool white fluorescent lamps— should be used to enhance cool colors such as green, blue, purple, etc.
2. Warm white light is preferred for sources used at low light levels and, conversely, cool sources for high light levels. For example, in an elegant restaurant where a low lighting level is designed for a relaxed atmosphere, warm light sources, such as incandescent or warm white fluorescent lamps, are the proper choice. On the other hand, in a fast food restaurant, a cooler light source is preferred to speed customer turnover.
3. Cooler white light sources should be used for high lighting levels. For example, cool white fluorescent lamps should be used in research laboratories, where the high lighting level is required for close observation.
4. When the color rendering is important, high color rendering index (CRI) or color preference index (CPI) light sources should be selected.

The success of lighting in a building is not just to provide lighting at the proper level, but also to provide color harmony resulting from the coordinated efforts of the architect, interior designer, and lighting engineer.

14.3 PHYSICS OF LIGHT

14.3.1 Photometric Units

Although the International System of units, abbreviated SI, is accepted worldwide as a standard for calculations of illumination, the traditional units of quantities (foot, pound, second) have not been phased out in the United States. Accordingly, photometric quantities introduced in this section will be given in both sets of units. The basic photometric quantities are given in Table 14–3.

14.3.2 Energy of Light

Light energy can be converted from other forms of energy, such as heat, electrical energy, and chemical energy, although it is normally converted from electrical energy because that conversion process is most efficient. Light energy is not normally calculated in lighting design, since light energy quickly degrades to heat energy and thus cannot be stored.

14.3.3 Power and Intensity of Light

Power is the rate of consuming energy. In lighting, power is the luminous flux emitted by a light source in a unit of time. The unit of lighting power is the lumen. If the conversion of electrical energy to light energy has no loss, then one electrical watt converts to 683 lumens of a single wavelength of green light. If the conversion is into white light, the conversion is only about 200 lumens per watt. Chapter 15 discusses the efficacy, or luminous conversion efficiency of var-

TABLE 14–3

Basic photometric quantities

Quantity	Symbol	Units and Abbreviations	
		Conventional Units	SI Units
Energy (luminous energy)	Q	Lumen-hours (lm-hr)	Lumen-hours (lm-hr)
Power (luminous flux)	F	Lumen (lm)	Lumen (lm)
Intensity (candlepower)	I	Candela (cd)*	Candela (cd)
Illumination (illuminance)	E	Foot-candle (fc)*	Lux (lx)
Brightness (luminance)	L	Foot-lambert (fL)*	Candela/m^2 (cd/m^2)
		Candela/in.^{2} (cd/in.^{2})	

*The use of these units is being phased out gradually in the United States.

ious electrical light sources, which varies widely from 10 to 150 lumens per watt.

A lumen is defined as the lighting power emitted within a unit solid angle (one steradian) by a point source having a uniform luminous intensity of one candela. A candela is a unit of intensity of light. In layman's terms, one candela is illustrated as the light intensity produced by a candle. This is a useful analogy, but is not a scientific definition, since a candle can be of any size and material. In scientific terms, a candle is defined as an international candle with luminance equal to that of a black body at the freezing point of platinum. By definition, one candela is one lumen per steradian. A sphere consists of 4π (12.57) steradians. Thus, a candle (candlepower) is equivalent to 12.57 lumens.

14.3.4 Illumination

Illumination is luminous power per unit area. The quantity is analogous to Btu/hr/area in thermody-namics and watts/area in electrical power. Although illumination is replaced by illuminance in engineering terminology, it is still used widely in practice. In conventional units, illuminance is in footcandles (fc); in SI units, it is in lux (lx). By definition, illuminance is expressed as:

$$E = \frac{\text{Luminous Flux}}{\text{Area}} = \frac{F}{A} \qquad (14\text{-}2)$$

where E = Illuminance, fc (or lux in SI units)
F = Light flux (power), lumen
A = Area, sq ft (or square meter in SI units)

Since one square meter is 10.76 times larger than one square foot, a surface with area one square foot illuminated to one footcandle will have 10.76 times as much illuminance than a surface with area one square meter illuminated to one lux. Thus, one footcandle is equal to 10.76 lux (approximately 10.8 lux, but should not be taken as equal to 10 lux as many believe it to be).

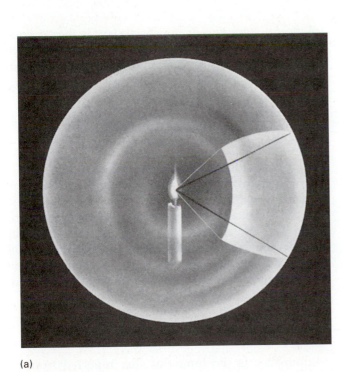

(a)

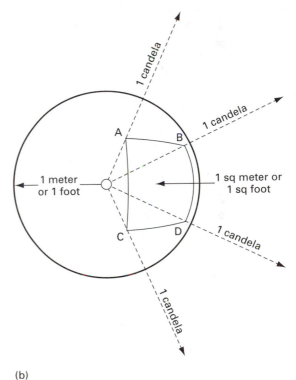

(b)

■   FIGURE 14–2

(a) Illustration of a one candela-source at the center of a clear sphere. (b) The illuminance at any point on the sphere is one lux (one lumen per square meter) when the radius is one meter, or one footcandle (one lumen per square foot) when the radius is one foot. The solid angle subtended by the area ABCD is one steradian. The flux density is therefore one lumen per steradian, which corresponds to a luminous intensity of one candela, as originally assumed. The sphere has a total area of 12.57 (4π) square meters or square feet, and there is a luminous flux of one lumen falling on each square meter or square foot. Thus, the source provides a total of 12.57 lumens.

The relationship between lighting power and illuminance is best illustrated in Figure 14–2(a), which depicts the light flux radiated from a point source, and in Figure 14–2(b) which illustrates the relationship between intensity (candela), power (lumen) and illuminance (footcandle or lux).

Commonly encountered illumination levels in nature and in buildings are given in Table 14–4. Chapter 16 will introduce the recommended method for determining the levels of illumination in various interior spaces.

14.3.5 Luminous Intensity

The light power in a given direction is defined as the luminous flux per unit solid angle, or the luminous flux on a surface normal to that direction, divided by the solid angle in steradians ($I = d\Phi/d\omega$). This definition applies strictly to a point source. In practice, however, light emitted from a larger light source, such as an incandescent lighting fixture whose dimensions are negligible in comparison with the distance from which the illuminance is measured, may also apply. Luminous intensity is measured in candela or candlepower.

Inverse-square law   The inverse-square law states that the illumination E at a point on a surface varies directly with the intensity I of the source and inversely with the square of the distance d between the source and the point. If the surface at the point is normal to the direction of the incident light, the law may be expressed as

$$E = \frac{I}{d^2} \qquad (14\text{–}3)$$

TABLE 14–4

Illumination encountered in nature and in buildings

At the Surround of	Nominal Level, fc
Starlight	0.0002
Moonlight	0.02
Street light	1.5–10.0
Daylight:	
In shade (outdoors)	100–1000
In direct sunlight	5000–10,000
Transient space (corridor, etc.)	5–10
Office lighting	50–70
Classrooms (kindergarten to college)	20–70
Drafting rooms	70–150
Sports facilities (schools to professional facilities)	30–300

where E is in footcandles when d is in feet and E is in lux when d is in meters

I is in candlepower when d is in feet and I is in candelas when d is in meters

and d is the distance between the light source and the surface perpendicular to the light source

This equation can be directly applied in lighting design calculations if the distance from the light source to the task surface is at least five times the maximum dimension of the source as viewed from the point on the surface.

Cosine law   The Lambert cosine law states that the illuminance on any surface varies with the cosine of the angle of incidence. The angle of incidence (θ) is the angle between the normal to the surface and the direction of the incident light. The inverse-square law and the cosine law can be combined into the form

$$E = \frac{I}{d^2} \times \cos\theta \qquad (14\text{–}4)$$

A useful extension of the cosine law is the "cosine-cubed" equation. By substituting h for the distance directly under the light source in lieu of the straight-line distance d, Equation (14–4) may be rewritten as

$$E = \frac{I \cos^3\theta}{h^2} \qquad (14\text{–}5)$$

Figure 14–3 illustrates the cosine law.

The inverse-square law shows how the same quantity of light flux is distributed over a greater area as the distance from the source to the surface is increased. The Lambert cosine law shows that the light flux striking a surface at angles other than the normal to the surface is distributed over a greater area. The cosine-cubed law explains the transformation from the one formula to the other.

14.3.6 Luminance and Brightness

Luminance is the luminous flux reflected from or transmitted through a surface. It is equal to the illuminance on the surface multiplied by the reflectance or transmittance factors of the surface. Luminance is perceived by our eyes as a sensation of brightness that is affected in part by the measurable luminance and in part by the state of adaptation of the eye. For example, if our eye has become adapted to an environment with very low illuminance, such as in a

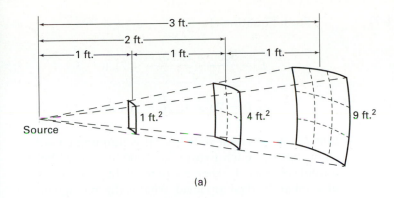

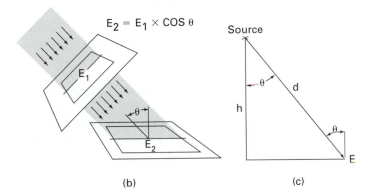

(b)   (c)

■ **FIGURE 14–3**
Illustrations of the inverse square law.

movie theater, a very dim light will appear to be bright. On the other hand, in daylight, a bright automobile headlight will appear dim. Thus, brightness is not synonymous with luminance, although the two terms have often been used interchangeably in common practice. The brightness of an object or a surface will appear less bright if the object or surface is in a bright environment and more bright if it is in a dim environment. To minimize the sensation of brightness, the designer should avoid locating the light source within the normal line of sight and provide less contrast surrounding the source.

Luminance is a directional quantity. Depending on the viewing angle and characteristics of the surface, luminance is expressed as lumens/steradian or candelas/area. Mathematically, the derivation of luminance is very complex. Luminance is expressed in the following units:

One luminance (L) = 1 lumen/m^2
One foot-lambert (fL) = 1 lumen/ft^2
One foot-lambert (fL) = 3.4 cd/m^2

Table 14–5 provides typical luminance values for natural and artificial light sources.

14.3.7 Luminous Exitance (Formerly Luminous Emittance)

For a perfectly or fairly diffused surface, the difficulty of measuring luminance can be resolved by using the quantity *luminous exitance*, which is defined as the density of luminous flux leaving a surface, whether the surface is opaque or translucent. Luminous exitance is in units of lumens/area and can be easily measured by a simple light (illuminance) meter in accordance with either of the following formulas:

For a reflecting surface, $M = \rho \times E$ (14–6)

TABLE 14–5

Approximate luminance values for various light sources

Light Source	Typical Luminance (fL)	Typical Luminance (cd/m²)
Sun (as observed from earth)	450,000,000	1,540,000,000
Moon (as observed from earth)	2400	8000
Snow in sunlight	9000	31,000
Overcast sky	600	2000
Candle flame	2900	10,000
Filament lamp (60 watts, inside frosted)	8800	30,000
40-watt fluorescent (cool white) lamp	5000	17,000

For a transmitting surface, $M = \tau \times E$ (14–7)

where M = Luminous exitance, lumens/area (ft^2 or m^2)
ρ = Reflectance of the surface, percent
τ = Transmittance of the surface, percent
E = Illuminance reaching the surface (fc or lux, depending on units in which area is expressed)

EXAMPLES

Example 1   If the lighting power reaching a surface of 10 square feet is 900 lumens, what is the average illuminance in footcandles or lux?
From Equation (14–1),

$$E = \frac{F}{A} = 900/10 = 90 \text{ footcandles, or}$$

$$E = \text{fc} \times 10.76 = 90 \times 10.76 = 960 \text{ lux}$$

Example 2   A lighting fixture has an intensity of 9000 candelas directly below the fixture. What is the illuminance on a table 10 feet below? 20 feet below?
From Equation (14–3),

$$E = \frac{I}{d^2} = \frac{9000}{10 \times 10} = 90 \text{ footcandles}$$

$$= \frac{9000}{20 \times 20} = 22.5 \text{ footcandles.}$$

Example 3   If a spotlight with 5000 candelas at the center is aimed at a painting on the wall 5 ft from the light (h), and the angle Φ is 45°, what is the illuminance level at the center of the painting?

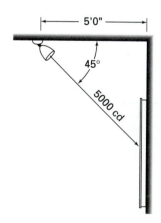

From Equation (14–5)

$$E = \frac{5000 \times (0.707)^3}{5^2} = 71 \text{ fc}$$

or $E = 70 \times 10.76 = 764 \text{ lux}$

Example 4   A room 10 ft $\times$ 20 ft is illuminated with eight fixtures. Each fixture has 3000 lumens of light output (power). If 70 percent of the light power can be utilized at the desktop level, what is the average illuminance at the desktop?
From Equation (14–2),

$$E = \frac{(8)\ (3000)\ (70\%)}{10 \times 20} = \frac{16,800 \text{ lumens}}{200 \text{ ft}^2} = 84 \text{ fc}$$

Example 5   If the luminance L of an object is 1000 cd/m^2 with a surrounding background luminance of 50 cd/m^2, what is the luminance contrast? If the luminances of the object and its background are reversed, what is the luminance contrast?[1]
From Equation (14–1), $C = |L_t - L_b|/L_b = (1000 - 50)/50 = 19$. If the luminances are reversed, $C = (50 - 1000)/1000 = |(-)0.95| = 0.95$.

14.4 LIGHT CONTROLS

The lighting in a space may be too bright or may come from the wrong direction, causing visual discomfort or inefficient utilization. Or the light may be the wrong color, causing poor color discrimination. For all of these reasons, light must be controlled.

14.4.1 Means of Controlling Light

Light travels in clean air without bending or notable loss until it is intercepted by another medium, which will either reflect, absorb, transmit, refract, diffuse, or polarize the light. These characteristics of varying materials are utilized as a method of controlling light to achieve better lighting. Six means of control normally employed in illumination design are illustrated in Figure 14–4.

---

[1]Luminous contrast varies considerably, depending on whether the background is darker or brighter. One alternative convention is to express the equation as $C = (L_g - L_e)/L_g$, where L_g is greater luminance and L_e is lesser luminance. With this convention, the luminous contrast will always be smaller than unity.

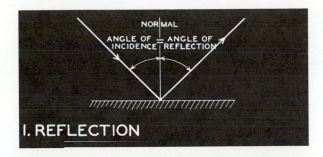

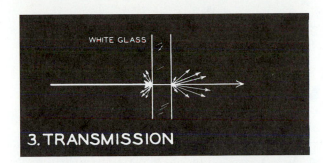

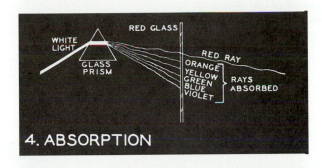

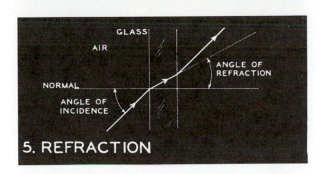

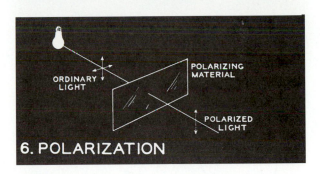

■ FIGURE 14–4
Six means of controlling light. (Courtesy: General Electric, Nela Park, OH.)

Reflection Light is reflected from the surface of a material. If the surface is shiny or specular, such as the surface of a mirror, then reflected light will follow the law of reflection, according to which the angle of reflection is equal to the angle of incidence, as illustrated in Figure 14–5. Even a transparent material reflects some light from its surface.

Diffusion When the surface is not shiny or matte, then the reflected light will be diffused. The diffused light may be directional or totally nondirectional, as a "cosine distribution."

Transmission When the material is transparent (clear glass), spread (etched glass), or totally diffused (white glass), light will pass through it in a controlled mode.

Absorption Light is absorbed when it is directed to an opaque material or passed through a transparent or translucent material. There will be light loss in either case. The amount of light absorbed is the balance of the incident light that is reflected or transmitted.

Refraction The property of a material to change the direction of light at the interface between two different materials, such as air and glass, is called refraction. Refraction is the most effective means of controlling light and is commonly used by lighting designers. Snell's law of refraction is

$$n_1 \times \sin i = n_2 \times \sin r \qquad (14\text{–}8)$$

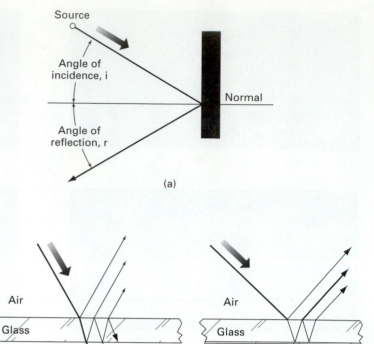

(a)

FIGURE 14–5

(a) The law of reflection states that the angle of incidence, *i*, is equal to the angle of reflection, *r*, from a specular surface.
(b) Reflection from a transparent medium reflects from the front as well as the rear surface.
(c) Reflection from a glass mirror with the rear surface coated is primarily from the rear, but a small portion is also reflected from the clear front surface. The images reflected are offset by the index of refraction of the glass medium. (Source: *IESNA Lighting Handbook*.)

(b)

(c)

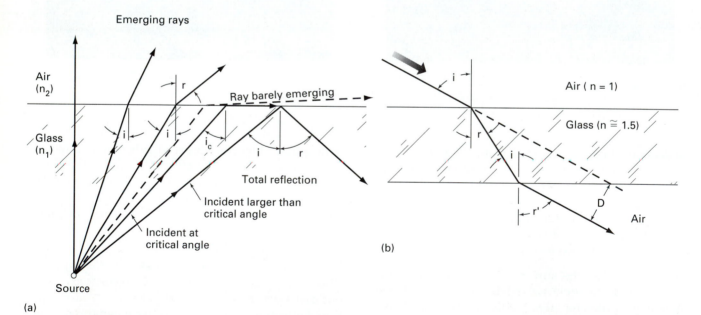

(a)

(b)

FIGURE 14–6

(a) Total reflection occurs when sin *r* = 1. The critical angle at which light is totally refracted (reflected) varies with the medium. The critical angle between glass (*n* = 1.5) and air (*n* = 1) is 41.8 degrees.
(b) Refraction of light rays at a plane surface causes bending of the incident rays and displacement of the emerging rays. A ray passing from a lighter to a denser medium bends toward the normal, while a ray passing from a denser to a lighter medium bends away from the normal. (Source: *IESNA Lighting Handbook*.)

where n_1 = the index of refraction of
the first medium
i = the angle the incident light ray forms
with the normal to the surface
n_2 = the index of refraction of
the second medium
r = the angle the refracted light ray
forms with the normal to the surface.

When the first medium is air ($n_1 = 1$), equation 14–8 can be simplified to

$$\sin r = \sin i \times n_1/n_2 = \sin i/n_2 \quad (14–9)$$

Figure 14–6 illustrates the refraction of a light ray from air to glass and back to air. If the angle of the light ray in the glass is such that the sine of the refracted angle is equal to unity, light will be trapped (confined) within the glass. The angle of incidence is then called the *critical angle* (θ_c) between glass and air. This phenomenon is the basic principle of fiber optics.

Polarization   Light travels at high speed with waves vibrating in all planes at right angles to the direction of travel. Polarization is the phenomenon wherein the waves vibrate only in one plane. A polarizing material (filter) is called a *polarizer*. When light passes through two polarizers in tandem, but with their optical axes oriented at 90°, the light will be totally polarized. One of the more popular uses of polarizers is in polarized sunglasses, which are recognized as the most effective means of reducing glare from the sun. If one were to place two polarized sunglasses at right angles to each other, one would find that the sunlight was nearly 100 percent filtered. Polarization is used in lighting controls in the form of multilayered polarizing lenses (diffusers). Figure 14–7 illustrates the principle of polarization.

14.4.2 Application of Light Control Techniques

Any combination of the six methods of light control can be incorporated into lighting design. Popular geometric forms for lighting controls are illustrated in Figures 14–8 through 14–15.

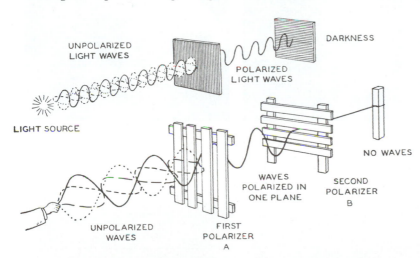

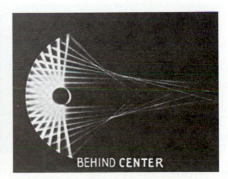

FIGURE 14–7
A light wave traveling in space may be compared to random waves traveling along a rope when one end of it is shaken. The waves vibrate in all planes at right angles to the direction of travel. The first polarizer (A) allows only the vertical vibrations to go through, thus polarizing the waves. The second polarizer (B) stops the vertical waves, so none of the polarized waves can go beyond, resulting in total darkness. (Source: IESNA, *Laboratory Activities with Light.*)

FIGURE 14–8
The circular reflector is used to reflect light to produce either spreadout or concentrated light patterns, depending on the position of the lamp in relation to the center of the circle. (Courtesy: General Electric, *Fundamentals of Light and Lighting.*)

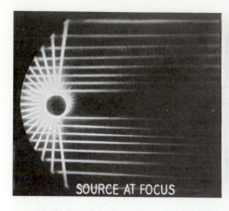
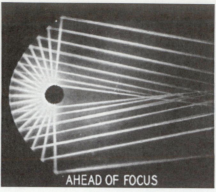

■ FIGURE 14–9

A parabola has one focal point. When the lamp is located at the focus, all reflected light beams will be parallel to each other. Parabolic reflectors are used primarily for spotlight applications. By varying the position of the lamp, the distribution may be a very narrow spotlight to a very wide floodlight. (Courtesy: General Electric, *Fundamentals of Light and Lighting.*)

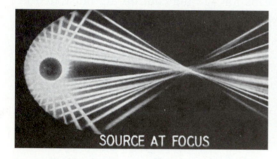

■ FIGURE 14–10

An elliptical reflector is used primarily to focus the light beam through the second focal point, allowing all light beams to pass through a pinhole. (Courtesy: General Electric, *Fundamentals of Light and Lighting.*)

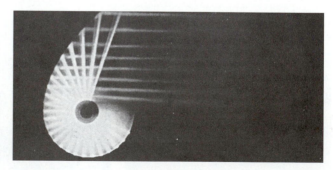

■ FIGURE 14–11

Reflection with compound circular-parabolic reflectors. (Courtesy: General Electric, *Fundamentals of Light and Lighting.*)

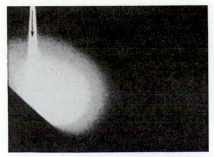
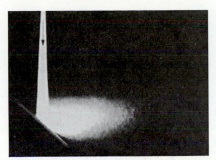
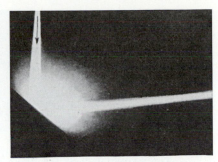

Diffuse reflection

- mat paint
- plaster
- lime stone
- terra cotta

Spread reflection

- aluminum paint
- oxidized aluminum

Diffuse-specular reflection

- glossy enamel
- glossy paper

■ **FIGURE 14–12**
Characteristics of diffuse materials. (Courtesy: General Electric, *Fundamentals of Light and Lighting*.)

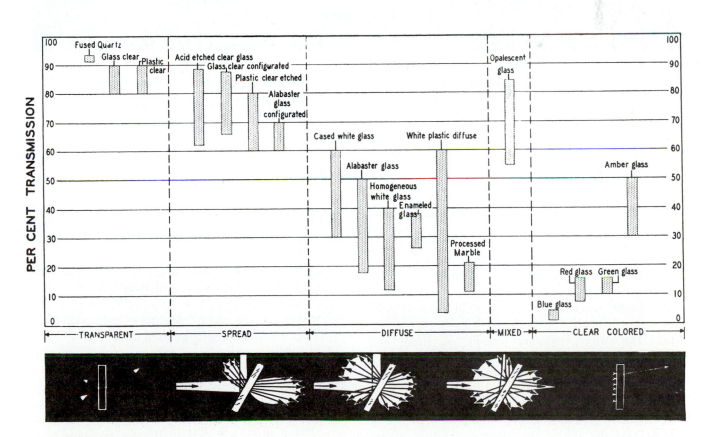

■ **FIGURE 14–13**
Characteristics of transmission materials. (Courtesy: General Electric, *Fundamentals of Light and Lighting*.)

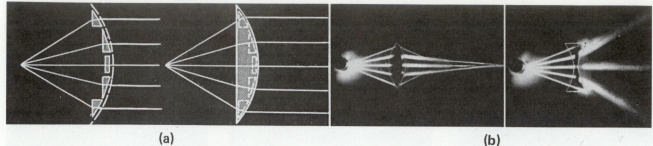

(a)

The refraction light through a series of tiny prisms and a single convex lens is identical.

(b)

The refraction of light through double convex and concave lens.

■ **FIGURE 14–14**

Refraction of light through lenses and prisms. (a) Refraction of light through a series of tiny prisms and refraction through a single convex lens are identical. (b) Refraction of light through double convex and concave lens. (Courtesy: General Electric, *Fundamentals of Light and Lighting.*)

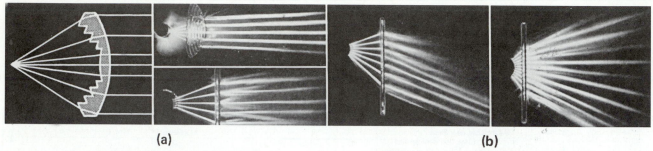

(a)

Construction of convex Fresnel lens and prismatic lens plate.

(b)

Lens plates may provide a symmetrical light pattern.

■ **FIGURE 14–15**

Refraction of light through a Fresnel lens and plates. (a) Construction of convex Fresnel lens and prismatic lens plate. (b) Lens plates may provide a symmetrical light pattern. (Courtesy: General Electric, *Fundamentals of Light and Lighting.*)

QUESTIONS

14.1 Natural light is radiant energy from the sun. The visible spectrum ranges from (   ) to (   ) nanometers.

14.2 Light generated from electrical light sources (lamps) is radiant energy. (True) (False).

14.3 The three primary colors of light are (   ).

14.4 The three subtractive primaries of light are (   ).

14.5 What is the unit of measure used for illuminance, or illumination level?

14.6 Daylight under an overcast sky is about (   ) foot-candles.

14.7 Daylight under sunlight is about (   ) foot-candles.

14.8 Color temperature is a way to define the heat generated by various types of lamps. (True) (False).

14.9 What is the approximate color temperature of a gas-filled incandescent lamp? A fluorescent lamp?

14.10 What is the unit of measure of light power in a conventional unit? In an SI unit?

14.11 What are the normal human responses to various colors?

14.12 What is the meaning of luminous intensity?

14.13 A higher luminance contrast is desired for viewing a fast-moving object, such as a baseball, but is not important for a stationary task. (True) (False).

14.14 Color difference also improves visual acuity even if the objects and its background are at the same luminance level or zero contrast. (True) (False)

14.15 Does a high luminance contrast provide more visual comfort?

14.16 What are glare, direct glare, reflected glare, discomfort glare, and disability glare?

14.17 The sun's ray reaching the earth has a balanced energy level of all wavelengths. (True) (False)

14.18 What wavelength is most sensitive to human eyes?

14.19 White light is a mixture of colored lights. (True) (False)

14.20 What are the basic characteristics of color?

14.21 If 56 lumens reaches a surface of 10 sq ft, what is the illumination level in foot-candles and in lux?

14.22 What is the recommended illuminance level for office occupancies?

14.23 When is inverse square law and cosine law used?

14.24 A surface two feet from a light source is illuminated by a light source having a candlepower of 20 candela. What will be the illuminance on the source?

14.25 If the light source in Question 14.24 is at a 30-degree angle with the normal of the surface, what is the illuminance?

14.26 What are the means of light controls?

14.27 Light travels in a straight line in a single medium, such as air, water or glass. (True) (False)

14.28 Light always bends the same way between media. (True) (False)

14.29 What is the refracted angle inside the glass if light enters from air into a glass at 35° from the normal axis and the index of refraction of glass is 1.4?

14.30 Refraction is the most effective means of light control. (True) (False)

14.31 The reflected light beams from a parabolic reflector are parallel to each other. (True) (False)

14.32 What is the primary use of an elliptical reflector?

14.33 What is a Fresnel lens?

14.34 Describe the light control principles used for the following light sources or lighting fixtures:
   a. A fixture with a diffusing glass globe.
   b. A metal can with an open bottom. The inside surfaces of the fixture is painted black.
   c. A ceiling-mounted fixture with translucent plastic diffusers at the bottom.
   d. A ceiling-mounted fixture with Fresnel (prismatic) control lens.
   e. A suspended fixture with opaque bottom and open top (indirect distribution).

14.35 A book is printed in black ink on buff paper. The luminance measured on the paper is 70 cd/m^2 and the prints are 5 cd/m^2. What is the contrast of the reading task (print and its background)?

14.36 A book is 0.75 sq. ft. in area and the average lighting power, directly and indirectly, reaching a 3-sq-ft desk is 75 lumens. What is the illuminance on the book? On the desk?

14.37 A lighting system consists of 10 spotlights, of which one is located directly above a task and the other 9 are aimed elsewhere. The reflected illuminance from these 9 fixtures is calculated to be 3 foot-candles on the task. What is the illuminance on the task if the candlepower at the center of the spotlight is 2500 candela and the spotlight is 10 feet above the task?

14.38 What is the luminous exitance of the task if its reflectance is 30%?

REFERENCES

1. Illuminating Engineering; Society of North America (IESNA), *Lighting Handbook: Reference and Application, 8th ed.* New York, NY, 1994.

2. General Electric, Lighting Company, *Fundamentals of Light and Lighting,* Nela Park, OH: 1960.

3. General Electric, Lighting Business Group, *Light and Color,* Nela Park, OH: 1995.

4. General Electric, Lighting Company, *Specifying Light and Color,* Nela Park, OH: 1995.

5. Minolta Corporation, *Precise Color Communication,* Ramsey, NJ: 1995.

6. IESNA, *IES Education Series (Introductory),* New York: 1995.

7. IESNA, *IES Education Series (Intermediate),* New York: 1995.

8. Joseph B. Murdoch, *Illumination Engineering,* New York: MacMillan, 1985.

9. Ronald H. Helms and M. Clay Belcher, *Lighting for Energy-Efficient Luminous Environments,* Englewood, Cliffs, NJ: Prentice Hall, 1991.

10. M. David Egan, *Concepts in Architectural Lighting,* New York: McGraw Hill, 1983.

15

LIGHTING EQUIPMENT AND SYSTEMS

LIGHT MAY ORIGINATE IN MANY WAYS—FROM SO-lar energy (daylight), from combustion and chemical reactions, and from the conversion of electrical energy. Of all the light sources, daylight is both most plentiful and free of charge. However, daylight is not available at night and fluctuates widely during the day, sometimes being too bright for visual comfort or too hot to stay under for very long. Still, when properly controlled, e.g., with sunglasses and air conditioning, it is the most economical of all sources of light.

In buildings, electrical lighting has become the sole source of light at night and a supplemental source during the day. Frequently, electrical lighting must be used even if there is plenty of daylight—for example, during a visual presentation in a classroom or conference room, and during many sporting events, when daylight from low angles may cause blinding glare to the players or the spectators.

The development of light sources through human history may be represented by the growth of a tree with branches, leaves, and flowers standing for different events and inventions. Figure 15–1 illustrates the chronological development of light sources and the relationship between them. All modern light sources applicable to the interior lighting of buildings are universally electric lights, while combustion-type light sources, such as gaslights, are limited to decorative use only.

The fundamental component of lighting equipment is the light source, commonly called a lamp. The assembly that holds a lamp or lamps together to provide lighting is the luminaire, commonly called a lighting fixture. Although discouraged in academic circles, the term *lighting fixture* is more accepted than *luminaire* by the general public and among professionals in the design and construction industries. Thus, the two terms are used interchangeably in this chapter, and similarly for *light sources* and *lamps*.

Lighting fixtures must be designed for a particular type of lamp and are usually not suitable for other types of lamp. One reason for this is the difference in configuration of the lamp holder (socket). Another reason is the amount of heat generated, which affects the material of the fixture, the ventilation, and the physical clearance between the lamp and its surrounding surfaces. Thus, lighting fixtures and lamps must be compatible and within the power (wattage) limit of the fixture.

Lighting fixtures are important elements of interior design, since they are prominently displayed in the space. In addition to lighting performance, size, texture, shape, and color must be considered when selecting a fixture. To strive for a successful lighting design, the architect, interior designer, and lighting engineer must be knowledgeable about both light sources and the performance of various fixtures. This chapter presents the fundamentals of lamps, fixtures, and lighting systems.

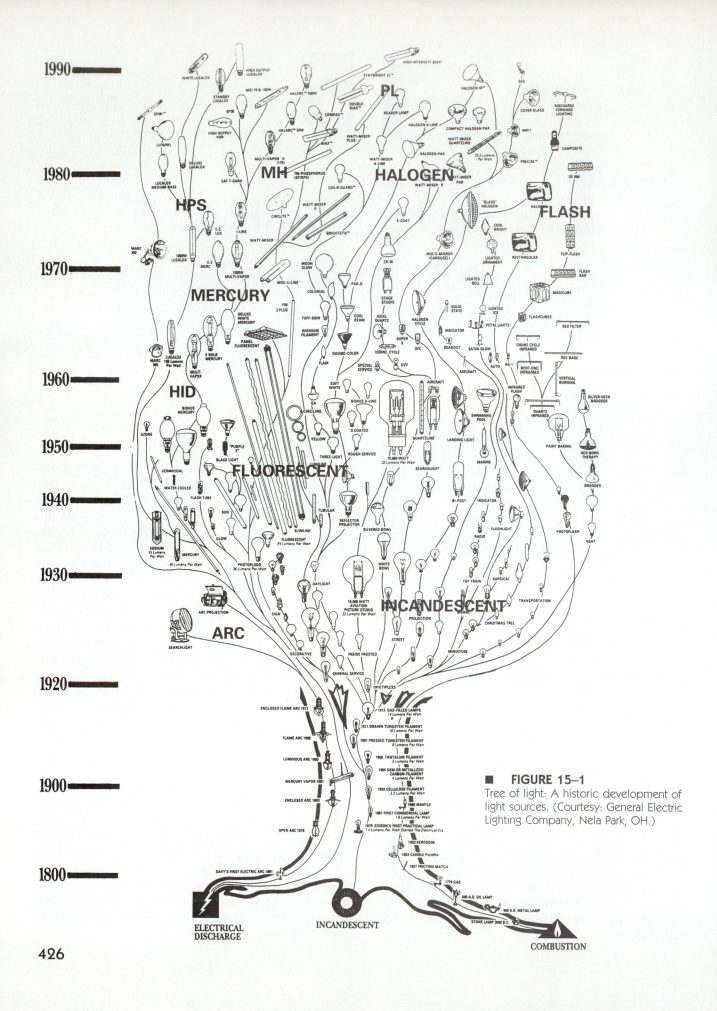

■ **FIGURE 15–1**
Tree of light: A historic development of light sources. (Courtesy: General Electric Lighting Company, Nela Park, OH.)

Lighting accounts for about 25 percent of all electrical energy consumed in the United States. For each kWh of electrical energy consumed, a corresponding amount of pollutant is being released into the atmosphere. The impact of energy consumption on the global environment is addressed in Chapter 1. Lighting design professionals have a responsibility to minimize energy consumption by using efficient light sources and equipment, while creating an environment that is aesthetically pleasing and conducive to higher productivity.

15.1 ELECTRICAL LIGHT SOURCES

Electrical light sources are called lamps. Although thousands of lamps are made for diversified applications, they can be grouped into four major classes, based on their operating principles.

Incandescent lamps employ the principle of converting electrical energy into heat at a temperature that causes the filament of the lamp to become incandescent (red or white hot). The process closely resembles the heating of a black body discussed in Chapter 14.

Fluorescent lamps contain mercury vapor. When proper voltage is applied, an electric arc is produced between the opposing electrodes, generating some visible, but mostly invisible, ultraviolet radiation. The ultraviolet radiation excites the phosphor coating on the inside of the bulb, thereby emitting visible light.

High-intensity discharge (HID) lamps produce high-intensity light within an inner arc tube contained in an outer bulb. The metallic gas within the arc tube may be mercury, sodium, or a combination of other metallic vapors. The outer bulb may be clear or coated with phosphor. HID lamps are classified as mercury, metal halide, and high-pressure sodium types.

Miscellaneous lamps include a wide variety of lamps operating on different principles. Although they have limited application in buildings, future breakthroughs in technology and production may provide a new dimension to the world of architecture and environmental design. Some of the promising new types of lamp are the following:

1. *Short-arc lamps*, or compact-arc lamps, such as the Xenon family of lamps, produce light in small arc tubes and are the closest thing to a true point source of high luminance. They are used primarily as searchlights, in projectors, and in optical instruments.

2. *Low-pressure sodium lamps (LPSs)* are monochromatic lamps in the yellow region of the spectrum (589 to 589.6 nm). The efficacy of an LPS lamp is as high as 180 lumens per watt, but the lamp has limited applications due to its color. Typical applications are along highways and in storage yards.

3. *Electroluminescent lamps* emit light by the direct excitation of phosphor from an alternating current. Therefore, they can be made in any shape, size, and form. Electroluminescent lamps can produce different colors by the mixing of phosphors. Although extremely efficient at about 200 lumens per watt, their use is limited to signs and decorative applications.

4. *Electrodeless lamps* are gaseous lamps excited by means of electromagnetic or microwave energy without the use of electrodes. These types of lamps have a promising future in building lighting applications through the use of specially designed fixtures or the use of "light pipes" and will be discussed later in the chapter.

15.2 FACTORS TO CONSIDER IN SELECTING LIGHT SOURCES AND EQUIPMENT

There are many factors to consider in selecting lamps of all varieties.

15.2.1 Light Output

Light power, expressed in lumens, is defined as follows:

- *Initial lumens:* the initial rated light power output.
- *Average lumens:* the average of the initial lumens output and the lumens output at the end of the rated life of the lamp.
- *Mean lumens:* the lumens output at 40 percent of the rated life of the lamp.
- *Beam lumens:* the initial lumens output within the central beam of a floodlight, usually defined to exclude light intensity less than 10 percent of the maximum intensity of the light.

15.2.2 Intensity

Light intensity is expressed in candelas at various angles from the lamp or fixture. The data are usually provided by manufacturers in the form of candlepower distribution curves.

15.2.3 Luminous Efficacy

Luminous efficacy or, simply, efficacy is defined as the light output per unit of electrical power (watts) input, or lumens/watt (lm/w). Theoretically, 1 watt of electrical power can be converted to 683 lumens of monochromic green light, or about 200 lumens of white light of equal energy level among all visual spectrum wave lengths. With this as reference, the 10- to 25-lm/w efficacy of incandescent lamps is far short of ideal. The efficacy of a lamp should include the power consumed by its accessories, that is:

- The *lamp efficacy*, in lumens/watt, for lamps only, or
- The *net lamp efficacy* in lumens/watt, of the lamps and accessories (ballast), for electrical discharge types (fluorescent and HID) of lamps.

15.2.4 Luminaire Efficiency

The luminaire efficiency is a measure of the total light power output in lumens versus the total light power input of all lamps in the luminaire. It is expressed in percent. Luminaire efficiency is a good measure for comparing luminaires of similar candlepower distribution characteristics, but is not necessarily a measure of how well the light power is being utilized. For example, if one must illuminate a painting on a wall, a bare bulb fixture, which is 100 percent efficient, suspended in front of the painting will not be as good as a spotlight that is only 60 percent efficient. Chapter 16 will introduce the concept of the coefficient of utilization (CU) in lighting design calculations.

15.2.5 Rated Lamp Life

The rated life of a lamp is defined as the time elapsed when only 50 percent of a group of lamps remain burning. The rated life follows closely the mortality curve of most statistics for a large number of subjects. Figure 15–2 shows the mortality curve of incandescent lamps.

Example   If the rated life of one type of incandescent lamp is 750 hours, what would be the expected percentage of survival for a large number of lamps installed in a building after 500 hours of use?

Solution. The hours of use of the lamp are ⅔, or 67 percent of the rated life. From Figure 15–2, 92 percent of the lamps would likely survive.

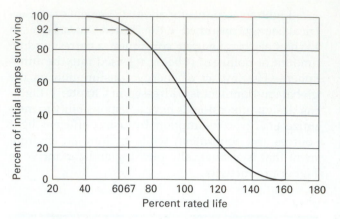

■ FIGURE 15–2
Range of typical mortality or life expectancy curves for incandescent lamps.

15.2.6 Lumen Depreciation

Light output depreciates with time. The loss of light, known as lumen depreciation, may be as high as 20 to 30 percent of a lamp's rated life. This characteristic must be taken into consideration in illumination design. Typical lumen depreciation characteristics for various lamps are listed in Table 15–1.

15.2.7 Color Temperature

Color temperature is the color the lamp appears, expressed in black body temperature. Chapter 14 provides the background for the concept of color temperature. Table 15–2 lists the color temperature characteristics of some typical lamps.

TABLE 15–1

Depreciation characteristic of lamps
(in percent of the initial light output)

	Approximate Light Output (%)	
Type of Lamp	@ 50% of Life	@ 100% of Life
Incandescent,		
General-service type	90	82
Tungsten halogen	97	92
Fluorescent,		
Light loading (low brightness)	92	90
Medium loading (standard)	85	82
High loading (high output)	75	65
High-intensity discharge,		
Mercury (H)	77	60
Metal halide (MH)	70	65
High-pressure sodium (HPS)	90	70

TABLE 15–2
Color temperature of typical light sources

Light Source	Color Temperature Range °K
Incandescent,	
60 watt	2500–2700
100 watt	2700–2900
500 watt	2900–3100
Halogen, tungsten	3000–3200
Fluorescent,	
Warm white	2900–3000
Cool white	4000–4500
Daylight	6000–6500
Mercury,	
Clear	5500–5800
Improved	4400–4500
Metal halide,	
Clear	3700–3800
Coated	3200–4000
High-pressure sodium,	
Normal	2000–2100
Color improved	3000–4000
Low-pressure sodium	1700–1800
Photoflood lamps	3200
Photoflash lamps	5500

15.2.8 Color-Rendering Index

The concept of the color-rendering index (CRI) was introduced in Chapter 14. It is useful to measure how well a lamp renders color, compared to a reference light source of the same color temperature range. An incandescent lamp with color temperature at 3000°K has been selected by the CIE as the reference source (CRI = 100) for comparison with all other lamps having color temperature below 5000°K. This selection of CRI = 100 does not mean that that particular incandescent lamp can render the true colors of every object or material. Such a lamp is selected only because it has a smooth profile of spectral energy distribution (S.E.D.) without extremely narrow energy bands.

15.2.9 Color Preference Index

The color preference index (CPI) was introduced in Chapter 14. It is useful in expressing the color quality of a light source on a preferential basis, such as red meat, green vegetables, blue sky, and pink complexions.

15.2.10 Flicker and Stroboscopic Effect

Theoretically, the cyclic flow of a 60-Hz current through a lamp can have light fluctuations 120 times per second. This is called *flicker*. Because of the light retention characteristic of incandescent filaments and the phosphor coatings of HID lamps, flicker is not normally perceivable except in noncoated HID lamps. When rapidly moving objects are observed under clear HID lamps, the object may appear to be at a standstill or moving at lower harmonic frequencies. This is called the *stroboscopic effect*. It can be eliminated or minimized by the use of lead-lag ballasts for multiple lamps and the wiring of fixtures on alternate three-phase circuits.

When the electrical power system frequency is lower than 60 Hz, flicker and the stroboscopic effect will be exaggerated. Conversely, at higher frequencies, flicker will not be perceived.

15.2.11 Brightness

Physically, small light sources of high intensity such as incandescent lamps, are excellent for light control, but can be too bright for visual comfort. The location and aiming of the fixtures must therefore be carefully selected. Uncomfortable glare can be minimized by methods discussed in Chapter 14.

15.2.12 Intensity Control

Light intensity can be controlled by multilevel switching or by dimming. Incandescent lamps can easily be dimmed by the use of autotransformers or solid-state dimmers. Fluorescent and HID lamps can be dimmed by special ballasts and circuitry, but at a considerably higher cost. (See Chapter 17 for the effect of dimming on typical residential or public spaces.)

15.2.13 Accessories

Among the many accessories that should be considered before selecting light sources are ballasts, starters, and dimmers.

15.3 INCANDESCENT LIGHT SOURCES

Edison's first successful lamp, using carbon filament in a vacuum, produced 1.4 lumens per watt. Since then, incandescent lamps have improved dramatically, using a tungsten filament in an inert gas-filled bulb. Incandescent lamps are easily controlled, both in intensity and in direction; however, they are the least energy-efficient lamps. Following are some characteristics of incandescent lamps.

15.3.1 Shape and Base

Incandescent lamps come with a variety of bases and in numerous shapes. (See Figure 15–3.) Lamps are made from a fraction of a watt to 5000 watts.

15.3.2 Size Designation

The physical size of the bulb is described in units of ⅛″ increments. Size 1 means that the bulb is ⅛″ in diameter at its widest part. The lamp prefix provides an indication of the shape of the bulb. Thus, P–25 indicates a pear-shaped bulb 3⅛″ in diameter, and T–8 denotes a tubular-shaped bulb 1″ in diameter.

15.3.3 Rated Life

Incandescent lamps are short lived. General-service types are normally rated for 750 hours as a bare lamp. (See Figure 15–2 for the range of a typical mortality curve for lamps.) When the lamp is installed in a fixture, its real life is even shorter. Halogen and quartz lamps have an improved life from 1000 to 4000 hours. However, they are still inferior compared with the 10,000 to 20,000 hours of rated life for fluorescent and HID lamps.

15.3.4 Efficacy

Efficacy is luminous efficiency in unit of lumens per watt (lm/w). Incandescent lamps are among the poorest lamps in efficacy. They range from 10 lm/w for 40-watt lamps to 25 lm/w for 1500-watt lamps. Compared with the 80- to 120-lm/w efficacy of fluorescent and HID lamps, the 10- to 25-lm/w efficacy of incandescent lamps renders them only about one-fifth as efficient. In other words, an incandescent lamp requires about five times as much electrical power as a fluorescent or HID lamp to produce the same amount of illumination. The use of incandescent lamps should thus be limited to applications that do not require high illuminance.

15.3.5 Depreciation of Light Output

Standard incandescent lamps have a moderate rate of depreciation of their light output. At the rated life of the lamp, the lumen output is about 82 percent of its initial output. (See Figure 15–4.) Halogen lamps, on the other hand, are extremely good in maintaining their light output, having practically no depreciation at all.

15.3.6 Effect of Variation in Line Voltage

Incandescent lamps are normally rated at 115 volts. If the actual line voltage is higher or lower than the rated voltage, it will greatly affect the life, lumen output, and wattage draw of the lamp. Figure 15–5 shows the effect of voltage on the performance of an incandescent lamp. As an example, if the line voltage is 94 percent of 115 volts, or 108 volts, then the light output will drop to 80 percent of its rating, the power will drop to 90 percent, and the life of the lamp will increase to 220 percent.

15.3.7 Color Rendering and Color Preference

Incandescent lamps have excellent color rendering and color preference. Indeed, they are considered the best reference light source.

15.3.8 Major Types of Incandescent Lamps

All incandescent lamps use tungsten as their filament (see Figure 15–6). Tungsten can be designed to operate at temperatures ranging from 3800 to 5000°F. Higher operating temperatures will produce a whiter spectrum and higher energy conversion efficiency (efficacy), but a reduced life for the filament. The extremes of incandescent lamp design can best be illustrated by the performance of photoflood lamps: Whereas an incandescent lamp normally has an efficacy between 15 and 25 lm/w, a 500-watt R–40 photoflood lamp can produce 38,000 lm, or 76 lm/w, but has only four hours of rated life. Conversely, lamps can be designed to last essentially forever, but at extremely low efficacy.

There are numerous types of incandescent lamps. Following are some that are commonly used in building systems.

General service type   These lamps are the standard for use in buildings. They are designed to have relatively good efficacy (15–25 lm/w), a moderate rated life (750–1000 hours), and good color-rendering characteristics (95–100 CRI).

Rough and vibration services   These lamps are used in locations subject to rough handling or vibration, such as on machinery or in mechanical equipment. They are specially designed with heavy tungsten filaments and supports. Efficacy is between 10 and 19 lm/w.

Extended life service   These lamps are designed with heavy filaments to operate at considerably lower than

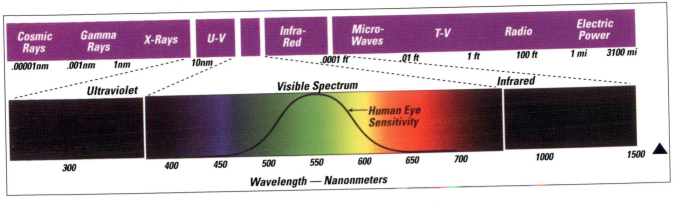

Plate 11 The electromagnetic spectrum. (Courtesy of General Electric Company)

(a)

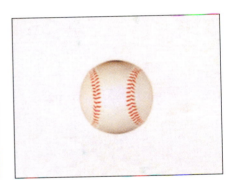

(b)

Plate 12 **(a)** The statement reads, "Contrast improves visual acuity and recognition." Without contrast, visual acuity is lost. Too much contrast may create visual discomfort. **(b)** Visual acuity is critical in high-speed sports. A white baseball in front of a dark background is easily visible, compared with a baseball on a white background. A baseball against a dark blue background shows up even more clearly because there is luminance contrast as well as a color difference.

Plate 13 Visual acuity is even more critical in tennis, where a ball is being hit at close range in random directions. The tennis court surface and the backdrops (vertical surfaces) should have high luminance contrast and color difference with the yellow or white tennis balls.

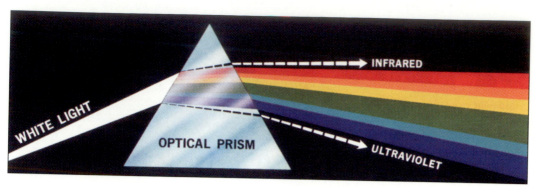

Plate 14 A prism can bend white light into its component colors. The effect of the prism is to bend shorter wavelengths more than longer wavelengths, separating them into distinctly identifiable bands of color. The colors can be recombined into a beam of white light by accurately orienting a properly designed second prism to intercept the dispersed colored light rays. Note that the prism is useful for separating infrared and ultraviolet rays, as well as visible rays. (Courtesy of General Electric Company)

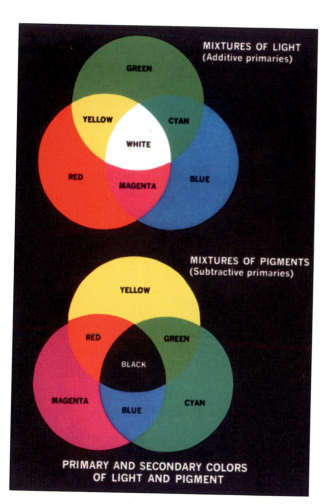

Plate 15 The primary colors of light are blue, green, and red. These are called *additive primaries*. Adding equal energy levels of the additive primaries produces white light. The primary colors of pigments are magenta, yellow, and cyan. These are called *subtractive primaries*. Adding equal amounts of subtractive primaries produces a black pigment. (Courtesy of General Electric Company)

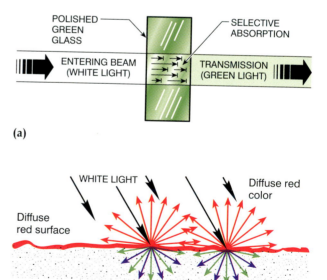

Plate 16 (a) White light passing through a green glass becomes green light because the green glass absorbs all color wavelengths except green. (b) A red painted or naturally red surface reflects only red wavelengths and absorbs all other wavelengths. (Courtesy of General Electric Company)

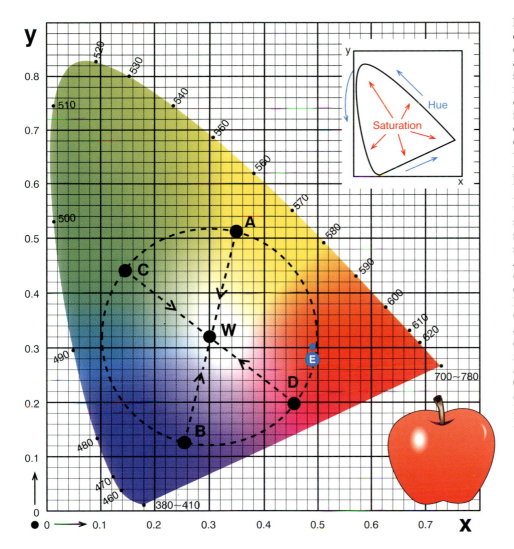

Plate 17 The CIE (x, y) chromaticity diagram is a diagram of the hue (color) and chroma (saturation) of spectrum colors on a selected neutral value (lightness-darkness plane). The diagram is plotted on the x- and y-coordinates with saturated colors at the perimeter. The white light at point W may be produced by mixing equal lumens of light sources A and B, or light sources C and D, or the combination of any other sources opposite point W. The specifications of the light sources represented at points A, B, C, D, and W can be read directly from the x- and y-coordinates. For example, light source D is a light with CIE chromaticity of $x = 0.46$ and $y = 0.2$. Point E on the chromaticity diagram represents the color of the apple with color values of $x = 0.49$ and $y = 0.30$. (Courtesy of Minolta Corporation, Instrument Systems Division, Ramsey, NJ)

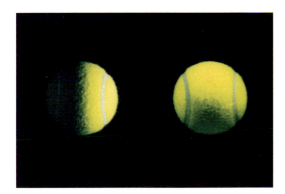

Plate 18 Light orientation is important to bring out the modeling (the 3-dimensional image) of an object. Shown at the left is a tennis ball illuminated from the right side only. The ball appears to be semispherical instead of a full sphere. When light comes from multiple directions with luminous differences between them, the 3-dimensional form of the ball is revealed, as shown in the photo on the right.

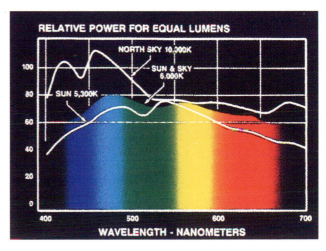

Plate 19 The distribution of light energy varies with the time of day. Shown are three variations: sun and sky at 6000K, sun at 5300K, and north sky at 10,000K. (Courtesy of General Electric Company)

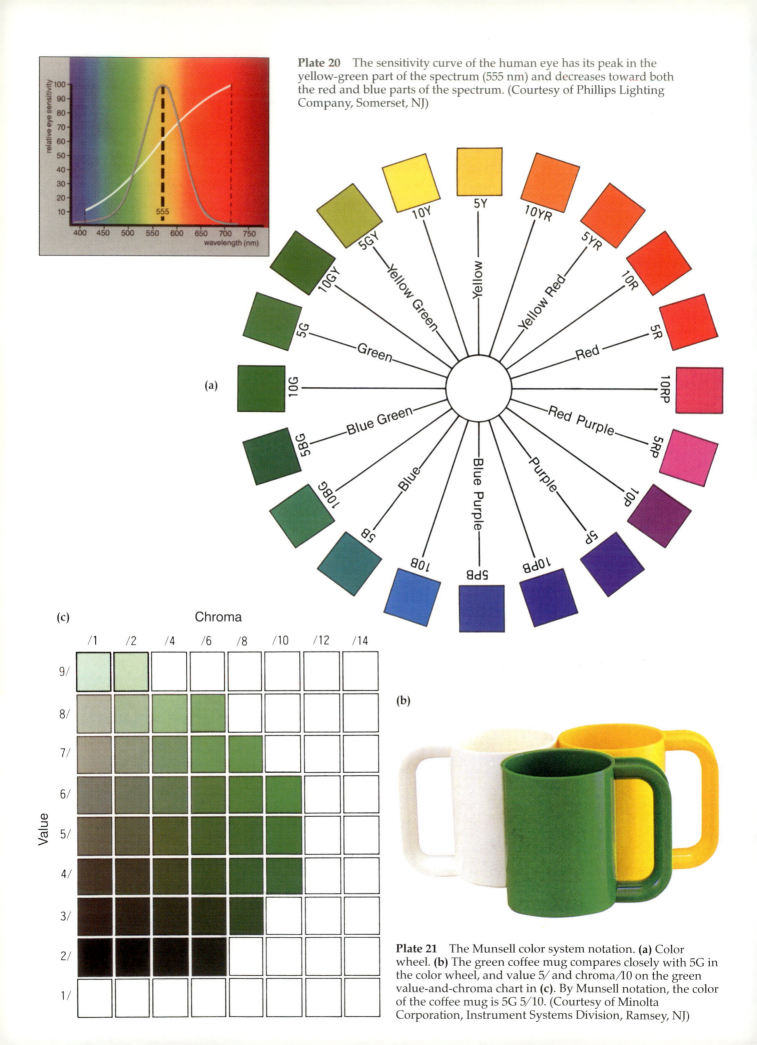

Plate 20 The sensitivity curve of the human eye has its peak in the yellow-green part of the spectrum (555 nm) and decreases toward both the red and blue parts of the spectrum. (Courtesy of Phillips Lighting Company, Somerset, NJ)

(a)

(b)

(c)

Chroma

Value

Plate 21 The Munsell color system notation. **(a)** Color wheel. **(b)** The green coffee mug compares closely with 5G in the color wheel, and value 5/ and chroma/10 on the green value-and-chroma chart in **(c)**. By Munsell notation, the color of the coffee mug is 5G 5/10. (Courtesy of Minolta Corporation, Instrument Systems Division, Ramsey, NJ)

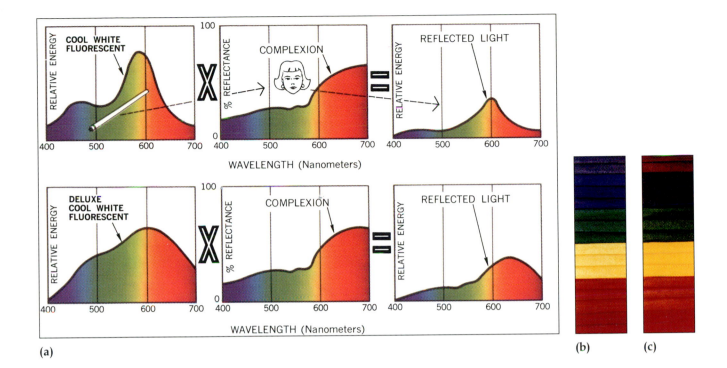

(a)

(b)   (c)

Plate 22 **(a)** How a light source affects color appearance. The illustration shows how spectral energy distributions (SEDs) of Cool White (CW) and Deluxe Cool White (CWX) lamps are modified by the reflectance characteristics of a typical human complexion. Under the deluxe lamp, much more red is reflected to the eye. The result is a healthier, more natural appearance. This is true even though CW and CWX lamps have approximately the same whiteness. The Deluxe Warm White lamp, which deemphasizes blues somewhat, is even more flattering to complexions. **(b)** The color of a multiple-color fabric as illuminated by a 150-watt incandescent flood lamp. **(c)** The color of the same fabric as in (b) but illuminated by a 50-watt, high CRI, high-pressure sodium lamp by Phillips. **(d)** The position of various "white" light sources are indicated on the CIE chromaticity diagram. They range from yellowish white to reddish white, but none is "pure (equal energy)" white. (Courtesy of General Electric Company)

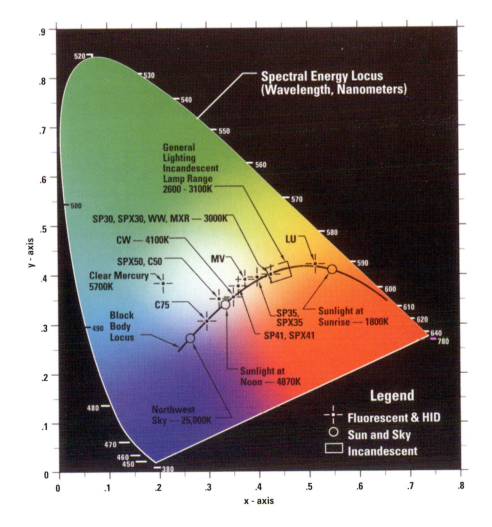

(d)

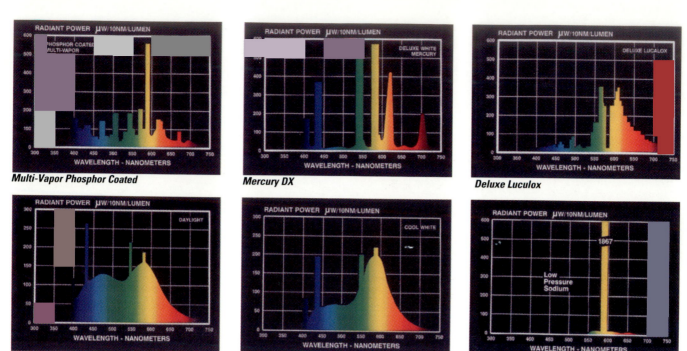

Multi-Vapor Phosphor Coated

Mercury DX

Deluxe Luculox

Daylight

Cool White

Low Pressure Sodium

(a)

(b)

Plate 23 **(a)** Spectral energy distribution (SED) of fluorescent and LPS lamps. [Note: Multi-vapor is metal halide, and Luculox is HPS of General Electric lamps.] **(b)** Effect of the light source on the complexion of a model. The pictures are intended, within the limits of modern high-speed printing, to give a good indication of the differences between SP and SPX lamps, at various chromaticities. [Note: SP and SPX are designations of General Electric. The number indicates the color temperature rating.] (Courtesy of General Electric Company)

(a)

(b)

(c)

(d)

(e)

(f)

Plate 24 Lighting systems and their effect on office environments. **(a)** *Indirect system:* High luminance on the ceiling; soft luminance contrast on furniture; low glare on the video display screen. **(b)** *Parabolic diffusers:* Uniform luminance; wall luminance varies; video display screen has more spill light. **(c)** *Prismatic lens:* Walls and desk are evenly illuminated; video display screen is washed out. **(d)** *High-cutoff louvers:* Walls are dark with pronounced scalloping; poor vertical illumination; less horizontal uniformity; video display screen has less spill light. **(e)** *Directional light produces shadows.* The degree of shadow (soft, harsh, strong) depends on the luminaire's orientation and distribution. **(f)** Any high luminance will appear on the video display screen and will follow the law of reflection. (Courtesy of Peerless Lighting Corp., Berkeley, CA)

10	30	65	70	75
20	50	70	75	85
65	75	80	85	90
10	30	50	75	80
10	20	30	55	70
15	25	40	60	75
10	20	40	65	75
0	35	60	75	80 - 95

Plate 25 Color chart showing base reflectance values of representative room surfaces. The value given under each color sample is the light reflectance value (LRV) of the surface expressed as a percentage. For example, the saturated red in the first row of the chart has an LRV of 10, or 10% of the incident light flux; the light pink in the first row has an LRV of 75 (75%). The base reflectance of white is between 80 and 95, because there are many shades of white ranging from off-white, to warm white, to high reflectance white, etc. In design applications, 80% LRV is normally selected for white surfaces to allow for aging and discoloration.

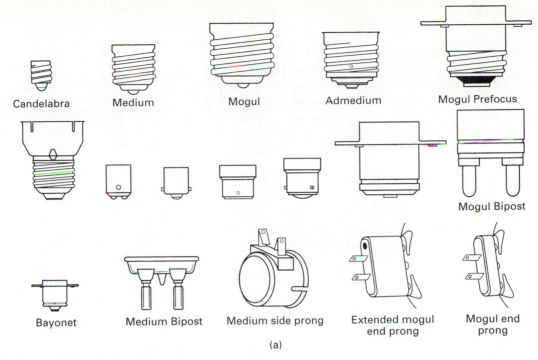

Candelabra   Medium   Mogul   Admedium   Mogul Prefocus

Mogul Bipost

Bayonet   Medium Bipost   Medium side prong   Extended mogul end prong   Mogul end prong

(a)

Bulb Shapes (Not Actual Sizes)

The size and shape of a bulb is designated by a letter or letters followed by a number. The letter indicates the shape of the bulb while the number indicates the diameter of the bulb in eighths of an inch. For example, "T-10" indicates a tubular shaped bulb having a diameter of 10/8 or 1¼ inches. The following illustrations show some of the more popular bulb shapes and sizes.

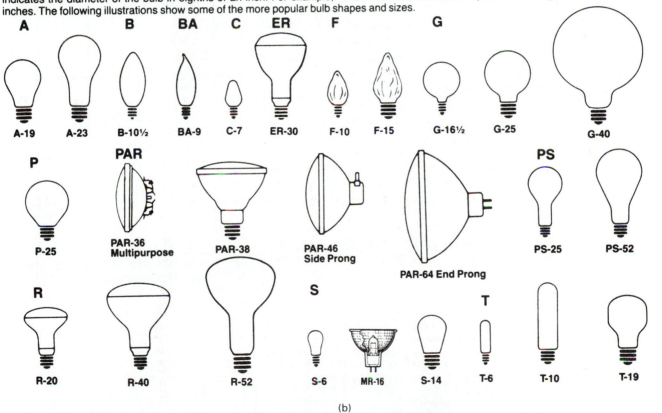

A   B   BA   C   ER   F   G

A-19   A-23   B-10½   BA-9   C-7   ER-30   F-10   F-15   G-16½   G-25   G-40

P   PAR   PS

P-25   PAR-36 Multipurpose   PAR-38   PAR-46 Side Prong   PAR-64 End Prong   PS-25   PS-52

R   S   T

R-20   R-40   R-52   S-6   MR-16   S-14   T-6   T-10   T-19

(b)

■ **FIGURE 15–3**

(a) Common incandescent lamp bases (from General Electric Company, Nela Park, OH).
(b) Common incandescent bulb shape designations (from G.E.):

A	Standard	GT	Globe tubular	R	Reflector
C	Cone	P	Pear	S	Straight inside
B	Decor	PAR	Parabolic	T	Tubular
F	Flame	PS	Pear, straight neck	MR	Minireflector
G	Globe	ER	Elliptical		

431

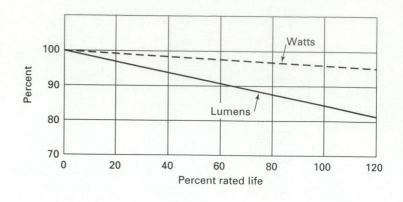

■ **FIGURE 15–4**
Depreciation of output of incandescent lamps. (From Sylvania.)

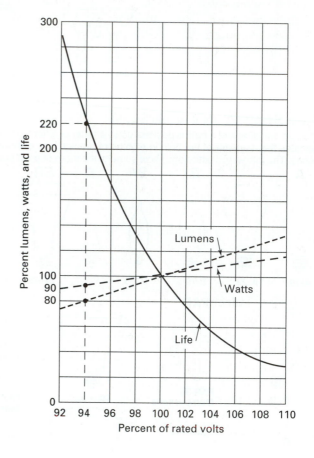

■ **FIGURE 15–5**
Effect of voltage on performance of incandescent lamp.

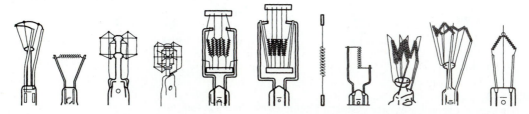

■ **FIGURE 15–6**
Filament coils and supports of incandescent lamps. (From General Electric Lighting Company, Nela Park, OH.)

3800°F and a rated life from 2500 to 10,000 hours. It should be expected that the longer the rated life, the lower will be the efficacy. A number of products claim to have 10,000 hours or more of lamp life. Most of these are constructed of heavier filaments with more filament supports, but operate at low temperatures and thus at extremely low light efficacy.

Dichroic reflector lamps These lamps transmit color selectively through a molecular layer of chemical coating, allowing only the desired wavelength of color to pass through. They are used to reduce the infrared wavelength, which causes heat. Typical applications of dichroic lamps are in retail merchandising displays or on art paintings, where heat in the light beam is substantially reduced. Life and efficacy are similar to those of standard lamps.

Krypton lamps These lamps are filled with the gas krypton and are designed for long life and good color rendition.

Tungsten-halogen lamps This type of lamp makes use of the halogen regenerative cycle to reduce blackening by redepositing the evaporated tungsten atoms on the filament. The bulb of a halogen lamp must be compact in size and thus is normally made with quartz to withstand extremely high temperatures. There are also halogen PAR lamps made of hard glass. Besides their compactness, halogen lamps are high in efficacy and low in lumen depreciation.

Miniature reflector (MR) lamps These lamps are a recent development. They are basically compact halogen lamps, such as the popular 50–100 watt MR-16 (2″-diameter) lamp. Miniature reflector (MR) lamps are high in efficacy, are easy to control, and have excellent color quality. MR lamps have become more popular in interior design applications, especially retail and residential applications and accent lighting in public and office spaces.

15.4 FLUORESCENT LIGHT SOURCES

Fluorescent lamps were developed in France in the early 1930s by André Claude, the inventor of the neon lamp, and were initially manufactured in the United States in 1939. During the second half of the 20th century, fluorescent lamps have improved manyfold in performance, efficacy (see Figure 15–7), color, life expectancy, and cost. They have become the chief light source in the world of lighting.

A fluorescent lamp contains electrodes at both ends of a tube that is filled with mercury vapor.

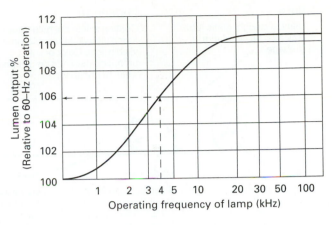

■ FIGURE 15–7
Percent efficacy of fluorescent lamps vs. operating frequency.

When an electric voltage is impressed between the electrodes, ultraviolet energy is generated and converted to visible energy by the phosphor coating on the inside of the bulb. The construction of a typical hot-cathode fluorescent lamp is shown in Figure 15–8(a).

The phosphor coating on the inside of a fluorescent lamp is a mixture of many chemicals that emit visible light when excited by the ultraviolet energy (at 253.7 nm) generated by the mercury vapor. Different phosphors emit different colors. The basic phosphor in white fluorescent lamps is calcium halophosphate, which emits light in the range of 350–750 nm with a peak energy at 610 nm. Other phosphors commonly used are cadmium borate (pink), calcium silicate (orange), calcium tungstate (blue), and zinc silicate (green).

By mixing phosphors in different proportions, hundreds of white fluorescent lamps are created, such as cool white, warm white, white, daylight, deluxe white, etc., and they vary slightly with each manufacturer. It is imperative that the lighting designer be knowledgeable about the S.E.D. characteristics of these variations to guide him or her in selecting a better light source. The following subsections present some characteristics of fluorescent lamps.

15.4.1 Shape and Size

Fluorescent lamps are generally tubular (T) in shape and are of varying lengths and diameters. A T–12 lamp is 1½″ in diameter. There are also U, circular, parallel lamp (PL), parallel lamp cluster (PLC) lamps. The designations of common fluorescent lamps, bases and sockets (lamp holders) are shown in Figure 15–9.

BULB

Usually straight glass tube. May also be circular or U-shaped.

PHOSPHOR

Coating inside the bulb transforms ultraviolet radiation into visible light. Color of light produced depends on composition of phosphor.

CATHODE

"Hot cathode" at each end of lamp is coated with emissive material which emits electrons. Usually made of coiled-coil or single-coil tungsten wire.

EXHAUST TUBE

Air is exhausted through this tube during manufacture and inert gas introduced into the bulb.

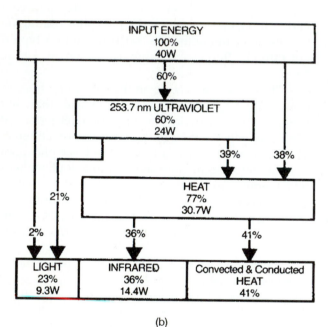

BASE

Several different types used to connect the lamp to the electric circuit and to support the lamp in the lampholder.

MERCURY

A minute quantity of liquid mercury is placed in the bulb to furnish mercury vapor.

GAS

Usually argon or a mixture of inert gasses at low pressure. Krypton is sometimes used.

STEM PRESS

The lead-in wires have an air tight seal here and are made of Dumet wire to assure about the same coefficient of expansion as the glass.

LEAD-IN WIRES

Connect to the base pins and carry the current to and from the cathodes and the mercury arc.

(a)

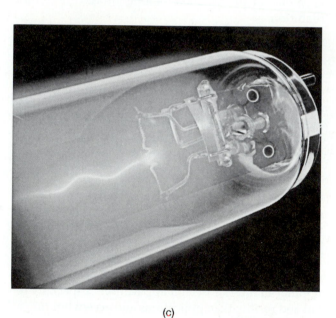

(b)

(c)

■ FIGURE 15–8

(a) Basic construction of a hot-cathode lamp (Courtesy: Osram-Sylvania, Danvers, MA.)
(b) Energy distribution of a typical 40-W cool white fluorescent lamp.
(c) Electric arc across the electrodes. (From General Electric.)
(d) Typical two-lamp series—sequence rapid start circuit.
(e) Preheat circuit with a glow switch. (From Philips Lighting Company, Somerset, NJ.)
(f) Typical lumen depreciation curves of fluorescent lamps.

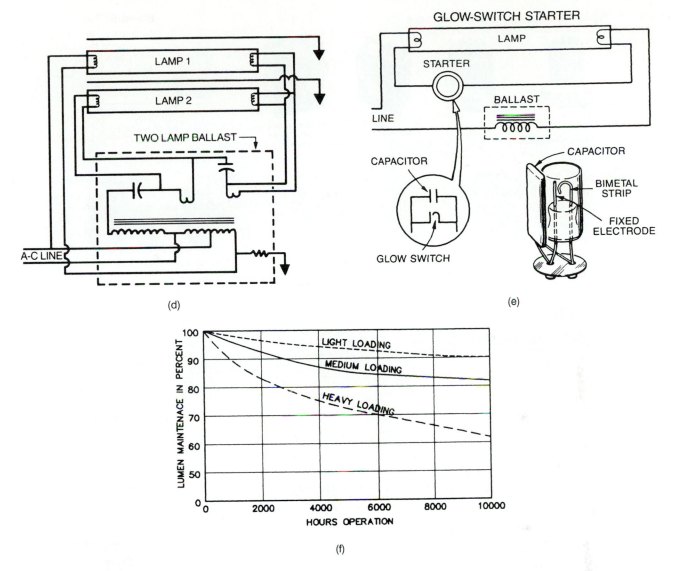

(d)

(e)

(f)

■ **FIGURE 15–8**
Continued

15.4.2 Rating

Fluorescent lamps are made with a fraction of a watt up to 215 watts.

15.4.3 Efficacy and Lumen Depreciation

One of the attractions of fluorescent lamps is their high efficacy. Even so, only about 25 percent of the total electrical power input is converted to visible energy. Depending on the type and rating, fluorescent lamps have a nominal efficacy from 50 to 80 lm/w. Figure 15–8(b) illustrates the distribution of energy of a typical 40-watt cool white rapid-start lamp. The lumens output of fluorescent lamps also depreciates substantially with time. For standard rapid-start lamps, the depreciation is about 20 percent, and for VHO lamps, the depreciation is up to 40 percent at the end of the rated life of the lamp.

Newer and more energy-efficient fluorescent lamps are filled with an argon-krypton gas mixture, rather than only argon gas. The lamp is coated with a transparent conductive coating to lower the starting voltage. The diameter of the popular T-12 lamp is reduced to T-8 to save about 50 percent of the volume of the gas. When the lamp is used in conjunction with an electronic ballast, the total power required for the lamp-ballast combination is reduced 30 to 40 percent, compared with the traditional design. This development has had a great impact on the demand for

Typical Lamp and Starter Holders

Rapid Start and Preheat

Rapid Start Lamps

14, 15, 20, 25, 30 & 40
Watt 2-Pin Starters

85, 90 & 100 Watt
2 & 4-Pin Starters

Starter Sockets

Butt-On Mounting

Surface Mounting Slimline

Slimline Lamps

Surface Mounting HO/VHO

Super-Hi Output/High Output

Surface Mounting PL* Sockets

Base Types (not Actual Sizes)

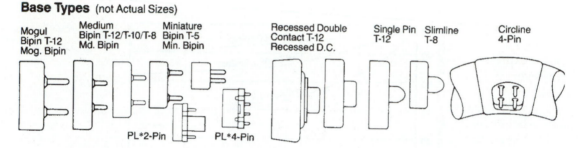

Mogul
Bipin T-12
Mog. Bipin

Medium
Bipin T-12/T-10/T-8
Md. Bipin

Miniature
Bipin T-5
Min. Bipin

PL*2-Pin   PL*4-Pin

Recessed Double
Contact T-12
Recessed D.C.

Single Pin
T-12

Slimline
T-8

Circline
4-Pin

Bulb Shapes (Not Actual Sizes)

The size and shape of a bulb is designated by a letter or letters followed by a number. The letter indicates the shape of the bulb while the number indicates the diameter of the bulb in eighths of an inch. For example, "T-12" indicates a tubular shaped bulb having a diameter of 12/8 or 1½ inches. The following illustrations show some of the more popular bulb shapes and sizes.

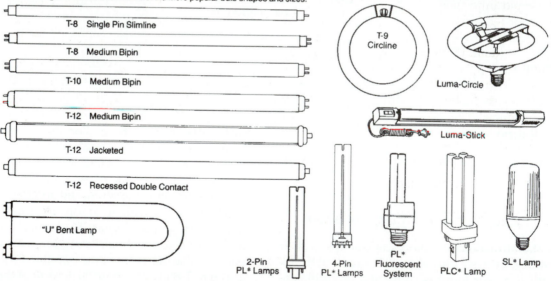

T-8   Single Pin Slimline

T-8   Medium Bipin

T-10   Medium Bipin

T-12   Medium Bipin

T-12   Jacketed

T-12   Recessed Double Contact

"U" Bent Lamp

2-Pin
PL* Lamps

4-Pin
PL* Lamps

PL*
Fluorescent
System

PLC* Lamp

SL* Lamp

T-9
Circline

Luma-Circle

Luma-Stick

■ **FIGURE 15–9**
Common fluorescent lamp holder, base, and bulb shapes. (Courtesy: General Electric Lighting Company, Nela Park, OH.)

power and energy consumption in the United States, since lighting accounts for about one-third of total U.S. electrical energy consumption, and fluorescent lamps account for two-thirds of that amount.

15.4.4 Rated Life

Fluorescent lamps have an excellent rated life. The popular types of fluorescents are rated between 5000 and 20,000 hours, and even longer. The life of the lamp is affected by the burning cycle, in terms of hours per start. At eight hours per start, rapid-start lamps have a rated life exceeding 24,000 hours. This is 30 times the rated life of general-service types of incandescent lamps.

15.4.5 Color Temperatures

The apparent color of the light generated by fluorescent lamps has many variations, ranging from warm (rich in yellow) to cool (rich in green). The apparent color may be expressed in terms of the color temperature. However, it varies among manufacturers. A representative color temperature rating of fluorescent lamps is given in Table 15–3, and the relative positions of these colors on the CIE chromaticity diagram are shown in Plate 22(d).

15.4.6 Color Rendering

With the rapid improvement of fluorescent technology, the color rendering indexes of some fluorescent lamps are nearly as good as that of incandescent lamps. While standard fluorescent lamps have ratings around 70, specially formulated triphosphor lamps have a CRI over 90 and a CPI up to 100. Plate 1(b) illustrates the color-rendering effect of several fluorescent lamps.

TABLE 15–3

Apparent color (color temperature) of fluorescent lamps

Lamp Designation	Color Temperature Range °K
Deluxe warm white	2900–3000
Warm white	3000–3100
White	3400–3500
Natural white	3500–3600
Cool white	4000–4500
Deluxe cool white	4500–5000
Daylight	6000–6500

15.4.7 Ballasts

Fluorescent lamps have so-called negative electrical resistance. That is, once an arc is struck across the lamp, the ionized mercury vapor becomes increasingly more conductive, and thus, more current will flow until the lamp is burnt out. For this reason, fluorescent and other HID lamps must be connected through a ballast, which serves both as a transformer to boost the voltage at the lamp terminals and as a choke to limit the maximum flow of current. There are four basic types of ballast:

1. *Magnetic type.* This is the conventional electromagnetic core-and-coil type of ballast operating at 60 Hz with secondary voltage between 200 and 700 volts, depending on the length of the lamp. A capacitor is provided in high power factor (HPF) models, and a thermal cutout is provided in P-rated models.

2. *Hybrid magnetic type.* This is basically a magnetic ballast, except that it has a built-in electronic switching device to save energy by disengaging the cathode current after the lamp or lamps are started.

3. *Hybrid electronic type.* This is a combination electronic and electromagnetic type of ballast that consists of an input electro-magnetic interference (EMI) filter, a rectifier to convert standard frequency (50–60 Hz) AC into DC, an inverter (oscillator) to convert DC to high frequency AC (20–30 kHz), a capacitor to correct the power factor, and an output transformer to provide the operating voltage and current limitation to the lamps.

4. *Electronic type.* This is the newest type of ballast, with all-electronic components. In addition to the rectifier and inverter, it incorporates a DC power preconditioner that provides power factor correction and a constant DC source to power the inverter. Models with integrated circuit and feedback controls can have additional control functions, such as manual and automatic dimming, detection of occupancy, and interfacing with the energy management system. The electronic ballast is quiet in operation, low in weight, and cool in temperature, and has the highest ballast efficiency factor (BEF). As the cost of an electronic ballast becomes more competitive, it will undoubtedly displace the electromagnetic type of ballast for all major lamp sizes in the 30- to 200-watt range. Small fluorescent lamps will continue to use the electromagnetic type for some years to come. Figure

15–10(a) shows the block diagram of an electronic ballast. Electronic ballasts should meet FCC standards on radio frequency interference (RFI) or EMI, IEEE/ANSI requirements on harmonics and UL requirements on safety. These requirements are too detailed to be discussed herein.

15.4.8 Ballast Factor

The ballast factor is the ratio of the light output produced by lamps operating on a commercial ballast to the light output of the same lamps operating on a standard reference ballast in the testing laboratory. If a ballast has a BF of 0.95, it means that the actual light output of the lamps connected with this commercial ballast will be only 95 percent of the manufacturer-published lumens output. In general, ballasts certified by the Certified Ballast Manufacturers (CBM) for 30–40-watt lamps must have a minimum BF factor of 92.5 percent. A medium value of BF = 95 percent is normally assumed. However, with the newer electronic ballasts, the BF is frequently more than 100 percent. The ballast factor can be expressed as

$$BF = F_a/F \qquad (15–1)$$

where: F_a = Actual light output of the lamps using the commercial, ballast, in lumens

F = Published light output listed in the lamp manufacturer's catalog (as tested with a reference ballast in the lab)

BF = Ballast factor, per unit

In other words, the actual light output F_a of the lamps used in the calculation of illuminance is

$$F_a = F \times BF \qquad (15–2)$$

15.4.9 Ballast Efficiency Factor

The ballast efficiency factor (BEF) is the ratio of the ballast factor to the power input to the specific ballast-and-lamp combination. The National Appliance Energy Conservation Act (NAECA) establishes the minimum energy efficiency standards for ballasts as

$$BEF = BF/P \qquad (15–3)$$

where BF = Ballast factor
P = Power input to the ballast-and-lamp combination, in watts

For example, NAECA specifies that the BEF shall not be lower than 1.805 for a single 40 w, 4-foot T-12 lamp ballast and 1.060 for a two-lamp ballast combination. Unfortunately, no correlation can be established between these BEF values.

15.4.10 Lamp Ballast Efficacy Factor (LBEF)

A more useful factor is the LBEF. It can be used for evaluating any lamp-ballast combination.

$$LBEF = F_a/P = F \times BF/P \qquad (15–4)$$

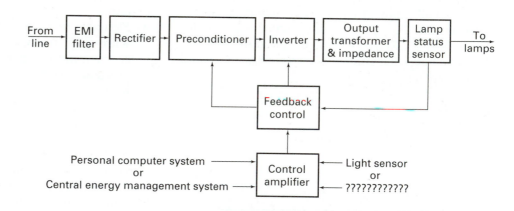

■ **FIGURE 15–10**
(a) The block diagram of an electronic ballast. (b) The cutaway interior view of an electronic ballast. (Courtesy: Advance Transformer Company, Rosemont, IL.)

where F_a = Actual light output of a lamp-ballast
combination, lumens

P = Lamp manufacturer's published
light output, lumens

LBEF = Lamp ballast efficacy factor, LM/w

For example:
A 4-foot T-8, 32-watt, R.S. lamp is rated for 2950 lumens. If the lamp is wired with an electronic ballast having a BF = 0.87 and a power input of 31-watt, then,

$$LBEF = F \times BF/P = (2950 \times 0.87)/31 = 82.8 \text{ Lm/w.}$$

This value can be compared with any other lamp-ballast combinations. A higher LBEF always represents a more efficient combination for the same size lamps or different size lamps.

15.4.11 Dimming of Fluorescent Lamps

Like incandescent lamps, fluorescent lamps can be dimmed. With the electro-magnetic type of fluorescents, the ballast contains a separate circuit for maintaining the voltage to the lamp electrodes, and the light output is reduced by "chopping" the voltage wave. For the electronic type, which operates at 20–30 kHz, light output is reduced by shortening the duration of the current wave.

15.4.12 High-Frequency Operation

A fluorescent lamp's efficacy increases with its frequency. As shown in Figure 15–11(c), the efficacy steadily increases to 111% from the standard utility frequency (50–60 Hz) to about 100 kHz. There is not much gain from 20 to 100 kHz. An aircraft's electrical system is generally designed for 400 Hz in order to reduce the weight of power equipment. At 400 Hz, fluorescent lights will gain about 6 percent more efficacy.

15.4.13 Types of Fluorescent Lamps

Fluorescent lamps may be divided into the hot-cathode and cold-cathode varieties. The major difference between the two is in the construction and operating temperature of their cathodes. Since most fluorescent lamps operate under the hot-cathode principle, their identification as such is often omitted. The following subsections briefly describe commercially available fluorescent lamps.

Preheat lamps   A manual or automatic starter switch is required in preheat lamps. These types of lamps are limited to small wattages. Lamps 30 watts or higher are not normally used in building systems. (See Figure 15–8(e) for the preheat circuit with a glow switch.)

Instant-start lamps   These lamps operate without starters. The ballast provides a high enough voltage at about 680 volts to strike the arc instantly. Instant-start lamps are also called "Slimline Lamps." They are available in 21- to 75-watt ratings.

Rapid-start lamps   Ballasts for rapid-start lamps have separate windings to provide continuous heating voltage for the lamp electrodes, as shown in the circuit diagram in Figure 15–8(d). The lamp starts in less than one second, nearly instantaneously. Following are a number of variations:

- Straight-tube rapid start (RS)
- U-tube rapid start (U/RS)
- Circline rapid start (T/C/RS)
- High-output rapid start (HO/RS)
- Very high-output rapid start (VHO/RS)

Compact lamps   Two kinds of compact fluorescent lamps can be used to replace low-efficacy incandescent lamps:

- *PL lamps* are single-ended fluorescent lamps. They are available from a few watts up to 18 watts. Due to their compact size, they are popular for illuminating general spaces that previously were lit by incandescent fixtures. PL lamps may last as long as 50,000 hours, compared with less than 2000 hours for incandescent lamps, and have an efficacy up to 50 lm/W compared with 15–20 lm/W for incandescent. Thus, they are over 300 percent more energy efficient. Figure 15–11(a) shows an Edison (medium-screw) base adapter used to replace incandescent lamps in an incandescent fixture.
- *SL Lamps* are specially designed to fit directly in an incandescent fixture. The lamp has a built-in ballast. Figure 15–11(b) shows the interior of an SL lamp.

Specialty Lamps   There are numerous varieties of specialty fluorescent lamps:

- *Black light lamps* produce energy in the near ultraviolet range.
- *UV lamps* produce ultraviolet energy below 320 nanometers for germicidal use.
- *Plant growth lamps* are designed especially to stimulate photosynthesis.

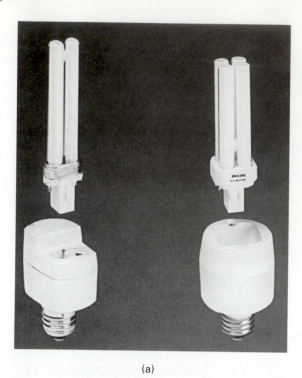

(a)   (b)

■ **FIGURE 15–11**
(a) PL lamp with an adapter base for replacing incandescent lamps. (b) The interior and exterior shape of an SL lamp. The ballast is located at the base of the lamp. The lamp shown is rated for 18 W and is intended to replace 60–75-W incandescent lamps. (Courtesy: Philips Lighting Company, Somerset, NJ.)

- *Cold-cathode lamps* are phosphor-coated lamps filled with mercury vapor and argon gas that operate at a voltage from 700 to 1000 volts. The cathodes are thimblelike in shape and can emit sufficient electrons at a much lower temperature than hot-cathode lamps. Cold-cathode lamps are generally small in diameter (T4 to T8) and can be bent into various shapes up to 20 ft in length. They have low efficacy (not much above half that of hot-cathode lamps) and a long rated life (over 40,000 hours). With different gas fills, they can also produce different colors. Cold-cathode lamps are used primarily for decorative lighting, such as for curved coves and in locations where replacing a lamp is difficult. Cold-cathode lamps are often used in lieu of neon lights in the interiors of buildings.

- *Neon lamps* are noncoated cold-cathode lamps operating at extremely high voltages (exceeding 5000 volts). The lamp tube is small in diameter and can easily be bent into any shape. Different gas fills generate different colors, e.g., neon emits red, and argon and mercury together emit blue and, combined with a blue-absorbing glass tube, will emit green. Other combinations of gas fills

and glass will result in various colors. Neon lights are used primarily for signs and decorative applications.

- *Subminiature lamps* constitute a family of tiny fluorescent lamps with a 7-mm (T2½) diameter with rating from 1 to 3 watts. They are used principally for backlighting of liquid crystal display signs or for lighting instruments.

- *Reflector lamps* have an internal reflector to cover up a portion of the bulb; thus, they reflect the light to the open aperture portion of the lamp at a higher intensity than the general-service–type lamps. Reflector lamps may be considered as a lighting fixture–lamp with built-in reflector. They are useful for display or in cove-lighting applications.

15.5 HIGH-INTENSITY DISCHARGE LIGHT SOURCES

High-intensity discharge (HID) lamps are a family of lamps that incorporate a high-pressure arc tube within the lamp envelope. The tube is filled with metallic gas, such as mercury, argon, sodium, etc.

When the gas is fully vaporized due to the flow of electrical current, the arc tube will have a high internal pressure of around 2 to 4 atmospheres (200 to 400 kilopascals). The concentrated light power generated in the arc tube has considerably high intensity (candela/sq in), from which the name of the lamp is derived. In the mercury type of HID lamp, the mercury gas generates a broader spectrum consisting of five principal visible spectral lines (404, 435, 546, 577, and 579 nm) in lieu of the single ultraviolet wavelength at 257 nm that simple fluorescent lamps produce. The envelope of HID lamps may be clear or coated with phosphor to improve color rendition.

15.5.1 Types of HID lamps

There are three major classes of HID lamps.

- *Mercury lamps* contain only mercury vapor and produce a predominantly bluish white light (5500–5800°K). Mercury lamps are used mostly for industrial and outdoor applications. Figure 15–12(a) shows the typical construction of a mercury lamp. Mercury lamps have an efficacy between 40 and 70 lm/W.
- *Metal halide lamps* contain mercury vapor and other halides to improve both their efficacy and their color-rendering characteristic. Typical halides include scandium, sodium oxide, dysprosium, indium oxide, and other rare earth iodides. These lamps have a color temperature around 3200°K to 4000°K, compared with 6000°K, for mercury lamps. Because they have an extremely high lamp efficacy—up to 120 lm/W, metal halide lamps are applicable to most indoor lighting needs. Figure 15–12(b) shows the construction of a metal halide lamp.
- *High-pressure sodium (HPS) lamps* contain xenon as a starting gas and an amalgam of sodium and mercury that is partially vaporized when the lamp attains its operating temperature. HPS lamps are constructed with two envelopes: an inner envelope (arc tube) made with material that has electrical resistance to sodium and an outer envelope designed to protect the arc tube. Some HPS lamps are also available with diffuse coatings on the outer tube. Figure 15–12(c) shows the construction of HPS lamps.

    Although HPS lamps radiate energy across the entire visible spectrum, yellow-orange wavelengths from 550 to 600 nanometers are most dominating. (See Plate 1 (a) for a typical spectral distribution.) HPS lamps may be described as golden white. Their color temperatures vary from 2000 to 4000°K. Increasing the sodium pressure increases the percentage of red radiation and thus improves color rendition, but at the sacrifice of life and efficacy. Figure 15–12(d) shows HPS lamps in sizes from 35 and 100 watts. At 80, these lamps have an amazing CRI. Plate 22(b) and (c) compares the color rendering of a special HPS lamp with that of an incandescent lamp.

    HPS lamps have the highest luminous efficacy of all metal halide lamps, ranging from 60 to 140 lm/W.

15.5.2 Characteristics of HID Lamps

HID lamps are similar to fluorescent lamps, except for their operating pressure, gas fill, and shape of the bulb. HIDs and fluorescents are all-electric discharge lamps. Following are some common characteristics of HID lamps.

Configuration and designation   The configurations of mercury and metal halide lamps are similar, but HPS lamps are more cylindrical. (See Figure 15–12(a–c).) The ANSI designation of mercury lamps starts with "H," MH lamps with "M" and HPS lamps with "S." Other numerals and letters in the designation denote the bulb size and wattage of the lamp. For example: MVR 150/C/vBu stands for multiple vapor, 150 watt, clear and design for vertical burning.

Performance   All HID lamps have high efficacy and long life, although they vary considerably in type and size. Table 15–4 shows the dimensions, rated life, initial lumens, efficacy (lm/W), and lamp lumen depreciation (see next subsection) of typical HID lamps.

Lamp lumen depreciation   HID lamps have a relatively higher rate of lamp lumen depreciation (LLD) than do fluorescent lamps; MH lamps have the highest rate, down to 60 percent at the end of their rated life. Economics will justify the replacement of these lamps prior to the end of their rated life, however. Figure 15–13 illustrates these characteristics.

Starting characteristics   HID lamps are not instant-start lamps. They all require time to reach their rated light output. Metal halide lamps require the longest time to start, and once turned off, they take a longer time to restart (restrike). In general, the starting time of HID lamps is from 2 to 5 minutes, and the restarting time is from 5 to 10 minutes. Thus, the

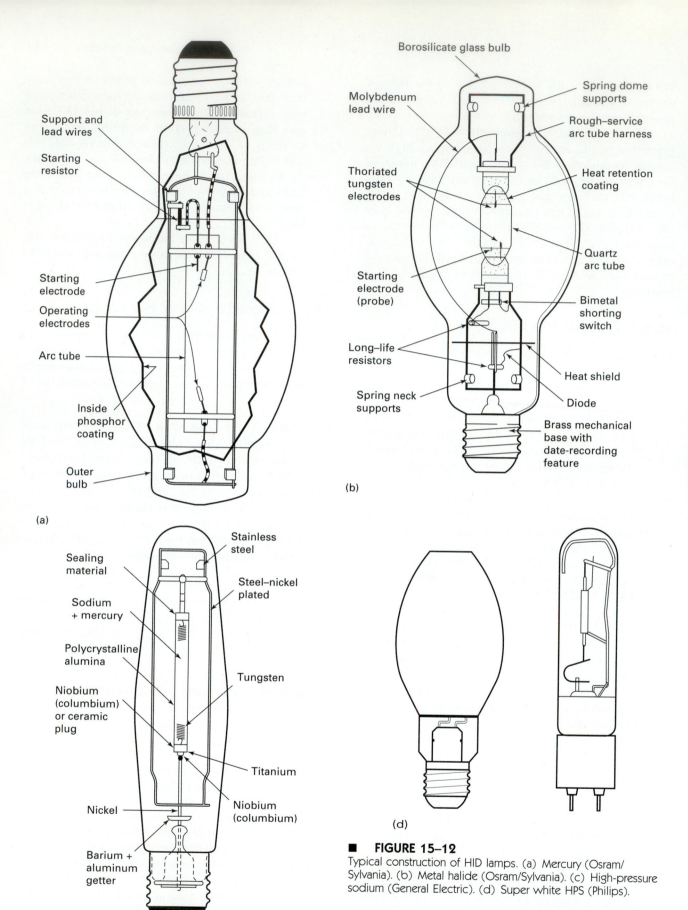

(a)

Support and lead wires
Starting resistor
Starting electrode
Operating electrodes
Arc tube
Inside phosphor coating
Outer bulb

(b)

Borosilicate glass bulb
Molybdenum lead wire
Thoriated tungsten electrodes
Starting electrode (probe)
Long–life resistors
Spring neck supports
Spring dome supports
Rough–service arc tube harness
Heat retention coating
Quartz arc tube
Bimetal shorting switch
Heat shield
Diode
Brass mechanical base with date-recording feature

(c)

Sealing material
Sodium + mercury
Polycrystalline alumina
Niobium (columbium) or ceramic plug
Nickel
Barium + aluminum getter
Stainless steel
Steel–nickel plated
Tungsten
Titanium
Niobium (columbium)

(d)

■ **FIGURE 15–12**
Typical construction of HID lamps. (a) Mercury (Osram/Sylvania). (b) Metal halide (Osram/Sylvania). (c) High-pressure sodium (General Electric). (d) Super white HPS (Philips).

442

TABLE 15–4

Typical characteristics of HID lamps

Watts	Outer Bulb Finish	ANSI Designation	Length (inches)	Rated Life (hours)	Initial Lumens[a]	Input Watts to Ballast	Lumens per Watt	LLD[b] (%)
Mercury Lamps								
100	Phos.	H38JA-100/DX	7½	24,000	4,425	120	33.9	69
175	Phos.	H39KC-175/DX	8⁵⁄₁₆	24,000	8,600	205	42.0	78
250	Phos.	H37KC-250/DX	8⁵⁄₁₆	24,000	12,775	285	44.9	76
400	Clear	H33CD-400	11½	24,000	21,000	450	46.7	80
400	Phos.	H33GL-400/DX	11½	24,000	23,125	450	51.4	71
1000	Clear	H36GV-1000	15⅜	24,000	56,150	1085	51.8	80
1000	Phos.	H36GW-1000/DX	15⅜	24,000	63,000	1085	58.1	71
Metal Halide Lamps								
175	Clear	M57PE-175	8⁵⁄₁₆	7,500	14,000	210	66.7	77
250	Clear	M58PG-250	8⁵⁄₁₆	10,000	20,500	290	70.7	76
400	Clear	M59PJ-400	11½	15,000	34,000	455	74.7	70
400	Phos.	M59PK-400	11½	15,000	34,000	455	74.7	68
1000	Clear	M47PA-1000	15⅜	10,000	110,000	1090	100.9	73
1000	Phos.	M47PB-1000	15⅜	10,000	105,000	1090	96.4	70
1500	Clear	M48PC-1500	15⅜	3,000	155,000	1610	96.3	—
High-Pressure Sodium (HPS) Lamps								
100	Clear	S54SB-100	7¾	24,000	9,500	130	73.1	—
150	Clear	S55SC-150	7¾	24,000	16,000	180	88.9	88
250	Clear	S50VA-250/S	9¾	24,000	30,000	295	101.7	88
400	Clear	S51WA-400	9¾	24,000	50,000	460	108.7	88
1000	Clear	S52XB-1000	15¹⁄₁₆	24,000	140,000	1085	129.0	90

[a]Lamps burning in the vertical position.

[b]Lamp lumen depreciation percentage of initial light output at 70% of rated life.

Source: Reprinted with permission from IES publications. The table is only a partial listing of HID lamps. Refer to the *IES Lighting Handbook*, 1984 reference volume, for a complete listing of all HID sources.

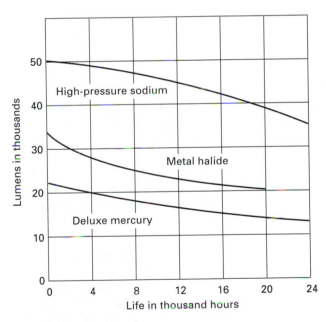

■ **FIGURE 15–13**

Lamp lumen depreciation (LLD) of typical 400-watt high-intensity discharge (HID) lamps: High-pressure sodium (HPS), metal halide, and deluxe mercury. (Reprinted with permission from IES publications.)

designer should be knowledgeable of the operation of the space when HID is used. HID lamps are not suitable for frequent on-and-off operations. When they are used in public spaces, such as an arena or a stadium, an auxiliary lighting system, such as fluorescent or incandescent lighting, should be provided to maintain minimum lighting during power interruption.

Operation   For continuous operation (24 hours/day and 7 days/week), the lamps should be turned off for about 10 minutes once a week. Otherwise, lamp life will be substantially shortened.

Equipment operating factor   The lumen output of HID lamps depends not only on the ballast, but also on the lamp operating position and other factors. The equipment operating factor (EOF) is defined as the ratio of the flux of an HID lamp-ballast-luminaire combination operating in a specific position to the lamp-luminaire combination operating with a reference ballast at the vertical position. The EOF includes the effect of the lamp position (or tilt) factor and the ballast factor. For determining initial illuminance in engineering calculations, the EOF should be applied to the manufacturer's published light output. For example, if the EOF of a 400-watt metal halide lamp-ballast-luminaire combination is given as 0.91, then the actual light output of this combination, using a 400-watt M59PK–400 metal halide lamp, is 34,000 lumens times 0.91, or only 30,940 lumens. See Chapter 16 for the use of lamp operating factor (LOF) instead of EOF.

15.6 MISCELLANEOUS LIGHT SOURCES

Many other light sources are used in lighting systems. The following are of particular interest.

15.6.1 Short Arc Lamps or Compact Arc Lamps

Short arc lamps produce high-intensity light from a small bulb, thus closely resembling a point source, which is important for critical light beam controls. These devices are HID lamps containing primarily xenon gas with mercury or argon as the starter. The lamps are available from 100 watts to 30,000 watts. Xenon lamps are used primarily as spotlights, in projectors, and as searchlights.

15.6.2 Low-Pressure Sodium Lamps

Low-pressure sodium (LPS) lamps produce a monochromic yellow-wavelength light at 589 nm. (See Plate 1a for the spectral energy distribution of typical LPS lamps.) The lamp is filled with sodium vapor and small amounts of argon, xenon, or helium as the starting gas. The arc tube temperature, which is critical, is optimal at 260°C. Deviating from this temperature will result in a great reduction in efficacy. For this reason, the outer bulb of the lamp is kept in a state of high vacuum, to retard heat transfer. When operating at the desired arc tube temperature, the lamp can produce up to 200 lm/W of light power, the highest of all light sources. LPS is not suitable for interior lighting applications in buildings. The principal applications of LPS lamps are highway exchanges, storage yards, and wherever color rendition is unimportant or a strong dominant color (yellow) is desired. For example, using LPS at highway exchanges provides a message of caution to motorists. On the other hand, LPS is not recommended for parking lots, because people are not flattered by black lips and pale clothing and because distorting the color of the cars makes them difficult to find.

15.6.3 Electrodeless Lamps

Electrodeless lamps include several new generation of lamps and promise to be the lamps of the future. They have a good color-rendering property, high efficacy, and up to 50,000 or more operating hours. Their light output will not depreciate, because they have no electrodes. They are ideal for installing in difficult-to-reach places. Two types of electrodeless lamps are:

- *Electromagnetic.* This type utilizes the principles of electromagnetism to excite the gas fill in the lamp. It consists of a magnetic core-and-coil assembly at the lamp's center, but external to the lamp envelope. The construction and operating principle of electromagnetic electrodless lamps are shown in Figures 15–14(a) and (b).
- *Microwave.* This type utilizes a concentrated microwave generator to direct microwaves to a glass bulb filled with sulfur gas. Depending on the microwave power, the light generated can be of very high intensity. Figures 15–14(c) and (d) illustrate the construction and operating principle of the microwave lamp. Because of its concentrated light power within a small bulb, this lamp is ideal for use with a light tube—a tubular luminaire with light originating from one end. By controlling refraction and reflection in the tube, light can be uniformly emitted throughout the entire length of the tube, which may stretch up to several hundred feet. This eliminates the need for multiple luminaires in the space to be lighted.

15.7 GENERAL COMPARISON OF LIGHT SOURCES

With the basic knowledge covered to this point, the reader should be able to make appropriate selections of light sources for various applications. Following is a summary of the factors that enter into these decisions:

1. *Efficacy.* Incandescent lamps—including the tungsten-halogen types—are the least energy-efficient light sources compared with fluorescent or HID lamps, which are three to five times as efficient.
2. *Spectrum power (energy) distribution.* No light source can produce equal power (energy) across the entire visible spectrum. Even daylight at

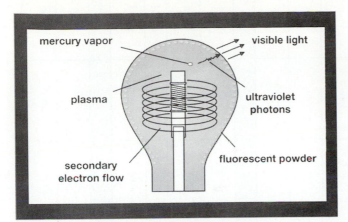

(a)

FIGURE 15–14

(a) and (b) illustrate the principle of light generation of an induction lamp, which has no direct electrical power connection, rather a power coupler, which generates high frequency. The high frequency induces an alternating magnetic field that excites the mercury gas so that it generates light in the same manner as a fluorescent lamp. (Courtesy: Philips Lighting Co.) (c) and (d) illustrate the principle of microwave light generation. The lamp is filled with sulfur gas, and when bombarded by concentrated microwaves, it generates high-intensity light. (Courtesy: Fusion Lighting, Inc., Rockville, MD.)

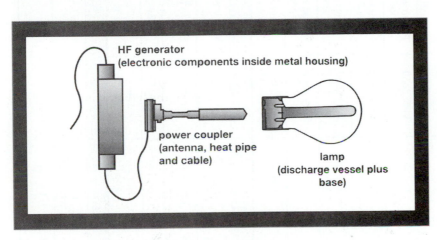

(b)

(c)

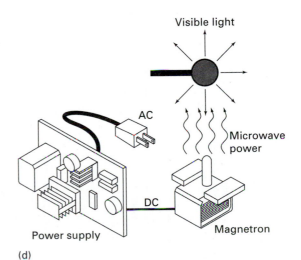

(d)

TABLE 15-5

General characteristics of commonly used light sources (reproduced with permission from IESNA)

General Characteristics of Commonly Used Light Sources

Light Source	Wattage Range	Efficacy (lm/W)	Life	Lumen Maintenance	Starting Time	Color Rendition	Ballast Required	Dimming Capability	Optical Control
Incandescent Filament	10 to 1500	Very low	Very low to low	Fair to good	Very good	Very good	No	Very good	Good
Tungsten-halogen	10 to 2000	Very low to low	Very low to low	Good to very good	Very good	Very good	No	Good	Very good
Low-Pressure Discharge									
Standard fluorescent	15 to 40	Low to good	Fair to very good	Fair to good	Good to very good	Low to very good	Yes	Good	Poor
Slimline fluorescent	20 to 75	Fair to good	Fair to good	Fair to good	Very good	Fair to very good	Yes	Low	Poor
High-output fluorescent	35 to 110	Fair to good	Fair to good	Fair to good	Very good	Fair to very good	Yes	Good	Poor
Very high-output fluorescent	38 to 215	Fair to good	Fair to good	Fair to good	Very good	Fair to very good	Yes	Good	Poor
Energy-saving fluorescent (T-12)	30 to 185	Fair to good	Fair to good	Fair to good	Very good	Low to very good	Yes	Low	Poor
High-efficacy fluorescent	18 to 40	Good	Good	Good	Very good	Good to very good	Yes	Fair	Poor
Compact fluorescent	5 to 40	Good	Fair to good	Good	Good to very good	Good to very good	Yes	Very low	Fair
High-Intensity Discharge									
Mercury	40 to 1000	Low to Fair	Good to very good	Very low to fair	Low	Very low to fair	Yes	Fair	Poor
Self-ballasted mercury	100 to 1500	Very low	Fair to very good	Low to fair	Fair	Low to fair	No	Very low	Poor
Metal halide	32 to 1500	Good	Low to Fair	Very low	Good	Low	Yes	Low	Good
High-pressure sodium	35 to 1000	Fair to good	Fair to very good	Fair to good	Fair	Low to good	Yes	Low	Good
Miscellaneous									
Low-pressure sodium	10 to 180	Fair to very good	Fair to good	Good to very good	Fair	Very low	Yes	Very low	Poor
Cold Cathode	10 t0 150	Low	Very good	Fair to good	Very good	Low to very good	Yes	Good	Poor

*See manufacturers' catalogs for specific data.

7500°K is richer in blue and poorer in red. Incandescent lamps are increasingly rich in the shorter spectral colors (orange and red), and fluorescent are rich in blue and green. Low-pressure sodium lamps generate only a very narrow band of yellow spectrum and lack all other colors. Typical spectrum energy distributions of selected light sources are shown in Plate 23a.

3. *Color rendering.* This is a very important aspect of lighting design. Plate 23b illustrates how a model appears under various light sources. The final color appearance is the combination of the spectrum energy on the life of the light source and the reflectivity of the model, as illustrated in Plate 22a. Many new light sources have excellent color-rendering qualities as well as high luminous efficacy. (See Plates 22b and 22c. The HPS lamp used in the illustration has a CRI of 80; the color rendition is favorably compared with that of an incandescent lamp with CRI taken as 100.)

4. *Color temperature.* The apparent color of white light sources, whether they are warm (yellowish, reddish) or cool (bluish, greenish), can be expressed by the black body temperature along the Planckian locus in the CIE chromaticity diagram. The characteristics of representative white light sources including that of incandescent, fluorescent lamps, sunlight and skylight are identified in Plate 22d.

5. *Rated life.* The lives of different light sources vary widely. In general, fluorescent and HID lamps last 10 to 20 times as long as incandescent lamps. This means that during the average life of a fluorescent or HID lamp, an incandescent lamp may have to be replaced 20 times.

6. *Cost of operation.* The energy savings and cost of maintenance of fluorescent and HID lamps are considerably lower than for incandescent lamps. The use of incandescent light sources should be limited to special-task illumination rather than general space illumination in most interior spaces, including the highly decorative commercial spaces.

The general characteristics of some currently popular light sources are summarized in Table 15–5 published by the Illuminating Engineering Society of North America. Table 15–6 is another valuable comparison of various types of lamps based on the efficacy, life, and typical applications. Both comparisons, of course, need to be updated periodically as technology advances.

15.8 LUMINAIRES

A luminaire, commonly called a lighting fixture, is a complete lighting unit that contains one or more lamps, structural supports, accessories, auxiliaries, wiring, and controls. Luminaires can be classified according to a number of constructional, physical, and photometric criteria, the most important of which are the following:

1. *The light source.* Whether the lamps within the luminaire are incandescent, halogen, fluorescent, mercury, metal halide, high-pressure sodium, low-pressure sodium, etc.
2. *The wattage.* Whether the luminaire is designed for a single or multiple number of lamps, and what is the maximum size of the luminaire(s), etc.
3. *The power supply.* Whether the luminaire is wired for 120, 220, 277, 380, or 480 volts and whether it is for AC or DC, etc.
4. *The application.* Whether the luminaire is for outdoor, indoor, underwater, underground, aviation, theater, roadway applications, for signs, etc.
5. *The construction.* Whether the luminaire is waterproof, weathertight, dust tight, explosion proof, bug tight, corrosion resistant, etc.
6. *The method of mounting.* Whether the luminaire is to be ceiling surface, laid in, suspension, bracket, wall, floor, pole mounted, etc.
7. *The control medium.* Whether the control is a prismatic lens, diffusing panel, or louver, and the specific cutoff angles, baffles, reflecting characters, etc.
8. *The ballast.* The type and number of ballasts, and the ballast factor (BF), ballast efficiency factor (BEF), etc.
9. *Special features.* Whether there is EMI shielding, a built-in emergency lamp, a built-in battery, etc.
10. *The light distribution.* Interior luminaires are classified by the IES into six categories and the CIE into five categories. The categories are based on the percentages and characteristics of the light flux distribution above and below the horizontal plane. Figure 15–15 illustrates the following six categories:
    - *Direct,* where 90 to 100 percent of the flux is directed downward.
    - *Semidirect,* where 60 to 90 percent of the flux is directed downward and the balance upward.
    - *General diffuse,* where between 40 and 60 percent of the flux is directed either upward or

TABLE 15–6

General comparison of different types of lamp: efficacy, life, and applications (courtesy of Phillips Lighting Company)

Category	Type	Typical Efficacy (Lm/W)	Rated Life (hours)	Characteristic Features	Typical Application
Incandescent Lamps	General Service and Reflector	5-22	750 to 2000	Easy to install, easy to use; many different versions; instant start; low cost; reflector lamps allow concentrated light beams	General lighting in the home, decorative lighting, localize lighting; accent and decorative lighting (reflector lamps)
	Halogen	12-36	2000 to 3000	Compact; high light output; white light; easy to install; long life compared with normal incandescent lamps	Accent lighting, floodlighting
Fluorescent Lamps	Tubular	80-100	12,000 to 24,000	Wide choice of light colors; high lighting levels possible; economical in use	All kinds of commercial and public buildings; streetlighting; home lighting
	Screw in base	50-60	9000 to 10,000	Energy/effective; direct replacement for incandescent lamps	Most appellations where incandescent lamps were used before
	Compact Fluorescent	27-80	9000 to 10,000	Compact; long life; energy/effective	To create a pleasant atmosphere in social areas, local lighting; signs, security, orientation lighting and general lighting
Gas-Discharge Lamps	Self-ballasted mercury	19-20	12,000 to 16,000	Long life; good color rendering; easy to install; better efficacy than incandescent lamps	Direct replacement for incandescent lamps; small industrial and public light projects; plant irradiation
	High-pressure mercury	50-60	12,000 to 24,000	High efficacy; long life; reasonable color quality	Residential area lighting; sports grounds; factory lighting
	Metal halide	80-115	10,000 to 20,000	Very high efficacy combined with excellent color rendering; long life	Floodlighting; especially for color TV; industrial lighting; road lighting, plant irradiation
	High-pressure sodium	90-140	10,000 to 24,000	Very high efficacy; extremely long life; good color rendering	Public lighting, floodlighting; industrial lighting; plant irradiation EL; direct replacement for mercury lamps
	Low-pressure sodium	130-180	14,000 to 18,000	Extremely high efficacy; very long life; high visual acuity; poor color rendering; monochromatic light	Many different application areas; wherever energy/cost effectiveness is important and color is not critical

Note: A Ballast factor(BF) must be applied to fluorescent and gas-discharge lamps to obtain the net lamp-ballast efficacy

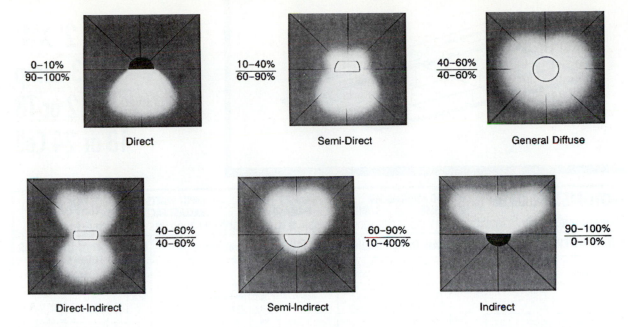

FIGURE 15–15
The IES and CIE classify luminaires in accordance with the percentage of the total luminaire output they emit above and below the horizontal. Within this percentage, there may be many variations, depending on the specific design of the luminaire.

downward and the distribution is more or less uniform in all directions.

- *Direct-Indirect*, where between 40 and 60 percent of the flux is directed either upward or downward and there is very little flux in the horizontal direction. The CIE considers the direct-indirect category as a part of the general diffuse type of luminaire. From a designer's point of view, these two types should be classified separately, as they produce distinct spatial relations, in terms of such properties as wall luminance, glare, and coefficient of utilization.
- *Semi-indirect*, where 60 to 90 percent of the flux is upward.
- *Indirect*, where 90 to 100 percent of the flux is upward.

Following are two examples of the general specification of a luminaire:

1. *Type RA.* The fixture shall be a ceiling lay-in type of fluorescent fixture for direct distribution. It shall contain three 34-W, T-8, warm white fluorescent lamps. The fixture shall have a 24-cell parabolic diffuser. The flux distribution shall have a batwing pattern and shall have a minimum coefficient of utilization of 75 percent, based on 80–50–20 ceiling-wall-floor reflectances and room cavity ratio (RCR) of 2.

2. *Type SB.* The fixture shall be a suspension-mounted fixture with a 12"-diameter opal glass globe. The fixture shall contain one 150-W general-service incandescent lamp. The hanger shall be of swivel construction with a 24" stem.

15.9 LUMINAIRES: PHOTOMETRIC DATA

Photometric data (figures 15–16 and 15–17) are the most important pieces of information required to determine the performance of a luminaire. They are the basic tool used in calculations of illumination. Photometric data are normally prepared by an independent testing laboratory or the laboratory of a qualified fixture manufacturer. They should contain much of the following information:

1. A description of the luminaire and the material out of which it is constructed.
2. The type, number, and rated lumens of the lamps used with the photometric test.
3. The candlepower distribution curve, including data. These are used to calculate the illuminance at a specific point on a surface from a single luminaire. For example, one may calculate the illuminance at a wall painting from a downlight or a wall washer. A downlight is a fixture with its lu-

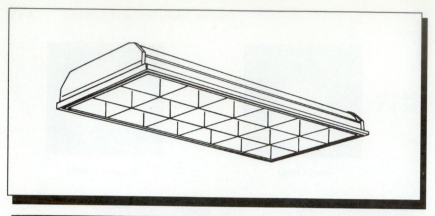

2' X 4' 3-Lamp T12 or T8 18 or 24 Cell

PHOTOMETRIC DATA

CATALOG # 2P3G340-36SL   LAMPS = F40 ES   INPUT WATTS = 118   LER = FP-41
TEST #15698   S/MH= 1.6   BALLAST = ESB   BALLAST FACTOR = .88

COMPARATIVE YEARLY LIGHTING ENERGY COST PER 1000 LUMENS = **$5.85** BASED ON 3000 HRS. AND $.08 PER KWH.

CANDLEPOWER

Angle	End	45	Cross
0	2210	2210	2210
5	2210	2212	2221
10	2172	2217	2260
15	2108	2206	2272
20	2038	2169	2241
25	1948	2102	2172
30	1852	2013	2205
35	1731	1917	2360
40	1593	1906	2127
45	1436	1809	1719
50	1255	1495	1135
55	1039	1022	706
60	777	575	417
65	343	255	172
70	54	66	49
75	22	20	22
80	10	9	9
85	4	4	3

MAINTAINED ILLUMINATION TABLE- Square Feet/Fixture*

- 80-50-20 Reflectances (Ceiling-Wall-Floor)
- LLF = 0.73 2650 Lumens/Lamp very clean
- Room width divided by room height = 5 or more, 2 or 1

Fixture Size & # of Lamps	Room Width / Room Height =	Approx. Area (sq. ft.) per Fixture				
		10 ft-c	30 ft-c	50 ft-c	70 ft-c	100 ft-c
2' X 4' 3 Lamp	5	–	143	86	61	43
	2	–	102	61	44	31
	1	–	76	45	32	–

*Observe Fixture S/MH Requirements for Specific Applications

AVERAGE LUMINANCE CD/SQ.M WITH 2650 LUMEN LAMPS

ANGLE	END	45°	CROSS
45	3244	4087	3884
55	2894	2846	1966
65	1297	964	650
75	136	123	136
85	73	73	55

TYPICAL V.C.P.'s

Room Size	Mounting Height Lengthwise		Crosswise	
	8.5	10	8.5	10
30x30	87	83	90	86
40x40	92	88	94	90
60x30	92	88	94	91
60x60	94	91	96	93
100x100	97	95	98	96

COEFFICIENT OF UTILIZATION

pfc	20							
pcc	80			70			50	
pw	70	50	30	70	50	30	50	30
RCR								
0	81	81	81	80	80	80	77	77
1	77	73	71	75	72	70	69	68
2	70	66	61	68	65	61	63	59
3	65	58	54	64	57	54	56	52
4	59	53	47	58	52	46	51	46
5	56	47	41	54	46	40	46	40
6	52	42	38	50	42	36	40	36
7	47	39	34	46	39	33	38	33
8	45	35	29	44	34	29	34	29
9	41	33	28	40	33	27	32	27
10	39	29	25	38	29	25	28	25

LIGHT DISTRIBUTION

DEGREES	LUMENS	% LAMP	% FIXTURE
0-30	1795	22.6	32.7
0-40	3041	38.3	55.5
0-60	5167	65.0	94.2
0-90	5483	69.0	100.0

LLF = .73 LLF = LIGHT LOSS FACTOR LLF = LDD X LLD X BF LDD = VERY CLEAN 0.94 CLEAN 0.90
LLD = 0.88 @ 40% RATED LAMP LIFE BF = 0.88 ESB BALLAST & 34W LAMP (RELAMP AT 70% LAMP LIFE)

PHOTOMETRIC DATA

CATALOG # 2P3GS332-36SL-1/3-EB   LAMPS = F32 T8   INPUT WATTS = 88   LER = FP-67
TEST #15694   S/MH= 1.7   BALLAST = ELECTRONIC   BALLAST FACTOR = .93

COMPARATIVE YEARLY LIGHTING ENERGY COST PER 1000 LUMENS = **$3.58** BASED ON 3000 HRS. AND $.08 PER KWH.

CANDLEPOWER

Angle	End	45	Cross
0	2520	2520	2520
5	2523	2527	2541
10	2476	2543	2608
15	2405	2544	2630
20	2323	2504	2587
25	2226	2429	2498
30	2107	2310	2426
35	1961	2188	2857
40	1801	2146	2909
45	1628	2238	2106
50	1424	1907	1211
55	1180	1205	743
60	880	636	460
65	375	276	200
70	65	75	56
75	26	23	26
80	11	10	11
85	4	4	3

MAINTAINED ILLUMINATION TABLE- Square Feet/Fixture*

- 80-50-20 Reflectances (Ceiling-Wall-Floor)
- LLF = 0.77 2900 Lumens/Lamp very clean
- Room width divided by room height = 5 or more, 2 or 1

Fixture Size & # of Lamps	Room Width / Room Height =	Approx. Area (sq. ft.) per Fixture				
		10 ft-c	30 ft-c	50 ft-c	70 ft-c	100 ft-c
2' X 4' 3 Lamp	5	–	–	105	75	53
	2	–	126	76	54	38
	1	–	93	56	40	–

*Observe Fixture S/MH Requirements for Specific Applications

AVERAGE LUMINANCE CD/SQ.M WITH 2900 LUMEN LAMPS

ANGLE	END	45°	CROSS
45	3678	5056	4758
55	3287	3356	2069
65	1418	1043	756
75	160	142	160
85	73	73	55

TYPICAL V.C.P.'s

Room Size	Mounting Height Lengthwise		Crosswise	
	8.5	10	8.5	10
30x30	86	81	89	84
40x40	90	86	92	88
60x30	91	87	93	89
60x60	93	90	95	92
100x100	96	94	97	95

COEFFICIENT OF UTILIZATION

pfc	20							
pcc	80			70			50	
pw	70	50	30	70	50	30	50	30
RCR								
0	86	86	86	84	84	84	81	81
1	81	79	76	80	77	75	73	71
2	75	69	66	73	68	65	67	63
3	69	63	57	68	61	56	59	56
4	64	56	51	63	55	50	54	48
5	58	51	45	57	50	44	47	44
6	55	46	40	53	45	40	44	39
7	51	41	35	50	40	34	40	34
8	46	38	32	46	38	32	36	32
9	44	34	28	42	34	28	34	28
10	40	32	27	40	32	27	30	26

LIGHT DISTRIBUTION

DEGREES	LUMENS	% LAMP	% FIXTURE
0-30	2061	23.7	32.3
0-40	3508	40.3	55.0
0-60	6015	69.1	94.4
0-90	6374	73.3	100.0

LLF = .77 LLF = LIGHT LOSS FACTOR LLF = LDD X LLD X BF LDD = VERY CLEAN 0.94 CLEAN 0.90
LLD = 0.88 @ 40% RATED LAMP LIFE BF = .93 ELECTRONIC BALLAST & T8 LAMP (RELAMP AT 70% LAMP LIFE)

■ **FIGURE 15–16**

Comprehensive photometric data sheet of a fluorescent fixture. This report contains photometrics of two different lamp–ballast configurations for the same fixture. The upper one uses (3) 40-watt, T-12, rapid start lamps with electromagnetic ballast having 0.88 ballast factor. The lower one uses (3) 32-watt, T-8 rapid start lamps with electronic ballast having 0.93 ballast factor. Note the substantial energy cost savings of the fixture using T-8 lamps with electronic ballast over the T-12 lamps with electromagnetic ballasts. (Courtesy: Thomas Industries, Tupelo, MS.)

minous flux primarily aimed downward to the floor, whereas a wall wash is aimed at the wall.

4. The lumens output within each zone of the distribution and the total lumens output of the entire luminaire.

5. The luminaire efficiency, that is, the total lumens output vs. the lamp lumens. Note that this is distinct from lamp efficacy.

6. The coefficient of utilization (CU). This number expresses the portion of lamp lumen that can be utilized on a working surface normally selected to be 30″ above the floor for horizontal illumination (although any height can be used). The CU may be considered the "net efficiency" of the selected luminaire for a particular installation. It depends on the following factors:

- *Reflectance of the room surfaces:* the ceiling, wall, floor, or the average of the floor and the desktop together.
- *Room dimensions and mounting:* the ceiling height, luminaire mounting height, and configuration and size of the room. The terms *room ratio* and *room index* are used to represent the overall effect of the room and luminaire relations.

The procedure for determining the CU of a design application is given in Chapter 16.

7. The maximum spacing-to–mounting height ratio. This concerns the uniformity of illumination on the selected working plane.

8. The visual comfort probability (VCP). This is a rating of the light system, expressed as the percentage of people who will find the system comfortable. The VCP is limited to office and school applications at selected positions in the room. It is not normally calculated.

9. The light loss factor (LLF). This is a modification factor expressing the depreciation of light output of the lighting system, including the luminaires and room conditions. It depends on the construction of the luminaire, the type of lamp used, and the cleanliness of the surroundings. Methods for determining the LLF are given in Chapter 16.

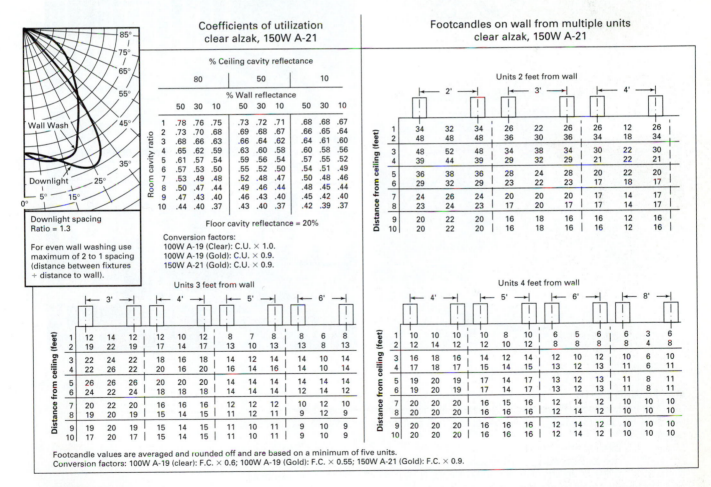

FIGURE 15–17
Typical photometric report for an incandescent luminaire. (Courtesy: Lightolier, Secaucus, NJ.)

Figures 15–16 and 15–17 give typical photometric data for luminaires. The CU table and candlepower distribution curve are the most important data in calculations of illumination.

15.10 GENERAL COMPARISON OF LIGHTING SYSTEMS

As a matter of general interest, several new lighting system concepts are shown in Figures 15–18 and 15–19. These include the light pipe, fiber optics cables and track lighting. Although track lighting is not a new concept, its use has been increasingly popular.

For each lighting system, thousands of luminaires are commercially available, and for each luminaire, there are many variations in terms of lamp type and size, and control medium, such as lens, louvers, baffles, etc. Therefore, it would be meaningless to illustrate just a few of these combinations. However, an overview of lighting systems can be given as follows.

A *direct system* is a lighting system that is based on the use of direct-distribution luminaires. It is the most effective system for horizontal illumination. However, there is the potential for both direct and reflected glare, the latter from visual situations, such as when reading or when viewing a visual display

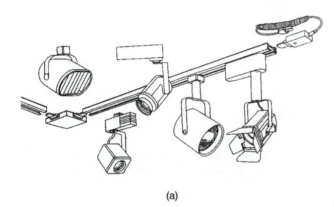

(a)

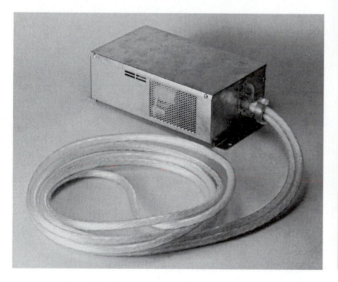

(b)

(c)

■ FIGURE 15–18
(a) Light track is a very flexible lighting system. Illustrated on the track are luminaires for: Compact fluorescent; MR 16; MR 16 projectors; PAR or R lamps; and MR 16 and electrical connections.
(Source: IES Lighting Handbook.)
(b) Light source and controls (c) Fiberoptic lighting system provides an attractive illuminated outline of an amphitheater in Lake Eola, FL. It is an effective way to decorate, and identify the structure and the facility.

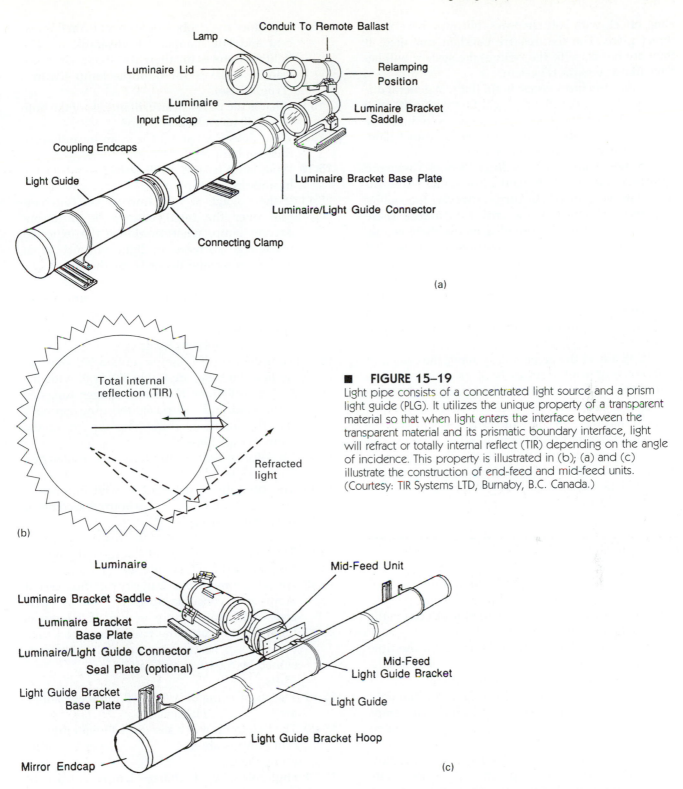

■ FIGURE 15–19
Light pipe consists of a concentrated light source and a prism light guide (PLG). It utilizes the unique property of a transparent material so that when light enters the interface between the transparent material and its prismatic boundary interface, light will refract or totally internal reflect (TIR) depending on the angle of incidence. This property is illustrated in (b); (a) and (c) illustrate the construction of end-feed and mid-feed units. (Courtesy: TIR Systems LTD, Burnaby, B.C. Canada.)

screen. There are many ways to overcome the glare, for example, by using large, low-brightness ceilings, sharp angle cutoffs, etc.

A *semidirect system* has small upward components. These will soften the high contrast between the ceiling and the luminaires and will improve the spatial brightness relations.

A *general diffuse system* uses luminaires with light flux more or less uniformly distributed in all directions. This system is likely to achieve more of a modeling effect, with soft shadows, but may have more direct glare. The fixtures are pendant mounted. If they are too close to the ceiling, the system will behave like a semidirect system.

A *direct-indirect system* is similar to a general diffuse system, except that the light flux within 10 to 30 degrees below the horizontal plane is reduced or shielded. This feature will greatly reduce direct glare, particularly on video display screens.

In *semi-indirect and indirect systems,* as more light flux is directed above the horizontal plane, direct glare is reduced. In turn, however, the ceiling brightness may become too great for comfort. These systems should not be installed unless there is adequate ceiling height and the fixtures are suspended within the recommendations of the fixture manufacturers. An indirect system is ideal for a video display space. If the overall illuminance level is not adequate for tasks such as close reading, the system may have to be supplemented with localized downlights.

With any of the preceding systems, the color and reflectance of wall surfaces is of great importance. Plate 25 is a chart of colored surfaces from which the reflectance value can be selected. As illustrated in Plate 24 the effects of color and reflectance of surfaces are quite different for different lighting systems. The proper selection of color will enhance the appearance of the spatial relations. The reflectance values are important data for designers and will be discussed in Chapters 16 and 17.

QUESTIONS

15.1 Daylight is plentiful and economical; however, it (   ) widely during the day.

15.2 Fluorescent lamps were first introduced in the (1930s) (1940s) (1950s).

15.3 The rated life of lamps is based on the time elapsed when only (   ) percent of a group of lamps still remain burning.

15.4 Light output depreciates with time. The light output of general-service incandescent lamps is about (   ) percent at their rated life.

15.5 Depending on the manufacturer's specifications, the color temperature of warm white fluorescent lamps is normally between (   ) and (   ) degrees Kelvin.

15.6 The CRI for a 3000°K incandescent lamp is given a rating of 100. Thus, it can reproduce the true color of any object. (True) (False)

15.7 To reduce the stroboscopic effect from fluorescent and HID lamps, it is desirable to wire lighting fixtures on alternate circuits or to use lead-lag ballasts for multiple lamp fixtures. (True) (False)

15.8 What is the physical size, in inches, of the bulb of a G-40 incandescent lamp?

15.9 What does PAR stand for in an incandescent reflector lamp?

15.10 What lamp base is a 150-watt general-service incandescent lamp?

15.11 If the voltage at the lamp socket is 6 percent over the rated voltage of an incandescent lamp, what would be the expected percentage increase in light output? What is the percentage decrease in the life of the lamp?

15.12 The efficacy of an incandescent lamp varies with the size of the lamp, ranging from approximately (   ) lumens/watt to (   ) lumens/watt.

15.13 Tungsten-halogen lamps contain halogen gas (iodine, fluorine, etc.), which reacts with the tungsten filament to form a halogen-tungsten compound and redeposits the tungsten back into the filament, prolonging the life of the lamp. (True) (False)

15.14 Miniature reflector lamps are compact halogen lamps. (True) (False)

15.15 The rated life of standard 40-watt fluorescent lamps is about (   ) hours, based on an 8-hour-per-start cycle.

15.16 What is the CRI rating of fluorescent lamps?

15.17 The color temperature of fluorescent lamps is fixed at 3500°K. (True) (False)

15.18 An electromagnetic ballast for fluorescent lamps may consume up to (   ) percent of the rated lamp power. If a lamp is rated for 40 watts, the ballast may consume (   ) watts; thus, the total power consumed by the lamp and ballast is (   ) watts.

15.19 The most commonly used fluorescent lamps are (   ) lamps.

15.20 The light output of a standard 40-watt fluorescent lamp is about (   ) percent of its input energy.

15.21 High-intensity discharge lamps consist of three major types:

15.22 What are the characteristics of LPS lamps?

15.23 The CRI of standard HPS lamps is very low. However, newly developed HPS lamps may have a CRI of (   ) or more.

15.24 The efficacy of HPS lamps is as high as (   ) lumens/watt, excluding the power loss of the ballast.

15.25 All incandescent lamps have the same color temperature. (True) (False).

15.26 All HID lamps require a ballast. (True) (False)

15.27 The ballast factor is used to modify the lumens output of a lamp-ballast combination. (True) (False)

15.28 The ballast factor is always smaller than 1. (True) (False)

15.29 What is BEF?

15.30 Can BEF be compared for different number and size lamps?

15.31 What is LBEF?

15.32 High-pressure sodium lamps are monochromic. (True) (False)

15.33 What is the typical efficacy of a 1000-W, clear MH lamp? (85) (100) (150) lumens/watt.

15.34 What is the depreciated lamp lumen of a 400-W, clear HPS lamp at 70% of its rated life? (65%) (70%) (73%)

15.35 Name two electrodeless lamps.

15.36 What is the difference between the IES and CIE lighting system classifications?

15.37 What are the two most valuable data on the performance of a luminaire in the luminaire's photometric report?

15.38 What is a light pipe and its applications?

16

CALCULATIONS OF ILLUMINATION

THE LIGHTING DESIGN PROCESS INVOLVES FOUR logical steps. The process starts with the analysis of visual tasks, followed by the determination of illumination requirements, engineering calculations, and ended with the equipment selection and system design. This chapter will focus on the first three steps with emphasis on engineering calculations, and Chapter 17 will focus on the equipment selection with emphasis on design.

16.1 QUANTITY AND QUALITY OF ILLUMINATION

Strictly speaking, quantity and quality of lighting are inseparable in that quantitative need is closely related with quality. High quality lighting can often compensate for the lack of quantity and conversely poor lighting often requires higher lighting quantity to overcome the loss of visual performance. For example, a light source aimed at a visual task, such as a book, but is not shielded from the normal viewing angle of the reader, will create glare or vailing reflection which washes out the reading task. At higher illuminance level, the glare may even cause discomfort to temporary blindness of the reader. In this case, the higher quantity of illumination even be counter productive.

Quality of lighting can be quantified or calculated; i.e., to be compared in numerical values. However, their calculations are usually very complex, thus they will only be briefly mentioned.

1. Glare may be evaluated in terms of:
   a) Disability glare factor (DGF). A numerical factor to measure the visibility of a task in a given lighting installation in comparison with its visibility under reference lighting conditions as a ratio of luminance contrast.
   b) Discomfort glare rating (DGR). A numerical assessment of the capacity of a number of sources of illuminance, such as luminaires, in a given visual environment.
2. Comfort may be expressed in terms of visual comfort probability (VCP), a rating of a lighting system expressed as a percent of people who, when viewing from a specific location and direction, will be expected to find it acceptable in terms of discomfort glare.
3. Equivalent sphere illumination (ESI). A measure of the equivalent illuminance in terms of a uniformly illuminated sphere with the task located at the center. This is often called ESI foot-candles or ESI lux.
4. Uniformity of illuminances on horizontal or vertical surfaces, as a ratio of minimum to average, average to minimum and maximum to minimum ratios.
5. Color rendering quality in terms presented in Chapters 14 and 15, such as color rendering index

(CRI), color preference index (CPI) and color temperatures.

16.2 DETERMINATION OF ILLUMINATION REQUIREMENTS

The illuminance selection procedure published by IESNA provides a rational basis for determining the illumination requirements for visual tasks expected in a space. The procedure is based on the assessment of the following factors:

1. *Type of visual activity.* Determine the visual tasks and the visual performance to be expected.
2. *The age of the observers.* Higher illuminance is required for older people. Unfortunately, in most design applications, the ages of the observers are not always known. So an educated guess may be required.
3. *Speed and accuracy.* The importance of speed and accuracy in the performance of visual tasks govern the criteria for higher or lower illuminance. For example, most sports require a high level of illumination due to their high-speed tasks (e.g., ball).
4. *Contrast of the task.* This is the background luminance on which the task is seen. For example, to detect black objects on a black background will require extremely high illuminance.

Determination of these factors in illuminance selection has been consolidated into the four selection steps as recommended by IESNA.

16.2.1 Step 1: Selection of the Visual Task

Determine the type of activity for which the illumination is required, for example, reading a printed book. At the same time establish the plane in which the visual task will be performed, such as on a desk 30 in. above the floor.

16.2.2 Step 2: Selection of the Illuminance Category

The IESNA and CIE have classified the illuminance into nine categories. These categories are described in Table 16–1. Categories A, B and C are for general lighting of a space; Categories D, E and F are for illuminance on the tasks; and Categories G, H and I are for illuminance on the task, but supplemented with local light sources specifically aimed at the tasks.

For convenience, IES also provided a comprehensive table listing of all commonly encountered interior spaces for commercial, institutional, residential, and public assemblies. An excerpt of this table is shown in Table 16–2. The readers will find a

---

TABLE 16–1

Illuminance categories and ranges of illuminance values for generic types of activities in interiors

Type of Activity	Illuminance Category	Ranges of Illuminances		Reference Work-Plane
		Lux*	Footcandles	
Public spaces with dark surroundings	A	20–30–50	2–3–5	General lighting throughout spaces
Simple orientation for short temporary visits	B	50–75–100	5–7.5–10	
Working spaces where visual tasks are only occasionally performed	C	100–150–200	10–15–20	
Performance of visual tasks of high contrast or large size	D	200–300–500	20–30–50	Illuminance on task
Performance of visual tasks of medium contrast or small size	E	500–750–1000	50–75–100	
Performance of visual tasks of low contrast or very small size	F	1000–1500–2000	100–150–200	
Performance of visual tasks of low contrast and very small size over a prolonged period	G	2000–3000–5000	200–300–500	Illuminance on task, obtained by a combination of general and local (supplementary lighting)
Performance of very prolonged and exacting visual task	H	5000–7500–10000	500–750–1000	
Performance of very special visual tasks of extremely low contrast and small size	I	10000–15000–20000	1000–1500–2000	

(Reproduced with permission from Figure 11–1, 8th edition, *IES Lighting HB*)

*The equivalence of lux to fc is 1 fc = 10.76 lux. Values are rounded.

complete listing of all spaces in the *IES Lighting Handbook*.

16.2.3 Step 3: Determine the Illuminance Range

Every illuminance category has a corresponding range of three illuminances—low, medium and high. For example, the range of illuminance recommended for illuminance category E is 500–750–1000 lux or 50–75–100 footcandles in approximation. The decision to choose any one of the three values depends on the evaluation of the room and occupant characteristics as determined in Step 4.

16.2.4 Step 4: Select the Target Illuminance

Target illuminance may be defined as the illuminance for which the lighting system is designed. The selection is aided by evaluation of several weighting factors, given in tables 16–3 and 16–4.

Example Select the illuminance value for the prescription counter of a pharmacy.

Step 1 The visual activity would likely be the frequent reading of doctors' prescription notes, normally a task of medium reflectance, and frequently hard to read.

Step 2 Select a category from Table 16–1, in this case, Category E.

Step 3 The illuminance range for Category E is 50-75-100 footcandles.

Step 4 Select and accumulate the weighting factors from Table 16–2.

Age (>55)	+1
Speed/accuracy (high)	+1
Reflectance	0
Total weighting factor	+2

Final selection: Select the high value of Category E, thus 1000 lux (100 fc).

TABLE 16–2
Recommended illuminance values of common task/areas

Auditoriums		Off-set Printing and Duplicating		Basketball (indoor)		
Assembly	C	area	D	Class I (4)		125
Social activity	B	Spaces with VDTs (1)		Class II		75
Stage/platform	E	*Reading*		Football (indoor)		
Banks		Xerograph, mimeograph	D	Class I (4)		200
Lobby-general	C	CRT screens	B	Class II		100
Writing area	D	#3 pencil and softer leads	E	Softball (outdoor)		
Teller's stations	E	#4 pencil and harder leads	F	Class II		50
Conference Rooms		Ball-point pen	D	Class III		30
Conferring	D	Reading mixed material	E	Soccer (*see* Football)		
Critical seeing	E	*Schools*		Tennis (outdoor & indoor)		
Corridors	C	Classrooms (*see* Reading)		Class I (4)		140
Drafting		Science laboratories	E	Class II		75
Low contrast	F	Shops (*see* IES handbook)		Class III		40
Blueprints	E	*Stairways and Corridors*	C			
Exhibition/Convention		*Residential Spaces*				
General	C	General lighting	B	*NOTES*		
Display	E	Entertainment	B	(1) Spaces with VDTs shall avoid spill		
Libraries		Passage areas	B	light and reflections on the screen. See		
Reading areas (*see* Reading)		Specific visual tasks		Section 5, Application Volume of IES		
Book stacks—active	D	Dining	C	Handbook.		
Card files	E	Grooming	D	(2) Illuminance for sports shall be for		
Audiovisual areas	D	Kitchen general	D	horizontal and vertical planes. See		
Merchandising Spaces		Kitchen counter	E	RP-6-88 IES Sports Lighting Manual		
Circulation	D	Kitchen range	E	(3) Class I—professional, international,		
Merchandise		Kitchen sink	E	national; Class II—college; Class III—		
Merchandise	E	Laundry	D	high school, etc.		
Offices		Music study (piano or organ)		(4) Updated to reflect current practices		
Accounting (*see* Reading)		Advanced scores	E	for TV requirements. Value shown is		
Audio-visual area		Sewing		for vertical illuminance.		
Conference (*see* Conference room)		Dark fabrics, low contrast	F			
Drafting (*see* Drafting)		Light to medium fabrics	E			
General and private offices	E	*Sports and Recreational* (2) (3)				
Libraries (*see* Libraries)		Baseball (infield/outfield)				
Lobbies, lounges and reception		Class I (4)	150/100			
areas	C	Class II	100/70			
Mail sorting	E	Class III	50/30			

(Excerpt from *IES Handbook* and current *Architectural/Engineering Practices*)

TABLE 16–3

Weighting factors to be considered in selecting specific illumance within ranges of values for each illuminance category

	a. For illuminance categories A through C		
	Weighting factor		
Room and occupant characteristics	*−1*	*0*	*+1*
Occupant ages	Under 40	40–55	Over 55
Room surface reflectances*	Greater than 70 percent	30 to 70 percent	Less than 30 percent
	b. For illuminance categories D through I		
	Weighting factor		
Task and worker characteristics	*−1*	*0*	*+1*
Worker's ages	Under 40	40–55	Over 55
Speed and/or accuracy[1]	Not important	Important	Critical
Reflectance of task background[2]	Greater than 70 percent	30 to 70 percent	Less than 30 percent

(Reproduced with permission from *IES Lighting Handbook*)

*Average weighted surface reflectances, including wall, floor and ceiling reflectances, if they encompass a large portion of the task area or visual surround. For instance, in an elevator lobby, where the ceiling height is 7.6 meters [25 feet], neither the task nor the visual surround encompass the ceiling, so only the floor and wall reflectances would be considered.

[1]In determining whether speed and/or accuracy is not important, important or critical, the following questions need to be answered: What are the time limitations? How important is it to perform the task rapidly? Will errors produce an unsafe condition or product? Will errors reduce productivity and be costly? For example, in reading for leisure there are no time limitations and it is not important to read rapidly. Errors will not be costly and will not be related to safety. Thus, speed and/or accuracy is not important. If however, a worker is involved in exacting work, accuracy is critical because of the close tolerances, and time is important because of production demands.

[2]The task background is that portion of the task upon which the meaningful visual display is exhibited. For example, on this page the meaningful visual display includes each letter which combines with other letters to form words and phrases. The display medium, or task background, is the paper, which has a reflectance of approximately 85 percent.

TABLE 16–4

Guidelines for selecting the illuminance level for weighting factor

Weighting Factor	Categories A–C (No Task Activity)	Categories D–I (Task Activity)
−3	xxx	Low
−2	Low	Low
−1	Low	Medium
0	Medium	Medium
1	High	Medium
2	High	High
3	xxx	High

From Table 16–4, select the high value of the range of illuminance in Table 16–1. Thus, the illuminance level selected is 100 footcandles.

16.3 LUMENS (ZONAL CAVITY) METHOD

The lumens, or zonal cavity, method is widely used for determining the average horizontal illuminance of a space. The method is simple to follow in princi-ple, but tedious to use in procedure. Introduced here are the complete procedures to be used for detailed manual calculations as well as the algorithm for com-puter programming. A simplified procedure is also introduced for preliminary design purposes.

The lumens method is based on the definition of illuminance, which is the luminous flux (lumens) in-cident on a unit area (sq m or sq ft). That is:

$$E = \text{luminous flux}/\text{Area} = F/A \qquad (14\text{–}2)$$

For example, if 1000 lumens of the luminous flux from all lighting fixtures falls, directly or indirectly, onto a 20-square m work plane, then the average il-luminance is $1000/20 = 50$ lux. Similarly, if the work plane area is 20 sq ft, then the average illuminance is $1000/20 = 50$ fc.

16.3.1 Initial Illuminance

The initial illuminance incident on a given work plane is:

$$E_i = \frac{F_a}{A} \qquad (16\text{–}1)$$

where E_i = initial illumance, fc (lux)
F_a = actual initial luminaires flux onto the work plane, lumens
A = area of the work plane, sq ft (sq m)

The actual initial luminous flux (F_a) reaching the work plane depends on many factors, including the luminous output of the lamps, the efficiency of the fixtures, the distribution pattern of the light flux, the location of the fixtures and the characteristics of the room such as the geometry and reflectance values of the surfaces in the rooms. The preceding formulas may be written as:

$$E_i = (\text{NF} \times \text{LPF} \times \text{LOF}) \times \text{CU}/A \quad (16\text{--}2)$$

where NF = Number of fixtures
LPF = Rated lamp lumens per fixture
= Number of lamps per fixture × *rated lumens* per lamp*
LOF = Lamp operating factor, a multiplier to modify the lumens output of each lamp under the conditions in which it is operated.
CU = Coefficient of utilization, a multiplier to account for the percentage of light flux that reaches the work plane.

Lamp operating factor The lamp manufacturer's published lumens output of a lamp is based on laboratory-controlled voltage and temperature conditions, a standard reactor ballast, and a vertical lamp burning position for HID lamps. All these conditions affect the initial light output of the lamp in the fixture under the installed conditions. Thus, a lamp operat-

*Rated lumens are the lumen output published by the lamp manufacturer under laboratory conditions.

ing factor (LOF) should be applied to reflect any deviations in the field:

$$\text{LOF} = \text{VF} \times \text{TF} \times \text{BF} \times \text{PF} \quad (16\text{--}3)$$

where VF = Voltage factor
TF = Temperature factor
BF = Ballast factor
PF = Position or tilt factor (for HID lamps)

The voltage factor (VF) is applied when the electrical system voltage is different from the voltage rating of the lamp. The actual voltage at the lamp socket, even with a well-designed electrical distribution system, is normally 3 to 5 percent lower than the system voltages, e.g., 115 volts on a 120-volt system, or 265 volts on a 277-volt system. Table 16–5 provides the approximate VF for lamps operating under different voltage conditions. VF is particularly significant for incandescent or halogen lamp applications. For example, at 96 percent of the rated lamp voltage, the light output of an incandescent lamp is only 88 percent of the published rating.

The temperature factor (TF) is applied when the fixture is operating at a different ambient temperature. Lamps and luminaires are normally tested at 25°C (77°F). Variations in temperature, within the range of those normally encountered in the interiors of buildings, have little effect on the light output of incandescent and HID lamp luminaires. There is more of an effect on air-handling types of fluorescent fixtures and outdoor mounted fixtures. Open fluorescent fixtures operating in freezing temperatures (even with special low-temperature ballasts) may produce only 80 percent of their rated lumens output. The manufacturer's photometric data should be consulted. For normal indoor applications, the temperature factor (TF) may be neglected, or TF = 1.

TABLE 16–5
Voltage factor (VF) for various types of lamps (in percent)[1]

Lamp Type	*Percent Deviation from Lamp Rated Voltage*					
	−5	−4	−3	−2	0	+2
Fluorescent (magnetic ballast)	95	96	97	98	100	102
Fluorescent (electronic ballast)[2]	—	—	—	—	100	—
Mercury (nonregulated ballast)	88	90	92	95	100	105
Incandescent (general service)	83	86	89	94	100	106
Incandescent (halogen)	80	84	88	92	100	108
Mercury (constant wattage)	97	98	99	99	100	102
Metal halide (reactor ballast)	91	93	95	97	100	103
High-pressure sodium	—	—	—	—	100	—

[1]Light output rounded to the nearest 0.5%.
[2]Obtain photometric data from fixture or ballast manufacturer for the true lumens output of the lamp ballast combinations.

The ballast factor (BF) is defined as the fraction of light output of a lamp or lamps using a specific commercial ballast, divided by the rated light output of the lamp or lamps using a standard reference ballast as in the laboratory. For electromagnetic ballasts, the BF is always lower than unity or 100 percent. However, for some electronic types of ballast, the BF may be higher than unity, say 120. In other words, in this latter case, the light output of the lamp-ballast combination is 20 percent more than the lamp manufacturer's rated output. For preliminary calculations of illuminance, the BF shown in Table 16–6 may be used.

The position factor (PF) is also called the tilt factor. The position of a lamp is critical for HID lamps, has a minor effect on incandescent lamps, and has no effect on fluorescent lamps. The lumens rating for HID lamps is normally based on a lamp mounted in the vertical position. Typical lumens outputs for HID lamps mounted at 30°, 45°, 60° and 90° from the vertical are 95, 90, 82 and 95 percent respectively. The manufacturer should be consulted for specific lamp performance characteristics.

Example   A fluorescent luminaire is designed to use lamps rated for 3000 lumens at 120 volts with a CBM-certified electromagnetic ballast. If the luminaire is installed in a building that has 115 volts at the lamp socket, what is the LOF of the installation? What is the actual light (lumens) output of each lamp?

From Table 16–5, the VF is 4 percent below the rated voltage or VF = 0.96

and from Table 16–6, the BF = 0.95
thus, LOF = 0.96 × 0.95   = 0.91
and, actual light output   = 3000 lumens × 0.91
                            = 2730 lumens

Coefficient of utilization   When a lighting system is turned on (energized) light power (flux) fills the

space. Depending on the flux distribution characteristics of the luminaires, some or all of the flux falls directly on the work plane and the rest will be escaped, absorbed or reflected from room surfaces. As a result, the flux emitted from the luminaires is only partially utilized. The coefficient of utilization (CU) is the multiplier that accounts for the fraction of the total flux that is being utilized at the work plane. That is:

$$CU = \text{Flux on the work plane}/ \text{Actual lamp flux} \qquad (16\text{–}4)$$

The lumens or zonal cavity method is based on the luminous radiative transfer theory, which holds that fluxes are interreflected until they reach equilibrium. This theory provides the engineering base for calculating the average illuminance level on the work plane. The method divides the room into three zones or cavities—ceiling, room and floor cavities—as illustrated in Figure 16–1.

There are four basic steps to follow in determining the CU of an installation.

Step 1: Calculate the Cavity Ratios   From the room dimensions and luminaire mounting, as illustrated in Figure 16–1, calculate the cavity ratios (CR) as follows.

$$\begin{aligned} CR &= 2.5 \times (\text{perimeter area of the cavity}/ \\ &\quad \text{floor area of the room}) \\ &= 2.5 \times (\text{perimeter} \times \text{cavity height})/ \\ &\quad \text{floor area of the room} \qquad (16\text{–}5) \\ &= 2.5 \times (\text{perimeter} \div \text{floor area of} \\ &\quad \text{the room}) \times \text{cavity height} \\ &= 2.5 \times \text{PAR} \times h \end{aligned}$$

TABLE 16–6

Typical ballast factor (BF) of electrical discharge lamps

Lamp-Ballast Combination	Typical BF
Fluorescent (30 watts and larger) with CBM-certified magnetic ballasts[1]	0.95
Fluorescent (below 30 watts) with CBM-certified magnetic ballasts[1]	0.85
Fluorescent with electronic ballasts (verify with ballast-lamp manufacturer)	0.8 to 1.2
HID lamps (mercury, metal halide, HPS)	EOF[2]

[1]CBM Certified Ballast Manufacturers.
[2]Luminaire manufacturer often provides a combined factor known as the *equipment operating factor* (EOF) for HID luminaires. This factor includes the TF, BF, and PF.

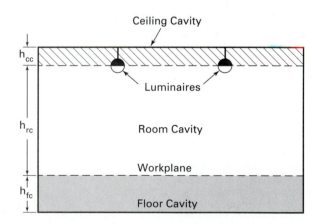

■   **FIGURE 16–1**
Terms of zonal cavity method.

where: h = cavity height, ft (m)
PAR = ratio of perimeter to floor area

$$= 2 \times (L + W)/(L \times W) \quad (16\text{–}5a)$$

for rectangular rooms

$$= 4/D \quad (16\text{–}5b)$$

for circular rooms

$$= 3.27/D \quad (16\text{–}5c)$$

for semicircular rooms

in which L = length of a rectangular room, ft (m)
W = width of a rectangular room, ft (m)
D = diameter of a circular room, ft (m)

Also, CCR = 2.5 PAR $\times h_{cc}$ (16–6a)
RCR = 2.5 PAR $\times h_{rc}$ (16–6b)
FCR = 2.5 PAR $\times h_{fc}$ (16–6c)

in which h_{cc} = ceiling cavity height, ft (m)
h_{rc} = room cavity height, ft (m)
h_{fc} = floor cavity height, ft (m)

Example   A room is 20 ft by 25 ft with h_{cc}, h_{rc} and h_{fc} to be 2.0, 7.0, and 2.5 ft, respectively. Find the cavity ratios.

From Equation 16–5a,
$$PAR = 2 \times (20 + 25)/20 \times 25 = 0.18$$

From Equation 16–6a, CCR = 2.5 $\times$ 0.18 $\times$ 2.0 = 0.9

From Equation 16–6b, RCR = 2.5 $\times$ 0.18 $\times$ 7.0 = 3.15

From Equation 16–6c, FCR = 2.5 $\times$ 0.18 $\times$ 2.5 = 1.13

Step 2: Select the Base Reflectances   The surface reflectances of the ceiling, walls, and floor greatly affect the coefficient of utilization of a lighting installation. Although the reflectance values of a space may not be known during the initial design of the space, the designer must make a reasonable assumption or an educated guess. In general, unless otherwise specified:

- As regards ceiling reflectance (R_c), assume a white ceiling having 70 to 80 percent base reflectance, unless otherwise given.
- For wall reflectance (R_w), assume 50-percent base reflectance for medium to light-colored walls, 20–30 percent for dark wood paneling, and 60–70 percent for white walls.
- With respect to floor reflectance (R_f), normally use 20-percent base reflectance for the combination of furniture and floor. Use 10 percent for dark floor finishes and 30 percent for light floor finishes.

Step 3: Determine the Effective Reflectances

- *Effective ceiling reflectance (ρ_{cc}).* As illustrated in Figure 16–1, when lighting fixtures are mounted a certain distance below the ceiling, the upper portion of the wall between the ceiling and the fixtures is, in effect, an extension of the ceiling. Thus, the effective reflectance of the ceiling is the combined reflectances of both the ceiling and the upper wall. For surface- or recess-mounted lighting installation (CCR = 0), ρ_{cc} is, of course, the same as the base ceiling reflectance. Table 16–7 gives effective ceiling or floor reflectances calculated from given base ceiling and floor reflectances.
- *Effective wall reflectance (ρ_w).* The effective wall reflectance is numerically equal to the base wall reflectance ($P_w = R_w$).
- *Effective floor reflectance (ρ_{fc}).* The effective floor reflectance for CU tables published by the lighting fixture manufacturers is standardized at 20 percent. Effective floor reflectances other than 20 percent shall be modified by a multiplier. The impact of the effective floor reflectance is less significant than that of the effective ceiling reflectance. In design practice, an effective floor reflectance in the 10-30 percent range is often selected. Table 16–8 provides the multipliers required from the values given for a 20 percent floor in a photometric report. Multiplier varies from 1.1 for 30-percent effective floor reflectance for large spaces to 1.0 for zero-percent reflectance for small spaces. For extremely light colored floors, the multiplier may be extrapolated from these data.

Step 4: Determine the CU   Use the manufacturer's furnished photometric data and the previously calculated values RCC, ρ_{cc}, ρ_w and ρ_{fc} to determine the CU.

Photometrics normally include data on the CU of the luminaire. The latter data are presented in a tubular format so that the designer can choose between the room cavity ratio (RCR), effective ceiling reflectances (80, 70 and 50 percent), wall reflectances (70, 50 and 30 percent) and effective floor reflectance (20 percent) only. A typical CU table for a fluorescent luminaire is shown in Figure 16–2.

Example   Determine the CU for the luminaire shown in Figure 16–2 installed in a room with effective floor reflectance (ρ_f) of 20 percent, effective ceiling reflectance (ρ_c), of 70 percent, and wall reflectance (ρ_w) of 30 percent. The room configuration is such that its RCR is 2.5.

From Figure 16–2, the CU for 70 percent ceiling, 30 percent wall, and RCR = 2 is 61 percent; with

TABLE 16–7

Selection of effective ceiling or floor reflectances from base ceiling or floor reflectances

Per Cent Base† Reflectance	80										70									
Per Cent Wall Reflectance	90	80	70	60	50	40	30	20	10	0	90	80	70	60	50	40	30	20	10	0
Cavity Ratio																				
0.2	79	78	78	77	77	76	76	75	74	72	70	69	68	68	67	67	66	66	65	64
0.4	79	77	76	75	74	73	72	71	70	68	69	68	67	66	65	64	63	62	61	58
0.6	78	76	75	73	71	70	68	66	65	63	69	67	65	64	63	61	59	58	57	54
0.8	78	75	73	71	69	67	65	63	61	57	68	66	64	62	60	58	56	55	53	50
1.0	77	74	72	69	67	65	62	60	57	55	68	65	62	60	58	55	53	52	50	47
1.2	76	73	70	67	64	61	58	55	53	51	67	64	61	59	57	54	50	48	46	44
1.4	76	72	68	65	62	59	55	53	50	48	67	63	60	58	55	51	47	45	44	41
1.6	75	71	67	63	60	57	53	50	47	44	67	62	59	56	53	47	45	43	41	38
1.8	75	70	66	62	58	54	50	47	44	41	66	61	58	54	51	46	42	40	38	35
2.0	74	69	64	60	56	52	48	45	41	38	66	60	56	52	49	45	40	38	36	33
2.2	74	68	63	58	54	49	45	42	38	35	66	60	55	51	48	43	38	36	34	32
2.4	73	67	61	56	52	47	43	40	36	33	65	60	54	50	46	41	37	35	32	30
2.6	73	66	60	55	50	45	41	38	34	31	65	59	54	49	45	40	35	33	30	28
2.8	73	65	59	53	48	43	39	36	32	29	65	59	53	48	43	38	33	30	28	26
3.0	72	65	58	52	47	42	37	34	30	27	64	58	52	47	42	37	32	29	27	24
3.2	72	65	57	51	45	40	35	33	28	25	64	58	51	46	40	36	31	28	25	23
3.4	71	64	56	49	44	39	34	32	27	24	64	57	50	45	39	35	29	27	24	22
3.6	71	63	54	48	43	38	32	30	25	23	63	56	49	44	38	33	28	25	22	20
3.8	70	62	53	47	41	36	31	28	24	22	63	56	49	43	37	32	27	24	21	19
4.0	70	61	53	46	40	35	30	26	22	20	63	55	48	42	36	31	26	23	20	17
4.2	69	60	52	45	39	34	29	25	21	18	62	55	47	41	35	30	25	22	19	16
4.4	69	60	51	44	38	33	28	24	20	17	62	54	46	40	34	29	24	21	18	15
4.6	69	59	50	43	37	32	27	23	19	15	62	53	45	39	33	28	24	21	17	14
4.8	68	58	49	42	36	31	26	22	18	14	62	53	45	38	32	27	23	20	16	13
5.0	68	58	48	41	35	30	25	21	18	14	61	52	44	36	31	26	22	19	16	12
6.0	66	55	44	38	31	27	22	19	15	10	60	51	41	35	28	24	19	16	13	09
7.0	64	53	41	35	28	24	19	16	12	07	58	48	38	32	26	22	17	14	11	06
8.0	62	50	38	32	25	21	17	14	11	05	57	46	35	29	23	19	15	13	10	05
9.0	61	49	36	30	23	19	15	13	10	04	56	45	33	27	21	18	14	12	09	04
10.0	59	46	33	27	21	18	14	11	08	03	55	43	31	25	19	16	12	10	08	03

Per Cent Base† Reflectance	60										50									
Per Cent Wall Reflectance	90	80	70	60	50	40	30	20	10	0	90	80	70	60	50	40	30	20	10	0
Cavity Ratio																				
0.2	60	59	59	59	58	57	56	56	55	53	50	50	49	49	48	48	47	46	46	44
0.4	60	59	59	58	57	55	54	53	52	50	50	49	48	48	47	46	45	45	44	42
0.6	60	58	57	56	55	53	51	51	50	46	50	48	47	46	45	44	43	42	41	38
0.8	59	57	56	55	54	51	48	47	46	43	50	48	47	45	44	42	40	39	38	36
1.0	59	57	55	53	51	48	45	44	43	41	50	48	46	44	43	41	38	37	36	34
1.2	59	56	54	51	49	46	44	42	40	38	50	47	45	43	41	39	36	35	34	29
1.4	59	56	53	49	47	44	41	39	38	36	50	47	45	42	40	38	35	34	32	27
1.6	59	55	52	48	45	42	39	37	35	33	50	47	44	41	39	36	33	32	30	26
1.8	58	55	51	47	44	40	37	35	33	31	50	46	43	40	38	35	31	30	28	25
2.0	58	54	50	46	43	39	35	33	31	29	50	46	43	40	37	34	30	28	26	24
2.2	58	53	49	45	42	37	34	31	29	28	50	46	42	38	36	33	29	27	24	22
2.4	58	53	48	44	41	36	32	30	27	26	50	46	42	37	35	31	27	25	23	21
2.6	58	53	48	43	39	35	31	28	26	24	50	46	41	37	34	30	26	23	21	20
2.8	58	53	47	43	38	34	29	27	24	22	50	46	41	36	33	29	25	22	20	19
3.0	57	52	46	42	37	32	28	25	23	20	50	45	40	36	32	28	24	21	19	17
3.2	57	51	45	41	36	31	27	23	22	18	50	44	39	35	31	27	23	20	18	16
3.4	57	51	45	40	35	30	26	23	20	17	50	44	39	35	30	26	22	19	17	15
3.6	57	50	44	39	34	29	25	22	19	16	50	44	39	34	29	25	21	18	16	14
3.8	57	50	43	38	33	29	24	21	19	15	50	44	38	34	29	25	21	17	15	13
4.0	57	49	42	37	32	28	23	20	18	14	50	44	38	33	28	24	20	17	15	12
4.2	56	49	42	37	32	27	22	19	17	14	50	43	37	32	28	24	20	17	14	12
4.4	56	49	42	36	31	27	22	19	16	13	50	43	37	32	27	23	19	16	13	11
4.6	56	49	41	35	30	26	21	18	16	13	50	43	36	31	26	22	18	15	13	10
4.8	56	48	41	34	29	25	21	18	15	12	50	43	36	31	26	22	18	15	12	09
5.0	56	48	40	34	28	24	20	17	14	11	50	42	35	30	25	21	17	14	12	09
6.0	55	45	37	31	25	21	17	14	11	07	50	42	34	29	23	19	15	13	10	06
7.0	54	43	35	30	24	20	15	12	09	05	49	41	32	27	21	18	14	11	08	05
8.0	53	42	33	28	22	18	14	11	08	04	49	40	30	25	19	16	12	10	07	03
9.0	52	40	31	26	20	16	12	10	07	03	48	39	29	24	18	15	11	09	07	03
10.0	51	39	29	24	18	15	11	09	07	02	47	37	27	22	17	14	10	08	06	02

* Values in this table are based on a length to width ratio of 1.6.
† Ceiling, floor or floor of cavity.

Per Cent Base† Reflectance	40										30									
Per Cent Wall Reflectance	90	80	70	60	50	40	30	20	10	0	90	80	70	60	50	40	30	20	10	0
Cavity Ratio																				
0.2	40	40	39	39	39	38	38	37	36	36	31	31	30	30	29	29	29	28	28	27
0.4	41	40	39	39	38	37	36	35	34	34	31	31	30	30	29	28	28	27	26	25
0.6	41	40	39	38	37	36	34	33	32	31	32	31	30	29	28	27	26	26	25	23
0.8	41	40	38	37	36	35	33	32	31	29	32	31	30	29	28	26	25	25	23	22
1.0	42	40	38	37	35	33	32	31	29	27	33	32	30	29	27	25	24	23	22	20
1.2	42	40	38	36	34	32	30	29	27	25	33	32	30	28	27	25	23	22	21	19
1.4	42	39	37	35	33	31	29	27	25	23	34	32	30	28	26	24	22	21	19	18
1.6	42	39	37	35	32	30	27	25	23	22	34	33	29	27	25	23	22	20	18	17
1.8	42	39	36	34	31	29	26	24	22	21	35	33	29	27	25	23	21	19	17	16
2.0	42	39	36	34	31	28	25	23	21	19	35	33	29	26	24	22	20	18	16	14
2.2	42	39	36	33	30	27	24	22	19	18	36	32	29	26	24	22	19	17	15	13
2.4	43	39	35	33	29	27	24	21	18	17	36	32	29	26	24	22	19	16	14	12
2.6	43	39	35	32	29	26	23	20	17	15	36	32	29	25	23	21	18	16	14	12
2.8	43	39	35	32	28	25	22	19	16	14	37	33	29	25	23	21	17	15	13	11
3.0	43	39	35	31	27	24	21	18	16	13	37	33	29	25	22	20	17	15	12	10
3.2	43	39	35	31	27	23	20	71	15	13	37	33	29	25	22	19	16	14	12	10
3.4	43	39	34	30	26	23	20	17	14	12	37	33	29	25	22	19	16	14	11	09
3.6	44	39	34	30	26	22	19	16	14	11	38	33	29	24	21	18	15	13	10	09
3.8	44	38	33	29	25	22	18	16	13	10	38	33	28	24	21	18	15	13	10	08
4.0	44	38	33	29	25	21	18	15	12	10	38	33	28	24	21	18	14	12	09	07
4.2	44	38	33	29	24	21	17	15	12	10	38	33	28	24	20	17	14	12	09	07
4.4	44	38	33	28	24	20	17	14	11	09	39	33	28	24	20	17	14	11	09	06
4.6	44	38	32	28	23	19	16	14	11	08	39	33	28	24	20	17	13	10	08	06
4.8	44	38	32	27	22	19	16	13	10	08	39	33	28	24	20	17	13	10	08	05
5.0	45	38	31	27	22	19	15	13	10	07	39	33	28	24	19	16	13	10	08	05
6.0	44	37	30	25	20	17	13	11	08	05	39	33	27	23	18	15	11	09	06	04
7.0	44	36	29	24	19	16	12	10	07	04	40	33	26	22	17	14	10	08	05	03
8.0	44	35	28	23	18	15	11	09	06	03	40	33	26	21	16	13	09	07	04	02
9.0	44	35	26	21	16	13	10	08	05	02	40	33	25	20	15	12	09	07	04	02
10.0	43	34	25	20	15	12	08	07	05	02	40	32	24	19	14	11	08	06	03	01

Per Cent Base† Reflectance	20										10									
Per Cent Wall Reflectance	90	80	70	60	50	40	30	20	10	0	90	80	70	60	50	40	30	20	10	0
Cavity Ratio																				
0.2	21	20	20	20	20	20	19	19	19	17	11	11	11	10	10	10	10	09	09	09
0.4	22	21	20	20	20	19	19	18	18	16	12	11	11	11	11	10	10	09	09	08
0.6	23	21	21	20	19	19	18	18	17	15	13	13	12	11	11	10	10	09	08	08
0.8	24	22	21	20	19	19	18	17	16	14	15	14	13	12	11	10	10	09	08	07
1.0	25	23	22	20	19	18	17	16	15	13	16	14	13	12	12	11	10	09	08	07
1.2	25	23	22	20	19	17	17	16	14	12	17	15	14	13	12	11	10	09	07	06
1.4	26	24	22	20	18	17	16	15	13	12	18	16	14	13	12	11	10	09	07	06
1.6	26	24	22	20	18	17	16	15	13	11	19	17	15	14	12	11	09	08	07	06
1.8	27	25	23	20	18	17	15	14	12	10	19	17	15	14	13	11	09	08	06	05
2.0	28	25	23	20	18	16	15	13	11	09	20	18	16	14	13	11	09	08	06	05
2.2	28	25	23	20	18	16	14	12	10	09	21	19	16	14	13	11	09	07	06	05
2.4	29	26	23	20	18	16	14	12	10	08	22	19	17	15	13	11	09	07	06	05
2.6	29	26	23	20	18	16	14	11	09	08	23	20	17	15	13	11	09	07	06	04
2.8	30	27	23	20	18	15	13	11	09	07	23	20	18	16	13	11	09	07	05	03
3.0	30	27	23	20	17	15	13	11	09	07	24	21	18	16	13	11	09	07	05	03
3.2	31	27	23	20	17	15	12	11	09	06	25	21	18	16	13	11	09	07	05	03
3.4	31	27	23	20	17	15	12	10	08	06	26	22	18	16	13	11	09	07	05	03
3.6	32	27	23	20	17	15	12	10	08	05	26	22	19	16	13	11	09	06	04	03
3.8	32	28	23	20	17	15	12	10	07	05	27	23	19	17	14	11	09	06	04	02
4.0	33	28	23	20	17	14	11	09	07	05	27	23	20	17	14	11	09	06	04	02
4.2	33	28	23	20	17	14	11	09	07	04	28	24	20	17	14	11	09	06	04	02
4.4	34	28	24	20	17	14	11	09	07	04	28	24	20	17	14	11	08	06	04	02
4.6	34	29	24	20	17	14	11	09	07	04	29	25	20	17	14	11	08	06	04	02
4.8	35	29	24	20	17	13	10	08	06	04	29	25	20	17	14	11	08	06	04	02
5.0	35	29	24	20	16	13	10	08	06	04	30	25	20	17	14	11	08	06	04	02
6.0	36	30	24	20	16	13	10	08	05	02	31	26	21	18	14	11	08	06	03	01
7.0	36	30	24	20	15	12	09	07	04	02	32	27	21	17	13	11	08	06	03	01
8.0	37	30	23	19	15	12	08	06	03	01	33	27	21	17	13	10	07	05	03	01
9.0	37	29	23	19	14	11	08	06	03	01	34	28	21	17	13	10	07	05	02	01
10.0	37	29	22	18	13	10	07	05	03	01	34	28	21	17	12	10	07	05	02	01

* Values in this table are based on a length to width ratio of 1.6.
† Ceiling, floor or floor of cavity.

Reprinted with permission from *Lighting Handbook* of Illuminating Engineering Society of North America, New York, NY.

TABLE 16–8

Multiplier for other than 20 percent effective floor cavity reflectance

% Effective Ceiling Cavity Reflectance, ρ_{cc}	80				70				50			30			10		
% Wall Reflectance, ρ_w	70	50	30	10	70	50	30	10	50	30	10	50	30	10	50	30	10
For 30-Percent Effective Floor Cavity Reflectance (20 Percent = 1.00)																	
Room Cavity Ratio																	
1	1.092	1.082	1.075	1.068	1.077	1.070	1.064	1.059	1.049	1.044	1.040	1.028	1.026	1.023	1.012	1.010	1.008
2	1.079	1.066	1.055	1.047	1.068	1.057	1.048	1.039	1.041	1.033	1.027	1.026	1.021	1.017	1.013	1.010	1.006
3	1.070	1.054	1.042	1.033	1.061	1.048	1.037	1.028	1.034	1.027	1.020	1.024	1.017	1.012	1.014	1.009	1.005
4	1.062	1.045	1.033	1.024	1.055	1.040	1.029	1.021	1.030	1.022	1.015	1.022	1.015	1.010	1.014	1.009	1.004
5	1.056	1.038	1.026	1.018	1.050	1.034	1.024	1.015	1.027	1.018	1.012	1.020	1.013	1.008	1.014	1.009	1.004
6	1.052	1.033	1.021	1.014	1.047	1.030	1.020	1.012	1.024	1.015	1.009	1.019	1.012	1.006	1.014	1.008	1.003
7	1.047	1.029	1.018	1.011	1.043	1.026	1.017	1.009	1.022	1.013	1.007	1.018	1.010	1.005	1.014	1.008	1.003
8	1.044	1.026	1.015	1.009	1.040	1.024	1.015	1.007	1.020	1.012	1.006	1.017	1.009	1.004	1.013	1.007	1.003
9	1.040	1.024	1.014	1.007	1.037	1.022	1.014	1.006	1.019	1.011	1.005	1.016	1.009	1.004	1.013	1.007	1.002
10	1.037	1.022	1.012	1.006	1.034	1.020	1.012	1.005	1.017	1.010	1.004	1.015	1.009	1.003	1.013	1.007	1.002
For 10-Percent Effective Floor Cavity Reflectance (20 Percent = 1.00)																	
Room Cavity Ratio																	
1	.923	.929	.935	.940	.933	.939	.943	.948	.956	.960	.963	.973	.976	.979	.989	.991	.993
2	.931	.942	.950	.958	.940	.949	.957	.963	.962	.968	.974	.976	.980	.985	.988	.991	.995
3	.939	.951	.961	.969	.945	.957	.966	.973	.967	.975	.981	.978	.983	.988	.988	.992	.996
4	.944	.958	.969	.978	.950	.963	.973	.980	.972	.980	.986	.980	.986	.991	.987	.992	.996
5	.949	.964	.976	.983	.954	.968	.978	.985	.975	.983	.989	.981	.988	.993	.987	.992	.997
6	.953	.969	.980	.986	.958	.972	.982	.989	.977	.985	.992	.982	.989	.995	.987	.993	.997
7	.957	.973	.983	.991	.961	.975	.985	.991	.979	.987	.994	.983	.990	.996	.987	.993	.998
8	.960	.976	.986	.993	.963	.977	.987	.993	.981	.988	.995	.984	.991	.997	.987	.994	.998
9	.963	.978	.987	.994	.965	.979	.989	.994	.983	.990	.996	.985	.992	.998	.988	.994	.999
10	.965	.980	.989	.995	.967	.981	.990	.995	.984	.991	.997	.986	.993	.998	.988	.994	.999
For 0-Percent Effective Floor Cavity Reflectance (20 Percent = 1.00)																	
Room Cavity Ratio																	
1	.859	.870	.879	.886	.873	.884	.893	.901	.916	.923	.929	.948	.954	.960	.979	.983	.987
2	.871	.887	.903	.919	.886	.902	.916	.928	.926	.938	.949	.954	.963	.971	.978	.983	.991
3	.882	.904	.915	.942	.898	.918	.934	.947	.936	.950	.964	.958	.969	.979	.976	.984	.993
4	.893	.919	.941	.958	.908	.930	.948	.961	.945	.961	.974	.961	.974	.984	.975	.985	.994
5	.903	.931	.953	.969	.914	.939	.958	.970	.951	.967	.980	.964	.977	.988	.975	.985	.995
6	.911	.940	.961	.976	.920	.945	.965	.977	.955	.972	.985	.966	.979	.991	.975	.986	.996
7	.917	.947	.967	.981	.924	.950	.970	.982	.959	.975	.988	.968	.981	.993	.975	.987	.997
8	.922	.953	.971	.985	.929	.955	.975	.986	.963	.978	.991	.970	.983	.995	.976	.988	.998
9	.928	.958	.975	.988	.933	.959	.980	.989	.966	.980	.993	.971	.985	.996	.976	.988	.998
10	.933	.962	.979	.991	.937	.963	.983	.992	.969	.982	.995	.973	.987	.997	.977	.989	.999

Reprinted with permission from *Lighting Handbook* of Illuminating Engineering Society of North America, New York, NY.

RCR = 3, the CU is 54 percent. Thus, by interpolation, with RCR = 2.5, the CU is 57.5 percent.

Example Using the same fluorescent fixture just described, calculate the CU of a lighting design for a small office space based on the following data:

- Room dimensions: Length, 20 ft; width, 15 ft; ceiling height, 9 ft
- Room finishes: Base ceiling = 80 percent; wall = 60 percent; base floor = 20 percent

- Work plane: 2′6″ above floor (desk height)
- Luminaire: 1′6″ suspended below ceiling

Calculations

- From Equation 16–5: PAR = 2 × (20 + 15)/(20 × 15) = 70/300 = 0.23
- From Figure 16–1: Cavity heights, h_{cc} = 1.5 ft; h_{rc} = 5.0 ft; h_{fc} = 2.5 ft
- From Equation 16–6a: CCR = 2.5 × 0.23 × 1.5 = 0.87 (use 0.9)
- From Equation 16–6b: RCR = 2.5 × 0.23 × 5.0 = 2.87 (use 2.9 or even 3.0)

- From Equation 16–6c: FCR = 2.5 × 0.23 × 2.5 = 1.4
- From Table 16–7: Effective ceiling reflectance (ρ_{cc}) for 80 percent ceiling, 60 percent wall, and CCR of 0.9 is 70 percent.
- From Table 16–7: Effective floor reflectance (ρ_{fc}) for 20 percent floor, 60 percent wall, and FCR of 1.4 is still 20 percent (there is no change).
- From Figure 16–2, determine the CU based on the following data:

ρ_{cc} = 52 percent (interpolating between 70 percent and 50 percent)

ρ_w = 60 percent

ρ_{fc} = 20 percent (Note: ρ_{fc} and R_f are identical in this case. Otherwise, a multiplier determined from Table 16–8 shall be applied).

CU = 58 percent (Since the manufacturer's CU table is normally abbreviated, much visual interpolation between the rows and columns has to be made.)

Figure 16–3 shows how the preceding calculation is arrived at using the manufacturer's data.

Simplified method for determining the coefficient of utilization The previous example demonstrates the tedious and time-consuming nature of the manual procedure for calculating the CU. To save time and effort one may use a computer program. For preliminary design purposes, the designer may bypass some of the time-consuming steps by choosing effective reflectances at the outset, rather than base reflectances. With this method, CU can be found directly from the

pfc	20								
pcc	80			70			52	50	
pw	70	50	30	70	50	30	60	50	30
RCR									
0	81	81	81	80	80	80		77	77
1	77	73	71	75	72	70		69	68
2	70	66	61	68	65	61		63	59
3	65	58	54	64	57	54	●	56	52
4	59	53	47	58	52	46		51	46
5	56	47	41	54	46	41	**55**	46	40
6	52	42	38	50	42	36		40	36
7	47	39	34	46	39	33		38	33
8	45	35	29	44	34	29		34	29
9	41	33	28	40	33	27		32	27
10	39	29	25	38	29	25		28	25

■ **FIGURE 16–3**
Exercise in visual interpolation from tabulated data.

photometric data, although it will always be higher than when it is determined by any other method.

Example For the same design problem given in example 16–4 assume that the effective reflectances are 80 percent, 60 percent and 20 percent for the ceiling, walls and floors respectively. Then the CU value is found to be 61.5 percent (interpolating between 65 and 58) in lieu of the 58 percent calculated before. There is about 6 percent discrepancy between the results. This is, of course, a significant difference which cannot be justified in final design decisions. As a rule, a more conservative selection of the reflectance values will minimize this discrepancy. For example, if ρ_w = 50 then CU will be 58.

Coefficient of utilization of generic luminaires The CU of the luminaire should be provided by the manufacturer based on certified laboratory tests, preferably by an independent testing laboratory. Experience indicates that the CU of a higher-quality product may be considerably greater than that of a lower-quality product due to a number of factors, such as the effective reflectance of luminaire surfaces, the positioning of the lamps, the geometry of the housing, the efficacy of the diffusing media, and the power loss of the ballast. This variation may result in the need for more or less luminaires to be installed in an identical space. It is, therefore, extremely important to use the specific laboratory testing data of the selected luminaire prior to final design calculations.

For preliminary design and evaluation purposes, using the CU values of generic luminaire types published in the *Illumination Engineering Society Handbook* can provide valuable comparison between different types of luminaires. These are shown in Figure 16–4.

Two fixtures may appear to be similar, but have widely different CU values. For example, with

(Text continues page 471.)

COEFFICIENT OF UTILIZATION

pfc	20								
pcc	80			(70)			50		
pw	70	50	30	70	50	(30)	50	30	
RCR									
0	81	81	81	80	80	80	77	77	
1	77	73	71	75	72	70	69	68	
(2)	70	66	61	68	65	(61)	63	59	
(3)	65	58	54	64	57	(54)	56	52	
4	59	53	47	58	52	46	51	46	
5	56	47	41	54	46	41	46	40	
6	52	42	38	50	42	36	40	36	
7	47	39	34	46	39	33	38	33	
8	45	35	29	44	34	29	34	29	
9	41	33	28	40	33	27	32	27	
10	39	29	25	38	29	25	28	25	

■ **FIGURE 16–2**
Typical data on the CU of a luminaire, provided by luminaire manufacturers.

Typical Luminaire	Typical Intensity Distribution and Per Cent Lamp Lumens			ρcc →	80			70			50			30			10			0	WDRC	ρcc →
				ρw →	50	30	10	50	30	10	50	30	10	50	30	10	50	30	10	0		ρw →
		Maint. Cat.	SC	RCR ↓					Coefficients of Utilization for 20 Per Cent Effective Floor Cavity Reflectance (ρFC = 20)													RCR ↓
1	V	1.5		0	.87	.87	.87	.81	.81	.81	.70	.70	.70	.59	.59	.59	.49	.49	.49	.45		0
				1	.71	.66	.62	.65	.61	.58	.55	.52	.49	.46	.44	.42	.38	.36	.34	.30	.368	1
				2	.60	.53	.48	.55	.50	.45	.47	.42	.38	.39	.35	.32	.31	.29	.26	.23	.279	2
	35½% ↕			3	.52	.44	.38	.48	.41	.36	.40	.35	.31	.33	.29	.26	.27	.24	.21	.18	.227	3
				4	.45	.37	.32	.42	.35	.29	.35	.30	.25	.29	.25	.21	.23	.20	.17	.14	.192	4
	45% ↕			5	.40	.32	.27	.37	.30	.25	.31	.25	.21	.26	.21	.18	.21	.17	.14	.12	.166	5
				6	.35	.28	.23	.33	.26	.21	.28	.22	.18	.23	.19	.15	.19	.15	.12	.10	.146	6
				7	.32	.25	.19	.29	.23	.18	.25	.20	.16	.21	.16	.13	.17	.13	.11	.09	.130	7
				8	.29	.22	.17	.27	.20	.16	.23	.17	.14	.19	.15	.12	.15	.12	.09	.07	.117	8
Pendant diffusing sphere with incandescent lamp				9	.26	.19	.15	.24	.18	.14	.21	.16	.12	.17	.13	.10	.14	.11	.08	.07	.107	9
				10	.24	.17	.13	.22	.16	.12	.19	.14	.11	.16	.12	.09	.13	.10	.08	.06	.098	10
2	II	N.A.		0	.83	.83	.83	.72	.72	.72	.50	.50	.50	.30	.30	.30	.12	.12	.12	.03		0
				1	.72	.69	.66	.62	.60	.57	.43	.42	.40	.26	.25	.25	.10	.10	.10	.03	.018	1
				2	.63	.58	.54	.54	.50	.47	.38	.35	.33	.23	.22	.20	.09	.09	.08	.02	.015	2
				3	.55	.49	.45	.47	.43	.39	.33	.30	.28	.20	.19	.17	.08	.07	.07	.02	.013	3
	83% ↑			4	.48	.42	.37	.42	.37	.33	.29	.26	.23	.18	.16	.15	.07	.06	.06	.02	.012	4
				5	.43	.36	.32	.37	.32	.28	.26	.23	.20	.16	.14	.12	.06	.06	.05	.01	.011	5
				6	.38	.32	.27	.33	.28	.24	.23	.20	.17	.14	.12	.11	.06	.05	.04	.01	.010	6
	3½% ↓			7	.34	.28	.23	.30	.24	.21	.21	.17	.15	.13	.11	.09	.05	.04	.04	.01	.009	7
				8	.31	.25	.20	.27	.21	.18	.19	.15	.13	.12	.10	.08	.05	.04	.03	.01	.008	8
Concentric ring unit with incandescent silvered-bowl lamp				9	.28	.22	.18	.24	.19	.16	.17	.14	.11	.10	.09	.07	.04	.03	.03	.01	.008	9
				10	.25	.20	.16	.22	.17	.14	.16	.12	.10	.10	.08	.06	.04	.03	.03	.01	.007	10
3	IV	1.3		0	.99	.99	.99	.97	.97	.97	.93	.93	.93	.89	.89	.89	.85	.85	.85	.83		0
				1	.87	.84	.81	.85	.82	.79	.82	.79	.77	.79	.76	.74	.76	.74	.72	.71	.323	1
	0% ↑			2	.76	.70	.65	.74	.69	.65	.71	.67	.63	.69	.65	.62	.66	.63	.60	.59	.311	2
				3	.66	.59	.54	.65	.59	.53	.62	.57	.53	.60	.56	.52	.58	.54	.51	.49	.288	3
				4	.58	.51	.45	.57	.50	.45	.55	.49	.44	.53	.48	.44	.51	.47	.43	.41	.264	4
				5	.52	.44	.39	.51	.44	.38	.49	.43	.38	.47	.42	.37	.46	.41	.37	.35	.241	5
	83½% ↓			6	.46	.39	.33	.46	.38	.33	.44	.38	.33	.43	.37	.33	.41	.36	.32	.31	.221	6
				7	.42	.34	.29	.41	.34	.29	.40	.33	.29	.39	.33	.29	.38	.32	.28	.27	.203	7
				8	.38	.31	.26	.37	.31	.26	.36	.30	.26	.35	.30	.25	.34	.29	.25	.24	.187	8
Porcelain-enameled ventilated standard dome with incandescent lamp				9	.35	.28	.23	.34	.28	.23	.33	.27	.23	.32	.27	.23	.32	.26	.23	.21	.173	9
				10	.32	.25	.21	.32	.25	.21	.31	.25	.21	.30	.24	.21	.29	.24	.20	.19	.161	10
4	IV	0.5		0	.82	.82	.82	.80	.80	.80	.76	.76	.76	.73	.73	.73	.70	.70	.70	.69		0
				1	.78	.77	.75	.76	.75	.74	.74	.73	.72	.71	.70	.70	.69	.68	.68	.67	.051	1
	0% ↑			2	.74	.72	.71	.73	.71	.70	.71	.70	.68	.69	.68	.67	.67	.66	.66	.65	.050	2
				3	.71	.69	.67	.71	.68	.67	.69	.67	.66	.67	.66	.65	.66	.65	.64	.63	.049	3
				4	.69	.66	.64	.68	.66	.64	.67	.65	.63	.66	.64	.63	.64	.63	.62	.61	.048	4
				5	.67	.64	.62	.66	.63	.62	.65	.63	.61	.64	.62	.61	.63	.61	.60	.59	.047	5
				6	.64	.62	.60	.64	.61	.60	.63	.61	.59	.62	.60	.59	.61	.60	.59	.58	.045	6
	68½% ↓			7	.63	.60	.58	.62	.60	.58	.61	.59	.57	.61	.59	.57	.60	.58	.57	.56	.044	7
				8	.61	.58	.56	.60	.58	.56	.60	.58	.56	.59	.57	.56	.59	.57	.56	.55	.043	8
Recessed baffled downlight, 140 mm (5 ½″) diameter aperture—150-PAR/FL lamp				9	.59	.56	.55	.59	.56	.55	.58	.56	.54	.58	.56	.54	.57	.55	.54	.54	.042	9
				10	.58	.55	.53	.57	.55	.53	.57	.55	.53	.56	.54	.53	.56	.54	.53	.52	.041	10
5	IV	0.5		0	1.01	1.01	1.01	.99	.99	.99	.95	.95	.95	.91	.91	.91	.87	.87	.87	.85		0
				1	.96	.94	.93	.94	.93	.91	.91	.89	.88	.88	.87	.86	.85	.84	.83	.82	.085	1
	0% ↑			2	.91	.88	.86	.90	.87	.85	.87	.85	.83	.84	.83	.81	.82	.81	.80	.79	.084	2
				3	.87	.83	.81	.86	.83	.80	.83	.81	.79	.81	.79	.78	.80	.78	.77	.75	.082	3
				4	.83	.79	.76	.82	.79	.76	.80	.77	.75	.79	.76	.74	.77	.75	.73	.72	.080	4
				5	.79	.76	.73	.79	.75	.72	.77	.74	.72	.76	.73	.71	.75	.72	.71	.70	.078	5
	85% ↓			6	.76	.72	.70	.76	.72	.69	.74	.71	.69	.73	.71	.68	.72	.70	.68	.67	.076	6
				7	.73	.69	.67	.73	.69	.67	.72	.69	.66	.71	.68	.66	.70	.68	.66	.65	.073	7
				8	.71	.67	.64	.70	.67	.64	.69	.66	.64	.69	.66	.64	.68	.65	.63	.62	.071	8
Recessed baffled downlight, 140 mm (5½″) diameter aperture—75ER30 lamp				9	.68	.64	.62	.68	.64	.62	.67	.64	.62	.66	.63	.61	.66	.63	.61	.60	.069	9
				10	.66	.62	.60	.66	.62	.60	.65	.62	.59	.64	.61	.59	.64	.61	.59	.58	.067	10
6	IV	0.7		0	.52	.52	.52	.51	.51	.51	.48	.48	.48	.46	.46	.46	.45	.45	.45	.44		0
				1	.49	.48	.47	.48	.47	.46	.46	.45	.45	.44	.44	.43	.43	.43	.42	.41	.055	1
	0% ↑			2	.46	.44	.43	.45	.44	.43	.44	.43	.42	.43	.42	.41	.41	.41	.40	.39	.054	2
				3	.43	.41	.40	.43	.41	.40	.42	.40	.39	.41	.39	.38	.40	.39	.38	.37	.053	3
				4	.41	.39	.37	.41	.39	.37	.40	.38	.37	.39	.37	.36	.38	.37	.36	.35	.052	4
				5	.39	.37	.35	.39	.37	.35	.38	.36	.35	.37	.36	.34	.36	.35	.34	.34	.051	5
	43½% ↓			6	.37	.35	.33	.37	.35	.33	.36	.34	.33	.35	.34	.33	.35	.34	.33	.32	.049	6
				7	.35	.33	.31	.35	.33	.31	.34	.33	.31	.34	.32	.31	.33	.32	.31	.30	.048	7
				8	.34	.31	.30	.33	.31	.30	.33	.31	.30	.32	.31	.29	.32	.31	.29	.29	.046	8
EAR-38 lamp above 51 mm (2″) diameter aperture (increase efficiency to 54½% for 76 mm (3″) diameter aperture)*				9	.32	.30	.28	.32	.30	.28	.31	.30	.28	.31	.29	.28	.31	.29	.28	.28	.045	9
				10	.31	.28	.27	.31	.28	.27	.30	.28	.27	.30	.28	.27	.30	.28	.27	.26	.043	10

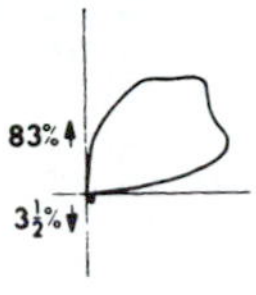

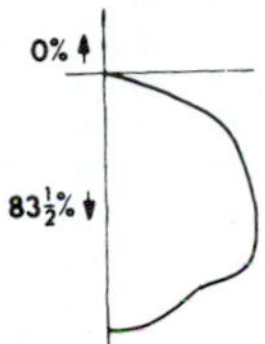

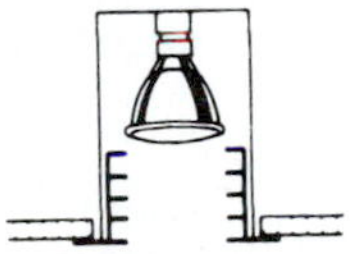

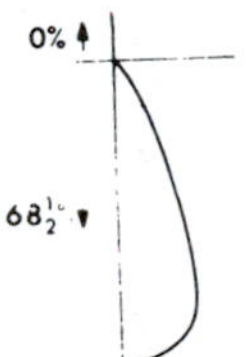

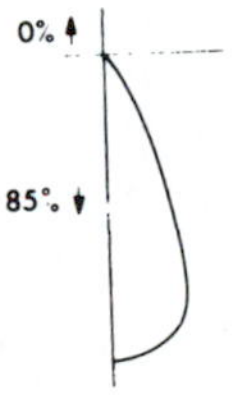

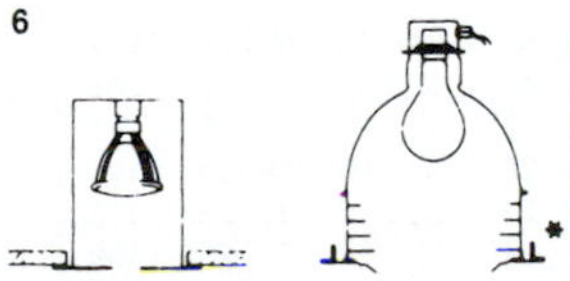

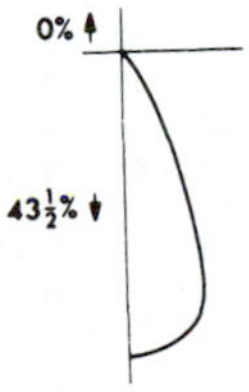

■ FIGURE 16–4

Coefficient of utilization (CU) of generic luminaires. (Courtesy: Illuminating Engineering Society of North America, New York, NY.)

Typical Luminaire	Maint. Cat.	SC	RCR ↓	ρcc→ 80 ρw→ 50	30	10	70 50	30	10	50 50	30	10	30 50	30	10	10 50	30	10	0	WDRC	RCR ↓
7 — R-40 flood without shielding (IV, 0%↑ 100%↓)	IV	0.8	0	1.19	1.19	1.19	1.16	1.16	1.16	1.11	1.11	1.11	1.06	1.06	1.06	1.02	1.02	1.02	1.00		
			1	1.08	1.05	1.03	1.06	1.03	1.01	1.02	1.00	.98	.98	.97	.95	.95	.93	.92	.90	.241	1
			2	.99	.94	.89	.97	.92	.88	.93	.90	.86	.90	.87	.84	.88	.85	.83	.81	.238	2
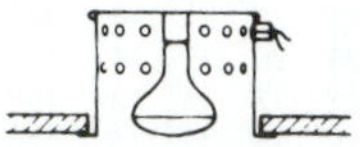			3	.90	.84	.79	.88	.83	.78	.86	.81	.77	.83	.79	.76	.81	.77	.74	.73	.227	3
			4	.82	.75	.70	.81	.75	.70	.79	.73	.69	.77	.72	.68	.75	.71	.67	.66	.215	4
			5	.76	.68	.63	.75	.68	.63	.73	.67	.62	.71	.66	.62	.69	.65	.61	.59	.202	5
			6	.70	.62	.57	.69	.62	.57	.67	.61	.57	.66	.60	.56	.64	.60	.56	.54	.191	6
			7	.65	.57	.52	.64	.57	.52	.62	.56	.52	.61	.56	.52	.60	.55	.51	.50	.180	7
			8	.60	.53	.48	.59	.53	.48	.58	.52	.48	.57	.52	.47	.56	.51	.47	.46	.169	8
			9	.56	.49	.44	.55	.49	.44	.54	.48	.44	.53	.48	.44	.52	.47	.44	.42	.160	9
			10	.52	.46	.41	.52	.45	.41	.51	.45	.41	.50	.45	.41	.49	.44	.41	.39	.152	10
8 — R-40 flood with specular anodized reflector skirt; 45° cutoff (IV, 0%↑ 85%↓)	IV	0.7	0	1.01	1.01	1.01	.99	.99	.99	.94	.94	.94	.90	.90	.90	.87	.87	.87	.85		
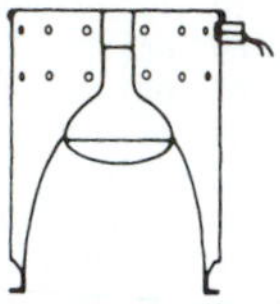			1	.95	.93	.91	.93	.91	.89	.89	.88	.87	.86	.85	.84	.83	.82	.82	.80	.115	1
			2	.89	.86	.83	.87	.84	.82	.85	.82	.80	.82	.80	.79	.80	.78	.77	.76	.115	2
			3	.83	.80	.77	.82	.79	.76	.80	.77	.75	.78	.76	.74	.76	.74	.72	.71	.113	3
			4	.79	.74	.71	.78	.74	.71	.76	.73	.70	.74	.71	.69	.73	.70	.68	.67	.110	4
			5	.74	.70	.67	.74	.69	.66	.72	.68	.66	.71	.68	.65	.69	.67	.65	.63	.107	5
			6	.70	.66	.62	.70	.65	.62	.68	.65	.62	.67	.64	.61	.66	.63	.61	.60	.104	6
			7	.67	.62	.59	.66	.62	.59	.65	.61	.58	.64	.61	.58	.63	.60	.58	.57	.100	7
			8	.63	.59	.56	.63	.58	.55	.62	.58	.55	.61	.58	.55	.60	.57	.55	.54	.097	8
			9	.60	.56	.53	.60	.56	.53	.59	.55	.52	.58	.55	.52	.58	.54	.52	.51	.094	9
			10	.57	.53	.50	.57	.53	.50	.56	.52	.50	.56	.52	.50	.55	.52	.49	.48	.091	10
9 — 2-lamp prismatic wraparound—see note 7 (V, 11½%↑ 58½%↓)	V	1.5/1.2	0	.81	.81	.81	.78	.78	.78	.72	.72	.72	.66	.66	.66	.61	.61	.61	.59		
			1	.71	.68	.66	.68	.66	.63	.63	.61	.59	.58	.57	.56	.54	.53	.52	.50	.223	1
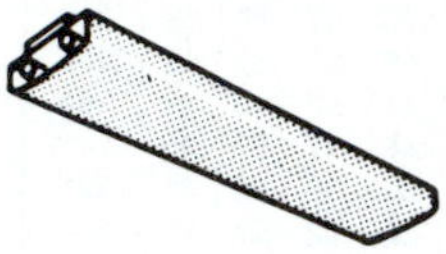			2	.63	.58	.55	.60	.56	.53	.56	.53	.50	.52	.50	.47	.48	.46	.45	.43	.201	2
	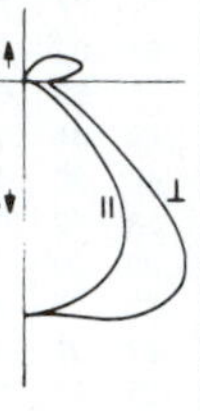		3	.56	.50	.46	.54	.49	.45	.50	.46	.43	.47	.43	.41	.43	.41	.39	.37	.183	3
			4	.50	.44	.40	.48	.43	.39	.45	.40	.37	.42	.38	.35	.39	.36	.34	.32	.167	4
			5	.45	.39	.34	.43	.38	.34	.40	.36	.32	.38	.34	.31	.35	.32	.30	.28	.153	5
			6	.40	.34	.30	.39	.34	.30	.37	.32	.28	.34	.30	.27	.32	.29	.26	.25	.142	6
			7	.37	.31	.27	.35	.30	.26	.33	.29	.25	.31	.27	.24	.30	.26	.23	.22	.131	7
			8	.33	.28	.24	.32	.27	.23	.30	.26	.23	.29	.25	.22	.27	.24	.21	.20	.122	8
			9	.31	.25	.21	.30	.25	.21	.28	.24	.20	.26	.23	.20	.25	.22	.19	.18	.114	9
			10	.28	.23	.19	.27	.22	.19	.26	.21	.18	.24	.21	.18	.23	.20	.17	.16	.107	10
10 — Prismatic bottom and sides, open top, 4-lamp suspended unit—see note 7 (VI, 33%↑ 50%↓)	VI	1.4/1.2	0	.91	.91	.91	.85	.85	.85	.74	.74	.74	.64	.64	.64	.54	.54	.54	.50		
			1	.80	.77	.74	.75	.72	.70	.65	.63	.61	.57	.55	.54	.49	.47	.47	.43	.179	1
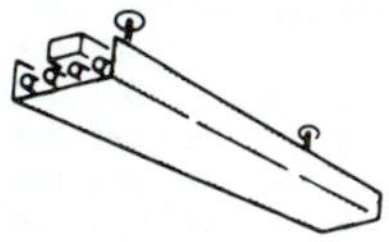			2	.70	.65	.61	.66	.62	.58	.58	.54	.52	.50	.48	.46	.43	.42	.40	.37	.166	2
	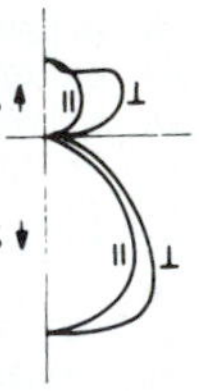		3	.62	.56	.51	.58	.53	.49	.51	.47	.44	.45	.42	.39	.39	.37	.35	.32	.153	3
			4	.55	.49	.44	.52	.46	.42	.46	.41	.38	.40	.37	.34	.35	.32	.30	.27	.140	4
			5	.50	.43	.38	.47	.41	.36	.41	.37	.33	.36	.33	.30	.32	.29	.26	.24	.129	5
			6	.45	.38	.33	.42	.36	.32	.37	.33	.29	.33	.29	.26	.29	.26	.23	.21	.119	6
			7	.40	.34	.29	.38	.32	.28	.34	.29	.26	.30	.26	.23	.26	.23	.21	.19	.111	7
			8	.37	.30	.26	.35	.29	.25	.31	.26	.23	.28	.24	.21	.24	.21	.19	.17	.103	8
			9	.34	.27	.23	.32	.26	.22	.29	.24	.21	.25	.22	.19	.22	.19	.17	.15	.096	9
			10	.31	.25	.21	.29	.24	.20	.26	.22	.19	.23	.20	.17	.21	.18	.15	.14	.090	10
11 — 2-lamp diffuse wraparound—see note 7 (V, 8%↑ 37½%↓)	V	1.3	0	.52	.52	.52	.50	.50	.50	.46	.46	.46	.43	.43	.43	.39	.39	.39	.38		
			1	.44	.42	.40	.42	.40	.39	.39	.37	.36	.36	.35	.33	.33	.32	.31	.30	.201	1
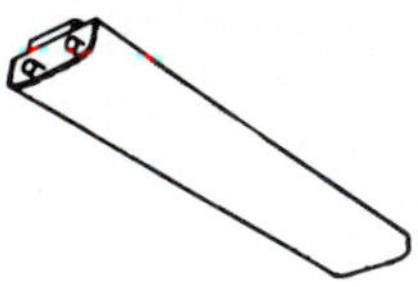			2	.38	.35	.32	.37	.33	.31	.34	.31	.29	.31	.29	.27	.28	.27	.25	.24	.171	2
	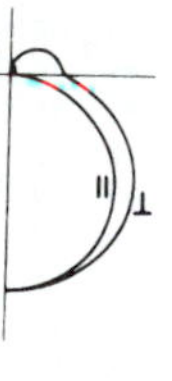		3	.33	.29	.26	.32	.28	.25	.29	.26	.24	.27	.25	.22	.25	.23	.21	.20	.149	3
			4	.29	.25	.22	.28	.24	.21	.26	.23	.20	.24	.21	.19	.22	.20	.18	.17	.132	4
			5	.26	.22	.19	.25	.21	.18	.23	.20	.17	.21	.18	.16	.20	.17	.15	.14	.117	5
			6	.23	.19	.16	.22	.18	.16	.21	.17	.15	.19	.16	.14	.18	.15	.13	.12	.106	6
			7	.21	.17	.14	.20	.16	.14	.19	.15	.13	.17	.15	.12	.16	.14	.12	.11	.096	7
			8	.19	.15	.12	.18	.15	.12	.17	.14	.12	.16	.13	.11	.15	.12	.11	.10	.088	8
			9	.17	.14	.11	.17	.13	.11	.16	.13	.10	.15	.12	.10	.14	.11	.09	.09	.081	9
			10	.16	.12	.10	.15	.12	.10	.14	.11	.09	.14	.11	.09	.13	.10	.09	.08	.075	10
12 — Fluorescent unit dropped diffuser, 4-lamp 610 mm (2') wide—see note 7 (V, 1%↑ 60½%↓)	V	1.2	0	.73	.73	.73	.71	.71	.71	.68	.68	.68	.65	.65	.65	.62	.62	.62	.60		
			1	.63	.60	.58	.62	.59	.57	.59	.57	.55	.56	.55	.53	.54	.53	.51	.50	.259	1
			2	.55	.51	.47	.54	.50	.46	.51	.48	.45	.49	.46	.44	.47	.45	.43	.42	.236	2
			3	.48	.43	.39	.47	.42	.39	.45	.41	.38	.43	.40	.37	.42	.39	.36	.35	.212	3
			4	.43	.37	.33	.42	.37	.33	.40	.36	.32	.39	.35	.32	.37	.34	.31	.30	.191	4
			5	.38	.33	.29	.37	.32	.28	.36	.31	.28	.35	.31	.28	.33	.30	.27	.26	.173	5
			6	.34	.29	.25	.34	.29	.25	.33	.28	.24	.31	.27	.24	.30	.27	.24	.23	.158	6
			7	.31	.26	.22	.31	.26	.22	.30	.25	.22	.29	.25	.21	.28	.24	.21	.20	.144	7
			8	.28	.23	.20	.28	.23	.20	.27	.23	.19	.26	.22	.19	.25	.22	.19	.18	.133	8
			9	.26	.21	.18	.26	.21	.18	.25	.21	.17	.24	.20	.17	.24	.20	.17	.16	.123	9
			10	.24	.19	.16	.24	.19	.16	.23	.19	.16	.22	.19	.16	.22	.18	.16	.15	.115	10

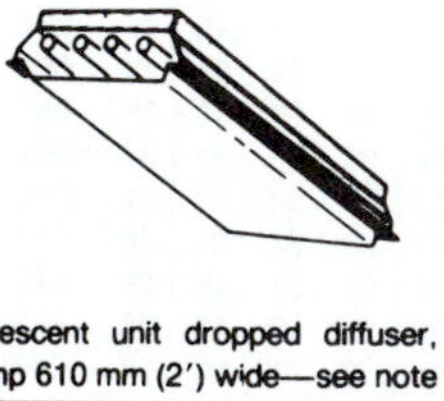

Coefficients of Utilization for 20 Per Cent Effective Floor Cavity Reflectance (ρFC = 20)

* Also, reflector downlight with baffles and inside frosted lamp.

■ **FIGURE 16–4**
Continued

Typical Luminaire	Typical Intensity Distribution and Per Cent Lamp Lumens	Maint. Cat.	SC	RCR	ρcc → 80			70			50			30			10			0	WDRC
					ρw → 50	30	10	50	30	10	50	30	10	50	30	10	50	30	10	0	
13 4-lamp, 610 mm (2') wide troffer with 45° white metal louver—see note 7	0% ↑ / 46% ↓	IV	0.9	0	.55	.55	.55	.54	.54	.54	.51	.51	.51	.49	.49	.49	.47	.47	.47	.46	
				1	.49	.48	.46	.48	.47	.46	.46	.45	.44	.45	.44	.43	.43	.42	.42	.41	.137
				2	.44	.42	.40	.43	.41	.39	.42	.40	.38	.40	.39	.37	.39	.38	.37	.36	.131
				3	.40	.37	.34	.39	.36	.34	.38	.36	.33	.37	.35	.33	.36	.34	.32	.32	.122
				4	.36	.33	.30	.36	.33	.30	.35	.32	.30	.34	.31	.29	.33	.31	.29	.28	.113
				5	.33	.30	.27	.33	.29	.27	.32	.29	.27	.31	.28	.26	.30	.28	.26	.25	.104
				6	.30	.27	.24	.30	.27	.24	.29	.26	.24	.29	.26	.24	.28	.25	.24	.23	.097
				7	.28	.25	.22	.28	.24	.22	.27	.24	.22	.26	.24	.22	.26	.23	.22	.21	.090
				8	.26	.23	.20	.26	.22	.20	.25	.22	.20	.25	.22	.20	.24	.22	.20	.19	.085
				9	.24	.21	.19	.24	.21	.19	.23	.20	.18	.23	.20	.18	.23	.20	.18	.18	.079
				10	.23	.19	.17	.22	.19	.17	.22	.19	.17	.22	.19	.17	.21	.19	.17	.16	.075
14 Bilateral batwing distribution—one-lamp, surface mounted fluorescent with prismatic wraparound lens	12% ↑ / 63½% ↓	V	N.A.	0	.87	.87	.87	.84	.84	.84	.77	.77	.77	.72	.72	.72	.66	.66	.66	.64	
				1	.75	.72	.69	.72	.69	.66	.67	.64	.62	.62	.60	.58	.57	.56	.54	.52	.296
				2	.65	.60	.56	.63	.58	.54	.58	.54	.51	.54	.51	.48	.50	.47	.45	.43	.261
				3	.57	.51	.46	.55	.49	.45	.51	.46	.42	.47	.43	.40	.44	.41	.38	.36	.232
				4	.50	.44	.39	.48	.42	.38	.45	.40	.36	.42	.38	.34	.39	.35	.32	.30	.209
				5	.45	.38	.33	.43	.37	.32	.40	.35	.31	.37	.33	.29	.35	.31	.28	.26	.189
				6	.40	.33	.28	.39	.32	.28	.36	.31	.26	.34	.29	.25	.31	.27	.24	.22	.172
				7	.36	.29	.25	.35	.29	.24	.32	.27	.23	.30	.26	.22	.28	.24	.21	.19	.158
				8	.33	.26	.22	.31	.25	.21	.29	.24	.20	.28	.23	.20	.26	.22	.19	.17	.146
				9	.30	.23	.19	.29	.23	.19	.27	.22	.18	.25	.21	.17	.24	.20	.17	.15	.135
				10	.27	.21	.17	.26	.21	.17	.25	.20	.16	.23	.19	.16	.22	.18	.15	.13	.126
15 Radial batwing distribution—4-lamp, 610 mm (2') wide fluorescent unit with flat prismatic lens—see note 7	0 / 59½ ↓	V	1.7	0	.71	.71	.71	.69	.69	.69	.66	.66	.66	.63	.63	.63	.61	.61	.61	.60	
				1	.62	.59	.57	.60	.58	.56	.58	.56	.54	.55	.54	.52	.53	.52	.51	.50	.251
				2	.53	.49	.46	.52	.48	.45	.50	.47	.44	.48	.45	.43	.46	.44	.42	.41	.237
				3	.46	.41	.37	.45	.41	.37	.44	.40	.36	.42	.39	.36	.40	.38	.35	.34	.216
				4	.41	.35	.31	.40	.35	.31	.38	.34	.30	.37	.33	.30	.36	.32	.30	.28	.196
				5	.36	.30	.26	.35	.30	.26	.34	.29	.26	.33	.29	.26	.32	.28	.25	.24	.178
				6	.32	.27	.23	.32	.26	.23	.31	.26	.22	.29	.25	.22	.29	.25	.22	.21	.162
				7	.29	.24	.20	.28	.23	.20	.28	.23	.19	.27	.22	.19	.26	.22	.19	.18	.149
				8	.26	.21	.17	.26	.21	.17	.25	.20	.17	.24	.20	.17	.24	.20	.17	.16	.137
				9	.24	.19	.15	.24	.19	.15	.23	.18	.15	.23	.18	.15	.22	.18	.15	.14	.127
				10	.22	.17	.14	.22	.17	.14	.21	.17	.14	.20	.16	.14	.20	.16	.14	.12	.118
16 Diffuse aluminum reflector with 35°CW shielding	17% ↑ / 66% ↓	II	1.5/1.3	0	.95	.95	.95	.91	.91	.91	.83	.83	.83	.76	.76	.76	.69	.69	.69	.66	
				1	.85	.82	.79	.81	.79	.76	.75	.73	.71	.69	.67	.66	.63	.62	.61	.58	.197
				2	.75	.71	.67	.72	.68	.65	.67	.63	.61	.62	.59	.57	.57	.55	.53	.51	.194
				3	.67	.61	.57	.65	.59	.55	.60	.56	.52	.55	.52	.49	.51	.49	.46	.44	.184
				4	.60	.54	.49	.58	.52	.48	.54	.49	.45	.50	.46	.43	.46	.43	.41	.39	.173
				5	.54	.47	.43	.52	.46	.42	.49	.43	.40	.45	.41	.38	.42	.39	.36	.34	.162
				6	.49	.42	.37	.47	.41	.37	.44	.39	.35	.41	.37	.33	.38	.35	.32	.30	.151
				7	.44	.38	.33	.43	.37	.32	.40	.35	.31	.38	.33	.30	.35	.31	.28	.27	.141
				8	.40	.34	.29	.39	.33	.29	.37	.31	.28	.34	.30	.27	.32	.28	.26	.24	.132
				9	.37	.31	.26	.36	.30	.26	.34	.29	.25	.32	.27	.24	.30	.26	.23	.21	.124
				10	.34	.28	.24	.33	.27	.23	.31	.26	.23	.29	.25	.22	.28	.24	.21	.19	.117
17 Porcelain-enameled reflector with 30°CW × 30°LW shielding	23½% ↑ / 57% ↓	II	1.0	0	.91	.91	.91	.86	.86	.86	.77	.77	.77	.68	.68	.68	.61	.61	.61	.57	
				1	.80	.77	.75	.76	.74	.71	.69	.67	.65	.62	.60	.59	.55	.54	.53	.50	.182
				2	.71	.67	.63	.68	.64	.60	.61	.58	.55	.55	.53	.51	.50	.48	.46	.43	.174
				3	.63	.58	.53	.60	.55	.51	.55	.51	.47	.50	.46	.44	.45	.42	.40	.38	.163
				4	.57	.51	.46	.54	.49	.44	.49	.45	.41	.45	.41	.38	.41	.38	.35	.33	.151
				5	.51	.45	.40	.49	.43	.39	.45	.40	.36	.41	.37	.34	.37	.34	.31	.29	.140
				6	.46	.40	.35	.44	.38	.34	.41	.36	.32	.37	.33	.30	.34	.30	.28	.26	.130
				7	.42	.36	.31	.40	.35	.30	.37	.32	.29	.34	.30	.27	.31	.28	.25	.23	.121
				8	.38	.32	.28	.37	.31	.27	.34	.29	.26	.31	.27	.24	.29	.25	.23	.21	.113
				9	.35	.29	.25	.34	.28	.25	.31	.27	.23	.29	.25	.22	.27	.23	.21	.19	.106
				10	.33	.27	.23	.31	.26	.22	.29	.24	.21	.27	.23	.20	.25	.21	.19	.17	.099
18 Enclosed reflector with an incandescent lamp	0% ↑ / 71½% ↓	V	1.4	0	.85	.85	.85	.83	.83	.83	.80	.80	.80	.76	.76	.76	.73	.73	.73	.72	
				1	.77	.75	.73	.76	.74	.72	.73	.71	.69	.70	.69	.67	.67	.66	.65	.64	.189
				2	.70	.66	.63	.68	.65	.62	.66	.63	.60	.64	.61	.59	.61	.60	.58	.56	.190
				3	.63	.58	.54	.62	.57	.54	.60	.56	.53	.58	.54	.52	.56	.53	.51	.50	.183
				4	.56	.51	.47	.56	.51	.47	.54	.50	.46	.52	.49	.46	.51	.48	.45	.44	.174
				5	.51	.46	.42	.50	.45	.41	.49	.44	.41	.48	.44	.40	.46	.43	.40	.39	.164
				6	.46	.41	.37	.46	.41	.37	.45	.40	.36	.43	.39	.36	.42	.39	.36	.34	.155
				7	.42	.37	.33	.42	.37	.33	.41	.36	.33	.40	.36	.32	.39	.35	.32	.31	.146
				8	.39	.33	.30	.38	.33	.29	.37	.33	.29	.37	.32	.29	.36	.32	.29	.28	.137
				9	.36	.30	.27	.35	.30	.27	.35	.30	.27	.34	.30	.26	.33	.29	.26	.25	.129
				10	.33	.28	.24	.33	.28	.24	.32	.27	.24	.31	.27	.24	.31	.27	.24	.23	.122

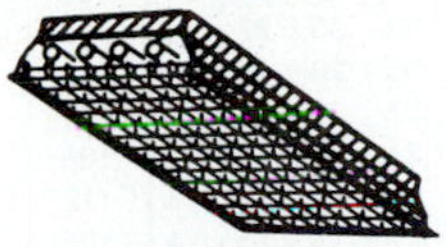

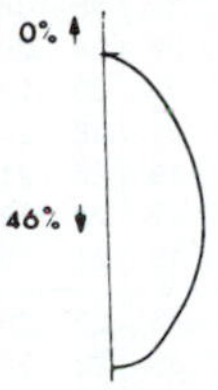

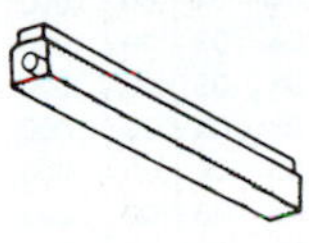

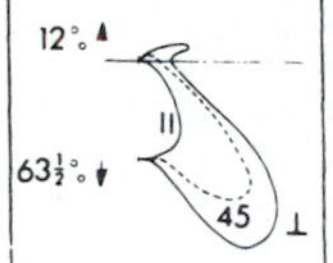

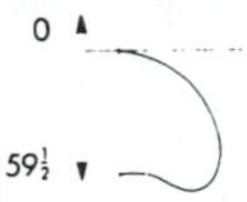

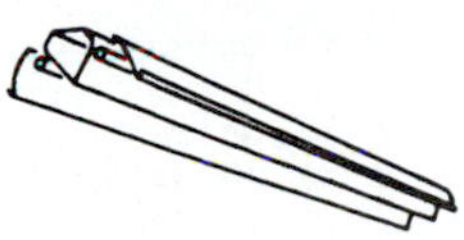

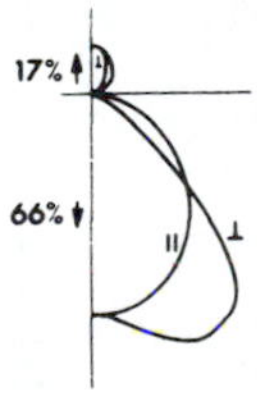

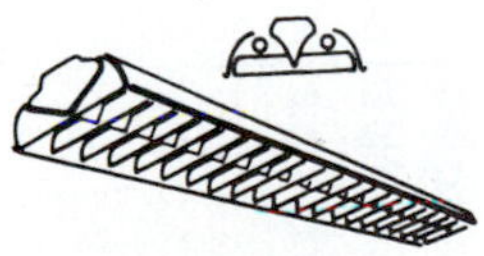

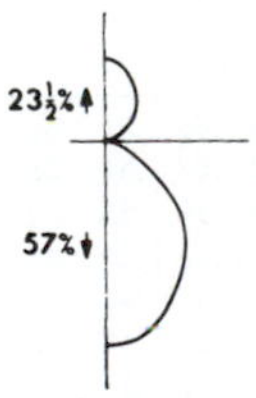

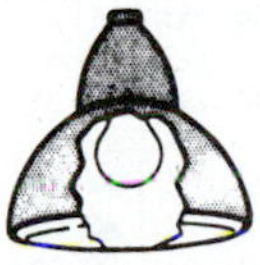

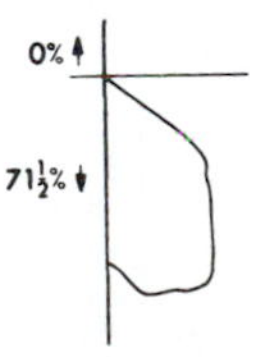

Coefficients of Utilization for 20 Per Cent Effective Floor Cavity Reflectance (ρFC = 20)

* Also, reflector downlight with baffles and inside frosted lamp.

■ FIGURE 16–4
Continued

Top Table — Luminaires 19–20

Typical Luminaire	Maint. Cat.	SC	RCR ↓	ρcc→ 80 (ρw 50)	80 (30)	80 (10)	70 (50)	70 (30)	70 (10)	50 (50)	50 (30)	50 (10)	30 (50)	30 (30)	30 (10)	10 (50)	10 (30)	10 (10)	0	WDRC	RCR ↓
19 Wide spread, recessed, small open bottom reflector with low wattage diffuse HID lamp (Dist. IV, 0% ↑, 56% ↓)	IV	1.7	0	.67	.67	.67	.65	.65	.65	.62	.62	.62	.60	.60	.60	.57	.57	.57	.56		0
	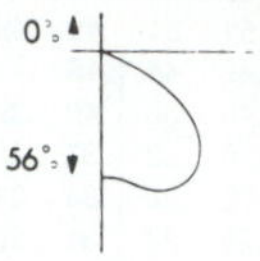		1	.60	.58	.56	.58	.57	.55	.56	.55	.53	.54	.53	.52	.52	.51	.50	.49	.177	1
			2	.53	.49	.46	.52	.48	.46	.50	.47	.45	.48	.46	.44	.46	.44	.43	.42	.179	2
			3	.46	.42	.39	.46	.42	.38	.44	.41	.38	.42	.40	.37	.41	.39	.37	.35	.172	3
			4	.41	.36	.33	.40	.36	.33	.39	.35	.32	.38	.34	.32	.37	.34	.31	.30	.161	4
			5	.37	.32	.28	.36	.31	.28	.35	.31	.28	.34	.30	.27	.33	.30	.27	.26	.150	5
			6	.33	.28	.24	.32	.28	.24	.31	.27	.24	.30	.27	.24	.30	.26	.24	.23	.139	6
			7	.30	.25	.21	.29	.25	.21	.28	.24	.21	.28	.24	.21	.27	.23	.21	.20	.129	7
			8	.27	.22	.19	.26	.22	.19	.26	.22	.19	.25	.21	.19	.24	.21	.19	.17	.120	8
			9	.25	.20	.17	.24	.20	.17	.24	.20	.17	.23	.19	.17	.22	.19	.17	.16	.112	9
			10	.22	.18	.15	.22	.18	.15	.22	.18	.15	.21	.18	.15	.21	.17	.15	.14	.105	10
20 Open top, indirect, reflector type unit with HID lamp (mult. by 0.9 for lens top) (Dist. VI, 78% ↑, 0% ↓)	VI	N.A.	0	.74	.74	.74	.63	.63	.63	.43	.43	.43	.25	.25	.25	.08	.08	.08	.00		0
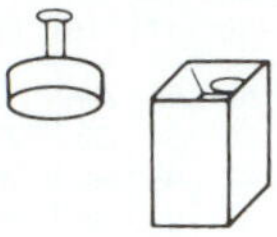	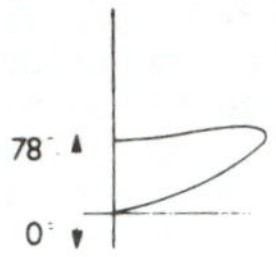		1	.64	.62	.59	.55	.53	.51	.38	.36	.35	.22	.21	.20	.07	.07	.07	.00	.000	1
			2	.56	.52	.48	.48	.45	.42	.33	.31	.29	.19	.18	.17	.06	.06	.06	.00	.000	2
			3	.49	.44	.40	.42	.38	.35	.29	.26	.24	.17	.15	.14	.05	.05	.05	.00	.000	3
			4	.43	.38	.34	.37	.33	.29	.26	.23	.20	.15	.13	.12	.05	.04	.04	.00	.000	4
			5	.38	.33	.28	.33	.28	.25	.23	.20	.17	.13	.12	.10	.04	.04	.03	.00	.000	5
			6	.34	.28	.24	.29	.25	.21	.20	.17	.15	.12	.10	.09	.04	.03	.03	.00	.000	6
			7	.31	.25	.21	.26	.22	.18	.18	.15	.13	.11	.09	.08	.03	.03	.03	.00	.000	7
			8	.28	.22	.18	.24	.19	.16	.16	.13	.11	.10	.08	.07	.03	.03	.02	.00	.000	8
			9	.25	.20	.16	.21	.17	.14	.15	.12	.10	.09	.07	.06	.03	.02	.02	.00	.000	9
			10	.23	.17	.14	.20	.15	.12	.14	.11	.09	.08	.06	.05	.03	.02	.02	.00	.000	10

Bottom Table — Luminaires 21–24

Coefficients of utilization for 20 Per Cent Effective Floor Cavity Reflectance, ρFC

Typical Luminaires	RCR ↓	ρcc→ 80 (ρw 50)	80 (30)	80 (10)	70 (50)	70 (30)	70 (10)	50 (50)	50 (30)	50 (10)	30 (50)	30 (30)	30 (10)	10 (50)	10 (30)	10 (10)	0
21 Single row fluorescent lamp cove without reflector, mult. by 0.93 for 2 rows and by 0.85 for 3 rows.	1	.42	.40	.39	.36	.35	.33	.25	.24	.23	Coves are not recommended for lighting areas having low reflectances.						
	2	.37	.34	.32	.32	.29	.27	.22	.20	.19							
	3	.32	.29	.26	.28	.25	.23	.19	.17	.16							
	4	.29	.25	.22	.25	.22	.19	.17	.15	.13							
	5	.25	.21	.18	.22	.19	.16	.15	.13	.11							
	6	.23	.19	.16	.20	.16	.14	.14	.12	.10							
	7	.20	.17	.14	.17	.14	.12	.12	.10	.09							
	8	.18	.15	.12	.16	.13	.10	.11	.09	.08							
	9	.17	.13	.10	.15	.11	.09	.10	.08	.07							
	10	.15	.12	.09	.13	.10	.08	.09	.07	.06							
22 Diffusing plastic or glass. 1) Ceiling efficiency ~60%; diffuser transmittance ~50%; diffuser reflectance ~40%. Cavity with minimum obstructions and painted with 80% reflectance paint—use ρc = 70. 2) For lower reflectance paint or obstructions—use ρc = 50. (ρcc from below ~65%) 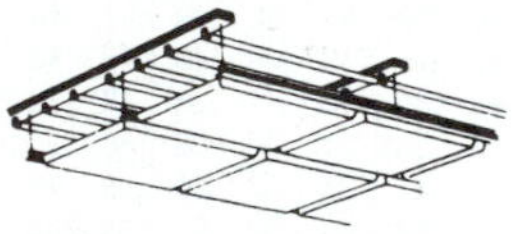	1				.60	.58	.56	.58	.56	.54							
	2				.53	.49	.45	.51	.47	.43							
	3				.47	.42	.37	.45	.41	.36							
	4				.41	.36	.32	.39	.35	.31							
	5				.37	.31	.27	.35	.30	.26							
	6				.33	.27	.23	.31	.26	.23							
	7				.29	.24	.20	.28	.23	.20							
	8				.26	.21	.18	.25	.20	.17							
	9				.23	.19	.15	.23	.18	.15							
	10				.21	.17	.13	.21	.16	.13							
23 Prismatic plastic or glass. 1) Ceiling efficiency ~67%; prismatic transmittance ~72%; prismatic reflectance ~18%. Cavity with minimum obstructions and painted with 80% reflectance paint—use ρc = 70. 2) For lower reflectance paint or obstructions—use ρc = 50. (ρcc from below ~60%) 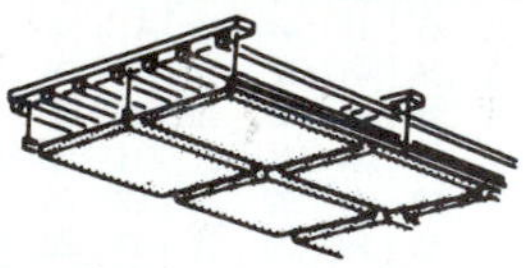	1				.71	.68	.66	.67	.66	.65	.65	.64	.62				
	2				.63	.60	.57	.61	.58	.55	.59	.56	.54				
	3				.57	.53	.49	.55	.52	.48	.54	.50	.47				
	4				.52	.47	.43	.50	.45	.42	.48	.44	.42				
	5				.46	.41	.37	.44	.40	.37	.43	.40	.36				
	6				.42	.37	.33	.41	.36	.32	.40	.35	.32				
	7				.38	.32	.29	.37	.31	.28	.36	.31	.28				
	8				.34	.28	.25	.33	.28	.25	.32	.28	.25				
	9				.30	.25	.22	.30	.25	.21	.29	.25	.21				
	10				.27	.23	.19	.27	.22	.19	.26	.22	.19				
24 Louvered ceiling. 1) Ceiling efficiency ~50%; 45° shielding opaque louvers of 80% reflectance. Cavity with minimum obstructions and painted with 80% reflectance paint—use ρc = 50. 2) For other conditions refer to Fig. 6-18. (ρcc from below ~45%) 	1							.51	.49	.48				.47	.46	.45	
	2							.46	.44	.42				.43	.42	.40	
	3							.42	.39	.37				.39	.38	.36	
	4							.38	.35	.33				.36	.34	.32	
	5							.35	.32	.29				.33	.31	.29	
	6							.32	.29	.26				.30	.28	.26	
	7							.29	.26	.23				.28	.25	.23	
	8							.27	.23	.21				.26	.23	.21	
	9							.24	.21	.19				.24	.21	.19	
	10							.22	.19	.17				.22	.19	.17	

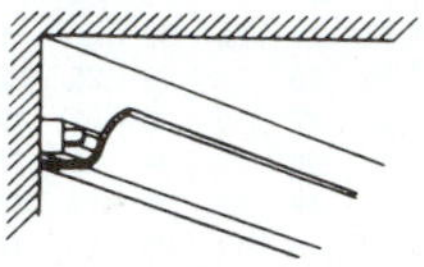

■ **FIGURE 16–4**
Continued

RC RW	80				70				50			0
	70	50	30	10	70	50	30	10	50	30	10	0
1	80	77	75	70	78	76	73	71	73	70	69	64
2	74	70	66	63	73	68	65	62	66	61	61	57
3	69	63	58	54	67	62	57	54	60	56	53	50
4	64	57	51	47	62	55	51	47	54	50	46	44
5	59	51	45	41	57	50	45	41	48	44	40	38
6	54	45	40	36	53	45	40	36	44	39	35	34
7	50	41	35	31	49	40	36	31	39	34	31	29
8	46	37	31	27	45	36	31	27	35	31	27	25
9	42	33	27	23	41	32	27	23	32	27	23	22
10	39	30	24	21	38	29	24	21	29	24	20	19

25

Typical CU for a 3-lamp fluorescent parabolic luminaire

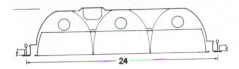

24

■ **FIGURE 16–4**
Continued

effective ceiling at 70 percent, wall at 30 percent, and RCR = 1, the CU for fixture type 7 is 1.03 (> 1.0) and for type 8 only 0.91. Type 7 is definitely more effective in this room configuration. However, for the same fixtures in another room with RCR = 10 (a very small room), the CU for type 7 is only 0.45; the CU for type 8 is 0.53, a more effective selection.

16.3.2 Maintained Illuminance

The light output of a lamp and of a luminaire installed in the field differs from that obtained in controlled laboratory conditions. The light output also depreciates with time. Lighting system design must therefore include a factor in the calculations to compensate for the anticipated losses. This factor is called the *light loss factor* (LLF), and the resulting illuminance is called the *maintained illuminance* (E_m, or simply, E). The relation between both of these and the initial illuminance is

$$E = E_i \times LLF$$
$$= (F_a \times CU/A) \times LLF \qquad (16\text{–}7)$$
$$= [(F \times LOF \times CU/A)] \times LLF$$

where E_i = Initial illuminance, fc (lux)
    F = Rated lamp lumens by published manufacturer
    F_a = Actual lumens produced by the lamps under field conditions
    E = Maintained illuminance, fc (lux)
    A = Area, sq ft (sq m)
    LOF = Lamp operating factor
    CU = Coefficient of utilization
    LLF = Light loss factor

Light loss factor Current practice for calculating light loss factor includes all causes affecting luminous output from field conditions to depreciation. More appropriately, these causes should be split into two groups: those that cause variations at the start of usage and those that cause luminous efficiency to drop

in time. The first group includes VF, TF, BF and PF which has been renamed the lamp operating factors (LOF). This applies to initial illuminance design calculations as discussed in section 16.3.1. The second group, which can be truly named as the light loss factor (LLF), shall include only the following factors.

$$LLF = LLD \times LDD \times LB \times RSDD \qquad (16\text{–}8)$$

where LLD = Lamp lumen depreciation factor
    LDD = Luminaire dirty depreciation factor
    LBO = Lamp burnout factor
    RSDD = Room surface dirty depreciation factor

Of these factors, only LLD and LDD are of importance. Lighting designers have often neglected LB and RSDD either as insignificant or unpredictable. When the environment is extremely dirty, such as in industries with heavy smokestacks, the designer should refer to the *IESNA Lighting Handbook* for recommended values on RSDD, otherwise, RSDD = 1 is always assumed. Unless the system is for large open spaces where group relamping* is the practice, the lamp burnout (BO) factor is also neglected, i.e., LBO = 1. Accordingly, the calculation for the LLF may be simplified to

$$LLF = LLD \times LDD \qquad (16\text{–}9)$$

The lamp lumen depreciation (LLD) value can be determined by selecting a percentage of the rated life of the lamp for which the lamps are expected to operate efficiently. Normally, the mean lumens or the lumens output at 70 percent of rated life is used. For example, if a fluorescent lamp is rated for 12,000 hours (based on 8-hours/start), the lumen's output at 8400 hours to

---

*Group relamping is a maintenance procedure when all lamps will be replaced at one time regardless of whether the lamps are still operating. This is usually done at 75 to 85 percent of the rated lamp life. The purpose is to save the cost of labor in hard-to-service locations.

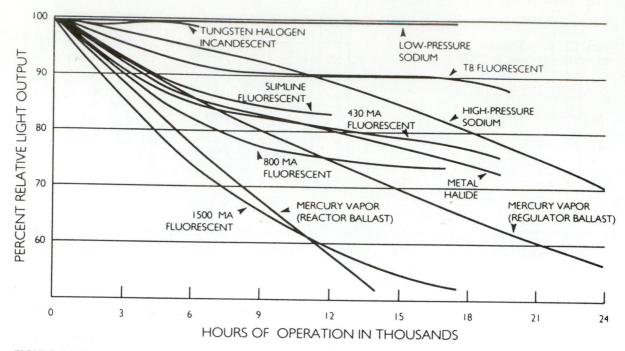

■ FIGURE 16–5

Lamp lumen depreciation (LLD) of typical light sources. LLD will vary between lamp sizes and models. Refer to manufacturer's publications for applicable values. (Courtesy: National Lighting Bureau, Washington DC.)

the rated lumen of the lamp in percent will determine the LLD factor. The LLD characteristics of various types of lamps are found in Chapter 15, but also can be found from Figure 16–5.

The luminaire dirty depreciation (LDD) values for various types of luminaires are shown in Figure 16–6. According to IESNA, there are six maintenance categories based on the construction of the luminaires of which five categories (I, II, III, IV, V) are included in this figure.

Atmospheric conditions are divided into five categories: very clean, clean, medium, dirty, and very dirty. For commercial and institutional lighting applications, Categories I and II are normally selected. The LDD also depends on how often the luminaire is being cleaned and on the maintenance program used. For design purposes, a six-month or one-year cleaning cycle is normally used.

Examples:

■ Determine the LLF for a lighting installation based on the following:

The luminaires are lay-in type fluorescent troffers with prismatic plastic lens. The lamps are 4-ft, T-8, 36-watt, rapid-start type (430 miliampere, light loading), and the rated life of the lamp is 20,000 hours. Building environment is very clean and the estimated cleaning cycle is 12 months after 10,000 hours of operation.

Answer: From Figure 16–5, the luminaire is top and bottom enclosed, the maintenance category of the luminaire is V. For a very clean environment and 12-month cleaning cycle, the estimated LDD of this installation is found to be 93 percent by interpolation.

From Figure 16–6 or from lamp manufacturer's data, the LLD is interpolated to be 81 percent, and

$$\text{LLF} = \text{LDD} \times \text{LLD} = 0.93 \times 0.81 = 0.75$$

■ Determine the LLF for another lighting installation based on the following:

The luminaires are suspension-mounted glass globes. The lamp is a general service type incandescent lamp with 1000-hr. rated life. The estimated normal usage is 4000 hours per year, and the scheduled cleaning cycle is 6 months. Building environment is medium (M).

Answer: From Figure 16–5, the maintenance category of the luminaire is V (totally enclosed). For medium environment and 6-month cleaning cycle, the LDD of this installation is 0.82. However, since the lamp has only a 1000 hour-rated life, and average lamp should have been burnt out and been replaced long before the scheduled cleaning time. It is reasonable to expect that the glass globe of the luminaire is cleaned each time

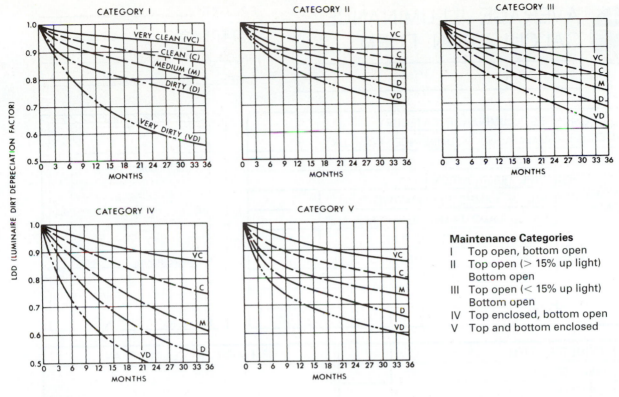

■ FIGURE 16–6
Luminaire dirt depreciation (LDD) factor for five luminaire categories. (Courtesy: Illuminating Engineering Society of North America, New York, NY; see IESHB for complete description of luminaire maintenance categories.)

the lamp is replaced. Therefore, the LDD value is taken at 3 months instead of 6 months, or LDD = 0.92.

From Figure 15–4, in Chapter 15, LLD at 100% of rated life is 0.85. Thus,

$$LLF = LDD \times LLD = 0.92 \times 0.85 = 0.78$$

■ Determine the LLF for the installation in the second example above with the lamp a tungsten halogen-type incandescent lamp.

Answer: The lumen's output of tungsten halogen type is nearly constant throughout its rated life, as demonstrated in Figure 16–5. Thus, LLD = 1.0, and the LLF for this installation is:

$$LLF = LDD \times LLD = 0.93 \times 1.0 = 0.93$$

16.3.3 Limitations and Applications of the Zonal Cavity Method

■ The illuminance calculated by the zonal cavity method is a representative average value only if luminaires are installed to meet manufacturers' recommended minimum mounting height and spacing. Even so, some variations are to be expected. Most likely the illuminance will be higher at the center of the space and lower near the walls.

■ The calculated illuminance is valid only under the conditions assumed for the calculation. It is entirely possible that the illuminance of a room with dark walnut paneling (15–20 percent reflectance) would be doubled if the wall were refinished in high-reflectance colors. This, of course, depends on the distribution characteristics of the luminaires. The effect of wall reflectance values is more pronounced for wide-distribution, direct or indirect lighting systems, including the diffuse (or direct-indirect) type. Ceiling reflectance is more critical for indirect systems and has little effect on narrow (spot) types of direct (down) lights. Calculated illuminance is for an assumed working plane, such as a desktop. Although design calculations normally select 30 inches as the working plane, this method will work well with the work plane at other heights.

■ The zonal cavity method may also be applied to determine the illuminance and luminance values on vertical surfaces by using the wall reflected radiation coefficient (WRRC), which can be

AVERAGE ILLUMINANCE CALCULATION FORM

PROJECT: _____

PROJECT NO: _____

CALCULATION BY: _____

DATE: _____ PAGE: _____

FOR ROOM	

ILLUMINANCE CRITERIA	IES ILLUMINANCE CATEGORY		
	MAINTAINED ILLUMINANCE, FC, (LUX)		
FIXTURE DATA	MFR/MODEL		
	TYPE DISTRIBUTION		
	NO. OF LAMPS PER FIXTURE		
	RATED LAMP LUMEN & WATTS/LAMP		
	LUMENS PER FIXTURE (LPF)		

ROOM DIMENSIONS	h		WIDTH(W)		LENGTH(L)	
ROOM CHARACTERS	h_{cc}		R_c		R_{w1}	
	h_{rc}		R_w		R_{w2}	
	h_{fc}		R_f		R_{w3}	

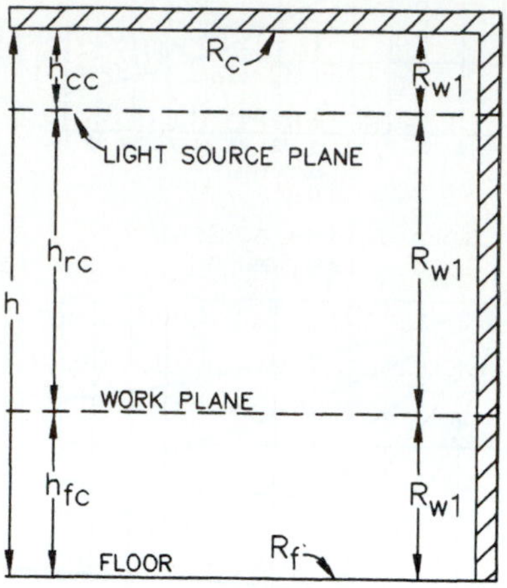

P	PERIMETER, FT(M):	
A	AREA, SF(SM):	
PAR	PERIMETER/AREA RATIO (P ÷ A)	
CCR	$2.5 \times$ PAR $\times h_{cc}$	
RCR	$2.5 \times$ PAR $\times h_{rc}$	
FCR	$2.5 \times$ PAR $\times h_{fc}$	
ρ_{cc}	FROM R_c & R_{w1} & CCR	
ρ_w	SAME AS R_w OR R_{w2}	
ρ_{fc}	FROM R_f & R_{w3} & FCR	
CU	FROM CU TABLE OF FIXTURE MFGR. INTERPOLATING BETWEEN RCR AND ρ_{cc}, ρ_w, ρ_{fc}	
LOF	BF – BALLAST FACTOR	
	VF – VOLTAGE FACTOR	
	OTHER	
LLF	LLD–LAMP LUMEN DEPREC.	
	LDD–LUMINAIRE DIRT DEPREC.	
	OTHER	

FLOOR OR CEILING PLAN

(USE SEPARATE DRAWINGS FOR ADDITIONAL LAYOUTS)

CALCULATIONS

MAINTAINED ILLUMINANCE

$$E = \frac{N \times (LPF \times LOF) \times CU \times LLF}{A}$$

INITIAL ILLUMINANCE

$$E_i = E \div LLF$$

CALCULATION & REMARKS:

* N – NUMBER OF FIXTURES

■ **FIGURE 16–7**

A typical average illuminance calculation form using the zonal cavity method. (Courtesy: William Tao & Associates, St. Louis, MO.)

found in the *IESNA Lighting Handbook*. Methods also are available for calculating illuminance in irregular-shaped spaces, illuminance affected by low partitions, and other applications. Refer to the *IES Lighting Education Manual* (ED-150) for these applications.

16.3.4 When to Use the Initial and Maintained Illuminances

Since light output of a lighting system depreciates with time, lighting system design must allow for such depreciation by a light-loss factor. The resulting illuminance is the maintained illuminance and is normally chosen as the design target, or target illuminance. This is not to say that the designer should overlook the initial performance of the system. In fact, many systems must be evaluated by its initial performance as well. For example, a professional sports lighting system which operates only infre-

quently, say only about 1000 hours a year using HD lamps with a rated life of 12,000 hours, must be evaluated by its initial performance. The initial illuminance criteria is also used to check the system performance during the commissioning of a new installation. There is no way to simulate the maintained or target illuminances ahead of time.

16.3.5 Zonal Cavity Design Form

A convenient design form developed by the authors is shown in Figure 16–7. It contains space for both input and output as well as a grid for a preliminary lighting layout or sketch. The space for lighting sketches is essential for comparing alternating designs in practice.

Work problem 16–1

Based on the fluorescent fixture shown in Figure 16–8, calculate the number of fixtures required for a classroom in a university.

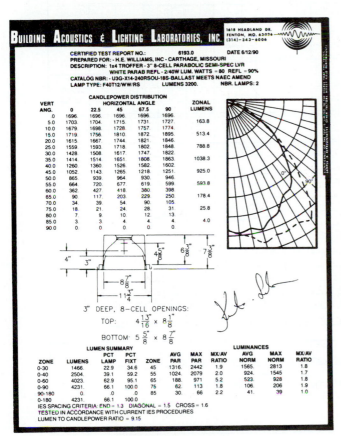

EFFICIENCY = 66.1%

Full Report Available

ZONAL CAVITY COEFFICIENTS OF UTILIZATION
EFFECTIVE FLOOR CAVITY REFLECTANCE = .20

CEILING	.80				.70				.50			.30		
WALL RCR	.70	.50	.30	.10	.70	.50	.30	.10	.50	.30	.10	.50	.30	.10
0	.79	.79	.79	.79	.77	.77	.77	.77	.73	.73	.73	.70	.70	.70
1	.74	.72	.70	.68	.72	.70	.68	.67	.68	.66	.65	.65	.64	.63
2	.69	.65	.62	.59	.68	.64	.61	.58	.62	.59	.57	.60	.58	.56
3	.64	.59	.55	.51	.63	.58	.54	.51	.56	.53	.50	.54	.52	.49
4	.60	.53	.49	.45	.58	.52	.48	.45	.51	.47	.44	.49	.46	.44
5	.55	.48	.43	.39	.54	.47	.43	.39	.46	.42	.39	.45	.41	.38
6	.51	.43	.38	.35	.50	.43	.38	.34	.42	.37	.34	.41	.37	.34
7	.47	.39	.34	.30	.46	.39	.34	.30	.38	.33	.30	.37	.33	.30
8	.43	.35	.30	.26	.42	.35	.30	.26	.34	.29	.26	.33	.29	.26
9	.40	.31	.26	.23	.39	.31	.26	.23	.30	.26	.23	.30	.25	.23
10	.37	.28	.23	.20	.36	.28	.23	.20	.27	.23	.20	.27	.23	.20

VISUAL COMFORT PROBABILITY
Reflectances = 80/50/20
Work Plane Illumination = 100 fc.

Room		Lengthwise				Crosswise			
		Mounting Height							
W	L	8.5	10	13	16	8.5	10	13	16
20	20	80	75	71	75	85	81	70	73
30	30	85	80	74	69	81	76	70	66
30	60	88	84	79	74	84	80	74	68
60	30	89	85	79	72	86	82	76	71
60	60	90	87	82	76	88	84	79	72

Calculated in accordance with RQQ2 1972

RCR

$$^{†}\,RCR = \frac{H \times 5 \times (L + W)}{L \times W}$$

H = DISTANCE FROM LUMINAIRE TO ILLUMINATED PLANE (TASK)
L = AREA LENGTH
W = AREA WIDTH

$$^{†}\,RCR = 2.5 \times PAR \times h_{re}$$

■ FIGURE 16–8
Photometric report of a fluorescent fixture. (Courtesy: H.E. Williams, Inc. Carthage, MO.)

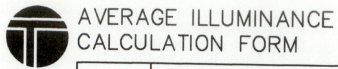

AVERAGE ILLUMINANCE CALCULATION FORM

FOR ROOM	DINING ROOM

ILLUMINANCE CRITERIA	IES ILLUMINANCE CATEGORY					E	
	MAINTAINED ILLUMINANCE, FC, (LUX)					50	
FIXTURE DATA	MFR/MODEL	WILLIAMS/SERIES U3G					
	TYPE DISTRIBUTION	DIRECT					
	NO. OF LAMPS PER FIXTURE				2		
	RATED LAMP LUMEN & WATTS/LAMP				3200/40		
	LUMENS PER FIXTURE (LPF)			6400Lms			
ROOM DIMENSIONS	h	9.5	WIDTH(W)	22	LENGTH(L)	40	
ROOM CHARACTERS	h_{cc}	0	R_C	0.7	R_{w1}	0.4	
	h_{rc}	7	R_W	0.4	R_{w2}	0.4	
	h_{fc}	2.5	R_f	0.2	R_{w3}	0.4	

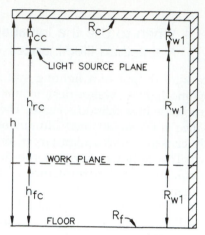

P	PERIMETER, FT(M):		124
A	AREA, SF(SM):		880
PAR	PERIMETER/AREA RATIO (P ÷ A)		0.14
CCR	2.5 x PAR x h_{cc}		0
RCR	2.5 x PAR x h_{rc}		2.45
FCR	2.5 x PAR x h_{fc}		0.80
ρ_{cc}	FROM R_c & R_{w1}		0.70
ρ_w	SAME AS R_w OR R_{w2}		0.40
ρ_{fc}	FROM R_f & R_{w3}		0.20
CU	FROM CU TABLE OF FIXTURE MFGR. INTERPOLATING BETWEEN RCR AND ρ_{cc}, ρ_w, ρ_{fc}		0.59
LOF	BF – BALLAST FACTOR	0.95	0.95
	VF – VOLTAGE FACTOR	1.0	
	OTHER	–	
LLF	LLD–LAMP LUMEN DEPREC.	0.80	0.70
	LDD–LUMINAIRE DIRT DEPREC.	0.80	
	OTHER	–	

FLOOR OR CEILING PLAN

40'

22'

(USE SEPARATE DRAWINGS FOR ADDITIONAL LAYOUTS)

CALCULATIONS

INITIAL ILLUMINANCE

$$E_i = \frac{(N \times LPF \times LOF) \times LLF \times CU}{A}$$

MAINTAINED ILLUMINANCE

$$E = E_i \times LLF$$

CALCULATION & REMARKS:

$$E_i = \frac{(N \times 6400 \times 0.95) \times 0.59 \times 0.70}{880} = 50 \text{ Lms.}$$

N = 17.5 FIXTURES (USE 18) @ 2 ROWS

* N – NUMBER OF FIXTURES

■ **ANSWERS TO WORK PROBLEM 16–1**

The design criteria and room data are as follows:

Room dimensions	40 ft × 22 ft
Ceiling height	8 to 11 ft, avg. 9.5 ft
Reflectances	$R_c = 80\%$, $R_W = 40\%$ avg., $R_f = 20\%$
Lamps	2 × 40 watt, T-12, fluorescent, 3,200 lumens/lamp
LLF	70%

Answer

1. Select the illuminance, $E = 50$ fc.
2. Calculate P, A, PAR, CCR, RCR and FCR
3. Determine $\rho_{cc} = R_c$; ρ_w; R_w; $\rho_{fc} = R_f$
4. Determine CU (see calculation form).
5. Estimate LLF = 0.70
6. Calculate number of fixtures, N = 17.5 (use 18)

Work problem 16–2

Determine the lamp size for a single incandescent lamp fixture with a diffused glass sphere (Luminaire Type 1) to be installed at the center of a circular dining room. Information on the room is as follows:

Room dimensions	12-ft diameter, 9-ft ceiling
Illumination maintained	15 fc
Room finishes	$R_c = 80\%$, $R_w = 70\%$, $R_f = 20\%$
Luminaire mounting	7 ft above floor

Answer

1. Use the Average Illuminance Calculation Form:
   - Enter maintenance (illuminance: 15 fc)
   - Enter fixture and room data
   - Calculate perimeter, $P = \pi D = 37.7$ ft
   - Calculate area, $A = \pi r^2 = 113$ sq ft
   - Calculate: PAR, CCR, RCR, FCR
   - Determine ρ_{cc} from Figure 16–4 with $R_c = .80$, $R_w = .70$, and CCR = 1.65; ρ_{cc} is interpolated to be = .66.
   - Determine CU from photometric data of Type 1 (Figure 16–4): ρ_{cc} of .66 (between 70 percent and 50 percent), RCR of 3.7 (between 3 and 4), and wall reflectance of 70 percent extrapolated beyond 50 percent; the interpolated and extrapolated value of CU is 0.47 (see calculation form).
   - Estimate the LLF to be 0.7 (LLD = 0.7, LDD = 1.0).

   - Calculate the lumens per fixture required = 5152 lumens
2. The lumen output of a 300-watt, ps-25 general-service type of incandescent lamp, found in the *IES Handbook* or in the lamp manufacturer data, is 6100 lumens. The output is about 20 percent higher than required. However, it would be appropriate to use a dimmer to vary the illuminance under different operating modes and to extend the life of the lamp.
3. If the wall were painted or covered with 30 percent wall covering, the ρ_{cc} will be reduced to 0.43 and the CU reduced to about 0.3. As a result, the lamp has to be enlarged to 500 watts, an increase of 66 percent (CU = 0.47 vs. 0.30). The impact of wall reflectance on illumination design, particularly in low-RCR spaces, is demonstrated.

16.4 POINT METHOD

The point method, also referred to as the point-by-point method, is based on the definition that the illuminance on a surface perpendicular to the light beam incident on it is inversely proportional to the distance from the light source to the surface (Equation 14–3). If the surface is not perpendicular to the light beam, then a cosine factor should be included (Equation 14–4). The latter equation can be rearranged as follows:

16.4.1 Initial Illuminance

The component of the initial illuminance on a horizontal plane is:

$$E_{ih} = \frac{I_a \times \cos \theta}{D^2} = \frac{I_a \times \cos \beta}{D^2} \qquad (16\text{--}10)$$
$$= \frac{I_a \times \cos^3 \theta}{H^3}$$

Refer to Figure 16–9. The component of the initial illuminance on a vertical plane is:

$$E_{iv} = \frac{I_a \times \sin \theta}{D^2} = \frac{I_a \times \cos \beta}{D^2} \qquad (16\text{--}11)$$
$$= \frac{I_a \times \cos^2 \theta \times \sin \theta}{H^2}$$

The intensity (I) in any direction from a luminaire is normally presented in a polar diagram in the photometric report provided by the luminaire manufacturer. The intensity values in candela (cd) are the ac-

AVERAGE ILLUMINANCE CALCULATION FORM

PROJECT: __RESIDENCE__

PROJECT NO: __WORK PROB. 16-2__

CALCULATION BY: __WKT__

DATE: __2/17__   PAGE: __1__

FOR ROOM	DINING ROOM

ILLUMINANCE CRITERIA	IES ILLUMINANCE CATEGORY					C
	MAINTAINED ILLUMINANCE, FC, (LUX)					15
FIXTURE DATA	MFR/MODEL	TYPE I, FIGURE 16-4				
	TYPE DISTRIBUTION	GENERAL DIFFUSE				
	NO. OF LAMPS PER FIXTURE				1	
	RATED LAMP LUMEN & WATTS/LAMP				TBD	
	LUMENS PER FIXTURE (LPF)				TBD	
ROOM DIMENSIONS	h	9	WIDTH(W)	–	LENGTH(L)	–
ROOM CHARACTERS	h_{cc}	2	R_c	0.8	R_{w1}	0.7
	h_{rc}	4.5	R_w	0.7	R_{w2}	0.7
	h_{fc}	4.5	R_f	0.2	R_{w3}	0.7

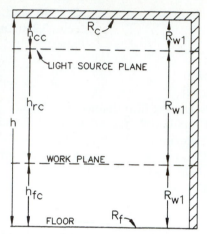

P	PERIMETER, FT(M):	37.7
A	AREA, SF(SM):	113
PAR	PERIMETER/AREA RATIO (P ÷ A)	0.33
CCR	2.5 x PAR x h_{cc}	1.65
RCR	2.5 x PAR x h_{rc}	3.7
FCR	2.5 x PAR x h_{fc}	2.1
ρ_{cc}	FROM R_c & R_{w1} & CCR	0.66
ρ_w	SAME AS R_w OR R_{w2}	0.70
ρ_{fc}	FROM R_f & R_{w3} & FCR	0.20
CU	FROM CU TABLE OF FIXTURE MFGR. INTERPOLATING BETWEEN RCR AND ρ_{cc}, ρ_w, ρ_{fc}	0.47

FLOOR OR CEILING PLAN

D = 12 ft

(USE SEPARATE DRAWINGS FOR ADDITIONAL LAYOUTS)

LOF	BF – BALLAST FACTOR	1.0	1.0
	VF – VOLTAGE FACTOR	1.0	
	OTHER	–	
LLF	LLD–LAMP LUMEN DEPREC.	0.7	0.7
	LDD–LUMINAIRE DIRT DEPREC.	1.0	
	OTHER	–	

CALCULATIONS

INITIAL ILLUMINANCE
$$E_i = \frac{(N \times LPF \times LOF) \times CU \times LLF}{A}$$

MAINTAINED ILLUMINANCE
$$E = E_i \times LLF$$

CALCULATION & REMARKS:
$$E_i = \frac{(1 \times 1.0 \times LPF) \times 0.47 \times 0.7}{113} = 15 \text{ Lms.}$$

LPF = 5152 Lumens

USE (1) PS-30 300W, GENERAL/SERVICE LAMP RATED FOR 6100 LUMENS

* N – NUMBER OF FIXTURES

■ **ANSWERS TO WORK PROBLEM 16-2**

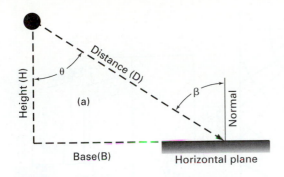

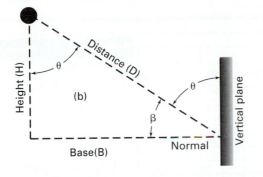

■ **FIGURE 16–9**
Trigonometric relationships applicable to the inverse square law. Light incident on (a) horizontal plane and (b) vertical plane.

tual intensity (I_a) of the luminaire including the temperature factor (TF), ballast factor (BF) and position factor (PF) except the voltage factor (VF). Thus, an LOF factor need not be included in the initial illuminance calculations when using the values from the photometric report.

16.4.2 Maintained illuminance

The component of the maintained illuminance on a horizontal plane is:

$$E_{mh} = E_{ih} \times LLF = \frac{I \times \cos^3 \times \theta}{H^2} \times LLF \quad (16\text{–}12)$$

The component of the maintained illuminance on a vertical plane is:

$$E_{mh} = E_{ih} \times LLF$$
$$= \frac{I \times \cos^2 \theta \times \sin \theta}{H^2} \times LLF \quad (16\text{–}13)$$

16.4.3 Typical Photometric Report

Photometric reports for point sources, such as down-lights, wall washers, and floodlights, usually include

a candlepower distribution curve in polar coordinates. In addition, the manufacturer may provide a handy chart depicting the illuminance levels at various locations (planes) away from the luminaire or group of luminaires. Figure 16–10 illustrates the construction and photometric report of a typical point source luminaire.

Example Based on the photometric report shown in Figure 16–10, calculate the illuminance on the wall at 60°, 45° and 30° using zero as the nadir (the point directly below the luminaire). The luminaire is installed 4 ft from the wall. The lamp is 150 watt R-40, and is rated for 1900 lumens.

From Equation 16–11, for the illuminance on a vertical plane,

$$E_{iv} = (I \times \sin \theta)/D^2$$

The calculations of illuminance can best be presented in tabular format, accompanied by a diagram (Figure 16–11):

ρ	θ	sin θ	D = 4/sin θ	I	E = I × sin θ/D²
ρ_1	60	0.87	$D_1 = 4.6$	$I_1 = 150$	$E_1 = \dfrac{150 \times .87}{(4.6)^2} = 6.2$
ρ_2	45	0.71	$D_2 = 5.6$	$I_2 = 250$	$E_2 = \dfrac{250 \times 0.71}{(5.6)^2} = 5.7$
ρ_3	30	0.50	$D_3 = 8.0$	$I_3 = 700$	$E_3 = \dfrac{700 \times 0.5}{8^2} = 5.5$

The foregoing example illustrates the steps used for determining the illuminance on a surface by the point method. To avoid this tedious work, the manufacturer often provides precalculated illuminance levels in tabulated format, as shown in Figure 16–10. For sports lighting applications, the precalculated illuminance levels may be given in the form of isolux (equal-illuminance) contours.

The illuminances calculated in the example are initial values. An LLF is applied to determine the maintained values.

The point method is commonly used for the following:

- Manual calculation of the illuminance at a point on a surface from single luminaire.
- Manual calculation of the illuminance at a point on a surface from multiple luminaires, provided that the calculations have taken into consideration the intricate angles of incidence between the luminaires and the point.
- Computerized calculation of the illuminance of a point on any surface from multiple sources at different angles, including the reflected flux from

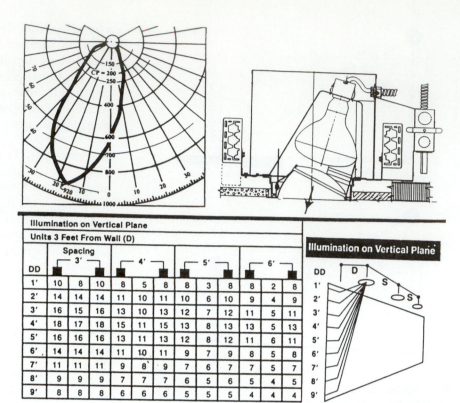

■ FIGURE 16–10
Typical photometric report of a luminaire with an asymmetrical candlepower distribution pattern. The report includes a precalculated illuminance table for luminaires shown at the indicated locations. (Courtesy: Halo Lighting Division, Cooper Industries, Elk Grove Village, IL.)

Illumination on Vertical Plane

Units 3 Feet From Wall (D)

DD	Spacing 3'			4'			5'			6'			DD
1'	10	8	10	8	5	8	8	3	8	8	2	8	1'
2'	14	14	14	11	10	11	10	6	10	9	4	9	2'
3'	16	15	16	13	10	13	12	7	12	11	5	11	3'
4'	18	17	18	15	11	15	13	8	13	13	5	13	4'
5'	16	16	16	13	11	13	12	8	12	11	6	11	5'
6'	14	14	14	11	10	11	9	7	9	8	5	8	6'
7'	11	11	11	9	8	9	7	6	7	7	5	7	7'
8'	9	9	9	7	7	7	6	5	6	5	4	5	8'
9'	8	8	8	6	6	6	5	5	5	4	4	4	9'

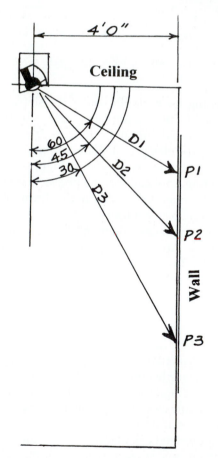

■ FIGURE 16–11
Section through the room showing location of the fixture in front of the wall.

the interior surfaces. The algorithm of the calculations is complex and includes thousands of calculations; thus, it can be done only by a computer. Computer programs are available from the Illuminating Engineering Society of North America, Consulting Engineers, lighting designers and most of the leading lighting fixture manufacturers.

■ Computerized calculations of sports lighting. The point method is the only method applicable to sports lighting design, which demands the accurate prediction of illuminances on the horizontal and vertical planes, as well as the uniformity of ratios. The design of sports lighting is very specialized.

Work problem 16–3

The following design is for a commercial carpet showroom that display carpets on the floor. From IES categories, it is determined that the medium value of category D is proper, i.e., 30 footcandles. The preliminary design determined to use two Type A indirect distribution luminaires (Figure 16–4, luminaire type 2) and six Type B direct distribution luminaires (Figure 16–4, luminaire type 8), as shown in the following sketch:

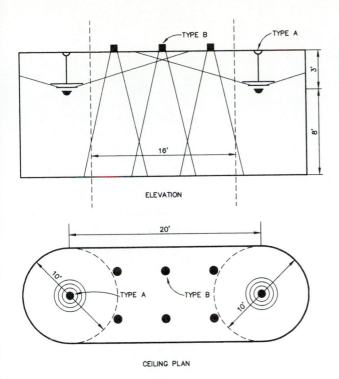

ELEVATION

16'

3'

8'

TYPE B   TYPE A

20'

10'

10'

TYPE A   TYPE B

CEILING PLAN

Determine the lamp lumen per fixture required, based on the following information:

1. Room dimensions: A rectangular room with semicircular ends, as shown in the drawing above.
2. Reflectances:
   - Ceiling, 80%
   - Wall, 50% average
   - Floor, 20% without carpets in place.
3. This work problem is designed to illustrate a number of variables encountered in lighting design applications:
   - The working plan is the floor, not the standard 30" level at desktops.
   - Although it is assumed that the floor reflectance is 20 percent, the actual floor reflectance could vary considerably depending on the reflectance of the carpets displayed in various parts of the room.
   - The room is not rectangular, but rather is a composite of a rectangular and a semicircular shape.
   - There are two types of luminaire to be used, each covering a different part of the room.
4. There is a number of approaches to take to tackle this design. One suggested way is to calculate the illuminance produced by the Type A and Type B luminaires separately and then superimpose the illuminance produced by both.

QUESTIONS

16.1 Name some illumination quality factors which can be expressed in numerical values.

16.2 What are the factors determining the selection of illuminance?

16.3 Name the steps recommended by IESNA for the selection of illuminance.

16.4 Select the maintained illuminance for a sewing factory.

16.5 State the reasons why initial illuminance has important design applications.

16.6 What is the recommended illuminance for professional football in an indoor stadium?

16.7 What is the lumen's output for a general service incandescent lamp if the voltage at the lamp socket is 5 percent below its rated voltage?

16.8 Is ballast factor always lower than 1.0?

16.9 What are two most important factors which make up the Lamp Operating Factor (LOF)?

16.10 What is the ceiling cavity ratio (CCR) for a ceiling recess-mounted fluorescent fixture? What is the effective ceiling reflectance if the base ceiling reflectance is 80 percent.

16.11 What is the simplified approach in zonal cavity method calculations? And what variance can be expected?

16.12 Determine the coefficient of utilization of a type 2 indirect distribution incandescent luminaire installed in a room with RCR = 2, $\rho_{cc} = 80\%$, $\rho_w = 50\%$, $\rho_{fc} = 20\%$?

16.13 For the same luminaire in 16.12, but is installed in a room with RCR = 10, $\rho_{cc} = 30\%$, $\rho_w = 30\%$ and $\rho_{fc} = 20\%$, what is the CU?

16.14 What is the maintenance category for a luminaire with open top and bottom and with 65 percent up-light?

16.15 If the luminaire is installed in an air-conditioned office building in a suburban environment, what is the appropriate LDD factor with a 12-month cleaning cycle?

16.16 A spotlight is aimed at the bulletin board on the wall. The light is 5 ft in front of the board. The center of the board is 10 ft below the light. If the maximum candlepower of the light is 10,000 candela based on rated lamp lumen, what is the illuminance at the center of the board?

16.17 What is the illuminance 4 feet below the ceiling for a wall washing installation, using the lighting fixtures illustrated in Figure 16–10? The fixtures are mounted 3 feet from the wall and spaced on 4-ft centers.

16.18 What are the principal applications of the point method?

17
LIGHTING DESIGN

LIGHT CAN AFFECT HUMAN BEHAVIOR, ENHANCE OR degrade a designed environment, improve or hinder human productivity, and light affects every minute of our lives.

Light can alter the character of a space, making it appear more spacious by lighting its walls, or more intimate by creating a barrier. Light can change the mood of people by the combination of color, brightness, contrast or simply by its presence. Light can originate from the ceiling, from the walls or even from the floor. Light can be displayed as a prominent feature in space or can be totally hidden through reflection or refraction techniques. Light is most versatile among all design elements.

Lighting design is an art as well as a science. The scientific aspects of lighting and lighting calculations have been introduced in Chapters 14, 15 and 16. This chapter will deal with the design process, with emphasis on the artistic aspect of lighting design.

17.1 LIGHTING DESIGN PROCESS

In the manner similar to that of other design processes, lighting design includes an orderly progression of design steps. The process may be divided into seven phases; the last three phases are actually an extension of the design phases.

1. Programming and conceptual formulation (P&C)
2. Schematic design or preliminary design (SC)
3. Design development (DD)
4. Construction documents (CD)
5. Bid negotiation (BN)
6. Construction administration (CA) and
7. Post-occupation evaluation (PE)

As always, there is an overlap between these phases and often in simple projects, several phases may be consolidated. *The Lighting Handbook* contains a complete description of the design issues to be considered. A flow diagram of the design process is shown in Figure 17–1. Notice that the flow diagram includes a return path between all phases, which shows that design is not always a one-way street.

17.2 DESIGN CONSIDERATIONS

17.2.1 Visual Performance Desired

A lighting design is a plan to achieve the visual performance desired for the visual task(s) for which the associated lighting system is to be used. Lighting also plays an important role in projecting spatial perception, activity, and a general atmosphere. For example, to create an impression of spaciousness, we adjust the luminance and distribution; of relaxation, we adjust

483

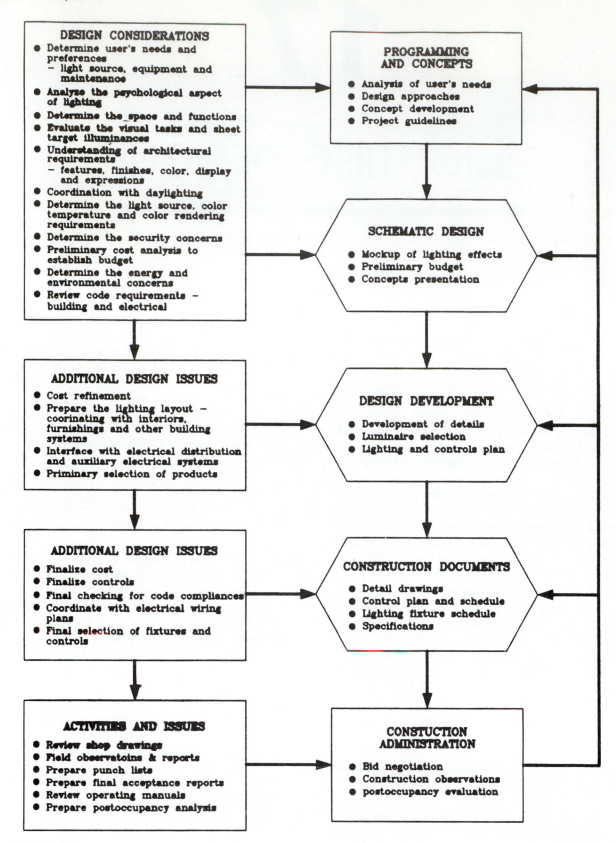

■ FIGURE 17–1

Lighting Design Flow Diagram and Check List. (Based on IESNA recommended procedure.)

degree of uniformity, distribution, and color; of privacy, we adjust luminance, degree of uniformity, and distribution; and of aesthetics, we adjust spatial relations, form, pattern and everything else.

17.2.2 Illuminance Selection

Chapter 16 gives calculations of illumination for visual task categories and the criteria to select the illuminance taking into account visual display, the age of the observers, the speed and accuracy required for the task, and the reflectance of the task and its background. Further refinement should take into account other influencing factors, such as the duration of the task, the interfacing of adjacent tasks in the space, and other spatial, architectural, structural, and mechanical relations.

Specific luminance ratios for various applications such as offices, educational facilities, institutions, industrial areas, and residences can be found in the *IES Lighting Handbook*. In general, a higher luminance ratio (contrast) is desired between the task and its immediate background for better visibility and visual acuity. (See Figure 17–2). However, a low luminance ratio is actually more desirable for visual comfort. Good lighting design tries to satisfy these conflicting criteria with appropriate solutions.

17.2.3 Visual Comfort

Visual comfort is achieved when there is no prolonged visual sensation due to excessively high luminances within the visual field. One measure of visual comfort is the *visual comfort probability* (VCP) of a lighted space due to either daylight or an interior lighting system. A VCP of 70 or higher is considered acceptable for visual comfort. Reflected glare can cause a loss of visibility. A lighting system with low reflected glare can be designed with the *equivalent spherical illumination* (ESI) method which requires the use of specialized computer software. In general, the reflected glare can be minimized by providing a proper surface texture for the task and by controlling the orientation of the luminous flux.

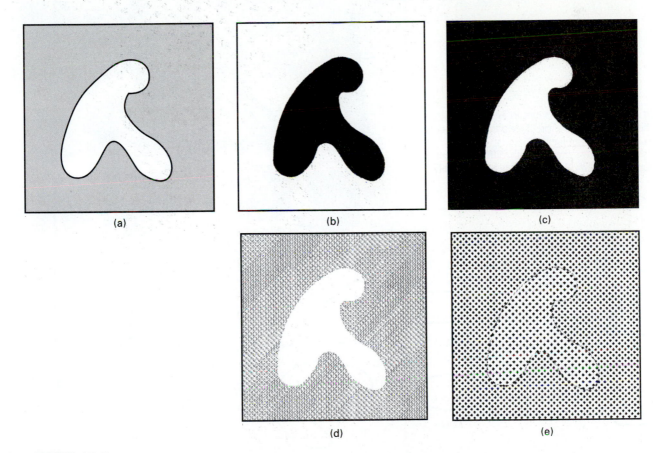

(a)   (b)   (c)

(d)   (e)

■ **FIGURE 17–2**
An object (task) is seen due to its color, shape and luminance against its background. When there is more contrast in luminance as in (b) or (c), the better the acuity. With less contrast as in (a) and (d), and less difference in texture (e), the poorer is the acuity.

17.2.4 Selection of Lighting System

Six general classes of lighting systems defined by IES are shown in Figure 15–16. A specific lighting system is selected based on the tasks to be performed, lighting distribution within the space, and orientation needed to provide visual comfort and acuity. Figure 17–3 illustrates the effect of the orientation a lighting system on 3-dimensional tasks and the resulting shadows.

17.2.5 Color

For a better understanding of color, chromaticity, color rendering, and the use of color, the lighting designer should refer to Section 5 of the reference volume of the *IES Handbook.* The terms chromaticity and color rendering are often mis-used. In simple terms, *chromaticity* refers to the color appearance of a light source, such as its color temperature, and *color rendering* refers to the ability of the light source to render colors of objects as one would expect them to appear at the same color temperature.

17.2.6 Lighting Layout

Perhaps the most important and most interesting aspect of lighting design is the layout of the lighting system in relation with the architectural expression of the space. This aspect of design will be emphasized in this chapter.

17.3 ARCHITECTURAL LIGHTING

17.3.1 Spatial Relations

Lighting must be coordinated with all other architectural elements in a space. Lighting elements may be concealed or exposed, prominently displayed or integrated with others. Basic lighting elements start with two-dimensional geometric shapes to three-dimensional forms. (See Figure 17–4 for the basic forms.) In most design applications, lighting fixtures are located on the ceiling plane either recess, regress, surface or suspension mounted. Lighting fix-

(a)   (b)

■ **FIGURE 17–3**
Light orientation is important in lighting design;
(a) harsh shadows are produced on the objects with uni-directional lighting.
(b) shadows are softened with diffused lighting.
(Courtesy: General Electric Co., Nela Park, OH)

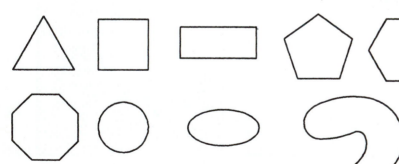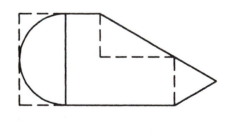

■ **FIGURE 17–4**
Spaces to be lighted may be any one of the basic geometric forms, such as squares, triangles, circles, etc. but also in irregular forms, such as free forms or a compound of several basic forms.

tures are also wall mounted or, in some designs, floor mounted.

17.3.2 Lighting Expressions

According to William Lam, the expression of a lighting system may be described as any of the following:

1. *Neutral.* The lighting system is de-emphasized and the lighting elements do not draw the special attention of the occupants of or visitors to the space.
2. *Expressive.* The lighting system is designed to harmonize, supplement, or enhance the architectural expressions in the space.
3. *Dominant.* The lighting system dominates the space, overpowering all the other elements.
4. *Confused (Disorganized).* The lighting system is disorderly, either in its configuration or in its relation to other elements in the space. In this case, lighting becomes a liability to the entire design.

For example, a small square luminaire installed at the center of the ceiling of a square room is obviously neutral, a large square could be expressive, and a square 2/3 of ceiling dimensions would be dominant. These designs are illustrated in Figure 17–5.

Although the classification of expressions is somewhat subjective, it is not difficult to reach consensus. No line can be drawn between these four expressions (neutral, expressive, dominant and confused). A design slightly different in dimensions, form, or location could easily be switched from one expression to the other. However, it is not difficult to agree on the expressions identified in Figure 17–6.

Figures 17–7, 17–8, and 17–9 illustrate lighting layouts for various space configurations. Some layouts can easily be identified as representing one expression class, while others are difficult to judge. A layout may be considered as one type, but will be totally different for another space occupancy. For example, a random layout as in Figure 17–8 (l) is considered "disorganized" in a classroom but will be

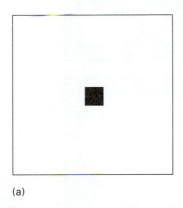 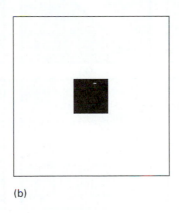 

(a)   (b)   (c)

■ **FIGURE 17–5**
Lighting expressions of lighting designs with a square fixture at the center of a square room: (a) neutral; (b) expressive; (c) dominant.

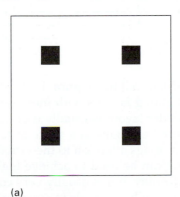 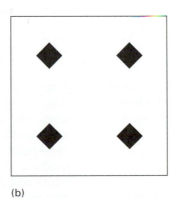 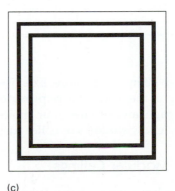

(a)   (b)   (c)

■ **FIGURE 17–6**
Lighting expression of sample lighting designs: (a) neutral; (b) confused; (c) dominant.

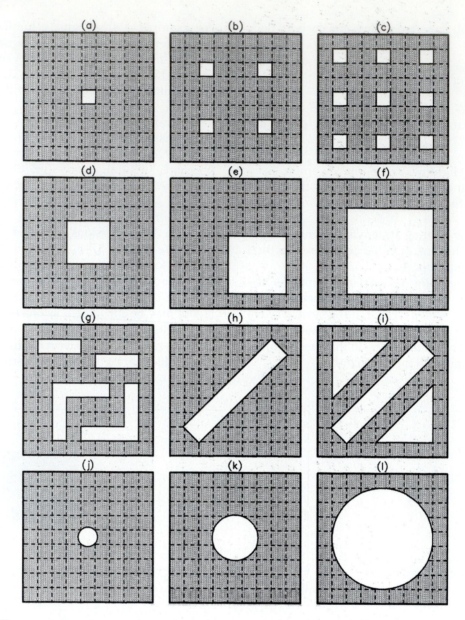

"expressive" in a retail store. Exercise your judgment by filling in the blank spaces provided at the bottom of each page, and discuss your choices with your instructor, fellow students or colleagues.

17.3.3 Modular Ceilings

Although less commonly used due to energy and cost considerations, some applications of lighting layouts require a modular ceiling. Figure 17–10 illustrates some typical lighting layouts with modular lighting units. The modular layout is usually used to supplement architectural elements in a space or to create a focal point that brings attention to the visual field. A modular ceiling can be used to achieve high illuminance and uniformity, but is usually very costly; it is also more aesthetically desirable to cover only a portion of the room.

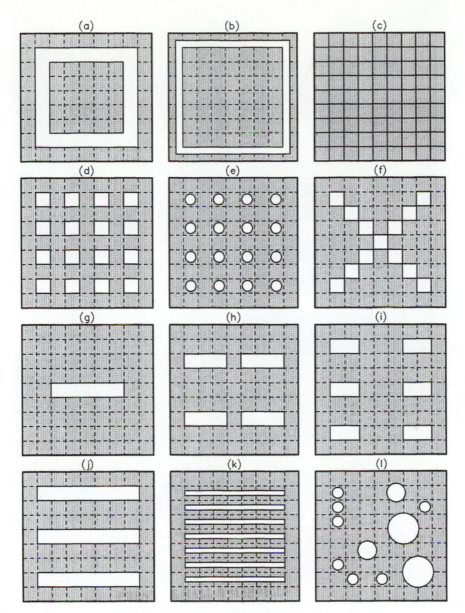

■ **FIGURE 17–8**
Exercise in evaluating lighting expressions in terms of form, pattern, orientation and locations.

(a) _____   (b) _____   (c) _____
(d) _____   (e) _____   (f) _____
(g) _____   (h) _____   (i) _____
(j) _____   (k) _____   (l) _____

17.3.4 Perspective

While geometric pattern of lighting fixtures on the ceiling is the starting point in most lighting design considerations, the designer must realize that lighting patterns will not be viewed in a two-dimensional plane. In fact, all architectural elements in a space, including lighting elements, are viewed in perspective. Often, a logical or attractive geometric pattern on the reflected ceiling plane may create a confused expression in a space. Figure 17–11 illustrates this problem.

17.4 LIGHTING LAYOUT

The number of luminaires determined by calculations of illuminance offers a basis for a lighting layout. Whether the objective is to provide uniform illumination throughout the entire space or only through a

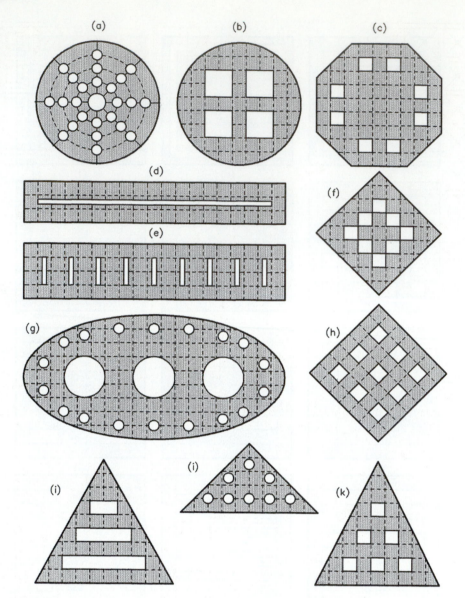

■ **FIGURE 17–9**
Exercise in evaluating lighting expressions in terms of form, pattern, orientation and locations.

(a) _____   (b) _____   (c) _____
(d) _____   (e) _____   (f) _____
(g) _____   (h) _____   (i) _____
(j) _____   (k) _____

portion of the space, a lighting layout can be developed to meet the objective. Two major criteria are used to achieve uniform illumination on the working plane.

17.4.1 Location of Luminaires

The working area to be illuminated is divided into as many unit areas as the number of luminaires to be installed. For example, if calculations show that the space requires eight luminaires, then the space is divided into eight unit areas, and one luminaire is located at the center of each unit area. The divided areas should be as symmetrical to the luminaire distribution pattern as is practical. Figures 17–11(a) and (b) illustrate the principle involved. If the objective is to illuminate only a portion of the space, then the working area to be illuminated is still divided into unit areas accordingly. Figures 17–11(c) and (d) illustrate this principle.

The general layout principle applies to all luminaires with symmetrical flux distribution. There

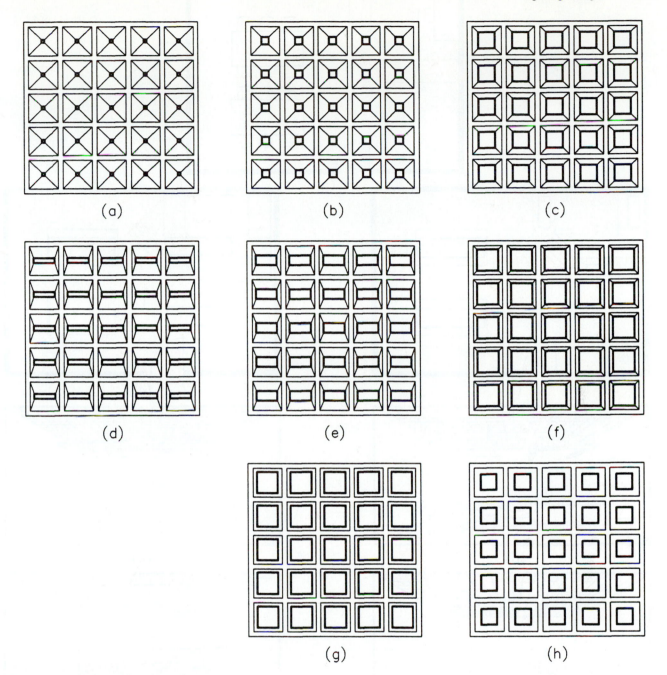

■ **FIGURE 17–10**
Representative modular ceiling lighting layouts. Modular ceiling does not have to cover an entire room. In fact, it is more aesthetically pleasing and economical to cover a portion of a room, or only the task area, when extremely low glare and uniformity are desired.

are, however, luminaires designed for asymmetrical distribution where the luminous flux (lumens) is stronger (higher) in one direction than the other, such as wall washers, or adjustable spot lights. In such cases, the location of luminaires is governed by their photometric distribution.

17.4.2 Spacing-to-Mounting Height Ratio

For uniform illumination, the spacing of the luminaire must not exceed the spacing-to-mounting height ratio (SMHR) recommended by the manufacturer. The problem associated with a design that ex-

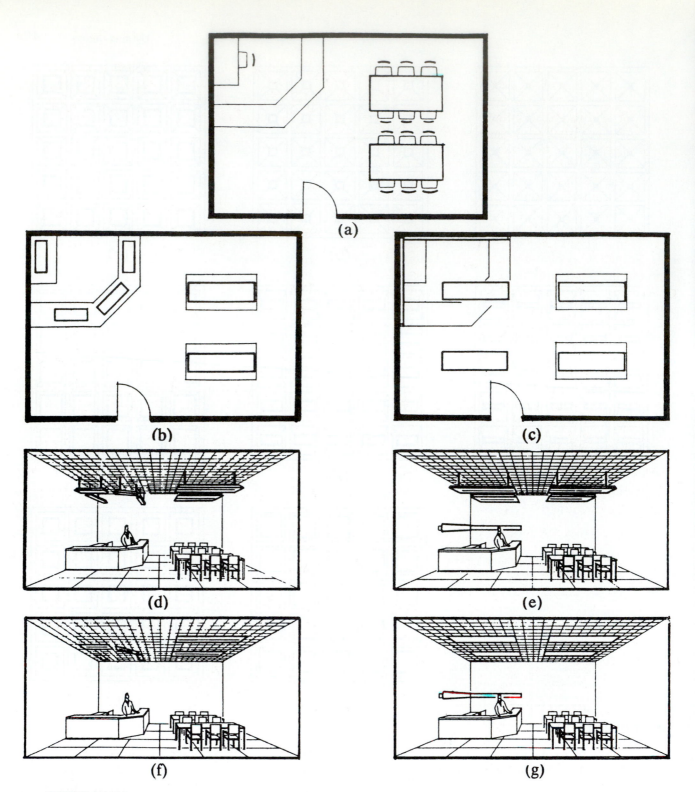

■ FIGURE 17–11

Evaluation of alternative lighting design in perspectives. (a) An architectural floor plan of a small reading room showing two reading tables, checkout counter and librarian's desk against the wall. (b) Design #1: Lighting layout follows furniture plan. (c) Design #2: Lighting layout follows room shape. (d) Perspective view of design #1 with suspension-mounted luminaires. The ceiling appears to be confused, cluttered (disorganized). (e) Perspective view of design #2 with suspended luminaires. The ceiling appears to be orderly, and too regimented (neutral). However, bracket mounted wall luminaire against the wall over the librarian's desk provides a luminous relief and practical solution to illumination need. (f) Perspective view of design #1 with ceiling recess-mounted luminaires softening ceiling clutter somewhat (still disorganized). (g) Perspective view of design #2 with ceiling recess mounted luminaires and bracket-mounted luminaires appear to be pleasant, neutral and expressive (a better solution). There are, of course, many others to be considered.

ceeds the recommended SMHR is illustrated in Figures 17–11 (e) and (f) which show a degree of darkness (lower illuminance) between the luminaires. In general, for a direct distribution type luminaire, the SMHR can be visually estimated by not exceeding the angles between the maximum and 50-percent values of the candlepower distribution curve provided in the photometric data. The SMHR is greater (more spread out) for semi-indirect and totally indirect luminaires. Narrow-beam direct distribution luminaires are normally selected to highlight a localized area, and selecting them to light a large space uniformly is improper.

Table 17–1 provides general guidelines for determining the maximum SMHR of different classes of luminaires. For example, the SMHR of a general diffuse class of luminaire should not exceed 1.0, and that of an indirect distribution luminaire can be approximately 1.25. In design applications, if the lumens output and class of distribution of luminaires are properly selected, the actual SMHR for most highly demanding visual performance spaces, such as classrooms, offices, and retail stores, is normally lower than the maximum SMHR required.

17.5 DAYLIGHT

When properly utilized daylight, can greatly reduce the energy required of interior lighting systems.

However, when improperly applied, it can impose an unnecessary cooling load in the space, thus increasing energy consumption.

The use of electric lighting systems is a relatively recent development in the history of the built environment. In early design, buildings were generally long and narrow, with their primary axis oriented east and west, and very large windows at both the north and south walls to take maximum advantage of natural light. The invention of electric lighting and mechanical systems freed designers from previous building forms, and daylight became less favored as electrical lighting grew more flexible and quality illumination could be provided to harmonize with a building's design. However, with energy becoming more expensive and scarce, daylight has emerged again as an effective means for reducing energy consumption in buildings.

Daylight increases the awareness of the occupants of a building by allowing a visual connection between the interior and the exterior environment of the building. The visual connection may have positive or negative effects on an occupant's anxiety and productivity, depending on the external environment.

17.5.1 Light and Thermal Balance

As indicated in Table 14–1, solar heat gain through a clear glass window can be as much as 250 btuh/sq ft

Table 17–1

Spacing–mounting height of luminaires

Mounting Height of Luminaires (above floor) except for Indirect and Semi-Indirect Luminaires, use Ceiling Height (above floor)	Suspension Distance For Indirect and Semi-Indirect Luminaires	(All dimensions in feet) MAXIMUM* SPACING DISTANCE BETWEEN LUMINAIRES							Distance from Walls All Types of Luminaires
		Indirect	Semi-Indirect	General Diffusing	Semi-Direct	Direct	Semi-Concentrating Direct	Concentrating Direct	
8	1-3	9.5	9.5	8	7	7	6.5	5	
9	1.5-3	10.5	10.5	9	8	8	7	5.5	⅓ Spacing
10	2-3	12	12	10	9	9	8	6	Distance if
11	2-3	13	13	11	10	10	9	6.5	desks or
12	2.5-4	14.5	14.5	12	11	11	9.5	7	work benches
13	3-4	15.5	15.5	13	12	12	10.5	8	are against
14	3-4	17	17	14	12.5	12.5	11	8.5	walls, other-
15	3-4	18	18	15	13.5	13.5	12	9	wise ½.
16	4-5	19	19	16	14.5	14.5	13	9.5	
18	4-5	22	22	18	16	16	14.5	11	
20 or more	4-6	24	24	20	18	18	16	12	

*The actual spacing is usually less than these maximum distances to suit bay or room dimensions or to provide adequate illumination. At an established mounting height, it is often necessary to reduce the spacing between luminaires or rows of them to provide specified footcandles. In such systems and particularly in small rooms, the utilization is reduced because the luminaires are closer to the walls. For example, in an 80-30-10 room with a room ratio of 0.6 and a flux ratio of 0.7, the utilization factor is .79 where the spacing-mounting height ratio is 1.0 (luminaires as far apart as they are above the work-plane). With the same room conditions but with luminaires spaced .4 as far apart as they are above the work-plane, the utilization factor is only .65.

From G. E. *Fundamentals of Light and Lighting.*

(2,800 KJ/m), with a luminous flux of 1000 to 10,000 lumens per sq ft. The extreme amount of heat gain is an asset during cold weather, but a liability during hot weather. With solar control devices and innovative architectural design, the effect of solar heat gain can be minimized while retaining the desirable level of daylight. In climates where substantial heating is required, the benefit of daylight can be evaluated between the solar heat gain and the conducted heat loss. The complexity of the lighting and thermal balance of building systems is addressed in sections 1.8 and 1.9 of Chapter 1.

In addition to direct sunlight, which varies with the sun's position and orientation, the sky provides an excellent source of relatively constant and uniform daylight. Three conditions must be taken into account designing daylight:

- Light from an overcast sky
- Light from a clear sky
- Light from a clear sky plus direct sunlight

Light reflected from the ground can also be an important source of daylight. Typically, 5 to 10 percent of daylight is ground-reflected light. On exposures not facing the sun, reflected daylight accounts for 50 percent of the possible daylight. Since ground-reflected light is normally directed at the ceiling, it is best utilized for lighting the interior part of a room.

The amount of daylight available is difficult to predict, as it varies from hour to hour. Thus, it must be supplemented by an equivalent interior lighting system. Veiling reflections from daylight can cause discomfort due to glare. A good design should provide glare control devices, such as blinds and exterior or interior shades.

17.5.2 Factors to Consider in Designing Daylight

In designing for daylight, one should take note of the following factors:

- Daylight is a dynamic source of light, varying both in position and intensity.
- External shading due to landscaping, the configuration of the building, and nearby structures must be considered.
- Proper interior and exterior controls of daylight should be provided.
- Attention must be paid to the glare factor due to windows and skylights.
- When combined with artificial light sources, daylight has an effect on colors in the interior of a building.

- The design must account for the interfacing of daylighting with interior lighting.

17.5.3 Rules of Thumb in Designing for Daylight

The following rules of thumb with respect to the features mentioned are applicable in designing for daylight:

- *Room geometry.* Minimize the room depth, and maximize the ceiling height.
- *Side lighting.* Orient workstations so that daylight will be on the left side of the worker (for a right-handed person).
- *Effective distance.* Daylight is usually assumed effective up to a depth 2.5 times the window height.
- *Solar gains.* Direct sunlight should be avoided.
- *Orientation.* South glazing provides the greatest amount of daylight. North glazing provides the most consistent illumination. East and west glazing have the greatest effect on the building cooling load. Large-area glazing may create overheating on the south side rooms in modern high-rise buildings when windows are prohibited from being opened.
- *Architectural elements.* Consider the use of overhangs, screens, louvers, and other control devices to manipulate the amount of daylight reflectance. Reflecting surfaces below the window level can increase daylight.
- *Window location.* Generally, the higher the window, the better the potential for daylight. Windows below the level of the visual task offer little illumination value.
- *Room finish.* Lighter room finishes are more effective. However, high-reflective white surfaces may cause discomfort due to glare.
- *Space planning.* Open spaces allow daylight to penetrate them.
- *Glazing area.* Glazed areas need not be greater than 25 percent of the floor area. Excessive glazing adds to heating and cooling loads and does not improve visual acuity.

17.6 DESIGN PRACTICE AND ALTERNATIVE SOLUTIONS

Although every designer should try to derive exactly one solution to a design problem, there is always more than one solution. The following exercise is intended to make this point.

Suppose you have been commissioned to design the lighting for an architectural or engineering office, as shown in the floor plan below. The office contains seven workstations, one reception desk, and one conference area (see Figure 17–12). The room data and design criteria are as follows:

- Room data:
  Dimensions   48' × 26'
  Ceiling   9'-6" height, 2' × 2' lay-in grid
  Ceiling reflectance   80% (base)
  Wall reflectance   50% (base = effective)
  Floor reflectance   20% (effective)
- Luminaire to be used: 2' × 2' parabolic, 3–20 watt (1500 lm), type 25 (see Figure 16–4).
- Design criteria: Maintained illuminance—50 fc minimum, LLF = 0.75

Refer to Figure 17–13.

1. Using the Average Illuminance Calculation Form, calculate the number of lighting fixtures required for uniform illumination over the workstation area only (36" × 26") The balance of the room is to be designed separately.
2. If the calculated number of fixtures includes a fraction, use the appropriate integral number of fixtures to best suit the workstation layout and the interior aesthetics. Lighting fixtures at each workstation should be in front of or above the drafting table whenever practical.
3. The central corridor space of the room is approximately one-third of the room width. Suppose the illuminance at the central corridor is not a concern, and make a modified layout to concentrate the lighting fixtures over the workstations by using slightly more than two-thirds of the number of fixtures calculated previously to fit the workstation pattern.

4. Make your own design for the reception and conference area. Select any luminaire(s) from Figure 16–4. Describe the design, and make a perspective of the area and viewed from the entrance door toward the conference table, as a visitor would see it.

Although this design exercise is only for a small architectural or engineering studio, it covers all aspects of the design practices. The exercise must address the following:

- Calculations of illumination for uniform lighting, nonuniform lighting, and lighting effectiveness
- Decisions on the layout of luminaires
- Coordination with ceiling grids
- Lighting aesthetics

The owning and operating costs of the design are more often than not the determining factors in the final selection of a system. However, in order to be realistic, data on costs must incorporate all tangible and intangible factors which vary widely among projects. These factors are too complex to be included here; nevertheless, all designers must be keenly aware of cost considerations in every design decision.

1. *Calculations*
   Using the Average Illuminance Calculation Form that follows, we find that for the zonal cavity method of calculation, the number of fixtures is 21.
2. *Uniform Layout*
   Based on the space configuration, the fixtures may be arranged in three rows of seven each. This configuration is obviously unsatisfactory to suit the workstation layout. Trials indicate that four rows of five each is most appropriate, even if this configuration is one fixture short of the number required. Thus, we have:
   - *Scheme A.* For uniformity within the entire area, the four rows of fixtures are more or less

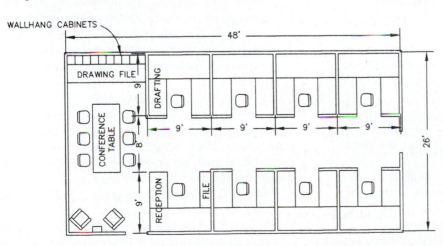

■ **FIGURE 17–12**
Floor plan of an architectual or engineering office.

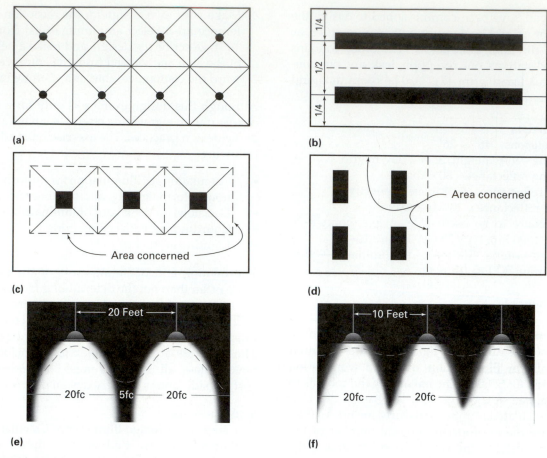

■ FIGURE 17–13

Layout of lighting fixtures for uniform lighting. (a), (b) The entire space is divided into unit areas for total uniformity. (c), (d) The fixtures are located in the area where uniformity is desired. (e), (f) Insufficient fixture spacing will create dark (shadow) areas.

evenly spaced on the ceiling in a north-south direction and at the front of each workstation in an east-west direction. This scheme provides a neutral expression and satisfactory illumination.

3. *Nonuniform Layouts*

   To achieve maximum lighting of task areas, illuminance levels at the central corridor may be reduced. Several schemes are possible:

   - *Scheme B.* The fixtures are moved within the workstation partition lines, with two fixtures in the front of each workstation and no fixtures in the corridor.
   - *Scheme C.* In lieu of locating the fixtures as in Scheme A, the second row of fixtures is moved closer to the first row, and the third row closer to the fourth row.
   - *Scheme D.* One 2′ × 4′ fixture is used to substitute for two 2′ × 2′ fixtures within the workstation areas, and three single 2′ × 2′ fixtures are used in the corridor area.
   - *Scheme E.* If the workstation partitions are

   higher than the 5-ft level within the 9′-6″ ceiling space, or if the partitions become full height to the ceiling, lighting fixtures must be moved within each workstation room.

   - *Scheme F.* The lighting fixtures in each workstation room in Scheme E are irregular in location and aesthetically not pleasing. This is because of the continuation of the grid in the ceiling. If the ceiling grids are installed on a room-by-room basis, the lighting within the individual rooms can be centered within each room.

4. *Lighting Plan for the Conference Area*

   There are of course, even more alternatives for illuminating the conference area than for the workstation area. With Scheme C as the base for the workstation area, several schemes are possible for the conference area. The schemes are:

   - *Scheme C1.* Continue the same 2′ × 2′ fixtures for the conference table area.
   - *Scheme C2.* Use the same 2′ × 2′ fixtures, except that the unit over the conference table is

AVERAGE ILLUMINANCE CALCULATION FORM

FOR ROOM	*Arch/Engr. office*

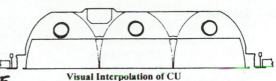

ILLUMINANCE CRITERIA	IES ILLUMINANCE CATEGORY					**E**
	MAINTAINED ILLUMINANCE, FC, (LUX)					**50**

Visual Interpolation of CU

* Pw of 40% is determined by the weighted average of N,E & S walls @ 50% and W wall @ 0%, or no reflectance at all.

\# CU is found by interpolating between Pw @40% and RCR @ 2.28.

	FIXTURE DATA					
	MFR/MODEL	*Type 25, chapter 16*				
	TYPE DISTRIBUTION	*Direct*				
	NO. OF LAMPS PER FIXTURE				**3**	
	RATED LAMP LUMEN & WATTS/LAMP				**1500/20W**	
	LUMENS PER FIXTURE (LPF)				**4500**	

ROOM DIMENSIONS	h	**9.5**	WIDTH(W)	**26**	LENGTH(L)	**36**
ROOM CHARACTERS	h_{cc}	**0**	R_C	**.8**	R_{w1}	**.5**
	h_{rc}	**7**	R_w	**.5**	R_{w2}	**.5**
	h_{fc}	**2.5**	R_f	**.2**	R_{w3}	**.5**

RC RW	80				*P_w=40*
	70	50	30	10	
1	80	77	75	70	
2	74	70	66	63	→ *CU=66*
3	69	63	58	54	*@ RCR*
4	64	57	51	47	*=2.28*
5	59	51	45	41	
6	54	45	40	36	
7	50	41	35	31	
8	46	37	31	27	
9	42	33	27	23	
10	39	30	24	21	

P	PERIMETER, FT(M): 2×36+2×26	**124**
A	AREA, SF(SM): 36 × 26	**936**
PAR	PERIMETER/AREA RATIO (P ÷ A)	**.13**
CCR	2.5 × PAR × h_{cc}	**0**
RCR	2.5 × PAR × h_{rc} (#)	**2.28**
FCR	2.5 × PAR × h_{fc}	**.81**
ρ_{cc}	FROM R_C & R_{w1} & CCR	**.80**
ρ_w	SAME AS R_w OR R_{w2} (✳)	**.40**
ρ_{fc}	FROM R_f & R_{w3} & FCR	**.20**
CU	FROM CU TABLE OF FIXTURE MFGR. INTERPOLATING BETWEEN RCR AND ρ_{cc}, ρ_w, ρ_{fc}	**.66**
LOF	BF – BALLAST FACTOR	**.95**	**.93**
	VF – VOLTAGE FACTOR	**.98**	
	OTHER	**–**	
LLF	LLD–LAMP LUMEN DEPREC.	**.85**	**.80**
	LDD–LUMINAIRE DIRT DEPREC.	**.95**	
	OTHER	**–**	

FLOOR OR CEILING PLAN

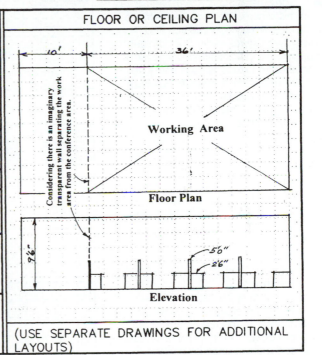

Considering there is an imaginary transparent wall separating the work area from the conference area.

Working Area

Floor Plan

Elevation

(USE SEPARATE DRAWINGS FOR ADDITIONAL LAYOUTS)

CALCULATIONS

MAINTAINED ILLUMINANCE

$$E = \frac{N \times (LPF \times LOF) \times CU \times LLF}{A}$$

INITIAL ILLUMINANCE

$$E_i = E \div LLF$$

CALCULATION & REMARKS:

$$E = 50 = \frac{N \times (4500 \times .93) \times 0.8 \times 0.66}{936}$$

$$N = 21$$

$$E_i = E/LLF = 50/0.8 = 62.5 \ fc$$

* N – NUMBER OF FIXTURES

■ Scheme A

For uniformity within the entire area, the four rows of fixtures are more or less spaced evenly on the ceiling in a N-S direction and at the front of each workstation in an E-W direction. Note that some of the lighting fixtures are located directly above the 5-ft-high partitions, which is acceptable. This scheme provides a neutral expression and satisfactory illumination.

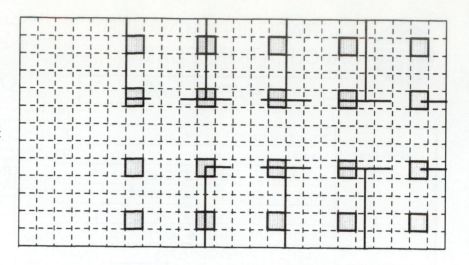

■ Scheme B

With this scheme, the fixtures are moved within the workstation partition lines, having two fixtures in the front of each workstation and no fixtures in the corridor. A total of 16 fixtures is used and a 20 percent reduction of fixtures. However, lighting in the workstations and the overall space is spotty. This schedule is not a desirable solution.

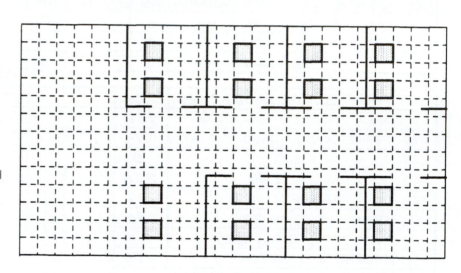

■ Scheme C

In lieu of locating the fixtures shown in Scheme A, the second row of fixtures is moved closer to the first row, and the third row closer to the fourth row. This reduces the illuminance level in the corridor and increases the illuminance level within the workstations to about 70 fc, which is a more desirable level for drafting tasks. This scheme is preferred.

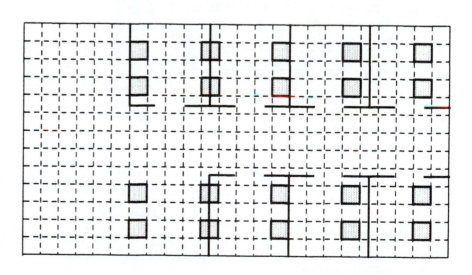

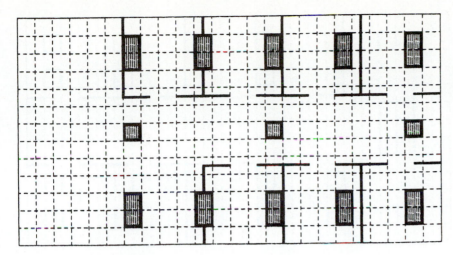

■ **Scheme D**

With this scheme, (one) 2′ × 4′ fixture is used to substitute for (two) 2′ × 2′ fixtures within the workstation areas, and (three) single 2′ × 2′ fixtures are used in the corridor area. Although this scheme uses more fixtures than Scheme B, it is lower in cost, since 2′ × 4′ fixtures are the most popular type used commercially, and thus more competitively priced.

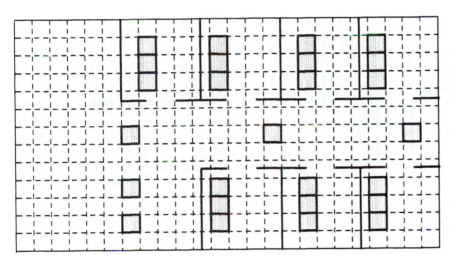

■ **Scheme E**

If the workstation partitions are higher than the 5-ft level within the 9′ 6″ ceiling space, or if the partitions become full height to the ceiling, lighting fixtures must be moved within each workstation room. In this case, the room cavity ratio of each workstation room will be greatly increased, and the CU values decreased, resulting in a need to add another fixture within each room.

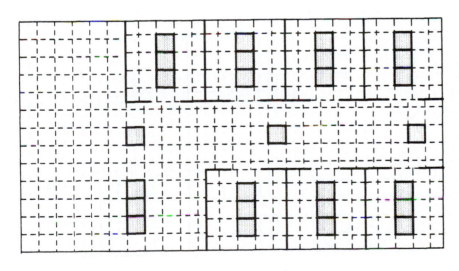

■ **Scheme F**

The lighting fixture locations within each workstation room shown in Scheme E are irregular in location and aesthetically not pleasing. This is because of the continuation of the ceiling grid in the ceiling. If the ceiling grids are installed on a room-by-room basis, lighting within the individual rooms can be centered within each room. This is a much better lighting design, although it will add ceiling cost.

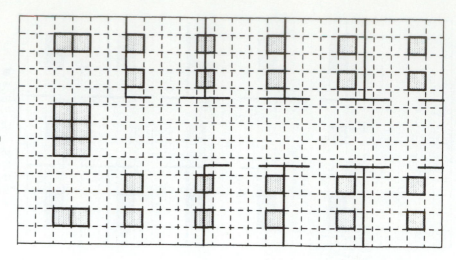

■ Scheme C1

Continuing the same 2' × 2' fixture for the conference table area. This scheme provides a consistent neutral expression in spatial relations. There is no visual focal point.

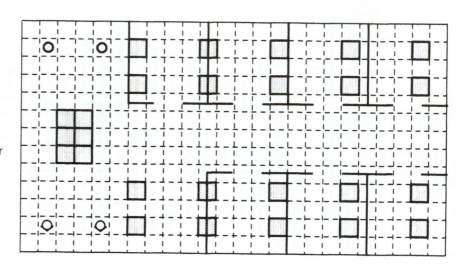

■ Scheme C2

Using the same 2' × 2' fixtures, except that the unit over the conference table is composed of (6) 2' × 2' fixtures suspended 3 ft. below the ceiling. The composite 4' × 6' unit may be selected for direct, indirect or direct-indirect distribution. The file storage and seating areas are illuminated with incandescent or panel fluorescent downlight or wall washers. The design provides good task illumination over the conference table, relaxed atmosphere at the seating area, and a balanced spatial relation at the storage area.

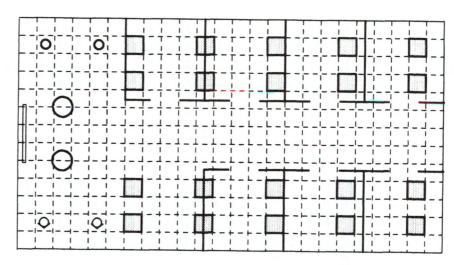

■ Scheme C3

As a totally different design, (two) decorative fixtures using either incandescent or panel fluorescent lamps are suspended over the conference table. The suspended fixtures are for direct distribution only, which will not provide sufficient illumination on the illustration board on the wall. A separate linear fluorescent unit is mounted over the board for vertical illumination. This scheme provides an expressive to dominant visual atmosphere in the space.

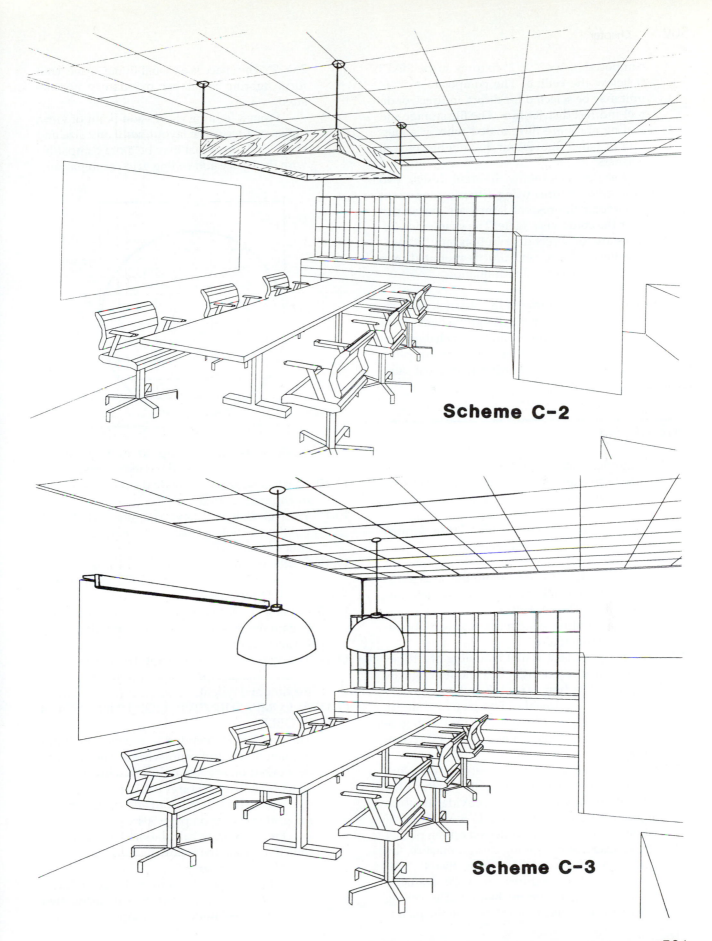

Scheme C-2

Scheme C-3

composed of six 2′ × 2′ fixtures suspended 3 ft below the ceiling. The composite 4′ × 6′ unit may be selected for direct, indirect, or direct-indirect distribution. The file storage and seating areas are illuminated with incandescent or panel fluorescent downlights, or wall washers.

- *Scheme C3.* As a totally different design, two decorative fixtures using either incandescent or circular fluorescent lamps are suspended over the conference table. The suspended fixtures are for direct distribution only. A separate linear fluorescent unit is mounted over the board for vertical illumination.

The preceding are just a few of the alternatives compatible with the basic lighting design (Scheme C) of the space. Lighting systems are indeed a challenge to designers. Creative ideas and close coordination among designers will be rewarded with great satisfaction.

QUESTIONS

17.1 A lighting system is an integral part of the architectural expression of the space, which may be described as _____, _____, _____, _____.

17.2 Illumination must be designed for uniformity on a horizontal plane. (True) (False)

17.3 For uniformity in horizontal illuminance, lighting fixtures must be mounted so as not to exceed the recommended space-to-mounting height ratio. (True) (False)

17.4 An ideal lighting system is a totally luminous ceiling. (True) (False)

17.5 To obtain the best visual modeling of a three-dimensional object, light flux should originate from one direction. (True) (False)

17.6 Lay out, on a reflected ceiling plan for uniform lighting, six round lighting fixtures with general diffuse distribution in a 15′ × 45′ rectangular space.

17.7 If the ceiling height is 10 feet and the work plane is 3 feet above the floor, do you think the preceding layout would provide uniform lighting? If not, what would you suggest? Show your design in a reflected ceiling plan.

17.8 As an exercise in lighting design practice, lay out on a reflected ceiling plan, five 4′ × 4′ square lighting fixtures in a 20′-diameter round room (see figure below). The ceiling grid for this round room shown on the plan is

2′ × 2′. The space is a casual living space. Uniform lighting is neither mandatory nor desired.

From a lighting expression point of view, make an alternative layout, with any size and shape of fixtures that may be more compatible with this space and ceiling grid combination.

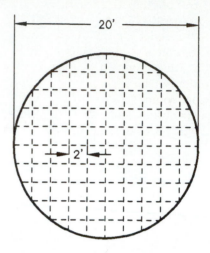

17.9 Daylight is effective up to a depth about _____ times the window height.

17.10 North glazing will provide the most consistent daylighting. (True) (False)

17.11 West glazing will have the most effect on a building's cooling load in the northern hemisphere. (True) (False)

17.12 South glazing may create overheating in spaces with southern exposure, even in winter time. (True) (False)

17.13 Glazing should be extended to the floor level to gain the most benefits from daylight. (True) (False)

17.14 The glazed area need not be greater than _____ percent of the floor area for maximum daylight.

17.15 A COMPREHENSIVE LIGHTING DESIGN PROBLEM

The problem is intended to practice the lighting layout, controls and wiring. The room to be worked on is a lecture room with auxiliary spaces. The architectural plan of the space including the furniture layout and the lay-in ceiling grid is shown on page 504.

From separate calculations, it was determined to install the following fixtures:

- Main lecture area:
  —(ten) type A, semi-indirect fixtures. Each fixture is 1-ft × 4-ft, using two 32W, T-8 fluorescent lamps.

—(six) type B, semi-indirect fixtures to be installed in front of the chalkboard. Each fixture is 1-ft × 4-ft using two 32W, T-8 fluorescent lamps. One of the two lamps provides downward asymmetrical distribution on the chalkboard which should be separately switched. See photometric drawing.

- Lab area: (2) type C, direct distribution fixtures. Each fixture contains (3) 32W, T-8 fluorescent lamps. It is desirable to switch the center lamps from one switch and the outer two lamps from another switch.

- Coat room: (1) type D, 2 ft × 2 ft, direct distribution fluorescent fixture using (4) 20W, T-8, fluorescent lamps.

- Closet: (1) type E, surface-mounted fluorescent fixture using (2) 40W, T-12 lamps.

Based on the calculated number of fixtures, make a lighting layout in these rooms. Keep in mind that:

- The ceiling grid is already determined.
- There are VDT workstations along the west wall. (Note: There are additional computers on student's desks. However, no special lighting provisions will need to be made.)

And as practice for electrical wiring, complete the branch circuit wiring for all lights in these rooms. Indicate the number of wires and the lamps or fixtures to be switched.

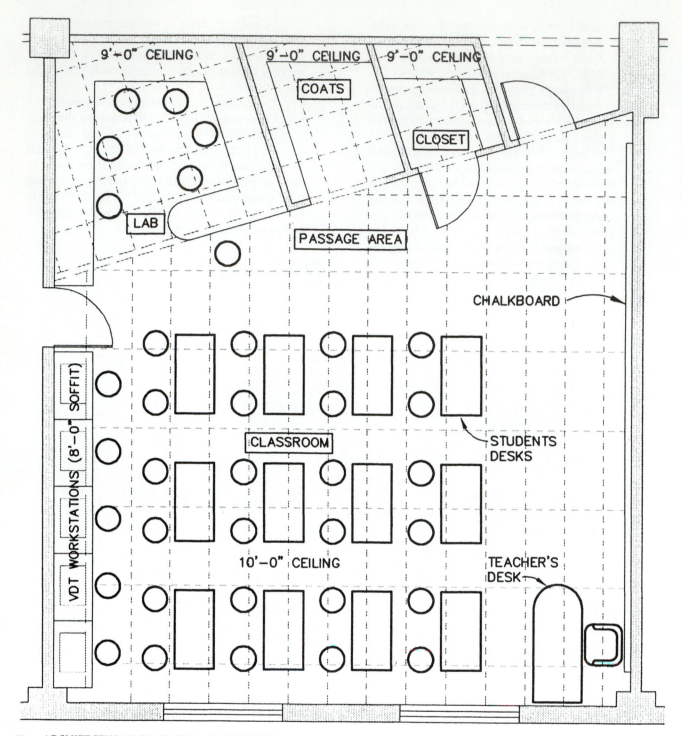

9'-0" CEILING   9'-0" CEILING   9'-0" CEILING

COATS

CLOSET

LAB

PASSAGE AREA

CHALKBOARD

VDT WORKSTATIONS (8'-0" SOFFIT)

CLASSROOM

STUDENTS DESKS

10'-0" CEILING

TEACHER'S DESK

■ **ARCHITECTURAL PLAN OF A CLASSROOM**
Showing architectural floor plan, furniture layout, ceiling heights and ceiling construction.

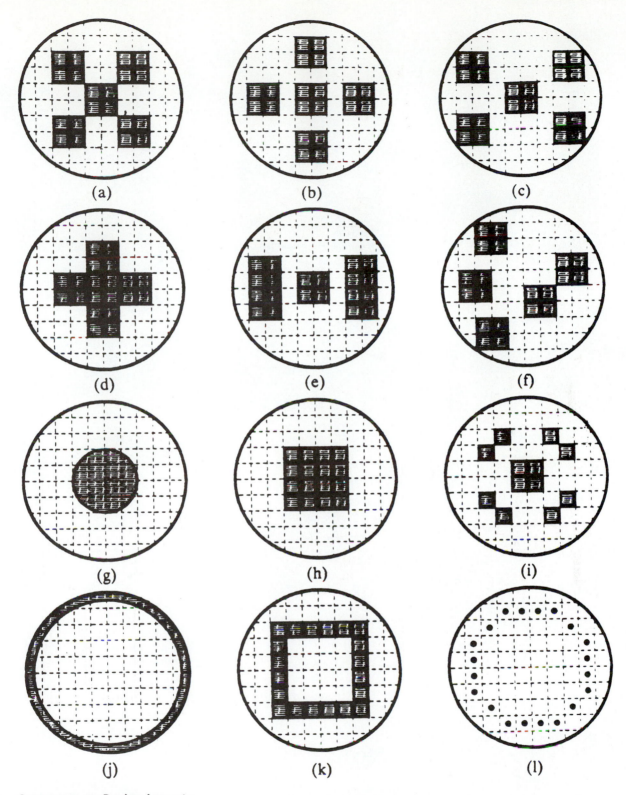

■ Comments on Design Layouts

(a) Neutral-Expressive (first choice)
(b) Neutral (second choice)
(c) Neutral-Disorganized (too spread out)
(d) Expressive-Dominant
(e) Neutral-Disorganized
(f) Expressive-Disorganized

(g) Neutral-Expressive
(h) Expressive-Dominant
(i) Dominant-Disorganized
(j) Neutral-Expressive
(k) Dominant
(l) Neutral-Disorganized (imperfect circle)

APPENDIX A

GLOSSARY OF TERMS, ACRONYMS, AND ABBREVIATIONS

THIS APPENDIX CONTAINS TERMS, ACRONYMS, AND abbreviations commonly used in the planning and design of mechanical, electrical, and illumination systems. For ease of understanding and quick reference, these terms are briefly explained rather than precisely defined. If a term is deemed essential for the basic understanding of the various systems, it will have been included in more detail in the text of the applicable chapters. If a term is only good to know as general or conversational knowledge, it may not be included in the text.

The terms are arranged alphabetically, with a letter in parentheses to identify their normal application in one or another of the various building systems. The identificatory letters are as follows:

(A)   Architectural
(AS)   Acoustics
(DP)   Data processing/computer
(E)   Electrical power systems
(FP)   Fire protection systems
(G)   General, applicable to all or most mechanical and/or electrical systems
(H)   Heating and air-conditioning systems
(I)   Illumination systems
(PS)   Plumbing and sanitation
(S)   Structural systems

(TC)   Telecommunications
(AX)   Auxiliary electrical systems

To find the exact location at which a term may be found in the text, search the index for chapter and page numbers.

Absolute humidity (H)   Weight of water vapor per unit volume of air-steam mixture. Usually expressed in grains per cubic foot.

Absolute pressure   The pressure above an absolute vacuum; the sum of the pressure shown by a gage and that indicated by a barometer.

Absolute temperature (H)   Temperature measured from absolute zero; on the Fahrenheit scale, the absolute temperature is °F + 459.6° and on the Centigrade scale °C + 273°.

Absorption (AS)   The ability of a material to absorb acoustical energy. Measured in sabins. The product of area (S) and absorption coefficient (α). Frequency sensitive.

Absorption coefficient (AS)   The fraction of sound energy impinging on a surface that is absorbed by that surface, usually denoted by α. Frequency sensitive.

Absorption refrigeration (H)   A process whereby a secondary fluid absorbs the refrigerant and, in doing so, gives up heat. Afterward, it releases the refrigerant, during which time it absorbs heat.

507

Acoustics (AS)   The science of sound, including its generation, transmission, and effects.

Acrylonitrile-butadiene-styrene, or ABS (PF)   A thermoplastic compound from which fittings, pipe, and tubing are made.

Active sludge (PF)   Sewage sediment, rich in destructive bacteria, that can be used to break down fresh sewage more quickly.

Adiabatic (H)   A change at constant total heat; an action or process during which no heat is added or subtracted.

Aeration (PF)   An artificial method in which water and air are brought into direct contact with each other to furnish oxygen to the water and to reduce obnoxious odors.

Aerobic (PF)   Living or active only in the presence of free oxygen (said of certain bacteria).

Airborne sound (AS)   Sound that is transmitted through air by a series of oscillating pressure fluctuations.

Air Change (PF)   The quantity of infiltration or ventilation air, in cubic feet per hour or minute, divided by the volume of the room.

Air Gap (PF)   The unobstructed vertical separation between the lowest opening from a pipe or faucet conveying water or waste to a plumbing fixture receptor.

Alternating current or AC (E)   Flow of electricity that cycles or alternates direction. The number of cycles per second is referred to as the frequency.

Ambient lighting (I)   Lighting throughout an area that produces general illumination.

Ambient noise (AS)   The sound associated with a given environment, usually a composite of many sources.

American Standard Code for Information Interchange, or ASCII (DP)   A system for referring to letters, numbers, and common symbols by code numbers. The numerical value assigned for each binary digit of an eight-digit byte is, from left to right, 128, 64, 32, 16, 8, 4, 2, and 1.

American Wire Gage, or AWG (E)   The standard system for measuring the size of wires in the United States.

Ampacity (E)   The current, in amperes, that a conductor can carry continuously under conditions of use without exceeding its temperature rating.

Analog (TC, DP)   Varying continuously (e.g., sound waves). Analog signals have their frequency and bandwidth measured in hertz. See Digital.

Angstrom (I, TC)   A unit of length, 0.1 nm or 10^{-10} m, often used to measure wavelength, but not part of the SI system of units. See nanometer.

Apparatus dew point   Dew point temperature of the air leaving the conditioning apparatus.

Atmospheric pressure   Pressure indicated by a barometer. The standard atmospheric pressure is 760 mm of mercury (29.921 in. of mercury) or 14.7 lb per sq in. at sea level.

Attenuation (AS)   Lessening or reduction, e.g., from 80 dB to 70 dB.

A-weighting (AS)   Prescribed frequency response defined by ANSI Standard S1.4-1971. Used to obtain a single number representing the sound pressure level of a noise in a manner approximating the response of the ear, by de-emphasizing the effects of the low and high frequencies. See dBA.

Background sound (AS)   Noise from all sources in an environment, exclusive of a specific sound of interest.

Backup (DP)   A copy of a program or document that you can use if the original is destroyed.

Ballast (I)   A device used with fluorescent and HID lamps to provide the necessary starting voltage and to limit the current during operation of the lamp.

Bandwidth (TC, DP)   The highest frequency that can be transmitted in an analog operation. Also (especially for digital systems), the information-carrying capacity of a system.

Baud (DP, TC)   Strictly speaking, the number of signal-level transitions per second in digital data. For some common coding schemes, this equals the number of bits per second, but it is not true for more complex coding, where the term is often misused. Telecommunication specialists prefer bits per second, which is less ambiguous.

Baud rate (DP)   The number of bits per second. Baud rates are most commonly used as a measure of how fast data are transmitted by a modem.

Beam angle (I)   The angle between the two directions for which the intensity is 50 percent of the maximum intensity, as measured in a plane through the nominal beam center line.

Bimetal (H)   Two metals of different coefficients of expansion welded together so that the piece will bend in one direction when heated and in the other when cooled. Thus, it can be used in opening and closing electrical circuits, as in thermostats.

Binary numbers (DP)   The base-2 numbering system of almost all computers, as opposed to the base-10 (decimal) numbers people use.

Binary system (DP)   A numbering system using 2 as base, as opposed to 10 as base in the decimal system. For example, 1 in binary system is 1; 10 is 2;

11 is 3; 100 is 4; 101 is 5; 110 is 6; 111 is 7; 1000 is 8; 1001 is 9; and 1110 is 10. The binary system is used in almost all computers.

Bit (DP) Abbreviated from "Binary Digit," the smallest possible unit of information. A bit represents one of two things: yes or no, on or off, or, as expressed in the binary numbers used in computers, 0 or 1.

Boiler (H) A closed vessel in which fuel is burned to generate steam. (A vessel that produces hot water should be called a hot-water heater.)

Boiler horsepower (H) The power required to evaporate 34.5 lb of water at 212°F per hour, or the equivalent of 33,475 Btu per hr.

Booting (DP) The starting up of a computer by loading an operating system into it. (The name *boot* comes from the idea that the operating system pulls itself up by its own bootstraps).

Branch circuit (E) The circuit conductors between the final overcurrent device protecting the circuit and the outlet(s).

British thermal unit, or Btu (G, H) The amount of heat required to raise the temperature of 1 lb (0.45 kg) of water 1 degree Fahrenheit (0.565°C).

Broadband (TC) In general, covering a wide range of frequencies. The broadband label is sometimes used for a network that carries many different services or for video transmission.

Buffer (DP) An area of memory or a separate memory cache that holds information until it is needed. Buffers are used to speed up printing, redraw the screen, etc.

Bug (DP) A mistake or unexpected occurrence in a piece of software (or, less commonly, a piece of hardware).

Bulletin board (bulletin board system), or BBS (DP) A computer dedicated to maintaining messages and software and making them available over phone lines at no charge. People upload (contribute) and download (gather) messages by calling the bulletin board from their own computers.

BX (E) An electrical cable comprised of a flexible metallic covering inside of which are two or more insulated wires for carrying electricity.

Byte (DP) A group of eight bits that the computer reads as a single letter, number, or symbol. For example, 01000101 is a typical byte of the binary system.

Calorie (H) In engineering, the large calorie is usually used, defined as one one-hundredth of the energy or heat required to raise the temperature of 1 kilogram of water from 0°C to 100°C; the small calorie is the heat required similarly to raise the temperature of 1 gram of water.

Candela (I) The unit of measurement of luminous intensity of a light source in a given direction.

Candlepower (I) Luminous intensity of a light source in a specific direction.

Candlepower distribution curve (I) A curve, generally polar, representing the variation in luminous intensity of a lamp or luminaire in a plane through the light center.

Cathode ray tube, or CRT (DP, TC) The display technology used on virtually all computer monitors and television sets.

Cathodic protection (PS) The control of the electrolytic corrosion of an underground or underwater metallic structure by the application of an electric current.

Central processing unit, or CPU (DP) The central part of a computer. The CPU includes the circuitry built around the CPU chip and mounted on the motherboard that actually performs the computer's calculations.

Chip (DP) A tiny piece of silicon or germanium impregnated with impurities in a pattern that creates different sorts of miniaturized computer circuits. A chip is the basic brain power controlling all electronic or computer equipment and systems.

Circuit breaker (E) A device designed to open and close a circuit manually or automatically on a predetermined overcurrent.

Clock rate (or clock speed) (DP) The operations of a computer are synchronized to a quartz crystal that pulses millions of times each second. These pulses determine things such as how often the screen is redrawing an image and how often the CPU accesses RAM or a hard disk. The frequency of the pulses—how often they occur—is measured in megahertz (millions of cycles per second) and is called the clock rate or clock speed.

Coefficient of performance In the heat pump, the ratio of the effect produced to the electrical input.

Coefficient of utilization or CU (I) The ratio of the luminous flux (lumens) from a luminaire, calculated as received on the workplane, to the luminous flux emitted by the luminaire's lamps alone.

Color rendering index, or CRI (I) Measure of the degree of color shift objects undergo when illuminated by a light source, as compared with a reference source, normally incandescent.

Color temperature (I) The absolute temperature of a blackbody radiator having a chromaticity equal to that of the light source.

Community antenna television, or CATV (TC) Cable television, a broadband transmission system generally using 75-ohm coaxial cable that simultaneously carries many frequency-divided TV channels.

Compact disk, or CD (DP, TC)   A 4.5-inch-diameter disk containing digital information that can be read by a laser source.

Compact disk read-only memory, or CD-ROM (DP, TC)   A 4.7-inch-diameter compact disk with read-only memory. It can store up to 660 megabytes of information.

Compiler (DP)   Software that implements a program by translating it all at once.

Computer-aided design, or CAD (DP)   Refers to both hardware and software.

CAD/CAM (DP)   See Computer-Aided Design and Computer-Aided Manufacturing.

Computer-aided manufacturing, or CAM (DP)   Computers and programs that run manufacturing machinery or even entire factories. (You seldom see the word CAM alone; it's usually combined with CAD.)

Condensate (H, PS)   Liquid formed by condensation.

Condenser (H)   Apparatus used to liquefy a gas.

Condensing unit (H)   An assembly attached to one base and including a refrigerating compressor, motor, condenser, receiver, and necessary accessories.

Conductivity (H)   The quantity of heat (usually Btu) transmitted per unit of time (usually 1 hour) from a unit surface (usually 1 sq ft) to an opposite unit of surface of one material per unit of thickness (usually 1 inch, but occasionally 1 foot) under a unit temperature differential (usually 1°F) between the surfaces.

Contrast (I)   Difference in brightness between an object and its background.

Convection (H)   The transfer of heat from one point to another within a fluid (such as air or water) by the mixing of one portion of the fluid with another. If the motion is due to differences in density, from temperature differences, the convection is natural; if the motion is imparted mechanically, it is forced convection.

Convector (H)   A heating unit containing heating elements that allow air to be heated through natural convection without using external power.

Cooling tower (H)   A device for cooling water by evaporation in the outside air. A natural-draft cooling tower is a cooling tower in which the airflow through it is due to the natural chimney effect; a mechanical-draft tower employs fans.

Coprocessor (DP)   A chip that specializes in mathematics, graphics, or some other specific kind of computation. When the CPU is handed the kind of job the coprocessor specializes in, it hands the job off to the coprocessor.

Condensate (G)   Water that has liquefied from steam or precipitated out from air.

Continuous load (E)   A load such that the maximum current is expected to continue for three hours or more.

Contrast (I)   The difference in brightness (luminance) between an object and its background.

Critical angle (I, TC)   The angle at which light undergoes total internal reflection.

Critical velocity (H)   The point above which streamline flow becomes turbulent.

dBA (TC)   Overall A-weighted sound pressure level expressed in decibels referenced to 20 micropascals. An A-weighting is a single number approximating the response of human hearing.

Debug (DP)   To search out bugs or defects in a piece of software and eliminate them.

Decay rate (AS)   The rate at which the sound pressure level (in dB) decreases when the source of the sound is eliminated.

Decibel or dB (AS, TC)   Ten times the logarithm to the base 10 of a quantity divided by a reference quantity; $dB = 10 * LOG_{10}(X/X_{ref})$.

Degree day, or DD (H)   The number of Fahrenheit degrees that the average outdoor temperature over a 24-hr period is less than 65°F.

Demand factor, or DF (E)   The ratio of the maximum demand to the total connected load of a system.

Demand load, or DL (E, H, PF)   The load, in appropriate units, such as kW, Btuh, gpm, etc., that an electrical or mechanical system encounters.

Density (G)   The weight of a unit of volume—usually, pounds per cubic foot.

Developed length (PF, H)   The length along the centerline of a pipe and its fittings.

Dewpoint or DP (G, H)   The temperature of a gas or liquid at which condensation or evaporation occurs.

Dielectric fitting (PS, H)   A fitting having insulating parts or material that prohibits the flow of electric current.

Diffuse sound field (AS)   A sound field in which the intensity of the sound is independent of its direction; an area over which the average rate of sound energy flow is equal in all directions.

Digital (DP, TC)   Expressed in binary code to represent information. See Analog.

Diode (DP, TC)   An electronic device that lets current flow in one direction. Semiconductor diodes used in fiber optics contain a junction between regions of different doping. (Doping means adding impurity elements to pure semi-conductors to improve their conductivity.) Diodes include light emitters (LEDs and laser diodes) and detectors (photodiodes).

Direct current, or DC (E)   Flow of electricity continuously in one direction from positive to negative.

Direct expansion (H)   An arrangement wherein a refrigerant expands in an evaporator in the airstream.

Direct glare (I)   Glare resulting from high luminances or insufficiently shielded light sources in the field of view.

Direct sound field (AS)   A sound field in which the energy arrives at the receiver in a direct path from the source, without any contribution from reflections.

Directivity index, or DI (AS)   A measure of the directionality of sound from a specific source, in decibels. The difference between the actual sound pressure level of the source and the sound pressure that would exist if the same source were a point source radiating spherically. The directivity index is sensitive to frequency.

Discomfort glare (I)   Glare producing discomfort. This type of glare does not necessarily interfere with visual performance or visibility.

Disconnecting means (E)   A device by which the conductors of a circuit can be disconnected from their source of supply.

Disk (DP)   A round, coated platter used to store information magnetically for computer processing. The two main types are floppy disks (3.5″ and 4.25″), which are removable and flexible, and hard disks, which are nonremovable and rigid.

Disk operating system, or DOS (DP)   The operating system used on IBM Personal Computers and compatible machines.

District heating (H)   The heating of a multiplicity of buildings from one central boiler plant, from which steam or hot water is piped to the buildings. District heating can be a private function, as in some housing projects, universities, hospitals, etc., or a public utility operation.

Diversity factor (H, E, PS)   The ratio of the sum of the individual maximum loads during a period to the simultaneous maximum loads of all the same units during the same period. Always unity or more.

Dot-matrix printer (DP)   A printer that forms characters out of a pattern of dots. Usually, each dot is made by a separate pin pushing an inked ribbon against the paper.

Dots per inch, or dpi (DP)   A measure of the resolution of a screen or printer; the number of dots in a line 1 inch long.

Downspout (PS)   The rain leader (pipe) from the roof to the means of disposal.

Drainage fixture unit, or DFU (PS)   A measure of probable discharge into the drainage system by various types of plumbing fixtures. By convention, 1 DFU is equivalent to 1 cf/min.

Dry-bulb temperature, or DB (H)   The temperature of air as measured by a thermometer.

Dynamic random access memory or DRAM (DP)   A memory chip that loses its memory when the computer is shut off.

Effluent (PS)   Treated or partially treated sewage flowing out of sewage treatment equipment.

Electronic data interchange, or EDI (TC)   A series of standards that provide for the exchange of data between computers over phone lines.

Electronic mail, or E-Mail (DP, TC)   Messages sent from computer to computer over phone lines or over a local area network (LAN).

Emissivity (H, I)   The ratio of radiant energy emitted by a body to that emitted by a perfect black body. A perfect black body has an emissivity of 1, a perfect reflector an emissivity of 0.

Energy management system, or EMS (H, E)   An automated control system designed to achieve higher energy efficiency or lower energy consumption in a building, an energy management system is not limited to the operation of HVAC, lighting, and other power systems.

Enthalpy (H)   For most engineering purposes, heat content or total heat, above some base temperature. Specific enthalpy is the ratio of the total heat to the weight of a substance.

Entropy (H)   The ratio of the heat added to a substance to the absolute temperature at which the heat is added. Specific entropy is the ratio of total heat to weight of the substance.

Equipment-grounding conductor (E)   The conductor used to connect the non-current-carrying metal parts of equipment, raceways, and other enclosures to the system, ground. (The equipment-grounding conductor is color coded green.)

Equivalent sphere illumination, or ESI (I)   The level of spherical illumination that would produce a task visibility equivalent to that produced by a specific lighting environment.

Ethernet (DP)   A relatively fast local area network (LAN) cabling system developed by Xerox. Ethernet components are also sold by other vendors.

Expansion joint (PS, H)   A joint designed to absorb longitudinal thermal expansion in the pipeline due to heat.

Expansion loop (PS, H)   A bend of large radius in a pipeline designed to absorb longitudinal thermal expansion in the line due to heat.

Face velocity (H)   The speed, in feet per minute, by which air leaves a register or a coil.

Fault (E)   A short circuit—either line to line or line to ground.

FAX (TC)   Facsimile—the transmitting of an image by means of telecommunication.

Feeder (E)   Circuit conductors between the service equipment and the final branch-circuit overcurrent device.

Feet of head   Pressure loss in psi divided by the factor 0.433.

Fiber optics, or FO (TC, DP, I)   A technology in which light is used to transmit information from one point to another. The transmitting medium is constructed of thin filaments (strands) of glass wires through which light beams are transmitted. For illumination purposes, the light may be one of any visible spectral wavelength. For data transmission, light must be of a single wavelength in order to be totally reflected (refracted) within the fiber. Light sources are usually generated by laser or LED.

Field angle (I)   The angle between the two directions for which the intensity of light is 10 percent of the maximum intensity, as measured in a plane through the nominal centerline of the light beam.

File server (DP)   A computer on a network that everyone on the network can access and get applications and documents from.

Finned tube, (H)   Tube or pipe containing fins used for heat transfer between water and air, usually by natural convection.

Fixture carrier (PS)   A metal unit designed to support an off-the-floor plumbing fixture.

Fixture unit, or FU (PF)   An index of the relative rate of flow of water to a fixture (water supply fixture unit, or WFU) or of sewage leaving a fixture (drainage fixture unit, or DFU).

Floppy disk (DP)   A flexible, removable disk (although the case in which the actual magnetic medium is housed may be hard, as it is on 3½″ floppies).

Floppy disk drive (DP)   A device for reading data from and writing data to floppy disks.

Flow rate (H, PF)   Cubic feet per minute (CFM) of air circulated in an air system or pounds of water per minute circulated through a hot-water heating system.

Flushometer valve (PS)   A device that discharges a predetermined quantity of water to fixtures for flushing purposes and is actuated by direct water pressure.

Font (DP, TC)   A collection of letters, numbers, punctuation marks, and symbols with an identifiable and consistent look.

Footcandle, or fc (I)   The illuminance on a surface 1 square foot in area on which there is a uniformly distributed flux of 1 lumen.

Foot Lambert, or fL (I)   A unit of luminance of a perfectly diffusing surface emitting or reflecting light at the rate of 1 lumen per square foot.

Formatting (DP)   All the characteristics of text other than the actual characters that make it up. Formatting includes things such as italics, boldface, type size, margins, line spacing, justification, and so on. Also, another term for initializing a disk.

Free field (AS)   A field free from boundaries that would otherwise tend to reflect sound.

Frequency (AS, E)   Number of complete oscillation cycles per unit of time. A unit of frequency often used is the hertz.

Function keys (DP)   Special keys labeled F1, F2, etc., on some extended keyboards.

Furnace (H)   Either the combustion space in a fuel-burning device or a direct-fired air heater. In the latter case, not to be confused with a boiler.

Fuse (E)   An overcurrent protective device with a circuit-opening fusible part that is heated and severed by the passage of overcurrent through it.

Galvanizing (PS)   A process wherein a surface of iron or steel is covered with a layer of zinc.

Gigabyte, (DP)   A measure of computer memory, disk space, and the like equal to 1024 megabytes (1,073,741,824 bytes), or about 179 million words. Sometimes a gigabyte is treated as an even billion bytes, but plainly, that is almost 74 million bytes short. Sometimes abbreviated gig (more often in speech than in writing).

Glare (I)   The sensation produced by luminance within the visual field that is sufficiently greater than the luminance to which the eyes are adapted. Glare may cause annoyance, discomfort, or loss of visual performance and visibility.

Ground (E)   A conducting connection, whether intentional or accidental, between electrical circuit or equipment and the earth.

Ground fault (circuit) interrupter, or GFI or GFCI (E)   A device that senses ground faults and reacts by opening the circuit.

Grounded conductor (E)   A system or circuit conductor that is intentionally grounded.

Grounding conductor (E)   A conductor used to connect equipment or the grounded circuit of a wiring system to a grounding electrode or electrodes.

Halon (FP)   A bromtrifluoromethane gas that is effective in extinguishing fires.

Handshake (DP)   What computers do when communicating, in order to establish a connection and agree on protocols for the transmission of data.

Hard disk drive (DP) A rigid, usually nonremovable disk or the disk drive that houses it. Hard disks store much more data and access the data much more quickly than floppy disks do.

Head end (TC) The central facility where signals are combined and distributed in an airborne sound (AS) system.

Heat pump (H) An all-electric heating/cooling device that takes energy for heating from outdoor air (or groundwater).

Hertz, or Hz (AS, DP, E) A measure of frequency (cycles per second).

High-density (or high-resolution) television, or HDTV (DP) Television with about double the resolution of present systems.

High-intensity discharge (HID) lamp (I) A lamp whose light source is mercury, metal halide, or high-pressure sodium.

High-pressure sodium (HPS) lamp (I) A HID lamp in which light is produced by radiation from sodium vapor.

Home run (E) The wiring run between the panel and the first outlet in the branch circuit. (Looking upstream, the wiring run between the last outlet and the panel.)

Horsepower (E, H, PF) A unit of power that equals 746 watts, or 1 hp = ¾ KW.

Humidity (H) Usually, water vapor mixed with dry air. Absolute humidity is the weight of water (or steam) per unit volume of the air-water mixture.

Icon (DP) A graphic symbol usually representing a file, folder, disk, or tool.

Illuminance (I) The density of the luminous flux incident on a surface; the quotient of the luminous flux divided by the area of the surface when the latter is uniformly illuminated.

Illuminance (lux or footcandle) meter (I) An instrument for measuring illuminance on a plane. Instruments that accurately respond to more than one spectral distribution are color corrected.

Incandescent lamp (I) A lamp in which light is produced by a filament heated to incandescence by an electric current.

Index of refraction (I, TC) The ratio of the speed of light in a vacuum to the speed of light in a material, usually abbreviated *n*.

Indirect lighting (I) Lighting by luminaires distributing 90 to 100 percent of the emitted light upward.

Input/output, or I/O (DP) Signal fed into and transmitted out of a circuit.

Integrated circuit, or IC (DP, TC) An electronic device that contains hundreds or thousands of separate components, such as transistors, resistors, switches, etc. Encapsulated in a plastic enclosure, it is also called a chip.

Integrated services digital network, or ISDN (TC) A digital standard calling for 144 kbit/sec transmission, corresponding to two 64-kbit/sec digital voice channels and one 16-kbit/sec data channel.

Intensity, or I (AS) The average rate of sound energy flow per unit of area in a direction perpendicular to the area.

Intensity, or I (I) The luminous flux per unit solid angle, expressed in lumens per steradian (lm/Sr) or candela.

Intensity level, or IL (AS) Ten times the logarithm of the ratio of the sound intensity to a reference sound intensity (I_{ref}) of 10^{-12} watt/M^2 (9.29 * 10^{-14} W/ft^2).

International organization for standardization, or ISO (G) An international organization to establish standards on scientific and technology quantities.

Internet (DP) A computer network that joins many government and private computers over phone lines. Internet was started in 1969 by the Defense Department, is managed by the National Science Foundation, and is now the most popular computer network in the world. Among the services is the World Wide Web (www).

Inverse square law (I) The law stating that the illuminance at a point on a surface varies directly with the intensity of a point source and inversely as the square of the distance between the source and the point.

Invert (PS) Lowest point on the interior of a horizontal pipe.

Isolux chart (I) A series of lines, plotted on any appropriate set of coordinates, each of which connects all the points on a surface having the same illumination.

Kilobyte (DP) A measure of computer memory, disk space, and the like equal to 1024 characters, or about 170 words. Abbreviated K.

Lamp efficacy (I) The ratio of lumens produced by a lamp to the watts consumed. Expressed as lumens per watt (LPW).

Lamp lumen depreciation, or LLD (I) Multiplicative factor in calculations of illumination for reduction in the light output of a lamp over a period of time.

Laser (TC) One of the wide range of devices that generate light by the principle. Laser light is directional, covers a narrow range of wavelengths, and is more coherent than ordinary light. Semiconductor diode lasers are the standard light sources in fiber optic systems.

Laser diode (DP, TC)   A semiconductor diode in which the injection of current carriers produces laser light by amplifying photons generates when holes and electrons recombine at the junction between *p*- and *n*-doped regions.

Latent heat (H)   Inherent heat in the form of fluid without a phase change.

Loudness (AS)   A subjective description of the level of sound. Typically, a 10-dB increase in sound pressure level is judged to be twice as loud as the level before the increase.

Light-emitting diode, or LED (AX, TC, DP)   A network that transmits data among many nodes in a small area (e.g., a building or campus).

Light loss factor, or LLF (I)   A factor used in calculating illuminance after a given period and under given conditions. (Formerly called *maintenance factor.*)

Liquid crystal display, or LCD (DP, TC)   A product that uses liquid crystals sealed in glass. The pixels contain individual transistors to generate alphanumeric and graphic images.

Local area network, or LAN (DP)   A network of computers and related devices that transmits data among many nodes in a relatively small area, such as one office or one building. See Wide area network.

Low-pressure sodium (LPS) Lamp, (I)   A discharge lamp in which a single wavelength of visible yellow light is produced by radiation of sodium vapor at a low pressure.

Lumen (I)   The unit of luminous flux; the luminous flux emitted within a unit solid angle (one steradian) by a point source having a uniform luminous intensity of one candela.

Luminaire (I)   A complete lighting unit consisting of a lamp or lamps together with the parts designed to distribute the light, to position and protect the lamps, and to connect the lamps to the power supply.

Luminaire dirt depreciation, or LDD (I)   Multiplicative factor used in calculations of illuminance for reduced illuminance due to dirt collecting on the luminaires.

Luminaire efficiency (I)   The ratio of the luminous flux (lumens) emitted by a luminaire to that emitted by the lamp or lamps used therein.

Luminance (I)   The amount of light reflected or transmitted by an object.

Lux (I)   The metric unit of illuminance. One lux is one lumen per square meter (lm/m^2).

Macro (DP)   A command that incorporates two or more other commands or actions. (The name comes from the idea that macro commands incorporate "micro" commands.)

Maintenance factor, or MF (I)   A factor used in calculating illuminance after a given period and under given conditions.

Makeup water (H, PS)   Water supplied to replenish that lost by leaks, evaporation, etc.

Masking (AS)   The rendering undetectable of one sound of interest by other sounds.

Mbh   Thousands of British thermal units (Btu) per hour (h).

MCM   Thousand circular mil (E); used to describe large wire sizes.

Meg (DP)   An abbreviation for *megabyte.*

Megabyte, or MB (DP)   A measure of computer memory, disk space, and the like equal to 1024K (1,048,576 bytes). Sometimes people try to equate a megabyte with an even million characters.

Megahertz, or MHz (DP, TC)   A million cycles, occurrences, alterations, or pulses per second. Used to describe the speed of computers' clock rates. Abbreviated MHz.

Memory (DP)   The retention of information electronically, on chips. There are two main types of memory: RAM, which is used for the short-term retention of information (that is, until the power is turned off), and ROM, which is used to store programs that are seldom, if ever, changed.

Menu (DP)   A list of commands to operate a computer. There are many types of menus, e.g., pop-up menus, submenus, etc.

Mercury lamp (I)   A high-intensity discharge (HID) lamp in which the major portion of the light is produced by radiation from mercury.

Metal halide (MH) lamp, (I)   A high-intensity discharge (HID) lamp in which the major portion of the light is produced by radiation of metal halides and their products of dissociation, possibly in combination with metallic vapors such as mercury.

Modulator-demodulator, or modem (DP)   A device that lets computers talk to each other over phone lines. Modems are used to send digital signals through telephone lines, to be converted back to analog signals at the receiving end.

Motherboard (DP)   The main board in a computer (or other computer device).

MS-DOS, (DP)   The original, and still the most popular, operating system used on IBM PCs and compatible computers. (The name stands for *Microsoft Disk Operating System.*) Also, sometimes called PC DOS.

Multimedia (DP, TC)   The combination of multiple media in an integrated system, such as audio, video, text, graphics, FAX, and telephone.

Multiplex (TC, DP)   To transmit two or more signals over a single channel.

Multiuser (DP)   Said of software or hardware that supports use by more than one person at one time.

Nadir (I)   Vertically downward directly below the luminaire or lamp; designated as 0°.

Nanometer (G, I, TC, AX)   A unit of length, 10^{-9} m. It is part of the SI system and has largely replaced the non-SI unit, "angstrom" (0.1 nm), in technical literature.

Nanosecond (G, DP, I, TC)   A billionth of a second. Used to measure the speed of memory chips, among other things. Abbreviated ns.

National Television Standard, or (TC)   The analog video broadcast standard used in North America and set by the National Television Standards Committee.

National Television Standards Committee, or NTSC (TC)   Committee that sets television transmission standards in North America at 30 frames per second and 525 horizontal lines per frame.

Network (DP, E)   Two or more computers (or other computer-related devices) connected to share information. Usually, the term refers to a local area network.

Network operating system, or NOS (DP)   A software program for a computer network that runs in a file server and controls access to files and other resources from multiple users.

Noise isolation class, or NIS (AS)   A single-number rating derived in the same manner as STC but based on NR (Noise Reduction) rather than TL (Sound Transmission Loss). It includes the acoustical absorption in the receiving room.

Noise reduction coefficient, or NRC (AS)   The average of the sound absorption coefficients in the 250-, 500-, 1000-, and 2000-Hz octave bands.

Noise reduction (AS)   The difference in decibels between the sound in one space and the sound in a second space attenuated by some intervening medium; e.g., from one room to another.

Octave (AS)   An interval between two frequencies having a ratio of two to one (2:1).

Octave band (AV)   A frequency band whose upper limit is twice the lower limit.

Off line (DP)   Said of things done while one is not actively connected to a computer or a network. For example, you might work on a message off line, then log onto an electronic mail system to send it. Opposite of on line.

Ohms Law (E)   The relationship between current and voltage in a circuit. The law states that current is proportional to voltage and inversely proportional to resistance. Algebraically, in DC circuits, $I = V / R$; in AC circuits, $I = V/Z$.

On line (DP)   On, or actively connected to, a computer or computer network. For example, on-line documentation appears on the screen rather than in a manual. Opposite of off line.

Open protocol (DP, TC)   A set of standard procedures that are agreed upon by all manufacturers so that data can be transmitted without obstructions.

Operating system (DP)   The basic software that controls a computer's operation.

Optic fiber cable, or OFC (TC, DP, I)   An information transmission medium consisting of a core of glass or plastic surrounded by a refractive cladding and a protective jacket. It transmits stranded and ribbon configurations. See Fiber optics.

Orifice (H, PS)   An opening. Commonly applied to discs placed in pipelines or radiator valves to reduce the flow of a fluid to a desired amount.

Overcurrent (E)   Any current in excess of the rated current of equipment or the ampacity of a conductor.

Overcurrent device (E)   A device, such as a fuse or a circuit breaker, designed to protect a circuit against excessive current by opening the circuit.

Panel or panelboard (E)   A box containing a group of overcurrent devices intended to supply branch circuits.

Parabolic reflector lamp, or PAR (I)   A lamp with internal reflector having the contour of a parabola to achieve beam control.

Pascal (Pa, AS)   Measure of pressure in the SI system ($= 1.45 \times 10^{-4}$ psi).

Personal computer, or PC (DP)   IBM machine introduced in 1981; also, its various clones.

Phase alternate line, or PAL (TC)   A television transmission standard used mostly in Europe; 25 frames per second and 625 lines per frame.

Picture element, or pixel (DP, TC)   Any of the little dots of light that make up the picture on a computer (or TV) screen. The more pixels there are in a given area—that is, the smaller and closer together they are—the higher is the resolution of the screen. Sometimes pixels are simply called dots. The number of pixels on a screen is usually expressed by the number of horizontal and vertical rows, such as 640×480, 1800×1280, etc.

Pitch (AS)   A subjective description of sound quality referring to the principal frequency content. The sound pressure level of a tone.

Point source (AS, I)   A source of essentially zero dimensions that radiates sound or light uniformly in all directions.

Polarization (I)   The process by which the transverse vibrations of light waves are oriented in a specific plane.

Polyvinyl chloride, or PVC (P)   An inert plastic material commonly used for pipes. It has high resistance to corrosion.

PostScript (DP)   A page-description programming language developed by Adobe and designed specifically to handle text and graphics and their placement on a page. Used primarily in laser printers and image setters.

Power factor, or PF (I)   The ratio of the working power to the apparent power of an AC circuit.

Private automatic branch exchange, or PABX (TC)   See Private Branch Exchange.

Private branch exchange, or PBX (TC, AX)   A private telephone system that usually interconnects with public telephone systems to serve a business or an organization. It can also provide access to a computer from a data terminal. PBX and PABX are used interchangeably.

Programmable ROM, or PROM (DP)   A read-only memory chip that can be changed with a special device.

Propeller fan (H)   A fan with airfoil blades that move the air in the general direction of the axis of the fan.

Protocol (DP, TC)   A set of standard procedures that control how information is transmitted between computer systems.

Pounds per square inch pressure, or psi (H, PF)   A unit measure of pressure, or force per unit area.

Psychrometer (H)   A device employing a wet-bulb and a dry-bulb thermometer to measure the humidity in the air.

Psychrometric chart (H)   A graph used in air conditioning and showing the properties of air-steam mixtures.

Raceway (E)   An enclosed channel of metal or nonmetal designed for holding wires or cables.

Radiation (H)   The transfer of energy in wave form from a hot body to a (relatively) cold body, independently of matter between the two bodies.

Radio frequency, or RF (TC)   Electromagnetic waves operating between 5 KHz and the MHz range. Although off-air (over-the-air) and satellite TV signals operate far above these frequencies (54 MHz and up to many GHz), these TV signals are frequently incorrectly referred to as RF signals in commercial applications.

RAM cache (DP)   An area of memory set aside to hold information recently read in from disk, so that if the information is needed again, it can be gotten from memory (which is much faster than getting it from disk).

RAM disk (DP)   A portion of memory set aside to act as a temporary disk.

Random access memory, or RAM (DP)   The part of a computer's memory used for the short-term retention of information (in other words, until the power is turned off). Programs and documents are stored in RAM while you use them. Actually, just about all kinds of memory are accessed randomly. See Read-Only Memory.

Random noise (AS)   Sound whose magnitude cannot be predicted at any time.

Rapid-start (RS) fluorescent lamp, (I)   A fluorescent lamp designed for operation with a ballast that starts the lamp light output within a split second.

Read-only memory, or ROM (DP)   The part of a computer's memory used to store programs that are seldom or never changed. A user can read information from ROM, but cannot write information to it. See Random Access Memory.

Reboot (DP)   Restart of a computer from scratch.

Reflectance (I)   The ratio of the reflected light to the incident light falling on a surface.

Reflected glare (I)   Glare resulting from specular reflections of high luminances in polished or glossy surfaces in the field of view.

Reflector (R) lamp (I)   A lamp with an internal reflector to redirect the light output to a controlled direction.

Refraction (I, TC)   The process by which the direction of a ray of light changes as the ray passes obliquely from one medium to another in which its speed is different.

Refrigerant (H)   A substance that absorbs heat while vaporizing and whose boiling point and other properties make it useful as a medium for refrigeration.

Relative humidity, or RH (H)   The ratio of the water vapor (by weight) in air to the water vapor saturated in air at the same pressure and temperature.

Relief vent (PS)   A vent designed to provide circulation of air between drainage and vent systems or to act as an auxiliary vent.

Residual pressure (FP)   Pressure remaining in a system while water is being discharged from outlets.

Resolution (DP)   The number of dots (or pixels) per square inch (or in any given area). The more dots there are, the higher is the resolution of the device.

Resonance (AS, G)   A state in which the forces of oscillation of a system occur at or near a natural frequency of the system.

Reverberant sound field (AS)   Sound that is reflected from the boundaries of and furnishings within an enclosed space. Excludes direct sound. (See Direct Sound Field.)

Reverberation (AS)   The persistence of sound in an enclosed space as a result of repeated reflection or scattering of the sound.

Reverberation time (AS)   The time required for the sound pressure level in a reverberant sound field to decay 60 dB after the source has been extinguished.

Riser diagram (E)   Electrical block-type diagram showing the connection of major items of equipment and components.

Romex (E)   One of several trade names for NEC-type nonmetallic sheathed flexible cable.

Romex (NM) cable, (E)   A cable composed of flexible plastic sheathing inside of which are two or more insulated wires for carrying electricity.

Room cavity ratio, or RCR (I)   A number indicating the proportions of a room cavity, calculated from the length, width, and height of the room.

R-value (H)   Resistance rating of thermal insulation.

Sabin (AS)   Unit of measure of acoustical absorption. Named after Wallace Clement Sabin, U.S. physicist.

Saturated air (H)   Air containing saturated water vapor, with both the air and the vapor at the same dry-bulb temperature.

Saturated pressure (H)   That pressure, for a given temperature, at which the vapor and the liquid phases of a substance can exist in stable equilibrium.

Saturated steam (H)   Steam at the boiling temperature corresponding to the pressure at which it exists. Dry saturated steam contains no water particles in suspension. Wet saturated steam does.

Seasonal Energy Efficiency Ratio (SEER)   Measures the efficiency of HVAC equipment on a seasonal, rather than a design-load, basis.

Semiconductor (DP, TC)   A material that has an electrical resistance somewhere in between a conductor (e.g., metal), and an insulator (e.g., plastic). Silicon and germanium are the two most commonly used semiconductors. The flow of electric current in a semiconductor can be changed by light or by electric or magnetic fields.

Sensible heat (H)   Heat that raises the air temperature.

Septic tank (PS)   A tank designed to separate solid waste from liquid waste.

Serial port (DP)   The jacks on the back of a PC into which you can plug printers, modems, etc. (*Serial* refers to the fact that data are transmitted through these ports serially, one bit after an-

other, rather than in parallel, several bits side by side.)

Service conductors (E)   The supply conductors that extend from the street main or from transformers to the service equipment of the premises supplied.

Shielded twisted pair, or STP (TC, DP, AX)   See TP.

Simplex (TC)   Single element (e.g., a simplex connector is a single-fiber connector).

Small computer system interface, or SCSI (DP)   An industry-standard interface for hard disks and other devices that allows for very fast transfer of information.

Software (DP)   The instructions that tell a computer what to do. Also called programs or, redundantly, software programs.

Sound power (AS)   The rate at which acoustic energy is radiated. Usually measured in watts.

Sound power level, or PWL (AS)   Ten times the logarithm of the ratio of sound power, in watts, to the reference level (W_{ref}) of 10^{-12} watt.

Sound pressure (AS)   The fluctuations of pressure about atmospheric pressure. Usually measured in micropascals (μPa).

Sound pressure level, or SPL (AS)   Ten times the logarithm of the ratio of the mean square pressure to the square of a reference pressure (p_{ref}) of 20 μPa.

Sound transmission class, or STC (AS)   A single-number rating system designed to provide a cursory estimate of the sound-insulating properties of a wall or partition.

Sound transmission loss, or STL (AS)   The difference in decibels of the sound pressure level on the receiver side of a partition or barrier from that on the source side, with the receiver side being free-field conditions.

Specific heat (H)   The heat absorbed by a unit weight of a substance per unit temperature rise of the substance.

Specular reflection (I)   Mirrorlike reflection.

Stack (PS)   The vertical main of a system of soil, waste, or vent piping extending through one or more stories.

Stack vent (PS)   The extension of a soil waste stack above the highest horizontal drain connected to the stack.

Storm sewer (PS)   A sewer used for conveying rain or surface or subsurface water.

Structure-borne sound (AS)   Sound transmitted through solid material by means of vibrations waves in the material.

Subcooling (H)   Cooling of a liquid refrigerant below the condensing temperature at constant pressure.

Subsoil drain (PS)   A drain that receives only subsurface or seepage water and conveys it to an approved place of disposal.

Sump pump (PS)   A mechanical device for removing liquid waste from a sump.

Supply fixture unit, or SFU (PS)   A measure of the probable hydraulic demand on the water supply by various types of plumbing fixtures.

Switchboard (E)   A large panel containing switches, overcurrent devices, buses, and, usually, instruments.

System disk (DP)   Any disk containing the system software a PC needs to begin operation.

Système International d'Unites (French), or International System of Units, or SI (G)   System of measurement based on the metric system (meter, gram, second); different from the traditional or British system (foot, pound, second). The United States agreed to convert to the SI system in Public Law 94-168, signed in 1975. However, the conversion has been slow.

System software (DP)   A catchall term for the basic programs that help computers work; system software includes operating systems, programming languages, certain utilities, and so on.

Task lighting (I)   Lighting directed to a specific surface or area that provides illumination for the performance of visual tasks.

Text file (DP)   An ASCII file—just characters and no formatting.

Thermocouple (H)   Two dissimilar metals joined together to produce an electromotive force that varies with the temperature. Used to measure temperature.

Total internal reflection (TC)   Total reflection of light back into a material when the light strikes the interface with another material having a lower refractive index at an angle below a critical value.

Tone (AS)   A sound having a pitch and capable of causing auditory sensation.

Transformer (E)   A device used to raise or lower electrical voltage by means of an electromagnetic core and windings.

Trap (PS)   A fitting or device designed to provide a liquid seal that will prevent a fluid from passing back to where it came from.

Tungsten-halogen lamp (I)   A gas-filled tungsten incandescent lamp containing a certain proportion of halogens.

Twisted pair, or TP (E, TC, DP)   A type of wire made of two insulated copper conductors twisted around each other to reduce induction (and thus interference) from one conductor to the other.

The twist, or lay, varies in length to reduce the potential for interference from signals between pairs in a multipair cable. The lay usually varies between 2 and 12 inches. Closer lay provides better attenuation between the conductors. TP cables are classified as unshielded (UTP) and shielded (STP), the latter of which includes a metal sheath surrounding the pairs within the protective jacket.

U Coefficient (H)   Rate of heat transmission in Btuh/sq ft-°F.

Ultrasonic (AS)   Sound above the audible range.

Uninterruptible power supply, or UPS (E, DP)   A power supply or system that provides a steady source of electrical power even when the normal (utility) power supply is interrupted. The system usually contains a storage battery floating on line with an inverter to convert battery power from DC to AC. Other UPSs may utilize a DC motor-driven AC generator set.

Unshielded twisted pair, or UTP (E, TC)   See Twisted pair.

Vacuum breaker (PS)   Check valve open to the atmosphere when the pressure in piping drops to atmospheric pressure.

Veiling reflections (I)   Regular reflections superimposed upon diffuse reflections from an object that partially or totally obscures the details to be seen by reducing the contrast. Veiling reflections are sometimes called reflected glare.

Ventilation (H)   The art or process of supplying outside (so-called fresh) air to or removing air from an enclosure.

Virtual memory (DP)   A technique that lets a computer treat part of a hard disk as if it were RAM.

Virus (DP)   A program that functions on your computer without your consent. A benign virus may do nothing more than duplicate itself, but some viruses are meant to destroy data.

Visual comfort probability, or VCP (I)   The rating of a lighting system, expressed as a percent of people who, when viewing from a specified location and in a specified direction, will be expected to find the system acceptable in terms of discomfort glare.

Visual display terminal, or VDT (G, DP)   A data terminal with a TV screen.

Voltage, or V (E)   The electric pressure in an electric circuit, expressed in volts.

Voltage drop, or VD (E)   The diminution of voltage around a circuit, including the wiring and loads. Must equal the supply voltage.

Voltage to ground (E)   For grounded circuits, the voltage between the ungrounded conductor and the ground; for ungrounded circuits, the greatest voltage between the given conductors.

Water hammer (PF)   Banging of pipes caused by the shock of closing faucets or other flow control devices.

Water hammer arrester (PS)   A device, other than an air chamber, designed to provide protection against excessive surge pressure.

Wavelength (electromagnetic) (I, TC, DP)   The distance between nodes of an electromagnetic wave, such as radio, TV, light, radar, etc.; given by L (length, m) $= 3 \times 10^8/f$, where f is the frequency, in hertz.

Wavelength (sound) (AS)   The distance between nodes in a sound wave; given by L (length, ft) $= 1130/f$ where f is the frequency, in hertz.

WB-wet-bulb (WB) temperature (H)   The temperature of the air as measured by a wet-bulb thermometer. Except when the air is saturated, the wet-bulb temperature is lower than the dry-bulb

temperature in inverse proportion to the humidity.

Wide area network, or WAN (DP)   A network of computers and related devices that transmit data among many nodes between buildings or cities. See Local Area Network.

Window (DP)   An enclosed area on the VDT or LCD screen that has a title bar (which one can use to drag the window around). Disks and folders open into windows, and documents appear in windows when recalled. Windows is also the name of an operating system.

Zonal cavity method (I)   A lighting design procedure used for predetermining the relation between the number and types of lamps or luminaires, the room characteristics, and the average illuminance on the work plane. The zonal cavity method takes into account both direct and reflected flux.

APPENDIX

B

Glossary of Technical Organizations

Abbreviations	Full Name
ACEC	American Consulting Engineers Council
AGA	American Gas Association
AGC	Associated General Contractors of America
AIA	American Institute of Architects
AIID	American Institute of Interior Designers
AIPE	American Institute of Plant Engineers
AMCA	Air Movement and Control Association, Inc.
ANSI	American National Standards Institute
ARI	Air Conditioning and Refrigeration Institute
ASCII	American Standard Code for Information Interchange
ASHRAE	American Society for Heating, Refrigerating and Air Conditioning Engineers
ASME	American Society of Mechanical Engineers
ASPE	American Society for Plumbing Engineers
ASSE	American Society of Sanitary Engineers
ASTM	American Society for Testing and Materials
AWWA	American Water Works Association
BICSI	Building Industry Consulting Service International
BOCA	Building Officials and Code Administration
BOMA	Building Owners and Managers Association, International
BRI	Building Research Institute
CSI	Construction Specification Institute
EJC	Engineers Joint Council
EPA	Environmental Protection Agency
FIA	Factory Insurance Association

FMS	Factory Mutual System
IBR	Institute of Boiler and Radiator Manufacturers
IEEE	Institute of Electrical and Electronics Engineers
IES	Illuminating Engineering Society
IESNA	Illuminating Engineering Society of North America
ISO	International Organization for Standards
NEC	National Electrical Code
NECA	National Electrical Contractors Association
NEMA	National Electrical Manufacturers Association
NFPA	National Fire Protection Association
NPC	National Plumbing Code
NSPE	National Society of Professional Engineers
SMACNA	Sheet Metal and Air Conditioning Contractors National Association
TIA/EIA	Telecommunication Industry Association (Formerly EIA/TIA)
UBC	Uniform Building Code
UFC	Uniform Fire Code
UL	Underwriter Laboratories, Inc.

APPENDIX
C

UNITS AND CONVERSION OF QUANTITIES

THIS APPENDIX CONTAINS THE COMMONLY ENcountered units of quantities and derived quantities in conventional (imperial or British) and SI units. Although the United States will eventually phase out the imperial system in favor of the International Standard (Système International, or SI) system, most U.S. construction measures are still expressed in the conventional system. For example, building dimensions are still given in feet and inches, pipe and conduit sizes are expressed in inches, etc. Thus, a conversion table between the two systems is needed until such time that only the SI system is in use.

This appendix provides an abbreviated list of units and conversion factors between the conventional and the SI units. For more complete information on units, the reader is referred to the following publications:

- AIA: *Metric Building and Construction Guide*
- ASTM: *Standard Practice for Use of the International System of Units*

For convenience, this appendix also includes information on the standard exponents and Greek letters.

C.1 GREEK LETTERS

A	α	Alpha	E	ϵ	Epsilon	I	ι	Iota	N	ν	Nu	P	ρ	Rho	Φ	ϕ	Phi
B	β	Beta	Z	ζ	Zeta	K	κ	Kappa	Ξ	ξ	Xi	Σ	$\sigma\ \varsigma$	Sigma	X	χ	Chi
Γ	γ	Gamma	H	η	Eta	Λ	λ	Lambda	O	o	Omicron	T	τ	Tau	Ψ	ψ	Psi
Δ	δ	Delta	Θ	$\vartheta\ \theta$	Theta	M	μ	Mu	Π	π	Pi	Υ	υ	Upsilon	Ω	ω	Omega

C.2 EXPONENTS

Prefix	Symbol	Factor by Which the Unit is Multiplied
exa	E	$1{,}000{,}000{,}000{,}000{,}000{,}000 = 10^{18}$
peta	P	$1{,}000{,}000{,}000{,}000{,}000 = 10^{15}$
tera	T	$1{,}000{,}000{,}000{,}000 = 10^{12}$
giga	G	$1{,}000{,}000{,}000 = 10^{9}$
mega	M	$1{,}000{,}000 = 10^{6}$
kilo	k	$1{,}000 = 10^{3}$
hecto	h	$100 = 10^{2}$
deka	da	$10 = 10^{1}$
deci	d	$0.1 = 10^{-1}$
centi	c	$0.01 = 10^{-2}$
milli	m	$0.001 = 10^{-3}$
micro	μ	$0.000{,}001 = 10^{-6}$
nano	n	$0.000{,}000{,}001 = 10^{-9}$
pico	p	$0.000{,}000{,}000{,}001 = 10^{-12}$
femto	f	$0.000{,}000{,}000{,}000{,}001 = 10^{-15}$
atto	a	$0.000{,}000{,}000{,}000{,}000{,}001 = 10^{-18}$

C.3 PARTIAL LIST OF UNITS AND ABBREVIATIONS

A	ampere	kV	kilovolt
atm	atmosphere	kVA	kilovolt-ampere
bps	bits per second	kVAr	reactive kilovolt-ampere
Cd	candela	kW	kilowatt
°C	degree Celsius	kWh	kilowatt-hour
cal	calorie	L	luminance
cgs	centimeter-gram-second (system)	lm	lumen
cm	centimeter	lx	lux
cp	candlepower	m	meter
CRI	color-rendering index	m^2	square meter
CU	coefficient of utilization	mA	milliampere
dB	decibel	MHz	megahertz
emf	electromotive force	min	minute (time)
°F	degree Fahrenheit	mm	milimeter
fc	footcandle	mph	mile per hour
ft	foot	nm	nanometer
ft^2	square foot	ns	nanosecond
fl	foot Lambert	R	reflectance factor
h	hour	rad	radian
hp	horsepower	sec	second (time)
Hz	hertz	sq	square
in	inch	sr	steradian
in^2	square inch	V	volt
J	joule	VA	volt-ampere
K	kelvin	Var	reactive volt-ampere
kcal	kilocalorie	W	watt
kg	kilogram	μA	microampere
kHz	kilohertz	μV	microvolt
km	kilometer	μW	microwatt
km/sec	kilometer per second		

C.4 CONVERSION OF ILLUMINATION VALUES

	m						m				
	cm						cm				
	kcd/m²						kcd/m²				
	cd/m²						cd/m²				
lx*	fc	fL	cd/in²	in	ft	lx*	fc	fL	cd/in²	in	ft
1	.09	.29	.65	.39	3.3	500	46.5	146.0	322.5	196.9	1641
2	.19	.58	1.29	.79	6.6	510	47.4	148.9	329.0	200.8	1673
3	.28	.88	1.94	1.18	9.8	520	48.3	151.8	335.4	204.7	1706
4	.37	1.17	2.58	1.57	13.1	530	49.2	154.7	341.9	208.7	1739
5	.47	1.46	3.23	1.97	16.4	540	50.2	157.6	348.3	212.6	1772
6	.56	1.75	3.87	2.36	19.7	550	51.1	160.5	354.8	216.5	1805
7	.65	2.04	4.52	2.76	23.0	560	52.0	163.5	361.2	220.5	1837
8	.74	2.34	5.16	3.15	26.2	570	53.0	166.4	367.7	224.4	1870
9	.84	2.63	5.81	3.54	29.5	580	53.9	169.3	374.1	228.3	1903
						590	54.8	172.2	380.6	232.3	1936
100	9.3	29.2	64.5	39.4	328	600	55.7	175.1	387.0	236.2	1969
110	10.2	32.1	71.0	43.3	361	610	56.7	178.1	393.5	240.2	2001
120	11.1	35.0	77.4	47.2	394	620	57.6	181.0	399.9	244.1	2034
130	12.1	37.9	83.9	51.2	427	630	58.5	183.9	406.4	248.0	2067
140	13.0	40.9	90.3	55.1	459	640	59.5	186.8	412.8	252.0	2100
150	13.9	43.8	96.8	59.1	492	650	60.4	189.7	419.3	255.9	2133
160	14.9	46.7	103.2	63.0	525	660	61.3	192.7	425.7	259.8	2165
170	15.8	49.6	109.7	66.9	558	670	62.2	195.6	432.2	263.8	2198
180	16.7	52.5	116.1	70.9	591	680	63.2	198.5	438.6	267.7	2231
190	17.7	55.5	122.6	74.8	623	690	64.1	201.4	445.1	271.7	2264
200	18.6	58.4	129.0	78.7	656	700	65.0	204.3	451.5	275.6	2297
210	19.5	61.3	135.5	82.7	689	710	66.0	207.2	458.0	279.5	2330
220	20.4	64.2	141.9	86.6	722	720	66.9	210.2	464.4	283.5	2362
230	21.4	67.1	148.4	90.6	755	730	67.8	213.1	470.9	287.4	2395
240	22.3	70.1	154.8	94.5	787	740	68.7	216.0	477.3	291.3	2428
250	23.2	73.0	161.3	98.4	820	750	69.7	218.9	483.8	295.3	2461
260	24.2	75.9	167.7	102.4	853	760	70.6	221.8	490.2	299.2	2494
270	25.1	78.8	174.2	106.3	886	770	71.5	224.8	496.7	303.1	2526
280	26.0	81.7	180.6	110.2	919	780	72.5	227.7	503.1	307.1	2559
290	26.9	84.7	187.1	114.2	951	790	73.4	230.6	509.6	311.0	2592
300	27.9	87.6	193.5	118.1	984	800	74.3	233.5	516.0	315.0	2625
310	28.8	90.5	200.0	122.0	1017	810	75.2	236.4	522.5	318.9	2658
320	29.7	93.4	206.4	126.0	1050	820	76.2	239.4	528.9	322.8	2690
330	30.7	96.3	212.9	130.0	1083	830	77.1	242.3	535.4	326.8	2723
340	31.6	99.2	219.3	133.9	1116	840	78.0	245.2	541.8	330.7	2756
350	32.5	102.2	225.8	137.8	1148	850	79.0	248.1	548.3	334.6	2789
360	33.4	105.8	232.2	141.7	1181	860	79.9	251.0	554.7	338.6	2822
370	34.4	108.0	238.7	145.7	1214	870	80.8	254.0	561.2	342.5	2854
380	35.3	110.9	245.1	149.6	1247	880	81.8	256.9	567.6	346.5	2887
390	36.2	113.8	251.6	153.5	1280	890	82.7	259.8	574.1	350.4	2920
400	37.2	116.8	258.0	157.5	1312	900	83.6	262.7	580.5	354.3	2953
410	38.1	119.7	264.5	161.4	1345	910	84.5	265.6	587.0	358.3	2986
420	39.0	122.6	270.9	165.4	1378	920	85.5	268.5	593.4	362.2	3019
430	39.9	125.5	277.4	169.3	1411	930	86.4	271.5	600.0	366.1	3051
440	40.0	128.4	283.8	173.2	1444	940	87.3	274.4	606.3	370.1	3084
450	41.8	131.4	290.3	177.2	1476	950	88.3	277.3	612.8	374.0	3117
460	42.7	134.3	296.7	181.1	1509	960	89.2	280.2	619.2	378.0	3150
470	43.7	137.2	303.2	185.0	1542	970	90.1	283.1	625.7	381.9	3183
480	44.6	140.1	309.6	189.0	1575	980	91.0	286.1	632.1	385.8	3215
490	45.5	143.0	316.1	192.9	1608	990	92.0	289.0	638.6	389.8	3248

Source: *IES Lighting Ready Reference.*

*Also useful for converting from ft² to m².

C.5 CONVERSION OF GENERAL, MECHANICAL AND ELECTRICAL UNITS

Multiply	By	To Obtain
Angstrom units	3.939×10^{-9}	Inches
Acre	0.4047	Hect-acres
Acre	43,560	Square feet
Atmospheres	760.0	Millimeters of mercury (at 32F)
Atmospheres	29.921	Inches of mercury (at 32F)
Atmospheres	33.97	Feet of water (at 62F)
Atmospheres	10,333.0	Kilograms per square meter
Atmospheres	14.697	Pounds per square inch (psi)
Barrels (oil)	42	Gallons
Boiler horsepower	33,475	Btu per hour (Btuh)
Btu	.252	Calories (large)
Btu	252	Calories (small)
Btu	778	Foot pounds
Btu	1.055	Kilo-Joules (KJ)
Btu	.000293	Kilowatt hour (kWh)
Btu per pound (Btu/lb)	2.326	Joules per gram (J/g)
Btu per hour (Btuh)	.252	Calories (large) per hour
Btu per hour per square foot per F (Btu/h · ft² · F)	.488	Calories (small) per hour per square centimeter per C (cal/h · cm² · c)
Btu-inch per square foot per hour per F (Btu-in/ft² · h · F)	1.24	Calories (small) per centimeter per sq. cm. per hr. per C (cal-cm/cm² · h · c)
Calories (large)	3.968	British thermal units (Btu)
Calories (large)	1.1619	Watt-hours (Wh)
Calories (large) per hour	3.968	Btu per hour
Centimeters	0.3937	Inches
Centimeters of mercury	136.0	Kilograms per square meter
Centimeters of mercury	0.1934	Pounds per square inch
Cubic centimeters	0.06102	Cubic inches
Cubic feet	.028317	Cubic meters
Cubic feet	1728.0	Cubic inches
Cubic feet	7.48052	Gallons
Cubic feet	28.32	Liters
Cubic feet of water	62.37	Pounds (at 60F)
Cubic feet per minute (cfm)	472.0	Cubic centimeters per sec. (cm²/s)
Cubic feet per minute	0.4720	Liters per second (L/S)
Cubic inches	16.39	Cubic centimeters
Cubic meters	35.3145	Cubic feet
Feet	30.48	Centimeters
Feet of water	62.37	Pounds per square foot
Feet of water	0.4335	Pounds per square inch
Feet per minute (fpm)	0.5080	Centimeters per second (cm/s)
Feet per minute	0.3048	Meters per minute
Feet per second (fps)	30.48	Centimeters per second (cm/s)
Feet per second (fps)	18.29	Meters per minute
Gallons (U.S.)	0.1337	Cubic feet
Gallons (U.S.)	231.0	Cubic inches
Gallons (U.S.)	3.7853	Liters
Gallons of water	8.3453	Pounds of water (at 60F)
Gallons per minute	0.06308	Liters per second
Grains	.0648	Grams
Horsepower	33,000.0	Foot pounds per minute
Horsepower	2546.0	British thermal units per hour
Horsepower	42.42	British thermal units per minute
Horsepower	0.7457	Kilowatts
Horsepower (boiler)	33,475	British thermal units per hour
Inches	2.540	Centimeters
Inches of mercury (at 62F)	13.57	Inches of water (at 62F)
Inches of mercury (at 62F)	.4912	Pounds per square inch
Inches of water (at 62F)	.07355	Inches of mercury
Inches of water (at 62F)	25.40	Kilograms per square meter
Inches of water (at 62F)	.03613	Pounds per square inch
Inches of water (at 62F)	5.202	Pounds per square foot
Kilograms	2.20462	Pounds
Kilogram-calories	3.968	British thermal units
Kilograms per sq. cm.	14.220	Pounds per square inch
Kilograms per sq. meter	.2048	Pounds per square foot

Kilometers	.62137	Miles
Kilowatts	1.341	Horsepower
Kilowatt-hours	3415.0	British thermal units
Kilowatt-hours	860.5	Kilogram-calories
Latent heat of ice	143.33	British thermal units per pound
Liters	0.03531	Cubic feet
Liters	61.02	Cubic inches
Liters	0.2642	Gallons
Meters	3.28083	Feet
Meters	39.37	Inches
Meters per minute	3.281	Feet per minute
Meters per minute	0.05468	Feet per second
Meters per second	3.281	Feet per second
Miles	1.60935	Kilometers
Miles per hour (mph)	1.61	Kilometer per hour (Km/h)
Pounds	7000.0	Grains
Pounds	0.45359	Kilograms
Pounds of water (at 60F)	27.68	Cubic inches
Pounds of water evaporated from and at 212F	970.4	British thermal units
Pounds per square foot (psf)	4.883	Kilograms per square meter
Pounds per square inch (psi)	2.309	Feet of water (at 62F)
Pounds per square inch	6.895	Kilo pascal (KPa)
Pounds per square inch	.0703	Kilograms per square centimeter
Square feet	929.0	Square centimeters
Square inches	6.452	Square centimeters
Square meters	10.765	Square feet
Temperature (C) + 273	1	Absolute temperature (C)
Temperature (C) + 17.78	1.8	Temperature (F)
Temperature (F) + 460	1	Absolute temperature (F)
Temperature (F) − 32	5/9, (0.556)	Temperature (C)
Tons of refrigeration (Ton)	12,000.0	British thermal units per hour (Btuh)
Tons of refrigeration	200.0	British thermal units per minute (Btu/min)
Tons of refrigeration	50.4	Calories per minute (cal/min.)
Watts	3.415	British thermal units per hour (Btuh)
Watts	0.01434	Kilogram-calories per min. (KCal/min.)
Watt-hours (Wh)	3.415	British thermal units (Btu)
Watt-hours	0.8605	Kilogram-calories (Kg/cal.)

INDEX